RUIN PROBABILITIES

Second Edition

ADVANCED SERIES ON STATISTICAL SCIENCE & APPLIED PROBABILITY

Editor: Ole E. Barndorff-Nielsen

Advanced Series on

Statistical Science &

Applied Probability

Vol. 14

RUIN PROBABILITIES

Second Edition

Søren Asmussen

Aarhus University, Denmark

Hansjörg Albrecher

University of Lausanne, Switzerland

 **World Scientific**

NEW JERSEY · LONDON · SINGAPORE · BEIJING · SHANGHAI · HONG KONG · TAIPEI · CHENNAI

Published by

World Scientific Publishing Co. Pte. Ltd.

5 Toh Tuck Link, Singapore 596224

USA office: 27 Warren Street, Suite 401-402, Hackensack, NJ 07601

UK office: 57 Shelton Street, Covent Garden, London WC2H 9HE

Library of Congress Cataloging-in-Publication Data
Asmussen, Søren.
    Ruin probabilities / by Søren Asmussen & Hansjörg Albrecher
    p. cm. -- (Advanced series on statistical science and applied probability ; v. 14)
    Includes bibliographical references and index.
    ISBN-13: 978-981-4282-52-9 (hardcover : alk. paper)
    ISBN-10: 981-4282-52-9 (hardcover : alk. paper)
    1. Insurance--Mathematics. 2. Risk. I. Albrecher, Hansjörg. II. Title.
    HG8781.A83 2010
    368'.01--dc22

                                                        2010023280

British Library Cataloguing-in-Publication Data
A catalogue record for this book is available from the British Library.

Printed in Singapore by World Scientific Printers.

Contents

Preface

This book is a second edition of the book of the same title by the first author which was published in 2000. The subject of ruin probabilities and related topics has since then undergone a considerable development, not to say boom. This much expanded and revised second edition aims at covering a substantial part of these developments as well as the classical topics.

Risk theory in general and ruin probabilities in particular are traditionally considered as part of insurance mathematics, and has been an active area of research from the days of Lundberg all the way up to today. One reason for writing this book is a feeling that the area has in recent years achieved a considerable mathematical maturity, which has in particular removed one of the standard criticisms of the area, namely that it can only say something about very simple models and questions. Although in insurance practice, usually simpler (and coarser) risk measures like Value-at-Risk are used, it is widely believed that the thinking advocated by ruin theory is still important for modern risk management. For instance, in times of market-consistent valuation principles, the role of the time diversification effect of insurance portfolios, which is one of the core elements of ruin theory, should not be forgotten. In addition, ruin theory has fruitful methodological links and applications to other fields of applied probability, like queueing theory and mathematical finance (pricing of barrier options, credit products etc.). Apart from these remarks, we have deliberately stayed away from discussing the practical relevance of the theory; if the formulations occasionally give a different impression, it is not by intention. Thus, the book is basically mathematical in its flavor.

The present second edition is more than 50% longer than the first and has more than double the number of references. The longer parts of the new material, reflecting subareas that have been particularly active in the last decade, are collected in Chapters XI–XIV, which treat Lévy processes, Gerber-Shiu functions, dependence and stochastic control, respectively. Shorter additions include more about martingales and generators (II.4), various versions in Chapter VIII of models with level dependence, e.g. tax or stochastic investments, Erlangization (IX.8), statistical techniques for distinguishing between light and heavy tails (X.6), more material on discrete-time risk models (XVI.1) and recent advances in simulation techniques scattered in Chapter XV. In addition, there are amendments and updates at a large number of places.

A book like this can be organized in many ways. One is by model, another by method. The present book is somewhere between these two possibilities. Chapters IV–VIII introduce some of the main models and give a first derivation of some of their properties. Chapters IX–XV then go into more depth with some of the special approaches for analyzing specific models and add a number of results on the models in Chapters IV–VIII. Chapters II and III are essentially methodological in flavor.

Here is a suggestion on how to get started with the book. For a brief orientation, read first Chapter I, continue with II.1–3 to see some of the simplest ruin calculations, the first part of III.5 (to understand the Pollaczeck-Khinchine formula in IV.2 more properly), and then, to get acquainted with the classical theory of the Cramér-Lundberg model, IV.1–5, V.4a, VIII.1, IX.1–3 and X.1–2. For a second reading, incorporate II.4, III.1–3, IV.8–9, V.1–2, V.5, VII.1–3, VIII.2, X.3–4, XII.1–2, XIII.1-2 and XV.1–3. The rest is up to your specific interests. Enjoy!

The symbols used for the quantities appearing in the book differ among the disciplines. We chose to use those that are common in the queueing community, partly also to be in line with the first edition. We apologize for the confusion this may cause for readers who are used to other symbols. In a book project like this it is impossible to avoid conflicts of notation in the sense that the same symbol may be used for different quantities. We hope that it will always be clear from the context what the notation refers to. In addition, we have collected a number of conventions, abbreviations and symbols after this Preface.

We have tried to be fairly exhaustive in citing references close to the text, but it is obvious that such a system involves a number of inconsistencies and omissions, for which we apologize to the reader and to the authors of the many papers that ought to have been on the list.

We intend to keep a list of misprints and remarks posted on the web page

http://www.hec.unil.ch/halbrecher/rp2.html

and we are grateful to get relevant material sent by email to

hansjoerg.albrecher@unil.ch

Finally, we would like to thank Corina Constantinescu, Hans Gerber, Peter Glynn, Dominik Kortschak, Ronnie Loeffen, Stefan Thonhauser and Hailiang Yang for discussions and proofreading parts of the manuscript, and Dominik Kortschak for help with some figures and general LaTeX issues.

Most of all, we would like to thank our wives May Lise and Renate for their support and patience during the writing of this book.

Aarhus and Lausanne, May 2010

Søren Asmussen   Hansjörg Albrecher

PREFACE

Notation and conventions

Numbering and reference system

The chapter number is specified only when it is not the current one. Thus Proposition 4.2, formula (5.3) or Section 3 of Chapter VI are referred to as Proposition VI.4.2, formula VI.(5.3) and Section VI.3 (or just VI.3), respectively, in all other chapters whereas in VI we just write Proposition 4.2, formula (5.3) or Section 3. References like Proposition A.4, (A.29) refer to the Appendix.

Throughout the book, [APQ] refers to the first author's earlier book *Applied Probability and Queues*, reference [69].

Abbreviations

a.s. almost surely

c.d.f. cumulative distribution function $\mathbb{P}(X \leq x)$

c.g.f. cumulant generating function, i.e. $\log \widehat{B}[s]$ where $\widehat{B}[s]$ is the m.g.f.

IDE integro-differential equation

i.i.d. independent identically distributed

i.o. infinitely often

l.h.s. left-hand side (of equation)

m.g.f. moment generating function, see under $\widehat{B}[s]$ below.

ODE ordinary differential equation

r.h.s. right-hand side (of equation)

r.v. random variable

s.c.v. squared coefficient of variation, $\mathbb{E}X^2/(\mathbb{E}X)^2$.

w.r.t. with respect to

w.p. with probability

Mathematical notation

$\mathbb{P}$ probability.

$\mathbb{E}$ expectation.

$\sim$ Used in asymptotic relations to indicate that the ratio between two expressions is 1 in the limit. E.g. $n! \sim \sqrt{2\pi}\, n^{n+1/2}\mathrm{e}^{-n}$, $n \to \infty$.

$\approx$ A different type of asymptotics: less precise, say a heuristic approximation, or a more precise one like $\mathrm{e}^h \approx 1 + h + h^2/2$, $h \to 0$.

$\overset{\log}{\sim}$ Used in asymptotic relations to indicate that the ratio between logarithms of two expressions is 1.

$\prec_{\mathrm{st}}$ stochastic order.

$\prec_{\mathrm{cx}}$ convex order.

$\prec_{\mathrm{icx}}$ increasing convex order (i.e. stop-loss order).

$\prec_{\mathrm{sm}}$ supermodular order.

$\Re(s)$ the real part of a complex number s.

— The same symbol B is used for a probability measure $B(\mathrm{d}x) = \mathbb{P}(X \in \mathrm{d}x)$ and its c.d.f. $B(x) = \mathbb{P}(X \le x) = \int_{-\infty}^{x} B(\mathrm{d}y)$.

$\widehat{B}[r]$ the m.g.f. $\int_{-\infty}^{\infty} \mathrm{e}^{rx} B(\mathrm{d}x)$ of the distribution B. If, as for typical claim size distributions, B is concentrated on $[0, \infty)$, $\widehat{B}[r]$ is always defined if $\Re(r) \le 0$ and sometimes in a larger strip (for example, if $\overline{B}(x) \sim c\mathrm{e}^{-\delta x}$, then for $\Re(r) < \delta$). The Laplace-Stieltjes transform is $\widehat{B}[-s]$.

$\overline{B}(x)$ the tail $1 - B(x) = \mathbb{P}(X > x)$ of B.

$\lambda(x)$ the failure rate of the distribution B, i.e. $\lambda(x) = b(x)/\overline{B}(x)$.

$\|G\|$ the total mass (variation) of a (signed) measure G. In particular, for a probability distribution $\|G\| = 1$, and for a defective probability distribution $\|G\| < 1$.

μ_B the mean $\mathbb{E}X = \int xB(\mathrm{d}x)$ of B.

$\mu_B^{(n)}$ the nth moment $\mathbb{E}X^n = \int x^n B(\mathrm{d}x)$ of B.

$I(A)$ the indicator function of the event A.

$\mathbb{E}[X; A]$ means $\mathbb{E}[XI(A)]$.

$\square$ marks the end of a proof, an example or a remark.

X_{t-} the left limit $\lim_{s \uparrow t} X_s$, i.e. the value just before t.

C^n the space of functions that are n times continuously differentiable.

$D[0, \infty)$ the space of $\mathbb{R}$-valued functions which are right-continuous and have left limits. *Unless otherwise stated, all stochastic processes considered in this book are assumed to have sample paths in this space.* Usually, the processes we consider are piecewise continuous, i.e. only have finitely many jumps in each finite interval. Then the assumption of D-paths just means that we use the convention that the value at each jump epoch is the right limit rather than the left limit.

In the French-inspired literature, often the term 'cadlag' (continues a droite avec limites a gauche) is used for the D-property.

$N(\mu, \sigma^2)$ the normal distribution with mean μ and variance σ^2.

Matrices and vectors

are denoted by bold letters. Usually, matrices have uppercase Roman or Greek letters like $\boldsymbol{T}$, $\boldsymbol{\Lambda}$, row vectors have lowercase Greek letters like $\boldsymbol{\alpha}$, $\boldsymbol{\pi}$, and column vectors have lowercase Roman letters like $\boldsymbol{t}$, $\boldsymbol{a}$. In particular:

$\boldsymbol{I}$ is the identity matrix.

$\boldsymbol{e}$ is the column vector with all entries equal to 1.

$\boldsymbol{e}_i$ is the ith unit column vector, i.e. the ith entry is 1 and all other 0.

(the dimension is usually clear from the context and left unspecified in the notation). For a given set $x_1, \ldots, x_p$ of numbers,

$(x_i)_{\mathrm{diag}}$ denotes the diagonal matrix with the x_i on the diagonal

$(x_i)_{\mathrm{row}}$ denotes the row vector with the x_i as components

$(x_i)_{\mathrm{col}}$ denotes the column vector with the x_i as components.

T is the transposition operator acting on vectors or matrices. E.g, the ith unit row vector is $\boldsymbol{e}_i^\mathsf{T}$.

Special notation for risk processes

A_t the total claims up to time t.

β the arrival intensity (when the arrival process is Poisson). Notation like β_i and $\beta(t)$ in Chapter VII has a similar, though slightly more complicated, intensity interpretation.

B the claim size distribution. Notation like B_i and $B^{(t)}$ in Chapter VII has a similar, though slightly more complicated, interpretation.

B_0 the stationary excess (integrated tail) distribution of B.

B_t Brownian motion.

δ the rate parameter of B for the exponential case $\overline{B}(x) = \mathrm{e}^{-\delta x}$. Alternatively, at some places the discount rate for time-dependent considerations.

η the safety loading, cf. I.1.

γ The adjustment coefficient (for negative drift, the positive solution of $\kappa(\gamma) = 0$).

γ_δ The adjustment coefficient in a time-dependent context (for negative drift, the smallest positive solution of $\kappa(\alpha) = \delta$).

$\kappa(\alpha)$ the c.g.f. of the increment distribution in the context of discrete-time random walks; for continuous-time processes with stationary and independent increments, $\kappa(\alpha)$ is the Lévy exponent as defined in III.(3.5). For Markov additive processes, the corresponding extension is discussed in II.4 and VI.3b. In Section XVI.1, an adaptation for processes with dependent increments is used.

$m(u)$ the Gerber-Shiu function for initial capital u.

ν The Lévy measure. At some places also the rate parameter of an exponential distribution.

ϕ Depending on the context it is sometimes used as a symbol for a m.g.f. and sometimes as the survival probability $\phi(u) = 1 - \psi(u)$.

$\psi(u)$ the ruin probability for initial capital u.

$\psi(u, T)$ the ruin probability up to time T for initial capital u.

$\widehat{\psi}[-s]$ the Laplace transform $\int_0^\infty \mathrm{e}^{-su} \psi(u)\, \mathrm{d}u$ of the ruin probability.

R_t the risk reserve process at time t.

ρ the net amount $\beta \mu_B$ of claims per unit time, or quantities with a similar time average interpretation, cf. I.1.

ρ_δ The absolute value of the largest non-positive solution of the time-dependent Lundberg equation (for negative drift, $\kappa(-\rho_\delta) = \delta$).

S_t the claim surplus process (i.e. aggregate loss process) $u - R_t$ at time t.

$\tau(u)$ the time of ruin for initial capital u.

u usually the initial capital (in the chapter on control, u denotes the control strategy).

W_t sometimes an alternative notation for B_t, i.e. Brownian motion.

$\xi(u)$ the deficit at ruin of the process R_t starting in u (or, equivalently, the overshoot over u of S_t).

$\mathbb{P}_L$, $\mathbb{E}_L$ the probability measure and its corresponding expectation corresponding to the exponential change of measure given by Lundberg conjugation, cf. e.g. IV.5, VII.5.

Chapter I

Introduction

1 The risk process

In this chapter, we give a very brief summary of some of the models, results and topics to be studied in the rest of the book, and some terminology is introduced.

A *risk reserve process* $\{R_t\}_{t\geq 0}$, as defined in broad terms, is a model for the time evolution of the reserves of an insurance company. We denote throughout the initial reserve by $u = R_0$. The probability $\psi(u)$ of ultimate ruin is the probability that the reserve ever drops below zero,

$$\psi(u) \;=\; \mathbb{P}\Big(\inf_{t\geq 0} R_t < 0\Big) \;=\; \mathbb{P}\Big(\inf_{t\geq 0} R_t < 0 \,\Big|\, R_0 = u\Big).$$

The probability of ruin before time T is

$$\psi(u,T) \;=\; \mathbb{P}\Big(\inf_{0\leq t\leq T} R_t < 0\Big).$$

We also refer to $\psi(u)$ and $\psi(u,T)$ as ruin probabilities with infinite horizon and finite horizon, respectively. *They are the main topics of study of the present book.*

For mathematical purposes, it is frequently more convenient to work with the *claim surplus process* (also called *aggregate loss process*) $\{S_t\}_{t\geq 0}$ defined by $S_t = u - R_t$. Letting

$$\tau(u) \;=\; \inf\{t \geq 0 : R_t < 0\} \;=\; \inf\{t \geq 0 : S_t > u\}, \tag{1.1}$$

$$M \;=\; \sup_{0\leq t<\infty} S_t, \qquad M_T \;=\; \sup_{0\leq t\leq T} S_t, \tag{1.2}$$

be the time to ruin and the maxima with infinite and finite horizon, respectively, the ruin probabilities can then alternatively be written as

$$\psi(u) \quad = \quad \mathbb{P}(\tau(u) < \infty) \ = \ \mathbb{P}(M > u), \tag{1.3}$$

$$\psi(u,T) \quad = \quad \mathbb{P}(\tau(u) \leq T) \ = \ \mathbb{P}(M_T > u). \tag{1.4}$$

So far we have not imposed any assumptions on the risk reserve process. However, the following set-up will cover a main part of the book:

- There are only finitely many claims in finite time intervals. That is, the number N_t of arrivals in $[0,t]$ is finite. We denote the interarrival times of claims by $T_2, T_3, \ldots$ and T_1 is the time of the first claim. Thus, the time of arrival of the nth claim is $\sigma_n = T_1 + \cdots + T_n$, and $N_t = \min\{n \geq 0 : \sigma_{n+1} > t\} = \max\{n \geq 0 : \sigma_n \leq t\}$.

- The size of the nth claim is denoted by U_n.

- Premiums flow in at rate p, say, per unit time.

Putting things together, we see that

$$R_t \quad = \quad u + pt - \sum_{k=1}^{N_t} U_k, \quad S_t = \sum_{k=1}^{N_t} U_k - pt. \tag{1.5}$$

The sample paths of $\{R_t\}$ and $\{S_t\}$ and the connection between the two processes are illustrated in Fig. I.1.

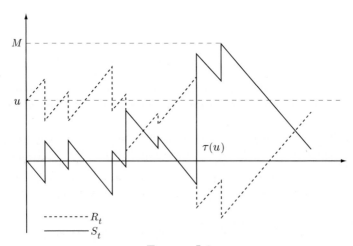

FIGURE I.1

Note that it is a matter of taste (or mathematical convenience) whether one allows $\{R_t\}$ and/or $\{S_t\}$ to continue its evolution after the time $\tau(u)$ of ruin. Thus, for example, one could well replace R_t by $R_{t\wedge\tau(u)}$ or $R_{t\wedge\tau(u)} \vee 0$. For the purpose of studying ruin probabilities this distinction is, of course, immaterial.

Some main examples of models not incorporated in the above set-up are:

- Models which are non-homogeneous in space, for example with a premium depending on the reserve (i.e on Fig. I.1 the slope of $\{R_t\}$ should depend also on the level). We study this case in Chapter VIII.

- Brownian motion or more general diffusions. Traditionally, Brownian motion has mainly been used as an approximation to the risk process rather than as a model of intrinsic merit and we look at this in Chapter V. However, since any modeling involves some approximative assumptions, it has (partly inspired from the modeling in mathematical finance) become more and more common to use Brownian motion as an intrinsically reasonable model.

- General Lévy processes (defined as continuous time processes with stationary independent increments) where the jump component has infinite Lévy measure, allowing a countable infinity of jumps on Fig. I.1. We treat Lévy processes in Chapter XI.

The models we consider will often have the property that there exists a constant ρ such that

$$\frac{1}{t}\sum_{k=1}^{N_t} U_k \stackrel{\text{a.s.}}{\to} \rho, \quad t \to \infty. \tag{1.6}$$

The interpretation of ρ is as the average amount of claim per unit time. A further basic quantity is the *safety loading* (or the *security loading*) η defined as the relative amount by which the premium rate p exceeds ρ,

$$\eta = \frac{p - \rho}{\rho}.$$

It is sometimes stated in the theoretical literature that the typical values of the safety loading η are relatively small, say $10\% - 20\%$; we shall, however, not discuss whether this actually corresponds to practice. It would appear obvious, however, that the insurance company should try to ensure $\eta > 0$, and in fact:

Proposition 1.1 *Assume that* (1.6) *holds. If* $\eta < 0$, *then* $M = \infty$ *a.s. and hence* $\psi(u) = 1$ *for all* u. *If* $\eta > 0$, *then* $M < \infty$ *a.s. and hence* $\psi(u) < 1$ *for all sufficiently large* u.

Proof. It follows from (1.6) that

$$\frac{S_t}{t} = \frac{\sum_{k=1}^{N_t} U_k - pt}{t} \overset{\text{a.s.}}{\to} \rho - p, \quad t \to \infty.$$

If $\eta < 0$, then this limit is > 0 which implies $S_t \overset{\text{a.s.}}{\to} \infty$ and hence $M = \infty$ a.s. If $\eta > 0$, then similarly $\lim S_t/t < 0$, $S_t \overset{\text{a.s.}}{\to} -\infty$, $M < \infty$ a.s. $\square$

In concrete models, we obtain typically a somewhat stronger conclusion, namely that $M = \infty$ a.s., $\psi(u) = 1$ for all u holds also when $\eta = 0$, and that $\psi(u) < 1$ for *all* $u > 0$ when $\eta > 0$. However, this needs to be verified in each separate case.

The simplest concrete example (to be studied in Chapter IV) is the *Cramér-Lundberg* or *compound Poisson* model, where $\{N_t\}$ is a Poisson process with rate β (say) and $U_1, U_2, \ldots$ are i.i.d. and independent of $\{N_t\}$. Here it is easy to see that $\rho = \beta \mathbb{E} U$ (on the average, β claims arrive per unit time and the mean of a single claim is $\mathbb{E} U$) and that also

$$\lim_{t \to \infty} \mathbb{E} \frac{1}{t} \sum_{k=1}^{N_t} U_k = \rho. \tag{1.7}$$

Again, (1.7) is a property which we will typically encounter. However, not all models considered in the literature have this feature:

Example 1.2 (COX PROCESSES) Here $\{N_t\}$ is a Poisson process with *random rate* $\beta(t)$ (say) at time t. If $U_1, U_2, \ldots$ are i.i.d. and independent of $\{(\beta(t), N_t)\}$, it is not too difficult to show that ρ as defined by (1.6) is given by

$$\rho = \mathbb{E} U \cdot \lim_{t \to \infty} \frac{1}{t} \int_0^t \beta(s) \, \mathrm{d}s$$

(provided the limit exists). Thus ρ may well be random for such processes, namely, if $\{\beta(t)\}$ is non-ergodic. The simplest example is $\beta(t) \equiv V$ where V is a r.v. This case is referred to as the *mixed Poisson process*, with the most notable special case being V having a Gamma distribution, corresponding to the *Pólya process*. $\square$

We shall only encounter a few instances of a Cox process, in connection with risk processes in a Markovian or periodic environment (Chapter VII), and here (1.6), (1.7) hold with ρ constant.

Proposition 1.3 *Assume $p \neq 1$ and define $\widetilde{R}_t = R_{t/p}$. Then the connection between the ruin probabilities for the given risk process $\{R_t\}$ and those $\widetilde{\psi}(u)$, $\widetilde{\psi}(u, T)$ for $\{\widetilde{R}_t\}$ is given by*

$$\psi(u) = \widetilde{\psi}(u), \quad \psi(u, T) = \widetilde{\psi}(u, Tp).$$

The proof is trivial. Since $\{\widetilde{R}_t\}$ has premium rate 1, the role of the result is to justify taking $p = 1$, which is feasible since in most cases the process $\{\widetilde{R}_t\}$ has a similar structure as $\{R_t\}$ (for example, the claim arrivals are Poisson or renewal at the same time). Note that when $p = 1$, the assumption $\eta > 0$ is equivalent to $\rho < 1$; in a number of models, we shall be able to identify ρ with the traffic intensity of an associated queue, and in fact $\rho < 1$ is the fundamental assumption of queueing theory ensuring steady-state behavior (existence of a limiting stationary distribution).

Notes and references The study of ruin probabilities, often referred to as *collective risk theory* or just *risk theory*, was largely initiated in Sweden in the first half of the century. Some of the main general ideas were laid down by Lundberg [614], while the first mathematically substantial results were given in Lundberg [615] and Cramér [265]; another important early Swedish work is Täcklind [826]. The Swedish school was pioneering not only in risk theory, but also in probability and applied probability as a whole; in particular, many results and methods in random walk theory originate from there and the area was ahead of related ones like queueing theory.

Some early surveys are given in Cramér [265], Segerdahl [792] and Philipson [699]. Some main later textbooks are (in alphabetical order) Bühlmann [208], Dickson [309], Daykin, Pentikäinen & Pesonen [279], De Vylder [300], Gerber [398], Grandell [429], Rolski, Schmidli, Schmidt & Teugels [746] and Seal [784, 788]. Besides in standard journals in probability and applied probability, the research literature is often published in journals like *Astin Bulletin, Insurance: Mathematics and Economics*, the *North American Actuarial Journal*, the *Scandinavian Actuarial Journal* and *Mitteilungen der Schweizerischen Aktuarvereinigung*. Note that the latter has recently been merged with *Blätter der Deutschen Gesellschaft für Versicherungs- und Finanzmathematik* and a number of further Actuarial Bulletins of European countries into *The European Actuarial Journal*.

The term *risk theory* is often interpreted in a broader sense than as just to comprise the study of ruin probabilities. An idea of the additional topics and problems one may incorporate under risk theory can be obtained from the survey paper [665] by Norberg; see also Chapter XVI. In the even more general area of non-life insurance mathematics, some main texts (typically incorporating some ruin theory but emphasizing the topic to a varying degree) are Bowers *et al.* [195], Bühlmann [208], Daykin *et al.* [279], Embrechts *et al.* [349], Heilmann [458], Hipp & Michel [468], Kaas *et al.* [515], Klugman, Panjer & Willmot [536], Mikosch [638], Schmidt [782], Straub [818], Sundt

[820] and Taylor [840]. Note that life insurance (e.g. Gerber [402]) has a rather different flavor, and we do not get near to the topic anywhere in this book.

Cox processes are treated extensively in Grandell [429]. For mixed Poisson processes and Pólya processes, see e.g. the recent survey by Grandell [431] and references therein.

2 Claim size distributions

This section contains a brief survey of some of the most popular classes of distributions B which have been used to model the claims $U_1, U_2, \ldots$ We roughly classify these into two groups, *light-tailed distributions* (sometimes the term 'Cramér-type conditions' is used), and *heavy-tailed distributions*. Here light-tailed means that the tail $\overline{B}(x) = 1 - B(x)$ satisfies $\overline{B}(x) = \mathrm{O}(\mathrm{e}^{-sx})$ for some $s > 0$. Equivalently, the m.g.f. $\widehat{B}[s]$ is finite for some $s > 0$. In contrast, B is *heavy-tailed* if $\widehat{B}[s] = \infty$ for all $s > 0$, but different more restrictive definitions are often used: subexponential, regularly varying (see below) or even regularly varying with infinite variance. On the more heuristical side, one could mention also the folklore in actuarial practice to consider B heavy-tailed if '20% of the claims account for more than 80% of the total claims', i.e. if

$$\frac{1}{\mu_B} \int_{b_{0.2}}^{\infty} x\, B(\mathrm{d}x) \;\geq\; 0.8,$$

where $\overline{B}(b_{0.2}) = 0.2$ and μ_B is the mean of B.

2a Light-tailed distributions

Example 2.1 (THE EXPONENTIAL DISTRIBUTION) Here the density is

$$b(x) = \delta \mathrm{e}^{-\delta x}. \tag{2.1}$$

The parameter δ is referred to as the *rate* or the *intensity*, and can also be interpreted as the (constant) failure rate $b(x)/\overline{B}(x)$.

As in a number of other applied probability areas, the exponential distribution is by far the simplest to deal with in risk theory as well. In particular, for the compound Poisson model with exponential claim sizes the ruin probability $\psi(u)$ can be found in closed form. The crucial feature is the *lack of memory*: if U is exponential with rate δ, then the conditional distribution of $U - x$ given $U > x$ is again exponential with rate δ (this is essentially equivalent to the failure rate being constant). For example in the compound Poisson model, a simple stopping time argument shows that this implies that the conditional distribution of the overshoot $S_{\tau(u)} - u$ at the time of ruin given $\tau(u)$ is again exponential with rate δ, a fact which turns out to contain considerable information. □

Example 2.2 (THE GAMMA DISTRIBUTION) The gamma distribution with parameters p, δ has density

$$b(x) = \frac{\delta^p}{\Gamma(p)} x^{p-1} e^{-\delta x} \tag{2.2}$$

and m.g.f.

$$\widehat{B}[s] = \left(\frac{\delta}{\delta - s}\right)^p, \quad s < \delta.$$

The mean $\mathbb{E}U$ is p/δ and the variance $\mathbb{V}ar\, U$ is p/δ^2. In particular, the squared coefficient of variation (s.c.v.)

$$\frac{\mathbb{V}ar\, U}{(\mathbb{E}U)^2} = \frac{1}{p}$$

is < 1 for $p > 1$, > 1 for $p < 1$ and $= 1$ for $p = 1$ (the exponential case).

The exact form of the tail $\overline{B}(x)$ is given by the incomplete Gamma function $\Gamma(x; p)$,

$$\overline{B}(x) = \frac{\Gamma(\delta x; p)}{\Gamma(p)} \quad \text{where} \quad \Gamma(x; p) = \int_x^\infty t^{p-1} e^{-t} \, dt.$$

Asymptotically, one has

$$\overline{B}(x) \sim \frac{\delta^{p-1}}{\Gamma(p)} x^{p-1} e^{-\delta x}.$$

In the sense of the theory of infinitely divisible distributions, the Gamma density (2.2) can be considered as the pth power of the exponential density (2.1) (or the $1/p$th root if $p < 1$). In particular, if p is integer and U has the gamma distribution p, δ, then $U \overset{\mathscr{D}}{=} X_1 + \cdots + X_p$ where $X_1, X_2, \ldots$ are i.i.d. and exponential with rate δ. This special case is referred to as the *Erlang distribution with p stages*, or just the Erlang(p) distribution. An appealing feature is its simple connection to the Poisson process: $\overline{B}(x) = \mathbb{P}(U_1 + \cdots + U_p > x)$ is the probability of at most $p - 1$ Poisson events in $[0, x]$ so that

$$\overline{B}(x) = \sum_{i=0}^{p-1} e^{-\delta x} \frac{(\delta x)^i}{i!}.$$

In the present text, we develop computationally tractable results mainly for the Erlang case (i.e. $p \in \mathbb{N}$). Ruin probabilities for the general case have been studied, among others, by Grandell & Segerdahl [433] and Thorin [847]. □

Example 2.3 (THE HYPEREXPONENTIAL DISTRIBUTION) This is defined as a finite mixture of exponential distributions,

$$b(x) = \sum_{i=1}^{p} \alpha_i \delta_i e^{-\delta_i x} \tag{2.3}$$

where $\sum_{1}^{p} \alpha_i = 1$, $0 \le \alpha_i \le 1$, $i = 1, \ldots, p$. An important property of the hyperexponential distribution is that its s.c.v. is > 1.

If $\alpha_i \in \mathbb{R}$, then one speaks of the distribution as a *combination of exponentials* and this class is dense in the set of all distributions on the positive halfline. □

Example 2.4 (PHASE-TYPE DISTRIBUTIONS) A phase-type distribution is the distribution of the absorption time in a Markov process with finitely many states, of which one is absorbing and the rest transient. Important special cases are the exponential, the Erlang and the hyperexponential distributions. *This class of distributions plays a major role in this book as one within computationally tractable exact forms of the ruin probability $\psi(u)$ can be obtained.*

The parameters of a phase-type distribution are the set E of transient states, the restriction $\boldsymbol{T}$ of the intensity matrix of the Markov process to E and the row vector $\boldsymbol{\alpha} = (\alpha_i)_{i \in E}$ of initial probabilities. The density and c.d.f. are

$$b(x) = \boldsymbol{\alpha} e^{\boldsymbol{T} x} \boldsymbol{t}, \quad \text{resp. } B(x) = \boldsymbol{\alpha} e^{\boldsymbol{T} x} \boldsymbol{e}, \qquad x \ge 0,$$

where $\boldsymbol{t} = \boldsymbol{T} \boldsymbol{e}$ and $\boldsymbol{e} = (1 \ldots 1)^{\mathsf{T}}$ is the column vector with 1 at all entries. The couple $(\boldsymbol{\alpha}, \boldsymbol{T})$ or sometimes the triple $(E, \boldsymbol{\alpha}, \boldsymbol{T})$ is called the *representation*. We give a more comprehensive treatment in IX.1 and defer further details to Chapter IX. □

Example 2.5 (DISTRIBUTIONS WITH RATIONAL TRANSFORMS) A distribution B has a rational m.g.f. (or, equivalently, a rational Laplace transform) if $\widehat{B}[r] = p(r)/q(r)$ with $p(r)$ and $q(r)$ being polynomials of finite degree. An equivalent characterization is that the density $b(x)$ is the solution of a homogeneous ordinary differential equation with constant coefficients

$$b^{(q)}(x) + d_{q-1} b^{(q-1)}(x) + \cdots + d_0 = 0; \quad d_j \in \mathbb{R}, d_0 \ne 0,$$

where one of the initial conditions is determined by $\int_0^\infty b(x) \, \mathrm{d}x = 1$. Consequently the density $b(x)$ has one of the forms

$$b(x) = \sum_{j=0}^{q} c_j x^j e^{\eta_j x}, \tag{2.4}$$

$$b(x) = \sum_{j=0}^{q_1} c_j x^j e^{\eta_j x} + \sum_{j=0}^{q_2} d_j x^j \cos(a_j x) e^{\delta_j x} + \sum_{j=0}^{q_3} e_j x^j \sin(b_j x) e^{\epsilon_j x}, \tag{2.5}$$

where the parameters in (2.4) are possibly complex-valued but the parameters in (2.5) are real-valued. This class of distributions is popular in the literature on both risk theory and queues, but often the attention is restricted to the class of phase-type distributions, which is slightly smaller but more amenable to probabilistic reasoning. We give some theory for matrix-exponential distributions in IX.6. □

Example 2.6 (DISTRIBUTIONS WITH BOUNDED SUPPORT) This example (i.e. there exists a $x_0 < \infty$ such that $\overline{B}(x) = 0$ for $x \geq x_0$, $\overline{B}(x) > 0$ for $x < x_0$) is of course a trivial instance of a light-tailed distribution. However, it is notable from a practical point of view because of reinsurance: if excess-of-loss reinsurance has been arranged with retention level x_0, then the claim size which is relevant from the point of view of the insurance company itself is $U \wedge x_0$ rather than U (the excess $(U - x_0)^+$ is covered by the reinsurer). See XVI.6. □

2b Heavy-tailed distributions

Example 2.7 (THE WEIBULL DISTRIBUTION) This distribution originates from reliability theory. Here failure rates $\delta(x) = b(x)/\overline{B}(x)$ play an important role, the exponential distribution representing the simplest example since here $\delta(x)$ is constant. However, in practice one may observe that $\delta(x)$ is either decreasing or increasing and may try to model smooth (increasing or decreasing) deviations from constancy by $\delta(x) = dx^{r-1}$ $(0 < r < \infty)$. Writing $c = d/r$, we obtain the Weibull distribution

$$\overline{B}(x) = e^{-cx^r}, \quad b(x) = crx^{r-1}e^{-cx^r}, \tag{2.6}$$

which is heavy-tailed when $0 < r < 1$. All moments are finite. Another interpretation is that it is the distribution of $X^{1/r}$, where X is exponential with parameter c. □

Example 2.8 (THE LOGNORMAL DISTRIBUTION) The *lognormal distribution* with parameters σ^2, μ is defined as the distribution of e^V where $V \sim N(\mu, \sigma^2)$, or equivalently as the distribution of $e^{\sigma W + \mu}$ where $W \sim N(0, 1)$. It follows that the density is

$$
\begin{aligned}
b(x) &= \frac{d}{dx} \Phi\left(\frac{\log x - \mu}{\sigma}\right) = \frac{1}{\sigma x} \varphi\left(\frac{\log x - \mu}{\sigma}\right) \\
&= \frac{1}{x\sigma\sqrt{2\pi}} \exp\left\{-\frac{1}{2}\left(\frac{\log x - \mu}{\sigma}\right)^2\right\}.
\end{aligned}
\tag{2.7}
$$

Asymptotically, the tail is

$$\overline{B}(x) \sim \frac{\sigma}{\log x \sqrt{2\pi}} \exp\left\{-\frac{1}{2}\left(\frac{\log x - \mu}{\sigma}\right)^2\right\}, \tag{2.8}$$

which is heavier than the one of the Weibull distribution. The lognormal distribution has moments of all orders. In particular, the mean is $e^{\mu+\sigma^2/2}$ and the second moment is $e^{2\mu+2\sigma^2}$.   □

Example 2.9 (THE PARETO DISTRIBUTION) Here the essence is that the tail $\overline{B}(x)$ decreases like a power of x. There are various variants of the definition around, the simplest one being

$$\overline{B}(x) = x^{-\alpha}, \quad x \geq 1, \tag{2.9}$$

which can be interpreted as the distribution of e^X for an exponential r.v. X with parameter α. Another variant is often referred to as US-Pareto and defined by

$$\overline{B}(x) = \frac{a^\alpha}{(a+x)^\alpha}, \quad b(x) = \frac{\alpha a^\alpha}{(a+x)^{\alpha+1}}, \quad x \geq 0, \tag{2.10}$$

for some $a > 0$. The pth moment is finite if and only if $p < \alpha - 1$.

The Laplace-Stieltjes transform of the Pareto distribution defined in (2.9) can be expressed through the incomplete Gamma function by

$$\widehat{B}[-s] = \int_1^\infty e^{-sx} \frac{\alpha}{x^{\alpha+1}} \, dx = \alpha \, s^\alpha \Gamma(-\alpha, s).$$

Similarly, the Laplace-Stieltjes transform of the US Pareto distribution is $\widehat{B}[-s] = \alpha \, (as)^\alpha e^{as} \Gamma(-\alpha, as)$. These relatively simple expressions have not always been noted.

Abate, Choudhury & Whitt [1] introduced a somewhat related class of random variables called Pareto mixture of exponentials, which are products of Pareto and exponential r.v.'s and lead to quite explicit Laplace-Stieltjes transforms.   □

Example 2.10 (THE LOGGAMMA DISTRIBUTION) The *loggamma distribution with parameters* p, δ is defined as the distribution of e^V where V has the gamma density (2.2). The density is

$$b(x) = \frac{\delta^p (\log x)^{p-1}}{x^{\delta+1} \Gamma(p)}. \tag{2.11}$$

The pth moment is finite if $p < \delta$ and infinite if $p > \delta$. For $p = 1$, the loggamma distribution is a Pareto distribution.   □

Example 2.11 (DISTRIBUTIONS WITH REGULARLY VARYING TAILS) The tail $\overline{B}(x)$ of a distribution B is said to be *regularly varying with index* α if

$$\overline{B}(x) \sim \frac{L(x)}{x^\alpha}, \quad x \to \infty, \tag{2.12}$$

where $L(x)$ is *slowly varying*, i.e. satisfies $L(xt)/L(x) \to 1$ as $x \to \infty$ (any L having a limit in $(0, \infty)$ is slowly varying; another standard example is $(\log x)^\eta$). Thus, examples of distributions with regularly varying tails are the Pareto distribution (2.10) (here $L(x) \to 1$), the loggamma distribution (with index δ) and a Pareto mixture of exponentials. □

Example 2.12 (THE SUBEXPONENTIAL CLASS OF DISTRIBUTIONS) We say that a distribution B is subexponential if

$$\lim_{x \to \infty} \frac{\overline{B^{*2}}(x)}{\overline{B}(x)} = 2. \tag{2.13}$$

It can be proved (see X.1) that any distribution with a regularly varying tail is subexponential. Also, for example the lognormal distribution is subexponential (but not regularly varying), though the proof of this is non-trivial, and so is the Weibull distribution with $0 < r < 1$. Thus, the subexponential class of distributions provide a convenient framework for studying large classes of heavy-tailed distributions. We return to a closer study in X.1. □

When studying ruin probabilities, it will be seen that we obtain completely different results depending on whether the claim size distribution is exponentially bounded or heavy-tailed. From a practical point of view, this phenomenon represents one of the true controversies of the area. Namely, the knowledge of the claim size distribution will typically be based upon statistical data, and based upon such information it seems questionable to extrapolate to tail behavior. However, one may argue that this difficulty is not restricted to ruin probability theory alone. Similar discussion applies to the distribution of the accumulated claims (XVI.2) or even to completely different applied probability areas like extreme value theory: if we are using a Gaussian process to predict extreme value behavior, we may know that such a process (with a covariance function estimated from data) is a reasonable description of the behavior of the system under study in typical conditions, but can never be sure whether this is also so for atypical levels for which far less detailed statistical information is available. We give some discussion on standard methods to distinguish between light and heavy tails in Chapter X.

3 The arrival process

For the purpose of modeling a risk process, the claim size distribution represents of course only one aspect (though a major one). At least as important is the specification of the structure of the point process $\{N_t\}$ of claim arrivals and its possible dependence with the claims.

By far the most prominent case is the compound Poisson (Cramér-Lundberg) model where $\{N_t\}$ is Poisson and independent of the claim sizes $U_1, U_2, \ldots$ The reason is in part mathematical since this model is the easiest to analyze, but the model also admits a natural interpretation: a large portfolio of insurance holders, which each have a (time-homogeneous) small rate of experiencing a claim, gives rise to an arrival process which is very close to a Poisson process, in just the same way as the Poisson process arises in telephone traffic (a large number of subscribers each calling with a small rate), radioactive decay (a huge number of atoms each splitting with a tiny rate) and many other applications. The compound Poisson model is studied in detail in Chapters IV, V (and, with the extension to premiums depending on the reserve, in Chapter VIII).

To the authors' knowledge, not so many detailed studies of the goodness-of-fit of the Poisson model in insurance are available. Some of them have concentrated on the marginal distribution of N_T (say $T = $ one year), found the Poisson distribution to be inadequate and suggested various other univariate distributions as alternatives, e.g. the negative binomial distribution. The difficulty in such an approach lies in that it may be difficult or even impossible to imbed such a distribution into the continuous set-up of $\{N_t\}$ evolving over time, and also that the ruin problem may be hard to analyze. Nevertheless, getting away from the simple Poisson process seems a crucial step in making the model more realistic, in particular to allow for certain inhomogeneities.

Historically, the first extension to be studied in detail was $\{N_t\}$ to be renewal (the interarrival times $T_1, T_2, \ldots$ are i.i.d. but with a general not necessarily exponential distribution). This model, to be studied in Chapter VI, has some mathematically appealing random walk features, which facilitate the analysis. However, it is more questionable whether it provides a model with a similar intuitive content as the Poisson model. One could possibly argue that renewal models are a compromise between choosing a tractable model and taking into account statistical information that may indicate that exponential interarrival time distributions do not calibrate given data well enough. Of course, one is then still left to believe in the independence assumption and – with the introduced memory between claims – one has to be aware that the resulting model is for most applications to be seen as an interpolation rather than a causal model.

A more appealing way to allow for inhomogeneity is by means of an intensity $\beta(t)$ fluctuating over time. An obvious example is $\beta(t)$ depending on the time of the year (the season), so that $\beta(t)$ is a periodic function of t; we study this case in VII.6. Another one is Cox processes, where $\{\beta(t)\}_{t \geq 0}$ is an arbitrary stochastic process. In order to prove reasonably substantial and interesting results, Cox processes are, however, too general and one needs to specialize to more concrete assumptions. The one we focus on (Chapter VII) is a Markovian environment: the environmental conditions are described by a finite Markov process $\{J_t\}_{t \geq 0}$, such that $\beta(t) = \beta_i$ when $J_t = i$. I.e. with a common term $\{N_t\}$ is a *Markov-modulated Poisson process*; its basic feature is to allow more variation (bursty arrivals) than inherent in the simple Poisson process. This model can be intuitively understood in some simple cases like $\{J_t\}$ describing weather conditions in car insurance, epidemics in life insurance etc. In others, it may be used in a purely descriptive way when it is empirically observed that the claim arrivals are more bursty than allowed for by the simple Poisson process.

Mathematically, the periodic and the Markov-modulated models also have attractive features. The point of view we take here is Markov-dependent random walks in continuous time (Markov additive processes), see III.4. This applies also to the case where the claim size distribution depends on the time of the year or the environment (VII.6), and which seems well motivated from a practical point of view as well.

4   A summary of main results and methods

4a   Duality with other applied probability models

Risk theory may be viewed as one of many applied probability areas, others being branching processes, genetics models, queueing theory, dam/storage processes, reliability, interacting particle systems, stochastic differential equations, time series and Gaussian processes, extreme value theory, stochastic geometry, point processes and so on. Some of these have a certain resemblance in flavor and methodology, others are quite different.

The ones which appear most related to risk theory are queueing theory and dam/storage processes. In fact, it is a recurrent theme of this book to stress this connection which was often neglected in the early specialized literature on risk theory. Mathematically, the classical result is that the ruin probabilities for the compound Poisson model are related to the workload (virtual waiting time) process $\{V_t\}_{t \geq 0}$ of an initially empty M/G/1 queue by means of

$$\psi(u, T) = \mathbb{P}(V_T > u), \quad \psi(u) = \mathbb{P}(V > u), \qquad (4.1)$$

where V is the limit in distribution of V_t as $t \to \infty$. The M/G/1 workload process $\{V_t\}$ may also be seen as one of the simplest storage models, with Poisson arrivals and constant release rule $p(x) \equiv 1$. A general release rule $p(x)$ means that $\{V_t\}$ decreases according to the differential equation $\dot{V} = -p(V)$ in between jumps, and here (4.1) holds as well provided the risk process has a premium rule depending on the reserve, $\dot{R} = p(R)$ in between jumps. Similarly, ruin probabilities for risk processes with an input process which is renewal, Markov-modulated or periodic can be related to queues with similar characteristics. Thus, it is desirable to have a set of formulas like (4.1) permitting to translate freely between risk theory and the queueing/storage setting. In Chapter VIII we will also see a direct and natural link between the maximum workload of an M/G/1 queue and the ruin probability in a compound Poisson risk model in terms of excursions. In general, methods or modeling ideas developed in one area often have relevance for the other one as well.

A stochastic process $\{V_t\}$ is said to be in the *steady state* if it is strictly stationary (in the Markov case, this amounts to V_0 having the stationary distribution of $\{V_t\}$), and the limit $t \to \infty$ is the *steady-state limit*. The study of the steady state is by far the most dominant topic of queueing and storage theory, and a lot of information on steady-state r.v.'s like V is available. It should be noted, however, that quite often the emphasis is on computing expected values like $\mathbb{E}V$. In the setting of (4.1), this gives only $\int_0^\infty \psi(u)\mathrm{d}u$ which is of limited intrinsic interest. Thus, the two areas, though overlapping, have to some extent a different flavor.

A prototype of the duality results in this book is Theorem III.2.1, which gives a sample path version of (4.1) in the setting of a general premium rule $p(x)$: the events $\{V_T > u\}$ and $\{\tau(u) \le T\}$ coincide when the risk process and the storage process are coupled in a suitable way (via time-reversion). The infinite horizon (steady state) case is covered by letting $T \to \infty$. The fact that Theorem III.2.1 is a sample path relation should be stressed: in this way the approach also applies to models having supplementary r.v.'s like the environmental process $\{J_t\}$ in a Markov-modulated setting.

4b   Exact solutions

Of course, the ideal is to be able to come up with closed form solutions for the ruin probabilities $\psi(u)$, $\psi(u,T)$. The cases where this is possible are basically the following for the infinite horizon ruin probability $\psi(u)$:

- The compound Poisson model with constant premium rate $p = 1$ and exponential claim size distribution B, $\overline{B}(x) = \mathrm{e}^{-\delta x}$. Here $\psi(u) = \rho \mathrm{e}^{-\gamma u}$ where β is the arrival intensity, $\rho = \beta/\delta$ and $\gamma = \delta - \beta$.

- The compound Poisson model with constant premium rate $p = 1$ and B being phase-type with just a few phases. Here $\psi(u)$ is given in terms of a matrix-exponential function (Corollary IX.3.1), which can be expanded into a sum of exponential terms by diagonalization (see, e.g., Example IX.3.2). The qualifier 'with just a few phases' refers to the fact that the diagonalization has to be carried out numerically in higher dimensions.

- The compound Poisson model with a claim size distribution degenerate at one point, see Corollary IV.3.7.

- The compound Poisson model with some rather special heavy-tailed claim size distributions, see Boxma & Cohen [193] and Abate & Whitt [3].

- The compound Poisson model with premium rate $p(x)$ depending on the reserve and exponential claim size distribution B. Here $\psi(u)$ is explicit provided that, as is typically the case, the functions

$$\omega(x) = \int_0^x \frac{1}{p(y)} \, \mathrm{d}y, \quad \int_0^\infty \frac{1}{p(x)} e^{\beta\omega(x) - \delta x} \, \mathrm{d}x$$

can be written in closed form, see Corollary VIII.1.9.

- The compound Poisson model with a piecewise constant premium rule $p(x)$ and B being phase-type with just a few phases, see IX.7.

- Renewal models with exponential claim sizes, see Theorem VI.2.2.

- Renewal model variants of the above cases for which the interclaim time is phase-type with just a few phases.

- Any Lévy model where the risk reserve process (not the claim surplus process!) is downward skipfree (Theorem XI.2.3). This includes Brownian motion.

- Any Lévy model for which the scale function is explicitly available, see XI.3 (for an early example cf. Furrer [381]).

A further notable fact (see again XVI.1) is the explicit form of the ruin probability when $\{R_t\}$ is a diffusion with infinitesimal drift and variance $\mu(x), \sigma^2(x)$:

$$\psi(u) = \frac{\int_u^\infty \exp\left\{-\int_0^x 2\mu(y)/\sigma^2(y) \, \mathrm{d}y\right\} \, \mathrm{d}x}{\int_0^\infty \exp\left\{-\int_0^x 2\mu(y)/\sigma^2(y) \, \mathrm{d}y\right\} \, \mathrm{d}x} = 1 - \frac{S(u)}{S(\infty)} \quad (4.2)$$

where

$$S(u) = \int_0^u \exp\left\{-\int_0^x 2\mu(y)/\sigma^2(y) \, \mathrm{d}y\right\} \, \mathrm{d}x$$

is the natural scale.

The finite horizon ruin probability $\psi(u, T)$ is explicit for Brownian motion (III.1) and the compound Poisson model with exponential claim size distribution (V.1). Later in the book a number of further rather specific cases will be discussed for which explicit expressions exist.

4c   Numerical methods

Next to a closed-form solution, the second best alternative is a numerical procedure which allows to calculate the exact values of the ruin probabilities. Here are some of the main approaches:

Laplace transform inversion Often, it is easier to find the Laplace transforms

$$\widehat{\psi}[-s] = \int_0^\infty \mathrm{e}^{-su}\psi(u)\,\mathrm{d}u\,, \quad \widehat{\psi}[-s,-\omega] = \int_0^\infty \int_0^\infty \mathrm{e}^{-su-\omega T}\psi(u,T)\,\mathrm{d}u\,\mathrm{d}T$$

in closed form rather than the ruin probabilities $\psi(u)$, $\psi(u,T)$ themselves. In that case $\psi(u)$, $\psi(u,T)$ can be calculated numerically by some method for transform inversion, say the fast Fourier transform (FFT) as implemented in Grübel [438] for infinite horizon ruin probabilities for the renewal model. We do not discuss Laplace transform inversion much; other relevant references are Abate & Whitt [2], Embrechts, Grübel & Pitts [346] and Grübel & Hermesmeier [439]; see also Albrecher, Avram & Kortschak [14] and the Bibliographical Notes in [746, p. 191].

Matrix-analytic methods This approach is relevant when the claim size distribution is of phase-type (or matrix-exponential), and in quite a few cases (Chapter IX), $\psi(u)$ is then given in terms of a matrix-exponential function $\mathrm{e}^{\boldsymbol{U}u}$ (here $\boldsymbol{U}$ is some suitable matrix) which can be computed by diagonalization, as the solution of linear differential equations or by some series expansion (not necessarily the straightforward $\sum_0^\infty \boldsymbol{U}^n u/n!$ one!). In the compound Poisson model with $p = 1$, $\boldsymbol{U}$ is explicit in terms of the model parameters, whereas for the renewal arrival model and the Markovian environment model $\boldsymbol{U}$ has to be calculated numerically, either as the iterative solution of a fixed point problem or by finding the diagonal form in terms of the complex roots to certain transcendental equations.

Differential- and integral equations The idea here is to express $\psi(u)$ or $\psi(u,T)$ as the solution to a differential- or integral equation, and carry out the solution by some standard numerical method. One example where this is feasible is the renewal equation for $\psi(u)$ (Corollary IV.3.3) in the compound Poisson model which is an integral equation of Volterra type. The method is, however, restricted to models that have a certain degree of Markovian structure in which case conditioning (or applying the more formal tool of generators, see II.4a) leads to equations that often involve both differential and integral terms. We will discuss cases where this approach can even lead to explicit solutions (see e.g. IX.7 and XII.3c). In many more cases, numerical solution methods are applicable, although the initial or boundary conditions can be a challenge.

If an integral equation is available, it is often possible to define a contractive integral operator and to identify the ruin probability as its fixed point, in which case the ruin probability can be approximated by iterated application of the integral operator to some starting function. The resulting high-dimensional integral can then be calculated by standard Monte Carlo and Quasi-Monte Carlo techniques (see e.g. Albrecher *et al.* [31, 24]). In comparison to the alternative of direct simulation of the risk process (as discussed in Section 4g), this technique often has significant computational advantages over the latter.

4d Approximations

The Cramér-Lundberg approximation This is one of the most celebrated results of risk theory (and probability theory as a whole). For the compound Poisson model with $p = 1$ and claim size distribution B with moment generating function (m.g.f.) $\widehat{B}[s]$, it states that

$$\psi(u) \sim Ce^{-\gamma u}, \quad u \to \infty, \tag{4.3}$$

where $C = (1 - \rho)/(\beta\widehat{B}'[\gamma] - 1)$ and $\gamma > 0$ is the solution of the Lundberg equation

$$\beta(\widehat{B}[\gamma] - 1) - \gamma = 0, \tag{4.4}$$

which can equivalently be written as

$$\widehat{B}[\gamma] = 1 + \frac{\gamma}{\beta}. \tag{4.5}$$

It is rather standard to call γ the *adjustment coefficient* but a variety of other terms are also frequently encountered (and often the notation R instead of γ is used in the literature). The Cramér-Lundberg approximation is renowned not only for its mathematical beauty but also for being very precise, often for *all* $u \geq 0$ and not just for large u. It has generalizations to other Lévy models, to the models with renewal arrivals, a Markovian environment or periodically varying parameters. However, in such cases the evaluation of C is more cumbersome. In fact, when the claim size distribution is of phase-type, the exact solution is as easy to compute as the Cramér-Lundberg approximation in some of these models.

The shape of the l.h.s. of equation (4.4) and its extensions to other models lie at the heart of ruin theory. Its level sets (not only the one at 0) reveal a lot of (in particular asymptotic) properties of ruin-related quantities and will play an important role in this book.

Diffusion approximations Here the idea is simply to approximate the risk process by a Brownian motion (or a more general diffusion) by fitting the first and second moments, and to use the fact that first passage probabilities are more readily calculated for diffusions than for the risk process itself. Diffusion approximations are easy to calculate, but typically not very precise in their first naive implementation. However, incorporating correction terms may change the picture dramatically. In particular, corrected diffusion approximations (see V.6) are by far the best one can do in terms of finite horizon ruin probabilities $\psi(u, T)$.

Large claims approximations In order for the Cramér-Lundberg approximation to be valid, the claim size distribution should have an exponentially decreasing tail $\overline{B}(x)$. In the case of heavy-tailed distributions, other approaches are thus required. Approximations for $\psi(u)$ as well as for $\psi(u, T)$ for large u are available in most of the models we discuss. For example, for the compound Poisson model under certain assumptions on B

$$\psi(u) \ \sim \ \frac{\rho}{\mu_B(1 - \rho)} \int_u^\infty \overline{B}(x) \, \mathrm{d}x, \quad u \to \infty. \tag{4.6}$$

In fact, in some cases the results are even more complete than for light tails. See Chapter X.

This list of approximations does by no means exhaust the topic; some further possibilities are surveyed in IV.7 and V.2.

4e   Bounds and inequalities

The outstanding result in the area is Lundberg's inequality

$$\psi(u) \ \leq \ \mathrm{e}^{-\gamma u}. \tag{4.7}$$

Compared to the Cramér-Lundberg approximation (4.3), it has the advantage of not involving approximations and also, as a general rule, of being somewhat easier to generalize beyond the compound Poisson setting. We return to various extensions and sharpenings of Lundberg's inequality (finite horizon versions, lower bounds etc.) at various places and in various settings.

When comparing different risk models, it is a general principle that *adding random variation to a model increases the risk*. For example, one expects a model with a deterministic claim size distribution B, say degenerate at m, to have smaller ruin probabilities than when B is non-degenerate with the same mean m. This is proved for the compound Poisson model in IV.8 (see also further ordering results for dependent risks in Section XIII.8). Empirical evidence shows that the general principle holds in a broad variety of settings, though precise mathematical results are not always available.

4f Statistical methods

Any of the approaches and results above assume that the parameters of the model are completely known. In practice, they have however to be estimated from data, obtained say by observing the risk process in $[0, T]$. This procedure in itself is fairly straightforward; e.g., in the compound Poisson model, it splits up into the estimation of the Poisson intensity (the estimator is $\widehat{\beta} = N_T/T$) and of the parameter(s) of the claim size distribution, which is a standard statistical problem since the claim sizes $U_1, \ldots, U_{N_T}$ are i.i.d. given N_T. However, the difficulty comes in when drawing inference about the ruin probabilities. How do we produce a confidence interval? And, more importantly, can we trust the confidence intervals for the large values of u which are of interest? In the present authors' opinion, this is extrapolation from data due to the extreme sensitivity of the ruin probabilities to the tail of the claim size distribution in particular (in contrast, fitting a parametric model to $U_1, \ldots, U_{N_T}$ may be viewed as an interpolation or smoothing of the histogram). For example, one may question whether it is possible to distinguish between claim size distributions which are heavy-tailed and those that have an exponentially decaying tail. This issue will be further discussed in Section X.6.

4g Simulation

The development of modern computers has made simulation a popular experimental tool in all branches of applied probability and statistics, and of course the method is relevant in risk theory as well. Simulation may be used just to get some vague insight in the process under study: simulate one or several sample paths, and look at them to see whether they exhibit the expected behavior or some surprises come up. However, the more typical situation is to perform a Monte Carlo experiment to estimate probabilities (or expectations or distributions) which are not analytically available. For example, this is a straightforward way to estimate finite horizon ruin probabilities.

The infinite horizon case presents a difficulty, because it appears to require an infinitely long simulation. Truncation to a finite horizon (or above a certain surplus level) has been used, but is not very satisfying. Still, good methods exist in a number of models and are based upon representing the ruin probability $\psi(u)$

as the expected value of a r.v. (or a functional of the expectation of a set of r.v's) which can be generated by simulation. The problem is entirely analogous to estimating steady-state characteristics by simulation in queueing/storage theory, and in fact methods from that area can often be used in risk theory as well. We look at a variety of such methods in Chapter XV, and also discuss how to develop methods which are efficient in terms of producing a small variance for a fixed simulation budget. A main problem is that ruin is typically a rare event (i.e. having small probability) and that therefore it is expensive or even infeasible in terms of computer time to obtain reasonably precise estimates of the ruin probability through naive simulation. Variance reduction techniques to improve the situation are discussed in Chapter XV.

Chapter II

Martingales and simple ruin calculations

1 Wald martingales

A random walk in discrete time is defined as $R_n = R_0 + Y_1 + \cdots + Y_n$ where the Y_i are i.i.d., with common distribution F (say). Here F is a general probability distribution on $\mathbb{R}$ (the special case of F being concentrated on $\{-1, 1\}$ is often referred to as *simple random walk* or *Bernoulli random walk*). Most often in the probability literature, $R_0 = 0$, but since we are here thinking of the random walk as a model for the risk reserve, we often allow $R_0 = u > 0$. Denote by $\widehat{F}[\alpha] = \mathbb{E}e^{\alpha Y_1}$ the m.g.f. (moment generating function) of F and let $\kappa(\alpha) = \log \widehat{F}[\alpha]$ be the c.g.f. (cumulant generating function).

Theorem 1.1 *Let $R_n = R_0 + Y_1 + \cdots + Y_n$ be a random walk. Then for any α such that $\widehat{F}[\alpha] < \infty$, the sequence*

$$\frac{e^{\alpha(Y_1 + \cdots + Y_n)}}{\widehat{F}[\alpha]^n} = e^{\alpha(Y_1 + \cdots + Y_n) - n\kappa(\alpha)} \tag{1.1}$$

is a martingale.

Proof. Denote by M_n the expression (1.1). Then

$$
\begin{aligned}
\mathbb{E}\big[M_{n+1} \,\big|\, Y_1, \ldots, Y_n\big] &= \frac{e^{\alpha(Y_1 + \cdots + Y_n)}}{\widehat{F}[\alpha]^n} \mathbb{E}\big[e^{\alpha Y_{n+1}} / \widehat{F}[\alpha] \,\big|\, Y_1, \ldots, Y_n\big] \\
&= M_n \mathbb{E}e^{\alpha Y_{n+1}} / \widehat{F}[\alpha] = M_n. \qquad \square
\end{aligned}
$$

The martingale in Theorem 1.1 is denoted the *Wald martingale*. The main application is optional stopping, i.e. exploiting the identity

$$\mathbb{E}e^{\alpha(Y_1+\cdots+Y_\tau)-\tau\kappa(\alpha)} = 1 \qquad (1.2)$$

where $\tau < \infty$ is a finite stopping time. A sufficient condition for (1.2) is that

$$\mathbb{E}\sup_{n\le\tau} e^{\alpha(Y_1+\cdots+Y_n)-n\kappa(\alpha)} < \infty.$$

For a necessary and sufficient condition, see III.1.4.

The Wald martingale generalizes to a Lévy process $\{X_t\}$, defined as a continuous time process with stationary and independent increments. The traditional formal definition is that $\{X_t\}$ is $\mathbb{R}$-valued with the increments

$$X_{t_1} - X_{t_0}, X_{t_2} - X_{t_1}, \ldots, X_{t_n} - X_{t_{n-1}}$$

being independent whenever $t_0 < t_1 < \ldots < t_n$ and with $X_{t_i} - X_{t_{i-1}}$ having distribution depending only on $t_i - t_{i-1}$. An equivalent characterization is $\{X_t\}$ being Markov with state space $\mathbb{R}$ and

$$\mathbb{E}\big[f(X_{t+s} - X_t)\,\big|\,\mathscr{F}_t\big] = \mathbb{E}_0 f(X_s), \qquad (1.3)$$

where $\mathbb{E}_x$ refers to the case $X_0 = x$. Note that the structure of such a process admits a complete description: $\{X_t\}$ can be written as the independent sum of a pure drift $\{\mu t\}$, a Brownian component $\{B_t\}$ (scaled by a variance constant σ) and a pure jump process $\{M_t\}$,

$$X_t = X_0 + \mu t + \sigma B_t + M_t. \qquad (1.4)$$

More precisely, the pure jump process is given by its Lévy measure $\nu(\mathrm{d}x)$, a positive measure on $\mathbb{R}$ with the properties

$$\int_{-\epsilon}^{\epsilon} x^2\nu(\mathrm{d}x) < \infty, \qquad \int_{\{x:|x|>\epsilon\}} \nu(\mathrm{d}x) < \infty \qquad (1.5)$$

for some (and then all) $\epsilon > 0$. Roughly, the interpretation is that the rate of a jump of size x is $\nu(\mathrm{d}x)$ (if $\int_{-\epsilon}^{\epsilon} |x|\nu(\mathrm{d}x) = \infty$, this description needs some amendments). The simplest case is $\beta = \|\nu\| < \infty$, which corresponds to the compound Poisson case: here jumps of $\{M_t\}$ occur at rate β and have distribution $B = \nu/\beta$ (in particular, the claim surplus process for the compound Poisson risk model, with premium rate p, corresponds to a Lévy process with $\mu = -p$, $\sigma^2 = 0$ and $\nu = \beta B$). A general jump process can be thought of as limit of compound Poisson processes with drift by considering a sequence $\nu^{(n)}$

of bounded measures with $\nu^{(n)} \uparrow \nu$. These issues are discussed in more detail in XI.1. For the moment, it suffices to have Brownian motion (possibly with a non-zero drift μ and a general variance constant σ^2) in mind as a second main example besides the compound Poisson model. We have:

Theorem 1.2 *Let* $\{X_t\}$ *be a Lévy process and* $\alpha \in \mathbb{R}$. *Then* $\mathbb{E}e^{\alpha X_t}$ *is either finite for all* $t > 0$ *or infinite for all* $t > 0$. *In the first case,* $\mathbb{E}e^{\alpha X_t} = e^{t\kappa(\alpha)}$ *for some* $\kappa(\alpha) \in \mathbb{R}$, *and the process*

$$e^{\alpha X_t - t\kappa(\alpha)} \tag{1.6}$$

is a martingale.

Proof. The first part is easily seen to hold with $\kappa(\alpha) = \log \mathbb{E}e^{\alpha(X_1 - X_0)}$. For the second, denote by M_t the expression (1.6) and let $\{\mathscr{F}_t\}$ be the natural filtration of $\{X_t\}$. Then

$$\begin{aligned}
\mathbb{E}[M_{t+s} \mid \mathscr{F}_t] &= M_t \, \mathbb{E}\big[e^{\alpha(X_{t+s} - X_t) - s\kappa(\alpha)} \mid \mathscr{F}_t\big] \\
&= M_t \, \mathbb{E}e^{\alpha(X_{t+s} - X_t) - s\kappa(\alpha)} = M_t \,.
\end{aligned}$$

$\square$

A sufficient condition for optional stopping, i.e. $\mathbb{E}M_\tau = 1$, is $\mathbb{E}\sup_{t \leq \tau} M_t < \infty$. For a necessary and sufficient condition, see again III.1.4.

For Brownian motion with drift μ and variance constant σ^2, X_1 is $N(\mu, \sigma^2)$. Therefore $\mathbb{E}e^{\alpha X_1} = e^{\alpha\mu + \alpha^2\sigma^2/2}$, so that

$$\kappa(\alpha) = \alpha\mu + \alpha^2\sigma^2/2 \,. \tag{1.7}$$

For the Cramér-Lundberg process R_t with premium rate p, Poisson parameter β and claim size distribution B, it is shown in IV.1 that

$$\kappa(\alpha) = \beta\big(\widehat{B}[-\alpha] - 1\big) + \alpha p \,. \tag{1.8}$$

For the claim surplus process $S_t = u - R_t$,

$$\kappa(\alpha) = \beta\big(\widehat{B}[\alpha] - 1\big) - \alpha p \,. \tag{1.9}$$

2 Gambler's ruin. Two-sided ruin. Brownian motion

The first solution of a ruin problem appears to be that of de Moivre (1711) for the *gambler's ruin* problem, which is a two-sided ruin problem. That is, starting from $u \in [0, a]$ we define the two-barrier ruin probability $\psi_a(u)$ as the probability of being ruined before the reserve reaches level a. I.e.

$$\psi_a(u) \; = \; \mathbb{P}\big(\tau(u, a) = \tau(u)\big) \; = \; 1 - \mathbb{P}\big(\tau(u, a) = \tau_+(a)\big),$$

where[1]

$$\tau(u) \; = \; \inf \{t \ge 0 : R_t \le 0\}, \quad \tau_+(a) \; = \; \inf \{t \ge 0 : R_t \ge a\},$$

$$\tau(u, a) \; = \; \tau(u) \wedge \tau_+(a) \; = \; \inf \{t \ge 0 : R_t \le 0 \text{ or } R_t \ge a\}.$$

Note that $\mathbb{P}(\tau(u, a) < \infty) = 1$, because the interval $[0, a]$ is finite. Besides its intrinsic interest, $\psi_a(u)$ can also be a useful vehicle for computing $\psi(u)$ by letting $a \to \infty$.

De Moivre considered a Bernoulli random walk, defined as $R_0 = u$ with $u \in \{0, 1, \ldots, a\}$, $R_n = u + Y_1 + \cdots + Y_n$ where $Y_1, Y_2, \ldots$ are i.i.d. and $\{-1, 1\}$-valued with $\mathbb{P}(Y_k = 1) = \theta$. His result was:

Proposition 2.1 *For a Bernoulli random walk with $\theta \ne 1/2$,*

$$\psi_a(u) = \frac{\left(\dfrac{1-\theta}{\theta}\right)^a - \left(\dfrac{1-\theta}{\theta}\right)^u}{\left(\dfrac{1-\theta}{\theta}\right)^a - 1}, \qquad a = u, u+1, \ldots. \tag{2.1}$$

If $\theta = 1/2$, then $\psi_a(u) = \dfrac{a - u}{a}$.

We give two proofs, one elementary but difficult to generalize to other models, and the other more advanced but applicable also in some other settings.

Proof 1. Conditioning upon Y_1 yields immediately the recursion

$$\begin{aligned}
\psi_a(1) &= 1 - \theta + \theta\psi_a(2), \\
\psi_a(2) &= (1 - \theta)\psi_a(1) + \theta\psi_a(3), \\
&\;\;\vdots \\
\psi_a(a - 2) &= (1 - \theta)\psi_a(a - 3) + \theta\psi_a(a - 1), \\
\psi_a(a - 1) &= (1 - \theta)\psi_a(a - 2),
\end{aligned}$$

---

[1]Note that the definition of $\tau(u)$ differs from the rest of the book where we use $\tau(u) = \inf \{t \ge 0 : R_t < 0\}$ (sharp inequality); in most cases, either this makes no difference ($\mathbb{P}(R_{\tau(u)} = 0) = 0$) or it is trivial to translate from one set-up to the other, as e.g. in the Bernoulli random walk example below.

and insertion shows that (2.1) is the solution satisfying the obvious boundary conditions $\psi_a(0) = 1$, $\psi_a(a) = 0$.[2] $\square$

The second proof uses martingales. We remark here as a matter of terminology that whereas our general definition of the Lundberg coefficient γ in Chapter I is as the non-zero solution of $\log \mathrm{e}^{\gamma S_1} = 0$ where $S_t = R_0 - R_t$ is the claim surplus, we work here directly with the reserve process R_t, so that for our Bernoulli random walk the Lundberg coefficient is the non-zero solution of $\log \mathrm{e}^{-\gamma(R_1 - R_0)} = 0$, i.e. $\widehat{F}[-\gamma] = 1$ where $\widehat{F}[s] = \theta \mathrm{e}^s + (1-\theta)\mathrm{e}^{-s}$. In view of the discrete nature of a Bernoulli random walk, we write $z = \mathrm{e}^{-\gamma}$. The Lundberg equation is then

$$1 = \widehat{F}[-\gamma] = \theta z + (1 - \theta)\frac{1}{z}$$

with solution $z = (1 - \theta)/\theta$.

Similar remarks apply when computing the Lundberg coefficient for Brownian motion below.

Proof 2. Wald's exponential martingale with $\alpha = -\gamma$ is just $\{\mathrm{e}^{-\gamma R_t}\} = \{z^{R_n}\}$. By optional stopping,

$$\begin{aligned}
z^u &= \mathbb{E}z^{R_0} = \mathbb{E}z^{R_{\tau(u,a)}} \\
&= z^0 \mathbb{P}\left(R_{\tau(u,a)} = 0\right) + z^a \mathbb{P}\left(R_{\tau(u,a)} = a\right) \\
&= z^0 \psi_a(u) + z^a\left(1 - \psi_a(u)\right),
\end{aligned} \qquad (2.2)$$

and solving for $\psi_a(u)$ yields $\psi_a(u) = (z^a - z^u)/(z^a - 1)$.

If $\theta = 1/2$, (2.2) is trivial ($z = 1$). However, $\{R_n\}$ is then itself a martingale and we get in a similar manner

$$u = \mathbb{E}R_0 = \mathbb{E}R_{\tau_{u,a}} = 0 \cdot \psi_a(u) + a(1 - \psi_a(u)), \quad \psi_a(u) = \frac{a - u}{a}.$$

$\square$

We note that if $\theta \leq 1/2$, then a Bernoulli random walk hits 0 w.p. 1 so $\psi(u) = 1$ for $u \geq 0$. In contrast:

Corollary 2.2 *For a Bernoulli random walk with $\theta > 1/2$,*

$$\psi(u) = \left(\frac{1 - \theta}{\theta}\right)^u. \qquad (2.3)$$

If $\theta \leq 1/2$, then $\psi(u) = 1$.

---

[2] Alternatively, a constructive solution of the difference equation $\psi_a(x) = (1 - \theta)\psi_a(x - 1) + \theta\psi_a(x + 1)$ is to substitute r^x, leading to the two choices $r = 1$ and $r = (1 - \theta)/\theta$ and the result then follows from their linear combination determined by the boundary conditions.

Proof. Let $a \to \infty$ in (2.1).   □

Now let us turn to Brownian motion.

Proposition 2.3 *Let $\{R_t\}$ be Brownian motion starting from $u \geq 0$ and with drift μ and variance constant σ^2. Then for $\mu \neq 0$,*

$$\psi_a(u) = \frac{e^{-2\mu a/\sigma^2} - e^{-2\mu u/\sigma^2}}{e^{-2\mu a/\sigma^2} - 1}. \tag{2.4}$$

If $\mu = 0$, then $\psi_a(u) = \dfrac{a - u}{a}$.

Proof. By (1.7), the Lundberg equation $\kappa(-\gamma) = 0$ is $\gamma^2\sigma^2/2 - \gamma\mu = 0$ with solution $\gamma = 2\mu/\sigma^2$. Applying optional stopping to the exponential martingale $\{e^{-\gamma R_t}\}$ yields

$$e^{-\gamma u} = \mathbb{E}e^{-\gamma R_0} = \mathbb{E}e^{-\gamma R_{\tau(u,a)}} = e^0\psi_a(u) + e^{-\gamma a}(1 - \psi_a(u)),$$

and solving for $\psi_a(u)$ yields $\psi_a(u) = (e^{-\gamma a} - e^{-\gamma u})/(e^{-\gamma a} - 1)$ for $\mu \neq 0$.

If $\mu = 0$, $\{R_t\}$ is itself a martingale and just the same calculation as in the proof of Proposition 2.1 yields $\psi_a(u) = (a - u)/a$.   □

Corollary 2.4 *For Brownian motion starting in $u \geq 0$ with drift $\mu > 0$ and variance constant σ^2,*

$$\psi(u) = e^{-2\mu u/\sigma^2}. \tag{2.5}$$

If $\mu \leq 0$, then $\psi(u) = 1$.

Proof. Let $a \to \infty$ in (2.4).   □

The reason that the calculations work out so smoothly for Bernoulli random walks and Brownian motion is the skip-free nature of the paths, implying $R_{\tau(u,a)} = a$ on $\{\tau(u, a) = \tau_+(a)\}$ and similarly for the boundary 0. For most standard risk processes, the paths are upwards skip-free but not downwards, and thus one encounters the problem of controlling the undershoot under level 0. Here is one more case where this is feasible:

Example 2.5 Consider a compound Poisson model with exponential claims (with rate, say, δ). That is,

$$R_t = u + t - \sum_{i=1}^{N_t} U_i$$

where N is a Poisson(β) process and $\mathbb{P}(U_i > x) = \mathrm{e}^{-\delta x}$. Now consider $R_{\tau(u)}$, assume $R_{\tau(u)-} = x > 0$, and let $Z = -R_{\tau(u)} + x$ be the size of the claim leading to ruin. The available information on Z is that its distribution is that of a claim size U given $U > x$, and thus by the memoryless property of the exponential distribution, the conditional distribution of Z is again just exponential with rate δ. Hence

$$
\begin{aligned}
\mathrm{e}^{-\gamma u} &= \mathbb{E}\mathrm{e}^{-\gamma R_0} \\
&= \mathbb{E}\big[\mathrm{e}^{-\gamma R_{\tau(u,a)}} \,\big|\, R_{\tau(u,a)} \le 0\big]\mathbb{P}(R_{\tau(u,a)} \le 0) + \mathrm{e}^{-\gamma a}\,\mathbb{P}(R_{\tau(u,a)} = a) \\
&= \frac{\delta}{\delta - \gamma}\,\mathbb{P}(R_{\tau(u,a)} \le 0) + \mathrm{e}^{-\gamma a}\,\mathbb{P}(R_{\tau(u,a)} = a) \\
&= \frac{\delta}{\delta - \gamma}\,\psi_a(u) + \mathrm{e}^{-\gamma a}\big(1 - \psi_a(u)\big).
\end{aligned}
$$

It follows from (1.8) that $\gamma = \delta - \beta$, from which we obtain

$$
\psi_a(u) = \frac{\mathrm{e}^{-\gamma u} - \mathrm{e}^{-\gamma a}}{\delta/\beta - \mathrm{e}^{-\gamma a}}\,. \tag{2.6}
$$

Again, letting $a \to \infty$ yields the classical expression $\rho\,\mathrm{e}^{-\gamma u}$ for $\psi(u)$ where $\rho = \beta/\delta$ (cf. I.4b), valid if $\rho < 1$ (otherwise, $\psi(u) = 1$). $\qquad\square$

However, passing to even more general cases the method quickly becomes unfeasible (see, however, IX.5a). It may then be easier to first compute the one-barrier ruin probability $\psi(u)$:

Proposition 2.6 *If the paths of $\{R_t\}$ are upward skip-free and $\psi(a) < 1$,*

$$
\psi_a(u) = \frac{\psi(u) - \psi(a)}{1 - \psi(a)}, \quad 0 \le u \le a. \tag{2.7}
$$

Proof. By the upward skip-free property,

$$
\psi(u) = \psi_a(u) + \big(1 - \psi_a(u)\big)\,\psi(a).
$$

If $\psi(a) < 1$, this immediately yields (2.7). $\qquad\square$

We will meet this argument again in VIII.1a for computing ruin probabilities for a two-step premium function.

Let us now return to Bernoulli random walk and Brownian motion and consider finite horizon ruin probabilities. For the symmetric (drift 0) case these are easily computable by means of the reflection principle:

Proposition 2.7 *For Brownian motion starting in $u \geq 0$ with drift 0 and variance constant σ^2,*

$$\psi(u, T) \;=\; \mathbb{P}(\tau(u) \leq T) \;=\; 2\Phi\left(\frac{-u}{\sigma\sqrt{T}}\right). \qquad (2.8)$$

Proof. In terms of the claim surplus process $\{S_t\} = \{u - R_t\}$, we have $\psi(u, T) = \mathbb{P}(M_T \geq u)$ where $M_T = \max_{0 \leq t \leq T} S_t$. Here $\{S_t\}$ is Brownian motion with drift 0 (starting from 0), in particular symmetric so that from time $\tau(u)$ (when the level is u) it is equally likely to go to levels $< u$ and levels $> u$ in time $T - \tau(u)$. Hence $\mathbb{P}(M_T \geq u, S_T < u) = \mathbb{P}(M_T \geq u, S_T > u)$, and one gets

$$
\begin{aligned}
\mathbb{P}(M_T \geq u) \;&=\; \mathbb{P}(S_T \geq u) + \mathbb{P}(S_T < u, M_T \geq u) \\
&=\; \mathbb{P}(S_T \geq u) + \mathbb{P}(S_T > u, M_T \geq u) \\
&=\; \mathbb{P}(S_T \geq u) + \mathbb{P}(S_T > u) \qquad (2.9) \\
&=\; 2\mathbb{P}(S_T > u).
\end{aligned}
$$

$\square$

Small modifications also apply to Bernoulli random walks:

Proposition 2.8 *For the Bernoulli random walk with $\theta = 1/2$,*

$$\psi(u, T) \;=\; \mathbb{P}(S_T = u) + 2\,\mathbb{P}(S_T > u), \qquad (2.10)$$

whenever u, T are integer-valued and non-negative. Here

$$
\mathbb{P}(S_T = v) \;=\;
\begin{cases}
2^{-T}\dbinom{T}{(v+T)/2}, & v = -T, -T+1, \ldots, T - 2, T, \\
\\
0, & \text{otherwise.}
\end{cases}
$$

Proof. The argument leading to (2.9) goes through unchanged, and (2.10) is the same as (2.9). The expression for $\mathbb{P}(S_T = v)$ is just a standard formula for the binomial distribution. $\square$

Notes and references   All material of the present section is standard. A classical reference for further aspects of Bernoulli random walks is Feller [361]. For generalizations of Proposition 2.6 to Markov-modulated models, see Asmussen & Perry [95]. Further early references on two-barrier ruin problems include Dickson & Gray [312, 313].

3 Further simple martingale calculations

We consider the claim surplus process $\{S_t\}$ of a general risk process. As usual, the time to ruin $\tau(u)$ is $\inf\{t \geq 0 : S_t > u\}$, and the ruin probabilities are

$$\psi(u) = \mathbb{P}(\tau(u) < \infty), \quad \psi(u, T) = \mathbb{P}(\tau(u) \leq T).$$

Our first result is a representation formula for $\psi(u)$ obtained by using the martingale optional stopping theorem. Let $\xi(u) = S_{\tau(u)} - u$ denote the overshoot.

Proposition 3.1 *Assume that* (a) *for some* $\gamma > 0$, $\{e^{\gamma S_t}\}_{t \geq 0}$ *is a martingale,* (b) $S_t \overset{\text{a.s.}}{\to} -\infty$ *on* $\{\tau(u) = \infty\}$. *Then*

$$\psi(u) = \frac{e^{-\gamma u}}{\mathbb{E}\left[e^{\gamma \xi(u)} \mid \tau(u) < \infty\right]}, \quad u \geq 0. \tag{3.1}$$

Proof. We shall use optional stopping at time $\tau(u) \wedge T$. [3] We get

$$
\begin{aligned}
1 &= \mathbb{E}e^{\gamma S_0} = \mathbb{E}e^{\gamma S_{\tau(u) \wedge T}} \\
&= \mathbb{E}\left[e^{\gamma S_{\tau(u)}}; \tau(u) \leq T\right] + \mathbb{E}\left[e^{\gamma S_T}; \tau(u) > T\right].
\end{aligned}
\tag{3.2}
$$

As $T \to \infty$, the second term converges to 0 by (b) and dominated convergence ($e^{\gamma S_T} \leq e^{\gamma u}$ on $\{\tau(u) > T\}$), and in the limit (3.2) takes the form

$$
\begin{aligned}
1 &= \mathbb{E}\left[e^{\gamma S_{\tau(u)}}; \tau(u) < \infty\right] + 0 \\
&= e^{\gamma u}\mathbb{E}\left[e^{\gamma \xi(u)}; \tau(u) < \infty\right] = e^{\gamma u}\mathbb{E}\left[e^{\gamma \xi(u)} \mid \tau(u) < \infty\right]\psi(u).
\end{aligned}
$$

$\square$

Example 3.2 Consider the compound Poisson model with Poisson arrival rate β, claim size distribution B and $\rho = \beta\mu_B < 1$. Thus

$$S_t = \sum_{i=1}^{N_t} U_i - t,$$

where $\{N_t\}$ is a Poisson process with rate β and the U_i are i.i.d. with common distribution B (and independent of $\{N_t\}$).

Condition (a) of Proposition 3.1 is satisfied, if a positive solution $\gamma > 0$ to the Lundberg equation (1.9) (i.e. an adjustment coefficient) exists. Property (b) follows from $\rho < 1$ and the law of large numbers (see Proposition IV.1.2(c)).

$\square$

---

[3] We cannot use the stopping time $\tau(u)$ directly because $\mathbb{P}(\tau(u) = \infty) > 0$ and also because the conditions of the optional stopping theorem present a problem; however, using $\tau(u) \wedge T$ invokes no problems because $\tau(u) \wedge T$ is bounded by T.

Example 3.3 Assume that $\{R_t\}$ is Brownian motion with variance constant σ^2 and drift $\mu > 0$. Then $\{S_t\}$ is Brownian motion with variance constant σ^2 and drift $-\mu < 0$. Since the positive solution to the Lundberg equation (1.7) is $\gamma = 2\mu/\sigma^2$, the conditions of Proposition 3.1 are satisfied. $\qquad\square$

Corollary 3.4 (LUNDBERG'S INEQUALITY) *Under the conditions of Proposition 3.1, $\psi(u) \le e^{-\gamma u}$ for all $u \ge 0$.*

Proof. Just note that $\xi(u) \ge 0$ a.s. $\qquad\square$

We also retrieve again the exact expression of Section I.4b for exponential claims:

Corollary 3.5 *For the compound Poisson model with B exponential, $\overline{B}(x) = e^{-\delta x}$, and $\rho = \beta/\delta < 1$, the ruin probability is $\psi(u) = \rho e^{-\gamma u}$ where $\gamma = \delta - \beta$.*

Proof. As before, from (1.9) it is immediately seen that $\gamma = \delta - \beta$. Now at the time $\tau(u)$ of ruin, $\{S_t\}$ upcrosses level u by making a jump. By the memoryless property of the exponential distribution, the conditional distribution of the overshoot $\xi(u)$ is again just exponential with rate δ. Thus

$$\mathbb{E}\big[e^{\gamma \xi(u)} \,\big|\, \tau(u) < \infty\big] = \int_0^\infty e^{\gamma x} \delta e^{-\delta x} \,\mathrm{d}x = \int_0^\infty \delta e^{-\beta x} \,\mathrm{d}x = \frac{\delta}{\beta} = \frac{1}{\rho}.$$

$\qquad\square$

If $\{R_t\}$ is Brownian motion with variance constant σ^2 and drift $\mu > 0$, then $\xi(u) = 0$ by continuity of Brownian motion and $\psi(u) = e^{-2\mu u/\sigma^2}$, which reconfirms Corollary 2.4.

Notes and references   The first use of martingales in risk theory is due to Gerber [397], and is further exploited in his book [398]. More recent references are Dassios & Embrechts [273], Grandell [429, 430], Embrechts, Grandell & Schmidli [345], Delbaen & Haezendonck [287] and Schmidli [770, 780].

4   More advanced martingales

4a   Generators. The Dynkin martingale

Assume that a stochastic process $\{R_t\}$ is a Markov process and write $\mathbb{P}_u$ and $\mathbb{E}_u$ for the case $R_0 = u$. In loose terms, the *generator* $\mathscr{A}$ is then an operator in an appropriate function space such that

$$\frac{\mathrm{d}}{\mathrm{d}t}\mathbb{E}_u f(R_t)\bigg|_{t=0} = \mathscr{A}f(u)$$

or equivalently

$$\lim_{h\downarrow 0} \frac{\mathbb{E}_u f(R_h) - f(u)}{h} = \mathscr{A}f(u) \qquad (4.1)$$

for a sufficiently rich class of functions f.

Historically, there have, however, been several different ways to make this loose definition precise, and in particular, one will find many definitions of the domain $\mathscr{D}(\mathscr{A})$ on which $\mathscr{A}$ is defined. For example, some older definitions require (4.1) to hold uniformly or locally uniformly in u. The most standard current definition is in terms of martingales: $f \in \mathscr{D}(\mathscr{A})$ with $\mathscr{A}f = g$ (g can be shown to be unique up to some null set complications (cf. Davis [278, p. 32]) if and only if

$$f(R_t) - \int_0^t g(R_s)\, \mathrm{d}s \qquad (4.2)$$

is a local martingale[4], commonly referred to as the *Dynkin martingale*. The motivation relating to (4.1) is loosely the following. Denote by M_t the expression (4.2), assume it is a martingale (and not just a local martingale) and that the function $s \to \mathbb{E}_u \mathscr{A}f(R_s)$ is sufficiently well-behaved at $s = 0$, say continuous and bounded. Then

$$\begin{aligned}
\mathbb{E}_u f(R_h) - f(u) &= \mathbb{E}_u[M_h - M_0] + \mathbb{E}_u \int_0^h \mathscr{A}f(R_s)\, \mathrm{d}s \\
&= 0 + h\mathscr{A}f(u) + \mathrm{o}(h)
\end{aligned}$$

so that (4.1) holds.

Example 4.1 Assume that $\{R_t\}$ is Brownian motion with drift μ and variance constant σ^2. Let further $f \in C^3$ have compact support. Then if V is a standard normal r.v., we have[5]

$$\begin{aligned}
\mathbb{E}_u f(R_h) &= \mathbb{E}f(u + \mu h + \sqrt{h}\sigma V) \\
&= \mathbb{E}\big[f(u) + f'(u)[\mu h + \sqrt{h}\sigma V] + f''(u)[\mu h + \sqrt{h}\sigma V]^2/2 + \mathrm{O}(h^2)\big] \\
&= f(u) + f'(u)\mu h + f''(u)h\sigma^2\, \mathbb{E}V^2/2 + \mathrm{O}(h^{3/2}).
\end{aligned}$$

Thus (4.1) holds with

$$\mathscr{A}f = \mu f' + (\sigma^2/2)\, f''.$$

---

[4]Strictly speaking, a local martingale w.r.t. $\mathbb{P}_x$ for all x. For ease of exposition, we omit such specification here and in the following.

[5]This calculation is of course a heuristic derivation of Itô's formula. In its full generality Itô's formula permits to weaken the assumption on f to $f \in C^2$.

It is less clear how much one can relax the assumptions on f to still get a local martingale and we will not go into this issue here. Nevertheless, it is clear that for a suitable class of twice differentiable functions f, one should have $f \in \mathscr{D}(\mathscr{A})$ and that $\mathscr{A}f = \mu f' + (\sigma^2/2) f''$.

The operator sending a twice differentiable function f to $\mu f' + (\sigma^2/2) f''$ is often called the *differential operator* of the Brownian motion. Similarly, a diffusion process with local drift $\mu(u)$ and local variance $\sigma^2(u)$ has differential operator $\mu(u)f'(u) + (\sigma^2(u)/2) f''(u)$. $\quad\square$

Example 4.2 Consider the Cramér-Lundberg model with parameters β, B and let U be a generic claim. Then, conditioning on whether or not a claim occurs in $(0, h)$, we have under suitable conditions on f

$$
\begin{aligned}
\mathbb{E}_u f(R_h) &= (1 - \beta h)f(u + h) + \beta h \, \mathbb{E}f\big(u - U + \mathrm{O}(h)\big) + \mathrm{o}(h) \\
&= f(u) - \beta f(u)h + hf'(u) + \beta h \int_0^\infty f(u - x) \, B(\mathrm{d}x) + \mathrm{o}(h).
\end{aligned}
$$

Thus, it is clear that for a suitable class of differentiable functions f, one should have $f \in \mathscr{D}(\mathscr{A})$ and that

$$
\mathscr{A}f(u) = -\beta f(u) + f'(u) + \beta \int_0^\infty f(u - x) \, B(\mathrm{d}x).
$$

$\quad\square$

A function f such that $f(R_t)$ is a martingale is called *harmonic*. Obviously, in view of (4.1) a harmonic function will have the property $f \in \mathscr{D}(\mathscr{A})$ and $\mathscr{A}f \equiv 0$.

Example 4.3 As a simple example of the relevance of harmonic functions for ruin theory, consider the Brownian setting of Example 4.1 and take $f(u) = \psi(u)$ (the ruin probability), with the convention $\psi(u) = 1$ for $u \leq 0$. For $u > 0$, $\mathbb{P}(\tau(u) \leq h) = \mathrm{o}(h)$ [6] and so by boundedness of $\psi(u)$ and the Markov property of R_t,

$$
\psi(u) = \mathbb{E}_u\big[\psi(R_h); \tau(u) > h\big] + \mathbb{E}_u\big[I(\tau(u) \leq h)\big] = \mathbb{E}_u\psi(R_h) + \mathrm{o}(h).
$$

The same holds for $u < 0$, and so (omitting the details for $u = 0$) we conclude that $\psi(u)$ is harmonic. Combining this with the remarks of Example 4.1, we conclude that

$$
\mu\psi'(u) + \psi''(u)\sigma^2/2 = 0.
$$

---

[6]This for instance follows from III.(1.8).

The general solution of this differential equation is $C_+ e^{\lambda_+ u} + C_- e^{\lambda_- u}$, where $\lambda_\pm$ are the solutions of the quadratic equation $0 = \lambda\mu + \lambda^2/2$, i.e. $\lambda_+ = 0$ and $\lambda_- = -2\mu/\sigma^2$. If $\mu > 0$, we have $\psi(u) \to 0$ as $u \to \infty$, and so $C_+ = 0$. Together with the boundary condition $\psi(0) = 1$ (due to the oscillation) we arrive at $\psi(u) = e^{-2\mu u/\sigma^2}$, as found in Corollary 2.4 by different means.

The method employed may be seen as a continuous time analogue of Proof 1 of Theorem 2.1. □

Notes and references Ethier & Kurtz [359] is a good reference for the modern approach to generators. Further references are Davis [278] and Rolski *et al.* [746].

4b Diffusions and two-sided ruin

One of the major areas where generators play a main role is diffusion processes. Thus let $\{R_t\}$ be a diffusion on $[0, \infty)$ with drift $\mu(x)$ and variance $\sigma^2(x)$ at x. We assume that $\mu(x)$ and $\sigma^2(x)$ are continuous with $\sigma^2(x) > 0$ for $x > 0$. Thus, close to x, $\{R_t\}$ behaves as Brownian motion with drift $\mu = \mu(x)$ and variance $\sigma^2 = \sigma^2(x)$. We can define a 'local' adjustment coefficient $\gamma(x) = -2\mu(x)/\sigma^2(x)$ for the locally approximating Brownian motion. Let

$$s(y) = e^{\int_0^y \gamma(x)\,dx}, \quad S(x) = \int_0^x s(y)\,dy, \quad S(\infty) = \int_0^\infty s(y)\,dy. \tag{4.3}$$

The following result gives a complete solution of the ruin problem for the diffusion subject to the assumption that $S(x)$, as defined in (4.3) with 0 as lower limit of integration, is finite for all $x > 0$. If this assumption fails, the behavior at the boundary 0 is more complicated and it may happen, e.g., that $\psi(u)$, as defined above as the probability of actually hitting 0, is zero for all $u > 0$ but that nevertheless $R_t \overset{\text{a.s.}}{\to} 0$ (the problem leads into the complicated area of boundary classification of diffusions, see e.g. Breiman [199] or Karlin & Taylor [522, p. 226]).

Theorem 4.4 *Consider a diffusion process $\{R_t\}$ on $[0, \infty)$, such that the drift $\mu(x)$ and the variance $\sigma^2(x)$ are continuous functions of x and that $\sigma^2(x) > 0$ for $x > 0$. Assume further that $S(x)$ as defined in (4.3) is finite for all $x > 0$. If*

$$S(\infty) < \infty, \tag{4.4}$$

then $0 < \psi(u) < 1$ for all $u > 0$ and

$$\psi(u) = 1 - \frac{S(u)}{S(\infty)}.$$

Conversely, if (4.4) fails, then $\psi(u) = 1$ for all $u > 0$.

Lemma 4.5 *Let $0 < b \leq u \leq a$ and let $\psi_{a,b}(u)$ be the probability that $\{R_t\}$ hits b before a starting from u. Then*

$$\psi_{a,b}(u) = \frac{S(a) - S(u)}{S(a) - S(b)}. \tag{4.5}$$

Proof. Recall that under mild conditions on q, $\mathbb{E}_u q(R_h) = q(u) + \mathscr{A}q(u)h + o(h)$, where

$$\mathscr{A}q(u) = \frac{\sigma^2(u)}{2} q''(u) + \mu(u) q'(u).$$

If $b < u < a$, the probability of ruin or hitting the upper barrier a before h is of order $o(h)$, so that

$$\psi_{a,b}(u) = \mathbb{E}_u \psi_{a,b}(R_h) + o(h) = \psi_{a,b}(u) + \mathscr{A}\psi_{a,b}(u)h + o(h),$$

i.e $\mathscr{A}\psi_{a,b}(u) = 0$. Using $s'(x)/s(x) = -2\mu(x)/\sigma^2(x)$, elementary calculus shows that we can rewrite $\mathscr{A}$ as

$$\mathscr{A}q(u) = \frac{1}{2}\sigma^2(u)s(u)\frac{\mathrm{d}}{\mathrm{d}u}\left[\frac{q'(u)}{s(u)}\right]. \tag{4.6}$$

Hence $\mathscr{A}\psi_{a,b}(u) = 0$ implies that $\psi'_{a,b}(u)/s(u)$ is constant, i.e. $\psi_{a,b}(u) = \alpha + \beta S(u)$. The obvious boundary conditions $\psi_{a,b}(b) = 1$, $\psi_{a,b}(a) = 0$ then yield the result. □

Proof of Theorem 4.4. Letting $b \downarrow 0$ in (4.5) yields $\psi_a(u) = 1 - S(u)/S(a)$. Letting $a \uparrow \infty$ and considering the cases $S(\infty) = \infty$, $S(\infty) < \infty$ separately completes the proof. □

Notes and references   A good introduction to diffusions is in Karlin & Taylor [522]; see in particular pp. 191–195 for material related to Theorem 4.4. In view of (4.5), the function $S(x)$ is referred to as the *natural scale* in the general theory of diffusions (in case of integrability problems at 0, one works instead with a lower limit $\epsilon > 0$ of integration in (4.3)). Another basic quantity is the *speed measure* M, defined by the density $1/(\sigma^2(u)s(u))$ showing up in (4.6). For related arguments concerning the local adjustment coefficient, see VIII.3.

4c   The Kella-Whitt martingale

Example 4.6   As a motivating example, consider the surplus process $\{S_t\}$ of a Cramér-Lundberg model with a claim size distribution B that is a mixture of two exponential distributions. To be definite, let the density be

$$b(x) = \frac{1}{2}3e^{-3x} + \frac{1}{2}7e^{-7x}$$

and let the Poisson parameter be $\beta = 3$. Then

$$\kappa(\alpha) = 3\left(\frac{1}{2}\frac{3}{3-\alpha} + \frac{1}{2}\frac{7}{7-\alpha} - 1\right) - \alpha. \tag{4.7}$$

Thus a solution of the Lundberg equation $\kappa(\alpha) = 0$ must satisfy

$$0 = \frac{9}{2}(7-\alpha) + \frac{21}{2}(3-\alpha) - 3(3-\alpha)(7-\alpha) - \alpha(3-\alpha)(7-\alpha).$$

This is a cubic equation, and roots are easily seen to be 0, $\gamma = 1$, $\gamma^* = 6$; that the Lundberg coefficient γ must be 1 and not 6 follows since $\mathbb{E}e^{\alpha S_1}$ is only finite if $\alpha < 3$. Consider the problem of two-sided exit of $\{S_t\}$ from $[-a, b]$ with $a, b > 0$. Let p_0 be the probability of exit at the lower barrier and p_3, p_7 the probabilities of exit at the upper barrier as result of a jump which is exponential with rate 3 and 7, respectively. For brevity, write $\tau = \tau(u, a)$. Now S_τ equals 0 for lower exit and $b + V_3$ for upper exit as result of an exponential(3) jump (where V_3 is again exponential(3) due to the lack-of-memory property of the exponential distribution). Defining V_7 similarly, we get by optional stopping of $\{e^{\gamma S_t}\}$

$$1 = e^{\gamma S_0} = \mathbb{E}e^{\gamma S_\tau} = p_0 e^{-\gamma a} + p_3 e^{\gamma b}\frac{3}{3-\gamma} + p_7 e^{\gamma b}\frac{7}{7-\gamma},$$

i.e.

$$6 = 6 p_0 e^{-a} + 9 p_3 e^b + 7 p_7 e^b. \tag{4.8}$$

Also obviously $1 = p_0 + p_3 + p_7$. To get the missing third equation, it is tempting to formally proceed with γ^* as with γ, which would give

$$1 = e^{\gamma^* S_0} = \mathbb{E}e^{\gamma^* S_\tau} = p_0 e^{-\gamma^* a} + p_3 e^{\gamma^* b}\frac{3}{3-\gamma^*} + p_7 e^{\gamma^* b}\frac{7}{7-\gamma^*}, \tag{4.9}$$

i.e.

$$1 = p_0 e^{-6a} + e^{6b}(-p_3 + 7p_7). \tag{4.10}$$

The problem is, however, that $\mathbb{E}e^{\gamma^* S_t} = \infty$ so $\{e^{\gamma^* S_t}\}$ is not a martingale (or for that sake a local martingale), so we are missing a rigorous justification for (4.9). □

We will see in Example 4.9 that the solution is nevertheless correct. For this and other purposes, we will exploit a martingale introduced by Kella & Whitt [527]. Let $\{R_t\}$ be a Lévy process with Lévy exponent $\kappa(\alpha)$. The Wald martingale is then $M_t = e^{\alpha R_t - t\kappa(\alpha)}$. The Kella-Whitt martingale is a stochastic integral w.r.t. $\{M_t\}$ and has a somewhat different range of applications; in particular, it allows for a more direct study of aspects of reflected Lévy processes.

Theorem 4.7 *Let $\{R_t\}$ be a Lévy process with Lévy exponent $\kappa(\alpha)$, let*

$$Y_t = \int_0^t dY_s^c + \sum_{0 \leq s \leq t} \Delta Y_s$$

be an adapted process of locally bounded variation with continuous part $\{Y_t^c\}$, D-paths and jumps $\Delta Y_s = Y_s - Y_{s-}$, and define $Z_t = x + R_t + Y_t$. For each t, let K_t be the r.v.

$$K_t = \int_0^t e^{\alpha Z_s} dM_s$$

$$= \kappa(\alpha) \int_0^t e^{\alpha Z_s} ds + e^{\alpha x} - e^{\alpha Z_t} + \alpha \int_0^t e^{\alpha Z_s} dY_s^c + \sum_{0 \leq s \leq t} e^{\alpha Z_s}(1 - e^{-\alpha \Delta Y_s}).$$

Then $\{K_t\}$ is a local martingale whenever $\kappa(\alpha)$ is well-defined.

Proof. Let $B_t = e^{\alpha Y_t + t\kappa(\alpha)}$. Then, by the general theory of stochastic integration, $K_t^* = \int_0^t B_{s-} dM_s$ is a local martingale. Using the formula for integration by parts (see e.g. Protter [718, p.60] for a version sufficiently general to deal with the present case) yields

$$M_t B_t - M_0 B_0 = \int_0^t M_{s-} dB_s + K_t^* + \sum_{0 \leq s \leq t} \Delta M_s \Delta B_s.$$

Inserting

$$\sum_{0 \leq s \leq t} \Delta M_s \Delta B_s = \int_0^t \Delta M_s dB_s = \int_0^t (M_s - M_{s-}) dB_s,$$

it follows that

$$-K_t^* = \int_0^t M_s dB_s + M_0 B_0 - M_t B_t. \tag{4.11}$$

Using $M_s B_s = e^{\alpha Z_s}$ and $dB_s = B_s(\alpha \, dY_s^c + \kappa(\alpha)ds + 1 - e^{-\alpha \Delta Y_s})$ shows that the r.h.s. of (4.11) reduces to K_t. $\qquad\square$

For practical purposes, the main application is optional stopping which is often verified via the following standard result from martingale theory:

Theorem 4.8 *If for a given t we have $\mathbb{E} \sup_{s \leq t} |K_s| < \infty$, then $\{K_t\}$ is a proper martingale. Further, let τ be a stopping time such that $\mathbb{E} \sup_{t \leq \tau} |K_t| < \infty$. Then $\mathbb{E} K_\tau = \mathbb{E} K_0 = 0$.*

Example 4.9 Consider again the mixed exponential setting of Example 4.6. Take $Y_t \equiv 0$. Thus we have simply

$$K_t = \kappa(\alpha) \int_0^t e^{\alpha S_s} \, ds + 1 - e^{\alpha S_t}$$

whenever $\alpha < 3$. Noting that $-a \leq S_s < b$, we get

$$|K_s| \leq |\kappa(\alpha)| s e^{|\alpha| \max(a,b)} + 1 + e^{|\alpha|(b+V_3)} + e^{|\alpha|(b+V_7)}.$$

Since $\mathbb{E}\tau < \infty$, this shows the conditions of Theorem 4.8 for $|\alpha| < 3$ and gives as in (4.9) that

$$0 = \kappa(\alpha)\phi(\alpha) + 1 - p_0 e^{-\alpha a} - p_3 e^{\alpha b} \frac{3}{3-\alpha} - p_7 e^{-\alpha b} \frac{7}{7-\alpha}, \tag{4.12}$$

where $\phi(\alpha) = \mathbb{E} \int_0^t e^{\alpha R_s} \, ds$. The same bound as used above easily gives that $\phi(\alpha)$ is defined for *all* $\alpha \in \mathbb{R}$, not only for $|\alpha| < 3$, and is an analytic function of α. Since everything else in (4.12) is analytic in the region $\Omega = \mathbb{R}\backslash\{3,7\}$ (think of $\kappa(s)$ as the r.h.s. of (4.7), i.e. the analytic continuation of $\log \mathbb{E}e^{sS_1}$), we conclude that (4.12) holds in the whole of Ω. Taking $\alpha = \gamma^* = 6$ we get the desired rigorous proof of (4.9). $\qquad\square$

The picture that emerges is that all roots of the analytic continuation of $\log \mathbb{E}e^{s(R_1-R_0)}$ may enter in ruin formulas, and we will see several examples of this, in particular when phase-type distributions are involved (see e.g. XI.5 for a much more elaborate version of Example 4.9).

The problem with the Wald martingale is of course that one only gets (4.9) for 0 and γ, and this is not enough to do an analytic continuation. Other values of α require conditional expectations of $e^{\alpha \tau}$ given the type of exit, and we are not aware of how to approach these via the Wald martingale. However, Example 4.9 shows how to solve the problem via the Kella-Whitt martingale.

Notes and references Optional stopping of the Kella-Whitt martingale is further discussed in Asmussen & Kella [85], and Markov-modulated versions of the martingale are in Asmussen & Kella [84]. A variety of different applications are surveyed in [APQ, Ch. IX.3].

Chapter III

Further general tools and results

1 Likelihood ratios and change of measure

We consider stochastic processes $\{X_t\}$ with a Polish state space E and paths in the Skorohod space $D_E = D_E[0, \infty)$, which we equip with the natural filtration $\{\mathscr{F}_t\}_{t \geq 0}$ and the Borel σ-field $\mathscr{F}$. Two such processes may be represented by probability measures $\mathbb{P}$, $\widetilde{\mathbb{P}}$ on $(D_E, \mathscr{F})$, and in analogy with the theory of measures on finite-dimensional spaces one could study conditions for the Radon-Nikodym derivative $\mathrm{d}\widetilde{\mathbb{P}}/\mathrm{d}\mathbb{P}$ to exist. However, as shown by the following example this set-up is too restrictive: typically[1], the parameters of the two processes can be reconstructed from a single infinite path, and $\mathbb{P}$, $\widetilde{\mathbb{P}}$ are then singular (concentrated on two disjoint measurable sets).

Example 1.1 Let $\mathbb{P}$, $\widetilde{\mathbb{P}}$ correspond to the claim surplus process of two compound Poisson risk processes with Poisson rates β, $\widetilde{\beta}$ and claim size distributions B, $\widetilde{B}$. The number $N_t^{(\epsilon)}$ of jumps $> \epsilon$ before time t is a (measurable) r.v. on $(D_E, \mathscr{F})$, hence so is $N_t = \lim_{\epsilon \downarrow 0} N_t^{(\epsilon)}$. Thus the sets

$$S = \left\{ \lim_{t \to \infty} \frac{N_t}{t} = \beta \right\}, \quad \widetilde{S} = \left\{ \lim_{t \to \infty} \frac{N_t}{t} = \widetilde{\beta} \right\}$$

are both in $\mathscr{F}$. But if $\beta \neq \widetilde{\beta}$, then S and $\widetilde{S}$ are disjoint, and by the law of

---

[1]Though not always: it is not difficult to construct a counterexample say in terms of transient Markov processes.

large numbers for the Poisson process, $\mathbb{P}(S) = \widetilde{\mathbb{P}}(\widetilde{S}) = 1$. A somewhat similar argument gives singularity when $B \neq \widetilde{B}$. □

The interesting concept is therefore to look for absolute continuity only on finite time intervals (possibly random, cf. Theorem 1.3 below). I.e. we look for a process $\{L_t\}$ (the likelihood ratio process) such that

$$\widetilde{\mathbb{P}}(A) = \mathbb{E}[L_t; A], \quad A \in \mathscr{F}_t, \tag{1.1}$$

(i.e. the restriction of $\widetilde{\mathbb{P}}$ to $(D_E, \mathscr{F}_t)$ is absolutely continuous w.r.t. the restriction of $\mathbb{P}$ to $(D_E, \mathscr{F}_t)$).

The following result gives the connection to martingales.

Proposition 1.2 *Let $\{\mathscr{F}_t\}_{t\geq 0}$ be the natural filtration on D_E, $\mathscr{F}$ the Borel σ-field and $\mathbb{P}$ a given probability measure on $(D_E, \mathscr{F})$.*
(i) If $\{L_t\}_{t\geq 0}$ is a non-negative martingale w.r.t. $(\{\mathscr{F}_t\}, \mathbb{P})$ such that $\mathbb{E}L_t = 1$, then there exists a unique probability measure $\widetilde{\mathbb{P}}$ on $\mathscr{F}$ such that (1.1) holds.
(ii) Conversely, if for some probability measure $\widetilde{\mathbb{P}}$ and some $\{\mathscr{F}_t\}$-adapted process $\{L_t\}_{t\geq 0}$ (1.1) holds, then $\{L_t\}$ is a non-negative martingale w.r.t. $(\{\mathscr{F}_t\}, \mathbb{P})$ such that $\mathbb{E}L_t = 1$.

Proof. Under the assumptions of (i), define $\widetilde{\mathbb{P}}$ by $\widetilde{\mathbb{P}}_t(A) = \mathbb{E}[L_t; A]$, $A \in \mathscr{F}_t$. Then $L_t \geq 0$ and $\mathbb{E}L_t = 1$ ensure that $\widetilde{\mathbb{P}}_t$ is a probability measure on $(D_E, \mathscr{F}_t)$. Let $s < t$, $A \in \mathscr{F}_s$. Then

$$\begin{aligned}
\widetilde{\mathbb{P}}_t(A) &= \mathbb{E}[L_t; A] = \mathbb{E}\,\mathbb{E}\big[L_t I(A) \,\big|\, \mathscr{F}_s\big] = \mathbb{E}\, I(A)\mathbb{E}[L_t | \mathscr{F}_s] \\
&= \mathbb{E}\, I(A)L_s = \widetilde{\mathbb{P}}_s(A),
\end{aligned}$$

using the martingale property in the fourth step. Hence the family $\{\widetilde{\mathbb{P}}_t\}_{t\geq 0}$ is consistent and hence extendable to a probability measure $\widetilde{\mathbb{P}}$ on $(D_E, \mathscr{F})$ such that $\widetilde{\mathbb{P}}(A) = \widetilde{\mathbb{P}}_t(A)$, $A \in \mathscr{F}_t$. This proves (i).

Conversely, under the assumptions of (ii) we have for $A \in \mathscr{F}_s$ and $s < t$ that $A \in \mathscr{F}_t$ as well and hence $\mathbb{E}[L_s; A] = \mathbb{E}[L_t; A]$. The truth of this for all $A \in \mathscr{F}_s$ implies that $\mathbb{E}[L_t | \mathscr{F}_s] = L_s$ and the martingale property. Finally, $\mathbb{E}L_t = 1$ follows by taking $A = D_E$ in (1.1) and non-negativity by letting $A = \{L_t < 0\}$. Then $\widetilde{\mathbb{P}}(A) = \mathbb{E}[L_t; L_t < 0]$ can only be non-negative if $\mathbb{P}(A) = 0$. □

The following likelihood ratio identity (typically with τ being the time $\tau(u)$ to ruin) is a fundamental tool throughout the book:

Theorem 1.3 *Let $\{L_t\}$, $\widetilde{\mathbb{P}}$ be as in Proposition 1.2(i). If τ is a stopping time and $G \in \mathscr{F}_\tau$, $G \subseteq \{\tau < \infty\}$, then*

$$\mathbb{P}(G) = \widetilde{\mathbb{E}}\left[\frac{1}{L_\tau}; G\right]. \tag{1.2}$$

Further, if Z is a r.v. which is $\mathscr{F}_\tau$-measurable and 0 on the set $\{\tau = \infty\}$, then

$$\mathbb{E}Z = \widetilde{\mathbb{E}}[Z/L_\tau]. \tag{1.3}$$

Proof. Assume first $G \subseteq \{\tau \leq T\}$ for some fixed deterministic $T < \infty$. By the martingale property, we have $\mathbb{E}[L_T|\mathscr{F}_\tau] = L_\tau$ on $\{\tau \leq T\}$. Hence

$$\widetilde{\mathbb{E}}\left[\frac{1}{L_\tau}; G\right] = \mathbb{E}\left[\frac{L_T}{L_\tau}; G\right] = \mathbb{E}\left[\frac{1}{L_\tau}I(G)\mathbb{E}[L_T|\mathscr{F}_\tau]\right]$$

$$= \mathbb{E}\left[\frac{1}{L_\tau}I(G)L_\tau\right] = \mathbb{P}(G). \tag{1.4}$$

In the general case, applying (1.4) to $G \cap \{\tau \leq T\}$ we get

$$\mathbb{P}(G \cap \{\tau \leq T\}) = \widetilde{\mathbb{E}}\left[\frac{1}{L_\tau}; G \cap \{\tau \leq T\}\right].$$

Since everything is non-negative, both sides are increasing in T, and letting $T \to \infty$, (1.2) follows by monotone convergence.

For (1.3), just use standard measure-theoretic arguments to extend from indicators to r.v.'s. $\square$

A main example of change of measure is to take $\{L_t\}$ as the Wald martingale $e^{\theta X_t - t\kappa(\theta)}$ in the case where $\{X_t\}$ is a random walk or a Lévy process. We then write $\mathbb{P}_\theta$ rather than $\widetilde{\mathbb{P}}$, and talk about *exponential change of measure* or *exponential tilting*. We will see in Section 3 that $\{X_t\}$ remains a random walk/Lévy process under $\mathbb{P}_\theta$, only with changed parameters. A first elegant application of the change-of-measure technique is the following observation:

Corollary 1.4 *The necessary and sufficient condition for optional stopping of the Wald martingale, i.e. $1 = \mathbb{E}e^{\theta X_\tau - \tau\kappa(\theta)}$ for a given stopping time with $\tau < \infty$, is that $\mathbb{P}_\theta(\tau < \infty) = 1$.*

Proof. Let

$$Z = L_\tau = e^{\theta X_\tau - \tau\kappa(\theta)} = e^{\theta X_\tau - \tau\kappa(\theta)}I(\tau < \infty).$$

Then the assertion means $\mathbb{E}Z = 1$, whereas (1.3) gives $\mathbb{E}Z = \widetilde{\mathbb{P}}(\tau < \infty)$. $\square$

From Theorem 1.3 we obtain a likelihood ratio representation of the ruin probability $\psi(u)$ parallel to the martingale representation II.(3.1) in Proposition II.3.1:

Corollary 1.5 *Under condition* (a) *of Proposition* II.3.1,

$$\psi(u) = e^{-\gamma u} \mathbb{E}_\gamma [e^{-\gamma \xi(u)}; \tau(u) < \infty]. \tag{1.5}$$

Proof. Letting $G = \{\tau(u) < \infty\}$, we have $\mathbb{P}(G) = \psi(u)$. Now just rewrite the r.h.s. of (1.2) by noting that

$$\frac{1}{L_{\tau(u)}} = e^{-\gamma S_{\tau(u)}} = e^{-\gamma u} e^{-\gamma \xi(u)}. \qquad \square$$

The advantage of (1.5) compared to II.(3.1) is that it seems in general easier to deal with the (unconditional) expectation $\mathbb{E}_\gamma [e^{-\gamma \xi(u)}; \tau(u) < \infty]$ occurring there than with the (conditional) expectation $\mathbb{E}[e^{\gamma \xi(u)} | \tau(u) < \infty]$ in II.(3.1). The crucial step is to obtain information on the process evolving according to $\widetilde{\mathbb{P}}$, and this problem will now be studied, first in the Markov case and next (Sections 3, 4) for processes with some random-walk-like structure.

As another simple application of the change-of-measure technique, we shall establish a formula for the finite-time ruin probability of Brownian motion:

Corollary 1.6 *Let* $\{R_t\}$ *be Brownian motion with drift* μ *and variance constant 1. Then the density and c.d.f. of* $\tau(u)$ *are*

$$\mathbb{P}_\mu(\tau(u) \in dT) = \frac{u}{\sqrt{2\pi}} T^{-3/2} \exp\left\{-\frac{(u - T\mu)^2}{2T}\right\}, \tag{1.6}$$

$$\mathbb{P}_\mu(\tau(u) \le T) = 1 - \Phi\left(\frac{u}{\sqrt{T}} - \mu\sqrt{T}\right) + e^{2\mu u} \Phi\left(-\frac{u}{\sqrt{T}} - \mu\sqrt{T}\right). \tag{1.7}$$

For a general variance constant σ^2, *one furthermore obtains*

$$\mathbb{P}(\tau(u) \le T) = \Phi\left(\frac{-u + \mu T}{\sigma\sqrt{T}}\right) + e^{2\mu u / \sigma^2} \Phi\left(\frac{-u - \mu\sqrt{T}}{\sigma\sqrt{T}}\right). \tag{1.8}$$

Proof. Consider first the case $\sigma^2 = 1$. For $\mu = 0$, (1.7) is the same as II.(2.8), and (1.6) follows then by straightforward differentiation. For $\mu \neq 0$, the ratio $d\mathbb{P}_\mu / d\mathbb{P}_0$ of densities of S_t is $e^{\mu S_t - t\mu^2 / 2}$, since $\kappa(\theta) = -\theta\mu + \theta^2 \sigma^2 / 2$. Hence

$$\mathbb{P}_\mu(\tau(u) \in dT) = \mathbb{E}_0\left[e^{\mu S_{\tau(u)} - \tau(u)\mu^2 / 2}; \tau(u) \in dT\right]$$

$$= e^{\mu u - T\mu^2 / 2} \mathbb{P}_0\left(\tau(u) \in dT\right)$$

$$= e^{\mu u - T\mu^2 / 2} \frac{u}{\sqrt{2\pi}} T^{-3/2} \exp\left\{-\frac{1}{2} \cdot \frac{u^2}{T}\right\},$$

which is the same as (1.6). (1.7) then follows by checking that the derivative of the r.h.s. is (1.6) and that the value at 0 is 0.

For a general σ^2, a completely analogous calculation can be done, but the analogue of (1.6) is more tedious to write down, so we omit the details. $\square$

The same argument as used for Corollary 1.6 also applies to Bernoulli random walk with $\theta \neq 1/2$, but we again omit the details.

Consider next a (time-homogeneous) Markov process $\{X_t\}$ with state space E, say, in continuous time (the discrete time case is parallel but slightly simpler). In the context of ruin probabilities, one would typically have $X_t = R_t$, $X_t = S_t$, $X_t = (J_t, R_t)$ or $X_t = (J_t, S_t)$, where $\{R_t\}$ is the risk reserve process, $\{S_t\} = \{u - R_t\}$ the claim surplus process and $\{J_t\}$ a process of supplementary variables possibly needed to make the process Markovian. A change of measure is performed by finding a process $\{L_t\}$ which is a martingale w.r.t. each $\mathbb{P}_x$, is non-negative and has $\mathbb{E}_x L_t = 1$ for all x, t. The problem is thus to investigate which characteristics of $\{X_t\}$ and $\{L_t\}$ ensure a given set of properties of the changed probability measure.

First we ask when the Markov property is preserved. To this end, we need the concept of a *multiplicative functional*. For the definition, we assume for simplicity that $\{X_t\}$ has D-paths, is Markov w.r.t. the natural filtration $\{\mathscr{F}_t\}$ on D_E and define $\{L_t\}$ to be a multiplicative functional if $\{L_t\}$ is adapted to $\{\mathscr{F}_t\}$, non-negative and

$$L_{t+s} = L_t \cdot (L_s \circ \theta_t) \tag{1.9}$$

$\mathbb{P}_x$-a.s. for all x, s, t, where θ_t is the shift operator. The precise meaning of this is the following: being $\mathscr{F}_t$-measurable, L_t has the form

$$L_t = \varphi_t\big(\{X_u\}_{0 \le u \le t}\big)$$

for some mapping $\varphi_t : D_E[0,t] \to [0, \infty)$, and then

$$L_s \circ \theta_t = \varphi_s\big(\{X_{t+u}\}_{0 \le u \le s}\big).$$

Theorem 1.7 *Let $\{X_t\}$ be Markov w.r.t. the natural filtration $\{\mathscr{F}_t\}$ on D_E, let $\{L_t\}$ be a non-negative martingale with $\mathbb{E}_x L_t = 1$ for all x, t and let $\widetilde{\mathbb{P}}_x$ be the probability measure given by $\widetilde{\mathbb{P}}_x(A) = \mathbb{E}_x[L_t; A]$. Then the family $\{\widetilde{\mathbb{P}}_x\}_{x \in E}$ defines a time-homogeneous Markov process if and only if $\{L_t\}$ is a multiplicative functional.*

Proof. Since both sides of (1.9) are $\mathscr{F}_{t+s}$ measurable, (1.9) is equivalent to

$$\mathbb{E}_x[L_{t+s} V_{t+s}] = \mathbb{E}_x\big[L_t \cdot (L_s \circ \theta_t) V_{t+s}\big] \tag{1.10}$$

for any $\mathscr{F}_{t+s}$-measurable r.v. V_{t+s}, which in turn is the same as

$$\mathbb{E}_x\big[L_{t+s}Z_t \cdot (Y_s \circ \theta_t)\big] \;=\; \mathbb{E}_x\big[L_t \cdot (L_s \circ \theta_t)Z_t \cdot (Y_s \circ \theta_t)\big] \qquad (1.11)$$

for any $\mathscr{F}_t$-measurable Z_t and any $\mathscr{F}_s$-measurable Y_s. Indeed, since $Z_t \cdot (Y_s \circ \theta_t)$ is $\mathscr{F}_{t+s}$-measurable, (1.10) implies (1.11). The converse follows since the class of r.v.'s of the form $Z_t \cdot (Y_s \circ \theta_t)$ comprises all r.v.'s of the form $\prod_1^n f_i(X_{t(i)})$ with all $t(i) \le t + s$.

Similarly, the Markov property can be written

$$\widetilde{\mathbb{E}}_x[Y_s \circ \theta_t | \mathscr{F}_t] \;=\; \widetilde{\mathbb{E}}_{X_t} Y_s, \quad t < s,$$

for any $\mathscr{F}_s$-measurable r.v. Y_s, which is the same as

$$\widetilde{\mathbb{E}}_x\big[Z_t(Y_s \circ \theta_t)\big] \;=\; \widetilde{\mathbb{E}}_x\big[Z_t \widetilde{\mathbb{E}}_{X_t} Y_s\big]$$

for any $\mathscr{F}_t$-measurable r.v. Z_t. By definition of $\widetilde{\mathbb{P}}_x$, this in turn means

$$\mathbb{E}_x\big[L_{t+s}Z_t(Y_s \circ \theta_t)\big] \;=\; \mathbb{E}_x\big[L_t Z_t \mathbb{E}_{X_t}[L_s Y_s]\big],$$

or, since $\mathbb{E}_{X_t}[L_s Y_s] = \mathbb{E}\big[(Y_s \circ \theta_t)(L_s \circ \theta_t) \,\big|\, \mathscr{F}_t\big]$,

$$\mathbb{E}_x\big[L_{t+s}Z_t(Y_s \circ \theta_t)\big] \;=\; \mathbb{E}_x\big[L_t Z_t(Y_s \circ \theta_t)(L_s \circ \theta_t)\big], \qquad (1.12)$$

which is the same as (1.11). □

Remark 1.8 For $\big\{\widetilde{\mathbb{P}}_x\big\}_{x \in E}$ to define a time-homogeneous Markov process, it suffices to assume that $\{L_t\}$ is a multiplicative functional with $\mathbb{E}_x L_t = 1$ for all x, t. Indeed, then

$$\mathbb{E}[L_{t+s} | \mathscr{F}_t] \;=\; L_t \mathbb{E}[L_s \circ \theta_t | \mathscr{F}_t] \;=\; L_t \mathbb{E}_{X_t} L_s \;=\; L_t,$$

(using the Markov property in the second step) so that the martingale property is automatic. □

Notes and references The results of the present section are essentially known in a very general Markov process formulation, see Dynkin [338] and Kunita [562]. A more elementary version along the lines of Theorem 1.7 can be found in Küchler & Sørensen [561], with a proof somewhat different from the present one. A further relevant reference is Barndorff-Nielsen & Shiryaev [139].

2 Duality with other applied probability models

In this section, we shall establish a general connection between ruin probabilities and certain stochastic processes which occur for example as models for queueing and storage. The formulation has applications to virtually all the risk models studied in this book.

The result is a sample path relation, and thus for the moment no parametric assumptions (on say the structure of the arrival process) are needed. We work on a finite time interval $[0, T]$ in the following set-up (which can be much generalized):

The risk process $\{R_t\}_{0 \leq t \leq T}$ has arrivals at epochs $\sigma_1, \ldots, \sigma_N$, $0 \leq \sigma_1 \leq \ldots \leq \sigma_N \leq T$. The corresponding claim sizes are $U_1, \ldots, U_N$. In between jumps, the premium rate is $p(r) > 0$ when the reserve is r (i.e., $\dot{R} = p(R)$). Thus

$$R_t = R_0 + \int_0^t p(R_s)\, \mathrm{d}s - A_t \quad \text{where} \quad A_t = \sum_{k:\, \sigma_k \leq t} U_k. \tag{2.1}$$

The initial condition is arbitrary, $R_0 = u$ (say), and the time to ruin is $\tau(u) = \inf\{t \geq 0 : R_t < 0\}$.

The storage process $\{V_t\}_{0 \leq t \leq T}$ is essentially defined by time-reversion, reflection at zero and initial condition $V_0 = 0$. More precisely, the arrival epochs are $\sigma_1^*, \ldots, \sigma_N^*$ where $\sigma_k^* = T - \sigma_{N-k+1}$, and just after time σ_k^* $\{V_t\}$ makes an upwards jump of size $U_k^* = U_{N-k+1}$. In between jumps, $\{V_t\}$ decreases at rate $p(r)$ when $V_t = r$ (i.e., $\dot{V} = -p(V)$). That is, instead of (2.1) we have

$$V_t = A_t^* - \int_0^t p(V_s)\, \mathrm{d}s \quad \text{where} \quad A_t^* = \sum_{k:\, \sigma_k^* < t} U_k^* = A_T - A_{T-t},$$

and we use the convention $p(0) = 0$ to make zero a reflecting barrier (when hitting 0, $\{V_t\}$ remains at 0 until the next arrival).

Note that these definitions make $\{R_t\}$ right-continuous (as standard) and $\{V_t\}$ left-continuous. The sample path relation between these two processes is illustrated in Fig. III.1.

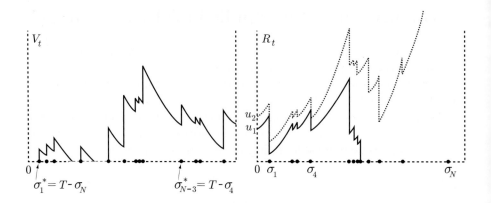

FIGURE III.1

Define $\tau(u) = \inf\{t \geq 0 : R_t < 0\}$ ($\tau(u) = \infty$ if $R_t \geq 0$ for all $t \leq T$) and let

$$\psi(u, T) = \mathbb{P}\left(\inf_{0 \leq t \leq T} R_t < 0\right) = \mathbb{P}(\tau(u) \leq T)$$

be the finite-time ruin probability.

Theorem 2.1 *The events* $\{\tau(u) \leq T\}$ *and* $\{V_T > u\}$ *coincide. In particular,*

$$\psi(u, T) = \mathbb{P}(V_T > u). \tag{2.2}$$

Proof. Let $r_t^{(u)}$ denote the solution of $\dot{R} = p(R)$ subject to $r_0^{(u)} = u$. Then $r_t^{(u)} > r_t^{(v)}$ for all t when $u > v$.

Suppose first $V_T > u$ (this situation corresponds to the solid path of $\{R_t\}$ in Fig. III.1 with $R_0 = u = u_1$). Then

$$V_{\sigma_N^*} = r_{\sigma_1}^{(V_T)} - U_1 > r_{\sigma_1}^{(u)} - U_1 = R_{\sigma_1}.$$

If $V_{\sigma_N^*} > 0$, we can repeat the argument and get $V_{\sigma_{N-1}^*} > R_{\sigma_2}$ and so on. Hence if n satisfies $V_{\sigma_{N-n+1}^*} = 0$ (such an n exists, if nothing else $n = N$), we have $R_{\sigma_n} < 0$ so that indeed $\tau(u) \leq T$.

Suppose next $V_T \leq u$ (this situation corresponds to the dotted path of $\{R_t\}$ in Fig. III.1 with $R_0 = u = u_2$). Then similarly

$$V_{\sigma_N^*} = r_{\sigma_1}^{(V_T)} - U_1 \leq r_{\sigma_1}^{(u)} - U_1 = R_{\sigma_1}, \quad V_{\sigma_{N-1}^*} \leq R_{\sigma_2},$$

and so on. Hence $R_{\sigma_n} \geq 0$ for all $n \leq N$, and since ruin can only occur at the times of claims, we have $\tau(u) > T$. $\square$

A basic example is when $\{R_t\}$ is the risk reserve process corresponding to claims arriving at Poisson rate β and being i.i.d. with distribution B, and a general premium rule $p(r)$ when the reserve is r. Then the time reversibility of the Poisson process ensures that $\{A_t\}$ and $\{A_t^*\}$ have the same distribution (for finite-dimensional distributions, the distinction between right- and left continuity is immaterial because the probability of a Poisson arrival at any fixed time t is zero). Thus we may think of $\{V_t\}$ as having compound Poisson input and being defined for all $t < \infty$. Historically, this represents a model for storage, say of water in a dam though other interpretations like the amount of goods stored are also possible. The arrival epochs correspond to rainfalls, and in between rainfalls water is released at rate $p(r)$ when V_t (the *content*) is r. We get:

Corollary 2.2 *Consider the compound Poisson risk model with a general premium rule $p(r)$. Then the storage process $\{V_t\}$ has a proper limit in distribution as $t \to \infty$, say V, if and only if $\psi(u) < 1$ for all u, and then*

$$\psi(u) \;=\; \mathbb{P}(V > u). \tag{2.3}$$

Proof. Let $T \to \infty$ in (2.2). $\qquad\qquad\qquad\qquad\qquad\qquad\qquad\qquad\square$

Consider now a compound Poisson risk model with constant premium rate 1 and claim arrival rate β. Then a direct relationship can be obtained between the survival probability in the risk model and the maximum workload $V_{\max}(u)$ of a busy period in an M/G/1 queue with arrival rate β.

Theorem 2.3 *Under the safety loading condition $\eta > 0$, we have for every $u \geq 0$*

$$\mathbb{P}(V_{\max} > u) = \frac{1}{\beta} \frac{\mathrm{d}}{\mathrm{d}u} \log \phi(u), \tag{2.4}$$

where $\phi(u) = 1 - \psi(u)$ is the survival probability.

Proof. The risk process R_t starting in u can only survive if after each claim that occurs at some running maximum $s > u$, the level s will be reached again before ruin occurs. This is equivalent to the statement that the maximum workload $V_{\max}$ of a busy period in an M/G/1 queue (with traffic intensity $\rho < 1$) does not exceed s. As we are only concerned about eventual survival, we can cut out every such 'surviving' excursion away from the running maximum of the risk process and only consider those claims at the running maximum which lead to a downward excursion causing ruin before recovering to level s. The instantaneous probability of having a claim of the latter type is

$$\beta \, \mathrm{d}t \, \mathbb{P}(V_{\max} > s) = \beta \, \mathrm{d}s \, \mathbb{P}(V_{\max} > s),$$

since at the running maximum we have $ds = dt$. Consequently, the survival probability $\phi(u)$ can simply be interpreted as the probability to have zero events during $[u, \infty)$ of an inhomogeneous Poisson process with rate $\beta(s) = \beta \, \mathbb{P}(V_{\max} > s)$ (which constitutes a thinning of the original Poisson process). This finally implies

$$\phi(u) \;=\; \exp\Big(-\int_u^\infty \beta(s)\,ds\Big) \;=\; \exp\Big(-\beta \int_u^\infty \mathbb{P}(V_{\max} > s)\,ds\Big). \qquad (2.5)$$

$\square$

The relation (2.4) also follows from a combination of (2.3) and the identity

$$\mathbb{P}(V_{\max} > u) = \frac{1}{\beta} \frac{d}{du} \log \mathbb{P}(V < u)$$

for an M/G/1 queue (see for instance Cohen [249, p. 618]), but the above proof establishes a direct and self-contained link between $V_{\max}$ and $\phi(u)$ that will also be useful later on (cf. VIII.4).

Notes and references   Two main references on storage processes are Harrison & Resnick [451] and Brockwell, Resnick & Tweedie [203]. Theorem 2.1 and its proof is from Asmussen & Schock Petersen [104], Corollary 2.2 from Harrison & Resnick [452]. The results can be viewed as special cases of *Siegmund duality*, see Siegmund [808]. Some further more general references are Asmussen [63] and Asmussen & Sigman [105].

Theorem 2.3 is due to Albrecher, Borst, Boxma & Resing [16].

Historically, the connection between risk theory and other applied probability areas appears first to have been noted by Prabhu [711] in a queueing context. It is a standard tool today, but one may feel that the interaction between the different areas was surprisingly limited in the first decades after the appearance of [711].

3   Random walks in discrete or continuous time

Consider a random walk $X_n = X_0 + Y_1 + \cdots + Y_n$ in discrete time where the Y_i are i.i.d., with common distribution F.

For discrete time random walks, there is an analogue of Theorem 2.1 in terms of *Lindley processes*. For a given i.i.d. $\mathbb{R}$-valued sequence $Z_1, Z_2, \ldots$, the Lindley process $W_0, W_1, W_2, \ldots$ generated by $Z_1, Z_2, \ldots$ is defined by assigning W_0 some arbitrary value ≥ 0 and letting

$$W_{n+1} \;=\; (W_n + Z_{n+1})^+. \qquad (3.1)$$

Thus $\{W_n\}_{n=0,1,\ldots}$ evolves as a random walk with increments $Z_1, Z_2, \ldots$ as long as the random walk only takes non-negative values, and is reset to 0 once the random walk hits $(-\infty, 0)$. I.e., $\{W_n\}_{n=0,1,\ldots}$ can be viewed as the reflected version of the random walk with increments $Z_1, Z_2, \ldots$ In particular, if $W_0 = 0$ then

$$W_N = Z_1 + \cdots + Z_N - \min_{n=0,1,\ldots,N}(Z_1 + \cdots + Z_n) \qquad (3.2)$$

(for a rigorous proof, just verify that the r.h.s. of (3.2) satisfies the same recursion as in (3.1)).

Theorem 3.1 *Let* $\tau(u) = \inf\{n : u + Y_1 + \cdots + Y_n < 0\}$. *Let further N be fixed and let* $W_0, W_1, \ldots, W_N$ *be the Lindley process generated by $Z_1 = -Y_N$, $Z_2 = -Y_{N-1}, \ldots, Z_N = -Y_1$ according to $W_0 = 0$. Then the events $\{\tau(u) \leq N\}$ and $\{W_N > u\}$ coincide.*

Proof. By (3.2),

$$
\begin{aligned}
W_N &= -Y_N - \cdots - Y_1 - \min_{n=0,1,\ldots,N}(-Y_N - \cdots - Y_{N-n+1}) \\
&= -\min_{n=0,1,\ldots,N}(Y_1 + \cdots + Y_{N-n}) = -\min_{n=0,1,\ldots,N}(Y_1 + \cdots + Y_n).
\end{aligned}
$$

From this the result immediately follows. $\qquad\square$

Corollary 3.2 *The following assertions are equivalent:*
(a) $\psi(u) = \mathbb{P}(\tau(u) < \infty) < 1$ *for all* $u \geq 0$;
(b) $\psi(u) = \mathbb{P}(\tau(u) < \infty) \to 0$ *as* $u \to \infty$;
(c) *The Lindley process $\{W_N\}$ generated by $Z_1 = -Y_1$, $Z_2 = -Y_2$, $\ldots$ has a proper limit W in distribution as $n \to \infty$;*
(d) $m = \inf_{n=0,1,\ldots}(Y_1 + \cdots + Y_n) > -\infty$ *a.s.;*
(e) $Y_1 + \cdots + Y_n \not\to -\infty$ *a.s.*
In that case, $W \overset{\mathcal{D}}{=} -m$ *and* $\mathbb{P}(W > u) = \mathbb{P}(-m > u) = \psi(u)$.

Proof. Since $(Y_N, \ldots, Y_1)$ has the same distribution as $(Y_1, \ldots, Y_N)$, the Lindley processes in Corollary 3.2 and Theorem 3.1 have the same distribution for $n = 0, 1, \ldots, N$. Thus the assertion of Theorem 3.1 is equivalent to

$$W_N \overset{\mathcal{D}}{=} M_N = \sup_{n=0,1,\ldots,N}(Z_1 + \cdots + Z_n)$$

so that $W_N \overset{\mathcal{D}}{\to} M = \sup_{n=0,1,\ldots}(Z_1 + \cdots + Z_n) = -m$ and $\mathbb{P}(W > u) = \mathbb{P}(M > u) = \psi(u)$. By Kolmogorov's 0-1 law, either $M = \infty$ a.s. or $M < \infty$ a.s. Combining these facts gives easily the equivalence of (a)–(d).

Clearly, (d) $\Rightarrow$ (e). The converse follows from general random walk theory since it is standard that $\limsup(Y_1 + \cdots + Y_n) = \infty$ when $Y_1 + \cdots + Y_n \not\to -\infty$. $\qquad\square$

By the law of large numbers, a sufficient condition for (e) is that $\mathbb{E}Y$ is well-defined and ≥ 0. In general, the condition

$$\sum_{n=1}^{\infty} \frac{1}{n} \mathbb{P}(Y_1 + \cdots + Y_n < 0) < \infty$$

is known to be necessary and sufficient ([APQ, p. 231]) but appears to be rather intractable.

Remark 3.3 The i.i.d. assumption on the $Z_1, \ldots, Z_N$ (or, equivalently, on the $Y_1, \ldots, Y_N$) in Theorem 3.1 is actually not necessary — the result is a sample path relation as is Theorem 2.1. Similarly, there is a more general version of Corollary 3.2. One then assumes Y_n to be a stationary sequence, w.l.o.g. doubly infinite $(n = 0, \pm 1, \pm 2, \ldots)$ and defines $Z_n = -Y_{-n}$. □

Next consider change of measure via likelihood ratios.

For a random walk, a Markovian change of measure as in Theorem 1.7 does not necessarily lead to a random walk: if, e.g., F has a strictly positive density and $\widetilde{\mathbb{P}}_x$ corresponds to a Markov chain such that the density of X_1 given $X_0 = x$ is also strictly positive, then the restrictions of $\mathbb{P}_x, \widetilde{\mathbb{P}}_x$ to $\mathscr{F}_n$ are equivalent (have the same null sets) so that the likelihood ratio L_n exists. The following result gives the necessary and sufficient condition for $\{L_n\}$ to define a new random walk:

Proposition 3.4 *Let $\{L_n\}$ be a multiplicative functional of a random walk with $\mathbb{E}_x L_n = 1$ for all n and x. Then the change of measure in Theorem 1.7 corresponds to a new random walk if and only if*

$$L_n = h(Y_1) \cdots h(Y_n) \tag{3.3}$$

$\mathbb{P}_x$-a.s. for some function h with $\mathbb{E}h(Y) = 1$. In that case, the changed increment distribution is $\widetilde{F}(x) = \mathbb{E}[h(Y); Y \leq x]$.

Proof. If (3.3) holds, then

$$\widetilde{\mathbb{E}}_x \prod_{i=1}^{n} f_i(Y_i) = \mathbb{E}_x \prod_{i=1}^{n} f_i(Y_i) h(Y_i)$$

$$= \prod_{i=1}^{n} \mathbb{E} f_i(Y_i) h(Y_i) = \prod_{i=1}^{n} \int f_i(y) \widetilde{F}(dy)$$

from which the random walk property is immediate with the asserted form of $\widetilde{F}$. Conversely, the random walk property implies $\widetilde{\mathbb{E}}_x f(Y_1) = \widetilde{\mathbb{E}}_0 f(Y_1)$. Since L_1 has the form $g(X_0, Y_1)$, this means $\mathbb{E}[g(x, Y)f(Y)] = \mathbb{E}[g(0, Y)f(Y)]$ for all f and x, implying $g(x, Y) = h(Y)$ a.s. where $h(y) = g(0, y)$. In particular, (3.3) holds for $n = 1$. For $n = 2$, we get

$$L_2 = L_1(L_1 \circ \theta_1) = h(Y_1)g(X_1, Y_2) = h(Y_1)h(Y_2),$$

and so on for $n = 3, 4, \ldots$. □

A particular important example is exponential change of measure ($h(y) = \mathrm{e}^{\alpha y - \kappa(\alpha)}$ where $\kappa(\alpha) = \log \widehat{F}[\alpha]$ is the c.g.f. of F). The corresponding likelihood ratio is

$$L_n = \exp\{\alpha(Y_1 + \cdots + Y_n) - n\kappa(\alpha)\}. \tag{3.4}$$

Thus $\{L_n\}$ is the Wald martingale, cf. II.1. We get:

Corollary 3.5 *Consider a random walk and an α such that*

$$\kappa(\alpha) = \log \widehat{F}[\alpha] = \log \mathbb{E}\mathrm{e}^{\alpha Y}$$

is finite, and define L_n by (3.4). Then the change of measure in Theorem 1.7 corresponds to a new random walk with changed increment distribution

$$\widetilde{F}(x) = \mathrm{e}^{-\kappa(\alpha)} \int_{-\infty}^{x} \mathrm{e}^{\alpha y} F(\mathrm{d}y).$$

Discrete time random walks have classical applications in queueing theory via the Lindley process representation of the waiting time, see Chapter VI. In risk theory, they arise as models for the reserve or claim surplus at a discrete sequence of instants, say the beginning of each month or year, or imbedded into continuous time processes, say by recording the reserve or claim surplus just before or just after claims (see Chapter VI for some fundamental examples). However, the tradition in the area is to use continuous time models.

Now consider reflected versions of Lévy processes (cf. II.1).

First assume in the setting of Section that $\{R_t\}$ is the risk reserve process for the compound Poisson risk model with constant premium rate $p(r) \equiv 1$. Then the storage process $\{V_t\}$ has constant release rate 1, i.e. has upwards jumps governed by B at the epochs of a Poisson process with rate β and decreases linearly at rate 1 in between jumps. A different interpretation is as the *workload* or *virtual waiting time* process in an M/G/1 queue, defined as a system with a single server working at a unit rate, having Poisson arrivals with rate β and distribution B of the service times of the arriving customers. Here 'workload' refers to the fact that we can interpret V_t as the amount of time the server will have to work until the system is empty provided no new customers arrive; virtual waiting time refers to V_t being the amount of time a customer would have to wait before starting service if he arrived at time t (this interpretation requires FIFO = First In First Out queueing discipline: the customers are served in the order of arrival).

Corollary 3.6 *In the compound Poisson risk model with constant premium rate $p(r) \equiv 1$, $\psi(u, T) = \mathbb{P}(V_T > u)$, where V_T is the virtual waiting time at time T in an initially empty $M/G/1$ queue with the same arrival rate β and the service times having the same distribution B as the claims in the risk process. Furthermore, $V_T \xrightarrow{\mathscr{D}} V$ for some r.v. $V \in [0, \infty]$, and $\psi(u) = \mathbb{P}(V > u)$.*

[The condition for $V < \infty$ a.s. is easily seen to be $\beta \mu_B < 1$, cf. Chapter IV.]

Processes with a more complicated path structure like Brownian motion or jump processes with unbounded Lévy measure are not covered by Section 2, and the reflected version is then defined by means of the abstract reflection operator as in (3.2),

$$W_T = X_T - \min_{0 \le t \le T} X_t$$

(assuming $W_0 = X_0 = 0$ for simplicity).

Proposition 3.7 *If $\{X_t\}$ is a Lévy process of the form $X_t = X_0 + \mu t + \sigma B_t + M_t$ as in II.(1.4), then*

$$\mathbb{E}e^{\alpha(X_t - X_0)} = \mathbb{E}_0 e^{\alpha X_t} = e^{t\kappa(\alpha)}, \qquad (3.5)$$

where

$$\kappa(\alpha) = \alpha\mu + \alpha^2 \sigma^2/2 + \int_{-\infty}^{\infty} (e^{\alpha x} - 1)\nu(\mathrm{d}x) \qquad (3.6)$$

provided the Lévy measure of the jump part $\{M_t\}$ satisfies $\int_{-\epsilon}^{\epsilon} |x|\, \nu(\mathrm{d}x) < \infty$.

Proof. This is basically an easy application of formulas II.(1.7), II.(1.8). To repeat: by standard formulas for the normal distribution,

$$\mathbb{E}e^{\alpha(\mu t + \sigma B_t)} = e^{t\{\alpha\mu + \alpha^2\sigma^2/2\}}.$$

By explicit calculation, we show in the compound Poisson case ($\|\nu\| < \infty$) in Proposition IV.1.1 that

$$\mathbb{E}e^{\alpha M_t} = \exp\left\{\int_{-\infty}^{\infty} (e^{\alpha x} - 1)\nu(\mathrm{d}x)\right\}.$$

In the general case, use the representation as limit of compound Poisson processes.   □

Note that (3.6) is the Lévy-Khinchine representation of the c.g.f. of an infinitely divisible distribution (see, e.g., Chung [246]). This is of course no coincidence since the distribution of $X_1 - X_0$ is necessarily infinitely divisible when $\{X_t\}$ has stationary independent increments.

Theorem 3.8 *Assume that $\{X_t\}$ is a Lévy process with $\int_{-\epsilon}^{\epsilon} |x|\,\nu(\mathrm{d}x) < \infty$, and that $\{L_t\}$ is a non-negative multiplicative functional of the form $L_t = g(t, X_t - X_0)$ with $\mathbb{E}_x L_t = 1$ for all x, t. Then the Markov process given by Theorem 1.7 is again a Lévy process. In particular, if $L_t = \mathrm{e}^{\theta(X_t - X_0) - t\kappa(\theta)}$, then the changed parameters in the representation (1.4) are*

$$\widetilde{\mu} = \mu + \theta\sigma^2, \quad \widetilde{\sigma}^2 = \sigma^2, \quad \widetilde{\nu}(\mathrm{d}x) = \mathrm{e}^{\theta x}\nu(\mathrm{d}x).$$

Proof. For the first statement, we use the characterization (1.3) and get

$$
\begin{aligned}
\widetilde{\mathbb{E}}\big[f(X_{t+s} - X_t)\,\big|\,\mathscr{F}_t\big] &= \mathbb{E}\big[f(X_{t+s} - X_t)L_s \circ \theta_t\,\big|\,\mathscr{F}_t\big] \\
&= \mathbb{E}\big[f(X_{t+s} - X_t)g(s, X_{t+s} - X_t)\,\big|\,\mathscr{F}_t\big] \\
&= \mathbb{E}_0 f(X_s)g(s, X_s) = \mathbb{E}_0 f(X_s)L_s \\
&= \widetilde{\mathbb{E}}_0 f(X_s).
\end{aligned}
$$

For the second, let $\mathrm{e}^{\widetilde{\kappa}(\alpha)} = \widetilde{\mathbb{E}}_0 \mathrm{e}^{\alpha X_1}$. Then

$$
\begin{aligned}
\mathrm{e}^{\widetilde{\kappa}(\alpha)} &= \mathbb{E}_0\big[L_1 \mathrm{e}^{\alpha X_1}\big] = \mathrm{e}^{-\kappa(\theta)}\mathbb{E}_0\big[\mathrm{e}^{(\alpha+\theta)X_1}\big] = \mathrm{e}^{\kappa(\alpha+\theta)-\kappa(\theta)}, \\
\widetilde{\kappa}(\alpha) &= \kappa(\alpha+\theta) - \kappa(\theta) \\
&= \alpha\mu + \big((\alpha+\theta)^2 - \theta^2\big)\sigma^2/2 + \int_{-\infty}^{\infty}\big(\mathrm{e}^{(\alpha+\theta)x} - \mathrm{e}^{\theta x}\big)\,\nu(\mathrm{d}x) \\
&= \alpha(\mu + \theta\sigma^2) + \alpha^2\sigma^2/2 + \int_{-\infty}^{\infty}(\mathrm{e}^{\alpha x} - 1)\mathrm{e}^{\theta x}\nu(\mathrm{d}x).
\end{aligned}
$$

    □

Remark 3.9 If $X_0 = 0$, then the martingale $\big\{\mathrm{e}^{\theta X(t) - t\kappa(\theta)}\big\}$ is the continuous time analogue of the Wald martingale (3.4).   □

Example 3.10 Let X_t be the claim surplus process of a compound Poisson risk process with Poisson rate β and claim size distribution B, corresponding to $\mu = -1$, $\sigma = 0$, $\nu(\mathrm{d}x) = \beta B(\mathrm{d}x)$. Then we can write

$$\widetilde{\nu}(\mathrm{d}x) = \beta\mathrm{e}^{\theta x}B(\mathrm{d}x) = \widetilde{\beta}\widetilde{B}(\mathrm{d}x), \quad \text{where } \widetilde{\beta} = \beta\widehat{B}[\theta], \quad \widetilde{B}(\mathrm{d}x) = \frac{\mathrm{e}^{\theta x}}{\widehat{B}[\theta]}B(\mathrm{d}x).$$

Thus (since $\widetilde{\mu} = \mu = -1$, $\widetilde{\sigma} = \sigma = 0$) *the changed process is the claim surplus process of another compound Poisson risk process with Poisson rate $\widetilde{\beta}$ and claim size distribution $\widetilde{B}$.* □

Example 3.11 For an example of a likelihood ratio not covered by Theorem 3.8, let the given Markov process (specified by the $\mathbb{P}_x$) be the claim surplus process of a compound Poisson risk process with Poisson rate β and claim size distribution B, and let the $\widetilde{\mathbb{P}}_x$ refer to the claim surplus process of another compound Poisson risk process with Poisson rate $\widetilde{\beta} = \beta$ and claim size distribution $\widetilde{B} \neq B$. Recalling that $\sigma_1, \sigma_2, \ldots$ are the arrival times and $U_1, U_2, \ldots$ the corresponding claim sizes, it is then easily seen that

$$L_t = \prod_{i:\ \sigma_i \le t} \frac{\mathrm{d}\widetilde{B}}{\mathrm{d}B}(U_i)$$

whenever the Radon-Nikodym derivative $\mathrm{d}\widetilde{B}/\mathrm{d}B$ exists (e.g. $\mathrm{d}\widetilde{B}/\mathrm{d}B = \widetilde{b}/b$ when $B, \widetilde{B}$ have densities $b, \widetilde{b}$ with $b(x) > 0$ for all x such that $\widetilde{b}(x) > 0$). □

4 Markov additive processes

A Markov additive process, abbreviated as MAP in this section[2], is defined as a bivariate Markov process $\{X_t\} = \{(J_t, S_t)\}$, where $\{J_t\}$ is a Markov process with state space E (say) and the increments of $\{S_t\}$ are governed by $\{J_t\}$ in the sense that

$$\mathbb{E}\big[f(S_{t+s} - S_t)g(J_{t+s}) \,\big|\, \mathscr{F}_t\big] = \mathbb{E}_{J_t,0}\big[f(S_s)g(J_s)\big]. \qquad (4.1)$$

For shorthand, we write $\mathbb{P}_i, \mathbb{E}_i$ instead of $\mathbb{P}_{i,0}, \mathbb{E}_{i,0}$ in the following.

As for processes with stationary independent increments, the structure of MAP's is completely understood when E is finite:

In discrete time, a MAP is specified by the measure-valued matrix (kernel) $\boldsymbol{F}(\mathrm{d}x)$ whose ijth element is the defective probability distribution $F_{ij}(\mathrm{d}x) = \mathbb{P}_i(J_1 = j, Y_1 \in \mathrm{d}x)$ where $Y_n = S_n - S_{n-1}$. An alternative description is in terms of the transition matrix $\boldsymbol{P} = (p_{ij})_{i,j \in E}$ (here $p_{ij} = \mathbb{P}_i(J_1 = j)$) and the probability measures

$$H_{ij}(\mathrm{d}x) = \mathbb{P}\big(Y_1 \in \mathrm{d}x \,\big|\, J_0 = i, J_1 = j\big) = \frac{F_{ij}(\mathrm{d}x)}{p_{ij}}.$$

---

[2] and only there; one reason is that in parts of the applied probability literature, MAP stands for the Markovian arrival process discussed below.

In simulation language, this means that the MAP can be simulated by first simulating the Markov chain $\{J_n\}$ and next the $Y_1, Y_2, \ldots$ by generating Y_n according to H_{ij} when $J_{n-1} = i$, $J_n = j$.

If all F_{ij} are concentrated on $(0, \infty)$, a MAP is the same as a semi-Markov or Markov renewal process, with the Y_n being interpreted as interarrival times.

In continuous time (assuming D-paths), $\{J_t\}$ is specified by its intensity matrix $\mathbf{\Lambda} = (\lambda_{ij})_{i,j \in E}$. On an interval $[t, t+s)$ where $J_t \equiv i$, $\{S_t\}$ evolves like a Lévy process and the parameters μ_i, σ_i^2, $\nu_i(dx)$ in II.(1.4) depending on i. In addition, a jump of $\{J_t\}$ from i to $j \neq i$ has probability q_{ij} of giving rise to a jump of $\{S_t\}$ at the same time, the distribution of which has some distribution B_{ij}. [That a process with this description is a MAP is obvious; the converse requires a proof, which we omit and refer to Neveu [663] or Çinlar [247].]

If E is infinite a MAP may be much more complicated. As an example, let $\{J_t\}$ be standard Brownian motion on the line. Then a Markov additive process can be defined by letting

$$S_t = \lim_{\epsilon \downarrow 0} \frac{1}{2\epsilon t} \int_0^t I\big(|J_s| \leq \epsilon\big) \, ds$$

be the local time at 0 up to time t.

As a generalization of the m.g.f., consider the matrix $\widehat{\boldsymbol{F}}_t[\alpha]$ with ijth element $\mathbb{E}_i\big[e^{\alpha S_t}; J_t = j\big]$.

Proposition 4.1 *For a MAP in discrete time and with E finite, $\widehat{\boldsymbol{F}}_n[\alpha] = \widehat{\boldsymbol{F}}[\alpha]^n$ where*

$$\widehat{\boldsymbol{F}}[\alpha] = \widehat{\boldsymbol{F}}_1[\alpha] = \big(\mathbb{E}_i[e^{\alpha S_1}; J_1 = j]\big)_{i,j \in E} = \big(\widehat{F}_{ij}[\alpha]\big)_{i,j \in E} = \big(p_{ij}\widehat{H}_{ij}[\alpha]\big)_{i,j \in E}.$$

Proof. Conditioning upon (J_n, S_n) yields

$$\mathbb{E}_i[e^{\alpha S_{n+1}}; J_{n+1} = j] = \sum_{k \in E} \mathbb{E}_i[e^{\alpha S_n}; J_n = k] \, \mathbb{E}_k[e^{\alpha Y_1}; J_1 = j],$$

which in matrix formulation is the same as $\widehat{\boldsymbol{F}}_{n+1}[\alpha] = \widehat{\boldsymbol{F}}_n[\alpha]\widehat{\boldsymbol{F}}[\alpha]$. □

Proposition 4.2 *Let E be finite and consider a continuous time Markov additive process with parameters $\mathbf{\Lambda}$, μ_i, σ_i^2, $\nu_i(dx)$ $(i \in E)$, q_{ij}, B_{ij} $(i, j \in E)$ and*

$S_0 = 0$. *Then the matrix* $\widehat{\boldsymbol{F}}_t[\alpha]$ *with* ijth *element* $\mathbb{E}_i\left[e^{\alpha S_t}; J_t = j\right]$ *is given by* $e^{t\boldsymbol{K}[\alpha]}$, *where*

$$\boldsymbol{K}[\alpha] = \boldsymbol{\Lambda} + \left(\kappa^{(i)}(\alpha)\right)_{\text{diag}} + \left(\lambda_{ij}q_{ij}(\widehat{B}_{ij}[\alpha] - 1)\right),$$

$$\kappa^{(i)}(\alpha) = \alpha\mu_i + \alpha^2\sigma_i^2/2 + \int_{-\infty}^{\infty} (e^{\alpha x} - 1)\nu_i(dx).$$

Proof. Let $\left\{S_t^{(i)}\right\}$ be a Lévy process with parameters μ_i, σ_i^2, $\nu_i(dx)$. Then, up to $o(h)$ terms,

$$\mathbb{E}_i\left[e^{\alpha S_{t+h}}; J_{t+h} = j\right]$$
$$= (1 + \lambda_{jj}h)\mathbb{E}_i\left[e^{\alpha S_t}; J_t = j\right]\mathbb{E}_j e^{S_h^{(j)}}$$
$$+ \sum_{k \neq j}\lambda_{kj}h\mathbb{E}_i\left[e^{\alpha S_t}; J_t = k\right]\left\{1 - q_{kj} + q_{kj}\widehat{B}_{kj}[\alpha]\right\}$$
$$= \mathbb{E}_i\left[e^{\alpha S_t}; J_t = j\right]\left(1 + h\kappa^{(j)}(\alpha)\right)$$
$$+ h\sum_{k \in E}\mathbb{E}_i\left[e^{\alpha S_t}; J_t = k\right]\left\{\lambda_{kj} + \lambda_{kj}q_{kj}(\widehat{B}_{kj}[\alpha] - 1)\right\}$$

(recall that $q_{jj} = 0$). In matrix formulation, this means that

$$\widehat{\boldsymbol{F}}_{t+h}[\alpha] = \widehat{\boldsymbol{F}}_t[\alpha]\left(\boldsymbol{I} + h\left(\kappa^{(i)}(\alpha)\right)_{\text{diag}} + h\boldsymbol{\Lambda} + h\left(\lambda_{ij}q_{ij}(\widehat{B}_{ij}[\alpha] - 1)\right)\right),$$
$$\widehat{\boldsymbol{F}}_t'[\alpha] = \widehat{\boldsymbol{F}}_t[\alpha]\boldsymbol{K},$$

which in conjunction with $\widehat{\boldsymbol{F}}_0[\alpha] = \boldsymbol{I}$ implies $\widehat{\boldsymbol{F}}_t[\alpha] = e^{t\boldsymbol{K}[\alpha]}$ according to the standard solution formula for systems of linear differential equations. $\square$

In the following, assume that the Markov chain/process $\{J_t\}$ is ergodic. By Perron-Frobenius theory (see A.4c), we infer that in the discrete time case the matrix $\widehat{\boldsymbol{F}}[\alpha]$ has a real eigenvalue $\kappa(\alpha)$ with maximal absolute value and that in the continuous time case $\boldsymbol{K}[\alpha]$ has a real eigenvalue $\kappa(\alpha)$ with maximal real part. The corresponding left and right eigenvectors $\boldsymbol{\nu}^{(\alpha)}$, $\boldsymbol{h}^{(\alpha)}$ may be chosen with strictly positive components. Since $\boldsymbol{\nu}^{(\alpha)}$, $\boldsymbol{h}^{(\alpha)}$ are only given up to a constants, we are free to impose two normalizations, and we shall take

$$\boldsymbol{\nu}^{(\alpha)}\boldsymbol{h}^{(\alpha)} = 1, \quad \boldsymbol{\pi}\boldsymbol{h}^{(\alpha)} = 1,$$

where $\boldsymbol{\pi} = \boldsymbol{\nu}^{(0)}$ is the stationary distribution. Then $\boldsymbol{h}^{(0)} = \boldsymbol{e}$.

The function $\kappa(\alpha)$ plays in many respects the same role as the cumulant g.f. of a random walk, as will be seen from the following results. In particular, its derivatives are 'asymptotic cumulants', cf. Corollary 4.7, and appropriate generalizations of the Wald martingale (and the associated change of measure) can be defined in terms of $\kappa(\alpha)$ (and $\boldsymbol{h}^{(\alpha)}$), cf. Proposition 4.4.

Corollary 4.3 $\mathbb{E}_i\left[e^{\alpha S_t}; J_t = j\right] \sim h_i^{(\alpha)}\nu_j^{(\alpha)}e^{t\kappa(\alpha)}$.

Proof. By Perron-Frobenius theory (see A.4c). □

We also get an analogue of the Wald martingale for random walks:

Proposition 4.4 $\mathbb{E}_i e^{\alpha S_t}h_{J_t}^{(\alpha)} = h_i^{(\alpha)}e^{t\kappa(\alpha)}$. *Furthermore,*

$$\left\{e^{\alpha S_t - t\kappa(\alpha)}h_{J_t}^{(\alpha)}\right\}_{t\geq 0}$$

is a martingale.

Proof. For the first assertion, just note that

$$\mathbb{E}_i e^{\alpha S_t}h_{J_t}^{(\alpha)} = \boldsymbol{e}_i^{\mathsf{T}}\widehat{\boldsymbol{F}}_t[\alpha]\boldsymbol{h}^{(\alpha)} = \boldsymbol{e}_i^{\mathsf{T}}e^{t\boldsymbol{K}[\alpha]}\boldsymbol{h}^{(\alpha)} = \boldsymbol{e}_i^{\mathsf{T}}e^{t\kappa(\alpha)}\boldsymbol{h}^{(\alpha)} = e^{t\kappa(\alpha)}h_i^{(\alpha)}.$$

It then follows that

$$
\begin{aligned}
&\mathbb{E}\left[e^{\alpha S_{t+v} - (t+v)\kappa(\alpha)}h_{J_{t+v}}^{(\alpha)} \mid \mathscr{F}_t\right] \\
&= e^{\alpha S_t - t\kappa(\alpha)}\mathbb{E}\left[e^{\alpha(S_{t+v} - S_t) - v\kappa(\alpha)}h_{J_{t+v}}^{(\alpha)} \mid \mathscr{F}_t\right] \\
&= e^{\alpha S_t - t\kappa(\alpha)}\mathbb{E}_{J_t}\left[e^{\alpha S_v - v\kappa(\alpha)}h_{J_v}^{(\alpha)}\right] = e^{\alpha S_t - t\kappa(\alpha)}h_{J_t}^{(\alpha)}.
\end{aligned}
$$

□

Let $\boldsymbol{k}^{(\alpha)}$ denote the derivative of $\boldsymbol{h}^{(\alpha)}$ w.r.t. α, and write $\boldsymbol{k} = \boldsymbol{k}^{(0)}$.

Corollary 4.5 $\mathbb{E}_i S_t = t\kappa'(0) + k_i - \mathbb{E}_i k_{J_t} = t\kappa'(0) + k_i - \boldsymbol{e}_i^{\mathsf{T}}e^{\boldsymbol{\Lambda}t}\boldsymbol{k}$.

Proof. By differentiation in Proposition 4.4,

$$\mathbb{E}_i\left[S_t e^{\alpha S_t}h_{J_t}^{(\alpha)} + e^{\alpha S_t}k_{J_t}^{(\alpha)}\right] = e^{t\kappa(\alpha)}\left(k_i^{(\alpha)} + t\kappa'(\alpha)h_i^{(\alpha)}\right). \tag{4.2}$$

Let $\alpha = 0$ and recall that $\boldsymbol{h}^{(0)} = \boldsymbol{e}$ so that $h_i^{(0)} = h_{J_t}^{(0)} = 1$. □

The argument is slightly heuristic (e.g., the existence of exponential moments is assumed) but can be made rigorous by passing to characteristic functions. In the same way, one obtains a generalization of Wald's identity $\mathbb{E}S_\tau = \mathbb{E}\tau \cdot \mathbb{E}S_1$ for a random walk:

Corollary 4.6 *For any stopping time τ with finite mean,*

$$\mathbb{E}_i S_\tau = \kappa'(0)\mathbb{E}\tau + k_i - \mathbb{E}_i k_{J_\tau}.$$

Corollary 4.7 *No matter the initial distribution $\boldsymbol{\nu}$ of J_0,*

$$\lim_{t\to\infty} \frac{\mathbb{E}_\nu S_t}{t} = \kappa'(0), \quad \lim_{t\to\infty} \frac{\mathbb{V}\mathrm{ar}_\nu S_t}{t} = \kappa''(0).$$

Proof. The first assertion is immediate by dividing by t in Corollary 4.5. For the second, we differentiate (4.2) to get

$$\mathbb{E}_i \left[S_t^2 e^{\alpha S_t} h_{J_t}^{(\alpha)} + 2 S_t e^{\alpha S_t} k_{J_t}^{(\alpha)} + e^{\alpha S_t} k_{J_t}^{(\alpha)'} \right]$$
$$= e^{t\kappa(\alpha)} \left(k_i^{(\alpha)'} + t\kappa'(\alpha) k_i^{(\alpha)} + t\{\kappa''(\alpha) h_i^{(\alpha)} + t\kappa'(\alpha)^2 h_i^{(\alpha)} + \kappa'(\alpha) k_i^{(\alpha)}\} \right).$$

Multiplying by ν_i, summing and letting $\alpha = 0$ yields

$$\mathbb{E}_\nu \left[S_t^2 + 2 S_t k_{J_t} \right] = t^2 \kappa'(0)^2 + 2t\kappa'(0)\boldsymbol{\nu}\boldsymbol{k} + t\kappa''(0) + O(1).$$

Squaring in Corollary 4.5 yields

$$[\mathbb{E}_\nu S_t]^2 = t^2 \kappa'(0)^2 + 2t\kappa'(0)\boldsymbol{\nu}\boldsymbol{k} - 2t\kappa'(0)\mathbb{E}_\nu k_{J_t} + O(1).$$

Since it is easily seen by an asymptotic independence argument that $\mathbb{E}_\nu [S_t k_{J_t}] = t\kappa'(0)\mathbb{E}_\nu k_{J_t} + O(1)$, subtraction yields $\mathbb{V}\mathrm{ar}_\nu S_t = t\kappa''(0) + O(1)$. $\square$

Remark 4.8 Also for E being infinite (possibly uncountable), $\mathbb{E}e^{\alpha S_t}$ typically grows asymptotically exponential with a rate $\kappa(\alpha)$ independent of the initial condition (i.e., the distribution of J_0). More precisely, there is typically a function $h = h^{(\alpha)}$ on E and a $\kappa(\alpha)$ such that

$$\mathbb{E}_x e^{\alpha S_t - t\kappa(\alpha)} \to h(x), \quad t \to \infty,$$

for all $x \in E$. From (4.1) one then (at least heuristically) obtains

$$\begin{aligned}
h(x) &= \lim_{v\to\infty} \mathbb{E}_x e^{\alpha S_v - v\kappa(\alpha)} \\
&= \lim_{v\to\infty} \mathbb{E}_x \left[e^{\alpha S_t - t\kappa(\alpha)} \mathbb{E}_{J_t} e^{\alpha S_{v-t} - (v-t)\kappa(\alpha)} \right] \\
&= \mathbb{E}_x e^{\alpha S_t - t\kappa(\alpha)} h(J_t).
\end{aligned}$$

It then follows as in the proof of Proposition 4.4 that

$$\left\{ \frac{h(J_t)}{h(J_0)} e^{\alpha S_t - t\kappa(\alpha)} \right\}_{t\geq 0} \tag{4.3}$$

is a martingale. In view of this discussion, we take the martingale property as our basic condition below (though this is automatic in the finite case). An example beyond the finite case occurs for periodic risk processes in VII.6, where $\{J_t\}$ is deterministic period motion on $E = [0,1)$ (i.e., $J_t = (s+t) \bmod 1$ $\mathbb{P}_s$-a.s. for $s \in E$). □

Remark 4.9 The condition that (4.3) is a martingale can be expressed via the generator $\mathscr{A}$ (cf. II.4a) of $\{X_t\} = \{(J_t, S_t)\}$ as follows. Given a function h on E, let $h_\alpha(i,s) = e^{\alpha s} h(i)$. We then want to determine h and $\kappa(\alpha)$ such that $\mathbb{E}_i e^{\alpha S_t} h(J_t) = e^{t\kappa(\alpha)} h(i)$. For t small, this leads to

$$h(i) + t\mathscr{A}h_\alpha(i,0) = h(i)\big(1 + t\kappa(\alpha)\big),$$

i.e.

$$\mathscr{A}h_\alpha(i,0) = \kappa(\alpha)h(i).$$

We shall not exploit this approach systematically; see, however, VI.3b and Remark VII.6.5. □

Proposition 4.10 *Let $\big\{(J_t, S_t)\big\}$ be a MAP and let θ be such that*

$$\{L_t\}_{t\geq 0} = \Big\{\frac{h(J_t)}{h(J_0)} e^{\theta S_t - t\kappa(\theta)}\Big\}_{t\geq 0}$$

is a $\mathbb{P}_x$-martingale for each $x \in E$. Then $\{L_t\}$ is a multiplicative functional, and the family $\big\{\widetilde{\mathbb{P}}_x\big\}_{x\in E}$ given by Theorem 1.7 defines a new MAP.

Proof. That $\{L_t\}$ is a multiplicative functional follows from

$$L_s \circ \theta_t = \frac{h(J_{t+s})}{h(J_t)} e^{\theta(S_{t+s} - S_t) - s\kappa(\theta)}.$$

The proof that we have a MAP is contained in the proof of Theorem 4.11 below in the finite case. In the infinite case, one can directly verify that (4.1) holds for the $\widetilde{\mathbb{P}}_x$. We omit the details. □

Theorem 4.11 *Consider the irreducible case with E finite. Then the MAP in Proposition 4.10 is given by*

$$\widetilde{\boldsymbol{P}} = e^{-\kappa(\theta)} \boldsymbol{\Delta}_{h^{(\theta)}}^{-1} \widehat{\boldsymbol{F}}[\theta] \boldsymbol{\Delta}_{h^{(\theta)}}, \qquad \widetilde{H}_{ij}(\mathrm{d}x) = \frac{e^{\theta x}}{\widehat{H}_{ij}[\theta]} H_{ij}(\mathrm{d}x)$$

in the discrete time case, and by

$$\widetilde{\boldsymbol{\Lambda}} = \boldsymbol{\Delta}_{\boldsymbol{h}^{(\theta)}}^{-1} \boldsymbol{K}[\theta] \boldsymbol{\Delta}_{\boldsymbol{h}^{(\theta)}} - \kappa(\theta)\boldsymbol{I}, \quad \widetilde{\mu}_i = \mu_i + \theta\sigma_i^2, \quad \widetilde{\sigma}_i^2 = \sigma_i^2,$$

$$\widetilde{\nu}_i(\mathrm{d}x) = \mathrm{e}^{\theta x}\nu_i(\mathrm{d}x), \quad \widetilde{q}_{ij} = \frac{q_{ij}\widehat{B}_{ij}[\theta]}{1 + q_{ij}\big(\widehat{B}_{ij}[\theta] - 1\big)}, \quad \widetilde{B}_{ij}(\mathrm{d}x) = \frac{\mathrm{e}^{\theta x}}{\widehat{B}_{ij}[\theta]}B_{ij}(\mathrm{d}x)$$

in the continuous time case. Here $\boldsymbol{\Delta}_{\boldsymbol{h}^{(\theta)}}$ *is the diagonal matrix with the* $h_i^{(\theta)}$ *on the diagonal. In particular, if* $\nu_i(\mathrm{d}x)$ *is compound Poisson,* $\nu_i(\mathrm{d}x) = \beta_i B_i(\mathrm{d}x)$ *with* $\beta_i < \infty$ *and* B_i *a probability measure, then also* $\widetilde{\nu}_i(\mathrm{d}x)$ *is compound Poisson with*

$$\widetilde{\beta}_i = \beta_i\widehat{B}_i[\theta], \quad \widetilde{B}_i(\mathrm{d}x) = \frac{\mathrm{e}^{\theta x}}{\widehat{B}_i[\theta]}B_i(\mathrm{d}x).$$

Remark 4.12 The expression for $\widetilde{\boldsymbol{\Lambda}}$ means

$$\widetilde{\lambda}_{ij} = \frac{h_j^{(\theta)}}{h_i^{(\theta)}}\lambda_{ij}\Big[1 + q_{ij}\big(\widehat{B}_{ij}[\theta] - 1\big)\Big], \quad i \neq j. \tag{4.4}$$

In particular, this gives a direct verification that $\widetilde{\boldsymbol{\Lambda}}$ is an intensity matrix: the off-diagonal elements are non-negative because $\lambda_{ij} \geq 0$, $0 \leq q_{ij} \leq 1$ and $\widehat{B}_{ij}[\theta] > 0$. That the rows sum to 1 follows from

$$\begin{aligned}\widetilde{\boldsymbol{\Lambda}}\boldsymbol{e} &= \boldsymbol{\Delta}_{\boldsymbol{h}^{(\theta)}}^{-1} \boldsymbol{K}[\theta]\boldsymbol{h}^{(\theta)} - \kappa(\theta)\boldsymbol{e} = \kappa(\theta)\boldsymbol{\Delta}_{\boldsymbol{h}^{(\theta)}}^{-1}\boldsymbol{h}^{(\theta)} - \kappa(\theta)\boldsymbol{e} \\ &= \kappa(\theta)\boldsymbol{e} - \kappa(\theta)\boldsymbol{e} = 0\,.\end{aligned}$$

That $0 \leq \widetilde{q}_{ij} \leq 1$ follows from the inequality

$$\frac{qb}{1 + q(b - 1)} \leq 1, \quad 0 \leq q \leq 1, \ 0 < b < \infty\,. \qquad \square$$

Proof of Theorem 4.11. First note that the ijth element of $\widehat{\widetilde{\boldsymbol{F}}}_t[\alpha]$ is

$$\widetilde{\mathbb{E}}_i[\mathrm{e}^{\alpha S_t}; J_t = j] = \mathbb{E}_i[L_t\mathrm{e}^{\alpha S_t}; J_t = j] = \frac{h_j^{(\theta)}}{h_i^{(\theta)}}\mathrm{e}^{-t\kappa(\theta)}\mathbb{E}_i[\mathrm{e}^{(\alpha+\theta)S_t}; J_t = j].$$

In matrix notation, this means that

$$\widehat{\widetilde{\boldsymbol{F}}}_t[\alpha] = \mathrm{e}^{-t\kappa(\theta)}\boldsymbol{\Delta}_{\boldsymbol{h}^{(\theta)}}^{-1}\widehat{\boldsymbol{F}}_t[\alpha + \theta]\boldsymbol{\Delta}_{\boldsymbol{h}^{(\theta)}}\,. \tag{4.5}$$

Consider first the discrete time case. Here the stated formula for $\widetilde{\boldsymbol{P}}$ follows immediately by letting $t = 1$, $\alpha = 0$ in (4.5). Further

$$
\begin{aligned}
\widetilde{F}_{ij}(\mathrm{d}x) &= \widetilde{\mathbb{P}}_i(Y_1 \in \mathrm{d}x, J_1 = j) = \mathbb{E}_i[L_t; Y_1 \in \mathrm{d}x, J_1 = j] \\
&= \frac{h_j^{(\theta)}}{h_i^{(\theta)}} e^{\theta x - \kappa(\theta)} \mathbb{P}_i(Y_1 \in \mathrm{d}x, J_1 = j) = \frac{h_j^{(\theta)}}{h_i^{(\theta)}} e^{\theta x - \kappa(\theta)} F_{ij}(\mathrm{d}x).
\end{aligned}
$$

This shows that $\widetilde{F}_{ij}$ is absolutely continuous w.r.t. F_{ij} with a density proportional to $e^{\theta x}$. Hence the same is true for $\widetilde{H}_{ij}$ and H_{ij}; since $\widetilde{H}_{ij}$, H_{ij} are probability measures, it follows that indeed the normalizing constant is $\widehat{H}_{ij}[\theta]$.

Similarly, in continuous time (4.5) yields

$$
e^{t\widetilde{\boldsymbol{K}}[\alpha]} = \boldsymbol{\Delta}_{\boldsymbol{h}^{(\theta)}}^{-1} e^{t(\boldsymbol{K}[\alpha + \theta] - \kappa(\theta)\boldsymbol{I})} \boldsymbol{\Delta}_{\boldsymbol{h}^{(\theta)}}.
$$

By a general formula (A.13) for matrix-exponentials, this implies

$$
\widetilde{\boldsymbol{K}}[\alpha] = \boldsymbol{\Delta}_{\boldsymbol{h}^{(\theta)}}^{-1}(\boldsymbol{K}[\alpha + \theta] - \kappa(\theta)\boldsymbol{I})\boldsymbol{\Delta}_{\boldsymbol{h}^{(\theta)}} = \boldsymbol{\Delta}_{\boldsymbol{h}^{(\theta)}}^{-1} \boldsymbol{K}[\alpha + \theta]\boldsymbol{\Delta}_{\boldsymbol{h}^{(\theta)}} - \kappa(\theta)\boldsymbol{I}.
$$

Letting $\alpha = 0$ yields the stated expression for $\widetilde{\boldsymbol{\Lambda}}$.

Now we can write

$$
\begin{aligned}
\widetilde{\boldsymbol{K}}[\alpha] &= \widetilde{\boldsymbol{\Lambda}} + \boldsymbol{\Delta}_{\boldsymbol{h}^{(\theta)}}^{-1}\big(\boldsymbol{K}[\alpha + \theta] - \boldsymbol{K}[\theta]\big)\boldsymbol{\Delta}_{\boldsymbol{h}^{(\theta)}} \\
&= \widetilde{\boldsymbol{\Lambda}} + \big(\kappa^{(i)}(\alpha + \theta) - \kappa^{(i)}(\theta)\big)_{\mathrm{diag}} + \left(\frac{h_j^{(\theta)}}{h_i^{(\theta)}}\lambda_{ij}q_{ij}\big(\widehat{B}_{ij}[\alpha + \theta] - \widehat{B}_{ij}[\theta]\big)\right).
\end{aligned}
$$

That $\kappa^{(i)}(\alpha + \theta) - \kappa^{(i)}(\theta)$ corresponds to the stated parameters $\widetilde{\mu}_i, \widetilde{\sigma}_i^2, \widetilde{\nu}_i(\mathrm{d}x)$ of a Lévy process follows from Theorem 3.8. Finally note that by (4.4),

$$
\begin{aligned}
\frac{h_j^{(\theta)}}{h_i^{(\theta)}}\lambda_{ij}q_{ij}\big(\widehat{B}_{ij}[\alpha + \theta] - \widehat{B}_{ij}[\theta]\big) &= \frac{h_j^{(\theta)}}{h_i^{(\theta)}}\lambda_{ij}q_{ij}\widehat{B}_{ij}[\theta]\big(\widehat{\widetilde{B}}_{ij}[\alpha] - 1\big) \\
&= \widetilde{\lambda}_{ij}\widetilde{q}_{ij}\big(\widehat{\widetilde{B}}_{ij}[\alpha] - 1\big).
\end{aligned}
$$

$\square$

Notes and references The earliest paper on treatment of MAP's in the present spirit we know of is Nagaev [654]. Much of the pioneering was done in the sixties in papers like Keilson & Wishart [524, 525, 526] and Miller [642, 643, 644] in discrete time; the literature on the continuous time case tends more to deal with special cases. Though the literature on MAP's is extensive, there is, however, hardly a single comprehensive treatment; an extensive bibliography on aspects of the theory can be found in Asmussen [58].

Conditions for analogues of Corollary 4.3 for an infinite E are given by Ney & Nummelin [657]. For the Wald identity in Corollary 4.6, see also Fuh & Lai [380] and Moustakides [651]. The closest reference on exponential families of random walks on a Markov chain we know of within the more statistical oriented literature is Höglund [477], which, however, is slightly less general than the present setting.

5   The ladder height distribution

We consider the claim surplus process $\{S_t\}$ of a risk process with jumps U_i, interclaim times $T_i > 0$ and premium rate 1 (but note that no independence or Poisson assumptions are made). As usual, $\tau(u) = \inf\{t > 0 : S_t > u\}$ is the time to ruin. In the particular case $u = 0$, write $\tau_+ = \tau(0)$ and define the associated *ladder height* S_{τ_+} and *ladder height distribution* by

$$G_+(x) \;=\; \mathbb{P}(S_{\tau_+} \le x) \;=\; \mathbb{P}(S_{\tau_+} \le x, \tau_+ < \infty).$$

Note that G_+ is concentrated on $(0, \infty)$, i.e. has no mass on $(-\infty, 0]$, and is typically defective,

$$\|G_+\| \;=\; G_+(\infty) \;=\; \mathbb{P}(\tau_+ < \infty) \;=\; \psi(0) < 1$$

when $\eta > 0$ (there is positive probability that $\{S_t\}$ will never come above level 0).

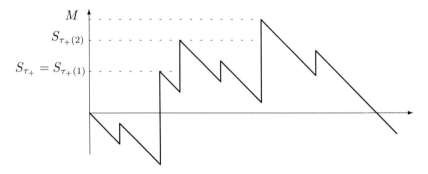

FIGURE III.2

The term *ladder height* is motivated from the shape of the process $\{M_t\}$ of relative maxima, see Fig. III.2. The first ladder step is precisely S_{τ_+}, and the maximum M is the total height of the ladder, i.e. the sum of all the ladder steps (if $\eta > 0$, there are only finitely many). In Fig. III.2, the second ladder point is $S_{\tau_+(2)}$ where $\tau_+(2)$ is the time of the next relative maximum after $\tau_+(1) = \tau_+$, the second ladder height (step) is $S_{\tau_+(2)} - S_{\tau_+(1)}$ and so on. In simple cases like the compound Poisson model, the ladder heights are i.i.d., a fact which turns out to be extremely useful. In other cases like the Markovian environment model, they have a semi-Markov structure (but in complete generality, the dependence structure seems too complicated to be useful). In any case, at present we concentrate on the first ladder height. The main result of this section is Theorem 5.5 below, which gives an explicit expression for G_+ in a very general setting, where basically only stationarity is assumed.

To illustrate the ideas, we shall first consider the compound Poisson model in the notation of Example II.3.2. Recall that $\overline{B}(x) = 1 - B(x)$ denotes the tail of B.

Theorem 5.1 *For the compound Poisson model with* $\rho = \beta \mu_B < 1$, G_+ *is given by the defective density* $g_+(x) = \beta \overline{B}(x) = \rho b_0(x)$ *on* $(0, \infty)$. *Here* $b_0(x) = \overline{B}(x)/\mu_B$.

For the proof of Theorem 5.1, define the *pre-τ_+-occupation measure* R_+ by

$$R_+(A) = \mathbb{E} \int_0^\infty I(S_t \in A, \tau_+ > t)\, dt = \mathbb{E} \int_0^{\tau_+} I(S_t \in A)\, dt.$$

The interpretation of $R_+(A)$ is as the expected time $\{S_t\}$ spends in the set A before τ_+. Thus, R_+ is concentrated on $(-\infty, 0]$, i.e., has no mass on $(0, \infty)$. Also, by approximation with step functions, it follows that for $g \geq 0$ measurable,

$$\int_{-\infty}^0 g(y) R_+(dy) = \mathbb{E} \int_0^{\tau_+} g(S_t)\, dt. \tag{5.1}$$

Lemma 5.2 R_+ *is the restriction of the Lebesgue measure to* $(-\infty, 0]$.

Proof. Let T be fixed and define $S_t^* = S_T - S_{T-t}$, $0 \leq t \leq T$. That is, $\{S_t^*\}_{0 \leq t \leq T}$ is constructed from $\{S_t\}_{0 \leq t \leq T}$ by time-reversion and hence, since the distribution of the Poisson process is invariant under time reversion, has the same distribution as $\{S_t\}_{0 \leq t \leq T}$, see Fig. III.3. Thus,

$$\mathbb{P}(S_T \in A, \tau_+ > T)$$
$$= \mathbb{P}(S_T \in A, S_t \leq 0, 0 \leq t \leq T)$$
$$= \mathbb{P}(S_T^* \in A, S_T^* \leq S_{T-t}^*, 0 \leq t \leq T)$$
$$= \mathbb{P}(S_T^* \in A, S_T^* \leq S_t^*, 0 \leq t \leq T)$$
$$= \mathbb{P}(S_T \in A, S_T \leq S_t, 0 \leq t \leq T). \tag{5.2}$$

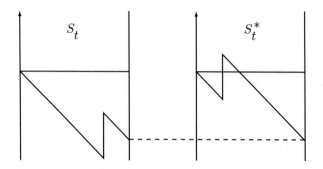

FIGURE III.3(A): $\tau_+ > t$

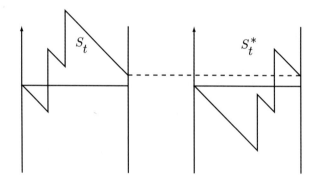

FIGURE III.3(B): $\tau_+ \le t$

Integrating w.r.t. dT, it follows that $R_+(A)$ is the expected time when S_T is in A and at a minimum at the same time. But since $S_t \to -\infty$ a.s., this is just the Lebesgue measure of A, cf. Fig. III.4 where the bold lines correspond to minimal values.　　□

Lemma 5.3 G_+ *is the restriction of* $\beta R_+ * B$ *to* $(0, \infty)$*. That is, for* $A \subseteq (0, \infty)$,

$$G_+(A) \;=\; \beta \int_{-\infty}^{0} B(A - y) R_+(dy) \,.$$

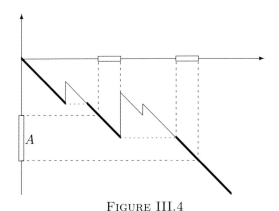

FIGURE III.4

Proof. A jump of $\{S_t\}$ at time t and of size U contributes to the event $\{S_{\tau_+} \in A\}$ precisely when $\tau_+ \geq t, U + S_{t-} \in A$. The probability of this given $\{S_u\}_{u<t}$ is $B(A - S_{t-})I(\tau_+ \geq t)$, and since the jump rate is β, we get

$$
\begin{aligned}
G_+(A) &= \int_0^\infty \beta \, dt \, \mathbb{E}\big[B(A - S_{t-}); \tau_+ \geq t\big] \\
&= \beta \int_0^\infty \mathbb{E}\big[B(A - S_t); \tau_+ > t\big] \, dt \\
&= \beta \, \mathbb{E} \int_0^{\tau_+} g(S_t) \, dt = \beta \int_{-\infty}^0 g(y) R_+(dy)
\end{aligned}
$$

where $g(y) = B(A - y)$ (here we used the fact that the probability of a jump at t is zero in the second step, and (5.1) in the last). $\qquad \square$

Proof of Theorem 5.1. With $r_+(y) = I(y < 0)$ denoting the density of R_+, Lemma 5.3 yields

$$
g_+(x) = \beta \int_0^\infty r_+(x - z) \, B(dz) = \beta \int_0^\infty I(x < z) \, B(dz) = \beta \overline{B}(x).
$$

$\qquad \square$

Generalizing the set-up, we consider the claim surplus process $\{S_t^*\}_{t \geq 0}$ of a risk reserve process in a very general set-up, assuming basically stationarity in time and space,

$$
\{S_{t+s}^* - S_s^*\}_{t \geq 0} \overset{\mathscr{D}}{=} \{S_t^*\}_{t \geq 0} \tag{5.3}
$$

for all $s \geq 0$. The sample path structure is assumed to be as for the compound Poisson case: $\{S_t^*\}$ is generated from interclaim times T_k^* and claim sizes U_k^*

according to premium 1 per unit time, i.e.

$$S_t^* = \sum_{k=1}^{N_t^*} U_k^* - t \quad \text{where} \quad N_t^* = \max\{k = 0, 1, \ldots : T_1^* + \cdots + T_k^* \le t\}.$$

The first ladder epoch τ_+^* is defined as $\inf\{t > 0 : S_t^* > 0\}$ and the corresponding ladder height distribution is

$$G_+^*(A) \;=\; \mathbb{P}\big(S_{\tau_+^*}^* \in A\big) \;=\; \mathbb{P}\big(S_{\tau_+^*}^* \in A, \tau_+^* < \infty\big).$$

The traditional representation of the input sequence $\{(T_k^*, U_k^*)\}_{k=1,2,\ldots}$ is as a marked point process $\mathscr{M}^*$, i.e. as a point process on $[0, \infty) \times (0, \infty)$. The points in the plane (marked by $\times$ on Fig. III.5) are (σ_k^*, U_k^*) $(k = 1, 2, \ldots)$ where $\sigma_k^* = T_1^* + \cdots + T_k^*$, the first component representing time (the arrival time σ_k^*) and the second the mark (the claim size U_k^*). The marked point process $\mathscr{M}^* \circ \theta_s$ shifted by s is defined the obvious way, cf. Fig. III.5 (the points in the plane are $(\sigma_k^* - s, U_k^*)$ for those k for which $\sigma_k^* - s \ge 0$). We call $\mathscr{M}^*$ *stationary* if $\mathscr{M}^* \circ \theta_s$ has the same distribution as $\mathscr{M}^*$ for all $s \ge 0$; obviously, this is equivalent to the risk process $\{S_t^*\}$ being stationary in the sense of (5.3). In the stationary case, we define the arrival rate as $\beta = \mathbb{E}\#\{k : \sigma_k^* \in [0, h]\}/h$ (by stationarity, this does not depend on h).

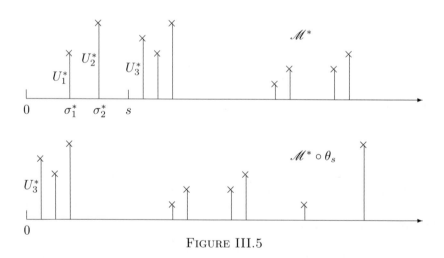

FIGURE III.5

Given a stationary marked point process $\mathscr{M}^*$, we define its *Palm version* $\mathscr{M}$ as a marked point process having the conditional distribution of $\mathscr{M}^*$ given an arrival at time 0, i.e. $\sigma_1^* = 0$. We represent $\mathscr{M}$ by the sequence $(T_k, U_k)_{k=1,2,\ldots}$ where $T_1 = 0$, and let $T = T_2$ denote the first proper interarrival time. The two fundamental formulas connecting $\mathscr{M}^*$ and $\mathscr{M}$ are

$$\mathbb{E}\varphi(\mathscr{M}) \;=\; \frac{1}{\beta h}\mathbb{E}\sum_{k:\,\sigma_k^* \in [0,h]} \varphi(\mathscr{M}^* \circ \theta_{\sigma_k^*}), \qquad (5.4)$$

$$\mathbb{E}\varphi(\mathscr{M}^*) \;=\; \frac{1}{\mathbb{E}T}\mathbb{E}\int_0^T \varphi(\mathscr{M} \circ \theta_t)\,\mathrm{d}t,$$

where T is the first arrival time > 0 of $\mathscr{M}$ and $h > 0$ an arbitrary constant (in the literature, most often one takes $h = 1$). As above, the r.h.s. of (5.4) does not depend on h; letting $h \downarrow 0$, βh becomes the approximate probability $\mathbb{P}(\sigma_1^* \leq h)$ of an arrival in $[0, h]$ and the sum approximately $\varphi(\mathscr{M}^*)I(\sigma_1 \leq h)$. This more or less gives a proof that indeed (5.4) represents the conditional distribution of $\mathscr{M}^*$ given $\sigma_1^* = 0$. Note also that (again by stationarity) the Palm distribution also represents the conditional distribution of $\mathscr{M}^* \circ \theta_t$ given an arrival at time t. See, e.g., Sigman [812] or [APQ, VII.6] for these and further aspects of Palm theory.

Example 5.4 Consider a finite Markov additive process (cf. Section 4) which has pure jump structure corresponding to $\mu_i = \sigma_i^2 = 0$, $\nu_i(\mathrm{d}x) = \beta_i B_i(\mathrm{d}x)$. Assume $\{J_t\}$ irreducible so that a stationary distribution $\boldsymbol{\pi} = (\pi_i)_{i\in E}$ exists.

Interpreting jump times as arrival times and jump sizes as marks, we get a marked point process generated by Poisson arrivals at rate β_i and mark distribution B_i when $J_t = i$, and by some additional arrivals which occur w.p. q_{ij} when $\{J_t\}$ jumps from i to j and have mark distribution B_{ij}. A stationary marked point process $\mathscr{M}^*$ is obtained by assigning J_0 distribution $\boldsymbol{\pi}$. If $J_{t-} = i$, an arrival for $\mathscr{M}^*$ occurs before time $t + \mathrm{d}t$ w.p.

$$\mathrm{d}t\Big\{\beta_i + \sum_{j\neq i}\lambda_{ij}q_{ij}\Big\}.$$

Thus the arrival rate for $\mathscr{M}^*$ is

$$\beta \;=\; \sum_{i\in E}\pi_i\Big\{\beta_i + \sum_{j\neq i}\lambda_{ij}q_{ij}\Big\}.$$

Given that an arrival occurs at time t, the probability α_{ij} of $J_{t-} = i$, $J_t = j$ is $\pi_i\beta_i/\beta$ for $i = j$ and $\pi_i\lambda_{ij}q_{ij}/\beta$ for $i \neq j$. It follows that we can describe the Palm version $\mathscr{M}$ as follows. First choose (J_{0-}, J_0) w.p. α_{ij} for (i, j) and let the initial mark U_1 have distribution B_i when $i = j$ and B_{ij} otherwise. After that, let the arrivals and their marks be generated by $\{J_t\}$ starting from $J_0 = j$.

Note in particular that the Palm distribution of the mark size (i.e., the distribution of U_1) is the mixture

$$B = \sum_{i \in E} \left\{ \alpha_{ii} B_i + \sum_{j \neq i} \alpha_{ij} B_{ij} \right\} = \sum_{i \in E} \frac{\pi_i}{\beta} \left\{ \beta_i B_i + \sum_{j \neq i} \lambda_{ij} q_{ij} B_{ij} \right\}.$$

□

Theorem 5.5 *Consider a general stationary claim surplus process $\{S_t^*\}_{t \geq 0}$, let U_0 be a r.v. having the Palm distribution of the claim size and $F(x) = \mathbb{P}(U_0 \leq x)$ its distribution. Assume that $S_t^* \to -\infty$ a.s. and that $\rho = \beta \mathbb{E} U_0 < 1$. Then the ladder height distribution G_+^* is given by the (defective) density $g_+^*(x) = \beta \overline{F}(x)$.*

Before giving the proof, we note:

Corollary 5.6 *Under the assumptions of Theorem 5.5, the ruin probability $\psi^*(0)$ with initial reserve $u = 0$ is $\rho = \beta \mathbb{E} U_0$.*

This follows by noting that

$$\psi^*(0) = \|G_+^*\| = \int_0^\infty g_+^*(x)\,dx = \beta \int_0^\infty \overline{F}(x)\,dx = \beta \mathbb{E} U_0.$$

By (5.4),

$$\psi^*(0) = \mathbb{E} \sum_{k:\, \sigma_k^* \in [0,1]} U_k^*;$$

here the r.h.s. has a very simple interpretation as the average amount of claims received per unit time. The result is notable by giving an explicit expression for ruin in great generality and by only depending on the parameters of the model through the arrival rate β and the average (in the Palm sense) claim size $\mathbb{E} U_0$. The last property is referred to as *insensitivity* in the applied probability literature.

Proof of Theorem 5.5. A standard argument for stationary processes ([199, p. 105]) shows that one can assume w.l.o.g. that $\mathcal{M}^*$ and $\mathcal{M}$ have doubly infinite time (i.e., are point processes on $(-\infty, \infty) \times (0, \infty)$). We then represent $\mathcal{M}$ by the mark (claim size) U_0 of the arrival at time 0, the arrival times $0 < \sigma_1 < \sigma_2 < \ldots$ in $(0, \infty)$ and the arrival times $0 > \sigma_{-1} > \sigma_{-2} > \ldots$ in $(-\infty, 0)$; the mark at time σ_k is denoted by U_k.

Let $p(t)$ be the conditional probability that $S_{\tau_+}^* \in A, \tau_+ = t$ given the event A_t that an arrival at t occurs. Then clearly

$$G_+^*(A) = \mathbb{P}(S_{\tau_+}^* \in A) = \int_0^\infty p(t)\beta\,dt.$$

Consider a process $\{\check{S}_t\}_{t\geq 0}$, which makes an upwards jump at time $-\sigma_{-k}$ ($k = 1, 2, \ldots$), moves down linearly at a unit rate in between jumps and starts from $\check{S}_0 = U_0$.

Now conditionally upon A_t, $\{S_u^*\}_{0\leq u\leq t}$ is distributed as a process $\{\widetilde{S}_u^*\}_{0\leq u\leq t}$ where a claim arrives at time t and has size U_0, and the kth preceding claim arrives at time $t - \sigma_{-k}$ and has size U_{-k}. The sample path relation between $\{\widetilde{S}_u^*\}$ and $\{\check{S}_u\}$ amounts to $\check{S}_u = \widetilde{S}_t^* - \widetilde{S}_{t-u-}^*$ (left limit) when $0 \leq u \leq t$ and is illustrated on Fig. III.6. It follows that for $A \subseteq (0, \infty)$

$$
\begin{aligned}
p(t) &= \mathbb{P}\big(S_t^* \in A, S_u^* \leq 0, 0 < u < t \,\big|\, A_t\big) \\
&= \mathbb{P}\big(S_t^* \in A, S_{u-}^* \leq 0, 0 < u < t \,\big|\, A_t\big) \\
&= \mathbb{P}\big(\widetilde{S}_t^* \in A, \widetilde{S}_{u-}^* \leq 0, 0 < u < t\big) \\
&= \mathbb{P}\big(\check{S}_t \in A, \check{S}_t \leq \check{S}_{t-u}, 0 < u < t\big) \\
&= \mathbb{P}\big(\check{S}_t \in A, \check{S}_t \leq \check{S}_u, 0 < u < t\big) \\
&= \mathbb{P}\big(\check{S}_t \in A; \check{M}_t\big),
\end{aligned}
$$

where $\check{M}_t = \{\check{S}_t \leq \check{S}_u, 0 < u < t\}$ is the event that $\{\check{S}_u\}$ has a relative minimum at t. In Fig. III.6, time instants corresponding to such minimal values have been marked with bold lines in the path of $\{\check{S}_t\}$, and we let $L(dy)$ be the random measure $L(A) = \int_0^\infty I(\check{S}_t \in A; \check{M}_t)\,dt$.

Since $\check{S}_0 = U_0$, the support of L has right endpoint U_0, and since by assumption $S_t^* \to -\infty$ a.s., $t \to \infty$, the left endpoint of the support is $-\infty$. A sample path inspection just as in the proof of Lemma 5.2 therefore immediately shows that $L(dy)$ is Lebesgue measure on $(-\infty, U_0]$, cf. Fig. III.6 where the boxes on the time axis correspond to time intervals where $\{\check{S}_t\}$ is at a minimum belonging to A and split A into pieces corresponding to segments where $\{\check{S}_u\}$ is at a relative minimum. Thus,

$$
\begin{aligned}
G_+^*(A) &= \beta \int_0^\infty \mathbb{P}(\check{S}_t \in A; M_t)\,dt = \beta \mathbb{E} L(A) \\
&= \beta \mathbb{E} \int_{-\infty}^\infty I(U_0 > y) I(y \in A)\,dy = \beta \int_A \mathbb{P}(U_0 > y)\,dy \\
&= \beta \int_A \overline{F}(y)\,dy.
\end{aligned}
$$

$\square$

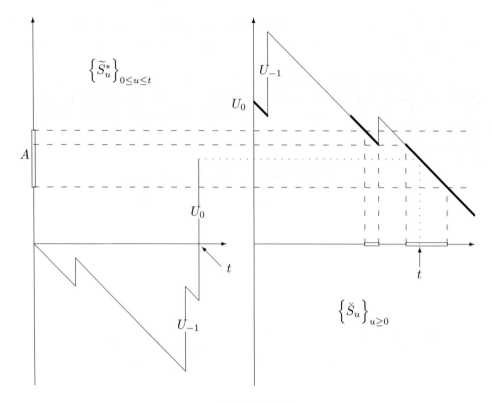

$$\text{Figure III.6}$$

Notes and references Theorem 5.5 is due to Schmidt & co-workers [102, 372, 648] (a special case of the result appears in Proposition VII.2.1). A further relevant reference related to Corollary 5.6 is Björk & Grandell [170].

Two alternative somewhat simpler approaches to prove Theorem 5.1 will be given in Chapter IV (after Theorem IV.2.1 and in Remark IV.3.6).

Chapter IV

The compound Poisson model

We consider throughout this chapter a risk reserve process $\{R_t\}_{t\geq 0}$ in the terminology and notation of Chapter I, and assume that

- $\{N_t\}_{t\geq 0}$ is a Poisson process with rate β.

- the claim sizes $U_1, U_2, \ldots$ are i.i.d. with common distribution B, say, and independent of $\{N_t\}$.

- the premium rate is $p = 1$.

Thus, $\{R_t\}$ and the associated claims surplus process $\{S_t\}$ are given by

$$R_t = u + t - \sum_{i=1}^{N_t} U_i, \quad S_t = u - R_t = \sum_{i=1}^{N_t} U_i - t.$$

An important omission of the discussion in this chapter is the numerical evaluation of the ruin probability. Some possibilities are numerical Laplace transform inversion via Corollary 3.4 below, exact matrix-exponential solutions under the assumption that B is phase-type (see further IX.3), Panjer's recursion (Corollary XVI.2.6) and simulation methods (Chapter XV). For finite horizon ruin probabilities, see Chapter V.

It is worth mentioning that much of the analysis of this chapter can be carried over in a direct way to more general Lévy processes, see Chapter XI.

1   Introduction

For later reference, we shall start by giving the basic formulas for moments, cumulants, m.g.f.'s etc. of the claim surplus $S_t = u - R_t$. Write

$$\mu_B^{(n)} = \mathbb{E}U^n, \quad \mu_B = \mu_B^{(1)} = \mathbb{E}U, \quad \rho = \beta\mu_B = 1/(1+\eta) \ .$$

Proposition 1.1 (a) $\mathbb{E}S_t = t(\beta\mu_B - 1) = t(\rho - 1)$;
(b) $\mathbb{V}ar\, S_t = t\beta\mu_B^{(2)}$;
(c) $\mathbb{E}e^{rS_t} = e^{t\kappa(r)}$ where $\kappa(r) = \beta\big(\widehat{B}[r] - 1\big) - r$;
(d) The kth cumulant of S_t is $t\beta\mu_B^{(k)}$ for $k \geq 2$.

Proof. It was noted in Chapter I that $\rho - 1$ is the expected claim surplus per unit time, and this immediately yields (a). A more formal proof goes as follows:

$$
\begin{aligned}
\mathbb{E}S_t &= \mathbb{E}\sum_{k=1}^{N_t} U_k - t = \mathbb{E}\,\mathbb{E}\Big[\sum_{k=1}^{N_t} U_k \,\Big|\, N_t\Big] - t \\
&= \mathbb{E}[N_t \mu_B] - t = \beta t \mu_B - t = t(\rho - 1).
\end{aligned}
$$

The same method yields also the variance as

$$
\begin{aligned}
\mathbb{V}ar\, S_t &= \mathbb{V}ar \sum_{k=1}^{N_t} U_k = \mathbb{V}ar\,\mathbb{E}\Big[\sum_{k=1}^{N_t} U_k \,\Big|\, N_t\Big] + \mathbb{E}\,\mathbb{V}ar\Big[\sum_{k=1}^{N_t} U_k \,\Big|\, N_t\Big] \\
&= \mathbb{V}ar\,[N_t \mu_B] + \mathbb{E}[N_t \mathbb{V}ar\, U] = t\beta\mu_B^2 + t\beta\mathbb{V}ar\, U = t\beta\mu_B^{(2)}.
\end{aligned}
$$

For (c), we get

$$
\begin{aligned}
\mathbb{E}e^{rS_t} &= e^{-rt} \sum_{k=0}^{\infty} \mathbb{E}e^{r(U_1+\cdots+U_k)}\mathbb{P}(N_t = k) \\
&= e^{-rt} \sum_{k=0}^{\infty} \widehat{B}[r]^k \cdot e^{-\beta t}\frac{(\beta t)^k}{k!} = \exp\{-rt - \beta t + \widehat{B}[r]\beta t\} = e^{t\kappa(r)}.
\end{aligned}
$$

Finally, for (d) just note that the kth cumulant of S_t is $t\kappa^{(k)}(0)$, where $\kappa^{(k)}(0)$ is the kth derivative of κ at 0, and that $\widehat{B}^{(k)}[0] = \mu_B^{(k)}$.   □

The linear way the index t enters in the formulas in Proposition 1.1 is the same as if $\{S_t\}$ was a random walk indexed by $t = 0, 1, 2, \ldots$ The connections to random walks are in fact fundamental, and there are at least two ways to exploit this:

Recalling that σ_k is the time of the kth claim, we have $S_{\sigma_k} - S_{\sigma_{k-1}} = U_k - T_k$, where T_k is the time between the kth and the $(k-1)$th claim. Obviously, the $U_k - T_k$ are i.i.d. so that $\{S_{\sigma_k}\}$ is a random walk with mean

$$\mathbb{E}U - \mathbb{E}T = \mathbb{E}U - \frac{1}{\beta} = \frac{\beta\mathbb{E}U - 1}{\beta} = -\eta\mu_B$$

where η is the safety loading. In this way, we get a discrete time random walk imbedded in the claim surplus process $\{S_t\}$, which is often used in the literature for obtaining information about $\{S_t\}$ and the ruin probabilities. For example, obviously $\psi(u) = \mathbb{P}(\max_k S_{\sigma_k} > u)$. We return to this approach in Chapter VI.

The point of view in the present chapter is, however, rather to view $\{S_t\}$ directly as a random walk in continuous time, meaning that the increments are stationary and independent, cf. III.3, so we have a Lévy process. Here is one immediate application:

Proposition 1.2 (DRIFT AND OSCILLATION)
(a) *No matter the value of η, $S_t/t \overset{\text{a.s.}}{\to} \rho - 1$ as $t \to \infty$;*
(b) *If $\eta < 0$, then $S_t \overset{\text{a.s.}}{\to} \infty$;*
(c) *If $\eta > 0$, then $S_t \overset{\text{a.s.}}{\to} -\infty$;*
(d) *If $\eta = 0$, then $\liminf_{t\to\infty} S_t = -\infty$, $\limsup_{t\to\infty} S_t = \infty$.*

For the proof, we need the following lemma:

Lemma 1.3 *If $nh \le t \le (n+1)h$, then*

$$S_{nh} - h \le S_t \le S_{(n+1)h} + h.$$

Proof. We first note that for $u, v \ge 0$,

$$S_{u+v} \ge S_u - v.$$

Indeed, $S_{u+v} - S_u$ attains its minimal value when there are no arrivals in $(u, u+v]$, and the value is then precisely v. In particular, if $t = nh + v$ with $0 \le v \le h$, then

$$S_t \ge S_{nh} - v \ge S_{nh} - h.$$

The inequality on the right in (1.3) is proved similarly. $\square$

Proof of Proposition 1.2. For any fixed h, $\{S_{nh}\}_{n=0,1,\dots}$ is a discrete time random walk, and hence by the strong law of large numbers, $S_{nh}/n \overset{\text{a.s.}}{\to} \mathbb{E}S_h = h(\rho - 1)$.

Thus using Lemma 1.3, we get

$$
\begin{aligned}
\liminf_{t\to\infty} \frac{S_t}{t} &= \liminf_{n\to\infty}\ \inf_{nh\le t\le (n+1)h} \frac{S_t}{t} \\
&\ge \frac{1}{h}\liminf_{n\to\infty} \frac{S_{nh}-h}{n} = \frac{1}{h}\mathbb{E}S_h = \rho-1.
\end{aligned}
$$

A similar argument for $\limsup$ proves (a), and (b), (c) are immediate consequences of (a). Part (d) follows by a (slightly more intricate) general random walk result ([APQ, pp. 224–225]) stating that $\liminf_{n\to\infty} S_{nh} = -\infty$, $\limsup_{n\to\infty} S_{nh} = \infty$ (Lemma 1.3 is not needed for (d)).   □

Corollary 1.4 *The ruin probability $\psi(u)$ is 1 for all u when $\eta \le 0$, and < 1 for all u when $\eta > 0$.*

Proof. The case of $\eta \le 0$ is immediate since then $M = \infty$ by Proposition 1.2. If $\eta > 0$, it suffices to prove $\psi(0) = \mathbb{P}(M > 0) < 1$. However, if $\mathbb{P}(M > 0) = 1$, then $\{S_t\}$ upcrosses level 0 a.s. at least once. Considering the next downcrossing (which occurs w.p. 1 since $S_t \to -\infty$) and repeating the argument, it is seen that upcrossing occurs at least twice, hence by induction i.o. This contradicts $S_t \to -\infty$.   □

There is also a central limit version of Proposition 1.2:

Proposition 1.5 *The limiting distribution of $\big(S_t - t(\rho-1)\big)/\sqrt{t}$ as $t \to \infty$ is normal with mean zero and variance $\beta\mu_B^{(2)}$.*

Proof. Since $\{S_t\}_{t\ge0}$ is a Lévy process (a random walk in continuous time), $\{S_{nh}\}_{n=0,1,\dots}$ is a discrete-time random walk for any $h > 0$, and hence it follows from standard central limit theory and the expression $\mathbb{V}ar(S_t) = t\beta\mu_B^{(2)}$ (Proposition 1.1(b)) that the assertion holds as $t \to \infty$ through values of the form $t = 0, h, 2h \dots$. The general case now follows either by another easy application of Lemma 1.3, or by a general result on discrete skeletons ([APQ, p. 415]).   □

Remark 1.6 Often it is of interest to consider *size-fluctuations*, where the size of the portfolio at time t is $M(t)$. Assuming that each risk generates claims at Poisson intensity β and pays premium 1 per unit time, this case can be reduced to the compound Poisson model by an easy operational time transformation $T^{-1}(t)$ where $T(s) = \beta \int_0^s M(t)\mathrm{d}t$ (this was already pointed out by Lundberg [614], see also [11] for an overview).   □

Notes and references   All material of the present section is standard.

2 The Pollaczeck-Khinchine formula

The time to ruin $\tau(u)$ was already defined in Chapter I as $\inf\{t > 0 : S_t > u\}$, and we shall here exploit the decomposition of the maximum M as sum of ladder heights, cf. Fig. III.2. We assume throughout $\eta > 0$ or, equivalently, $\rho < 1$.

It is crucial to note that for the compound Poisson model, *the ladder heights are i.i.d.* This follows simply by noting that the process repeats itself after reaching a relative maximum. The decomposition of M as a sum of ladder heights now yields:

Theorem 2.1 *The distribution of M is* $(1 - \|G_+\|)\sum_{n=0}^{\infty} G_+^{*n}$, *where G_+ is given by the defective density* $g_+(x) = \beta\overline{B}(x) = \rho b_0(x)$ *on* $(0, \infty)$. *Here $b_0(x) = \overline{B}(x)/\mu_B$.*

The formula for g_+ was already obtained in Theorem III.5.1, but before showing the rest of Theorem 2.1, we give an alternative argument which is short and intuitive, but also slightly heuristical:

Proof of $g_+(x) = \beta\overline{B}(x)$: Assume B has a density b. Note that if there is a claim arrival before dt, then $S_{\tau_+} \in (u, u + du]$ occurs precisely when the claim has size u. Hence the contribution to $g_+(u)$ from this event is $b(u)\beta\,dt$. If there are no claim arrivals before dt, consider the process $\{\widetilde{S}_t\}_{t\geq 0}$ where $\widetilde{S}_t = S_{t+dt} - S_{dt} = S_{t+dt} + dt$. For $S_{\tau_+} \in (u, u + du]$ to occur, $\widetilde{S}$ must either have its first ladder point equal to $u + dt$ or $v \in (0, dt]$, and in the latter case the process starting from v must have its first ladder point equal to $u + v$, i.e. the probability is $\int_0^{dt} g_+(v)g_+(u + v)\,dv$. Collecting all first order terms, it follows that,

$$
\begin{aligned}
g_+(u) &= b(u)\beta\,dt + (1 - \beta\,dt)\big(g_+(u + dt) + g_+(0)g_+(u)\,dt\big) + o(dt) \\
&= b(u)\beta\,dt + (1 - \beta\,dt)\big(g_+(u) + g_+'(u)\,dt + g_+(0)g_+(u)\,dt\big) + o(dt) \\
&= g_+(u) + dt\big(-\beta g_+(u) + g_+'(u) + \beta g_+(0)g_+(u) + \beta b(u)\big) + o(dt), \\
g_+'(u) &= \big(\beta - g_+(0)\big)g_+(u) - \beta b(u). \tag{2.1}
\end{aligned}
$$

Integrating from 0 to x gives

$$
g_+(x) = g_+(0) + \big(\beta - g_+(0)\big)\mathbb{P}(S_{\tau_+} \leq x, \tau_+ < \infty) - \beta B(u).
$$

Letting $x \to \infty$ and assuming (heuristical but reasonable!) that then $g_+(x) \to 0$, we get

$$
0 = g_+(0) + \big(\beta - g_+(0)\big)\mathbb{P}(\tau_+ < \infty) - \beta = -\big(\beta - g_+(0)\big)\mathbb{P}(\tau_+ = \infty).
$$

Since $\mathbb{P}(\tau_+ = \infty) > 0$ because of the assumption of a positive loading, we therefore have $g_+(0) = \beta$. Thus (2.1) simply means $g'_+(u) = -\beta b(u)$, and the solution satisfying $g_+(0) = \beta$ is $g_+(u) = \beta \overline{B}(u)$. $\qquad \square$

Proof of Theorem 2.1. The probability that M is attained in precisely n ladder steps and does not exceed x is $G_+^{*n}(x)(1 - \|G_+\|)$ (the parenthesis gives the probability that there are no further ladder steps after the nth). Summing over n, the formula for the distribution of M follows. $\qquad \square$

Alternatively, we may view the ladder heights as a terminating renewal process and M becomes then the lifetime.

Combined with $\psi(u) = \mathbb{P}(M > u)$, Theorem 2.1 provides a representation formula for $\psi(u)$, which we henceforth refer to as the *Pollaczeck-Khinchine formula*. Note that the integrated tail distribution B_0 with density b_0 is familiar from renewal theory as the limiting stationary distribution of the overshoot (*forward recurrence time*), see [APQ, V.3–4] or A.1e. Thus, we can rewrite the Pollaczeck-Khinchine formula as

$$\psi(u) = \mathbb{P}(M > u) = (1 - \rho) \sum_{n=1}^{\infty} \rho^n \overline{B}_0^{*n}(u), \qquad (2.2)$$

representing the distribution of M as a geometric compound.

As a vehicle for computing $\psi(u)$, (2.2) is not entirely satisfying because of the infinite sum of convolution powers, but we shall be able to extract substantial information from the formula, nevertheless.

The following result generalizes the fact that the conditional distribution of the deficit $S_{\tau(0)}$ just after ruin given that ruin occurs (i.e., that $\tau(0) < \infty$) is B_0: taking $y = 0$ shows that the conditional distribution of the risk reserve immediately before ruin (i.e. $-S_{\tau(0)-}$) is again B_0, and we further get information about the joint conditional distribution of this quantity and the deficit. Note that this distribution is the same as the limiting joint distribution of the age and excess life in a renewal process governed by B, cf. Theorem A1.5.

Theorem 2.2 *The joint distribution of* $(-S_{\tau(0)-}, S_{\tau(0)})$ *is given by the following four equivalent statements:*

(a) $\mathbb{P}\left(-S_{\tau(0)-} > x, \, S_{\tau(0)} > y; \, \tau(0) < \infty\right) = \beta \int_{x+y}^{\infty} \overline{B}(z)\,\mathrm{d}z;$

(b) *the joint distribution of* $(-S_{\tau(0)-}, S_{\tau(0)})$ *given* $\tau(0) < \infty$ *is the same as the distribution of* $(VW, (1 - V)W)$ *where* V, W *are independent,* V *is uniform on* $(0, 1)$ *and* W *has distribution* F_W *given by* $\mathrm{d}F_W/\mathrm{d}B(x) = x/\mu_B;$

(c) *the marginal distribution of* $-S_{\tau(0)-}$ *is* B_0, *and the conditional distribution of* $S_{\tau(0)}$ *given* $-S_{\tau(0)-} = y$ *is the overshoot distribution* $B_0^{(y)}$ *given by* $\overline{B_0^{(y)}}(z) =$

$\overline{B_0}(y + z)/\overline{B_0}(y)$;

(d) *the marginal distribution of $S_{\tau(0)-}$ is B_0, and the conditional distribution of $-S_{\tau(0)-}$ given $S_{\tau(0)-} = z$ is $B_0^{(z)}$.*

The proof is given in V.2 and it gives an alternative derivation of the distribution of the deficit $S_{\tau(0)}$.

Notes and references   The Pollaczeck-Khinchine formula is standard in queueing theory, see for example [APQ], Feller [362] or Wolff [894]. The proof of Theorem III.5.1 is traditionally carried out for the imbedded discrete time random walk, where it requires slightly more calculation. As shown in Theorem III.5.5, the form of G_+ is surprisingly insensitive to the form of $\{S_t\}$ and holds in a certain general marked point process set-up. However, in this setting there is no decomposition of M as a sum of i.i.d. ladder heights so that the results do not appear too useful for estimating $\psi(u)$ for $u > 0$.

Theorem 2.2(a) is from Dufresne & Gerber [333]. Again, there is a general marked point process version, cf. Asmussen & Schmidt [103]. For the study of the joint distribution of the surplus $S_{\tau(u)-}$ just before ruin and the deficit $S_{\tau(u)}$ at ruin, see Schmidli [773] and references therein. In Chapter XII these results will be generalized in various directions.

In risk theory literature, the Pollaczeck-Khinchine formula is often referred to as *Beekman's convolution formula*, cf. Beekman [152, 153].

3   Special cases of the Pollaczeck-Khinchine formula

The model and notation is the same as in the preceding sections. We assume $\eta > 0$ throughout.

3a   The ruin probability when the initial reserve is zero

The case $u = 0$ is remarkable by giving a formula for $\psi(u)$ which depends on the claim size distribution only through its mean:

Corollary 3.1 $\psi(0) = \rho = \beta\mu_B = \dfrac{1}{1 + \eta}.$

Proof   Recall that $\tau_+ = \tau(0)$ and note that

$$\psi(0) = \mathbb{P}(\tau_+ < \infty) = \|G_+\| = \beta \int_0^\infty \overline{B}(x)\,\mathrm{d}x = \beta\mu_B.$$

$\square$

Notes and references   The fact that $\psi(u)$ only depends on B through μ_B is often referred to as an *insensitivity property*. As shown in III.6, the formula for $\psi(0)$ holds in a more general setting; a slightly modified version also holds for certain two-sided jumps, cf. Section XII.4. A further relevant reference is Björk & Grandell [170].

3b   Exponential claims

Corollary 3.2 *If B is exponential with rate δ, then $\psi(u) = \rho e^{-(\delta-\beta)u}$.*

Proof   The distribution B_0 of the ascending ladder height (given that it is defined) is the distribution of the overshoot of $\{S_t\}$ at time τ_+ over level 0. But claims are exponential, hence without memory, and so this overshoot has the same distribution as the claims themselves. I.e., B_0 is exponential with rate δ and the result can now be proved from the Pollaczeck-Khinchine formula by elementary calculations. Thus, B_0^{*n} is the Erlang distribution with n phases and the density of M at $x > 0$ is

$$(1-\rho)\sum_{n=1}^{\infty}\rho^n\frac{\delta^n x^{n-1}}{(n-1)!}e^{-\delta x} = (1-\rho)\rho\delta e^{-\delta(1-\rho)x} = \rho(\delta-\beta)e^{-(\delta-\beta)x}.$$

Integrating from u to ∞, the result follows. Alternatively, use Laplace transforms.

The result can, however, also be seen probabilistically without summing infinite series. Let $\lambda(x)$ be the failure rate of M at $x > 0$. For a failure at x, the current ladder step must terminate which occurs at rate δ and there must be no further ones which occurs w.p. $1 - \rho$. Thus $\lambda(x) = \delta(1-\rho) = \delta - \beta$ so that the conditional distribution of M given $M > 0$ is exponential with rate $\delta - \beta$ and

$$\psi(u) = \mathbb{P}(M > u) = \mathbb{P}(M > 0)\,\mathbb{P}\big(M > u \,\big|\, M > 0\big) = \rho e^{-(\delta-\beta)u}.$$

$\square$

In IX.3, we show that expressions for $\psi(u)$ which are explicit (up to matrix exponentials) come out in a similar way also when B is phase-type. E.g. (Example IX.3.2), if $\beta = 3$ and B is a mixture of two exponential distributions with rates 3 and 7, and weights $1/2$ for each, then

$$\psi(u) = \frac{24}{35}e^{-u} + \frac{1}{35}e^{-6u}. \tag{3.1}$$

For heavy-tailed B, we use the Pollaczeck-Khinchine formula in Chapter X to show that

$$\psi(u) \sim \frac{\rho}{1-\rho}\overline{B}_0(u), \quad u \to \infty.$$

Notes and references Corollary 3.2 is one of the main classical early results in the area. A variety of proofs are available. We mention in particular the following: (a) check that $\psi(u) = \rho \, e^{-(\delta - \beta)u}$ is the solution of the renewal equation (3.2) below; (b) use stopped martingales, cf. II.3.

3c Some classical analytical results

Recall the notation $\overline{G}_+(u) = \int_u^\infty G_+(\mathrm{d}x)$.

Corollary 3.3 *The ruin probability $\psi(u)$ satisfies the defective renewal equation*

$$\psi(u) \;=\; \overline{G}_+(u) + G_+ * \psi(u) \;=\; \beta \int_u^\infty \overline{B}(y) \, \mathrm{d}y + \int_0^u \psi(u - y) \beta \overline{B}(y) \, \mathrm{d}y. \quad (3.2)$$

Equivalently, the survival probability $\phi(u) = 1 - \psi(u)$ satisfies the defective renewal equation

$$\phi(u) \;=\; 1 - \rho + G_+ * \phi(u) \;=\; 1 - \rho + \int_0^u \phi(u - y) \beta \overline{B}(y) \, \mathrm{d}y. \quad (3.3)$$

Proof Write $\psi(u)$ as

$$\mathbb{P}(M > u) \;=\; \mathbb{P}(S_{\tau_+} > u, \tau_+ < \infty) + \mathbb{P}(M > u, S_{\tau_+} \le u, \tau_+ < \infty).$$

Then the first term on the r.h.s. is $\overline{G}_+(u)$, and conditioning upon $S_{\tau_+} = y$ yields

$$\mathbb{P}(M > u, S_{\tau_+} \le u, \tau_+ < \infty)$$
$$= \int_0^u \mathbb{P}(M > u - y) G_+(\mathrm{d}y) \;=\; \int_0^u \psi(u - y) G_+(\mathrm{d}y).$$

For the last identity in (3.2), just insert the explicit form of G_+. The case of (3.3) is similar (equivalently, (3.3) can be derived by elementary algebra from (3.2)). □

Corollary 3.4 *The Laplace transform of the ruin probability is*

$$\int_0^\infty e^{-su} \psi(u) \mathrm{d}u \;=\; \frac{\beta - \beta \widehat{B}[-s] - \rho s}{s(\beta - s - \beta \widehat{B}[-s])}. \quad (3.4)$$

Proof. We first find the m.g.f. $\widehat{B}_0$ of B_0 as

$$\widehat{B}_0[r] \;=\; \int_0^\infty e^{ru} \frac{\overline{B}(u)}{\mu_B} \mathrm{d}u \;=\; \int_0^\infty \frac{e^{ru} - 1}{r\mu_B} B(\mathrm{d}u) \;=\; \frac{\widehat{B}[r] - 1}{r\mu_B}. \quad (3.5)$$

Hence

$$\mathbb{E}e^{rM} = (1-\rho)\sum_{n=0}^{\infty} \rho^n \widehat{B}_0[r]^n = \frac{1-\rho}{1-\rho\widehat{B}_0[r]} = \frac{(1-\rho)r}{r+\beta-\beta\widehat{B}[r]}, \quad (3.6)$$

$$\int_0^{\infty} e^{-su}\psi(u)du = \int_0^{\infty} e^{-su}\mathbb{P}(M>u)du = \frac{1}{s}\left(1-\mathbb{E}e^{-sM}\right)$$

$$= \frac{1}{s}\left(1 + \frac{(1-\rho)s}{\beta-s-\beta\widehat{B}[-s]}\right),$$

which is the same as (3.4). $\square$

Corollary 3.5 *The first two moments of M are*

$$\mathbb{E}M = \int_0^{\infty}\psi(u)\,du = \frac{\rho\mu_B^{(2)}}{2(1-\rho)\mu_B}, \quad \mathbb{E}M^2 = \frac{\rho\mu_B^{(3)}}{3(1-\rho)\mu_B} + \frac{\beta^2\mu_B^{(2)2}}{2(1-\rho)^2}. \quad (3.7)$$

Proof. This can be shown, for example, by analytical manipulations (L'Hôpital's rule) from (3.6). We omit the details (see, e.g., [APQ, p. 237]). $\square$

Remark 3.6 As mentioned in the Notes below, one can also derive (3.2) by analytical techniques. At the same time, (3.4) follows from (3.2) directly by taking Laplace transforms and noting that the Laplace transform of a convolution of two functions is the product of their Laplace transforms. The Laplace transform of the survival probability $\phi(u)$ correspondingly is

$$\widehat{\phi}[-s] = \int_0^{\infty} e^{-su}\phi(u)du = \frac{1-\rho}{s-\beta(1-\widehat{B}[-s])}.$$

This can now be used to provide yet another more analytical proof of the ladder height density for a compound Poisson process. From $\phi(u) = \mathbb{P}(M \leq u)$ (or from (3.5)) one sees that

$$\mathbb{E}e^{-sM} = \phi(0) + \int_0^{\infty} e^{-su}\phi'(u)\,du = \frac{(1-\rho)s}{s-\beta(1-\widehat{B}[-s])}.$$

On the other hand, as a sum of i.i.d. ladder heights, M is a geometric compound with $\mathbb{E}(e^{-sM}) = \mathbb{E}\big((\widehat{G^+}[-s]/\rho)^N\big)$ and N is geometric$(1-\rho)$, leading to $\mathbb{E}e^{-sM} = (1-\rho)/\big(1-\widehat{G^+}[-s]\big)$. A comparison of those two representations for $\mathbb{E}e^{-sM}$ now gives $\widehat{G^+}[-s] = \beta\left(1-\widehat{B}[-s]\right)/s$, so that $g_+(x) = \rho\,b_0(x)$. $\square$

Notes and references Corollary 3.3 is standard, see e.g. [APQ, pp. 144–145] or Feller [362]. The approach there is to condition upon the first claim occurring at time t and having size x, which yields the survival probability as

$$\phi(u) = \int_0^\infty \beta e^{-\beta t} dt \int_0^{u+t} \phi(u + t - x) B(dx),$$

from which (3.3) can be derived by elementary but tedious manipulations (in Section XII.3 a formal procedure will be discussed that is applicable in much more general models). Of course, it is not surprising that such arguments are more cumbersome since the ladder height representation is not used.

Also (3.6) and Corollary 3.5 can be found in virtually any queueing book. In fact, either of these sets of formulas are what many authors call the Pollaczeck-Khinchine formula.

In view of (3.4), numerical inversion of the Laplace transform is one of the classical approaches for computing ruin probabilities, see e.g. Abate & Whitt [2], Embrechts, Grübel & Pitts [346], Grübel [438], Thorin & Wikstad [848] and Albrecher, Avram & Kortschak [14] (see also the Bibliographical Notes in [746, p. 191]).

3d Deterministic claims

Corollary 3.7 *If B is degenerate at μ, then*

$$\psi(u) = 1 - (1 - \rho) \sum_{k=0}^{\lfloor u/\mu \rfloor} e^{-\rho(k - u/\mu)} \frac{[\rho(k - u/\mu)]^k}{k!}.$$

Proof. By replacing $\{S_t\}$ by $\{S_{t\mu}/\mu\}$ if necessary, we may assume $\mu = 1$ so that the stated formula in terms of the survival probability $\phi(u) = 1 - \psi(u)$ takes the form

$$\phi(u) = (1 - \beta) \sum_{k=0}^{\lfloor u \rfloor} e^{-\beta(k-u)} \frac{[\beta(k-u)]^k}{k!}. \tag{3.8}$$

The renewal equation (3.3) for $\phi(u)$ means

$$
\begin{aligned}
\phi(u) &= 1 - \beta + \int_0^{1 \wedge u} \phi(u - y) \beta I(0 \le y \le 1) \, dy \\
&= 1 - \beta + \int_{u - 1 \wedge u}^u \phi(y) \beta I(0 \le u - y \le 1) \, dy \\
&= \begin{cases} 1 - \beta + \beta \displaystyle\int_0^u \phi(y) \, dy, & 0 \le u \le 1, \\[2mm] 1 - \beta + \beta \displaystyle\int_{u-1}^u \phi(y) \, dy, & 1 \le u < \infty. \end{cases}
\end{aligned}
$$

For $0 \leq u \leq 1$, differentiation yields $\phi'(u) = \beta\phi(u)$ which together with the boundary condition $\phi(0) = 1 - \beta$ yields $\phi(u) = (1 - \beta)e^{\beta u}$ so that (3.8) follows for $0 \leq u \leq 1$. Assume (3.8) shown for $n - 1 \leq u \leq n$ and let $\widetilde{\phi}(u)$ denote the r.h.s. of (3.8). For $n \leq u \leq n+1$, differentiation yields $\phi'(u) = \beta\phi(u) - \beta\phi(u-1)$,

$$
\begin{aligned}
\widetilde{\phi}'(u) &= \frac{\mathrm{d}}{\mathrm{d}u}(1 - \beta) \sum_{k=0}^{n} e^{-\beta(k-u)} \frac{\left[\beta(k - u)\right]^{k}}{k!} \\
&= (1 - \beta)\beta e^{\beta u} + (1 - \beta) \sum_{k=1}^{n} \beta e^{-\beta(k-u)} \frac{\left[\beta(k - u)\right]^{k}}{k!} \\
&\quad - (1 - \beta) \sum_{k=1}^{n} e^{-\beta(k-u)} \frac{\beta\left[\beta(k - u)\right]^{k-1}}{(k - 1)!} \\
&= \beta\widetilde{\phi}(u) - \beta(1 - \beta) \sum_{k=0}^{n-1} e^{-\beta(k-u+1)} \frac{\left[\beta(k - u + 1)\right]^{k}}{k!} \\
&= \beta\widetilde{\phi}(u) - \beta\widetilde{\phi}(u - 1) .
\end{aligned}
$$

Since $\phi(n) = \widetilde{\phi}(n)$ by the induction hypothesis, it follows that $\phi(u) = \widetilde{\phi}(u)$ for $n \leq u \leq n + 1$. $\qquad\square$

Notes and references   Corollary 3.7 is identical to the formula for the M/D/1 waiting time distribution derived by Erlang [356]. See also Iversen & Staalhagen [496] for a discussion of computational aspects and further references.

4   Change of measure via exponential families

If X is a random variable with c.d.f. F and c.g.f.

$$
\kappa(\alpha) = \log \mathbb{E}e^{\alpha X} = \log \int_{-\infty}^{\infty} e^{\alpha x} F(\mathrm{d}x) = \log \widehat{F}[\alpha],
$$

the standard definition of the exponential family $\{F_\theta\}$ generated by F is

$$
F_\theta(\mathrm{d}x) = e^{\theta x - \kappa(\theta)} F(\mathrm{d}x), \tag{4.1}
$$

or equivalently, in terms of the c.g.f. of F_θ,

$$
\kappa_\theta(\alpha) = \kappa(\alpha + \theta) - \kappa(\theta). \tag{4.2}
$$

(Here θ is any such number such that $\kappa(\theta)$ is well-defined.)

The adaptation of this construction to Lévy processes (such as $\{S_t\}$) has been carried out in III.3, but will now be repeated for the sake of self-containedness. We could first tentatively consider the claim surplus $X = S_t$ for a single t, say $t = 1$: recall from Proposition 1.1 that $\kappa(\alpha) = \beta\big(\widehat{B}[\alpha] - 1\big) - \alpha$, and define κ_θ by (4.2). The question then naturally arises whether κ_θ is the c.g.f. corresponding to a compound Poisson risk process in the sense that for a suitable arrival intensity β_θ and a suitable claim size distribution B_θ we have

$$\kappa_\theta(\alpha) \;=\; \kappa(\alpha + \theta) - \kappa(\theta) \;=\; \beta_\theta(\widehat{B}_\theta[\alpha] - 1) - \alpha. \tag{4.3}$$

The answer is yes: inserting in (4.2) shows that the solution is

$$\beta_\theta = \beta\widehat{B}[\theta], \qquad B_\theta(\mathrm{d}x) = \frac{\mathrm{e}^{\theta x}}{\widehat{B}[\theta]}B(\mathrm{d}x), \text{ or equivalently } \widehat{B}_\theta[\alpha] = \frac{\widehat{B}[\alpha + \theta]}{\widehat{B}[\theta]}. \tag{4.4}$$

Repeating for $t \neq 1$, we just have to multiply (4.3) by t, and thus (4.4) works as well. Formalizing this for the purpose of studying the whole process $\{S_t\}$, we set up

Definition 4.1 *Let $\mathbb{P}$ be the probability measure on $D[0, \infty)$ governing a given compound Poisson risk process with arrival intensity β and claim size distribution B, and define β_θ, B_θ by (4.4). Then $\mathbb{P}_\theta$ denotes the probability measure governing the compound Poisson risk process with arrival intensity β_θ and claim size distribution B_θ; the corresponding expectation operator is $\mathbb{E}_\theta$.*

The following result (Proposition 4.2, with T taking the role of n) is the analogue of the expression

$$\exp\big\{\theta(x_1 + \cdots + x_n) - n\kappa(\theta)\big\} \tag{4.5}$$

for the density of n i.i.d. replications from F_θ (replace x by x_i in (4.1) and multiply from 1 to n).

Let $\mathscr{F}_T = \sigma(S_t : t \leq T)$ denote the σ-algebra spanned by the S_t, $t \leq T$, and $\mathbb{P}_\theta^{(T)}$ the restriction of $\mathbb{P}_\theta$ to $\mathscr{F}_T$.

Proposition 4.2 *For any fixed T, the $\mathbb{P}_\theta^{(T)}$ are mutually equivalent on $\mathscr{F}_T$, and*

$$\frac{\mathrm{d}\mathbb{P}_\theta^{(T)}}{\mathrm{d}\mathbb{P}^{(T)}} \;=\; \exp\big\{\theta S_T - T\kappa(\theta)\big\}.$$

That is, for $G \in \mathscr{F}_T$,

$$\mathbb{P}(G) \;=\; \mathbb{P}_0(G) \;=\; \mathbb{E}_\theta\big[\exp\big\{-\theta S_T + T\kappa(\theta)\big\}; G\big]. \tag{4.6}$$

Proof. We must prove that if Z is $\mathscr{F}_T$-measurable, then

$$\mathbb{E}_\theta Z = \mathbb{E}\left[Z e^{\theta S_T - T\kappa(\theta)}\right]. \tag{4.7}$$

By standard measure theory, it suffices to consider the case where Z is measurable w.r.t. $\mathscr{F}_T^{(n)} = \sigma\left(S_{kT/n} : k = 0, 1, \ldots, n\right)$ for a given n. But let $X_k = S_{kT/n} - S_{(k-1)T/n}$. Then the X_k are i.i.d. with common c.g.f. $T\kappa(\alpha)/n$, Z is measurable w.r.t. $\sigma(X_1, \ldots, X_n)$, and thus (4.7) follows by discrete exponential family theory, in particular the expression (4.5) for the density. The identity (4.6) now follows by taking $Z = e^{-\theta S_T + T\kappa(\theta)} I(G)$. $\square$

Theorem 4.3 *Let τ be any stopping time and let $G \in \mathscr{F}_\tau$, $G \subseteq \{\tau < \infty\}$. Then*

$$\mathbb{P}(G) = \mathbb{P}_0(G) = \mathbb{E}_\theta\left[\exp\{-\theta S_\tau + \tau\kappa(\theta)\}; G\right]. \tag{4.8}$$

Proof. We first note that for any fixed t,

$$\mathbb{E}_\theta e^{-\theta S_t + t\kappa(\theta)} = 1. \tag{4.9}$$

Now assume first that $G \subseteq \{\tau \leq T\}$ for some deterministic T. Then $G \in \mathscr{F}_T$, and hence (4.6) holds. Given $\mathscr{F}_\tau$, $t = T - \tau$ is deterministic. Thus by (4.9),

$$\mathbb{E}_\theta\left[\exp\{-\theta S_T + T\kappa(\theta)\} I(G) \,\middle|\, \mathscr{F}_\tau\right] = 1,$$

so that $\mathbb{P}Q$ equals

$$\begin{aligned}
& \mathbb{E}_\theta \, \mathbb{E}_\theta\left[\exp\{-\theta S_T + T\kappa(\theta)\} I(G) \,\middle|\, \mathscr{F}_\tau\right] \\
= \ & \mathbb{E}_\theta\left[\exp\{-\theta S_\tau + \tau\kappa(\theta)\} I(G) \mathbb{E}_\theta\left[\exp\{-\theta(S_T - S_\tau) + (T - \tau)\kappa(\theta)\} \,\middle|\, \mathscr{F}_\tau\right]\right] \\
= \ & \mathbb{E}_\theta\left[\exp\{-\theta S_\tau + \tau\kappa(\theta)\} I(G)\right].
\end{aligned}$$

Now consider a general G. Then $G_T = G \cap \{\tau \leq T\}$ satisfies $G_T \in \mathscr{F}_\tau$, $G_T \subseteq \{\tau \leq T\}$. Thus, according to what has just been proved, (4.8) holds with G replaced by G_T. Letting $T \uparrow \infty$ and using monotone convergence then shows that (4.8) holds for G as well. $\square$

5 Lundberg conjugation

Being a c.g.f., $\kappa(\alpha)$ is a convex function of α. The behavior at zero is given by the first order Taylor expansion

$$\kappa(\alpha) \approx \kappa(0) + \kappa'(0)\alpha = 0 + \mathbb{E}S_1 \, \alpha = \alpha(\rho - 1) = -\frac{\eta}{1 + \eta}\alpha.$$

Thus, subject to the basic assumption $\eta > 0$ of a positive safety loading, the typical shape of κ is as in Fig. IV.1(a).

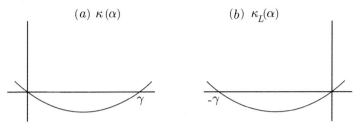

(a) $\kappa(\alpha)$ (b) $\kappa_L(\alpha)$

FIGURE IV.1

When the tail of the claim size distribution is exponentially bounded, then typically a $\gamma > 0$ satisfying

$$0 = \kappa(\gamma) = \beta\big(\widehat{B}[\gamma] - 1\big) - \gamma \tag{5.1}$$

exists. Equation (5.1) is known as the *Lundberg equation* and plays a fundamental role in risk theory; an equivalent version illustrated in Fig. IV.2 is

$$\widehat{B}[\gamma] = 1 + \frac{\gamma}{\beta}. \tag{5.2}$$

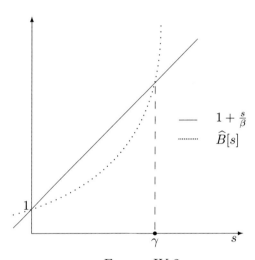

$$\underline{\hspace{1cm}} \quad 1 + \frac{s}{\beta}$$
$$\cdots\cdots\cdots \quad \widehat{B}[s]$$

FIGURE IV.2

As support for memory, we write $\mathbb{P}_L$ instead of $\mathbb{P}_\gamma$, β_L instead of β_γ and so on in the following. Note that

$$\kappa_L(\alpha) = \beta_L\big(\widehat{B}_L[\alpha] - 1\big) - \alpha = \kappa(\alpha + \gamma),$$

cf. Fig. IV.1(b). An established terminology is to call γ the *adjustment coefficient* but there are various alternatives around, e.g. *the Lundberg exponent*.

Example 5.1 Consider the case of exponential claims, $\widehat{B}[r] = \delta/(\delta - r)$. It is then readily seen that the non-zero solution of (5.1) (or (5.2)) is $\gamma = \delta - \beta$. Thus $\widehat{B}[\gamma] = \delta/\beta$, and (4.4) yields $\beta_L = \delta$ and that B_L is again exponential with rate $\delta_L = \beta$. Thus, Lundberg conjugation corresponds to interchanging the rates of the interarrival times and the claim sizes. $\square$

It is a crucial fact that when governed by $\mathbb{P}_L$, the claim surplus process has positive drift

$$\mathbb{E}_L S_1 = \kappa'_L(0) = \kappa'(\gamma) > 0, \qquad (5.3)$$

cf. Fig. IV.1(b). Taking $\tau = \tau(u)$, $G = \{\tau(u) < \infty\}$ in Theorem 4.3, we further note that (5.1) is precisely what is needed for one of the terms in the exponent to vanish so that Theorem 4.3 takes a particular simple form,

$$\psi(u) = \mathbb{P}(\tau(u) < \infty) = \mathbb{E}_L \left[\exp\left\{-\gamma S_{\tau(u)}\right\}; \tau(u) < \infty\right].$$

Letting $\xi(u) = S_{\tau(u)} - u$ be the overshoot and noting that $\mathbb{P}_L(\tau(u) < \infty) = 1$ by (5.3), we can rewrite this as

$$\psi(u) = e^{-\gamma u}\mathbb{E}_L e^{-\gamma \xi(u)}, \qquad (5.4)$$

see also III.(1.5).

Theorem 5.2 (LUNDBERG'S INEQUALITY) *For all $u \geq 0$, $\psi(u) \leq e^{-\gamma u}$.*

Proof. Just note that $\xi(u) \geq 0$ in (5.4). $\square$

Theorem 5.3 (THE CRAMÉR-LUNDBERG APPROXIMATION) $\psi(u) \sim Ce^{-\gamma u}$ *as* $u \to \infty$, *where*

$$C = \frac{1 - \rho}{\gamma \int_0^\infty x e^{\gamma x}\beta \overline{B}(x)\,dx} = \frac{1 - \rho}{\beta \widehat{B}'[\gamma] - 1}. \qquad (5.5)$$

Proof. By renewal theory, see A.1e, $\xi(u)$ has a limit $\xi(\infty)$ (in the sense of weak convergence w.r.t. $\mathbb{P}_L$) with density

$$\frac{1 - G_+^{(L)}(x)}{\mu_+^{(L)}} = \frac{\overline{G}_+^{(L)}(x)}{\mu_+^{(L)}},$$

where $G_+^{(L)}$ is the $\mathbb{P}_L-$ ascending ladder height distribution and $\mu_+^{(L)}$ its mean. Since $e^{-\gamma x}$ is continuous and bounded, we therefore have $\mathbb{E}_L e^{-\gamma \xi(u)} \to C$ where

$$\begin{aligned}
C &= \mathbb{E}_L e^{-\gamma \xi(\infty)} = \frac{1}{\mu_+^{(L)}}\int_0^\infty e^{-\gamma x}(1 - G_+^{(L)}(x))\,dx \\
&= \frac{1}{\gamma \mu_+^{(L)}}\int_0^\infty (1 - e^{-\gamma x})G_+^{(L)}(dx), \qquad (5.6)
\end{aligned}$$

and all that is needed to check is that (5.6) is the same as (5.5). To that end, take first $\theta = \gamma$, $\tau = \tau_+$, $G = \{S_{\tau_+} \in A\}$ in Theorem 4.3. Then

$$\mathbb{P}(S_{\tau_+} \in A) = \mathbb{E}_L \left[\exp\{-\gamma S_{\tau_+}\} ; S_{\tau_+} \in A\right],$$

which shows that

$$G_+^{(L)}(\mathrm{d}x) = \mathrm{e}^{\gamma x} G_+(\mathrm{d}x) = \mathrm{e}^{\gamma x} \beta \overline{B}(x) \, \mathrm{d}x. \tag{5.7}$$

In principle, this solves the problem of evaluating (5.6), but some tedious (though elementary) calculations remain to bring the expressions on a final form. Noting that $\|G_+^{(L)}\| = 1$ because of (5.3), we get

$$\int_0^\infty (1 - \mathrm{e}^{-\gamma x}) G_+^{(L)}(\mathrm{d}x) = 1 - \int_0^\infty \beta \overline{B}(x) \, \mathrm{d}x = 1 - \rho.$$

Using (5.7) yields

$$\mu_+^{(L)} = \beta \int_0^\infty x \mathrm{e}^{\gamma x} \overline{B}(x) \, \mathrm{d}x = \beta \varphi'(\gamma), \tag{5.8}$$

where

$$\varphi(\alpha) = \int_0^\infty \mathrm{e}^{\alpha x} \overline{B}(x) \, \mathrm{d}x = \frac{1}{\alpha} \left(\widehat{B}[\alpha] - 1\right) \tag{5.9}$$

so that

$$\varphi'(\gamma) = \frac{\gamma \widehat{B}'[\gamma] - \left(\widehat{B}[\gamma] - 1\right)}{\gamma^2} = \frac{\widehat{B}'[\gamma] - 1/\beta}{\gamma}$$

(using (5.1)) and

$$\gamma \mu_+^{(L)} = \gamma \beta \frac{\widehat{B}'[\gamma] - 1/\beta}{\gamma} = \beta \widehat{B}'[\gamma] - 1. \tag{5.10}$$

$\square$

Example 5.4 Consider first the exponential case $b(x) = \delta \mathrm{e}^{-\delta x}$. Then $\psi(u) = \rho \mathrm{e}^{-(\delta - \beta)u}$ where $\rho = \beta/\delta$. From this it follows, of course, that $\gamma = \delta - \beta$ (this was already found in Example 5.1 above) and that $C = \rho$. A direct proof of $C = \rho$ is of course easy:

$$\widehat{B}'[\gamma] = \frac{\mathrm{d}}{\mathrm{d}\gamma} \frac{\delta}{\delta - \gamma} = \frac{\delta}{(\delta - \gamma)^2} = \frac{\delta}{\beta^2},$$

$$C = \frac{1 - \rho}{\beta \widehat{B}'[\gamma] - 1} = \frac{1 - \rho}{\beta \delta/\beta^2 - 1} = \frac{1 - \rho}{1/\rho - 1} = \rho.$$

The accuracy of Lundberg's inequality in the exponential case thus depends on how close ρ is to one, or equivalently of how close the safety loading η is to zero. $\square$

Remark 5.5 Noting that

$$\rho_L - 1 \;=\; \beta_L \mu_{B_L} - 1 \;=\; \kappa_L'(0) \;=\; \kappa'(\gamma) = \beta\widehat{B}'[\gamma] - 1 \,,$$

we can rewrite the Cramér-Lundberg constant C in the nice symmetrical form

$$C \;=\; \frac{|\kappa'(0)|}{\kappa'(\gamma)} \;=\; \frac{1 - \rho}{\rho_L - 1}. \tag{5.11}$$

□

Remark 5.6 Let $\widehat{\psi}[-s] = \int_0^\infty \mathrm{e}^{-su}\psi(u)\,\mathrm{d}u$ denote the Laplace transform of the ruin probability. Obviously, the Laplace transform of $\psi(u)\mathrm{e}^{\gamma u}$ is then $\widehat{\psi}[-s + \gamma]$. Since from the damping property of Laplace transforms, for any function $f(u)$, $\lim_{u\to\infty} f(u) = \lim_{s\to 0} s\,\widehat{f}[-s]$, given that this limit exists, we can also determine the constant C in the Cramér-Lundberg approximation by

$$C = \lim_{s\to 0} s\,\widehat{\psi}[-s + \gamma],$$

which from (3.4) again gives (5.5). Although it looks tempting to use this procedure for determining C in more general models where γ exists and only the Laplace transform of ψ may be available explicitly, it is important to note that this procedure does not prove the Cramér-Lundberg approximation, but just gives the correct value of C in case the approximation holds (the approximation itself usually has to be established by other techniques and often only exists in a weaker logarithmic sense, cf. Chapter XIII).

For a related method to obtain the asymptotic behavior of $\psi(u)$ for regularly varying claims, see Chapter X.   □

In Chapter V, we shall need the following result which follows by a variant of the calculations in the proof of Theorem 5.3:

Lemma 5.7 *For $\alpha \neq \gamma$,* $\mathbb{E}_L \mathrm{e}^{-\alpha\xi(\infty)} \;=\; \dfrac{\gamma}{\alpha\kappa'(\gamma)}\left(1 - \beta\dfrac{\widehat{B}[\gamma - \alpha] - 1}{\gamma - \alpha}\right).$

Proof. Replacing γ by α in (5.6) and using (5.7), we obtain

$$\begin{aligned}
\mathbb{E}_L \mathrm{e}^{-\alpha\xi(\infty)} &= \frac{1}{\alpha\mu_+^{(L)}}\left(1 - \int_0^\infty \mathrm{e}^{(\gamma-\alpha)x}\beta\overline{B}(x)\,\mathrm{d}x\right) \\
&= \frac{1}{\alpha\mu_+^{(L)}}\left(1 - \beta\frac{\widehat{B}[\gamma - \alpha] - 1}{\gamma - \alpha}\right),
\end{aligned}$$

using integration by parts as in (3.5) in the last step. Inserting (5.10), the result follows. □

Notes and references The results of this section are classical, with Lundberg's inequality being given first in Lundberg [615] and the Cramér-Lundberg approximation in Cramér [265]. Therefore, extensions and generalizations are main topics in the area of ruin probabilities, and in particular numerous such results can be found later in this book; in particular, see Sections V.4, VI.3, VII.3, VII.6, XI.2, and XII.2–3.

The mathematical approach we have taken is more recent in risk theory (some of the classical ones can be found in the next subsection). The techniques are basically standard ones from sequential analysis, see for example Wald [869] and Siegmund [810].

5a Alternative proofs

For the sake of completeness, we shall here give some classical proofs, first one of Lundberg's inequality which is slightly longer but maybe also slightly more elementary:

Alternative proof of Lundberg's inequality. Let X be the value of $\{S_t\}$ just after the first claim, $F(x) = \mathbb{P}(X \leq x)$. Then, since X is the independent difference $U - T$ between an interarrival time T and a claim U,

$$\widehat{F}[\gamma] \;=\; \mathbb{E}e^{\gamma(U-T)} \;=\; \mathbb{E}e^{\gamma U} \cdot \mathbb{E}e^{-\gamma T} \;=\; \widehat{B}[\gamma]\frac{\beta}{\beta + \gamma} \;=\; 1,$$

where the last equality follows from $\kappa(\gamma) = 1$. Let $\psi^{(n)}(u)$ denote the probability of ruin after at most n claims. Conditioning upon the value x of X and considering the cases $x > u$ and $x \leq u$ separately yields

$$\psi^{(n+1)}(u) \;=\; \overline{F}(u) + \int_{-\infty}^{u} \psi^{(n)}(u - x)\, F(\mathrm{d}x).$$

We claim that this implies $\psi^{(n)}(u) \leq e^{-\gamma u}$, which completes the proof since $\psi(u) = \lim_{n\to\infty} \psi^{(n)}(u)$. Indeed, this is obvious for $n = 0$ since $\psi^{(0)}(u) = 0$. Assuming it proved for n, we get

$$
\begin{aligned}
\psi^{(n+1)}(u) \;&\leq\; \overline{F}(u) + \int_{-\infty}^{u} e^{-\gamma(u-x)}\, F(\mathrm{d}x) \\
&\leq\; e^{-\gamma u}\int_{u}^{\infty} e^{\gamma x}\, F(\mathrm{d}x) + \int_{-\infty}^{u} e^{-\gamma(u-x)}\, F(\mathrm{d}x) \\
&=\; e^{-\gamma u}\widehat{F}[\gamma] \;=\; e^{-\gamma u}.
\end{aligned}
$$

□

Of further proofs of Lundberg's inequality, we mention in particular the martingale approach, see III.1.

Next consider the Cramér-Lundberg approximation. Here the most standard proof is via the renewal equation in Corollary 3.3 (however, as will be seen, the calculations needed to identify the constant C are precisely the same as above):

Alternative proof of the Cramér-Lundberg's approximation. Recall from Corollary 3.3 that

$$\psi(u) = \beta \int_u^\infty \overline{B}(x)\,\mathrm{d}x + \int_0^u \psi(u-x)\beta\overline{B}(x)\,\mathrm{d}x.$$

Multiplying by $\mathrm{e}^{\gamma u}$ and letting

$$Z(u) = \mathrm{e}^{\gamma u}\psi(u), \quad z(u) = \mathrm{e}^{\gamma u}\beta \int_u^\infty \overline{B}(x)\,\mathrm{d}x, \quad F(\mathrm{d}x) = \mathrm{e}^{\gamma x}\beta\overline{B}(x)\,\mathrm{d}x,$$

we can rewrite this as

$$\begin{aligned}
Z(u) &= z(u) + \int_0^u \mathrm{e}^{\gamma(u-x)}\psi(u-x) \cdot \mathrm{e}^{\gamma x}\beta\overline{B}(x)\,\mathrm{d}x, \\
&= z(u) + \int_0^u Z(u-x)F(\mathrm{d}x),
\end{aligned}$$

i.e. $Z = z + F * Z$. Note that by (5.9) and the Lundberg equation, γ is precisely the correct exponent which will ensure that F is a proper distribution ($\|F\| = 1$). It is then a matter of routine to verify the conditions of the key renewal theorem (Proposition A1.1) to conclude that $Z(u)$ has the limit $C = \int_0^\infty z(x)\mathrm{d}x/\mu_F$, so that it only remains to check that C reduces to the expression given above. However, μ_F is immediately seen to be the same as $\mu_+^{(L)}$ calculated in (5.8), whereas

$$\begin{aligned}
\int_0^\infty z(u)\,\mathrm{d}u &= \int_0^\infty \beta\mathrm{e}^{\gamma u}\,\mathrm{d}u \int_u^\infty \overline{B}(x)\,\mathrm{d}x = \int_0^\infty \overline{B}(x)\,\mathrm{d}x \int_0^x \beta\mathrm{e}^{\gamma u}\,\mathrm{d}u \\
&= \int_0^\infty \overline{B}(x)\frac{\beta}{\gamma}\left(\mathrm{e}^{\gamma x} - 1\right)\,\mathrm{d}x = \frac{\beta}{\gamma}\left[\frac{1}{\gamma}(\widehat{B}[\gamma]-1) - \mu_B\right] \\
&= \frac{\beta}{\gamma}\left[\frac{1}{\beta} - \mu_B\right] = \frac{1-\rho}{\gamma},
\end{aligned}$$

using the Lundberg equation and the calculations in (5.9). Easy calculus now gives (5.5). □

Notes and references Another related, but slightly different proof of the Cramér-Lundberg's approximation in the spirit of Feller [362] that utilizes the Blackwell renewal theorem can be found in Albrecher & Teugels [36].

The asymptotic behavior of the ruin probability for heavy-tailed claims will be discussed in X.2.

6 Further topics related to the adjustment coefficient

6a On the existence of γ

In order that the adjustment coefficient γ exists, it is of course necessary that B is light-tailed in the sense of I.2a, i.e. that $\widehat{B}[\alpha] < \infty$ for some $\alpha > 0$. This excludes heavy-tailed distributions like the log-normal or Pareto, but may in many other cases not appear all that restrictive, and the following possibilities then occur:

1. $\widehat{B}[\alpha] < \infty$ for all $\alpha < \infty$.

2. There exists $\alpha^* < \infty$ such that $\widehat{B}[\alpha] < \infty$ for all $\alpha < \alpha^*$ and $\widehat{B}[\alpha] = \infty$ for all $\alpha \geq \alpha^*$.

3. There exists $\alpha^* < \infty$ such that $\widehat{B}[\alpha] < \infty$ for all $\alpha \leq \alpha^*$ and $\widehat{B}[\alpha] = \infty$ for all $\alpha > \alpha^*$.

In particular, monotone convergence yields $\widehat{B}[\alpha] \uparrow \infty$ as $\alpha \uparrow \infty$ in case 1, and $\widehat{B}[\alpha] \uparrow \infty$ as $\alpha \uparrow \alpha^*$ in case 2 (in exponential family theory, this is often referred to as the *steep* case). Thus the existence of γ is automatic in cases 1, 2; standard examples are distributions with finite support or tail satisfying $\overline{B}(x) = o(e^{-\alpha x})$ for all α in case 1, and phase-type or Gamma distributions in case 2. Case 3 may be felt to be rather atypical, but some non-pathological examples exist, for example the inverse Gaussian distribution (see Example 9.7 below for details). In case 3, γ exists provided $\widehat{B}[\alpha^*] \geq 1 + \alpha^*/\beta$ and not otherwise, that is, dependent on whether β is larger or smaller than the threshold value $\alpha^*/(\widehat{B}[\alpha^*] - 1)$.

Notes and references Ruin probabilities in case 3 with γ non-existent are studied, e.g., by Borovkov [182, p. 132] and Embrechts & Veraverbeke [353]. To the present authors' mind, this is a somewhat special situation and therefore not treated in this book.

6b Bounds and approximations for γ

Proposition 6.1 *If the adjustment coefficient exists, it can be bounded by*

$$\gamma \;<\; \frac{2(1 - \beta\mu_B)}{\beta\mu_B^{(2)}} \;=\; \frac{2\eta\mu_B}{\mu_B^{(2)}}.$$

Proof. From $U \geq 0$ it follows that $\widehat{B}[\alpha] = \mathbb{E}e^{\alpha U} > 1 + \mu_B\alpha + \mu_B^{(2)}\alpha^2/2$. Hence

$$1 \;=\; \frac{\beta\big(\widehat{B}[\gamma] - 1\big)}{\gamma} \;>\; \frac{\beta\big(\gamma\mu_B + \gamma^2\mu_B^{(2)}/2\big)}{\gamma} \;=\; \beta\mu_B + \frac{\beta\gamma\mu_B^{(2)}}{2}, \tag{6.1}$$

from which the results immediately follows. □

The upper bound in Proposition 6.1 is also an approximation for small safety loadings (heavy traffic, cf. Section 7c):

Proposition 6.2 *Let B be fixed but assume that $\beta = \beta(\eta)$ varies with the safety loading such that $\beta = \dfrac{1}{\mu_B(1 + \eta)}$. Then as $\eta \downarrow 0$,*

$$\gamma \;=\; \gamma(\eta) \;\sim\; \frac{2\eta\mu_B}{\mu_B^{(2)}}. \tag{6.2}$$

Further, the Cramér-Lundberg constant satisfies $C = C(\eta) \sim 1$.

Proof. Since $\psi(u) \to 1$ as $\eta \downarrow 0$, it follows from Lundberg's inequality that $\gamma \to 0$. Hence by Taylor expansion, the inequality in (6.1) is also an approximation so that

$$1 \;=\; \frac{\beta\big(\widehat{B}[\gamma] - 1\big)}{\gamma} \;\approx\; \frac{\beta\big(\gamma\mu_B + \gamma^2\mu_B^{(2)}/2\big)}{\gamma} \;=\; \rho + \frac{\beta\gamma\mu_B^{(2)}}{2},$$

$$\gamma \;\sim\; \frac{2(1 - \rho)}{\beta\mu_B^{(2)}} \;=\; \frac{2\eta\mu_B}{\mu_B^{(2)}}.$$

That $C \to 1$ easily follows from $\gamma \to 0$ and $C = \mathbb{E}_L e^{-\gamma\xi(\infty)}$ (in the limit, $\xi(\infty)$ is distributed as the overshoot corresponding to $\eta = 0$). For an alternative analytic proof, note that

$$C \;=\; \frac{1 - \rho}{\beta\widehat{B}'[\gamma] - 1} \;=\; \frac{\eta\mu_B}{\widehat{B}'[\gamma] - 1/\beta}$$

$$\approx\; \frac{\eta\mu_B}{\mu_B + \gamma\mu_B^{(2)} - \mu_B(1 + \eta)} \;=\; \frac{\eta}{\gamma\mu_B^{(2)}/\mu_B - \eta}$$

$$\approx\; \frac{\eta}{2\eta - \eta} \;=\; 1. \qquad\qquad\qquad □$$

Obviously, the approximation (6.2) is easier to calculate than γ itself. However, it needs to be used with caution say in Lundberg's inequality or the Cramér-Lundberg approximation, in particular when u is large.

6c A refinement of Lundberg's inequality

The following result gives a sharpening of Lundberg's inequality (because obviously $C_+ \leq 1$) as well as a supplementary lower bound:

Theorem 6.3 $C_- e^{-\gamma u} \leq \psi(u) \leq C_+ e^{-\gamma u}$ *where*

$$
C_- = \inf_{x \geq 0} \frac{\overline{B}(x)}{\int_x^\infty e^{\gamma(y-x)} B(\mathrm{d}y)}, \qquad C_+ = \sup_{x \geq 0} \frac{\overline{B}(x)}{\int_x^\infty e^{\gamma(y-x)} B(\mathrm{d}y)}.
$$

Proof. Let $H(\mathrm{d}t, \mathrm{d}x)$ be the $\mathbb{P}_L$-distribution of the time $\tau(u)$ of ruin and the reserve $u - S_{\tau(u)-}$ just before ruin. Given $\tau(u) = t$, $u - S_{\tau(u)-} = x$, a claim occurs at time t and has distribution $B_L(\mathrm{d}y)/\overline{B}_L(x)$, $y > x$. Hence

$$
\begin{aligned}
\mathbb{E}_L e^{-\gamma\xi(u)} &= \int_0^\infty \int_0^\infty H(\mathrm{d}t, \mathrm{d}x) \int_x^\infty e^{-\gamma(y-x)} \frac{B_L(\mathrm{d}y)}{\overline{B}_L(x)} \\
&= \int_0^\infty \int_0^\infty H(\mathrm{d}t, \mathrm{d}x) \frac{\int_x^\infty B(\mathrm{d}y)}{e^{-\gamma x} \overline{B}_L(x) \widehat{B}[\gamma]} \\
&= \int_0^\infty \int_0^\infty H(\mathrm{d}t, \mathrm{d}x) \frac{\overline{B}(x)}{\int_x^\infty e^{\gamma(y-x)} B(\mathrm{d}y)} \\
&\leq C_+ \int_0^\infty \int_0^\infty H(\mathrm{d}t, \mathrm{d}x) = C_+.
\end{aligned}
$$

The upper bound then follows from $\psi(u) = e^{-\gamma u} \mathbb{E}_L e^{-\xi(u)}$, and the proof of the lower bound is similar. $\qquad\square$

Example 6.4 If $\overline{B}(x) = e^{-\delta x}$, then an explicit calculation shows easily that

$$
\frac{\overline{B}(x)}{\int_x^\infty e^{\gamma(y-x)} B(\mathrm{d}y)} = \frac{e^{-\delta x}}{\int_x^\infty e^{(\delta-\beta)(y-x)} \delta e^{-\delta y} \,\mathrm{d}y} = \frac{\beta}{\delta} = \rho.
$$

Hence $C_- = C_+ = \rho$ so that the bounds in Theorem 6.3 collapse and yield the exact expression $\rho e^{-\gamma u}$ for $\psi(u)$. $\qquad\square$

The following concluding example illustrates a variety of the topics discussed above (though from a general point of view the calculations are deceivingly simple: typically, γ and other quantities will have to be calculated numerically).

Example 6.5 Assume as for (3.1) that $\beta = 3$ and

$$b(x) = \frac{1}{2} \cdot 3e^{-3x} + \frac{1}{2} \cdot 7e^{-7x},$$

and recall that the ruin probability is

$$\psi(u) = \frac{24}{35}e^{-u} + \frac{1}{35}e^{-6u}.$$

Since the dominant term is $24/35 \cdot e^{-u}$, it follows immediately that $\gamma = 1$ and $C = 24/35 = 0.686$ (also, bounding e^{-6u} by e^{-u} confirms Lundberg's inequality). For a direct verification, note that the Lundberg equation is

$$\gamma = \beta(\widehat{B}[\gamma] - 1) = 3\left(\frac{1}{2} \cdot \frac{3}{3-\gamma} + \frac{1}{2} \cdot \frac{7}{7-\gamma} - 1\right),$$

which after some elementary algebra leads to the cubic equation $2\gamma^3 - 14\gamma^2 + 12\gamma = 0$ with roots $0, 1, 6$. Thus indeed $\gamma = 1$ (6 is not in the domain of convergence of $\widehat{B}[\gamma]$ and therefore excluded). Further,

$$1 - \rho = 1 - \beta\mu_B = 1 - 3\left(\frac{1}{2} \cdot \frac{1}{3} + \frac{1}{2} \cdot \frac{1}{7}\right) = \frac{2}{7},$$

$$\widehat{B}'[\gamma] = \frac{1}{2} \cdot \frac{3}{(3-\alpha)^2} + \frac{1}{2} \cdot \frac{7}{(7-\alpha)^2}\bigg|_{\alpha=\gamma=1} = \frac{17}{36},$$

$$C = \frac{1-\rho}{\beta\widehat{B}'[\gamma] - 1} = \frac{\frac{2}{7}}{3 \cdot \frac{17}{36} - 1} = \frac{24}{35}.$$

For Theorem 6.3, note that the function

$$\frac{\displaystyle\int_u^\infty \left\{\frac{1}{2} \cdot 3e^{-3x} + \frac{1}{2} \cdot 7e^{-7x}\right\} dx}{\displaystyle\int_u^\infty e^{x-u}\left\{\frac{1}{2} \cdot 3e^{-3x} + \frac{1}{2} \cdot 7e^{-7x}\right\} dx} = \frac{3 + 3e^{-4u}}{9/2 + 7/2e^{-4u}}$$

attains its minimum $C_- = 2/3 = 0.667$ for $u = \infty$ and its maximum $C_+ = 3/4 = 0.750$ for $u = 0$, so that $0.667 \leq C \leq 0.750$ in accordance with $C = 0.686$. $\qquad\square$

Notes and references Theorem 6.3 is from Taylor [836]. Closely related results are given in a queueing setting in Kingman [535], Ross [748] and Rossberg & Siegel [749].

Some further references on variants and extensions of Lundberg's inequality are Kaas & Goovaerts [514], Willmot [886], Cai & Garrido [218], Dickson [306], Kalashnikov [516, 518] and Chadjiconstantinidis & Politis [227], all of which also go into aspects of the heavy-tailed case.

7 Various approximations for the ruin probability

7a The Beekman-Bowers approximation

The idea is to write $\psi(u)$ as $\mathbb{P}(M > u)$, fit a gamma distribution with parameters λ, δ to the distribution of M by matching the two first moments and use the incomplete gamma function approximation

$$\psi(u) \approx \int_u^\infty \frac{\delta^\lambda}{\Gamma(\lambda)} x^{\lambda-1} e^{-\delta x} \, \mathrm{d}x.$$

According to Corollary 3.5, this means that λ, δ are given by $\lambda/\delta = a_1$, $2\lambda/\delta^2 = a_2$

$$a_1 = \frac{\rho \mu_B^{(2)}}{2(1-\rho)\mu_B}, \quad a_2 = \frac{\rho \mu_B^{(3)}}{3(1-\rho)\mu_B} + \frac{\beta^2 \mu_B^{(2)^2}}{2(1-\rho)^2},$$

i.e. $\delta = 2a_1/a_2$, $\lambda = 2a_1^2/a_2$.

Notes and references The approximation was introduced by Beekman [151], with the present version suggested by Bowers in the discussion of [151].

7b De Vylder's approximation

Given a risk process with parameters β, B, $p = 1$, the idea is to approximate the ruin probability with the one for a different process with exponential claims, say with rate parameter $\widetilde{\delta}$, arrival intensity $\widetilde{\beta}$ and premium rate $\widetilde{p}$. In order to make the processes look as much as possible alike, we make the first three cumulants match, which according to Proposition 1.1 means

$$\frac{\widetilde{\beta}}{\widetilde{\delta}} - \widetilde{p} = \beta \mu_B - 1 = \rho - 1, \quad \frac{2\widetilde{\beta}}{\widetilde{\delta}^2} = \beta \mu_B^{(2)}, \quad \frac{6\widetilde{\beta}}{\widetilde{\delta}^3} = \beta \mu_B^{(3)}.$$

These three equations have solutions

$$\widetilde{\delta} = \frac{3\mu_B^{(2)}}{\mu_B^{(3)}}, \quad \widetilde{\beta} = \frac{9\beta \mu_B^{(2)^3}}{2\mu_B^{(3)^2}}, \quad \widetilde{p} = \frac{3\beta \mu_B^{(2)^2}}{2\mu_B^{(3)}} - \rho + 1. \tag{7.1}$$

Letting $\beta^* = \widetilde{\beta}/\widetilde{p}$, $\rho^* = \beta^*/\widetilde{\delta}$, the approximating risk process has ruin probability $\psi(u) = \rho^* e^{-(\widetilde{\delta}-\beta^*)u}$, cf. Proposition I.1.3 and Corollary 3.2, and hence the ruin probability approximation is

$$\psi(u) \approx \frac{\widetilde{\beta}}{\widetilde{p}\widetilde{\delta}} e^{-(\widetilde{\delta}-\widetilde{\beta}/\widetilde{p})u}. \tag{7.2}$$

Notes and references   The approximation (7.2) was suggested by De Vylder [299]. Though of course it is based upon purely empirical grounds, numerical evidence (e.g. Grandell [429, pp. 19–24], [432]) shows that it may produce surprisingly good results, in particular for light-tailed claim distributions. Extensions of this method to approximations with more general involved claim distributions are immediate, but there is a natural trade-off between complexity and accuracy of the approximation. For an investigation on the use of Coxian distributions of order two for the claim distribution of the approximating risk process, see Badescu & Stanford [120]. Due to its simplicity, the De Vylder approximation is also very popular for the study of effects of external mechanisms such as dividend payments and reinsurance on the probability of ruin (see for instance Beveridge, Dickson & Wu [161], Gerber, Shiu & Smith [413]).

A related procedure is to approximate $\psi(u)$ by a combination of two exponential terms, where one of them is the Cramér-Lundberg approximation (5.5) and the coefficient and exponent of the other are determined by matching $\mathbb{E}[M]$ and the mass of M in 0. This leads to the so-called *Tijms approximation*, see [852] and Lin & Willmot [892, Ch.8].

7c   The heavy traffic approximation

The term *heavy traffic* comes from queueing theory, but has an obvious interpretation also in risk theory: on the average, the premiums exceed only slightly the expected claims. That is, heavy traffic conditions mean that the safety loading η is positive but small, or equivalently that β is only slightly smaller than $\beta_{\max} = 1/\mu_B$. Mathematically, we shall represent this situation with a limit where $\beta \uparrow \beta_{\max}$ but B is fixed.

Proposition 7.1 *As $\beta \uparrow \beta_{\max}$, $(\beta_{\max} - \beta)M$ converges in distribution to the exponential distribution with rate $\delta = \dfrac{2\mu_B^2}{\mu_B^{(2)}}$.*

Proof. Note first that $1 - \rho = (\beta_{\max} - \beta)\mu_B$. Letting B_0 be the stationary excess life distribution, we have according to the Pollaczeck-Khinchine formula in the form (3.6) that

$$
\mathbb{E}e^{s(\beta_{\max}-\beta)M}
$$

$$
= \frac{1-\rho}{1-\rho\widehat{B}_0\big[s(\beta_{\max}-\beta)\big]} = \frac{1-\rho}{1-\rho+\rho\big\{1-\widehat{B}_0\big[s(\beta_{\max}-\beta)\big]\big\}}
$$

$$
\approx \frac{1-\rho}{1-\rho-\rho s(\beta_{\max}-\beta)\mu_{B_0}} \approx \frac{1-\rho}{1-\rho-s(\beta_{\max}-\beta)\mu_{B_0}}
$$

$$
= \frac{\mu_B}{\mu_B-s\mu_{B_0}} = \frac{\delta}{\delta-s},
$$

where $\delta = \mu_B/\mu_{B_0} = 2\mu_B^2/\mu_B^{(2)}$. $\qquad\square$

Corollary 7.2 *If $\beta \uparrow \beta_{\max}$, $u \to \infty$ in such a way that $(\beta_{\max} - \beta)u \to v$, then $\psi(u) \to e^{-\delta v}$.*

Proof. Write $\psi(u)$ as $\mathbb{P}\big((\beta_{\max} - \beta)M > (\beta_{\max} - \beta)u\big)$. $\qquad\square$

These results suggest the approximation

$$
\psi(u) \approx e^{-\delta(\beta_{\max}-\beta)u}. \tag{7.3}
$$

It is worth noting that this is essentially the same as the approximation

$$
\psi(u) \approx Ce^{-\gamma u} \approx e^{-2u\eta\mu_B/\mu_B^{(2)}} \tag{7.4}
$$

suggested by the Cramér-Lundberg approximation and Proposition 6.2. This follows since $\eta = 1/\rho - 1 \approx 1 - \rho$, and hence

$$
\delta(\beta_{\max} - \beta) = \frac{2\mu_B^2}{\mu_B^{(2)}} \cdot \frac{1-\rho}{\mu_B} \approx \frac{2\eta\mu_B}{\mu_B^{(2)}}.
$$

However, obviously Corollary 7.2 provides the better mathematical foundation.

Notes and references Heavy traffic limit theory for queues goes back to Kingman [534]. The present situation of Poisson arrivals is somewhat more elementary to deal with than the renewal case (see e.g. [APQ, X.7]). We return to heavy traffic from a different point of view (diffusion approximations) in Chapter V and give further references there. In the setting of risk theory, the first results of heavy traffic type seem to be due to Hadwiger [445].

Numerical evidence shows that the fit of (7.3) is reasonable for η being say 10–20% and u being small or moderate, while the approximation may be far off for large u.

7d   The light traffic approximation

As for heavy traffic, the term *light traffic* comes from queueing theory, but has an obvious interpretation also in risk theory: on the average, the premiums are much larger than the expected claims. That is, light traffic conditions mean that the safety loading η is positive and large, or equivalently that β is small compared to μ_B. Mathematically, we shall represent this situation with a limit where $\beta \downarrow 0$ but B is fixed.

Of course, in risk theory heavy traffic is most often argued to be the more typical case. However, light traffic is of some interest as a complement to heavy traffic, as well as it is needed for the interpolation approximation to be studied in the next subsection.

Proposition 7.3 *As $\beta \downarrow 0$,*

$$\psi(u) \; \approx \; \beta \int_u^\infty \overline{B}(x)\,\mathrm{d}x \; = \; \beta\mathbb{E}\big[U - u;\, U > u\big] \; = \; \beta\mathbb{E}(U - u)^+. \qquad (7.5)$$

Proof. According to the Pollaczeck-Khinchine formula,

$$\psi(u) \; = \; (1 - \rho)\sum_{n=1}^\infty \beta^n \mu_B^n \overline{B}_0^{*n}(u) \; \approx \; \sum_{n=1}^\infty \beta^n \mu_B^n \overline{B}_0^{*n}(u)\,.$$

Asymptotically, $\sum_{n=2}^\infty \cdots = O(\beta^2)$ so that only the first terms matters, and hence

$$\psi(u) \; \approx \; \beta\mu_B\overline{B}_0(u) \; = \; \beta \int_u^\infty \overline{B}(x)dx.$$

The alternative expressions in (7.5) follow by integration by parts.   □

Note that heuristically the light traffic approximation in Proposition 7.3 is the same which comes out by saying that basically ruin can only occur at the time T of the first claim, i.e. $\psi(u) \approx \mathbb{P}(U - T > u)$. Indeed, by monotone convergence

$$\mathbb{P}(U - T > u) \; = \; \int_0^\infty \overline{B}(x + u)\beta e^{-\beta x}\,\mathrm{d}x \; \approx \; \beta \int_u^\infty \overline{B}(x)\,\mathrm{d}x.$$

Notes and references   Light traffic limit theory for queues was initiated by Bloomfield & Cox [178]. For a more comprehensive treatment, see Daley & Rolski [270, 271], Asmussen [61] and references therein. Again, the Poisson case is much easier than the renewal case. Another way to understand that the present analysis is much simpler than in these references is the fact that in the queueing setting light traffic theory is much easier for virtual waiting times (the probability of the conditioning event $\{M > 0\}$ is explicit) than for actual waiting times, cf. Sigman [811]. Light traffic was first studied in risk theory in the first edition of this book.

7e Interpolating between light and heavy traffic

We shall now outline an idea of how the heavy and light traffic approximations can be combined. The crude idea of interpolating between light and heavy traffic leads to

$$
\begin{aligned}
\psi(u) &\approx \left(1 - \frac{\beta}{\beta_{\max}}\right) \lim_{\beta \downarrow 0} \psi(u) + \frac{\beta}{\beta_{\max}} \lim_{\beta \uparrow \beta_{\max}} \psi(u) \\
&= \left(1 - \frac{\beta}{\beta_{\max}}\right) \cdot 0 + \frac{\beta}{\beta_{\max}} \cdot 1 = \frac{\beta}{\beta_{\max}} = \rho,
\end{aligned}
$$

which is clearly useless. Instead, to get non-degenerate limits, we combine with our explicit knowledge of $\psi(u)$ for the exponential claim size distribution E with the same mean μ_B as the given one B, that is, with rate $1/\mu_B = \beta_{\max}$. Let $\widetilde{\psi}_{\mathrm{LT}}^{(B)}(u)$ denote the light traffic approximation given by Proposition 7.3 and use similar notation for $\psi^{(B)}(u) = \psi(u)$, $\psi^{(E)}(u) = \rho e^{-(\beta_{\max} - \beta)u}$, $\widetilde{\psi}_{\mathrm{LT}}^{(E)}(u)$, $\widetilde{\psi}_{\mathrm{HT}}^{(B)}(u)$, $\widetilde{\psi}_{\mathrm{HT}}^{(E)}(u)$. Substituting $v = u(\beta_{\max} - \beta)$, we see that the following limits exist:

$$
\lim_{\beta \uparrow \beta_{\max}} \frac{\widetilde{\psi}_{\mathrm{HT}}^{(B)}\left(\frac{v}{\beta_{\max} - \beta}\right)}{\widetilde{\psi}_{\mathrm{HT}}^{(E)}\left(\frac{v}{\beta_{\max} - \beta}\right)} = \frac{e^{-\delta v}}{e^{-2\mu_E^2/\mu_E^{(2)} \cdot v}} = e^{(1-\delta)v} = c_{\mathrm{HT}}(v) \quad (\text{say}),
$$

$$
\lim_{\beta \downarrow 0} \frac{\widetilde{\psi}_{\mathrm{LT}}^{(B)}\left(\frac{v}{\beta_{\max} - \beta}\right)}{\widetilde{\psi}_{\mathrm{LT}}^{(E)}\left(\frac{v}{\beta_{\max} - \beta}\right)} = \frac{\int_{v/\beta_{\max}}^{\infty} \overline{B}(x)\,\mathrm{d}x}{\int_{v/\beta_{\max}}^{\infty} e^{-\beta_{\max} x}\,\mathrm{d}x}
$$

$$
= \beta_{\max} e^{v} \int_{v/\beta_{\max}}^{\infty} \overline{B}(x)\,\mathrm{d}x = c_{\mathrm{LT}}(v) \quad (\text{say}),
$$

and the approximation we suggest is

$$
\begin{aligned}
\psi(u) &\approx \psi^{(E)}(u)\left(\left[1 - \frac{\beta}{\beta_{\max}}\right] c_{\mathrm{LT}}(u(\beta_{\max} - \beta)) + \frac{\beta}{\beta_{\max}} c_{\mathrm{HT}}(u(\beta_{\max} - \beta))\right) \\
&= \rho(1-\rho)\beta_{\max} \int_{u(1-\rho)}^{\infty} \overline{B}(x)\,\mathrm{d}x + \rho^2 e^{-\delta(\beta_{\max} - \beta)u}. \quad (7.6)
\end{aligned}
$$

The particular features of this approximation are that it is exact for the exponential distribution and asymptotically correct both in light and heavy traffic. Thus, even if the safety loading is not very small, one may hope that some correction of the heavy traffic approximation has been obtained.

Notes and references In the queueing setting, the idea of interpolating between light and heavy traffic is due to Burman & Smith [209, 210]. Another main queueing paper is Whitt [882], where further references can be found. The adaptation to risk theory is new; no empirical study of the fit of (7.6) is, however, available.

8   Comparing the risks of different claim size distributions

Given two claim size distributions $B^{(1)}$, $B^{(2)}$, we may ask which one carries the larger risk in the sense of larger values of the ruin probability $\psi^{(i)}(u)$ for a fixed value of β.

To this end, we shall need various ordering properties of distributions, for more detail and background on which we refer to Müller & Stoyan [653] or Shaked & Shantikumar [795].

Recall that $B^{(1)}$ is said to be *stochastically smaller* than $B^{(2)}$ (in symbols, $B^{(1)} \prec_{\text{st}} B^{(2)}$) if $\overline{B^{(1)}}(x) \leq \overline{B^{(2)}}(x)$ for all x; equivalent characterizations are $\int f \, dB^{(1)} \leq \int f \, dB^{(2)}$ for any non-decreasing function f, or the existence of random variables $U^{(1)}$, $U^{(2)}$ such that $U^{(1)}$ has distribution $B^{(1)}$, $U^{(2)}$ distribution $B^{(2)}$ and $U^{(1)} \leq U^{(2)}$ a.s.

A weaker concept is *increasing convex ordering*: $B^{(1)}$ is said to be smaller than $B^{(2)}$ in the increasing convex order (in symbols, $B^{(1)} \prec_{\text{icx}} B^{(2)}$) if

$$\int_x^\infty \overline{B^{(1)}}(y) \, dy \ \leq \ \int_x^\infty \overline{B^{(2)}}(y) \, dy \tag{8.1}$$

for all x; an equivalent characterization is $\int f \, dB^{(1)} \leq \int f \, dB^{(2)}$ for any non-decreasing convex function f. In the literature on risk theory, most often the term *stop-loss ordering* is used instead of increasing convex ordering because for a given distribution B, one can interpret $\int_x^\infty \overline{B}(y) \, dy$ as the net stop-loss premium in a stop-loss or excess-of-loss reinsurance arrangement with retention limit x, cf. XVI.4.

Finally, we have the *convex ordering*: $B^{(1)}$ is said to be convexly smaller than $B^{(2)}$ (in symbols, $B^{(1)} \prec_{\text{cx}} B^{(2)}$) if $\int f \, dB^{(1)} \leq \int f \, dB^{(2)}$ for any convex function f. Rather than measuring difference in size, this ordering measures difference in variability. In particular (consider the convex functions x and $-x$) the definition implies that $B^{(1)}$ and $B^{(2)}$ must have the same mean, whereas (consider x^2) $B^{(2)}$ has the larger variance. One can show that if $B^{(1)}$ and $B^{(2)}$ have the same mean and $B^{(1)} \prec_{\text{icx}} B^{(2)}$, this is equivalent to $B^{(1)} \prec_{\text{cx}} B^{(2)}$.

Proposition 8.1 *If $B^{(1)} \prec_{\text{st}} B^{(2)}$, then $\psi^{(1)}(u) \leq \psi^{(2)}(u)$ for all u.*

Proof. According to the above characterization of stochastic ordering, we can assume that $S_t^{(1)} \leq S_t^{(2)}$ for all t. In terms of the time to ruin, this implies $\tau^{(1)}(u) \geq \tau^{(2)}(u)$ for all u so that $\{\tau^{(1)}(u) < \infty\} \subseteq \{\tau^{(2)}(u) < \infty\}$. Taking probabilities, the proof is complete. $\square$

Of course, Proposition 8.1 is quite weak, and a particular deficit is that we cannot compare the risks of claim size distributions with the same mean: if $B^{(1)} \prec_{\mathrm{st}} B^{(2)}$ and $\mu_{B^{(1)}} = \mu_{B^{(2)}}$, then $B^{(1)} = B^{(2)}$. Here convex ordering is useful:

Proposition 8.2 *If* $B^{(1)} \prec_{\mathrm{icx}} B^{(2)}$ *and* $\mu_{B^{(1)}} = \mu_{B^{(2)}}$ *(i.e.* $B^{(1)} \prec_{\mathrm{cx}} B^{(2)}$)*, then* $\psi^{(1)}(u) \le \psi^{(2)}(u)$ *for all* u.

Proof. Since the means are equal, say to μ, we have

$$\overline{B_0^{(1)}}(x) = \frac{1}{\mu} \int_x^\infty \overline{B^{(1)}}(y) \, dy \le \frac{1}{\mu} \int_x^\infty \overline{B^{(2)}}(y) \, dy = \overline{B_0^{(2)}}(x). \qquad (8.2)$$

I.e., $B_0^{(1)} \prec_{\mathrm{st}} B_0^{(2)}$ which implies the same order relation for all convolution powers. Hence by the Pollaczeck-Khinchine formula

$$\psi^{(1)}(u) = (1-\rho) \sum_{n=1}^\infty \beta^n \mu^n \overline{B_0^{(1)*n}}(u) \le (1-\rho) \sum_{n=1}^\infty \beta^n \mu^n \overline{B_0^{(2)*n}}(u) = \psi^{(2)}(u).$$

$\square$

Remark 8.3 From the proof above it is clear that $\psi^{(1)}(u) \le \psi^{(2)}(u)$ for all u still holds if the assumption on the ordering of the claim size distribution is weakened to just ask for (8.2). Slightly more general, the ordering defined by

$$\frac{1}{\mu_{B^{(1)}}} \int_x^\infty \overline{B^{(1)}}(y) \, dy \le \frac{1}{\mu_{B^{(2)}}} \int_x^\infty \overline{B^{(2)}}(y) \, dy \quad \text{for all } x \ge 0,$$

is known as the *harmonic mean residual life* order and is sufficient for $\psi^{(1)}(u) \le \psi^{(2)}(u)$ to hold as long as $\beta_1 \mu_{B^{(1)}} \le \beta_2 \mu_{B^{(2)}}$. $\square$

A general picture that emerges from these results and numerical studies like in Example 8.6 below is that (in a rough formulation) *increased variation in B increases the risk* (assuming that we fix the mean). The problem is to specify what 'variation' means. A first attempt would of course be to identify 'variation' with variance. The heavy traffic approximation (7.4) certainly supports this view: noting that, with fixed mean, larger variance is paramount to larger second moment, it is seen that asymptotically in heavy traffic larger claim size variance leads to larger ruin probabilities. Proposition 8.2 provides another instance of this, and here is one more result of the same flavor:

Corollary 8.4 *Let D refer to the distribution degenerate at* μ_B. *Then* $\psi^{(D)}(u) \le \psi^{(B)}(u)$ *for all* u.

Proof. If f is convex, we have by Jensen's inequality that $\mathbb{E}f(U) \geq f(\mathbb{E}U)$. This implies that $D \prec_{\mathrm{cx}} B$ and we can apply Proposition 8.2. $\qquad\square$

A partial converse to Proposition 8.2 is the following:

Proposition 8.5 *If $\psi^{(1)}(u) \leq \psi^{(2)}(u)$ for all u and β, then $B^{(1)} \prec_{\mathrm{cx}} B^{(2)}$.*

Proof. Consider the light traffic approximation in Proposition 7.1. $\qquad\square$

We finally give a numerical example illustrating how differences in the claim size distribution B may lead to very different ruin probabilities even if we fix the mean $\mu = \mu_B$.

Example 8.6 Fix β at $1/1.1$ and μ_B at 1 so that the safety loading η is 10%, and consider the following claim size distributions:

B_1: the standard exponential distribution with density e^{-x};

B_2: the hyperexponential distribution with density $0.1\lambda_1 e^{-\lambda_1 x} + 0.9\lambda_2 e^{-\lambda_2 x}$ where $\lambda_1 = 0.1358$, $\lambda_2 = 3.4142$;

B_3: the Erlang distribution with density $4xe^{-2x}$.

B_4: the Pareto distribution with density $3/(1 + 2x)^{5/2}$.

Let u_α denote the α fractile of the ruin function, i.e. $\psi(u_\alpha) = \alpha$, and consider $\alpha = 5\%, 1\%, 0.1\%, 0.01\%$. One then obtains the following table:

	B_1	B_2	B_3	B_4
$u_{0.05}$	32	181	24	35
$u_{0.01}$	50	282	37	70
$u_{0.001}$	75	425	56	245
$u_{0.0001}$	100	568	74	1100

(the table was produced using simulation and the numbers are therefore subject to statistical uncertainty). Note to make the figures comparable, all distributions have mean 1. In terms of variances σ_k^2, we have

$$\sigma_3^2 = \frac{1}{2} < \sigma_1^2 = 1 < \sigma_2^2 = 10 < \sigma_4^2 = \infty$$

so that in this sense B_4 is the most variable. However, in comparison to B_2 the effect on the u_α does not show before $\alpha = 0.01\%$, which appears to be smaller than the range of interest in insurance risk (certainly not in queueing applications!), and this is presumably a consequence of a heavier tail rather than larger variance. For B_1, B_2, B_3 the comparison is as expected from the intuition concerning the variability of these distributions, with the hyperexponential distribution being more variable than the exponential distribution and the Erlang distribution less. $\qquad\square$

Notes and references Further relevant references are Goovaerts *et al.* [425], van Heerwarden [454], Klüppelberg [539], Pellerey [689] and (for the convex ordering) Makowski [623]. For the harmonic mean residual life order, see Michel [636] and Trufin, Albrecher & Denuit [854]. For relations between higher-order stop-loss orderings of claim size distributions and ruin probabilities see Cheng & Pai [236]. Tsai [856] considers orderings in the presence of perturbations. We return to ordering of ruin probabilities in a special problem in VII.4 and also in XIII.8.

For the situation that the claim size distribution and the Poisson parameter are unknown, but a sample of data points is available, Politis [709] considers the problem of semi-parametric estimation of ruin probabilities.

9 Sensitivity estimates

In a broad setting, sensitivity analysis (or pertubation analysis) deals with the calculation of the derivative (the gradient in higher dimensions) of a performance measure $s(\theta)$ of a stochastic or deterministic system, the behavior of which is governed by a parameter θ. A standard example from queueing theory is a queueing network, with θ the vector of service rates at different nodes and routing probabilities, and $s(\theta)$ the expected sojourn time of a customer in the network. In the present setting, $s(\theta)$ is of course the ruin probability $\psi = \psi(u)$ (with u fixed) and θ a set of parameters determining the arrival rate β, the premium rate p and the claim size distribution B. For example, we may be interested in $\partial \psi / \partial p$ for assessing the effects of a small change in the premium, or we may be interested in $\partial \psi / \partial \beta$ as a measure of the uncertainty on ψ if β is only approximatively known, say estimated from data.

Example 9.1 Consider the case of claims which are exponential with rate δ (the premium rate is one). Then $\psi = \frac{\beta}{\delta} e^{-(\delta - \beta)u}$, and hence

$$\frac{\partial \psi}{\partial \beta} = \frac{1}{\delta} e^{-(\delta - \beta)u} + \frac{\beta u}{\delta} e^{-(\delta - \beta)u} = \left(\frac{1}{\beta} + u \right) \psi(u),$$

which is of the order of magnitude $u\psi(u)$ for large u.

Assume for example that δ is known, while $\beta = \widehat{\beta}$ is an estimate, obtained say in the natural way as the empirical arrival rate N_t/t in $[0, t]$. Then if t is large, the distribution of $\widehat{\beta} - \beta$ is approximatively normal $N(0, \beta/t)$. Thus, if $\widehat{\psi} = \frac{\widehat{\beta}}{\delta} e^{-(\delta - \widehat{\beta})u}$, it follows that $\widehat{\psi} - \psi$ is approximatively normal $N(0, \sigma^2/t)$, where

$$\sigma^2 = \beta \left(\frac{\partial \psi}{\partial \beta} \right)^2 \sim \beta u^2 \psi^2.$$

In particular, the standard deviation on the normalized estimate $\widehat{\psi}/\psi$ (the relative error) is approximatively $\beta^{1/2}u$, i.e. increasing in u. Similar conclusions will be found below. □

Proposition 9.2 *Consider a risk process $\{R_t\}$ with a general constant premium rate p. Then*

$$\frac{\partial \psi}{\partial p} \;=\; -\beta \frac{\partial \psi}{\partial \beta},$$

where the partial derivatives are evaluated at $p = 1$.

Proof. This is an easy time transformation argument in a similar way as in Proposition I.1.3. Let $R_t^{(p)} = R_{t/p}$. Then the arrival rate $\beta^{(p)}$ for $\{R_t^{(p)}\}$ is β/p, and hence the effect of changing p from 1 to $1 + \Delta p$ corresponds to changing β to $\beta/(1 + \Delta p) \approx \beta(1 - \Delta p)$. Thus at $p = 1$,

$$\frac{\partial \psi}{\partial p} \;=\; \frac{\partial \beta}{\partial p}\frac{\partial \psi}{\partial \beta} \;=\; -\beta \frac{\partial \psi}{\partial \beta}.$$

□

As a consequence, it suffices to fix the premium at $p = 1$ and consider only the effects of changing β or/and B. In the case of the claim size distribution B, various parametric families of claim size distributions could be considered, but we shall concentrate on a special structure covering a number of important cases, namely that of a two-parameter exponential family of the form

$$B_{\theta,\zeta}(\mathrm{d}x) \;=\; \exp\{\theta x + \zeta t(x) - \omega(\theta,\zeta)\}\mu(\mathrm{d}x)\,, \quad x > 0 \qquad (9.1)$$

(see Remark 9.5 below for some discussion of this assumption).

Consider first the adjustment coefficient γ as function of β, θ, ζ, and write $\gamma_\beta = \partial\gamma/\partial\beta$ and so on. Similar notation for partial derivatives are used below, e.g. for the ruin probabilities $\psi = \psi(u)$ and the Cramér-Lundberg constant C.

Proposition 9.3

$$\gamma_\beta \;=\; \frac{\gamma}{\beta\big(1 - \omega_\theta(\theta + \gamma, \zeta)(\beta + \gamma)\big)}, \qquad (9.2)$$

$$\gamma_\theta \;=\; \frac{(\beta + \gamma)\big[\omega_\theta(\theta + \gamma, \zeta) - \omega_\theta(\theta, \zeta)\big]}{1 - (\beta + \gamma)\omega_\theta(\theta + \gamma, \zeta)}, \qquad (9.3)$$

$$\gamma_\zeta \;=\; \frac{(\beta + \gamma)\big[\omega_\zeta(\theta + \gamma, \zeta) - \omega_\zeta(\theta, \zeta)\big]}{1 - (\beta + \gamma)\omega_\theta(\theta + \gamma, \zeta)}. \qquad (9.4)$$

Proof. According to (9.8) below, we can rewrite the Lundberg equation as $\omega(\theta + \gamma, \zeta) - \omega(\theta, \zeta) = \log(1 + \gamma/\beta)$. Differentiating w.r.t. β yields

$$\omega_\theta(\theta + \gamma, \zeta)\gamma_\beta = \frac{1}{1 + \gamma/\beta}\left(\frac{\gamma_\beta}{\beta} - \frac{\gamma}{\beta^2}\right).$$

From this (9.2) follows by straightforward algebra, and the proofs of (9.3), (9.4) are similar. □

Now consider the ruin probability $\psi = \psi(u)$ itself. Of course, we cannot expect in general to find explicit expressions like in Example 9.1 or Proposition 9.3, but must look for approximations for the sensitivities ψ_β, ψ_θ, ψ_ζ. The most intuitive approach is to rely on the accuracy of the Cramér-Lundberg approximation, so that heuristically we obtain

$$\psi_\beta \approx \frac{\partial}{\partial\beta}Ce^{-\gamma u} = C_\beta e^{-\gamma u} - u\gamma_\beta Ce^{-\gamma u} \approx -u\gamma_\beta\psi \tag{9.5}$$

as $u \to \infty$. As will be seen below, this intuition is indeed correct. However, mathematically a proof is needed basically to show that two limits ($u \to \infty$ and the differentiation as limit of finite differences) are interchangeable.

Consider first the case of $\partial\psi/\partial\beta$:

Proposition 9.4 *As $u \to \infty$, it holds that*

$$\frac{\partial\psi}{\partial\beta} \sim ue^{-\gamma u}\frac{\gamma C^2}{\beta(1 - \rho)}.$$

Proof. We shall use the renewal equation (3.2) for $\psi(u)$,

$$\psi(u) = \beta\int_u^\infty \overline{B}(x)\,dx + \int_0^u \psi(u - x)\beta\overline{B}(x)\,dx. \tag{9.6}$$

Letting $\varphi = \partial\psi/\partial\beta$ and differentiating (9.6), we get

$$\varphi(u) = \int_u^\infty \overline{B}(x)\,dx + \int_0^u \psi(u - x)\overline{B}(x)\,dx + \int_0^u \varphi(u - x)\beta\overline{B}(x)\,dx.$$

Proceeding in a similar way as in the proof of the Cramér-Lundberg approximation based upon (9.6) (Section 5), we multiply by $e^{\gamma u}$ and let $Z(u) = e^{\gamma u}\varphi(u)$, $F(dx) = e^{\gamma x}\beta\overline{B}(x)dx$ and $z = z_1 + z_2$, where

$$z_1(u) = e^{\gamma u}\int_u^\infty \overline{B}(x)\,dx, \quad z_2(u) = e^{\gamma u}\int_0^u \psi(u - x)\overline{B}(x)\,dx.$$

Then $Z = z + F * Z$ and F is a proper probability distribution. By dominated convergence,

$$z_2(u) = \frac{1}{\beta} \int_0^u e^{\gamma(u-x)} \psi(u-x) \, F(\mathrm{d}x) \rightarrow \frac{1}{\beta} \int_0^\infty C \, F(\mathrm{d}x) = \frac{C}{\beta},$$

as $u \to \infty$, and also $z_1(u) \to 0$ because of $\widehat{B}[\gamma] < \infty$. Hence by a variant of the key renewal theorem (Proposition A1.2 of the Appendix), $Z(u)/u \to C/\beta\mu_F$ where μ_F is the mean of F. But from the proof of Theorem 5.3 (see in particular (5.10)), $\mu_F = (1-\rho)/C\gamma$. Combining these estimates, the proof is complete. $\square$

For the following, we note the formulas

$$\mathbb{E}_{\theta,\zeta} t(U) = \omega_\zeta(\theta,\zeta), \tag{9.7}$$

$$\mathbb{E}_{\theta,\zeta} e^{\alpha U} = \widehat{B}_{\theta,\zeta}[\alpha] = \exp\{\omega(\theta+\alpha,\zeta) - \omega(\theta,\zeta)\}, \tag{9.8}$$

$$\mathbb{E}_{\theta,\zeta} t(U) e^{\alpha U} = \omega_\zeta(\theta+\alpha,\zeta) \exp\{\omega(\theta+\alpha,\zeta) - \omega(\theta,\zeta)\} \tag{9.9}$$

which are well-known and easy to show (see e.g. Barndorff-Nielsen [136]). Further write

$$
\begin{aligned}
d_\theta &= \big[\omega_\theta(\theta+\gamma,\zeta) - \omega_\theta(\theta,\zeta)\big] \exp\{\omega(\theta+\gamma,\zeta) - \omega(\theta,\zeta)\} \\
&= \big[\omega_\theta(\theta+\gamma,\zeta) - \omega_\theta(\theta,\zeta)\big]\Big(1 + \frac{\gamma}{\beta}\Big), \\
d_\zeta &= \big[\omega_\zeta(\theta+\gamma,\zeta) - \omega_\zeta(\theta,\zeta)\big] \exp\{\omega(\theta+\gamma,\zeta) - \omega(\theta,\zeta)\} \\
&= \big[\omega_\zeta(\theta+\gamma,\zeta) - \omega_\zeta(\theta,\zeta)\big]\Big(1 + \frac{\gamma}{\beta}\Big).
\end{aligned}
$$

Proposition 9.5 *Assume that (9.1) holds. Then as $u \to \infty$,*

$$\frac{\partial\psi}{\partial\theta} \sim ue^{-\gamma u} \frac{\beta C^2 d_\theta}{1-\rho}, \qquad \frac{\partial\psi}{\partial\zeta} \sim ue^{-\gamma u} \frac{\beta C^2 d_\zeta}{1-\rho}. \tag{9.10}$$

Proof. By straightforward differentiation,

$$
\begin{aligned}
\frac{\partial \overline{B}(x)}{\partial \zeta} &= \frac{\partial}{\partial \zeta} \int_x^\infty \exp\{\theta y + \zeta t(y) - \omega(\theta,\zeta)\} \mu(\mathrm{d}y) \\
&= \int_x^\infty \big[t(y) - \omega_\zeta(\theta,\zeta)\big] B(\mathrm{d}y).
\end{aligned}
$$

Letting $\varphi = \partial\psi/\partial\zeta$, it thus follows from (9.6) that

$$\varphi(u) = e^{-\gamma u} z_1(u) + e^{-\gamma u} z_2(u) + \int_0^u \varphi(u-x)\beta\overline{B}(x) \, \mathrm{d}x,$$

where

$$z_1(u) = \beta e^{\gamma u} \int_u^\infty \int_x^\infty \left[t(y) - \omega_\zeta(\theta, \zeta) \right] B(\mathrm{d}y) \, \mathrm{d}x,$$

$$z_2(u) = e^{\gamma u} \int_0^u \psi(u - x) \beta \int_x^\infty \left[t(y) - \omega_\zeta(\theta, \zeta) \right] B(\mathrm{d}y) \, \mathrm{d}x.$$

Multiplying by $e^{\gamma u}$ and letting

$$Z(u) = e^{\gamma u} \varphi(u), \quad z = z_1 + z_2, \quad F(\mathrm{d}x) = e^{\gamma x} \beta \overline{B}(x) \mathrm{d}x,$$

this implies $Z = z + F * Z$. By dominated convergence and (9.7)-(9.9),

$$
\begin{aligned}
z_2(u) \quad &\to \quad \int_0^\infty C \cdot e^{\gamma x} \beta \int_x^\infty \left[t(y) - \omega_\zeta(\theta, \zeta) \right] B(\mathrm{d}y) \, \mathrm{d}x \\
&= \quad \beta C \int_0^\infty \left[t(y) - \omega_\zeta(\theta, \zeta) \right] \frac{1}{\gamma} (e^{\gamma y} - 1) \, B(\mathrm{d}y) \\
&= \quad \frac{\beta C}{\gamma} \int_0^\infty \left[t(y) - \omega_\zeta(\theta, \zeta) \right] e^{\gamma y} B(\mathrm{d}y) \\
&= \quad \frac{\beta C}{\gamma} d_\zeta
\end{aligned}
$$

as $u \to \infty$, and also $z_1(u) \to 0$ because of

$$\int_0^\infty e^{\gamma y} \left[t(y) - \omega_\zeta(\theta, \zeta) \right] B(\mathrm{d}y) \ < \ \infty.$$

Hence,

$$\frac{Z(u)}{u} \quad \to \quad \frac{\beta C}{\gamma \mu_F} d_\zeta,$$

from which the second assertion of (9.10) follows, and the proof of the first one is similar. $\qquad \square$

Example 9.6 Consider the gamma density

$$b(x) = \frac{\delta^\alpha}{\Gamma(\alpha)} x^{\alpha - 1} e^{-\delta x} = \exp\{-\delta x + \alpha \log x - (\log \Gamma(\alpha) - \alpha \log \delta)\} \cdot \frac{1}{x}.$$

Here (9.1) holds with

$$\mu(\mathrm{d}x) = x^{-1} \mathrm{d}x, \quad \theta = -\delta, \quad \zeta = \alpha, \quad t(x) = \log x,$$

$$w(\theta, \zeta) = \log \Gamma(\alpha) - \alpha \log \delta = \log \Gamma(\zeta) - \zeta \log(-\theta).$$

We get $\omega_\zeta(\theta,\zeta) = \Psi(\zeta) - \log(-\theta) = \Psi(\alpha) - \log\delta$ where $\Psi = \Gamma'/\Gamma$ is the Digamma function, $\omega_\theta(\theta,\zeta) = -\zeta/\theta = \alpha/\delta$. It follows after some elementary calculus that $\rho = \alpha\beta/\delta$ and, by inserting in the above formulas, that

$$C = \frac{1-\rho}{\alpha\beta\delta^\alpha/(\delta-\gamma)^{\alpha+1}-1},$$

$$d_\theta = \alpha\gamma\frac{\delta^{\alpha-1}}{(\delta-\gamma)^{\alpha+1}},$$

$$d_\zeta = \log\left(\frac{\delta}{\delta-\gamma}\right)\left(\frac{\delta}{\delta-\gamma}\right)^\alpha, \tag{9.11}$$

$$\gamma_\beta = \frac{\gamma^2-\delta\gamma}{\alpha\beta^2+\alpha\beta\gamma+\beta\gamma-\beta\delta}, \tag{9.12}$$

$$\gamma_\delta = -\gamma_\theta = -\frac{\alpha\beta\gamma+\alpha\gamma^2}{\delta^2-\delta\gamma-\alpha\beta\delta-\alpha\delta\gamma}, \tag{9.13}$$

$$\gamma_\alpha = \gamma_\zeta = \frac{(\beta\delta+\delta\gamma-\beta\gamma-\gamma^2)}{\delta-\gamma-\alpha\beta-\alpha\gamma}\log\left(\frac{\delta}{\delta-\gamma}\right). \tag{9.14}$$

Finally, (9.10) takes the form

$$\frac{\partial\psi}{\partial\delta} = -\frac{\partial\psi}{\partial\theta} \sim -ue^{-\gamma u}\frac{\beta C^2 d_\theta}{1-\rho}, \qquad \frac{\partial\psi}{\partial\alpha} = \frac{\partial\psi}{\partial\zeta} \sim ue^{-\gamma u}\frac{\beta C^2 d_\zeta}{1-\rho}.$$

$\square$

Example 9.7 Consider the inverse Gaussian density

$$b_{\xi,c}(x) = \frac{c}{\sqrt{2x^3\pi}}\exp\left\{\xi c - \frac{1}{2}\left(\frac{c^2}{x}+\xi^2 x\right)\right\}.$$

This has the form (9.1) with

$$\mu(dx) = \frac{1}{\sqrt{2x^3\pi}}dx, \quad \theta = -\frac{\xi^2}{2}, \quad \zeta = -\frac{c^2}{2}, \quad t(x) = \frac{1}{x},$$

$$\omega(\theta,\zeta) = -\xi c - \log c = -2\sqrt{(-\theta)(-\zeta)} - \frac{1}{2}\log(-\zeta) - \frac{1}{2}\log 2.$$

In particular, for $\alpha \le \alpha^* = \frac{\xi^2}{2}$

$$\widehat{B}_{\theta,\zeta}[\alpha] = \exp\{\omega(\theta+\alpha,\zeta)-\omega(\theta,\zeta)\} = \exp\{c(\xi-\sqrt{\xi^2-2\alpha})\}.$$

Thus the condition $\widehat{B}[\alpha^*] \geq 1 + \alpha^*/\beta$ of Section 6a needed for the existence of γ becomes $e^{\xi c} \geq 1 + \xi^2/2\beta$. Straightforward but tedious calculations, which we omit in part, further yield

$$
\begin{aligned}
\beta \widehat{B}'_{\theta,\zeta}[\gamma] - 1 &= \beta \exp\{c(\xi - \sqrt{\xi^2 - 2\gamma})\} \frac{1}{c\sqrt{\xi^2 - 2\gamma}} - 1 \\
&= \frac{\beta + \gamma}{c\sqrt{\xi^2 - 2\gamma}} - 1
\end{aligned}
$$

$$
\omega_\zeta(\theta, \zeta) = \sqrt{\frac{\theta}{\zeta} - \frac{1}{2\zeta}} = \frac{\xi}{c} + \frac{1}{c^2}, \quad \omega_\theta(\theta, \zeta) = \sqrt{\frac{\zeta}{\theta}} = \frac{c}{\xi},
$$

$$
\gamma_\beta = \frac{\gamma\sqrt{\xi^2 - 2\gamma}}{\beta\sqrt{\xi^2 - 2\gamma} - \beta c(\beta + \gamma)},
$$

$$
\gamma_\xi = -\xi\gamma_\theta = -c(\gamma + \beta)\frac{\xi - \sqrt{\xi^2 - 2\gamma}}{\sqrt{\xi^2 - 2\gamma} - c(\gamma + \beta)},
$$

$$
\gamma_c = -c\gamma_\zeta = -(\gamma + \beta)\frac{\xi^2 - 2\gamma - \xi\sqrt{\xi^2 - 2\gamma}}{\sqrt{\xi^2 - 2\gamma} - c(\gamma + \beta)},
$$

$$
d_\theta = c\left[\frac{1}{\sqrt{\xi^2 - 2\gamma}} - \frac{1}{\xi}\right]\left(1 - \frac{\gamma}{\beta}\right),
$$

$$
d_\zeta = \frac{1}{c}\left[\sqrt{\xi^2 - 2\gamma} - \xi\right]\left(1 - \frac{\gamma}{\beta}\right).
$$

Finally, (9.10) takes the form

$$
\frac{\partial \psi}{\partial \xi} = -\xi \frac{\partial \psi}{\partial \theta} \sim -\xi u e^{-\gamma u}\frac{\beta C^2 d_\theta}{1 - \rho}, \quad \frac{\partial \psi}{\partial c} = -c\frac{\partial \psi}{\partial \zeta} \sim -cu e^{-\gamma u}\frac{\beta C^2 d_\zeta}{1 - \rho}.
$$

$\square$

Remark 9.8 The specific form of (9.1) is motivated as follows. In general, the exponent of the density in an exponential family has the form $\theta_1 t_1(x) + \cdots + \theta_k t_k(x)$. Thus, we have assumed $k = 2$ and $t_1(x) = x$. That it is no restriction to assume one of the $t_i(x)$ to be linear follows since the whole set-up requires exponential moments to be finite (thus we can always extend the family if necessary by adding a term θx). That it is no restriction to assume $k \leq 2$ follows since if $k > 2$, we can just fix $k - 2$ of the parameters. Finally if $k = 1$, the exponent is either θx, in which case we can just let $t(x) = 0$, or $\zeta t(x)$, in which case the extension just described applies. $\square$

Notes and references The general area of sensitivity analysis (gradient estimation) is currently receiving considerable interest in queueing theory. However, the models there (e.g. queueing networks) are typically much more complicated than the one considered here, and hence explicit or asymptotic estimates are in general not possible. Thus, the main tool is simulation, for which we refer to XV.7 and references therein.

Comparatively less work seems to have been done in risk theory; thus, to our knowledge, the results presented here are new. Van Wouve *et al.* [861] consider a special problem related to reinsurance. For the study of perturbation via perturbed renewal equations, see Gyllenberg & Silvestrov [444].

10   Estimation of the adjustment coefficient

We consider a non-parametric set-up where β, B are assumed to be completely unknown, and we estimate γ by means of the empirical solution γ_T to the Lundberg equation. To this end, let

$$\beta_T = \frac{N_T}{T}, \quad \widehat{B}_T[\alpha] = \frac{1}{N_T} \sum_{i=1}^{N_T} e^{\alpha U_i}, \quad \kappa_T(\alpha) = \beta_T\big(\widehat{B}_T[\alpha] - 1\big) - \alpha,$$

and let γ_T be defined by $\kappa_T(\gamma_T) = 0$.

Note that if $N_T = 0$, then $\widehat{B}_T$ and hence γ_T is undefined. Also, if

$$\rho_T = \beta_T \frac{1}{N_T}(U_1 + \cdots + U_{N_T}) > 1,$$

then $\gamma_T < 0$. However, by the LLN both $\mathbb{P}(N_T = 0)$ and $\mathbb{P}(\rho_T > 1)$ converge to 0 as $T \to \infty$.

Theorem 10.1 *As $T \to \infty$, $\gamma_T \overset{\text{a.s.}}{\to} \gamma$. If furthermore $\widehat{B}[2\gamma] < \infty$, then*

$$\gamma_T - \gamma \approx N\Big(0, \frac{1}{T}\sigma_\gamma^2\Big),$$

where $\sigma_\gamma^2 = \beta\kappa(2\gamma)/\kappa'(\gamma)^2$.

For the proof, we need a lemma.

Lemma 10.2 *As $T \to \infty$,*

$$\widehat{B}_T[\gamma] \approx N\Big(\widehat{B}[\gamma], \frac{\widehat{B}[2\gamma] - \widehat{B}[\gamma]^2}{\beta T}\Big), \tag{10.1}$$

$$\kappa_T(\gamma) \approx N\Big(\frac{\kappa(2\gamma)}{T}\Big). \tag{10.2}$$

Proof. Since

$$\mathbb{V}ar(e^{\gamma U}) = \mathbb{E}e^{2\gamma U} - \left(\mathbb{E}e^{\gamma U}\right)^2 = \widehat{B}[2\gamma] - \widehat{B}[\gamma]^2,$$

we have

$$\frac{1}{n}\sum_{i=1}^{n} e^{\gamma U_i} \approx N\left(\widehat{B}[\gamma]\frac{\widehat{B}[2\gamma] - \widehat{B}[\gamma]^2}{n}\right).$$

Hence (10.1) follows from $N_T/T \stackrel{\text{a.s.}}{\to} \beta$ and Anscombe's theorem. More generally, since $N_T/T \approx N(\beta, \beta/T)$, it is easy to see that we can write

$$\left(\begin{array}{c}\beta_T \\ \widehat{B}_T[\gamma]\end{array}\right) \approx \left(\begin{array}{c}\beta \\ \widehat{B}[\gamma]\end{array}\right) + \frac{1}{\sqrt{T}}\left(\begin{array}{c}\sqrt{\beta}\,V_1 \\ \sqrt{\beta}\sqrt{\widehat{B}[2\gamma] - \widehat{B}[\gamma]^2}\,V_2\end{array}\right),$$

where V_1, V_2 are independent $N(0,1)$ r.v.'s. Hence

$$\begin{aligned}
\kappa_T(\gamma) &= \left(\beta + (\beta_T - \beta)\right)\left((\widehat{B}_T[\gamma] - \widehat{B}[\gamma]) + (\widehat{B}[\gamma] - 1)\right) - \gamma \\
&\approx \beta\left(\widehat{B}[\gamma] - 1\right) - \gamma + (\beta_T - \beta)\left(\widehat{B}[\gamma] - 1\right) + \beta\left(\widehat{B}_T[\gamma] - \widehat{B}[\gamma]\right) \\
&\approx 0 + \frac{1}{\sqrt{T}}\left\{\sqrt{\beta}(\widehat{B}[\gamma] - 1)V_1 + \sqrt{\beta}\sqrt{\widehat{B}[2\gamma] - \widehat{B}[\gamma]^2}\cdot V_2\right\} \\
&\stackrel{\mathscr{D}}{=} N\left(0, \frac{\beta}{T}\left\{(\widehat{B}[\gamma] - 1)^2 + \widehat{B}[2\gamma] - \widehat{B}[\gamma]^2\right\}\right) \\
&= N\left(\frac{\beta}{T}\left\{\widehat{B}[2\gamma] - \frac{2\gamma}{\beta} - 1\right\}\right)
\end{aligned}$$

which is the same as (10.2). □

Proof of Theorem 10.1. By the law of large numbers,

$$\beta_T \stackrel{\text{a.s.}}{\to} \beta, \quad \widehat{B}_T[\alpha] \stackrel{\text{a.s.}}{\to} \widehat{B}[\alpha], \quad \kappa_T(\alpha) \stackrel{\text{a.s.}}{\to} \kappa(\alpha).$$

Let $0 < \epsilon < \gamma$. Then

$$\kappa(\gamma - \epsilon) < 0 < \kappa(\gamma + \epsilon)$$

and hence

$$\kappa_T(\gamma - \epsilon) < 0 < \kappa_T(\gamma + \epsilon)$$

for all sufficiently large T. I.e., $\gamma_T \in (\gamma - \epsilon, \gamma + \epsilon)$ eventually, and the truth of this for all $\epsilon > 0$ implies $\gamma_T \stackrel{\text{a.s.}}{\to} \gamma$.

Now write

$$\kappa_T(\gamma_T) - \kappa_T(\gamma) = \kappa'_T(\gamma_T^*)(\gamma_T - \gamma), \tag{10.3}$$

where γ_T^* is some point between γ_T and γ. If $\gamma_T \in (\gamma - \epsilon, \gamma + \epsilon)$, we have

$$\kappa_T'(\gamma - \epsilon) < \kappa_T'(\gamma_T^*) < \kappa_T'(\gamma + \epsilon).$$

By the law of large numbers,

$$\widehat{B}_T'[\alpha] = \frac{1}{N_T} \sum_{i=1}^{N_T} U_i e^{\alpha U_i} \overset{\text{a.s.}}{\to} \mathbb{E} U e^{\alpha U} = \widehat{B}'[\alpha].$$

Hence $\kappa_T'(\alpha) \overset{\text{a.s.}}{\to} \kappa'(\alpha)$ for all α so that for all sufficiently large T

$$\kappa'(\gamma - \epsilon) < \kappa_T'(\gamma_T^*) < \kappa'(\gamma + \epsilon),$$

which implies $\kappa_T'(\gamma_T^*) \overset{\text{a.s.}}{\to} \kappa'(\gamma)$.

Combining (10.3) and Lemma 10.2, it follows that

$$\gamma_T - \gamma \approx \frac{\kappa_T(\gamma_T) - \kappa_T(\gamma)}{\kappa'(\gamma)} = -\frac{\kappa_T(\gamma)}{\kappa'(\gamma)}$$

$$\approx N\left(\frac{\kappa(2\gamma)}{T\kappa'(\gamma)}\right) = N\left(0, \sigma_\gamma^2/T\right).$$

$\square$

Theorem 10.1 can be used to obtain error bounds on the ruin probabilities when the parameters β, θ are estimated from data. To this end, first note that

$$e^{-\gamma_T u} \approx N\left(e^{-\gamma u}, u^2 e^{-2\gamma u} \sigma_\gamma^2/T\right).$$

Thus an asymptotic upper α confidence bound for $e^{-\gamma u}$ (and hence by Lundberg's inequality for $\psi(u)$) is

$$e^{-\gamma_T u} + \frac{f_\alpha}{\sqrt{T}} u\, e^{-\gamma u} \sigma_{\gamma;T}$$

where $\sigma_{\gamma;T}^2 = \beta_T \kappa_T(2\gamma_T)/\kappa_T'(\gamma_T)^2$ is the empirical estimate of σ_γ^2 and f_α satisfies $\Phi(-f_\alpha) = \alpha$ (e.g., $f_\alpha = 1.96$ if $\alpha = 2.5\%$).

Notes and references Theorem 10.1 is from Grandell [428]. A major restriction of the approach is the condition $\widehat{B}[2\gamma] < \infty$ which may be quite restrictive. For example, if B is exponential with rate δ so that $\gamma = \delta - \beta$, it means $2(\delta - \beta) < \delta$, i.e. $\delta < 2\beta$ or equivalently $\rho > 1/2$ or $\eta < 100\%$. For this reason, various alternatives have been developed. One (see Schmidli [771]) is to let $\{V_t\}$ be the workload process of an M/G/1 queue with the same arrival epochs as the risk process and service times $U_1, U_2, \ldots$, i.e. $V_t = S_t - \inf_{0 \le v \le t} S_v$. Letting

$$\omega_0 = 0, \quad \omega_n = \inf\{t > \omega_{n-1} : V_t = 0, V_s > 0 \text{ for some } t \in [\omega_{n-1}, t]\},$$

the nth busy cycle is then $[\omega_{n-1}, \omega_n)$, and the known fact that the

$$Y_n = \max_{t \in [\omega_{n-1}, \omega_n)} V_t$$

are i.i.d. with a tail of the form $\mathbb{P}(Y > y) \sim C_1 e^{-\gamma y}$ (see e.g. Asmussen [65]) can then be used to produce an estimate of γ. This approach in fact applies also for many models more general than the compound Poisson one.

Further work on estimation of γ with different methods can be found in Csörgő & Steinebach [268], Csörgő & Teugels [269], Deheuvels & Steinebach [285], Embrechts & Mikosch [348], Herkenrath [459], Hipp [464, 465], Frees [371], Mammitzsch [628], Brito & Freitas [202], Conti [256] and Pitts, Grübel & Embrechts [707].

Chapter V

The probability of ruin within finite time

This chapter is concerned with the finite time ruin probabilities

$$
\begin{aligned}
\psi(u, T) &= \mathbb{P}\big(\tau(u) \leq T\big) \\
&= \mathbb{P}\Big(\inf_{0 \leq t \leq T} R_t < 0 \,\big|\, R_0 = u\Big) \\
&= \mathbb{P}\Big(\sup_{0 \leq t \leq T} S_t > u\Big).
\end{aligned}
$$

Only the compound Poisson case is treated; generalizations to other models are either discussed in the Notes and References or in relevant chapters.

The notation is essentially as in Chapter IV. In particular, the premium rate is 1, the Poisson intensity is β and the claim size distribution is B with m.g.f. $\widehat{B}[\cdot]$ and mean μ_B. The safety loading is $\eta = 1/\rho - 1$ where $\rho = \beta\mu_B$. Unless otherwise stated, it is assumed that $\eta > 0$ and that the adjustment coefficient (Lundberg exponent) γ, defined as solution of $\kappa(\gamma) = 0$ where $\kappa(\alpha) = \beta(\widehat{B}[\alpha] - 1) - \alpha$, exists. Further let γ_m be the unique point in $(0, \gamma)$ where $\kappa(\alpha)$ attains its minimum value, see Fig. V.1 (the role of γ_y will be explained in Section 4b).

The claims surplus is $\{S_t\}$, the time of ruin is $\tau(u)$ and $\xi(u) = S_{\tau(u)} - u$ is the overshoot.

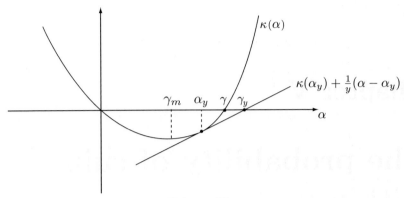

FIGURE V.1

1   Exponential claims

Proposition 1.1 *In the compound Poisson model with exponential claims with rate ν and safety loading $\eta > 0$, the conditional mean and variance of the time to ruin are given by*

$$\mathbb{E}\big[\tau(u)\,\big|\,\tau(u) < \infty\big] \;\; = \;\; \frac{\beta u + 1}{\nu - \beta}, \tag{1.1}$$

$$\mathbb{V}ar\big[\tau(u)\,\big|\,\tau(u) < \infty\big] \;\; = \;\; \frac{2\beta\nu u + \beta + \nu}{(\nu - \beta)^3}. \tag{1.2}$$

Proof. Let as in Example IV.5.1 $\mathbb{P}_L, \mathbb{E}_L$ refer to the exponentially tilted process with arrival intensity ν and exponential claims with rate β (thus, $\rho_L = \nu/\beta = 1/\rho > 1$). By the likelihood identity IV.(4.8), we have for $k = 1, 2$ that

$$\mathbb{E}\big[\tau(u)^k; \tau(u) < \infty\big] \;\; = \;\; \mathbb{E}_L \tau(u)^k e^{-\gamma S_{\tau(u)}} \;\; = \;\; e^{-\gamma u}\,\mathbb{E}_L e^{-\gamma \xi(u)}\,\mathbb{E}_L \tau(u)^k$$

$$= \;\; e^{-\gamma u}\frac{\beta}{\nu}\mathbb{E}_L \tau(u)^k \;\; = \;\; \psi(u)\mathbb{E}_L \tau(u)^k,$$

using that the overshoot $\xi(u)$ is exponential with rate β w.r.t. $\mathbb{P}_L$ and independent of $\tau(u)$. In particular,

$$\mathbb{E}\big[\tau(u)\,\big|\,\tau(u) < \infty\big] \;\; = \;\; \mathbb{E}_L \tau(u), \quad \mathbb{V}ar\big[\tau(u)\,\big|\,\tau(u) < \infty\big] \;\; = \;\; \mathbb{V}ar_L \tau(u).$$

For (1.1), we have by Wald's identity that (note that $\mathbb{E}_L S_t = t(\rho_L - 1)$)

$$\mathbb{E}_L S_{\tau(u)} \;\; = \;\; (\rho_L - 1)\mathbb{E}_L \tau(u),$$

$$\mathbb{E}_L \tau(u) \;\; = \;\; \frac{u + \mathbb{E}_L \xi(u)}{\rho_L - 1} \;\; = \;\; \frac{u + 1/\beta}{\nu/\beta - 1} \;\; = \;\; \frac{\beta u + 1}{\nu - \beta}.$$

For (1.2), Wald's second moment identity yields

$$\mathbb{E}_L \left(S_{\tau(u)} - (\rho_L - 1)\tau(u) \right)^2 = \sigma_L^2 \mathbb{E}_L \tau(u)$$

where $\sigma_L^2 = \kappa''(\gamma) = 2\nu/\beta^2$. Since $S_{\tau(u)}$ and $(\rho_L - 1)\tau(u)$ are independent with the same mean, the l.h.s. is

$$
\begin{aligned}
\mathbb{V}ar_L S_{\tau(u)} + \mathbb{V}ar_L((\rho_L - 1)\tau(u)) &= \mathbb{V}ar_L \xi(u) + (\rho_L - 1)^2 \mathbb{V}ar_L \tau(u) \\
&= \frac{1}{\beta^2} + \left(\frac{\nu}{\beta} - 1 \right)^2 \mathbb{V}ar_L \tau(u).
\end{aligned}
$$

Thus the l.h.s. of (1.2) is

$$\frac{\sigma_L^2 \mathbb{E}_L \tau(u) - 1/\beta^2}{(\nu/\beta - 1)^2} = \frac{2\nu(\beta u + 1)/(\nu - \beta) - 1}{(\nu - \beta)^2},$$

which is the same as the r.h.s. □

Proposition 1.2 *In the compound Poisson model with exponential claims with rate ν and safety loading $\eta > 0$, the Laplace transform of the time to ruin is given by*

$$\mathbb{E}e^{-\delta\tau(u)} = \mathbb{E}\left[e^{-\delta\tau(u)}; \tau(u) < \infty \right] = e^{-\rho_\delta u}\left(1 - \frac{\rho_\delta}{\nu} \right) \tag{1.3}$$

for $\delta \geq \kappa(\gamma_m) = 2\sqrt{\beta\nu} - \beta - \nu$, where

$$\rho_\delta = \frac{\nu - \beta - \delta + \sqrt{(\nu - \beta - \delta)^2 + 4\delta\nu}}{2}.$$

Proof. It is readily checked that $\gamma_m = \nu - \sqrt{\beta\nu}$ and hence that the value of $\kappa(\gamma_m)$ is as asserted.

Let $\rho_\delta > \gamma_m$ be determined by $\kappa(\rho_\delta) = \delta$. This means that $\beta(\nu/(\nu - \rho_\delta) - 1) - \rho_\delta = \delta$, which leads to the quadratic $\rho_\delta^2 + (\beta - \nu + \delta)\rho_\delta - \nu\delta = 0$ with solution ρ_δ (the sign of the square root is + because $\rho_\delta > 0$). But by the fundamental likelihood ratio identity (Theorem IV.4.3) we have

$$
\begin{aligned}
&\mathbb{E}\left[e^{-\delta\tau(u)}; \tau(u) < \infty \right] \\
&= \mathbb{E}_{\rho_\delta}\left[\exp\left\{ -\delta\tau(u) - \rho_\delta S_{\tau(u)} + \tau(u)\kappa(\rho_\delta) \right\}; \tau(u) < \infty \right] \\
&= e^{-\rho_\delta u}\mathbb{E}_{\rho_\delta}e^{-\rho_\delta\xi(u)} = e^{-\rho_\delta u}\frac{\nu_{\rho_\delta}}{\nu_{\rho_\delta} + \rho_\delta},
\end{aligned}
$$

where we used that $\mathbb{P}_{\rho_\delta}(\tau(u) < \infty) = 1$ because $\rho_\delta > \gamma_m$ and hence $\mathbb{E}_{\rho_\delta} S_1 = \kappa'(\rho_\delta) > 0$. Using $\nu_{\rho_\delta} = \nu - \rho_\delta$, the result follows. □

Note that it follows from Proposition 1.3 that we can write

$$\mathbb{E}e^{-\delta\tau(u)} = e^{-\rho_\delta u}\mathbb{E}e^{-\delta\tau(0)}. \tag{1.4}$$

The interpretation of this is that $\tau(u)$ can be written as the independent sum of $\tau(0)$ plus a r.v. $Y(u)$ belonging to a convolution semigroup. More precisely,

$$\tau(u) = \tau + \sum_{k=1}^{M(u)} \tau_k \tag{1.5}$$

where $\tau = \tau(0)$ is the length of the first ladder segment, $\tau_1, \tau_2, \ldots$ are the lengths of the ladder segments $2, 3, \ldots$, and $M(u) + 1$ is the index of the ladder segment corresponding to $\tau(u)$. Cf. Fig. V.2, where $Y_1, Y_2, \ldots$ are the ladder heights which form a terminating sequence of exponential r.v.'s with rate ν.

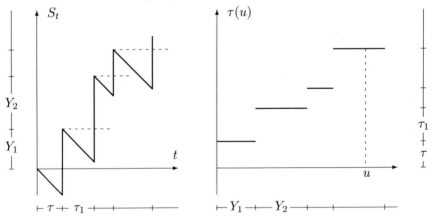

FIGURE V.2

For numerical purposes, the following formula is convenient by allowing $\psi(u, T)$ to be evaluated by numerical integration:

Proposition 1.3 *Assume that claims are exponential with rate $\nu = 1$. Then*

$$\psi(u, T) = \beta e^{-(1-\beta)u} - \frac{1}{\pi}\int_0^\pi \frac{f_1(\theta)f_2(\theta)}{f_3(\theta)}\,d\theta \tag{1.6}$$

where

$$
\begin{aligned}
f_1(\theta) &= \beta\exp\left\{2\sqrt{\beta T}\cos\theta - (1+\beta)T + u\left(\sqrt{\beta}\cos\theta - 1\right)\right\}, \\
f_2(\theta) &= \cos\left(u\sqrt{\beta}\sin\theta\right) - \cos\left(u\sqrt{\beta}\sin\theta + 2\theta\right), \\
f_3(\theta) &= 1 + \beta - 2\sqrt{\beta}\cos\theta.
\end{aligned}
$$

Note that the case $\nu \neq 1$ is easily reduced to the case $\nu = 1$ via the formula $\psi_{\beta,\nu}(u) = \psi_{\beta/\nu,1}(\nu u, \nu T)$.

Proof. We use the formula $\psi(u, T) = \mathbb{P}(V_T > u)$ where $\{V_t\}$ is the workload process in an initially empty M/M/1 queue with arrival rate β and service rate $\nu = 1$, cf. Corollary III.3.6. Let $\{Q_t\}$ be the queue length process of the queue (number in system, including the customer being currently served). If $Q_T = N > 0$, then $V_T = U_{1,T} + \cdots + U_{N,T}$, where $U_{1,T}$ is the residual service time of the customer being currently served and $U_{2,T}, \ldots, U_{N,T}$ the service times of the customers awaiting service. Since $U_{1,T}, U_{2,T}, \ldots, U_{N,T}$ are conditionally i.i.d. and exponential with rate $\nu = 1$, the conditional distribution of V_T given $Q_T = N$ is that of E_N where the r.v. E_N has an Erlang distribution with parameters $(N, 1)$, i.e. density $x^{N-1}e^{-x}/(N-1)!$. Hence

$$
\begin{aligned}
\psi(u, T) \;=\; \mathbb{P}(V_T > u) \;&=\; \sum_{N=1}^{\infty} \mathbb{P}(Q_T = N)\mathbb{P}(E_N > u) \\
&=\; \sum_{N=1}^{\infty} \mathbb{P}(Q_T = N) \sum_{k=0}^{N-1} e^{-u}\frac{u^k}{k!} \\
&=\; \sum_{k=0}^{\infty} e^{-u}\frac{u^k}{k!}\mathbb{P}(Q_T \geq k+1)\,.
\end{aligned}
\tag{1.7}
$$

For $j = 0, 1, 2, \ldots$, let (cf. [4])

$$
I_j(x) \;=\; \sum_{n=0}^{\infty} \frac{(x/2)^{2n+j}}{n!(n+j)!} \;=\; \frac{1}{\pi}\int_0^{\pi} e^{x\cos\theta}\cos j\theta \,\mathrm{d}\theta
\tag{1.8}
$$

denote the modified Bessel function of order j, let $I_{-j}(x) = I_j(x)$, and define $\iota_j = e^{-(1+\beta)T}\beta^{j/2}I_j(2\sqrt{\beta T})$. Then (see Prabhu [712, pp. 9–12], in particular equations (1.38), (1.44); similar formulas are in [APQ, pp. 87–89])

$$
\sum_{j=-\infty}^{\infty} \iota_j \;=\; 1,
$$

$$
\begin{aligned}
\mathbb{P}(Q_T \geq k+1) \;&=\; 1 - \sum_{j=-\infty}^{k} \iota_j + \beta^{k+1}\sum_{j=-\infty}^{-k-2} \iota_j \\
&=\; \beta^{k+1} + \sum_{j=k+1}^{\infty} \iota_j - \beta^{k+1}\sum_{j=-k-1}^{\infty} \iota_j\,.
\end{aligned}
$$

By Euler's formulas,

$$
\begin{aligned}
\sum_{j=k+1}^{\infty} \beta^{j/2} \cos(j\theta) &= \Re\Big[\sum_{j=k+1}^{\infty} \beta^{j/2} e^{ij\theta} \Big] = \Re\Big[\frac{\beta^{(k+1)/2} e^{i(k+1)\theta}}{\beta^{1/2} e^{i\theta} - 1} \Big] \\
&= \frac{\Re\big[\beta^{(k+1)/2} e^{i(k+1)\theta} \big(\beta^{1/2} e^{-i\theta} - 1 \big) \big]}{\big| \beta^{1/2} e^{i\theta} - 1 \big|^2} \\
&= \frac{\beta^{(k+1)/2} \big[\beta^{1/2} \cos(k\theta) - \cos\big((k+1)\theta\big) \big]}{f_3(\theta)},
\end{aligned}
$$

$$
\begin{aligned}
\beta^{k+1} \sum_{j=-k-1}^{\infty} \beta^{j/2} \cos(j\theta) &= \beta^{k+1} \Re\Big[\sum_{j=-k-1}^{\infty} \beta^{j/2} e^{ij\theta} \Big] = \Re \frac{\beta^{(k+1)/2} e^{-i(k+1)\theta}}{\beta^{1/2} e^{i\theta} - 1} \\
&= \frac{\Re\big[\beta^{(k+1)/2} e^{-i(k+1)\theta} \big(\beta^{1/2} e^{-i\theta} - 1 \big) \big]}{\big| \beta^{1/2} e^{i\theta} - 1 \big|^2} \\
&= \frac{\beta^{(k+1)/2} \big[\beta^{1/2} \cos\big((k+2)\theta\big) - \cos\big((k+1)\theta\big) \big]}{f_3(\theta)}.
\end{aligned}
$$

Hence the integral expression in (1.8) yields

$$
\begin{aligned}
&\mathbb{P}(Q_T \geq k+1) - \beta^{k+1} \\
&= e^{-(1+\beta)T} \frac{1}{\pi} \int_0^\pi e^{2\beta^{1/2} T \cos\theta} \cdot \frac{\beta^{(k+1)/2} \big[\beta^{1/2} \cos(k\theta) - \cos((k+2)\theta) \big]}{f_3(\theta)} \, d\theta.
\end{aligned}
$$

Since $\mathbb{P}(Q_\infty \geq k+1) = \beta^{k+1}$, it follows as in (1.7) that

$$
\psi(u) = \sum_{k=0}^{\infty} e^{-u} \frac{u^k}{k!} \beta^{k+1}.
$$

A further application of Euler's formulas yields

$$
\begin{aligned}
\sum_{k=0}^{\infty} \frac{u^k}{k!} \beta^{k/2} \cos\big((k+2)\theta\big) &= \Re\Big[e^{2i\theta} \sum_{k=0}^{\infty} \frac{(u\beta^{1/2} e^{i\theta})^k}{k!} \Big] = \Re e^{u\beta^{1/2} e^{i\theta} + 2i\theta} \\
&= e^{u\beta^{1/2} \cos\theta} \cos\big(u\beta^{1/2} \sin\theta + 2\theta\big),
\end{aligned}
$$

$$
\begin{aligned}
\sum_{k=0}^{\infty} \frac{u^k}{k!} \beta^{k/2} \cos(k\theta) &= \Re\Big[\sum_{k=0}^{\infty} \frac{(u\beta^{1/2} e^{i\theta})^k}{k!} \Big] = \Re e^{u\beta^{1/2} e^{i\theta}} \\
&= e^{u\beta^{1/2} \cos\theta} \cos\big(u\beta^{1/2} \sin\theta\big).
\end{aligned}
$$

The rest of the proof is easy algebra. □

Notes and references Proposition 1.3 was given in Asmussen [55] (as pointed out by Barndorff-Nielsen & Schmidli [138], there are several misprints in the formula there; however, the numerical examples in [55] are correct). Related formulas are in Takács [827]. Seal [785] gives a different numerical integration fomula for $1 - \psi(u, T)$ which, however, is numerically unstable for large T.

Alternatively, by using generators one can also represent $\psi(u, T)$ as the solution of the partial integro-differential equation

$$\frac{\partial \psi(u, T)}{\partial u} - \frac{\partial \psi(u, T)}{\partial T} - \beta \psi(u, T) + \beta \int_0^u \psi(u - y, T) \, \mathrm{d}B(y) + \beta\big(1 - B(u)\big) = 0$$

with boundary conditions $\lim_{u \to \infty} \psi(u, T) = 0$ for all $T > 0$ and $\psi(u, 0) = 0$ for all $u \geq 0$. For exponential claims this equation can be transformed into a second-order partial differential equation, which in Pervozvansky [694] was solved by Laplace transformation w.r.t. T and careful applications of the Cauchy residue theorem, resulting in an alternative integral representation of (1.6) in terms of trigonometric functions.

2 The ruin probability with no initial reserve

In this section, we are concerned with describing the distribution of the ruin time $\tau(0)$ in the case where the initial reserve is $u = 0$. We allow a general claim size distribution B and recall that we have the explicit formula $\psi(0) = \mathbb{P}\big(\tau(0) < \infty\big) = \rho$.

We first prove two classical formulas which are remarkable by showing that the ruin probabilities can be reconstructed from the distributions of the S_t, or, equivalently, from the accumulated claim distribution

$$F(x, t) = \mathbb{P}\Big(\sum_{i=1}^{N_t} U_i \leq x\Big)$$

(note that $\mathbb{P}(S_t \leq x) = F(x + t, t)$). The first formula, going back to Cramér, expresses $\psi(0, T)$ in terms of $F(\cdot, T)$, and the next one (often called Seal's formula but originating from Prabhu [711]) shows how to reduce the case $u \neq 0$ to this.

Theorem 2.1 $1 - \psi(0, T) = \dfrac{1}{T} \displaystyle\int_0^T F(x, T)\mathrm{d}x.$

Proof. For any $v \in [0, T]$, we define a new claim surplus process $\big\{S_t^{(v)}\big\}_{0 \leq t \leq T}$ by a 'cyclic translation', meaning that we interchange the two segments of the arrival process of $\{S_t\}_{0 \leq t \leq T}$ corresponding to the intervals $[0, v]$, resp. $[v, T]$. See Fig. V.3.

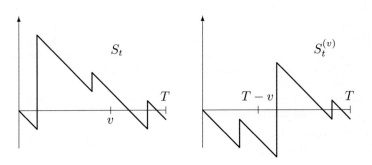

FIGURE V.3

In formulas,

$$S_t^{(v)} = \begin{cases} S_{t+v} - S_v & 0 \le t \le T - v \\ S_T - S_v + S_{t-T+v} & T - v \le t \le T \end{cases}.$$

Define

$$M(v,t) = \{S_t^{(v)} \le S_w^{(v)}, 0 \le w \le t\}$$

as the event that $\{S_w^{(v)}\}$ is at a minimum at time t. Then

$$1 - \psi(0,T) = \mathbb{P}\big(\tau(0) > T\big) = \mathbb{P}\big(M(0,T)\big)$$

$$= \frac{1}{T} \int_0^T \mathbb{P}\big(M(v,T)\big)\, dv = \frac{1}{T} \mathbb{E} \int_0^T I\big(M(v,T)\big)\, dv,$$

where the second equality follows from III.(5.2) with $A = (0, \infty)$, and the third from the obvious fact (exchangeability properties of the Poisson process) that $\{S_t^{(v)}\}$ has the same distribution as $S_t = \{S_t^{(0)}\}$ so that $\mathbb{P}\big(M(v,T)\big)$ does not depend on v.

Now consider the evaluation of $\int_0^T I\big(M(v,T)\big)\, dv$. Obviously, this integral is 0 if $S_T^{(v)} \equiv S_T > 0$. If $S_T < 0$, there exist v such that $M(v,T)$ occurs. For example, letting $\omega = \inf \{t > 0 : S_{t-} = \min_{0 \le w \le T} S_w\}$, we can take $v \in (\omega - \epsilon, \omega)$ for some small ϵ. We claim that if $M(0,T)$ occurs, then $M(v,T) = M(0,v)$. Indeed, we can write $M(v,T)$ as

$$\{S_T \le S_{t+v} - S_v, 0 \le t \le T - v\} \cap \{S_T \le S_T - S_v + S_{t-T+v}, T - v \le t \le T\}$$

$$= \{S_T \le S_t - S_v, v \le t \le T\} \cap \{S_T \le S_T - S_v + S_t, 0 \le t \le v\}$$

$$= \{S_T \le S_t - S_v, v \le t \le T\} \cap M(0,v) = M(0,v),$$

where the last equality follows from $S_T \leq S_t$ on $M(0,T)$ and $S_v \leq 0$ on $M(0,v)$.

It follows that if $M(0,T)$ occurs, then

$$\frac{1}{T} \int_0^T I\big(M(v,T)\big)\,\mathrm{d}v \;=\; \frac{1}{T} \int_0^T I\big(M(0,v)\big)\,\mathrm{d}v \;=\; -S_T$$

(note that the Lebesgue measure of the v for which $\{S_t\}$ is at a minimum at v is exactly $-S_T$ on $M(0,T)$). It is then clear from the cyclical nature of the problem that this holds irrespective of whether $M(0,T)$ occurs or not as long as $S_T < 0$. Hence

$$
\begin{aligned}
\frac{1}{T}\,\mathbb{E} & \int_0^T I\big(M(v,T)\big)\,\mathrm{d}v \\
&= \frac{1}{T}\mathbb{E}S_T^- \;=\; \frac{1}{T}\int_0^\infty \mathbb{P}(S_T \leq -x)\,\mathrm{d}x \\
&= \frac{1}{T}\int_0^T \mathbb{P}(S_T \leq -x)\,\mathrm{d}x \;=\; \frac{1}{T}\int_0^T \mathbb{P}\Big(\sum_{i=1}^{N_T} U_i \leq T - x\Big)\,\mathrm{d}x\,.
\end{aligned}
$$

$\square$

Let $f(\cdot,t)$ denote the density of $F(\cdot,t)$.

Theorem 2.2 $1 - \psi(u,T) \;=\; F(u+T,T) - \displaystyle\int_0^T \big(1 - \psi(0,T-t)\big)f(u+t,t)\,\mathrm{d}t.$

Proof. The event $\{S_T \leq u\} = \big\{\sum_1^{N_T} U_i \leq u + T\big\}$ can occur in two ways: either ruin does not occur in $[0,T]$, or it occurs, in which case there is a last time σ where S_t downcrosses level u, cf. Fig V.4.

Here $\sigma \in [t, t+\mathrm{d}t]$ occurs if and only if $S_t \in [u, u+\mathrm{d}t]$ and there is no upcrossing of level u after time t, which occurs w.p. $\psi(T-t)$. Hence

$$\mathbb{P}(S_T \leq u) \;=\; 1 - \psi(u,T) + \int_0^T \big(1 - \psi(0,T-t)\big)\mathbb{P}\big(S_t \in [u, u+\mathrm{d}t]\big)\,,$$

which is the same as the assertion of the theorem. $\square$

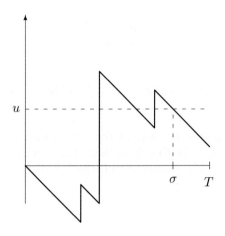

FIGURE V.4

The following representation of $\tau(0)$ will be used in the next section. The proof will be combined with the proof of Theorem IV.2.2.

Proposition 2.3 *Define* $\tau_-(z) = \inf\{t > 0 : S_t = -z\}$, $z > 0$. *Let Z be a r.v. which is independent of S_t and has the stationary excess distribution B_0. Then* $\mathbb{P}\big(\tau(0) \in \cdot \mid \tau(0) < \infty\big) = \mathbb{P}\big(\tau_-(Z) \in \cdot\big)$.

Proof of Theorem IV.2.2. For a fixed $T > 0$, define $S_t^* = S_T - S_{T-t-}$ and let

$$
\begin{aligned}
A(z,T) &= \big\{S_t < 0, \, 0 < t < T, \, S_{T-} = -z\big\}, \\
C(z,T) &= \big\{S_t > -z, \, 0 < t < T, \, S_{T-} = -z\big\}, \\
C^*(z,T) &= \big\{S_t^* > -z, \, 0 < t < T, \, S_{T-}^* = -z\big\}.
\end{aligned}
$$

Then

$$
\mathbb{P}\big(\tau(0) \in [T, T+\mathrm{d}T], \, -S_{\tau(0)-} \in [z, z+\mathrm{d}z]\big) = \mathbb{P}\big(A(z,T)\big)\beta\overline{B}(z)\,\mathrm{d}z\,\mathrm{d}T. \quad (2.1)
$$

But by sample path inspection (cf. Fig. V.5), $A(z,T) = C^*(z,T)$, and since $\{S_t\}_{0 \leq t \leq T}$, $\{S_t^*\}_{0 \leq t \leq T}$ have the same distribution, we therefore have $\mathbb{P}\big(A(z,T)\big) = \mathbb{P}\big(C(z;T)\big)$. Hence integrating (2.1) yields

$$
\begin{aligned}
\mathbb{P}\big(-S_{\tau(0)-} \in [z, z+\mathrm{d}z], \, \tau(0) < \infty\big) &= \beta\overline{B}(z)\,\mathrm{d}z \int_0^\infty \mathbb{P}\big(C(z,T)\big)\,\mathrm{d}T \\
&= \beta\overline{B}(z)\,\mathrm{d}z\,\mathbb{P}\big(\tau_-(z) < \infty\big) = \beta\overline{B}(z)\,\mathrm{d}z.
\end{aligned}
$$

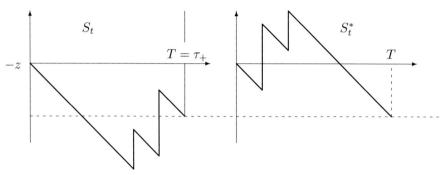

FIGURE V.5

Thus

$$\mathbb{P}\left(-S_{\tau(0)-} > x, \, S_{\tau(0)} > y; \, \tau(0) < \infty\right)$$

$$= \int_x^\infty \mathbb{P}\big(U > y + z \,\big|\, U > z\big)\, \mathbb{P}\big(-S_{\tau(0)-} \in [z, z+dz], \, \tau(0) < \infty\big)$$

$$= \int_x^\infty \frac{\overline{B}(y+z)}{\overline{B}(z)}\, \beta \overline{B}(z)\, dz \; = \; \beta \int_x^\infty \overline{B}(y+z)\, dz \; = \; \beta \int_{x+y}^\infty \overline{B}(z)\, dz \, ,$$

which is the assertion of Theorem IV.2.2. □

Proof of Proposition 2.3. It follows by division by

$$\mathbb{P}\big(S_{\tau(0)-} \in [z, z+dz], \, \tau(0) < \infty\big) \; = \; \beta \overline{B}(z)\, dz$$

in (2.1) that

$$\mathbb{P}\big(\tau(0) \in [T, T+dT] \,\big|\, S_{\tau(0)-} \in [z, z+dz], \, \tau(0) < \infty\big) \; = \; \mathbb{P}\big(C(z)\big)\, dT.$$

Hence

$$\mathbb{P}\big(\tau(0) \in [T, T+dT] \,\big|\, \tau(0) < \infty\big)$$

$$= \; dT \int_0^\infty \mathbb{P}\big(C(z)\big) \mathbb{P}\big(S_{\tau(0)-} \in [z, z+dz], \, \tau(0) < \infty\big)$$

$$= \; dT \int_0^\infty \mathbb{P}\big(C(z)\big) \mathbb{P}\big(Z \in [z, z+dz], \, \tau(0) < \infty\big)$$

$$= \; dT\, \mathbb{P}\big(\tau_-(Z) \in [T, T+dT]\big) \, .$$

□

Notes and references For Theorems 2.1, 2.2, see in addition to Prabhu [711] also Seal [784, 787]. Theorem 2.1 and the present proof is in the spirit of Ballot theorems, cf. Takács [827]; a martingale proof is in Delbaen & Haezendonck [287]. For related inequalities for positive u, see De Vylder & Goovaerts [304].

Proposition 2.3 was noted by Asmussen & Klüppelberg [86], who instead of the present direct proof gave two arguments, one based upon a result of Asmussen & Schmidt [103] generalizing Theorem III.5.5 and one upon excursion theory for Markov processes (see X.4a).

For discrete claim size distributions, Picard & Lefèvre [701] used generalized Appell polynomials to develop recursion formulae for finite time ruin probabilities, see also Rullière & Loisel [757]. This was later extended to more general set-ups including dependent claims, cf. for instance Ignatov & Kaishev [493, 494] and Lefèvre & Loisel [575]. Continuous versions of the discrete expressions of [701] are given in De Vylder & Goovaerts [303].

In the setting of general Lévy processes, some relevant references are Shtatland [798] and Gusak & Korolyuk [442].

3   Laplace transforms

As usual, $-\rho_\delta$ denotes the negative solution of the equation

$$\kappa(r) \;=\; \beta\big(\widehat{B}[r] - 1\big) - r \;=\; \delta. \tag{3.1}$$

Let $\tau_-(y)$ be defined as Proposition 2.3. Note that $\tau_-(y) < \infty$ a.s. because of $\eta > 0$.

Lemma 3.1 $\mathbb{E}e^{-\delta\tau_-(y)} = e^{-\rho_\delta y}$.

Proof. Optional stopping at $\tau_-(y) \wedge T$ of the martingale

$$\big\{e^{-\rho_\delta S_t - t\kappa(-\rho_\delta)}\big\} = \big\{e^{-\rho_\delta S_t - \delta t}\big\}$$

and letting $T \to \infty$ using dominated convergence yields $1 = e^{\rho_\delta y}\,\mathbb{E}e^{-\delta\tau_-(y)}$.   □

Let $g_\delta(x)$ be the density of the measure $\mathbb{E}\big[e^{-\delta\tau(0)};\, \tau(0) < \infty,\, \xi(0) \in dx\big]$ (recall that $\xi(0) = S_{\tau(0)}$).

Lemma 3.2 $g_\delta(x) \;=\; \beta\,e^{\rho_\delta x} \displaystyle\int_x^\infty e^{-\rho_\delta y}B(dy)$.

Proof. Let Z be the surplus $-S_{\tau(0)-}$ just before ruin. Then by Proposition 2.3,

$$\mathbb{E}\big[e^{-\delta\tau(0)} \,\big|\, \tau(0) < \infty,\, Z = y\big] \;=\; \mathbb{E}e^{-\delta\tau_-(y)} \;=\; e^{-\rho_\delta y}.$$

Further by Theorem IV.2.2

$$\mathbb{P}\big(Z \in [y, y + \mathrm{d}y], \, \xi(0) \in \mathrm{d}x\big) = \beta B(x + \mathrm{d}y) \, \mathrm{d}x$$

and hence

$$g_\delta(x) = \int_0^\infty e^{-\rho_\delta y} \beta B(x + \mathrm{d}y) = \beta \int_x^\infty e^{-\rho_\delta(y-x)} B(\mathrm{d}y). \qquad \square$$

Lemma 3.3 *For the Laplace transform* $\widehat{g}_\delta[-s] = \int_0^\infty e^{-sx} g_\delta(x) \, \mathrm{d}x$ *we have*

$$\widehat{g}_\delta[-s] = \frac{\kappa(-s) - s - \delta + \rho_\delta}{\rho_\delta - s}.$$

Proof.

$$\begin{aligned}
\widehat{g}_\delta[-s] &= \beta \int_0^\infty e^{x(-s+\rho_\delta)} \, \mathrm{d}x \int_x^\infty e^{-\rho_\delta y} B(\mathrm{d}y) \\
&= \beta \int_0^\infty e^{-\rho_\delta y} B(\mathrm{d}y) \int_0^y e^{x(-s+\rho_\delta)} \, \mathrm{d}x \\
&= \frac{\beta}{\rho_\delta - s} \int_0^\infty e^{-\rho_\delta y} B(\mathrm{d}y) [e^{y(\rho_\delta - s)} - 1] \\
&= \frac{\beta}{\rho_\delta - s} \big(\widehat{B}[-s] - \widehat{B}[-\rho_\delta]\big).
\end{aligned}$$

The result follows by inserting $\beta \widehat{B}[-s] = \kappa(-s) + \beta - s$ and $\kappa(-\rho_\delta) = \delta$. $\qquad \square$

Corollary 3.4 $\mathbb{E}[e^{-\delta\tau(0)}; \tau(0) < \infty] = 1 - \dfrac{\delta}{\rho_\delta}.$

Proof. Let $b = 0$. $\qquad \square$

Here is a classical result: the double Laplace transform of the ruin time $\tau(u)$:

Corollary 3.5 $\displaystyle\int_0^\infty e^{-su} \mathbb{E}\big[e^{-\delta\tau(u)}; \tau(u) < \infty\big] \, \mathrm{d}u = \dfrac{\kappa(-s)/s - \delta/\rho_\delta}{\kappa(-s) - \delta}.$

Proof. Define $Z_\delta(u) = \mathbb{E}\big[e^{-\delta\tau(u)}; \tau(u) < \infty\big]$. It is then easily seen that $Z_\delta(u)$ is the solution of the renewal equation $Z_\delta(u) = z_\delta(u) + \int_0^u Z_\delta(u - x) g_\delta(x) \, \mathrm{d}x$ where $z_\delta(u) = \int_u^\infty g_\delta(x) \mathrm{d}x$. Hence

$$\int_0^\infty e^{-su} \, \mathrm{d}u \, \mathbb{E}\big[e^{-\delta\tau(u)}; \tau(u) < \infty\big] = \widehat{Z}_\delta[-s] = \frac{\widehat{z}_\delta[-s]}{1 - \widehat{g}_\delta[-s]} = \frac{\widehat{g}_\delta[0] - \widehat{g}_\delta[-s]}{s(1 - \widehat{g}_\delta[-s])}.$$

Using Lemma 3.3, the result follows after simple algebra. □

Notes and references An explicit inversion of the double Laplace transform in Corollary 3.5 to obtain expressions for $\psi(u, t)$ in terms of infinite series can be found for claim size distributions of mixed Erlang type in Garcia [389] and Dickson & Willmot [322], see also Willmot & Woo [893] and Dickson [310]. For a power series expansion, see e.g. Usabel [859, 860]. An alternative very accurate numerical method is to randomize the time horizon T and exploit the resulting additional smoothness of the problem, in particular in a matrix-analytic framework, cf. Section IX.8.

In Chapter XII the results of this section will be extended in various directions in the context of Gerber-Shiu functions.

4   When does ruin occur?

For the general compound Poisson model, the known results are even less explicit than for the exponential claims case, and take basically the form of approximations and inequalities.

The first main result of the present section is that the value $u\, m_L$, where

$$m_L = \frac{1}{\kappa'(\gamma)} = \frac{1}{\beta \widehat{B}'[\gamma] - 1} = \frac{1}{\beta_L \mathbb{E}_L U - 1} = \frac{C}{1 - \rho},$$

is in some appropriate sense critical as the most 'likely' time of ruin (here C is the Cramér-Lundberg constant). Later results then deal with more precise and refined versions of this statement.

Theorem 4.1 *Assume $\eta > 0$. Then given $\tau(u) < \infty$, $\tau(u)/u \xrightarrow{\mathbb{P}} m_L$ as $u \to \infty$. That is, for any $\epsilon > 0$*

$$\mathbb{P}\left(\left| \frac{\tau(u)}{u} - m_L \right| > \epsilon \,\middle|\, \tau(u) < \infty \right) \to 0. \tag{4.1}$$

Further, for any m

$$\frac{\psi(u, mu)}{\psi(u)} \to \begin{cases} 0 & m < m_L \\ 1 & m > m_L. \end{cases} \tag{4.2}$$

For the proof, we need the following auxiliary result:

Proposition 4.2 *Assume $\eta < 0$, i.e. $\rho = \beta \mu_B > 1$. Then as $u \to \infty$,*

$$\frac{\tau(u)}{u} \xrightarrow{a.s.} m = \frac{1}{\rho - 1}, \qquad \frac{\mathbb{E}\tau(u)}{u} \to \frac{1}{\rho - 1}, \tag{4.3}$$

$$\frac{\tau(u) - mu}{\sqrt{u}} \xrightarrow{\mathscr{D}} N(0, \omega^2) \quad \text{where } \omega^2 = \beta \mu_B^{(2)} m^3. \tag{4.4}$$

Proof. The assumption $\eta < 0$ ensures that $\mathbb{P}(\tau(u) < \infty) = 1$ and $\tau(u) \stackrel{a.s.}{\to} \infty$. By Proposition IV.1.2, $S_t/t \stackrel{a.s.}{\to} 1/m$, and hence a.s.

$$m = \lim_{t \to \infty} \frac{t}{S_t} = \lim_{u \to \infty} \frac{\tau(u)}{S_{\tau}(u)} = \lim_{u \to \infty} \frac{\tau(u)}{u + \xi_{\tau(u)}} = \lim_{u \to \infty} \frac{\tau(u)}{u},$$

using $\xi_{\tau(u)} = o(u)$ a.s., cf. Proposition A1.6. This proves the first assertion of (4.3). For the second, note that by Wald's identity

$$u + \mathbb{E}\xi(u) = \mathbb{E}S_{\tau(u)} = \mathbb{E}\tau(u) \cdot \mathbb{E}S_1 = (\rho - 1)\mathbb{E}\tau(u)$$

and that $\mathbb{E}\xi(u)/u \to 0$, cf. again Proposition A1.6.

For (4.4), note first that (Proposition IV.1.5)

$$\frac{S_t - t/m}{\sqrt{t}} \stackrel{\mathscr{D}}{\to} N\left(0, \beta\mu_B^{(2)}\right).$$

According to Anscombe's theorem (e.g. Theorem 7.3.2 of [246]) and (4.3), the same conclusion holds with t replaced by $\tau(u)$. If $Z \sim N(0, 1)$, this can be rewritten as

$$\frac{u + \xi(u) - \tau(u)/m}{\sqrt{\tau(u)}} \approx \sqrt{\beta\mu_B^{(2)}}\, Z, \quad \text{implying}$$

$$\frac{\tau(u) - mu}{\sqrt{\tau(u)}} \approx -m\sqrt{\beta\mu_B^{(2)}}\, Z \stackrel{\mathscr{D}}{=} m\sqrt{\beta\mu_B^{(2)}}\, Z,$$

$$\frac{\tau(u) - mu}{\sqrt{u}} \approx m^{3/2}\sqrt{\beta\mu_B^{(2)}}\, Z = \omega Z.$$

$\square$

Proof of Theorem 4.1. The l.h.s. of (4.1) is

$$\frac{\mathbb{P}\left(\left|\frac{\tau(u)}{u} - m_L\right| > \epsilon, \tau(u) < \infty\right)}{\mathbb{P}(\tau(u) < \infty)}$$

$$= \frac{e^{-\gamma u}\mathbb{E}_L\left[e^{-\gamma\xi(u)}; \left|\frac{\tau(u)}{u} - m_L\right| > \epsilon, \tau(u) < \infty\right]}{\psi(u)}$$

$$\le \frac{e^{-\gamma u}\mathbb{P}_L\left(\left|\frac{\tau(u)}{u} - m_L\right| > \epsilon\right)}{O(e^{-\gamma u})}.$$

By Proposition 4.2, $\mathbb{P}_L(\cdot) \to 0$, proving (4.1), and (4.2) follows immediately from (4.1).

$\square$

Notes and references   Theorem 4.1 is standard, though it is not easy to attribute priority to any particular author. Thus, the result comes out not only by the present direct proof but also from any of the results in the following subsections.

For a study of the distribution of the number of claims until ruin, see Egidio dos Reis [340].

4a   Segerdahl's normal approximation

We shall now prove a classical result due to Segerdahl, which may be viewed both as a refinement of Theorem 4.1 (by considering $\psi(u, T)$ for T which are close to the critical value um_L), and as a time-dependent version of the Cramér-Lundberg approximation.

Corollary 4.3 (SEGERDAHL [791])   *Let C be the Cramér-Lundberg constant and define $\omega_L^2 = \beta_L \mathbb{E}_L U^2 m_L^3 = \beta \widehat{B}''[\gamma] m_L^3$ where $m_L = 1/(\rho_L - 1) = 1/(\beta \widehat{B}'[\gamma] - 1)$. Then for any y,*

$$\mathrm{e}^{\gamma u} \psi\big(u, um_L + y\omega_L \sqrt{u}\big) \;\to\; C\Phi(y). \tag{4.5}$$

For the proof, we need the following auxiliary result:

Proposition 4.4 (STAM'S LEMMA)   *If $\eta < 0$, then $\xi(u)$ and $\tau(u)$ are asymptotically independent in the sense that, letting Z be a $N(0, \omega^2)$ r.v. with ω^2 as in (4.4), one has*

$$\mathbb{E}f\big(\xi(u)\big)g\Big(\frac{\tau(u) - mu}{\sqrt{u}}\Big) \;\to\; \mathbb{E}f\big(\xi(\infty)\big) \cdot \mathbb{E}g(Z) \tag{4.6}$$

whenever f, g are continuous and bounded on $[0, \infty)$, resp. $(-\infty, \infty)$.

Proof. Define $u' = u - u^{1/4}$. Then the distribution of $\tau(u) - \tau(u')$ given $\mathscr{F}_{\tau(u')}$ is readily seen to be degenerate at zero if $S_{\tau(u')} > u$ and otherwise that of $\tau(v)$ with $v = u - S_{\tau(u')} = u^{1/4} - \xi(\tau(u'))$. Using (4.3), we get

$$
\begin{aligned}
\mathbb{E}\big[\tau(u) - \tau(u')\big] &= \mathbb{E}\big[\tau\big(u^{1/4} - \xi(u')\big); \xi(u') \le u^{1/4}\big] \\
&\le \mathbb{E}\tau(u^{1/4}) = \mathrm{O}(u^{1/4}),
\end{aligned}
$$

and thus in (4.6), we can replace $\tau(u)$ by $\tau(u')$. Let $h(u) = \mathbb{E}f\big(\xi(u)\big)$. Then $h(u) \to h(\infty) = \mathbb{E}f\big(\xi(\infty)\big)$, and similarly as above we get

$$
\begin{aligned}
&\mathbb{E}\big[f\big(\xi(u)\big) \,\big|\, \mathscr{F}_{\tau(u')}\big] \\
&= h\big(u^{1/4} - \xi(u')\big)I\big(\xi(u') \le u^{1/4}\big) + f\big(\xi(u') - u^{1/4}\big)I\big(\xi(u') > u^{1/4}\big) \\
&\xrightarrow{\;\mathbb{P}\;} h(\infty) + 0,
\end{aligned}
$$

using that $u^{1/4} - \xi(u') \overset{\mathbb{P}}{\to} \infty$ w.r.t. $\mathbb{P}$ because of $\xi(u') \overset{\mathscr{D}}{\to} \xi(\infty)$ (recall that $\eta < 0$). Hence

$$
\begin{aligned}
\mathbb{E}f\big(\xi(u)\big)g\Big(\frac{\tau(u') - mu}{\sqrt{u}}\Big) &= \mathbb{E}\Big[\mathbb{E}\big[f\big(\xi(u)\big)\,\big|\,\mathscr{F}_{\tau(u')}\big]g\Big(\frac{\tau(u') - mu}{\sqrt{u}}\Big)\Big] \\
&\sim h(\infty)\mathbb{E}g\Big(\frac{\tau(u') - mu}{\sqrt{u}}\Big) \sim h(\infty)\mathbb{E}g(Z)\,.
\end{aligned}
$$

$\square$

Proof of Corollary 4.3.

$$
\begin{aligned}
e^{\gamma u}\psi\big(u, um_L + y\omega_L\sqrt{u}\big) &= e^{\gamma u}\mathbb{P}\big(\tau(u) \le um_L + y\omega_L\sqrt{u}\big) \\
&= \mathbb{E}_L\big[e^{-\gamma\xi(u)};\,\tau(u) \le um_L + y\omega_L\sqrt{u}\big] \\
&\sim \mathbb{E}_L e^{-\gamma\xi(u)} \cdot \mathbb{P}_L\big(\tau(u) \le um_L + y\omega_L\sqrt{u}\big) \\
&\to C\Phi(y),
\end{aligned}
$$

where we used Stam's lemma in the third step and (4.4) in the last. $\square$

For practical purposes, Segerdahl's result suggests the approximation

$$
\psi(u, T) \approx Ce^{-\gamma u}\Phi\Big(\frac{T - um_L}{\omega_L\sqrt{u}}\Big). \tag{4.7}
$$

To arrive at this, just substitute $T = um_L + y\omega_L\sqrt{u}$ in (4.5) and solve for $y = y(T)$. The precise condition for (4.7) to be valid is that T varies with u in such a way that $y(T)$ has a limit in $(-\infty, \infty)$ as $u \to \infty$. Thus, in practice one would trust (4.7) whenever u is large and $|y(T)|$ moderate or small (numerical evidence presented in [55] indicates, however, that for the fit of (4.7) to be good, u needs to be very large).

A remarkably sharp and explicit asymptotic result in terms of the time horizon T is the following:

Theorem 4.5 *For every fixed $u \ge 0$, we have, as $T \to \infty$,*

$$
\psi(u) - \psi(u, T) \sim C_1\,e^{-\gamma_m u - |\kappa(\gamma_m)|T}\,T^{-3/2}\big(1 + H(u)\big), \tag{4.8}
$$

where $C_1 = \big(2\pi\beta B''[\gamma_m]\big)^{-1/2}\gamma_m^{-2}$ and $H(u)$ is a renewal function which satisfies $H(u) \sim 2\big(\beta B''[\gamma_m]\big)^{-1}u$ as $u \to \infty$.

The proof is quite involved and uses deep results from random walk theory; we refer to Teugels [841].

Notes and references   Corollary 4.3 is due to Segerdahl [791]. The present proof is basically that of Siegmund [806]; see also von Bahr [121] and Gut [443]. For refinements of Corollary 4.3 in terms of Edgeworth expansions, see Asmussen [55] and Malinovskii [625]. Cf. also Höglund [478].

4b   Gerber's time-dependent version of Lundberg's inequality

For $y > 0$, define α_y, γ_y by

$$\kappa'(\alpha_y) = \frac{1}{y}, \quad \gamma_y = \alpha_y - y\kappa(\alpha_y). \tag{4.9}$$

Note that $\alpha_y > \gamma_m$ and that $\gamma_y > \gamma$ (unless for the critical value $y = 1/m_L$), cf. Fig. V.1.

Theorem 4.6

$$\psi(u, yu) \;\leq\; e^{-\gamma_y u}, \quad y < \frac{1}{\kappa'(\gamma)}, \tag{4.10}$$

$$\psi(u) - \psi(u, yu) \;\leq\; e^{-\gamma_y u}, \quad y > \frac{1}{\kappa'(\gamma)}. \tag{4.11}$$

Proof. Consider first the case $y < 1/\kappa'(\gamma)$. Then $\kappa(\alpha_y) > 0$ (see Fig. V.1), and hence

$$
\begin{aligned}
\psi(u, yu) &= e^{-\alpha_y u}\mathbb{E}_{\alpha_y}\left[e^{-\alpha_y \xi(u) + \tau(u)\kappa(\alpha_y)}; \tau(u) \leq yu\right] \\
&\leq e^{-\alpha_y u}\mathbb{E}_{\alpha_y}\left[e^{\tau(u)\kappa(\alpha_y)}; \tau(u) \leq yu\right] \;\leq\; e^{-\alpha_y u + yu\kappa(\alpha_y)}.
\end{aligned}
$$

Similarly, if $y > 1/\kappa'(\gamma)$, we have $\kappa(\alpha_y) < 0$ and get

$$
\begin{aligned}
\psi(u) - \psi(u, yu) &= e^{-\alpha_y u}\mathbb{E}_{\alpha_y}\left[e^{-\alpha_y \xi(u) + \tau(u)\kappa(\alpha_y)}; yu < \tau(u) < \infty\right] \\
&\leq e^{-\alpha_y u}\mathbb{E}_{\alpha_y}\left[e^{\tau(u)\kappa(\alpha_y)}; yu < \tau(u) < \infty\right] \\
&\leq e^{-\alpha_y u + yu\kappa(\alpha_y)}.
\end{aligned}
$$

$\square$

Remark 4.7 It may appear that the proof uses considerably less information on α_y than is inherent in the definition (4.9). However, the point is that we want to select an α which produces the largest possible exponent in the inequalities. From the proof it is seen that this amounts to that α should maximize $\alpha - y\kappa(\alpha)$. Differentiating w.r.t. α, we arrive at the expression in (4.9). $\square$

In view of Theorem 4.6, γ_y is sometimes called the *time-dependent Lundberg exponent*.

An easy combination with the proof of Theorem IV.6.3 yields the following sharpening of (4.10):

Proposition 4.8 $\psi(u, yu) \leq C_+(\alpha_y)e^{-\gamma_y u}$ *where*

$$C_+(\alpha_y) = \sup_{x \geq 0} \frac{\overline{B}(x)}{\int_x^\infty e^{\alpha_y(y-x)}B(dy)}.$$

Notes and references Theorem 4.6 is due to Gerber [397], who used a martingale argument. For a different proof, see Martin-Löf [631]. Numerical comparisons are in Grandell [430]; the bound $e^{-\gamma_y u}$ turns out to be rather crude, which may be understood from Theorem 4.9 below, which shows that the correct rate of decay of $\psi(u, yu)$ is $e^{-\gamma_y u}/\sqrt{u}$.

Some further discussion is given in XVI.2, and generalizations to more general models are given in Chapter VII. Höglund [477] treats the renewal case.

4c Arfwedson's saddlepoint approximation

Our next objective is to strengthen the time-dependent Lundberg inequalities to approximations. As a motivation, it is instructive to reinspect the choice of the change of measure in the proof, i.e. the choice of α_y. For any $\alpha > \gamma_m$, Proposition 4.2 yields

$$\mathbb{E}_\alpha \tau(u) \sim \frac{u}{\kappa'_\alpha(0)} = \frac{u}{\kappa'(\alpha)}.$$

I.e., if we want $\mathbb{E}_\alpha \tau(u) \approx T$, then the relevant choice is precisely $\alpha = \alpha_y$ where $y = T/u$. We thereby obtain that T is 'in the center' of the $\mathbb{P}_\alpha$-distribution of $\tau(u)$. This idea is precisely what characterizes the saddlepoint method.

The traditional application of the saddlepoint method is to derive approximations, not inequalities, and in case of ruin probabilities the approach leads to the following result:

Theorem 4.9 *If* $y < 1/\kappa'(\gamma)$, *then the solution* $\widetilde{\alpha}_y < \alpha_y$ *of* $\kappa(\widetilde{\alpha}) = \kappa(\alpha_y)$ *is* < 0, *and*

$$\psi(u, yu) \sim \frac{\alpha_y - \widetilde{\alpha}_y}{\alpha_y |\widetilde{\alpha}_y| \sqrt{2\pi y \beta \widehat{B}''[\alpha_y]}} \cdot \frac{e^{-\gamma_y u}}{\sqrt{u}}, \quad u \to \infty. \tag{4.12}$$

If $y > 1/\kappa'(\gamma)$, then $\widetilde{\alpha}_y > 0$, and

$$\psi(u) - \psi(u, yu) \quad \sim \quad \frac{\alpha_y - \widetilde{\alpha}_y}{\alpha_y \widetilde{\alpha}_y \sqrt{2\pi y \beta \widehat{B}''[\alpha_y]}} \cdot \frac{e^{-\gamma_y u}}{\sqrt{u}}, \quad u \to \infty. \quad (4.13)$$

Proof. In view of Stam's lemma, the formula

$$\psi(u, yu) \; = \; e^{-\alpha_y u} \mathbb{E}_{\alpha_y} \left[e^{-\alpha_y \xi(u) + \tau(u)\kappa(\alpha_y)}; \tau(u) \le yu \right]$$

suggests heuristically that

$$\psi(u, yu) \; \approx \; e^{-\alpha_y u} \mathbb{E}_{\alpha_y} e^{-\alpha_y \xi(\infty)} \cdot \mathbb{E}_{\alpha_y} \left[e^{\tau(u)\kappa(\alpha_y)}; \tau(u) \le yu \right]. \quad (4.14)$$

Here the first expectation can be estimated similarly as in the proof of the Cramér–Lundberg's approximation in Chapter IV. Using Lemma IV.5.7 with $\mathbb{P}$ replaced by $\mathbb{P}_{\widetilde{\alpha}_y}$ and $\mathbb{P}_L$ by $\mathbb{P}_{\alpha_y}$, we have $\gamma_{\widetilde{\alpha}_y} = \alpha_y - \widetilde{\alpha}_y$ and get

$$
\begin{aligned}
\mathbb{E}_{\alpha_y} e^{-\alpha_y \xi(\infty)} \; &= \; \frac{\gamma_{\widetilde{\alpha}_y}}{\alpha_y \kappa'_{\widetilde{\alpha}_y}(\gamma_{\widetilde{\alpha}_y})} \left(1 - \beta_{\widetilde{\alpha}_y} \frac{\widehat{B}_{\widetilde{\alpha}_y}[\gamma_{\widetilde{\alpha}_y} - \alpha_y] - 1}{\gamma_{\widetilde{\alpha}_y} - \alpha_y} \right) \\
&= \; \frac{\alpha_y - \widetilde{\alpha}_y}{\alpha_y \kappa'(\gamma_{\widetilde{\alpha}_y} + \widetilde{\alpha}_y)} \left(1 + \beta \widehat{B}[\widetilde{\alpha}_y] \frac{\widehat{B}[\gamma_{\widetilde{\alpha}_y} - \alpha_y + \widetilde{\alpha}_y]/\widehat{B}[\widetilde{\alpha}_y] - 1}{\widetilde{\alpha}_y} \right) \\
&= \; \frac{\alpha_y - \widetilde{\alpha}_y}{\alpha_y \kappa'(\alpha_y)} \left(1 + \beta \frac{1 - \widehat{B}[\widetilde{\alpha}_y]}{\widetilde{\alpha}_y} \right) \\
&= \; \frac{y(\alpha_y - \widetilde{\alpha}_y)}{\alpha_y} \cdot \frac{\widetilde{\alpha}_y + \beta(1 - \widehat{B}[\widetilde{\alpha}_y])}{\widetilde{\alpha}_y} \\
&= \; \frac{-y(\alpha_y - \widetilde{\alpha}_y)\kappa(\widetilde{\alpha}_y)}{\widetilde{\alpha}_y} \; = \; \frac{y(\alpha_y - \widetilde{\alpha}_y)\kappa(\alpha_y)}{\alpha_y |\widetilde{\alpha}_y|}.
\end{aligned}
$$

For the second term in (4.14), it seems tempting to apply the normal approximation (4.4). Writing $\tau(u) \approx yu + u^{1/2}\omega V$, where V is normal$(0, 1)$ under $\mathbb{P}_{\alpha_y}$ and

$$\omega^2 \; = \; \beta_{\alpha_y} \mu_{B_{\alpha_y}}^{(2)} / (\rho_{\alpha_y} - 1)^3 \; = \; \beta \widehat{B}''[\alpha_y] / (\rho_{\alpha_y} - 1)^3 \; = \; y^3 \beta \widehat{B}''[\alpha_y],$$

we get heuristically that

$$
\begin{aligned}
\mathbb{E}_{\alpha_y} &\left[e^{\tau(u)\kappa(\alpha_y)}; \, \tau(u) \le yu \right] \\
&= e^{yu\kappa(\alpha_y)} \mathbb{E}_{\alpha_y} \left[e^{\kappa(\alpha_y)u^{1/2}\omega V}; V \le 0 \right] \\
&= e^{yu\kappa(\alpha_y)} \int_0^\infty e^{-\kappa(\alpha_y)u^{1/2}\omega x} \varphi(x) \, \mathrm{d}x \\
&= e^{yu\kappa(\alpha_y)} \frac{1}{\kappa(\alpha_y)u^{1/2}\omega} \int_0^\infty e^{-z} \varphi\big(z/(\kappa(\alpha_y)u^{1/2}\omega)\big) \, \mathrm{d}z \\
&\sim e^{yu\kappa(\alpha_y)} \frac{1}{\kappa(\alpha_y)u^{1/2}\omega} \int_0^\infty e^{-z} \cdot \frac{1}{\sqrt{2\pi}} \, \mathrm{d}z \\
&= e^{yu\kappa(\alpha_y)} \frac{1}{\kappa(\alpha_y)\sqrt{2\pi u\omega^2}}.
\end{aligned}
$$

Inserting these estimates in (4.14), (4.12) follows. The proof of (4.13) is completely similar. □

The difficulties in making the proof precise is in part to show (4.14) rigorously, and in part that for the final calculation one needs a sharpened version of the CLT for $\psi(u)$ (basically a local CLT with remainder term).

Example 4.10 Assume that $\overline{B}(x) = e^{-\nu x}$. Then $\kappa(\alpha) = \beta\big(\nu/(\nu - \alpha) - 1\big) - \alpha$, $\kappa'(\alpha) = \beta\alpha/(\nu - \alpha)^2 - 1$, and the equation $\kappa'(\alpha) = 1/y$ is easily seen to have solution

$$
\alpha_y = \nu - \sqrt{\frac{\beta\nu}{1 + 1/y}}
$$

(the sign of the square root is negative because the c.g.f. is undefined for $\alpha > \nu$). It follows that

$$
\nu_{\alpha_y} = \nu - \alpha_y = \sqrt{\frac{\beta\nu}{1 + 1/y}}, \quad \beta_{\alpha_y} = \beta + \alpha_y = \beta + \nu - \sqrt{\frac{\beta\nu}{1 + 1/y}},
$$

$$
\alpha_y - \widetilde{\alpha_y} = \beta_{\alpha_y} - \nu_{\alpha_y} = \beta + \nu - 2\sqrt{\frac{\beta\nu}{1 + 1/y}}, \quad \widetilde{\alpha_y} = \sqrt{\frac{\beta\nu}{1 + 1/y}} - \beta,
$$

$$
\widehat{B}''[\alpha_y] = \frac{2\nu}{(\nu - \alpha_y)^3} = \frac{2\nu^{1/2}(1 + y)^{3/2}}{\beta^{3/2}},
$$

and (4.12) gives the approximation

$$\frac{\beta^{1/4}\left(\beta + \nu - 2\sqrt{\dfrac{\beta\nu}{1 + 1/y}}\right)}{\left(\nu - \sqrt{\dfrac{\beta\nu}{1 + 1/y}}\right)\left(\beta - \sqrt{\dfrac{\beta\nu}{11/ + y}}\right)\nu^{1/4}(1 + 1/y)^{3/4}\sqrt{4\pi y}} \cdot \frac{e^{-\gamma_y u}}{\sqrt{u}}$$

for $\psi(u, yu)$ when $y < 1/\kappa'(\gamma) = \rho/1 - \rho$. □

Notes and references Theorem 4.9 is from Arfwedson [51]. A related result appears in Barndorff-Nielsen & Schmidli [138].

5   Diffusion approximations

The idea behind the diffusion approximation is to first approximate the claim surplus process by a Brownian motion with drift by matching the two first moments, and next to note that such an approximation in particular implies that the first passage probabilities are close.

The mathematical result behind is Donsker's theorem for a simple random walk $\{S_n^*\}_{n=0,1,\ldots}$ in discrete time: if $\mu = \mathbb{E}S_1^*$ is the drift and $\sigma^2 = \mathbb{V}ar(S_1^*)$ the variance, then

$$\left\{\frac{1}{\sigma\sqrt{c}}(S_{\lfloor tc\rfloor}^* - tc\mu)\right\}_{t\geq 0} \xrightarrow{\mathscr{D}} \{W_0(t)\}_{t\geq 0}, \quad c \to \infty, \tag{5.1}$$

where $\{W_\zeta(t)\}$ is Brownian motion with drift ζ and variance (diffusion constant) 1 (here $\xrightarrow{\mathscr{D}}$ refers to weak convergence in $D = D[0, \infty)$).

It is fairly straightforward to translate Donsker's theorem into a parallel statement for continuous time random walks (Lévy processes), of which a particular case is the claim surplus process (see the proof of Theorem 5.1 below). However, for the purpose of approximating ruin probabilities the centering around the mean (the $tc\mu$ term in (5.1)) is inconvenient. We want an approximation of the claim surplus process itself, and this can be obtained under the assumption that the safety loading η is small and positive. *This is the regime of the diffusion approximation* (note that this is just the same as for the heavy traffic approximation for infinite horizon ruin probabilities studied in IV.7c).

Mathematically, we shall represent this assumption on η by a family $\left\{S_t^{(p)}\right\}_{t\geq 0}$ of claim surplus processes indexed by the premium rate p, such that the claim size distribution B and the Poisson rate β are the same for all p (i.e., $S_t = \sum_{i=1}^{N_t} U_i - tp$), and consider the limit $p \downarrow \rho$, where ρ is the critical premium rate $\beta\mu_B$.

Theorem 5.1 *As $p \downarrow \rho$, we have*

$$\left\{ \frac{|\mu|}{\sigma^2} S^{(p)}_{t\sigma^2/\mu^2} \right\}_{t \geq 0} \quad \overset{\mathscr{D}}{\to} \quad \{W_{-1}(t)\}_{t \geq 0} \tag{5.2}$$

where $\mu = \mu_p = \rho - p$, $\sigma^2 = \beta \mu_B^{(2)}$.

Proof. The first step is to note that

$$\left\{ \frac{1}{\sigma\sqrt{c}} \left(S^{(p)}_{tc} - tc\mu_p \right) \right\} \;=\; \left\{ \frac{1}{\sigma\sqrt{c}} \left(S^{(\rho)}_{ct} - \rho ct \right) \right\} \overset{\mathscr{D}}{\to} \{W_0(t)\} \tag{5.3}$$

whenever $c = c_p \uparrow \infty$ as $p \downarrow \rho$. Indeed, this is an easy consequence of (5.1) with $S_n^* = S_n^{(p)}$ and the inequalities

$$S^{(\rho)}_{n/c} - \rho/c \;\leq\; S^{(\rho)}_t \;\leq\; S^{(\rho)}_{(n+1)/c} + \rho p/c, \quad n/c \leq t \leq (n+1)/c,$$

cf. Lemma IV.1.3.

Letting $c = \sigma^2/\mu_p^2$, (5.3) takes the form

$$\left\{ \frac{|\mu|}{\sigma^2} S^{(p)}_{t\sigma^2/\mu^2} + t \right\} \overset{\mathscr{D}}{\to} \{W_0(t)\},$$

$$\left\{ \frac{|\mu|}{\sigma^2} S^{(p)}_{t\sigma^2/\mu^2} \right\} \overset{\mathscr{D}}{\to} \{W_0(t) - t\} = \{W_{-1}(t)\}.$$

$\square$

Now let

$$\tau_p(u) \;=\; \inf\{t \geq 0 : S^{(p)}_t > u\}, \quad \tau_\zeta(u) \;=\; \inf\{t \geq 0 : W_\zeta(t) > u\}.$$

It is well-known (Corollary III.1.6 or [APQ, p. 263]) that the distribution $\text{IG}(\cdot; \zeta; u)$ of $\tau_\zeta(u)$ (often referred to as the *inverse Gaussian distribution*) is given by

$$\text{IG}(x; \zeta; u) \;=\; \mathbb{P}\big(\tau_\zeta(u) \leq x\big) \;=\; 1 - \Phi\left(\frac{u}{\sqrt{x}} - \zeta\sqrt{x} \right) + e^{2\zeta u}\Phi\left(-\frac{u}{\sqrt{x}} - \zeta\sqrt{x} \right). \tag{5.4}$$

Note that $\text{IG}(\cdot; \zeta; u)$ is defective when $\zeta < 0$.

Corollary 5.2 *As $p \downarrow \rho$,*

$$\psi_p\left(\frac{u\sigma^2}{|\mu|}, \frac{T\sigma^2}{\mu^2} \right) \;\to\; \text{IG}(T; -1; u).$$

Proof. Since $f \to \sup_{0 \le t \le T} f(t)$ is continuous on D a.e. w.r.t. any probability measure concentrated on the continuous functions, the continuous mapping theorem yields

$$\sup_{0 \le t \le T} \frac{|\mu|}{\sigma^2} S^{(p)}_{t\sigma^2/\mu^2} \xrightarrow{\mathscr{D}} \sup_{0 \le t \le T} W_{-1}(t).$$

Since the r.h.s. has a continuous distribution, this implies

$$\mathbb{P}\left(\sup_{0 \le t \le T} \frac{|\mu|}{\sigma^2} S^{(p)}_{t\sigma^2/\mu^2} > u\right) \to \mathbb{P}\left(\sup_{0 \le t \le T} W_{-1}(t) > u\right).$$

But the l.h.s. is $\psi_p\left(u\sigma^2/|\mu|, T\sigma^2/\mu^2\right)$, and the r.h.s. is $\mathrm{IG}(T; -1; u)$. □

For practical purposes, Corollary 5.2 suggests the approximation

$$\psi(u, T) \approx \mathrm{IG}(T\mu^2/\sigma^2; \, u|\mu|/\sigma^2). \tag{5.5}$$

Note that letting $T \to \infty$ in (5.5), we obtain formally the approximation

$$\psi(u) \approx \mathrm{IG}(\infty; \, u|\mu|/\sigma^2) = \mathrm{e}^{-2u|\mu|/\sigma^2}. \tag{5.6}$$

This is the same as the heavy-traffic approximation derived in IV.7c. However, since $\psi(u)$ has infinite horizon, the continuity argument above does not generalize immediately, and in fact some additional arguments are needed to justify (5.6) from Theorem 5.1. Because of the direct argument in Chapter IV, we omit the details; see Grandell [426], [427] or [APQ, pp. 196, 199].

Checks of the numerical fits of (5.5) and (5.6) are presented, e.g., in Asmussen [55]. The picture which emerges is that the approximations are not terribly precise, in particular for large u. In view of the excellent fit of the Cramér-Lundberg approximation, (5.6) therefore does not appear to be of much practical relevance for the compound Poisson model. However, for more general models it may be easier to generalize the diffusion approximation than the Cramér-Lundberg approximation; as an example of such a generalization we mention the paper [342] by Emanuel *et al.* on the premium rule involving interest. In contrast, the simplicity of (5.5) combined with the fact that finite horizon ruin probabilities are so hard to deal with even for the compound Poisson model makes this approximation more appealing. However, in the next subsection we shall derive a refinement of (5.5) for the compound Poisson model which does not require much more computation, and which is much more precise.

We conclude this section by giving a more general triangular array version of Theorem 5.1. The proof is a straightforward combination of the proof of Theorem 5.1 and Section VIII.6 of [APQ].

Theorem 5.3 *Consider a family* $\left\{S_t^{(\theta)}\right\}$ *of claim surplus processes indexed by a parameter* θ, *such that the Poisson rate* β_θ, *the claim size distribution* B_θ *and the premium rate* p_θ *depends on* θ. *Assume further that* $\beta_\theta \mu_{B_\theta} < p_\theta$, *that*

$$\beta_\theta \to \beta_{\theta_0}, \quad B_\theta \xrightarrow{\mathscr{D}} B_{\theta_0}, \quad p_\theta \to p_{\theta_0}, \quad p_\theta - \beta_\theta \mu_{B_\theta} \to 0,$$

as $\theta \to \theta_0$ *and that the* U^2 *are uniformly integrable w.r.t. the* B_θ. *Then as* $\theta \to \theta_0$, *we have*

$$\left\{ \frac{|\mu|}{\sigma^2} S_{t\sigma^2/\mu^2}^{(\theta)} \right\}_{t\geq 0} \xrightarrow{\mathscr{D}} \left\{W_{-1}(t)\right\}_{t\geq 0} \tag{5.7}$$

where $\mu = \mu_\theta = \rho_\theta - p_\theta = \beta_\theta \mu_{B_\theta} - p_\theta$, $\sigma^2 = \sigma_\theta^2 = \beta_\theta \mu_{B_\theta}^{(2)}$.

Notes and references   Diffusion approximations of random walks via Donsker's theorem is a classical topic of probability theory. See for example Billingsley [167]. The first application in risk theory is Iglehart [492], and two further standard references in the area are Grandell [426], [427]. All material of this section can be found in these references.

For claims with infinite variance, Furrer, Michna & Weron [383] suggested an approximation by a stable Lévy process rather than a Brownian motion. Further relevant references in this direction are Furrer [382], Boxma & Cohen [194] and Whitt [883].

6   Corrected diffusion approximations

The idea behind the simple diffusion approximation is to replace the risk process by a Brownian motion (by fitting the two first moments) and use the Brownian first passage probabilities as approximation for the ruin probabilities. Since Brownian motion is skip-free, this idea ignores (among other things) the presence of the overshoot $\xi(u)$, which we have seen to play an important role for example for the Cramér-Lundberg approximation. The objective of the corrected diffusion approximation is to take this and other deficits into consideration.

The set-up is the exponential family of compound risk processes with parameters β_θ, B_θ constructed in IV.4. However, whereas there we let the given risk process with safety loading $\eta > 0$ correspond to $\theta = 0$, it is more convenient here to use some value $\theta_0 < 0$ and let $\theta = 0$ correspond to $\eta = 0$ (zero drift); this is because in the regime of the diffusion approximation, η is close to zero, and we want to consider the limit $\eta \downarrow 0$ corresponding to $\theta_0 \uparrow 0$.

In terms of the given risk process with Poisson intensity β, claim size distribution B, $\kappa(\alpha) = \beta\big(\widehat{B}[\alpha] - 1\big) - \alpha$ and $\rho = \beta\mu_B < 1$, $\eta = 1/\rho - 1 > 0$, this means the following:

1. Determine $\gamma_m > 0$ by $\kappa'(\gamma_m) = 0$ and let $\theta_0 = -\gamma_m$.

2. Let $\mathbb{P}_0$ refer to the risk process with parameters

$$\beta_0 = \beta\widehat{B}[-\theta_0], \quad B_0(\mathrm{d}x) = \frac{\mathrm{e}^{-\theta_0 x}}{\widehat{B}[-\theta_0]}B(\mathrm{d}x).$$

Then $\mathbb{E}_0 U^k = \widehat{B}_0^{(k)}[0] = \widehat{B}^{(k)}[-\theta_0]/\widehat{B}[-\theta_0]$ and $\kappa_0(s) = \kappa(s-\theta_0) - \kappa(-\theta_0)$, $\kappa_0'(0) = 0$.

3. For each θ, let $\mathbb{P}_\theta$ refer to the risk process with parameters

$$\beta_\theta = \beta_0\widehat{B}_0[\theta] = \beta\widehat{B}[\theta-\theta_0], \quad B_\theta(\mathrm{d}x) = \frac{\mathrm{e}^{\theta x}}{\widehat{B}_0[\theta]}B_0(\mathrm{d}x) = \frac{\mathrm{e}^{(\theta-\theta_0)x}}{\widehat{B}[\theta-\theta_0]}B(\mathrm{d}x).$$

Then

$$\kappa_\theta(s) = \kappa_0(s+\theta) - \kappa_0(\theta) = \kappa(s+\theta-\theta_0) - \kappa(\theta-\theta_0)$$

and the given risk process corresponds to $\mathbb{P}_{\theta_0}$ where $\theta_0 = -\gamma_m$.

In this set-up, $\mathbb{P}_\theta(\tau(u) < \infty) = 1$ for $\theta \geq 0$, $\mathbb{P}_\theta(\tau(u) < \infty) < 1$ for $\theta < 0$, and we are studying $\psi(u, T) = \mathbb{P}_{\theta_0}(\tau(u) \leq T)$ for $\theta_0 < 0$, $\theta_0 \uparrow 0$.

Recall that $\mathrm{IG}(x; \zeta; u)$ denotes the distribution function of the passage time of Brownian motion $\{W_\zeta(t)\}$ with unit variance and drift ζ from level 0 to level $u > 0$. One has

$$\mathrm{IG}(x; \zeta; u) = \mathrm{IG}(x/u^2; \zeta u; 1). \tag{6.1}$$

The corrected diffusion approximation to be derived is

$$\psi(u, T) \approx \mathrm{IG}\left(\frac{T\nu_1}{u^2} + \frac{\nu_2}{u}; -\frac{\gamma u}{2}; 1 + \frac{\nu_2}{u}\right) \tag{6.2}$$

where as ususal $\gamma > 0$ is the adjustment coefficient for the given risk process, i.e. the solution of $\kappa(\gamma) = 0$, and

$$\nu_1 = \beta_0\mathbb{E}_0 U^2 = \beta\widehat{B}''[\gamma_m], \quad \nu_2 = \frac{\mathbb{E}_0 U^3}{3\mathbb{E}_0 U^2} = \frac{\widehat{B}'''[\gamma_m]}{3\widehat{B}''[\gamma_m]}.$$

Write the initial reserve u for the given risk process as $u = \zeta/\theta_0$ (note that $\zeta < 0$) and, for brevity, write $\tau = \tau(u)$, $\xi = \xi(u) = S_\tau - u$. The first step in the derivation is to note that

$$\mu = \kappa'(0) = \kappa_0'(\theta_0) \sim \theta_0\kappa_0''(0) = \theta_0\nu_1 = \frac{\zeta\nu_1}{u},$$

$$\mathbb{V}ar_{\theta_0}S_1 \sim \mathbb{V}ar_0 S_1 = \beta_0\mathbb{E}_0 U^2 = \nu_1, \quad \theta_0 \uparrow 0.$$

Theorem 5.3 applies and yields

$$\left\{ \frac{|\zeta|\nu_1}{u\nu_1} S_{t\nu_1 u^2/\zeta^2\nu_1^2} \right\}_{t\geq 0} \xrightarrow{\mathcal{D}} \{W_{-1}(t)\}_{t\geq 0}$$

which easily leads to

$$\left\{ \frac{1}{u\sqrt{\nu_1}} S_{tu^2} \right\}_{t\geq 0} \xrightarrow{\mathcal{D}} \{W_{\zeta\sqrt{\nu_1}}(t)\}_{t\geq 0},$$

$$\psi(u, tu^2) \to \mathrm{IG}\left(t; \zeta\sqrt{\nu_1}; \frac{1}{\sqrt{\nu_1}} \right) = \mathrm{IG}(t\nu_1; \zeta; 1).$$

Since

$$\int_0^\infty e^{-\alpha t} \mathrm{IG}(dt; \zeta, u) = e^{-uh(\lambda,\zeta)} \quad \text{where} \quad h(\lambda, \zeta) = \sqrt{2\lambda + \zeta^2} - \zeta, \qquad (6.3)$$

this implies (take $u = 1$)

$$\mathbb{E}_{\theta_0} \exp\{ -\lambda\nu_1\tau(u)/u^2 \} \to e^{-h(\lambda,\zeta)}. \qquad (6.4)$$

The idea of the proof is to improve upon this by an $O(u^{-1})$ term (in the following, $\approx$ means up to $o(u^{-1})$ terms):

Proposition 6.1 *As $u \to \infty$, $\theta_0 \uparrow 0$ in such a way that $\zeta = \theta_0 u$ is fixed, it holds for any fixed $\lambda > 0$ that*

$$\mathbb{E}_{\theta_0} \exp\{ -\lambda\nu_1\tau(u)/u^2 \} \approx \exp\{ -h(\lambda, -\gamma u/2)(1 + \nu_2/u) \}\left[1 + \frac{\lambda\nu_2}{u} \right]. \qquad (6.5)$$

Once this is established, we get by formal Laplace transform inversion that

$$\psi\left(u, \frac{tu^2}{\nu_1} \right) \approx \mathrm{IG}\left(t + \frac{\nu_2}{u}; -\frac{\gamma u}{2}; 1 + \frac{\nu_2}{u} \right).$$

Indeed, the r.h.s. is the c.d.f. of a (defective) r.v. distributed as $Z - \nu_2/u$ where Z has distribution $\mathrm{IG}(\cdot; -\gamma u/2; 1 + \nu_2/u)$. But the Laplace transform of such a r.v. is

$$\mathbb{E}e^{-\lambda Z}e^{\lambda\nu_2/u} \approx \mathbb{E}e^{-\lambda Z}[1 + \lambda\nu_2/u]$$

where the last expression coincides with the r.h.s. of (6.5) according to (6.3). To arrive at (6.2), just replace t by $T\nu_1/u^2$.

Note, however, that whereas the proof of Proposition 6.1 below is exact, the formal Laplace transform inversion is heuristic: an additional argument would be required to infer that the remainder term in (6.2) is indeed $o(u^{-1})$. The

justification for the procedure is the wonderful numerical fit which has been
found in numerical examples and which for a small or moderate safety loading
η is by far the best among the various available approximations [note, however,
that the saddlepoint approximation of Barndorff-Nielsen & Schmidli [138] is a
serious competitor and is in fact preferable if η is large]. A numerical illustration
is given in Fig. V.5, which is based upon exponential claims with mean $\mu_B = 1$.
The solid line represents the exact value, calculated using numerical integration
and Proposition 1.3, and the dotted line the corrected diffusion approximation
(6.2). In (1) and (2), we have $\rho = \beta = 0.7$, in (3) and (4), $\rho = 0.4$. The initial
reserve u has been selected such that the infinite horizon ruin probability $\psi(u)$
is 10% in (1) and (3), 1% in (2) and (4).

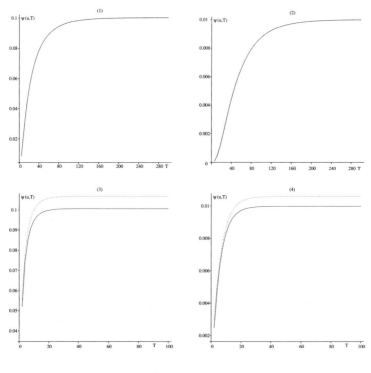

FIGURE V.5

It is seen that the numerical fit is extraordinary for $\rho = 0.7$. Note that the
ordinary diffusion approximation requires ρ to be close to 1 and $\psi(u)$ to be not
too small, and all of the numerical studies the authors knows of indicate that
its fit at $\rho = 0.7$ or at values of $\psi(u)$ like 1% is unsatisfying. Similarly, the fit
at $\rho = 0.4$ may not be outstanding but nevertheless, it gives the right order

of magnitude and the ordinary diffusion approximation hopelessly fails for this value of ρ. For further numerical illustrations, see Asmussen [55], Barndorff-Nielsen & Schmidli [138] and Asmussen & Højgaard [81].

The proof of Proposition 6.1 proceeds in several steps.

Lemma 6.2 $e^{-h(\lambda,\zeta)} \approx \mathbb{E}_{\theta_0} \exp\left\{h(\lambda,\zeta)\dfrac{\xi}{u} - \dfrac{\lambda\nu_1\tau}{u^2} - \dfrac{\nu_1\nu_2\tau}{2u^3}(\widetilde{\theta}^3 - \zeta^3)\right\}.$

Proof. For $\widetilde{\theta} \geq 0$,

$$1 = \mathbb{P}_{\widetilde{\theta}}(\tau < \infty) = \mathbb{E}_{\theta_0} \exp\left\{(\widetilde{\theta} - \theta_0)(u + \xi) - \tau\big(\kappa_0(\widetilde{\theta}) - \kappa_0(\theta_0)\big)\right\}.$$

Replacing $\widetilde{\theta}$ by $\widetilde{\theta}/u$ and θ_0 by ζ/u yields

$$e^{-(\widetilde{\theta}-\zeta)} = \mathbb{E}_{\theta_0}\exp\left\{(\widetilde{\theta} - \zeta)\xi/u - \tau\big(\kappa_0(\widetilde{\theta}/u) - \kappa_0(\zeta/u)\big)\right\}.$$

Let $\widetilde{\theta} = (2\lambda + \zeta^2)^{1/2} = h(\lambda,\zeta) + \zeta$ and note that

$$\kappa_0(\theta) = \frac{1}{2}\theta^2\beta_0\mathbb{E}_0 U^2 + \frac{1}{6}\theta^3\beta_0\mathbb{E}_0 U^3 + \cdots = \frac{\nu_1\theta^2}{2} + \frac{\nu_1\nu_2\theta^3}{2} + \cdots. \tag{6.6}$$

Using $\widetilde{\theta}^2 - \zeta^2 = 2\lambda$, the result follows. $\qquad\square$

Lemma 6.3 $\displaystyle\lim_{u\to\infty} \mathbb{E}_0\xi(u) = \mathbb{E}_0\xi(\infty) = \nu_2 = \dfrac{\mathbb{E}_0 U^3}{3\mathbb{E}_0 U^2}.$

Proof. By partial integration, the formulas

$$\mathbb{P}_0\big(\xi(0) > x\big) = \mathbb{P}_0(S_{\tau(0)} > x) = \frac{1}{\mathbb{E}_0 U}\int_x^\infty \mathbb{P}_0(U > y)\,\mathrm{d}y\,,$$

$$\mathbb{P}_0\big(\xi(\infty) > x\big) = \frac{1}{\mathbb{E}_0\xi(0)}\int_x^\infty \mathbb{P}_0\big(\xi(0) > y\big)\,\mathrm{d}y$$

imply

$$\mathbb{E}_0\xi(0)^k = \frac{\mathbb{E}_0 U^{k+1}}{(k+1)\mathbb{E}_0 U}\,, \qquad \mathbb{E}_0\xi(\infty)^k = \frac{\mathbb{E}_0\xi(0)^{k+1}}{(k+1)\mathbb{E}_0\xi(0)}\,.$$

$\square$

Lemma 6.4 $\mathbb{E}_{\theta_0}\exp\left\{-\dfrac{\nu_1\tau}{u^2}\right\}$

$$\approx \exp\left\{-h(\lambda,\zeta)\Big(1 + \frac{\nu_2}{u}\Big)\right\}\left\{1 + \frac{\nu_2}{2u}\Big[2\lambda + \zeta^2 - \frac{\zeta^3}{(2\lambda+\zeta)^{1/2}}\Big]\right\}.$$

Proof. It follows by a suitable variant of Stam's lemma (Proposition 4.4) that the r.h.s. in Lemma 6.2 behaves like

$$\mathbb{E}_{\theta_0} \exp\left\{-\frac{\lambda \nu_1 \tau}{u^2}\right\}\left[1 + h(\lambda, \zeta)\frac{\xi}{u} - \frac{\nu_1 \nu_2 \tau}{2u^3}(\widetilde{\theta}^3 - \zeta^3)\right]$$

$$\approx \mathbb{E}_{\theta_0} \exp\left\{-\frac{\lambda \nu_1 \tau}{u^2}\right\} + e^{-h(\lambda, \zeta)} h(\lambda, \zeta)\frac{\nu_2}{u}$$

$$- \frac{\nu_2}{2u}(\widetilde{\theta}^3 - \zeta^3)\mathbb{E}_{\theta_0}\left[\frac{\nu_1 \tau}{u^2} \exp\left\{-\frac{\lambda \nu_1 \tau}{u^2}\right\}\right]. \tag{6.7}$$

The last term is approximately

$$\frac{\nu_2}{2u}(\widetilde{\theta}^3 - \zeta^3)\frac{\mathrm{d}}{\mathrm{d}\lambda}e^{-h(\lambda, \zeta)}$$

$$= -\frac{\nu_2}{2u}\left[2\lambda + \zeta^2 - \frac{\zeta^3}{(2\lambda + \zeta)^{1/2}}\right]e^{-h(\lambda, \zeta)}$$

$$\approx -\frac{\nu_2}{2u}\left[2\lambda + \zeta^2 - \frac{\zeta^3}{(2\lambda + \zeta)^{1/2}}\right]\exp\left\{-h(\lambda, \zeta)\left(1 + \frac{\nu_2}{u}\right)\right\}.$$

The result follows by combining Lemma 6.2 and (6.7) and using

$$e^{-h(\lambda,\zeta)} - e^{-h(\lambda,\zeta)} h(\lambda, \zeta)\frac{\nu_2}{u} \approx \exp\left\{-h(\lambda, \zeta)\left(1 + \frac{\nu_2}{u}\right)\right\}.$$

$\square$

The last step is to replace $h(\lambda, \zeta)$ by $h(\lambda, -\gamma u/2)$. There are two reasons for this: in this way, we get the correct asymptotic exponential decay parameter γ in the approximation (6.2) for $\psi(u)$ (indeed, letting formally $T \to \infty$ yields $\psi(u) \approx C'e^{-\gamma u}$ where $C' = e^{-\gamma \nu_2}$); and the correction terms which need to be added cancels conveniently with some of the more complicated expressions in Lemma 6.4.

Lemma 6.5 $\exp\left\{-h(\lambda, \zeta)\left(1 + \frac{\nu_2}{u}\right)\right\}$

$$\approx \exp\left\{-h(\lambda, -\gamma u/2)\left(1 + \frac{\nu_2}{u}\right)\right\}\left\{1 - \frac{\nu_2}{2u}\left[\frac{\zeta^3}{\sqrt{2\lambda + \zeta^2}} - \zeta^2\right]\right\}.$$

Proof. Use first (6.6) and $\kappa_0(\theta_0) = \kappa_0(\gamma + \theta_0)$ to get

$$0 = \frac{\nu_1}{2}(\gamma^2 + 2\gamma\theta_0) + \frac{\nu_1 \nu_2}{2}(\gamma^3 + 3\gamma^2\theta_0 + 3\gamma\theta_0^2) + \mathrm{O}(u^{-4}),$$

i.e.

$$\frac{\gamma}{2} + \theta_0 = -\frac{\nu_2}{2}(\gamma^2 + 3\gamma\theta_0 + 3\theta_0^2) + \mathrm{O}(u^{-3}).$$

Thus $\gamma = -2\theta_0 + O(u^{-2})$, and inserting this and $\theta_0 = \zeta/u$ on the r.h.s. yields

$$\frac{\gamma}{2} + \theta_0 = -\frac{\nu_2\theta_0^2}{2} + O(u^{-3}) = -\frac{\nu_2\zeta^2}{2u^2} + O(u^{-3}).$$

Thus by Taylor expansion around $\zeta = \theta_0 u$, we get

$$h(\lambda, -\gamma u/2) \approx h(\lambda, \zeta) - \left[\frac{\zeta}{\sqrt{2\lambda + \zeta^2}} - 1\right](\gamma/2 + \theta_0)u,$$

$$\exp\left\{-h(\lambda, -\gamma u/2)\left(1 + \frac{\nu_2}{u}\right)\right\}$$

$$\approx \exp\left\{-h(\lambda, \zeta)\left(1 + \frac{\nu_2}{u}\right)\right\}\left\{1 - \left(1 + \frac{\nu_2}{u}\right)\left[\frac{\zeta}{\sqrt{2\lambda + \zeta^2}} - 1\right](\gamma/2 + \theta_0)u\right\}$$

$$\approx \exp\left\{-h(\lambda, \zeta)\left(1 + \frac{\nu_2}{u}\right)\right\}$$

$$\qquad - \left(1 + \frac{\nu_2}{u}\right)\left[\frac{\zeta}{\sqrt{2\lambda + \zeta^2}} - 1\right]\frac{\nu_2\zeta^2}{2u}\exp\{-h(\lambda, \zeta)\}$$

$$\approx \exp\left\{-h(\lambda, \zeta)\left(1 + \frac{\nu_2}{u}\right)\right\} - \frac{\nu_2}{2u}\left[\frac{\zeta^3}{\sqrt{2\lambda + \zeta^2}} - \zeta^2\right]\exp\{-h(\lambda, \zeta)\}.$$

$\square$

Proof of Proposition 6.1: Just insert Lemma 6.5 in Lemma 6.4. $\square$

Notes and references Corrected diffusion approximations were introduced by Siegmund [809] in a discrete random walk setting, with the translation to risk processes being carried out by Asmussen [55]; this case is in part simpler than the general random walk case because the ladder height distribution G_+ can be found explicitly (as ρB_0) which avoids the numerical integration involving characteristic functions which was used in [809] to determine the constants.

In Siegmund's book [810], the approach to the finite horizon case is in part different and uses local central limit theorems. The adaptation to risk theory has not been carried out.

The corrected diffusion approximation was extended to the renewal model in Asmussen & Højgaard [81], and to the Markov-modulated model of Chapter VII in Asmussen [58]; Fuh [379] considers the closely related case of discrete time Markov additive processes.

Hogan [473] considered a variant of the corrected diffusion approximation which does not require exponential moments. His ideas were adapted by Asmussen & Binswanger [72] to derive approximations for the infinite horizon ruin probability $\psi(u)$ when claims are heavy-tailed; the analogous analysis of finite horizon ruin probabilities $\psi(u, T)$ has not been carried out and seems non-trivial.

For corrected diffusion approximations with higher-order terms, see Blanchet & Glynn [174]; their results also cover some heavy-tailed cases.

7   How does ruin occur?

We saw in Section 4 that given that ruin occurs, the 'typical' value (say in sense of the conditional mean) was um_L, that is, the same as for the unconditional Lundberg process. We shall now generalize this question by asking what a sample path of the risk process looks like given it leads to ruin. The answer is similar: the process behaved as if it changed its whole distribution to $\mathbb{P}_L$, i.e. changed its arrival rate from β to β_L and its claim size distribution from B to B_L. Recall that $\mathscr{F}_{\tau(u)}$ is the stopping time σ-algebra carrying all relevant information about $\tau(u)$ and $\{S_t\}_{0 \le t \le \tau(u)}$. Define $\mathbb{P}^{(u)} = \mathbb{P}(\cdot|\tau(u) < \infty)$ as the distribution of the risk process given ruin with initial reserve u. We are concerned with describing the $\mathbb{P}^{(u)}$-distribution of $\{S_t\}_{0 \le t \le \tau(u)}$ (note that the behavior after $\tau(u)$ is trivial: by the strong Markov property, $\{S_{\tau(u)+t} - S_{\tau(u)}\}_{t \ge 0}$ is just an independent copy of $\{S_t\}_{t \ge 0}$).

Theorem 7.1 *Let* $\{F(u)\}_{u \ge 0}$ *be any family of events with* $F(u) \in \mathscr{F}_{\tau(u)}$ *and satifying* $\mathbb{P}_L F(u) \to 1$, $u \to \infty$. *Then also* $\mathbb{P}^{(u)} F(u) \to 1$.

Proof.

$$
\begin{aligned}
\mathbb{P}^{(u)} F(u)^c &= \frac{\mathbb{P}\big(F(u)^c; \tau(u) < \infty\big)}{\mathbb{P}\big(\tau(u) < \infty\big)} = \frac{\mathbb{E}_L\big[e^{-\gamma S_{\tau(u)}}; F(u)^c\big]}{\psi(u)} \\
&\le \frac{\mathbb{E}_L\big[e^{-\gamma u}; F(u)^c\big]}{\psi(u)} \sim \frac{e^{-\gamma u}\mathbb{P}_L F(u)^c}{Ce^{-\gamma u}} \to 0.
\end{aligned}
$$

$\square$

Corollary 7.2 *If* B *is exponential, then* $\mathbb{P}^{(u)}$ *and* $\mathbb{P}_L$ *coincide on*

$$
\mathscr{F}_{\tau(u)-} = \sigma\Big(\tau(u), \{S_t\}_{0 \le t < \tau(u)}\Big).
$$

Proof. Write $e^{-\gamma S_{\tau(u)}} = e^{-\gamma u}e^{-\gamma \xi(u)}$. In the exponential case, $\mathscr{F}_{\tau(u)-}$ and $\xi(u)$ are independent, so in the proof, the numerator becomes

$$
e^{-\gamma u}\mathbb{E}_L e^{-\gamma \xi(u)}\mathbb{P}_L(F(u)^c) = e^{-\gamma u}C\mathbb{P}_L(F(u)^c)
$$

when $F(u) \in \mathscr{F}_{\tau(u)-}$ and similarly the denominator is exactly equal to $Ce^{-\gamma u}$.
$\square$

Note that basically the difference between $\mathscr{F}_{\tau(u)}$ and $\mathscr{F}_{\tau(u)-}$ is that $\xi(u)$ is not $\mathscr{F}_{\tau(u)-}$-measurable. In fact, $\xi(u)$ is exponential with rate ν w.r.t. $\mathbb{P}^{(u)}$ and rate $\beta = \nu_L$ w.r.t. $\mathbb{P}_L$.

As example, we give a typical application of Theorem 7.1, stating roughly that under $\mathbb{P}^{(u)}$, the Poisson rate changes from β to β_L and the claim size distribution from B to B_L. Recall that $\beta_L = \beta \widehat{B}[\gamma]$ and $B_L(\mathrm{d}x) = \mathrm{e}^{\gamma x} B(\mathrm{d}x)/\widehat{B}[\gamma]$, and let $M(u)$ be the index of the claim leading to ruin (thus $\tau(u) = T_1 + T_2 + \cdots + T_{M(u)}$).

Corollary 7.3

$$\frac{1}{M(u)} \sum_{k=1}^{M(u)} I(T_k \leq x) \overset{\mathbb{P}^{(u)}}{\to} 1 - \mathrm{e}^{-\beta_L x},$$

$$\frac{1}{M(u)} \sum_{k=1}^{M(u)} I(U_k \leq x) \overset{\mathbb{P}^{(u)}}{\to} B_L(x).$$

Proof. For the first assertion, take

$$F(u) = \left\{ \left| \frac{1}{M(u)} \sum_{k=1}^{M(u)} I(T_k \leq x) - (1 - \mathrm{e}^{-\beta_L x}) \right| > \epsilon \right\}.$$

The proof of the second is similar. □

We finally consider the limiting joint distribution of

$$\zeta(u) = u - S_{\tau(u)-} = R_{\tau(u)-} \quad \text{and} \quad \xi(u) = S_{\tau(u)} - u = -R_{\tau(u)}$$

(the surplus prior to ruin, resp. the deficit at ruin).

Proposition 7.4 *Under the conditions of the Cramér-Lundberg approximation, $(\zeta(u), \xi(u))$ has a proper limit $(\zeta(\infty), \xi(\infty))$ as $u \to \infty$ in $\mathbb{P}^{(u)}$-distribution. The limiting distribution is given by*

$$\mathbb{P}(\zeta(\infty) \in \mathrm{d}x, \, \xi(\infty) \geq y) = \frac{\beta}{1 - \rho} \overline{B}(x + y)(\mathrm{e}^{\gamma x} - 1).$$

Proof. Define $Z(u) = \mathbb{P}(\zeta(u) \in \mathrm{d}x, \, \xi(u) \geq y)$. Consider first the case where $\zeta(u) \in \mathrm{d}x, \, \xi(u) \geq y$ occurs in the first ladder step, illustrated in Fig. V.6.

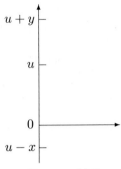

FIGURE V.6

For occurrence in the first ladder step, we need $x > u$ and $\zeta(0) = u - y$. Also, by Theorem IV.2.2 the (defective) density of $\zeta(0)$ is $\beta\overline{B}(z)$ and if $\zeta(0) = 0$, the available information on $\xi(0)$ given $\zeta(0) = z$ is that it has the distribution of a claim U conditioned to exceed z. Thus, taking $z = u - x$, we get the contribution to $Z(u)$ from the first ladder step as

$$\beta\overline{B}(y-u)\frac{\overline{B}(x+y)}{\overline{B}(y-u)}I(u < x) \;=\; \beta\overline{B}(x+y)I(u < x)\,.$$

If $\xi(0) = z < u$, everything repeats itself from z, and thus, since also $\xi(0)$ has density $\beta\overline{B}(z)$,

$$Z(u) \;=\; \beta\overline{B}(x+y)I(u < x) \;+\; \int_0^u Z(u-z)\beta\overline{B}(z)\,\mathrm{d}z\,.$$

This is a defective renewal equation, and since $\kappa(\gamma) = 0$ implies $\int_0^\infty e^{\gamma v}\beta\overline{B}(v)\,\mathrm{d}v = 1$, the usual exponential technique gives that $e^{\gamma u}Z(u)$ has the limit

$$\beta\overline{B}(x+y)\int_0^x e^{\gamma u}\,\mathrm{d}u \Big/ \int_0^u \beta\overline{B}(v)v e^{\gamma v}\,\mathrm{d}v \;=\; \beta\overline{B}(x+y)\frac{e^{\gamma x}-1}{\gamma}\frac{\gamma C}{1-\rho}\,,$$

where C is the Cramér-Lundberg constant (for the last equality, see the calculations around IV.(5.8) and IV.(5.11)). Thus $Z(u)/Ce^{-\gamma u}$ has the asserted limit which is what was to show.

That the limit is proper follows from

$$\int_0^\infty \frac{\beta}{1-\rho}\overline{B}(x)(e^{\gamma x}-1)\,\mathrm{d}x$$

$$= \frac{1}{1-\rho}\Big(\int_0^\infty \beta\overline{B}(x)e^{\gamma x}\,\mathrm{d}x \;-\; \rho\Big) \;=\; \frac{1}{1-\rho}\Big(\frac{\beta}{\gamma}\big(\widehat{B}[\gamma]-1\big) \;-\; \rho\Big)$$

$$= \frac{1}{1-\rho}(1-\rho) \;=\; 1\,. \qquad\qquad \square$$

Notes and references Proposition 7.4 can be found in Schmidli [773]. It will be shown in X.4 that in the heavy-tailed case, $\zeta(u)$ and $\xi(u)$ need to be scaled down before a conditional limit can be obtained.

The remaining results of the present section are part of a more general study carried out by the first author [54]. A somewhat similar study was carried out in the queueing setting by Anantharam [48], who also treated the heavy-tailed case; however, the queueing results are of a somewhat different type because of the presence of reflection at 0.

From a mathematical point of view, the subject treated in this section leads into the area of large deviations theory. This is currently a very active area of research, see further XIII.1.

Convergence properties of empirical finite-time ruin probabilities are investigated in Loisel, Mazza & Rullière [608].

Schmidli [781] derives some exact expressions for the distribution of the risk process given that ruin occurs. These may be seen as special h-transform calculations for piecewise deterministic Markov processes (recall that the distribution of a Markov chain or Markov process conditioned to hit a set is always a h-tranform, see e.g. Asmussen & Glynn [79, VI.7]).

Chapter VI

Renewal arrivals

1 Introduction

The basic assumption of this chapter states that the arrival epochs $\sigma_1, \sigma_2, \ldots$ of the risk process form a renewal process: letting $T_n = \sigma_n - \sigma_{n-1}$ ($T_1 = \sigma_1$), the T_n are independent, with the same distribution A (say) for $T_2, T_3, \ldots$ In the so-called *zero-delayed* case, the distribution A_1 of T_1 is A as well. A different important possibility is A_1 to be the stationary delay distribution A_0 with density $\overline{A}(x)/\mu_A$. Then the arrival process is stationary which could be a reasonable assumption in many cases (for these and further basic facts from renewal theory, see A.1).

We use much of the same notation as in Chapter I. Thus the premium rate is 1, the claim sizes $U_1, U_2, \ldots$ are i.i.d. with common distribution B, $\{S_t\}$ is the claim surplus process given by I.(1.5), with

$$N_t = \#\{n : \sigma_n \le t\}$$

the number of arrivals before t, and M is the maximum of $\{S_t\}$, $\tau(u)$ the time to ruin. The ruin probability corresponding to the zero-delayed case is denoted by $\psi(u)$, the one corresponding to the stationary case by $\psi^{(s)}(u)$, and the one corresponding to $T_1 = s$ by $\psi_s(u)$.

Proposition 1.1 *Define $\rho = \dfrac{\mu_B}{\mu_A}$. Then regardless of the distribution A_1 of T_1,*

$$\lim_{t \to \infty} \frac{S_t}{t} = \lim_{t \to \infty} \mathbb{E}\frac{S_t}{t} = \rho - 1, \tag{1.1}$$

$$\lim_{t \to \infty} \frac{\mathbb{V}ar(S_t)}{t} = \frac{\mu_B^2 \sigma_A^2 + \mu_A^2 \sigma_B^2}{\mu_A^3}. \tag{1.2}$$

Furthermore for any $a > 0$,

$$\lim_{t\to\infty} \mathbb{E}\left[S_{t+a} - S_t\right] = a(\rho - 1). \qquad (1.3)$$

Proof. Obviously,

$$\mathbb{E}S_t = \mathbb{E}\,\mathbb{E}\left[\sum_{i=1}^{N_t} U_i \,\Big|\, N_t\right] - t = \mathbb{E}N_t \cdot \mu_B - t.$$

However, by the elementary renewal theorem (cf. A.1) $\mathbb{E}N_t/t \to 1/\mu_A$. From this (1.1) follows, and (1.3) follows similarly by Blackwell's renewal theorem, stating that $\mathbb{E}[N_{t+a} - N_t] \to a/\mu_A$.

For (1.2), we get in the same way by using known facts about $\mathbb{E}N_t$ and $\mathbb{V}ar\,N_t$ that

$$
\begin{aligned}
\mathbb{V}ar(S_t) &= \mathbb{V}ar\,\mathbb{E}\left[\sum_{i=1}^{N_t} U_i \,\Big|\, N_t\right] + \mathbb{E}\,\mathbb{V}ar\left[\sum_{i=1}^{N_t} U_i \,\Big|\, N_t\right] \\
&= \mathbb{V}ar(\mu_B N_t) + \mathbb{E}(\sigma_B^2 N_t) \\
&= t\mu_B^2 \frac{\sigma_A^2}{\mu_A^3} + t\frac{\sigma_B^2}{\mu_A} + \mathrm{o}(t).
\end{aligned}
$$

$\square$

Of course, Proposition 1.1 gives the desired interpretation of the constant ρ as the expected claims per unit time. Thus, the definition $\eta = 1/\rho - 1$ of the safety loading appears reasonable here as well.

The renewal model is often referred to as the *Sparre Andersen process*, after E. Sparre Andersen whose 1959 paper [816] was the first to treat renewal assumptions in risk theory in more depth. The simplest case is of course the Poisson case where A and A_1 are both exponential with rate β. This has a direct physical interpretation (a large portfolio with claims arising with small rates and independently). Here are two special cases of the renewal model with a similar direct interpretation (see also the discussion in I.3):

Example 1.2 (DETERMINISTIC ARRIVALS) If A is degenerate, say at a, one could imagine that the claims are recorded only at discrete epochs (say each week or month) and thus each U_n is really the accumulated claims over a period of length a. $\square$

Example 1.3 (SWITCHED POISSON ARRIVALS) Assume that the process has a random environment with two states ON, OFF, such that no arrivals occur in the off state, but the arrival rate in the ON state is $\beta > 0$. If the environment is Markovian with transition rate λ from ON to OFF and μ from OFF to ON, the interarrival times become i.i.d. (an arrival occurs necessarily in the ON state, and then the whole process repeats itself). More precisely, A is phase-type (Example I.2.4) with phase space {ON,OFF}, initial vector (1 0) and phase generator

$$\begin{pmatrix} -\beta - \lambda & \lambda \\ \mu & -\mu \end{pmatrix}.$$ □

However, in general the mechanism generating a renewal arrival process appears much harder to interpret in the risk theory context and therefore the relevance of the model has been questioned repeatedly, see the discussion in I.3. Nevertheless we will present at least some basic features of the renewal model, if only for the mathematical elegance of the subject, the fundamental connections to the theory of queues and random walks, and for historical reasons.

The following representation of the ruin probability (already discussed in Section IV.1) will be a basic vehicle for studying the ruin probabilities:

Proposition 1.4 *The ruin probabilities for the zero-delayed case can be represented as* $\psi(u) = \mathbb{P}\big(M^{(d)} > u\big)$ *where* $M^{(d)} = \max\{S_n^{(d)} : n = 0, 1, \ldots\}$ *with* $\{S_n^{(d)}\}$ *a discrete time random walk with increments distributed as the independent difference* $U - T$ *between a claim* U *and an interarrival time* T.

Proof. The essence of the argument is that ruin can only occur at claim times. The values of the claim surplus process just after claims has the same distribution as $\{S_n^{(d)}\}$. Since the claim surplus process $\{S_t\}$ decreases in between arrival times, we have

$$\max_{0 \le t < \infty} S_t = \max_{n=0,1,\ldots} S_n^{(d)}.$$

From this the result immediately follows. □

For later use, we note that the ruin probabilities for the delayed case $T_1 = s$ can be expressed in terms of the ones for the zero-delayed case as

$$\psi_s(u) = \overline{B}(u+s) + \int_0^{u+s} \psi(u+s-y)B(dy). \tag{1.4}$$

Indeed, the first term represents the probability $\mathbb{P}(U_1 - s > u)$ of ruin at the time s of the first claim whereas the second is $\mathbb{P}\big(\tau(u) < \infty, U_1 - s \le u\big)$, as follows easily by noting that the evolution of the risk process after time s is that of a renewal risk model with initial reserve $U_1 - s$. For the stationary case, integrate (1.4) w.r.t. A_0.

2   Exponential claims.   The compound Poisson model with negative claims

We first consider a variant of the compound Poisson model obtained essentially by sign-reversion. That is, the claims and the premium rate are negative so that the risk reserve process, resp. the claim surplus process are given by

$$R_t^* \;=\; u + \sum_{i=1}^{N_t^*} U_i^* - t, \quad S_t^* \;=\; t - \sum_{i=1}^{N_t^*} U_i^*,$$

where $\{N_t^*\}$ is a Poisson process with rate β^* (say) and the U_i^* are independent of $\{N_t\}$ and i.i.d. with common distribution B^* (say) concentrated on $(0, \infty)$. This model is sometimes referred to as the *dual risk model* in the literature[1]. A typical sample path of $\{R_t^*\}$ is illustrated in Fig. VI.1.

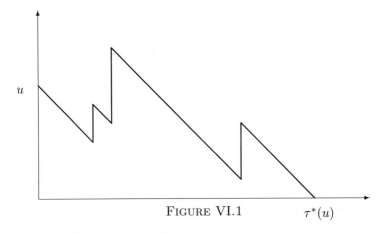

FIGURE VI.1 $\tau^*(u)$

One interpretation of the model is to have continuous expenses and events according to a Poisson process (e.g. innovations) which increase the value of the portfolio or company. Another interpretation is of course the workload in an M/G/1 queue in its first busy period.

Define the time of ruin $\tau^*(u) = \inf\{t > 0 : R_t^* < 0\}$. Using Lundberg conjugation, we shall be able to compute the ruin probability $\psi^*(u) = \mathbb{P}\big(\tau^*(u) < \infty\big)$ for this model very quickly. A simple sample path comparison will then provide us with the ruin probabilities for the renewal model with exponential claim size distribution.

---

[1]Although this terminology is not related to the other duality concepts used in this book.

Theorem 2.1 *If $\beta^*\mu_{B^*} \le 1$, then $\psi^*(u) = 1$ for all $u \ge 0$. If $\beta^*\mu_{B^*} > 1$, then $\psi^*(u) = e^{-\gamma u}$ where $\gamma > 0$ is the unique solution of*

$$0 = \kappa^*(-\gamma) = \beta^*\big(\widehat{B}^*[-\gamma] - 1\big) + \gamma. \tag{2.1}$$

[Note that $\kappa^*(\alpha) = \log \mathbb{E}e^{-\alpha S_1}$.]

Proof. Define

$$S_t^* = u - R_t^*, \quad \widetilde{S}_t = R_t^* - u = -S_t^*.$$

Then $\{\widetilde{S}_t\}$ is the claim surplus process of a standard compound Poisson risk process with parameters β^*, B^*. If $\beta^*\mu_{B^*} \le 1$, then by Proposition IV.1.2

$$\sup_{t \ge 0} S_t^* = -\inf_{t \ge 0} \widetilde{S}_t = \infty$$

and hence $\psi^*(u) = 1$ follows.

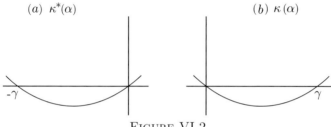

(a) $\kappa^*(\alpha)$ (b) $\kappa(\alpha)$

FIGURE VI.2

Assume now $\beta^*\mu_{B^*} > 1$. Then the function κ^* is defined on the whole of $(-\infty, 0)$ and has typically the shape on Fig. VI.2(a). Hence γ exists and is unique. Let

$$\beta = \beta^*\widehat{B}^*[-\gamma], \quad B(dx) = \frac{e^{-\gamma x}}{\widehat{B}^*[-\gamma]} B^*(dx),$$

and let $\{S_t\}$ be a compound Poisson risk process with parameters β, B. Then the c.g.f. of $\{S_t\}$ is $\kappa(\alpha) = \kappa^*(\alpha - \gamma)$, cf. Fig. VI.2(b), and the Lundberg conjugate of $\{S_t\}$ is $\{\widetilde{S}_t\}$ and vice versa. Define

$$\tau_-(u) = \inf\{t \ge 0 : S_t = -u\}, \quad \widetilde{\tau}_-(u) = \inf\{t \ge 0 : \widetilde{S}_t = -u = -u\}.$$

Since $\kappa'(0) < 0$, the safety loading of $\{S_t\}$ is > 0. Hence $\tau_-(u) < \infty$ a.s., and thus

$$1 = \mathbb{P}\big(\tau_-(u) < \infty\big) = \widetilde{\mathbb{E}}\big[e^{-\gamma \widetilde{S}_{\widetilde{\tau}_-(u)}}; \widetilde{\tau}_-(u) < \infty\big]$$
$$= e^{\gamma u}\mathbb{P}\big(\widetilde{\tau}_-(u) < \infty\big) = e^{\gamma u}\psi^*(u).$$

$\square$

Now return to the renewal model.

Theorem 2.2 *If B is exponential, with rate δ (say), and $\delta\mu_A > 1$, then $\psi(u) = \pi_+e^{-\gamma u}$ where $\gamma > 0$ is the unique solution of*

$$1 = \mathbb{E}e^{\gamma(U-T)} = \frac{\delta}{\delta - \gamma}\widehat{A}[-\gamma] \tag{2.2}$$

and $\pi_+ = 1 - \dfrac{\gamma}{\delta}$.

Proof. We can couple the renewal model $\{S_t\}$ and the compound Poisson model $\{S_t^*\}$ with negative claims in such a way that the interarrival times of $\{S_t^*\}$ are $T_0^*, T_1^* = U_1, T_2^* = U_2, \ldots$ Then $B^* = A$, $\beta^* = \delta$, and (2.1) means that $\delta(\widehat{A}[-\gamma] - 1) + \gamma = 0$ which is easily seen to be the same as (2.2).

Now the value of $\{S_t^*\}$ just before the nth claim is

$$T_0^* + T_1^* + \cdots + T_n^* - U_1^* - \cdots - U_n^*,$$

and from Fig. VI.1 it is seen that ruin is equivalent to one of these values being $> u$. Hence

$$
\begin{aligned}
M^* &= \max_{t \geq 0} S_t^* = \max_{n=0,1,\ldots} \left\{ T_0^* + T_1^* + \cdots + T_n^* - U_1^* - \cdots - U_n^* \right\} \\
&\overset{\mathscr{D}}{=} T_0^* + \max_{n=0,1,\ldots} \left\{ U_1 + \cdots + U_n - T_1 - \cdots - T_n \right\} \\
&\overset{\mathscr{D}}{=} T_0^* + M^{(d)}
\end{aligned}
$$

in the notation of Proposition 1.4.

Taking m.g.f.'s and noting that $\psi^*(u) = \mathbb{P}(M^* > u)$ so that Theorem 2.1 means that M^* is exponentially distributed with rate γ, we get

$$\mathbb{E}e^{\alpha M^{(d)}} = \frac{\mathbb{E}e^{\alpha M^*}}{\mathbb{E}e^{\alpha T_0^*}} = \frac{\gamma/(\gamma - \alpha)}{\delta/(\delta - \alpha)} = 1 - \pi_+ + \pi_+\frac{\gamma}{\gamma - \alpha}.$$

I.e. the distribution of $M^{(d)}$ is a mixture of an atom at zero and an exponential distribution with rate parameter γ with weights $1 - \pi_+$ and π_+, respectively. Hence $\mathbb{P}(M^{(d)} > u) = \pi_+e^{-\gamma u}$. $\qquad\square$

Remark 2.3 A variant of the last part of the proof, which has the advantage of avoiding transforms and leading up to the basic ideas of the study of the phase-type case in IX.4 goes as follows: define $\pi_+ = \mathbb{P}(M^{(d)} > 0)$ and consider $\{S_t^*\}$ only when the process is at a maximum value. According to Theorem 2.1, the failure rate of this process is γ. However, alternatively termination occurs at a jump time (having rate δ), with the probability that a particular jump time is not followed by any later maximum values being $1 - \pi_+$, and hence the failure rate is $\delta(1 - \pi_+)$. Putting this equal to γ, we see that $\gamma = \delta(1 - \pi_+)$ and hence $\pi_+ = 1 - \gamma/\delta$. However, consider instead the failure rate of $M^{(d)}$ and decompose $M^{(d)}$ into ladder steps as in III.6, IV.2. The probability that the first ladder step is finite is π_+. Furthermore, a ladder step is the overshoot of a claim size, hence exponential with rate δ. Thus a ladder step terminates at rate δ and is followed by one more with probability π_+. Hence the failure rate of $M^{(d)}$ is $\delta(1 - \pi_+) = \gamma$ and consequently $\mathbb{P}(M^{(d)} > u) = \mathbb{P}(M^{(d)} > 0)e^{-\gamma u} = \pi_+ e^{-\gamma u}$.

$\square$

3   Change of measure via exponential families

We shall discuss two points of view, the imbedded discrete time random walk and Markov additive processes.

3a   The imbedded random walk

The key steps have already been carried out in Corollary III.3.5, which states that for a given α, the relevant exponential change of measure corresponds to changing the distribution $F^{(d)}$ of $Y = U - T$ to

$$F_\alpha^{(d)}(x) \;=\; e^{-\kappa^{(d)}(\alpha)} \int_{-\infty}^x e^{\alpha y} F^{(d)}(dy)$$

where

$$\kappa^{(d)}(\alpha) \;=\; \log \widehat{F}^{(d)}[\alpha] \;=\; \log \widehat{B}[\alpha] + \log \widehat{A}[-\alpha]. \tag{3.1}$$

It only remains to note that this change of measure can be achieved by changing the interarrival distribution A and the claim distribution B to $A_\alpha^{(d)}$, resp. $B_\alpha^{(d)}$, where

$$A_\alpha^{(d)}(dt) \;=\; \frac{e^{-\alpha t}}{\widehat{A}[-\alpha]} A(dt), \quad B_\alpha^{(d)}(dx) \;=\; \frac{e^{\alpha x}}{\widehat{B}[\alpha]} B(dx).$$

This follows since, letting $\mathbb{P}_\alpha^{(d)}$ refer to the renewal risk model with these changed parameters, we have

$$
\begin{aligned}
\mathbb{E}_\alpha^{(d)} e^{\beta Y} &= \widehat{B}_\alpha^{(d)}[\beta] \widehat{A}_\alpha^{(d)}[-\beta] = \frac{\widehat{B}[\alpha + \beta]}{\widehat{B}[\alpha]} \cdot \frac{\widehat{A}[-\alpha - \beta]}{\widehat{A}[-\alpha]} \\
&= \frac{\widehat{F}^{(d)}[\alpha + \beta]}{\widehat{F}^{(d)}[\alpha]} = \widehat{F}_\alpha^{(d)}[\beta].
\end{aligned}
$$

Let
$$M(u) = \inf\{n = 1, 2, \ldots : S_n^{(d)} > u\}$$
be the number of claims leading to ruin and
$$\xi(u) = S_{\tau(u)} - u = S_{M(u)}^{(d)} - u$$
the overshoot, then we get:

Proposition 3.1 *For any α such that $\kappa^{(d)'}(\alpha) \geq 0$,*
$$\psi(u) = e^{-\alpha u}\mathbb{E}_\alpha^{(d)} e^{-\alpha\xi(u)+M(u)\kappa^{(d)}(\alpha)} .$$

Consider now the Lundberg case, i.e. let $\gamma > 0$ be the solution of $\kappa^{(d)}(\gamma) = 0$. We have the following versions of Lundberg's inequality and the Cramér-Lundberg approximation:

Theorem 3.2 *In the zero-delayed case,*
(a) $\psi(u) \leq e^{-\gamma u}$;
(b) $\psi(u) \sim Ce^{-\gamma u}$ where $C = \lim_{u\to\infty} \mathbb{E}_\gamma^{(d)} e^{-\gamma\xi(u)}$, *provided the distribution F of $U - T$ is non-lattice.*

Proof. Proposition 3.1 implies
$$\psi(u) = e^{-\alpha u}\mathbb{E}_\gamma^{(d)} e^{-\alpha\xi(u)} ,$$
and claim (a) follows immediately from this and $\xi(u) > 0$ a.s. For claim (b), just note that $F_\gamma^{(d)}$ is non-lattice when F is so. This is known to be sufficient for $\xi(0)$ to be non-lattice w.r.t. $\mathbb{P}_\gamma^{(d)}$ ([APQ, p. 222]) and thereby for $\xi(u)$ to converge in distribution, since $\mathbb{P}_\gamma^{(d)}(\tau(0) < \infty) = 1$ because of $\kappa^{(d)'}(\gamma) > 0$. $\qquad\square$

It should be noted that the computation of the Cramér-Lundberg constant C is much more complicated for the renewal case than for the compound Poisson case where $C = (1 - \rho)/(\beta\widehat{B}'[\gamma] - 1)$ is explicit given γ. In fact, in the easiest non-exponential case where B is phase-type, the evaluation of C is at the same level of difficulty as the evaluation of $\psi(u)$ in matrix-exponential form, cf. IX.4.

Corollary 3.3 *For the delayed case $T_1 = s$, $\psi_s(u) \sim C_s e^{-\gamma u}$ where $C_s = Ce^{-\gamma s}\widehat{B}[\gamma]$. For the stationary case, $\psi^{(s)}(u) \sim C^{(s)}e^{-\gamma u}$ where*
$$C^{(s)} = \frac{C(\widehat{B}[\gamma] - 1)}{\gamma\mu_A} .$$

Proof. Using (1.4), $\overline{B}(x) = \mathrm{o}(\mathrm{e}^{-\gamma x})$ and dominated convergence, we get

$$
\begin{aligned}
\mathrm{e}^{\gamma u}\psi_s(u) &= \mathrm{e}^{\gamma u}\overline{B}(u+s) + \int_0^{u+s} \mathrm{e}^{\gamma(y-s)}\mathrm{e}^{\gamma(u+s-y)}\psi(u+s-y)\,B(\mathrm{d}y) \\
&\to 0 + \int_0^\infty \mathrm{e}^{\gamma(y-s)}C\,B(\mathrm{d}y) = C_s.
\end{aligned}
$$

For the stationary case, another use of dominated convergence combined with $\widehat{A}_0[s] = (\widehat{A}[s] - 1)/s\mu_A$ yields

$$
\begin{aligned}
\mathrm{e}^{\gamma u}\psi^{(s)}(u) &= \int_0^\infty \mathrm{e}^{\gamma u}\psi_s(u)\,A_0(\mathrm{d}s) \to \int_0^\infty C\mathrm{e}^{-\gamma s}\widehat{B}[\gamma]\,A_0(\mathrm{d}s) \\
&= \frac{C\widehat{B}[\gamma]\big(\widehat{A}[-\gamma]-1\big)}{-\gamma\mu_A} = C^{(s)}.
\end{aligned}
$$

$\square$

Of course, a delayed version of Lundberg's inequality can be obtained in a similar manner. The expressions are slightly more complicated and we omit the details.

3b Markov additive representations

We take the Markov additive point of view of III.5. The underlying Markov process $\{J_t\}$ for the Markov additive process $\{X_t\} = \{(J_t, S_t)\}$ can be defined by taking J_t as the residual time until the next arrival. According to Remark III.4.9, we look for a function $h(s)$ and a κ (both depending on α) such that $\mathscr{G}h_\alpha(s,0) = \kappa(\alpha)h(s)$, where $\mathscr{G}$ is the generator of $\{X_t\} = \{(J_t, S_t)\}$ and $h_\alpha(s,y) = \mathrm{e}^{\alpha y}h(s)$. Let $\mathbb{P}_s, \mathbb{E}_s$ refer to the case $J_0 = s$. For $s > 0$,

$$
\mathbb{E}_s h_\alpha(J_{\mathrm{dt}}, S_{\mathrm{dt}}) = h(s - \mathrm{dt})\mathrm{e}^{-\alpha\,\mathrm{dt}} = h(s) - \mathrm{dt}\big(\alpha h(s) + h'(s)\big)
$$

so that $\mathscr{G}h_\alpha(s,0) = -\alpha h(s) - h'(s)$. Equating this to $\kappa h(s)$ and dividing by $h(s)$ yields $h'(s)/h(s) = -\alpha - \kappa$,

$$
h(s) = \mathrm{e}^{-(\alpha+\kappa(\alpha))s} \tag{3.2}
$$

(normalizing by $h(0) = 1$). To determine κ, we invoke the behavior at the boundary 0. Here

$$
1 = h_\alpha(0,0) = \mathbb{E}_0\big[h_\alpha(J_{\mathrm{dt}}, S_{\mathrm{dt}})\big] = \mathbb{E}\mathrm{e}^{\alpha U}h(T)
$$

means

$$
1 = \int_0^\infty \mathrm{e}^{\alpha y}B(\mathrm{d}y)\int_0^\infty h(s)A(\mathrm{d}s),
$$

i.e.

$$\widehat{B}[\alpha]\widehat{A}[-\alpha - \kappa(\alpha)] = 1. \tag{3.3}$$

As in III.5, we can now for each α define a new probability measure $\mathbb{P}_{\alpha;s}$ governing $\{(J_t, S_t)\}_{t \geq 0}$ by letting the likelihood ratio L_t restricted to $\mathscr{F}_t = \sigma\big((J_v, S_v) : 0 \leq v \leq t\big)$ be

$$L_t = \mathrm{e}^{\alpha S_t - t\kappa(\alpha)}\frac{h(J_t)}{h(s)} = \mathrm{e}^{\alpha S_t - t\kappa(\alpha)}\mathrm{e}^{-(\alpha + \kappa(\alpha))(J_t - s)}$$

where $\kappa(\alpha)$ is the solution of (3.3).

Proposition 3.4 *The probability measure $\mathbb{P}_{\alpha;s}$ is the probability measure governing a renewal risk process with $J_0 = s$ and the interarrival distribution A and the service time distribution B changed to A_α, resp. B_α where*

$$A_\alpha(\mathrm{d}t) = \frac{\mathrm{e}^{-(\alpha + \kappa(\alpha))t}}{\widehat{A}[-\alpha - \kappa(\alpha)]}A(\mathrm{d}t), \quad B_\alpha(\mathrm{d}x) = \frac{\mathrm{e}^{\alpha x}}{\widehat{B}[\alpha]}B(\mathrm{d}x).$$

Proof. $\mathbb{P}_{\alpha;s}(J_0 = s) = 1$ follows trivially from $L_0 = 1$. Further, since $J_{T_1} = J_s = T_2$,

$$\begin{aligned}
\mathbb{E}_{\alpha;s}\mathrm{e}^{\beta U_1 + \delta T_2} &= \mathbb{E}_s\big[\mathrm{e}^{\beta U_1 + \delta T_2}L_{T_1}\big] \\
&= \mathbb{E}_s\big[\mathrm{e}^{\beta U_1 + \delta T_2}\mathrm{e}^{\alpha(U_1 - s) - s\kappa(\alpha)}\mathrm{e}^{-(\alpha + \kappa(\alpha))(T_2 - s)}\big] \\
&= \widehat{B}[\alpha + \beta]\widehat{A}[\delta - \alpha - \kappa(\alpha)] = \frac{\widehat{B}[\alpha + \beta]}{\widehat{B}[\alpha]}\frac{\widehat{A}[\delta - \alpha - \kappa(\alpha)]}{\widehat{A}[-\alpha - \kappa(\alpha)]} \\
&= \widehat{B}_\alpha[\beta]\widehat{A}_\alpha[\delta],
\end{aligned}$$

which shows that U_1, T_2 are independent with distributions B_α, resp. A_α as asserted. An easy extension of the argument shows that $U_1, \ldots, U_n, T_2, \ldots, T_{n+1}$ are independent with distribution A_α for the T_k and B_α for the U_k. $\square$

Remark 3.5 For the compound Poisson case where A is exponential with rate β, (3.3) means $1 = \widehat{B}[\alpha]\beta/(\beta + \alpha + \kappa(\alpha))$, i.e. $\kappa(\alpha) = \beta(\widehat{B}[\alpha] - 1) - \alpha$ in agreement with Chapter IV. $\square$

Note that the changed distributions of A and B are in general not the same for $\mathbb{P}_{\alpha;s}$ and $\mathbb{P}_\alpha^{(d)}$. An important exception is, however, the determination of the adjustment coefficient γ where the defining equations $\kappa^{(d)}(\gamma) = 0$ and $\kappa(\gamma) = 0$ are the same.

The Markov additive point of view is relevant when studying problems which cannot be reduced to the imbedded random walk, say finite horizon ruin probabilities where the approach via the imbedded random walk yields results on the probability of ruin after N claims, not after time T. Using the Markov additive approach yields for example the following analogue of Theorem V.4.6:

Proposition 3.6 *Let $y < 1/\kappa'(\gamma)$, let $\alpha_y > 0$ be the solution of $\kappa'(\alpha_y) = 1/y$, and define $\gamma_y = \alpha_y - y\kappa(\alpha_y)$. Then*

$$\psi_s(u, yu) \leq \frac{\mathrm{e}^{-(\alpha_y + \kappa(\alpha_y))s}}{\widehat{A}\big[-\alpha_y - \kappa(\alpha_y)\big]} \mathrm{e}^{-\gamma_y u} = \mathrm{e}^{-(\alpha_y + \kappa(\alpha_y))s} \widehat{B}[\alpha_y] \mathrm{e}^{-\gamma_y u}.$$

In particular, for the zero-delayed case $\psi_s(u, yu) \leq \mathrm{e}^{-\gamma_y u}$.

Proof. As in the proof of Theorem V.4.6, it is easily seen that $\kappa(\alpha_y) > 0$. Let $M(u)$ be the number of claims leading to ruin. Then $J\big(\tau(u)\big) = T_{M(u)+1}$ and hence

$$\psi_s(u, yu) = \mathbb{E}_{\alpha_y;s}\left[\mathrm{e}^{-\alpha_y S_{\tau(u)} + \tau(u)\kappa(\alpha_y)} \frac{h(s)}{h(J_{\tau(u)})}; \tau(u) \leq yu\right]$$

$$\leq \mathrm{e}^{-\alpha_y u + yu\kappa(\alpha_y)} \mathbb{E}_{\alpha_y;s}\left[\frac{\mathrm{e}^{-(\alpha_y + \kappa(\alpha_y))s}}{\mathrm{e}^{-(\alpha_y + \kappa(\alpha_y))T_{M(u)+1}}}\right]$$

$$= \mathrm{e}^{-(\alpha_y + \kappa(\alpha_y))s} \mathrm{e}^{-\gamma_y u} \widehat{A}_{\alpha_y}\big[\alpha_y + \kappa(\alpha_y)\big],$$

which is the same as the asserted inequality for $\psi_s(u, yu)$. The claim for the zero-delayed case follows by integration w.r.t. $A(\mathrm{d}s)$. $\qquad\square$

Notes and references The approach via the imbedded random walk is standard, see e.g. [APQ]. The random walk interpretation also allows to translate general asymptotic results for finite time-horizon ruin probabilities of random walks to the corresponding renewal model, for instance the sharp time-horizon asymptotics of Veraverbeke & Teugels [864]. For the approach via Markov additive processes, see in particular Dassios & Embrechts [273] and Asmussen & Rubinstein [99].

4   The duality with queueing theory

We first review some basic facts about the GI/G/1 queue, defined as the single server queue with first in first out (FIFO; or FCFS = first come first served) queueing discipline and renewal interarrival times. Label the customers $1, 2, \ldots$ and assume that T_n is the time between the arrivals of customers $n - 1$ and n, and U_n the service time of customer n. The *actual waiting time* W_n of customer

n is defined as his time spent in queue (excluding the service time), that is, the time from which he arrives to the queue until he starts service. The *virtual waiting time* V_t at time t is the residual amount of work at time t, that is, the amount of time the server will have to work until the system is empty provided no new customers arrive (for this reason often the term *workload process* is used) or, equivalently, the waiting time a customer would have if he arrived at time t. Thus, since customer n arrives at time σ_n, we have

$$W_n = V_{\sigma_n -} \tag{4.1}$$

(left limit). The traffic intensity of the queue is $\rho = \mathbb{E}U/\mathbb{E}T$.

The following result shows that $\{W_n\}$ is a Lindley process in the sense of III.4:

Proposition 4.1 $W_{n+1} = (W_n + U_n - T_n)^+$.

Proof. The amount of residual work just before customer n arrives is $V_{\sigma_n -}$. It then jumps to $V_{\sigma_n -} + U_n$, whereas in $[\sigma_n, \sigma_{n+1}) = [\sigma_n, \sigma_n + T_n)$ the residual work decreases linearly until possibly zero is hit, in which case $\{V_t\}$ remains at zero until time σ_{n+1}. Thus $V_{\sigma_{n+1} -} = (W_n + U_n - T_n)^+$, and combining with (4.1), the proposition follows.   □

Applying Theorem III.3.1, we get:

Corollary 4.2 *Let* $M_n^{(d)} = \max_{k=0,\dots,n-1}\big(U_1 + \dots + U_k - T_1 - \dots T_k\big)$. *If* $W_1 = 0$, *then* $W_n \overset{\mathscr{D}}{=} M_n^{(d)}$.

The next result summarizes the fundamental duality relations between the steady-state behavior of the queue and the ruin probabilities (part (a) was essentially derived already in III.4):

Proposition 4.3 *Assume* $\eta > 0$ *or, equivalently,* $\rho < 1$. *Then:*
(a) *as* $n \to \infty$, W_n *converges in distribution to a random variable* W, *and we have*

$$\mathbb{P}(W > u) = \psi(u); \tag{4.2}$$

(b) *as* $t \to \infty$, V_t *converges in distribution to a random variable* V, *and we have*

$$\mathbb{P}(V > u) = \psi^{(s)}(u). \tag{4.3}$$

Proof. Part (a) is contained in Theorem III.3.1 and Corollary III.3.2, but we shall present a slightly different proof via the duality result given in Theorem III.2.1. Let the T there be the random time σ_N. Then $\mathbb{P}(\tau(u) \le T)$ is the probability $\psi^{(N)}(u)$ of ruin after at most N claims, and obviously $\psi(u) = \lim_{N \to \infty} \psi^{(N)}(u)$. Also $\{Z_t\}_{0 \le t \le T}$ evolves like the left-continuous version of the virtual waiting time process up to just before the Nth arrival, but interchanging the set $(T_1, \ldots, T_N)$ with $(T_N, \ldots, T_1)$ and similarly for the U_n. However, by an obvious reversibility argument this does not affect the distribution, and hence in particular Z_T is distributed as the virtual waiting time just before the Nth arrival, i.e. as W_N. It follows that $\mathbb{P}(W_N > u) = \psi^{(N)}(u)$ has the limit $\psi(u)$ for all u, which implies the convergence in distribution and (4.2).

For part (b), we let T be deterministic. Then the arrivals of $\{R_t\}$ in $[0, T]$ form a stationary renewal process with interarrival distribution A, hence (since the residual lifetime at 0 and the age at T have the same distribution, cf. A.1e) the same is true for the time-reversed point process which is the interarrival process for $\{Z_t\}_{0 \le t \le T}$. Thus as before, $\{Z_t\}_{0 \le t \le T}$ has the same distribution as the left-continuous version of the virtual waiting time process so that

$$\mathbb{P}^{(s)}(V_T > u) = \mathbb{P}^{(s)}\big(\tau(u) \le T\big), \tag{4.4}$$
$$\lim_{T \to \infty} \mathbb{P}^{(s)}(V_T > u) = \lim_{T \to \infty} \mathbb{P}^{(s)}\big(\tau(u) \le T\big) = \psi^{(s)}(u).$$

$\square$

It should be noted that this argument only establishes the convergence in distribution subject to certain initial conditions, namely $W_1 = 0$ in (a) and $V_0 = 0$, $T_1 \sim A_0$ in (b). In fact, convergence in distribution holds for arbitrary initial conditions, but this requires some additional arguments (involving regeneration at 0, but not difficult) that we omit.

Letting $n \to \infty$ in Corollary 4.2, we obtain:

Corollary 4.4 *The steady-state actual waiting time W has the same distribution as $M^{(d)}$.*

Corollary 4.5 (LINDLEY'S INTEGRAL EQUATION) *Let $F(x) = \mathbb{P}(U_1 - T_1 \le x)$, $K(x) = \mathbb{P}(W \le x)$. Then*

$$K(x) = \int_{-\infty}^{x} K(x - y) F(\mathrm{d}y), \quad x \ge 0. \tag{4.5}$$

Proof. Letting $n \to \infty$ in Proposition 4.1, we get $W \overset{\mathscr{D}}{=} (W + U^* - T^*)^+$, where U^*, T^* are independent and distributed as U_1, resp. T_1. Hence for $x \ge 0$, conditioning upon $U^* - T^* = y$ yields

$$K(x) = \mathbb{P}\big((W + U^* - T^*)^+ \le x\big) = \mathbb{P}(W + U^* - T^* \le x)$$
$$= \int_{-\infty}^{x} K(x - y) F(\mathrm{d}y)$$

($x \geq 0$ is crucial for the second equality!). □

Now return to the Poisson case. Then the corresponding queue is M/G/1, and we get:

Corollary 4.6 *For the M/G/1 queue with $\rho < 1$, the actual and the virtual waiting time have the same distribution in the steady state. That is, $W \overset{\mathscr{D}}{=} V$.*

Proof. For the Poisson case, the zero-delayed and the stationary renewal processes are identical. Hence $\psi(u) = \psi^{(s)}(u)$, implying $\mathbb{P}(W > u) = \mathbb{P}(V > u)$ for all u. □

Notes and references The GI/G/1 queue is a favorite of almost any queueing book (see e.g. Cohen [249] or [APQ, Ch. X]), despite the fact that the extension from M/G/1 is of equally doubtful relevance as we argued in Section 1 to be the case in risk theory. Some early classical papers are Smith [814] and Lindley [598].

Note that (4.5) looks like the convolution equation $K = F * K$ but is not the same (one would need (4.5) to hold for all $x \in \mathbb{R}$ and not just $x \geq 0$). Equation (4.5) is in fact a homogeneous Wiener-Hopf equation, see e.g. Asmussen [66] and references therein.

For some further explicit treatments beyond exponential claim sizes, see e.g. Malinovski [627] and Rongming & Haifeng [747].

The imbedded random walk approach also leads to a Pollaczek-Khinchine type formula in the renewal case, which will be exploited for the asymptotic behavior of the ruin probability with heavy-tailed claims in Section X.3. A detailed exposition of the compound geometric approach to renewal models is Willmot & Lin [892]. Whenever the interclaim time is phase-type, renewal models are a special case of Markov additive processes and hence we also refer to Chapter IX for related material.

Exploiting some links between wave governed random motions and the renewal risk process, Mazza & Rullière [632] give an algorithm for computing finite-time ruin probabilities for non-exponential interarrival times.

A number of further results on ruin-related quantities in renewal models will be discussed in XII.3.

Chapter VII

Risk theory in a Markovian environment

1 Model and examples

We assume that the arrivals form an inhomogeneous Poisson process, more precisely determined by a Markov process $\{J_t\}_{0 \leq t < \infty}$ with a finite state space E as follows:

- The arrival intensity is β_i when $J_t = i$;

- Claims arriving when $J_t = i$ have distribution B_i;

- The premium rate when $J_t = i$ is p_i.

Thus, $\{J_t\}$ describes the environmental conditions for the risk process. The intensity matrix governing $\{J_t\}$ is denoted by $\mathbf{\Lambda} = (\lambda_{ij})_{i,j \in E}$ and its stationary limiting distribution by $\boldsymbol{\pi}$; here $\boldsymbol{\pi}$ exists whenever $\mathbf{\Lambda}$ is irreducible which is assumed throughout, and can be computed as the positive solution of $\boldsymbol{\pi}\mathbf{\Lambda} = \mathbf{0}$, $\boldsymbol{\pi}\boldsymbol{e} = 1$ where $\boldsymbol{e}$ is the E-column-vector of ones. As in Chapter I, $\{S_t\}$ denotes the claim surplus process,

$$S_t = \sum_{i=1}^{N_t} U_i - \int_0^t p_{J_v}\, \mathrm{d}v,$$

and $\tau(u) = \inf\{t \geq 0 : S_t > u\}$, $M = \sup_{t \geq 0} S_t$. The ruin probabilities with initial environment i are

$$\psi_i(u) = \mathbb{P}_i\big(\tau(u) < \infty\big) = \mathbb{P}_i(M > u), \quad \psi_i(u,T) = \mathbb{P}_i\big(\tau(u) \leq T\big),$$

where as usual $\mathbb{P}_i$ refers to the case $J_0 = i$.

Unless otherwise stated, we shall assume that $p_i = 1$; this is no restriction when studying infinite horizon ruin probabilities, cf. the operational time argument given in Example 1.5 below.

We let

$$\rho_i = \beta_i \mu_{B_i}, \quad \rho = \sum_{i \in E} \pi_i \rho_i, \quad \eta = \frac{1 - \rho}{\rho}. \tag{1.1}$$

Then ρ_i is the average amount of claims received per unit time when the environment is in state i, and ρ is the overall average amount of claims per unit time, cf. Proposition 1.11 below.

An example of how such a mechanism could be relevant in risk theory follows.

Example 1.1 Consider car insurance, and assume that weather conditions play a major role for the occurrence of accidents. For example, we could distinguish between normal and icy road conditions, leading to E having two states n, i and corresponding arrival intensities β_n, β_i and claim size distributions B_n, B_i; one expects that $\beta_i > \beta_n$ and presumably also that $B_n \neq B_i$, meaning that accidents occurring during icy road conditions lead to claim amounts which are different from the normal ones. □

The versatility of the model in terms of incorporating (or at least approximating) many phenomena which look very different or more complicated at a first sight goes in fact much further (note that for the following discussion a basic knowledge of phase-type distributions is needed, cf. IX.1):

Example 1.2 (ALTERNATING RENEWAL ENVIRONMENT) The model of Example 1.1 implicitly assumes that the sojourn times of the environment in the normal and the icy states are exponential, with rates λ_{ni} and λ_{in}, respectively, which is clearly unrealistic. Thus, assume that the sojourn time in the icy state has a more general distribution $A^{(i)}$. According to Theorem A5.14, we can approximate $A^{(i)}$ with a phase-type distribution (cf. Example I.2.4) with representation $\left(E^{(i)}, \boldsymbol{\alpha}^{(i)}, \boldsymbol{T}^{(i)}\right)$, say. Assume similarly that the sojourn time in the normal state has distribution $A^{(n)}$ which we approximate with a phase-type distribution with representation $\left(E^{(n)}, \boldsymbol{\alpha}^{(n)}, \boldsymbol{T}^{(n)}\right)$, say. Then the state space for the environment is the disjoint union of $E^{(n)}$ and $E^{(i)}$, and we have $\beta_j = \beta_i$ when $j \in E^{(i)}$, $\beta_j = \beta_n$ when $j \in E^{(n)}$; in block-partitioned form, the intensity matrix is

$$\boldsymbol{\Lambda} = \begin{pmatrix} \boldsymbol{T}^{(n)} & \boldsymbol{t}^{(n)} \boldsymbol{\alpha}^{(i)} \\ \boldsymbol{t}^{(i)} \boldsymbol{\alpha}^{(n)} & \boldsymbol{T}^{(i)} \end{pmatrix},$$

where $\boldsymbol{t}^{(n)} = -\boldsymbol{T}^{(n)} \boldsymbol{e}$, $\boldsymbol{t}^{(i)} = -\boldsymbol{T}^{(i)} \boldsymbol{e}$ are the exit rates. □

Example 1.3 Consider again the alternating renewal model for car insurance in Example 1.2, but assume now that the arrival intensity changes during the icy period, say it is larger initially. One way to model this would be to take $A^{(i)}$ to be Coxian (cf. Example IX.1.4) with states $i_1, \ldots, i_q$ (visited in that order) and let $\beta_{i_1} > \ldots > \beta_{i_q}$. □

Example 1.4 (SEMI-MARKOVIAN ENVIRONMENT) Dependence between the length of an icy period and the following normal one (and vice versa) can be modelled by semi-Markov structure. This amounts to a family $\left(A^{(\eta)}\right)_{\eta \in H}$ of sojourn time distributions, such that a sojourn time of type η is followed by one of type ι w.p. $w_{\eta\iota}$ where $\boldsymbol{W} = (w_{\eta\iota})_{\eta,\iota \in H}$ is a transition matrix. Approximating each $A^{(\eta)}$ by a phase-type distribution with representation $\left(E^{(\eta)}, \boldsymbol{\alpha}^{(\eta)}, \boldsymbol{T}^{(\eta)}\right)$, say, the state space E for the environment is $\left\{(\eta, i) : \eta \in H, i \in E^{(\eta)}\right\}$, and

$$
\boldsymbol{\Lambda} = \begin{pmatrix}
\boldsymbol{T}^{(1)} + \omega_{11}\boldsymbol{t}^{(1)}\boldsymbol{\alpha}^{(1)} & \omega_{12}\boldsymbol{t}^{(1)}\boldsymbol{\alpha}^{(2)} & \cdots & \omega_{1q}\boldsymbol{t}^{(1)}\boldsymbol{\alpha}^{(q)} \\
\omega_{21}\boldsymbol{t}^{(2)}\boldsymbol{\alpha}^{(1)} & \boldsymbol{T}^{(2)} + \omega_{22}\boldsymbol{t}^{(2)}\boldsymbol{\alpha}^{(2)} & \cdots & \omega_{2q}\boldsymbol{t}^{(2)}\boldsymbol{\alpha}^{(q)} \\
\vdots & & \ddots & \vdots \\
\omega_{q1}\boldsymbol{t}^{(q)}\boldsymbol{\alpha}^{(1)} & \omega_{q2}\boldsymbol{t}^{(q)}\boldsymbol{\alpha}^{(2)} & \cdots & \boldsymbol{T}^{(q)} + \omega_{qq}\boldsymbol{t}^{(q)}\boldsymbol{\alpha}^{(q)}
\end{pmatrix}
$$

where $q = |H|$, $\boldsymbol{t}^{(\eta)} = -\boldsymbol{T}^{(\eta)}\boldsymbol{e}$. The simplest model for the arrival intensity amounts to $\beta_{\eta,j} = \beta_\eta$ depending only on η.

In the car insurance example, one could for example have $H = \{i_\ell, i_s, n_\ell, n_s\}$, such that the icy period is of two types (long and short) each with their sojourn time distribution $A^{(i_\ell)}$, resp. $A^{(i_s)}$, and similarly for the normal period. Then for example $w_{i_\ell n_s}$ is the probability that a long icy period is followed by a short normal one. □

Example 1.5 (MARKOV-MODULATED PREMIUMS) Returning for a short while to the case of general premium rates p_i depending on the environment i, let

$$
\theta(T) = \int_0^T p_{J_t}\, dt, \quad \widetilde{J}_t = J_{\theta^{-1}(t)}, \quad \widetilde{S}_t = S_{\theta^{-1}(t)}.
$$

Then (by standard operational time arguments) $\left\{\widetilde{S}_t\right\}$ is a risk process in a Markovian environment with unit premium rate, and $\widetilde{\psi}_i(u) = \psi_i(u)$. Indeed, the parameters are $\widetilde{\lambda}_{ij} = \lambda_{ij}/p_i$, $\widetilde{\beta}_i = \beta_i/p_i$. □

From now on, we assume again $p_i = 1$ so that the claim surplus is

$$
S_t = \sum_{i=1}^{N_t} U_i - t.
$$

We turn to some more mathematically oriented basic discussion. The key property for much of the analysis presented below is the following immediate observation:

Proposition 1.6 *The claim surplus process $\{S_t\}$ of a risk process in a Markovian environment is a Markov additive process corresponding to the parameters $\mu_i = -p_i$, $\sigma_i^2 = 0$, $\nu_i(dx) = \beta_i B_i(dx)$, $q_{ij} = 0$ in the notation of Chapter III.4.*

In particular, the Markov additive structure will be used for exponential change of measure and thereby versions of Lundberg's inequality and the Cramér-Lundberg approximation.

Next we note a semi-Markov structure of the arrival process:

Proposition 1.7 *The $\mathbb{P}_i$-distribution of T_1 is phase-type with representation $\left(e_i', \Lambda - (\beta_i)_{\mathrm{diag}}\right)$. More precisely,*

$$\mathbb{P}_i\left(T_1 \in dx,\ J_{T_1} = j\right) \;=\; \beta_j \cdot e_i' e^{(\Lambda - (\beta_i)_{\mathrm{diag}})x} e_j\, dx.$$

Proof. The result immediately follows by noting that T_1 is obtained as the lifelength of $\{J_t\}$ killed at the time of the first arrival and that the exit rate obviously is β_j in state j. $\square$

A remark which is fundamental for much of the intuition on the model consists in noting that to each risk process in a Markovian environment, one can associate in a natural way a standard Poisson one by averaging over the environment. More precisely, we put

$$\beta^* \;=\; \sum_{i \in E} \pi_i \beta_i, \quad B^* \;=\; \sum_{i \in E} \frac{\pi_i \beta_i}{\beta^*} B_i.$$

These parameters are the ones which the statistician would estimate if he ignored the presence of Markov-modulation:

Proposition 1.8 *As $t \to \infty$,*

$$\frac{N_t}{t} \overset{\text{a.s.}}{\to} \beta^*, \quad \frac{1}{N_t} \sum_{\ell=1}^{N_t} I(U_\ell \le x) \overset{\text{a.s.}}{\to} B^*(x).$$

Note that the last statement of the proposition just means that in the limit, the empirical distribution of the claims is B^*. Note also that (as the proof shows) $\pi_i \beta_i / \beta^*$ gives the proportion of the claims which are of type i (arrive in state i).

Proof. Let $t_i = \int_0^t I(J_s = i)\, ds$ be the time spent in state i up to time t and $N_t^{(i)}$ the number of claim arrivals in state i. Then it is standard that $t_i / t \overset{\text{a.s.}}{\to} \pi_i$ as

$t \to \infty$. However, given $\{J_t\}_{0 \le t < \infty}$, we may view $N_t^{(i)}$ as the number of events in a Poisson process where the accumulated intensity at time t is $\beta_i t_i$. Hence

$$\frac{N_t^{(i)}}{t_i} \overset{\text{a.s.}}{\to} \beta_i, \quad \frac{N_t^{(i)}}{t} \overset{\text{a.s.}}{\to} \pi_i \beta_i, \quad \frac{N_t}{t} = \sum_{i \in E} \frac{N_t^{(i)}}{t} \overset{\text{a.s.}}{\to} \beta^*.$$

Also, denoting the sizes of the claims arriving in state i by $U_1^{(i)}, U_2^{(i)}, \ldots$, the standard law of large numbers yields

$$\frac{1}{N} \sum_{k=1}^{N} I(U_k^{(i)} \le x) \overset{\text{a.s.}}{\to} B_i(x), \quad N \to \infty.$$

Hence

$$\frac{1}{N_t} \sum_{\ell=1}^{N_t} I(U_\ell \le x) = \frac{1}{N_t} \sum_{i \in E} \sum_{k=1}^{N_t^{(i)}} I(U_k^{(i)} \le x) \sim \sum_{i \in E} \frac{N_t^{(i)}}{N_t} B_i(x)$$

$$\sim \sum_{i \in E} \frac{t \pi_i \beta_i}{t \beta^*} B_i(x) = B^*(x).$$

$\square$

A different interpretation of B^* is as the Palm distribution of the claim size, cf. Example III.5.4.

The next result shows that we can think of the averaged compound Poisson risk model as the limit of the Markov-modulated one obtained by speeding up the Markov-modulation.

Proposition 1.9 *Consider a Markov-modulated risk process $\{S_t\}$ with parameters β_i, B_i, $\mathbf{\Lambda}$, and let $\{S_t^{(a)}\}$ refer to the one with parameters β_i, B_i, $a\mathbf{\Lambda}$, $\{S_t^*\}$ to the compound Poisson model with parameters β^*, B^*. Then $\{S_t^{(a)}\} \overset{\mathscr{D}}{\to} \{S_t^*\}$ in $D[0, \infty)$ as $a \to \infty$. In particular, $\psi_i^{(a)}(u) \to \psi^*(u)$ for all u and i.*

Proof. According to Proposition 1.7, the $\mathbb{P}_i$-distribution of T_1 in $\{S_t^{(a)}\}$ is phase-type with representation $(E, e_i', a\mathbf{\Lambda} - (\beta_i)_{\text{diag}})$. By Proposition A5.2, this converges to the exponential distribution with rate β^* as $a \to \infty$, and furthermore in the limit J_{T_1} has distribution $(\pi_i \beta_i / \beta^*)_{i \in E}$ and is independent of T_1. In particular, the limiting distribution of the first claim size U_1 is B^*. Conditioning upon $\mathscr{F}_{T_1}$ shows similarly that in the limit (T_2, U_2) are independent of $\mathscr{F}_{T_1}$, with T_2 being exponential with rate β^* and U_2 having distribution B^*. Continuing in this manner shows that the limiting distribution of $(T_n, U_n)_{n=1,2,\ldots}$ is as in $\{S_t^*\}$. From this the convergence in distribution follows by general facts on weak convergence in $D[0,\infty)$, which also yields $\psi^{(a)}(u,T) \to \psi^*(u,T)$ for all u and T. The fact that indeed $\psi_i^{(a)}(u) \to \psi^*(u)$ follows, e.g., from Theorem 3.2.1 of [370].   □

Example 1.10 Let

$$E = \{1,2\}, \qquad \Lambda = \begin{pmatrix} -a & a \\ a & -a \end{pmatrix}$$

$$\beta_1 = \frac{9}{2}, \qquad B_1 = \frac{3}{5}E_3 + \frac{2}{5}E_7,$$

$$\beta_2 = \frac{3}{2}, \qquad B_2 = \frac{1}{5}E_3 + \frac{4}{5}E_7,$$

where E_δ denotes the exponential distribution with intensity parameter δ and $a > 0$ is arbitrary. That is, we may imagine that we have two types of claims such that the claim size distributions are E_3 and E_7. Claims of type E_3 arrive with intensity $\frac{9}{2} \cdot \frac{3}{5} = \frac{27}{10}$ in state 1 and with intensity $\frac{3}{2} \cdot \frac{1}{5} = \frac{3}{10}$ in state 2, those of type E_7 with intensity $\frac{9}{2} \cdot \frac{2}{5} = \frac{9}{5}$ in state 1 and with intensity $\frac{3}{2} \cdot \frac{4}{5} = \frac{6}{5}$ in state 2. Thus, since E_3 is a more dangerous claim size distribution than E_7 (the mean is larger and the tail is heavier), state 1 appears as more dangerous than state 2, and in fact

$$\rho_1 = \beta_1\mu_{B_1} = \frac{9}{2}\left(\frac{3}{5}\cdot\frac{1}{3} + \frac{2}{5}\cdot\frac{1}{7}\right) = \frac{81}{70},$$

$$\rho_2 = \beta_1\mu_{B_2} = \frac{3}{2}\left(\frac{1}{5}\cdot\frac{1}{3} + \frac{4}{5}\cdot\frac{1}{7}\right) = \frac{19}{70}.$$

Thus in state 1 where $\rho_1 > 1$, the company even suffers an average loss, and (at least when a is small such that state changes of the environment are infrequent), the paths of the surplus process will exhibit the type of behavior in Fig. VII.1 with periods with positive drift alternating with periods with negative drift; the overall drift is negative since $\boldsymbol{\pi} = (\frac{1}{2} \ \frac{1}{2})$ so that $\rho = \pi_1\rho_1 + \pi_2\rho_2 = \frac{5}{7} < 1$. On Fig. VII.1, there are $p = 2$ background states of $\{J_t\}$, marked by thin, resp. thick, lines in the path of $\{S_t\}$.

Computing the parameters of the averaged compound Poisson model, we first get that

$$\beta^* = \frac{1}{2}\cdot\frac{9}{2} + \frac{1}{2}\cdot\frac{3}{2} = 3.$$

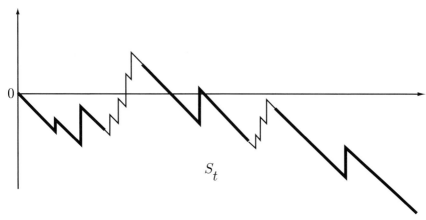

FIGURE VII.1

Thus, a fraction $\pi_1\beta_1/\beta^* = 3/4$ of the claims occurs in state 1 and the remaining fraction $1/4$ in state 2. Hence

$$B^* = \frac{3}{4}\left(\frac{3}{5}E_3 + \frac{2}{5}E_7\right) + \frac{1}{4}\left(\frac{1}{5}E_3 + \frac{4}{5}E_7\right) = \frac{1}{2}E_3 + \frac{1}{2}E_7.$$

That is, the averaged compound Poisson model is the same as in IV.(3.1). □

The definition (1.1) of the safety loading η is (as for the renewal model in Chapter VI) based upon an asymptotic consideration given by the following result:

Proposition 1.11 (a) $\mathbb{E}S_t/t \to \rho - 1$, $t \to \infty$;
(b) $S_t/t \to \rho - 1$ a.s., $t \to \infty$.

Proof. In the notation of Proposition 1.8, we have

$$\mathbb{E}\left[S_t + t \,\middle|\, (t_i)_{i\in E}\right] = \sum_{i\in E} t_i\beta_i\mu_{B_i} = \sum_{i\in E} t_i\rho_i.$$

Taking expectations and using the well-known fact $\mathbb{E}t_i/t \to \pi_i$ yields (a). For (b), note first that $\sum_1^N U_k^{(i)}/N \overset{\text{a.s.}}{\to} \mu_{B_i}$. Hence

$$\frac{S_t + t}{t} = \sum_{i\in E} \frac{N_t^{(i)}}{t} \cdot \frac{1}{N_t^{(i)}} \sum_{k=1}^{N_t^{(i)}} U_k^{(i)} \overset{\text{a.s.}}{\to} \sum_{i\in E} \pi_i\beta_i \cdot \mu_{B_i} = \rho.$$

□

Corollary 1.12 *If $\eta \leq 0$, then $M = \infty$ a.s., and hence $\psi_i(u) = 1$ for all i and u. If $\eta > 0$, then $M < \infty$ a.s., and $\psi_i(u) < 1$ for all i and u.*

Proof. The case $\eta < 0$ is trivial since then the a.s. limit $\rho - 1$ of S_t/t is > 0, and hence $M = \infty$. The case $\eta > 0$ is similarly easy. Now let $\eta = 0$, let some state i be fixed and define

$$\omega = \omega_1 = \inf \{t > 0 : J_{t-} \neq i, J_t = i\}, \quad \omega_2 = \inf \{t > \omega_1 : J_{t-} \neq i, J_t = i\},$$

$$X_1 = S_{\omega_1}, \quad X_2 = S_{\omega_2} - S_{\omega_1},$$

and so on. Then by standard Markov process formulas (e.g. [APQ, Th. II.4.2(i), p. 50]) $\mathbb{E}_i \omega_1 = -1/\pi_i \lambda_{ii}$ and

$$
\begin{aligned}
\mathbb{E}_i X_1 &= \mathbb{E}_i \int_0^\omega \beta_{J_t} \mu_{B_{J_t}} \, dt - \mathbb{E}_i \omega \\
&= \mathbb{E}_i \omega \cdot \left(\sum_{j \in E} \pi_j \beta_j \mu_{B_j} - 1 \right) = (\rho - 1) \mathbb{E}_i \omega = 0.
\end{aligned}
$$

Now obviously the ω_n form a renewal process, and hence $\omega_n/n \overset{\text{a.s.}}{\to} \mathbb{E}_i \omega$. Since the X_n are independent, with $X_2, X_3, \ldots$ having the $\mathbb{P}_i$-distribution of X, also

$$\frac{S_{\omega_n}}{n} = \frac{X_1 + \cdots + X_n}{n} \overset{\text{a.s.}}{\to} \mathbb{E}_i X = 0.$$

Thus $\{S_{\omega_n}\}$ is a discrete time random walk with mean zero, and hence oscillates between $-\infty$ and ∞ so that also here $M = \infty$. $\qquad\qquad\square$

Notes and references The Markov-modulated Poisson process has been very popular in queueing theory since the early 1980s, see the Notes to Section 7. In risk theory, some early studies are in Janssen & Reinhard [501, 730, 502], and a more comprehensive treatment in Asmussen [58]. The mainstream of the present chapter follows [58], with some important improvements being obtained in Asmussen [59] in the queueing setting and being implemented numerically in Asmussen & Rolski [97].

Statistical aspects are not treated here. See Meier [634] and Rydén [760, 761]. There seems still to be more to be done in this area, in particular in order to treat more than low-dimensional state spaces E.

Proposition 1.1 and the Corollary are standard. The proof of Proposition 1.1(b) is essentially the same as the proof of the strong law of large numbers for cumulative processes, see [APQ, p. 178] or A.1d.

2   The ladder height distribution

Our mathematical treatment of the ruin problem follows the model of Chapter IV for the simple compound Poisson model, and involves a version of the

Pollaczeck-Khinchine formula (see Proposition 2.2(a) below) where the ladder height distribution is evaluated by a time reversion argument.

Define the ladder epoch τ_+ by $\tau_+ = \inf\{t : S_t > 0\} = \tau(0)$, let

$$G_+(i, j; A) = \mathbb{P}_i\big(S_{\tau_+} \in A, J_{\tau_+} = j, \tau_+ < \infty\big)$$

and let $\boldsymbol{G}_+$ be the measure-valued matrix with ijth element $G_+(i, j; \cdot)$. The form of $\boldsymbol{G}_+$ turns out to be explicit (or at least computable), but is substantially more involved than for the compound Poisson case. However, by specializing results for general stationary risk processes (Theorem III.5.5; see also Example III.5.4) we obtain the following result, which represents a nice simplified form of the ladder height distribution $\boldsymbol{G}_+$ when taking certain averages: starting $\{J_t\}$ stationary, we get the same ladder height distribution as for the averaged compound Poisson model, cf. the definition of β^*, B^* in Section 1.

Proposition 2.1 $\boldsymbol{\pi}\boldsymbol{G}_+(\mathrm{d}y)\boldsymbol{e} = \beta^* \overline{B}^*(y)\mathrm{d}y$.

For measure-valued matrices, we define the convolution operation by the same rule as for multiplication of real-valued matrices, only with the product of real numbers replaced by convolution of measures. Thus, e.g., $\boldsymbol{G}_+^{*2}$ is the matrix whose ijth element is

$$\sum_{k \in E} G_+(i, k; \cdot) * G_+(k, j; \cdot).$$

Also, $\|\boldsymbol{G}_+\|$ denotes the matrix with ijth element

$$\big\|G_+(i, j; \cdot)\big\| = \int_0^\infty G_+(i, j; \mathrm{d}x).$$

Let further $\boldsymbol{R}$ denote the pre-τ_+ occupation kernel,

$$R(i, j; A) = \mathbb{E}_i \int_0^{\tau_+} I(S_t \in A, J_t = j)\,\mathrm{d}t,$$

and $\boldsymbol{S}(\mathrm{d}x)$ the measure-valued diagonal matrix with $\beta_i B_i(\mathrm{d}x)$ as ith diagonal element.

Proposition 2.2 (a) *The distribution of M is given by*

$$1 - \psi_i(u) = \mathbb{P}_i(M \le u) = \boldsymbol{e}_i^\mathsf{T} \sum_{n=0}^\infty \boldsymbol{G}_+^{*n}(u)\big(\boldsymbol{I} - \|\boldsymbol{G}_+\|\big)\boldsymbol{e}. \qquad (2.1)$$

(b) $\boldsymbol{G}_+(y, \infty) = \int_{-\infty}^0 \boldsymbol{R}(\mathrm{d}x)\boldsymbol{S}\big((y - x, \infty)\big)$. *That is, for $i, j \in E$,*

$$G_+\big(i, j; (y, \infty)\big) = \int_{-\infty}^0 R(i, j; \mathrm{d}x)\beta_j \overline{B}_j(y - x). \qquad (2.2)$$

Proof. The probability that there are n proper ladder steps not exceeding x and that the environment is j at the nth when we start from i is $e_i^\mathsf{T} G_+^{*n}(x) e_j$, and the probability that there are no further ladder steps starting from environment j is $e_j^\mathsf{T}(I - \|G_+\|)e$. From this (2.1) follows by summing over n and j. The proof of (2.2) is just the same as the proof of Lemma III.5.3. □

To make Proposition 2.2 useful, we need as in Chapters III, IV to bring R and G_+ on a more explicit form. To this end, we need to invoke the time-reversed version $\{J_t^*\}$ of $\{J_t\}$; the intensity matrix Λ^* has ijth element

$$\lambda_{ij}^* = \frac{\pi_j}{\pi_i}\lambda_{ji},$$

and we have

$$\mathbb{P}_i(J_T^* = j) = \frac{\pi_j}{\pi_i}\mathbb{P}_j(J_T = i). \tag{2.3}$$

We let $\{S_t^*\}$ be defined as $\{S_t\}$, only with $\{J_t\}$ replaced by $\{J_t^*\}$ (the β_i and B_i are the same), and let further $\{m_x\}$ be the E-valued process obtained by observing $\{J_t^*\}$ only when $\{S_t^*\}$ is at a minimum value. That is, $m_x = j$ when for some (necessarily unique) t we have $S_t^* = -x$, $J_t^* = j$, $S_t^* < S_u^*$ for $u < t$; see Figure VII.2 for an illustration in the case of $p = 2$ environmental states of $\{J_t\}$, marked by thin, resp. thick, lines in the path of $\{S_t\}$.

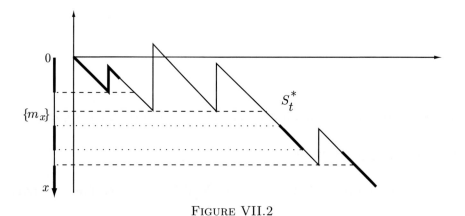

FIGURE VII.2

The following observation is immediate:

Proposition 2.3 *When $\eta > 0$, $\{m_x\}$ is a non-terminating Markov process on E, hence uniquely specified by its intensity matrix Q (say).*

Proposition 2.4 Q *satisfies the non-linear matrix equation* $Q = \varphi(Q)$ *where*

$$\varphi(Q) \;=\; \Lambda^* - (\beta_i)_{\mathrm{diag}} + \int_0^\infty S(\mathrm{d}x)\, e^{Qx},$$

and $S(\mathrm{d}x)$ *is the diagonal matrix with the* $\beta_i B_i(\mathrm{d}x)$ *on the diagonal. Furthermore, the sequence* $\{Q^{(n)}\}$ *defined by*

$$Q^{(0)} = \Lambda^* - (\beta_i)_{\mathrm{diag}}, \quad Q^{(n+1)} = \varphi\big(Q^{(n)}\big)$$

converges monotonically to Q.

Note that the integral in the definition of $\varphi(Q)$ is the matrix whose ith row is the ith row of

$$\beta_i \widehat{B}_i[Q] = \beta_i \int_0^\infty e^{Qx} B_i(\mathrm{d}x).$$

Proof. The argument relies on an interpretation in terms of excursions. An excursion of $\{S_t^*\}$ above level $-x$ starts at time t if $S_{t-}^* = -x$, $\{S_v^*\}$ is a minimum value at $v = t-$ and a jump (claim arrival) occurs at time t, and the excursion ends at time $s = \inf\{v > t : S_v^* = -x\}$. If there are no jumps in $(t, s]$, we say that the excursion has depth 0. Otherwise each jump at a minimum level during the excursion starts a subexcursion, and the excursion is said to have depth 1 if each of these subexcursions have depth 0. In general, we recursively define the depth of an excursion as 1 plus the maximal depth of a subexcursion. The definitions are illustrated on Figure VII.3 where there are three excursions of depth 1,0,2. For example the excursion of depth 2 has one subexcursion which is of depth 1, corresponding to two subexcursions of depth 0.

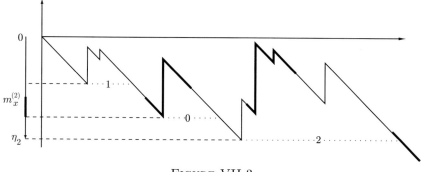

FIGURE VII.3

Let $p_{ij}^{(n)}$ be the probability that an excursion starting from $J_t^* = i$ has depth at most n and terminates at $J_s^* = j$ and p_{ij} the probability that an excursion starting from $J_t^* = i$ terminates at $J_s^* = j$. By considering minimum values within the excursion, it becomes clear that

$$p_{ij} = \int_0^\infty \left[e^{Qy} \right]_{ij} B_i(\mathrm{d}y) . \tag{2.4}$$

To show $Q = \varphi(Q)$, we first compute q_{ij} for $i \neq j$. Suppose $m_x = i$. Then a jump to j (i.e., $m_{x+dx} = j$) occurs in two ways, either due to a jump of $\{J_t^*\}$ which occurs with intensity λ_{ij}^*, or through an arrival starting an excursion terminating with $J_s^* = j$. It follows that $q_{ij} = \lambda_{ij}^* + \beta_i p_{ij}$. Similarly,

$$\mathbb{P}_i(m_h = i) = 1 + \lambda_{ii}^* h - \beta_i h + \beta_i h p_{ii} + o(h)$$

implies $q_{ii} = \lambda_{ii}^* - \beta_i + \beta_i p_{ii}$. Writing out in matrix notation, $Q = \varphi(Q)$ follows.

Now let $\{m_x^{(n)}\}$ be $\{m_x\}$ killed at the first time η_n (say) a subexcursion of depth at least n occurs. It is clear that $\{m_x^{(n)}\}$ is a terminating Markov process and that $\{m_x^{(0)}\}$ has subintensity matrix $\Lambda^* - (\beta_i)_{\mathrm{diag}} = Q^{(0)}$. The proof of $Q = \varphi(Q)$ then immediately carries over to show that the subintensity matrix of $\{m_x^{(1)}\}$ is $\varphi(Q^{(0)}) = Q^{(1)}$. Similarly by induction, the subintensity matrix of $\{m_x^{(n+1)}\}$ is $\varphi(Q^{(n)}) = Q^{(n+1)}$ which implies that

$$q_{ij}^{(n+1)} = \lambda_{ij}^* - \beta_i + \beta_i p_{ij}^{(n)} .$$

Now just note that $p_{ij}^{(n)} \uparrow p_{ij}$ and insert (2.4). □

Define a further kernel U by

$$U(i, j; A) = \int_{-A} \mathbb{P}_i(m_x = j) \, \mathrm{d}x = \int_{-A} e_i^\mathsf{T} e^{Qx} e_j \, \mathrm{d}x \tag{2.5}$$

(note that we use $-A = \{x : -x \in A\}$ on the r.h.s. of the definition to make U be concentrated on $(-\infty, 0)$).

Theorem 2.5 $R(i, j; A) = \dfrac{\pi_j}{\pi_i} U(j, i; A).$

Proof. We shall show that

$$\mathbb{P}_i\big(J_t = j, S_t \in A, \tau_+ > t \big) = \frac{\pi_j}{\pi_i} \mathbb{P}_j\big(J_t^* = i, S_t^* \in A, S_t^* < S_u^*, u < t \big), \tag{2.6}$$

from which the result immediately follows by integrating from 0 to ∞ w.r.t. $\mathrm{d}t$. To this end, consider stationary versions of $\{J_t\}$, $\{J_t^*\}$. We may then assume $J_u^* = J_{t-u}$, $S_u^* = S_t - S_{t-u}$, $0 \le u \le t$, and get

$$
\begin{aligned}
& \pi_i \mathbb{P}_i\big(J_t = j, S_t \in A, \tau_+ > t\big) \\
&= \mathbb{P}_\pi\big(J_t = j, J_0 = i, S_t \in A, S_u < 0, 0 < u < t\big) \\
&= \mathbb{P}_\pi\big(J_0^* = j, J_t^* = i, S_t^* \in A, S_t^* < S_{t-u}^*, 0 < u < t\big) \\
&= \pi_j \mathbb{P}_j\big(J_t^* = i, S_t^* \in A, S_t^* < S_u^*, 0 < u < t\big),
\end{aligned}
$$

and this immediately yields (2.6). $\qquad\square$

It is convenient at this stage to rewrite the above results in terms of the matrix $K = \Delta^{-1} Q^\mathsf{T} \Delta$, where Δ is the diagonal matrix with π on the diagonal:

Corollary 2.6 (a) $R(\mathrm{d}x) = e^{-Kx}\mathrm{d}x$, $x \le 0$;
(b) *for* $z \ge 0$, $G_+((z, \infty)) = \int_0^\infty e^{Kx} S\big((x + z, \infty)\big)\,\mathrm{d}x$;
(c) *the matrix* K *satisfies the non-linear matrix equation* $K = \varphi(K)$ *where*

$$
\varphi(K) = \Lambda - (\beta_i)_{\mathrm{diag}} + \int_0^\infty e^{Kx}\, S(\mathrm{d}x);
$$

(d) *the sequence* $\{K^{(n)}\}$ *defined by* $K^{(0)} = \Lambda - (\beta_i)_{\mathrm{diag}}$, $K^{(n+1)} = \varphi(K^{(n)})$ *converges monotonically to* K.

[The $\varphi(\cdot)$ here is of course not the same as in Proposition 2.4.]

From $Qe = 0$, it is readily checked that π is a left eigenvector of K corresponding to the eigenvalue 0 (when $\rho < 1$), and we let k be the corresponding right eigenvector normalized by $\pi k = 1$.

Remark 2.7 It is instructive to see how Proposition 2.1 can be rederived using the more detailed form of G_+ in Corollary 2.6(b): from $\pi K = 0$ we get

$$
\begin{aligned}
\pi G_+(\mathrm{d}y)e &= \int_0^\infty \pi e^{Kx}(\beta_i B_i(\mathrm{d}y + x))_{\mathrm{diag}}\,\mathrm{d}x \cdot e \\
&= \int_0^\infty \pi(\beta_i B_i(\mathrm{d}y + x))_{\mathrm{col}}\,\mathrm{d}x \\
&= \sum_{i \in E} \pi_i \beta_i \overline{B}_i(y)\,\mathrm{d}y = \beta^* \overline{B}^*(y)\mathrm{d}y.
\end{aligned}
$$

$\qquad\square$

Though maybe Corollary 2.6 is hardly all that explicit in general, we shall see that nevertheless we have enough information to derive, e.g., the Cramér-Lundberg approximation (Section 3), and to obtain a simple solution in the special case of phase-type claims (Chapter IX). As preparation, we shall give at this place some simple consequences of Corollary 2.6.

Lemma 2.8 $\left(I - \|G_+\|\right)e = (1 - \rho)k.$

Proof. Using Corollary 2.6(b) with $z = 0$, we get

$$\|G_+\| = \int_0^\infty e^{Kx} S\big((x, \infty)\big)\, dx. \tag{2.7}$$

In particular, multiplying by K and integrating by parts yields

$$K\|G_+\| = \int_0^\infty (e^{Kx} - I)S(dx)$$

$$= K - \Lambda + (\beta_i)_{\text{diag}} - \int_0^\infty S(dx) = K - \Lambda. \tag{2.8}$$

Let $L = (k\pi - K)^{-1}$. Then $(k\pi - K)k = k$ implies $Lk = k$. Now using (2.7), (2.8) and $\pi e^{Kx} = \pi$, we get

$$k\pi\|G_+\|e = k\int_0^\infty \pi S\big((x, \infty)\big)e\, dx = k(\pi_i \beta_i \mu_{B_i})_{\text{row}}e = \rho k,$$

$$K\|G_+\|e = Ke,$$

$$(k\pi - K)\big(I - \|G_+\|\big)e = k - Ke - \rho k + Ke = (1 - \rho)k.$$

Multiplying by L to the left, the proof is complete. □

Here is an alternative algorithm to the iteration scheme in Corollary 2.6 for computing K. Let $|A|$ denote the determinant of the matrix A and d the number of states in E.

Proposition 2.9 *The following assertions are equivalent:*
(a) *all d eigenvalues of K are distinct;*
(b) *there exist d distinct solutions $s_1, \ldots, s_d \in \{s \in C : \Re(s) \le 0\}$ of*

$$\left|\Lambda + (\beta_i(\widehat{B}_i[s] - 1))_{\text{diag}} - sI\right| = 0. \tag{2.9}$$

In that case, then $s_1, \ldots, s_d$ are precisely the eigenvalues of K, and the corresponding left row eigenvectors $a_1, \ldots, a_d$ can be computed by

$$a_i \left(\Lambda - (\beta_i(\widehat{B}_i[s_i] - 1))_{\text{diag}} - s_i I\right) = 0. \tag{2.10}$$

Thus,

$$K = \begin{pmatrix} a_1 \\ \vdots \\ a_d \end{pmatrix}^{-1} \begin{pmatrix} s_1 a_1 \\ \vdots \\ s_d a_d \end{pmatrix}. \tag{2.11}$$

Proof. Since K is similar to the subintensity matrix Q, all eigenvalues must indeed be in $\{s \in \mathbb{C} : \Re(s) \le 0\}$.

Assume $aK = sa$. Then multiplying $K = \varphi(K)$ by a to the left, we get

$$sa = a\Big(\Lambda - (\beta_i)_{\text{diag}} + \int_0^\infty e^{sx} S(\mathrm{d}x)\Big) = a\Big(\Lambda - (\beta_i)_{\text{diag}} + (\beta_i \widehat{B}_i[s])_{\text{diag}}\Big).$$

It follows that if (a) holds, then so does (b), and the eigenvalues and eigenvectors can be computed as asserted.

The proof that (b) implies (a) is more involved and omitted; see Asmussen [58]. □

In the computation of the Cramér-Lundberg constant C, we shall also need some formulas which are only valid if $\rho > 1$ instead of (as up to now) $\rho < 1$. Let M_+ denote the matrix with ijth entry

$$M_+(i,j) = \int_0^\infty x\, G_+(i,j;\mathrm{d}x).$$

Lemma 2.10 *Assume $\rho > 1$. Then $\|G_+\|$ is stochastic with invariant probability vector ζ_+ (say) proportional to $-\pi K$, $\zeta_+ = -\pi K/(-\pi Ke)$. Furthermore,*

$$-\pi K M_+ e = \rho - 1.$$

Proof. From $\rho > 1$ it follows that $S_t \overset{\text{a.s.}}{\to} \infty$ and hence $\|G_+\|$ is stochastic. That $-\pi K = -e^{\mathsf{T}} Q' \Delta$ is non-zero and has non-negative components follows since $-Qe$ has the same property for $\rho > 1$. Thus the formula for ζ_+ follows immediately by multiplying (2.8) by $-\pi$, which yields $-\pi K \|G_+\| = -\pi K$.

Further

$$M_+ = \int_0^\infty \mathrm{d}z \int_0^\infty e^{Kx} S\big((x+z,\infty)\big)\,\mathrm{d}x$$

$$= \int_0^\infty \mathrm{d}y \int_0^y e^{Kx}\,\mathrm{d}x\, S\big((y,\infty)\big)$$

$$= K^{-1} \int_0^\infty (e^{Ky} - I)\, S\big((y,\infty)\big)\,\mathrm{d}y,$$

$$-\pi K M_+ e = \pi \int_0^\infty \mathrm{d}y (I - e^{Ky})\, S\big((y,\infty)\big)e$$

$$= \pi(\beta_i \mu_{B_i})_{\text{diag}} e - \pi \|G_+\| e = \rho - 1$$

(since $\|\boldsymbol{G}_+\|$ being stochastic implies $\|\boldsymbol{G}_+\|\boldsymbol{e} = \boldsymbol{e}$).   □

Notes and references   The exposition follows Asmussen [59] closely (the proof of Proposition 2.4 is different). The problem of computing $\boldsymbol{G}_+$ may be viewed as a special case of Wiener-Hopf factorization for continuous-time random walks with Markov-dependent increments (Markov additive processes); the discrete-time case is surveyed in Asmussen [57] and references given therein.

3   Change of measure via exponential families

We first recall some notation and some results which were given in Chapter III in a more general Markov additive process context. Define $\boldsymbol{F}_t$ as the measure-valued matrix with ijth entry $F_t(i,j;x) = \mathbb{P}_i[S_t \le x; J_t = j]$, and $\widehat{\boldsymbol{F}}_t[s]$ as the matrix with ijth entry $\widehat{F}_t[i,j;s] = \mathbb{E}_i[e^{sS_t}; J_t = j]$ (thus, $\widehat{\boldsymbol{F}}[s]$ may be viewed as the matrix m.g.f. of $\boldsymbol{F}_t$ defined by entrywise integration). Define further

$$\boldsymbol{K}[\alpha] = \boldsymbol{\Lambda} + \left(\beta_i\big(\widehat{B}_i[\alpha] - 1\big)\right)_{\mathrm{diag}} - \alpha\boldsymbol{I}$$

(the matrix function $\boldsymbol{K}[\alpha]$ is of course not related to the matrix $\boldsymbol{K}$ of the preceding section). Then (Proposition III.4.2):

Proposition 3.1 $\widehat{\boldsymbol{F}}_t[\alpha] = e^{t\boldsymbol{K}[\alpha]}$.

It follows from III.4 that $\boldsymbol{K}[\alpha]$ has a simple and unique eigenvalue $\kappa(\alpha)$ with maximal real part, such that the corresponding left and right eigenvectors $\boldsymbol{\nu}^{(\alpha)}$, $\boldsymbol{h}^{(\alpha)}$ may be taken with strictly positive components. We shall use the normalization $\boldsymbol{\nu}^{(\alpha)}\boldsymbol{e} = \boldsymbol{\nu}^{(\alpha)}\boldsymbol{h}^{(\alpha)} = 1$. Note that since $\boldsymbol{K}[0] = \boldsymbol{\Lambda}$, we have $\boldsymbol{\nu}^{(0)} = \boldsymbol{\pi}$, $\boldsymbol{h}^{(0)} = \boldsymbol{e}$. The function $\kappa(\alpha)$ plays the role of an appropriate generalization of the c.g.f., see Theorem III.4.7.

Now consider some θ such that all $\widehat{B}_i[\theta]$ and hence $\kappa(\theta)$, $\boldsymbol{\nu}^{(\theta)}$, $\boldsymbol{h}^{(\theta)}$ etc. are well-defined. The aim is to define governing parameters $\beta_{\theta;i}$, $B_{\theta;i}$, $\boldsymbol{\Lambda}_\theta = (\lambda_{ij}^{(\theta)})_{i,j\in E}$ for a risk process, such that one can obtain suitable generalizations of the likelihood ratio identitites of Chapter III and thereby of Lundberg's inequality, the Cramér-Lundberg approximation etc.

According to Theorem III.4.11, the appropriate choice is

$$\beta_{\theta;i} = \beta_i\widehat{B}_i[\theta], \quad B_{\theta;i}(\mathrm{d}x) = \frac{e^{\theta x}}{\widehat{B}_i[\theta]}B_i(\mathrm{d}x),$$

$$\begin{aligned}
\boldsymbol{\Lambda}_\theta &= \boldsymbol{\Delta}_\theta^{-1}\boldsymbol{K}[\theta]\boldsymbol{\Delta}_\theta - \kappa(\theta)\boldsymbol{I} \\
&= \boldsymbol{\Delta}_\theta^{-1}\boldsymbol{\Lambda}\boldsymbol{\Delta}_\theta + \left(\beta_i(\widehat{B}_i[\theta] - 1)\right)_{\mathrm{diag}} - \big(\kappa(\theta) + \theta\big)\boldsymbol{I}
\end{aligned}$$

where $\boldsymbol{\Delta}_\theta$ is the diagonal matrix with $h_i^{(\theta)}$ as ith diagonal element. That is,

$$
\lambda_{ij}^{(\theta)} = \begin{cases} \lambda_{ij} \dfrac{h_j^{(\theta)}}{h_i^{(\theta)}} & i \neq j \\ \lambda_{ii} + \beta_i(\widehat{B}_i[\theta] - 1) - \kappa(\theta) - \theta & i = j \end{cases} .
$$

We recall that it was shown in III.4 that $\boldsymbol{\Lambda}_\theta$ is an intensity matrix, that $\mathbb{E}_i e^{\theta S_t} h_{J_t}^{(\theta)}$ $= e^{t\kappa(\theta)} h_i^{(\theta)}$ and that $\big\{ e^{\theta S_t - t\kappa(\theta)} h_{J_t}^{(\theta)} \big\}_{t \geq 0}$ is a martingale.

We let $\mathbb{P}_{\theta;i}$ be the governing probability measure for a risk process with parameters $\beta_{\theta;i}$, $B_{\theta;i}$, $\boldsymbol{\Lambda}_\theta$ and initial environment $J_0 = i$. Recall that if $\mathbb{P}_{\theta;i}^{(T)}$ is the restriction of $\mathbb{P}_{\theta;i}$ to $\mathscr{F}_T = \sigma\big\{ (S_t, J_t) : t \leq T \big\}$ and $\mathbb{P}_i^{(T)} = \mathbb{P}_{0;i}^{(T)}$, then $\mathbb{P}_{\theta;i}^{(T)}$ and $\mathbb{P}_i^{(T)}$ are equivalent for $T < \infty$. More generally, allowing T to be a stopping time, Theorem III.1.3 takes the following form:

Proposition 3.2 *Let τ be any stopping time and let $G \in \mathscr{F}_\tau$, $G \subseteq \{\tau < \infty\}$. Then*

$$
\mathbb{P}_i G = \mathbb{P}_{0;i} G = h_i^{(\theta)} \mathbb{E}_{\theta;i} \left[\frac{1}{h_{J_\tau}^{(\theta)}} \exp\{-\theta S_\tau + \tau\kappa(\theta)\} ; G \right]. \tag{3.1}
$$

Let $\widehat{\boldsymbol{F}}_{\theta;t}[s]$, $\kappa_\theta(s)$ and ρ_θ be defined the same way as $\widehat{\boldsymbol{F}}_t[s]$, $\kappa(s)$ and ρ, only with the original risk process replaced by the one with changed parameters.

Lemma 3.3 $\widehat{\boldsymbol{F}}_{\theta;t}[s] = e^{-t\kappa(\theta)} \boldsymbol{\Delta}^{-1} \widehat{\boldsymbol{F}}_t[s + \theta] \boldsymbol{\Delta}$.

Proof. Use III.(4.5). $\qquad\square$

Lemma 3.4 $\kappa_\theta(s) = \kappa(s + \theta) - \kappa(\theta)$. *In particular, $\rho_\theta > 1$ whenever $\kappa'(s) > 0$.*

Proof. The first formula follows by Lemma 3.3 and the second from $\rho_\theta = \kappa_\theta'(s)$. $\qquad\square$

Notes and references The exposition here and in the next two subsections (on likelihood ratio identities and Lundberg conjugation) follows Asmussen [58] closely (but is somewhat more self-contained).

3a Lundberg conjugation

Since the definition of $\kappa(s)$ is a direct extension of the definition for the Cramér-Lundberg model, the Lundberg equation is $\kappa(\gamma) = 0$. We assume that a solution $\gamma > 0$ exists and use notation like $\mathbb{P}_{L;i}$ instead of $\mathbb{P}_{\gamma;i}$; also, for brevity we write $\boldsymbol{h} = \boldsymbol{h}^{(\gamma)}$ and $\boldsymbol{\nu} = \boldsymbol{\nu}^{(\gamma)}$.

Substituting $\theta = \gamma$, $\tau = \tau(u)$, $G = \{\tau(u) < \infty\}$ in Proposition 3.2, letting $\xi(u) = S_{\tau(u)} - u$ be the overshoot and noting that $\mathbb{P}_{L;i}(\tau(u) < \infty) = 1$ by Lemma 3.4, we obtain:

Corollary 3.5

$$\psi_i(u, T) = h_i e^{-\gamma u} \mathbb{E}_{L,i}\left[\frac{e^{-\gamma \xi(u)}}{h_{J_{\tau(u)}}}; \tau(u) \le T\right], \qquad (3.2)$$

$$\psi_i(u) = h_i e^{-\gamma u} \mathbb{E}_{L,i} \frac{e^{-\gamma \xi(u)}}{h_{J_{\tau(u)}}}. \qquad (3.3)$$

Noting that $\xi(u) \ge 0$, (3.3) yields

Corollary 3.6 (LUNDBERG'S INEQUALITY) $\psi_i(u) \le \dfrac{h_i}{\min_{j \in E} h_j} e^{-\gamma u}$.

Assuming it has been shown that $C = \lim_{u \to \infty} \mathbb{E}_{L;i}[e^{-\gamma \xi(u)}/h_{J_{\tau(u)}}]$ exists and is independent of i (which is not too difficult, cf. the proof of Lemma 3.8 below), it also follows immediately that $\psi_i(u) \sim h_i C e^{-\gamma u}$. However, the calculation of C is non-trivial. Recall the definition of $\boldsymbol{G}_+$, $\boldsymbol{K}$, $\boldsymbol{k}$ from Section 2.

Theorem 3.7 (THE CRAMÉR-LUNDBERG APPROXIMATION) *In the light-tailed case, $\psi_i(u) \sim h_i C e^{-\gamma u}$, where*

$$C = \frac{1 - \rho}{(\rho_L - 1)} \boldsymbol{\nu k}. \qquad (3.4)$$

To calculate C, we need two lemmas. For the first, recall the definition of $\boldsymbol{\zeta}_+, \boldsymbol{M}_+$ in Lemma 2.10.

Lemma 3.8 *As $u \to \infty$, $\big(\xi(u), J_{\tau(u)}\big)$ converges in distribution w.r.t. $\mathbb{P}_{L;i}$, with the density $g_j(x)$ (say) of the limit $\big(\xi(\infty), J_{\tau(\infty)}\big)$ at $\xi(\infty) = x$, $J_{\tau(\infty)} = j$ being independent of i and given by*

$$g_j(x) = \frac{1}{\boldsymbol{\zeta}_+^L \boldsymbol{M}_+^L \boldsymbol{e}} \sum_{\ell \in E} \zeta_+^{L;\ell} G_+^L\big(\ell, j; (x, \infty)\big). $$

Proof. We shall need to invoke the concept of semi-regeneration, see A.1f. Interpreting the ladder points as semi-regeneration points (the types being the environmental states in which they occur), $\{(\xi(u), J_{\tau(u)})\}$ is semi-regenerative with the first semi-regeneration point being $(\xi(0), J_{\tau(0)}) = (S_{\tau_+}, J_{\tau_+})$. The formula for $g_j(x)$ now follows immediately from Proposition A1.7, noting that the non-lattice property is obvious because all $G_+^L(\ell, j; \cdot)$ have densities. $\qquad \square$

Lemma 3.9 $K^L = \Delta^{-1} K \Delta - \gamma I, \ \widehat{G}^L_+[-\gamma] = \Delta^{-1} \|G_+\| \Delta, \ \widehat{G}_+[\gamma] h = h.$

Proof. Appealing to the occupation measure interpretation of K, cf. Corollary 2.6, we get for $x < 0$ that

$$
\begin{aligned}
e_i^{\mathsf{T}} e^{-Kx} e_j \, \mathrm{d}x &= \int_0^{\infty} \mathbb{P}_i \big(S_t \in \mathrm{d}x, J_t = j, \tau_+ > t \big) \, \mathrm{d}t \\
&= \frac{h_i}{h_j} e^{-\gamma x} \int_0^{\infty} \mathbb{P}_{L;i} \big(S_t \in \mathrm{d}x, J_t = j, \tau_+ > t \big) \, \mathrm{d}t \\
&= \frac{h_i}{h_j} e^{-\gamma x} e'_i e^{-K^L x} e_j \, \mathrm{d}x,
\end{aligned}
$$

which is equivalent to the first statement of the lemma. The proof of the second is a similar but easier application of the basic likelihood ratio identity Proposition 3.2. In the same way we get $\widehat{G}_+[\gamma] = \Delta \|G^L_+\| \Delta^{-1}$, and since $\|G^L_+\| e = e$, it follows that

$$
\widehat{G}_+[\gamma] h = \Delta \|G^L_+\| \Delta^{-1} h = \Delta \|G^L_+\| e = \Delta e = h.
$$

$\square$

Proof of Theorem 3.7. Using Lemma 3.8, we get

$$
\begin{aligned}
\mathbb{E}_L \big[e^{-\gamma \xi(\infty)}; J_{\tau(\infty)} = j \big] &= \int_0^{\infty} e^{-\gamma x} g_j(x) \, \mathrm{d}x \\
&= \frac{1}{\zeta^L_+ M^L_+ e} \sum_{\ell \in E} \zeta^{L;\ell}_+ \int_0^{\infty} e^{-\gamma x} G^L_+ \big(\ell, j; (x, \infty) \big) \, \mathrm{d}x \\
&= \frac{1}{\zeta^L_+ M^L_+ e} \sum_{\ell \in E} \zeta^{L;\ell}_+ \int_0^{\infty} \frac{1}{\gamma} (1 - e^{-\gamma x}) G^L_+ (\ell, j; \mathrm{d}x) \\
&= \frac{1}{\gamma \zeta^L_+ M^L_+ e} \sum_{\ell \in E} \zeta^{L;\ell}_+ \big(\|G^L_+(\ell, j)\| - \widehat{G}^L_+[\ell, j; -\gamma] \big).
\end{aligned}
$$

In matrix formulation, this means that

$$
\begin{aligned}
C &= \mathbb{E}_{L;i} \frac{e^{-\gamma \xi(\infty)}}{h_{J_{\tau(\infty)}}} = \frac{1}{\gamma \zeta^L_+ M^L_+ e} \zeta^L_+ \big(\|G^L_+\| - \widehat{G}^L_+[-\gamma] \big) \Delta^{-1} e \\
&= \frac{1}{\gamma \zeta^L_+ M^L_+ e} \zeta^L_+ \big(I - \widehat{G}^L_+[-\gamma] \big) \Delta^{-1} e \\
&= \frac{1}{\gamma (\rho_L - 1)} (-\pi^L K^L) \big(I - \widehat{G}^L_+[-\gamma] \big) \Delta^{-1} e,
\end{aligned}
$$

using Lemma 2.10 for the last two equalities. Inserting first Lemma 3.9 and next Lemma 2.8, this becomes

$$\frac{1}{\gamma(\rho_L - 1)}\boldsymbol{\pi}^L\boldsymbol{\Delta}^{-1}(\gamma\boldsymbol{I} - \boldsymbol{K})(\boldsymbol{I} - \|\boldsymbol{G}_+\|)\boldsymbol{e}$$

$$= \frac{1 - \rho}{\gamma(\rho_L - 1)}\boldsymbol{\pi}^L\boldsymbol{\Delta}^{-1}(\gamma\boldsymbol{I} - \boldsymbol{K})\boldsymbol{k} = \frac{1 - \rho}{(\rho_L - 1)}\boldsymbol{\pi}^L\boldsymbol{\Delta}^{-1}\boldsymbol{k}.$$

Thus, to complete the proof it only remains to check that $\boldsymbol{\pi}^L = \boldsymbol{\nu}\boldsymbol{\Delta}$. The normalization $\boldsymbol{\nu}\boldsymbol{h} = 1$ ensures $\boldsymbol{\nu}\boldsymbol{\Delta}\boldsymbol{e} = 1$. Finally,

$$\boldsymbol{\nu}\boldsymbol{\Delta}\boldsymbol{\Lambda}_L = \boldsymbol{\nu}\boldsymbol{\Delta}\boldsymbol{\Delta}^{-1}\boldsymbol{K}[\gamma]\boldsymbol{\Delta} = 0$$

since by definition $\boldsymbol{\nu}\boldsymbol{K}[\gamma] = \kappa(\gamma)\boldsymbol{\nu} = 0$. $\square$

3b   Ramifications of Lundberg's inequality

We consider first the time-dependent version of Lundberg's inequality, cf. V.4. The idea is as there to substitute $T = yu$ in $\psi_i(u, T)$ and to replace the Lundberg exponent γ by $\gamma_y = \alpha_y - y\kappa(\alpha_y)$, where α_y is the unique solution of

$$\kappa'(\alpha_y) = \frac{1}{y}. \tag{3.5}$$

Graphically, the situation is just as in Fig. V.1. Thus, one always has $\gamma_y > \gamma$, whereas $\alpha_y > \gamma$, $\kappa(\alpha_y) > 0$ when $y < 1/\kappa'(\gamma)$, and $\alpha_y < \gamma$, $\kappa(\alpha_y) < 0$ when $y > 1/\kappa'(\gamma)$.

Theorem 3.10 *Let* $C_+^{(0)}(y) = \dfrac{1}{\min_{i \in E} h_i^{(\alpha_y)}}$. *Then*

$$\psi_i(u, yu) \leq C_+^{(0)}(y)h_i^{(\alpha_y)}e^{-\gamma_y u}, \quad y < \frac{1}{\kappa'(\gamma)}, \tag{3.6}$$

$$\psi_i(u) - \psi_i(u, yu) \leq C_+^{(0)}(y)h_i^{(\alpha_y)}e^{-\gamma_y u}, \quad y > \frac{1}{\kappa'(\gamma)}. \tag{3.7}$$

Proof. Consider first the case $y < 1/\kappa'(\gamma)$. Then, since $\kappa(\alpha_y) > 0$, (3.1) yields

$$
\begin{aligned}
& \psi_i(u, yu) \\
&= h_i^{(\alpha_y)} \mathbb{E}_{\alpha_y; i} \Big[\frac{1}{h_{J_{\tau(u)}}^{(\alpha_y)}} \exp\big\{ -\alpha_y S_{\tau(u)} + \tau(u)\kappa(\alpha_y) \big\} ; \tau(u) \le yu \Big] \\
&= h_i^{(\alpha_y)} \mathrm{e}^{-\alpha_y u} \mathbb{E}_{\alpha_y; i} \Big[\frac{1}{h_{J_{\tau(u)}}^{(\alpha_y)}} \exp\big\{ -\alpha_y \xi(u) + \tau(u)\kappa(\alpha_y) \big\} ; \tau(u) \le yu \Big] \\
&\le h_i^{(\alpha_y)} C_+^{(0)}(y) \mathrm{e}^{-\alpha_y u} \mathbb{E}_{\alpha_y; i} \big[\mathrm{e}^{\tau(u)\kappa(\alpha_y)}; \tau(u) \le yu \big] \\
&\le h_i^{(\alpha_y)} C_+^{(0)}(y) \mathrm{e}^{-\alpha_y u + yu\kappa(\alpha_y)}.
\end{aligned}
$$

Similarly, if $y > 1/\kappa'(\gamma)$, we have $\kappa(\alpha_y) < 0$ and get

$$
\begin{aligned}
& \psi_i(u) - \psi_i(u, yu) \\
&= h_i^{(\alpha_y)} \mathrm{e}^{-\alpha_y u} \mathbb{E}_{\alpha_y; i} \Big[\frac{1}{h_{J_{\tau(u)}}^{(\alpha_y)}} \exp\big\{ -\alpha_y \xi(u) + \tau(u)\kappa(\alpha_y) \big\} ; yu < \tau(u) < \infty \Big] \\
&\le h_i^{(\alpha_y)} C_+^{(0)}(y) \mathrm{e}^{-\alpha_y u} \mathbb{E}_{\alpha_y; i} \big[\mathrm{e}^{\tau(u)\kappa(\alpha_y)}; yu < \tau(u) < \infty \big] \\
&\le h_i^{(\alpha_y)} C_+^{(0)}(y) \mathrm{e}^{-\alpha_y u + yu\kappa(\alpha_y)}.
\end{aligned}
$$

$\square$

Note that the proof appears to use less information than is inherent in the definition (3.5). However, as in the classical case (3.5) will produce the maximal γ_y for which the argument works.

Our next objective is to improve upon the constant in front of $\mathrm{e}^{-\gamma u}$ in Lundberg's inequality as well as to supplement with a lower bound:

Theorem 3.11 *Let*

$$
\begin{aligned}
C_- &= \min_{j \in E} \frac{1}{h_j} \cdot \inf_{x \ge 0} \frac{\overline{B}_j(x)}{\int_x^\infty \mathrm{e}^{\gamma(y-x)} B_j(\mathrm{d}y)}, \\
C_+ &= \max_{j \in E} \frac{1}{h_j} \cdot \sup_{x \ge 0} \frac{\overline{B}_j(x)}{\int_x^\infty \mathrm{e}^{\gamma(y-x)} B_j(\mathrm{d}y)}.
\end{aligned}
\tag{3.8}
$$

Then for all $i \in E$ and all $u \ge 0$,

$$
C_- h_i \mathrm{e}^{-\gamma u} \le \psi_i(u) \le C_+ h_i \mathrm{e}^{-\gamma u}.
\tag{3.9}
$$

For the proof, we shall need the matrices $\boldsymbol{G}_+$ and $\boldsymbol{R}$ of Section 2. We further write $\overline{\boldsymbol{G}}(u)$ for the vector with ith component $\overline{G}_i(u) = \sum_{j \in E} G_+\big(i, j; (u, \infty)\big)$

and, for a vector $\boldsymbol{\varphi}(u) = (\varphi_i(u))_{i \in E}$ of functions, we let $\boldsymbol{G}_+ * \boldsymbol{\varphi}(u)$ be the vector with ith component

$$\sum_{j \in E} (G_+(i,j) * \varphi_j)(u) = \sum_{j \in E} \int_0^u \varphi_j(u-y) G_+(i,j;dy).$$

Lemma 3.12 *Assume* $\sup_{i,u} |\varphi_i^{(0)}(u)| < \infty$, *and define* $\boldsymbol{\varphi}^{(n+1)}(u) = \overline{\boldsymbol{G}}(u) + (\boldsymbol{G}_+ * \boldsymbol{\varphi}^{(n)})(u)$. *Then* $\varphi_i^{(n)}(u) \to \psi_i(u)$ *as* $n \to \infty$.

Proof. Write $\boldsymbol{U}_N = \sum_0^N \boldsymbol{G}_+^{*n}$, $\boldsymbol{U} = \boldsymbol{U}_\infty = \sum_0^\infty \boldsymbol{G}_+^{*n}$. Then iterating the defining equation $\boldsymbol{\varphi}^{(n+1)} = \overline{\boldsymbol{G}} + \boldsymbol{G}_+ * \boldsymbol{\varphi}^{(n)}$ we get

$$\boldsymbol{\varphi}^{(N+1)} = \boldsymbol{U}_N * \overline{\boldsymbol{G}} + \boldsymbol{G}_+^{*(N+1)} * \boldsymbol{\varphi}^{(0)}.$$

However, if $\tau_+(n)$ is the nth ladder epoch, we have

$$\left[\overline{\boldsymbol{G}^{*(N+1)}} * \boldsymbol{\varphi}^{(0)} \right]_i (u) \leq \sup_{i,u} |\varphi_i^{(0)}(u)| \mathbb{P}_i(\tau_+(N+1) < \infty) \to 0.$$

Hence $\lim \boldsymbol{\varphi}^{(n)}$ exists and equals $\boldsymbol{U} * \overline{\boldsymbol{G}}$.

To see that the ith component of $\boldsymbol{U} * \overline{\boldsymbol{G}}(u)$ equals $\psi_i(u)$, just note that the recursion $\boldsymbol{\varphi}^{(n+1)} = \overline{\boldsymbol{G}} + \boldsymbol{G}_+ * \boldsymbol{\varphi}^{(n)}$ holds for the particular case where $\varphi_i^{(n)}(u)$ is the probability of ruin after at most n ladder steps and that then obviously $\varphi_i^{(n)}(u) \to \psi_i(u)$, $n \to \infty$. □

Lemma 3.13 *For all* i *and* u,

$$C_- \sum_{j \in E} h_j \int_u^\infty e^{\gamma(y-u)} G_+(i,j;dy) \leq \overline{G}_i(u) \leq C_+ \sum_{j \in E} h_j \int_u^\infty e^{\gamma(y-u)} G_+(i,j;dy).$$

Proof. According to (2.2),

$$G_+(i,j;dy) = \beta_j \int_{-\infty}^0 B_j(dy-x) R(i,j;dx).$$

Thus

$$C_+ \sum_{j \in E} h_j \int_u^\infty e^{\gamma(y-u)} G_+(i,j;dy)$$

$$= C_+ \sum_{j \in E} \beta_j h_j \int_{-\infty}^0 R(i,j;dx) \int_u^\infty e^{\gamma(y-u)} B_j(dy - x)$$

$$= C_+ \sum_{j \in E} \beta_j h_j \int_{-\infty}^0 R(i,j;dx) \frac{\int_{u-x}^\infty e^{\gamma(y-u+x)} B_j(dy)}{\overline{B}_j(u-x)} \overline{B}_j(u-x)$$

$$\geq \sum_{j \in E} \beta_j \int_{-\infty}^0 R(i,j;dx) \overline{B}_j(u-x) = \overline{G}_i(u),$$

proving the upper inequality, and the proof of the lower one is similar. □

Proof of Theorem 3.11. Let first $\varphi_i^{(0)}(u) = C_- h_i e^{-\gamma u}$ in Lemma 3.13. We claim by induction that then $\varphi_i^{(n)}(u) \geq C_- h_i e^{-\gamma u}$ for all n, from which the lower inequality follows by letting $n \to \infty$. Indeed, this is obvious if $n = 0$, and assuming it shown for n, we get

$$\varphi_i^{(n+1)}(u) = \overline{G}_i(u) + \sum_{j \in E} \int_0^u \varphi_j^{(n)}(u-y) G_+(i,j;dy) \tag{3.10}$$

$$\geq C_- \sum_{j \in E} \int_u^\infty h_j e^{\gamma(y-u)} G_+(i,j;dy)$$

$$+ C_- \sum_{j \in E} \int_0^u h_j e^{\gamma(y-u)} G_+(i,j;dy)$$

$$= C_- e^{-\gamma u} \sum_{j \in E} \widehat{G}_+[i,j;\gamma] h_j = C_- e^{-\gamma u} h_i, \tag{3.11}$$

estimating the first term in (3.10) by Lemma 3.13 and the second by the induction hypothesis, and using Lemma 3.9 for the last equality in (3.11).

The proof of the upper inequality is similar, taking $\varphi_i^{(0)}(u) = 0$. □

Here is an estimate of the rate of convergence of the finite horizon ruin probabilities $\psi_i(u,T) = \mathbb{P}_i(\tau(u) \leq T)$ to $\psi_i(u)$ which is different from Theorem 3.10:

Theorem 3.14 *Let $\gamma_0 > 0$ be the solution of $\kappa'(\gamma_0) = 0$, let $C_+(\gamma_0)$ be as in (3.8) with γ replaced by γ_0 and h_i by $h_i^{(\gamma_0)}$, and let $\delta = e^{\kappa(\gamma_0)}$. Then*

$$0 \leq \psi_i(u) - \psi_i(u,T) \leq C_+(\gamma_0) h_i^{(\gamma_0)} e^{-\gamma_0 u} \delta^T. \tag{3.12}$$

Proof. We first note that just as in the proof of Theorem 3.11, it follows that

$$\psi_i(u) \leq C_-(\gamma_0) h_i^{(\gamma_0)} e^{-\gamma_0 u}. \tag{3.13}$$

Hence, letting $M_T = \max_{0 \leq t \leq T} S_t$, we have

$$
\begin{aligned}
\psi_i(u) - \psi_i(u, T) &= \mathbb{P}_i(M > u) - \mathbb{P}_i(M_T > u) = \mathbb{P}_i(M_T \leq u, M > u) \\
&= \mathbb{P}_i(S_T \leq u, M_T \leq u, M > u) \\
&= \mathbb{E}_i\big[\psi_{J_T}(u - S_T); M_T \leq u, S_T \leq u\big] \\
&\leq C_+(\gamma_0) e^{-\gamma_0 u} \mathbb{E}_i\big[h_{J_T}^{(\gamma_0)} e^{\gamma_0 S_T}\big] \\
&= C_+ h_i^{(\gamma_0)} e^{-\gamma_0 u} \delta^T .
\end{aligned}
$$

$\square$

Notes and references   The results and proofs are from Asmussen & Rolski [98]. Further related discussion is given in Grigelionis [435, 436].

Jasiulewicz [503] uses an integral equation approach to study the ruin probability in a Markov-modulated model with surplus-dependent premium rates, for approaches involving systems of IDEs see Siegl & Tichy [805] and Lu & Li [610]. For moments of discounted aggregate claims, see Kim & Kim [532]. Yin, Liu & Yang [906] deal with effects of state-space reduction of J_t on Lundberg-type bounds for the ruin probability. Zhu & Yang [921] investigate general regularity issues for ruin-related functions in a Markovian environment. For the stability of ruin probabilities w.r.t. parameter changes, see Enikeeva, Kalashnikov & Rusaityte [355]. Discrete-time models with Markovian environment are e.g. studied in Reinhard & Snoussi [731] and Wagner [866].

4   Comparisons with the compound Poisson model

4a   Ordering of the ruin functions

For two risk functions ψ', ψ'', we define the stochastic ordering by $\psi' \prec_{\text{st}} \psi''$ if

$$\psi'(u) \leq \psi''(u), \qquad u \geq 0. \tag{4.1}$$

Obviously, this corresponds to the usual stochastic ordering of the maxima M', M'' of the corresponding two claim surplus processes (note that $\psi'(u) = \mathbb{P}(M' > u)$, $\psi''(u) = \mathbb{P}(M'' > u)$).

Now consider the risk process in a Markovian environment and define $\psi_\pi(u) = \sum_{i \in E} \pi_i \psi_i(u)$. It was long conjectured that $\psi^* \prec_{\text{st}} \psi_\pi$, where $\psi^*(u)$ is the ruin probability for the averaged compound Poisson model defined in Section 1 and $\psi_{\boldsymbol{\pi}}$ is the one for the Markov-modulated one in the stationary case (the distribution of J_0 is $\boldsymbol{\pi}$). The motivation that such a result should be true came in part from numerical studies, in part from the folklore principle that any added stochastic variation increases the risk, and finally in part from queueing theory, where it has been observed repeatedly that Markov-modulation increases waiting times and in fact some partial results had been obtained. The results to be presented show that quite often this is so, but that in general the picture is more diverse.

The conditions which play a role in the following are:

$$\beta_1 \leq \beta_2 \ldots \leq \beta_p. \tag{4.2}$$

$$B_1 \prec_{\text{st}} B_2 \prec_{\text{st}} \ldots \prec_{\text{st}} B_p. \tag{4.3}$$

$$\textit{The Markov process } \{J_t\} \textit{ is stochastically monotone.} \tag{4.4}$$

To avoid trivialities, we also assume that there exist $i \neq j$ such that either $\beta_i < \beta_j$ or $B_i \neq B_j$. Occasionally we strengthen (4.3) to

$$B = B_i \quad \textit{does not depend on } i. \tag{4.5}$$

Note that whereas (4.2) alone just amounts to an ordering of the states, this is not the case for (4.3). For the notion of monotone Markov processes, we refer to Müller & Stoyan [653]; note that (4.4) is automatic in some simple examples like birth-death processes or $p = 2$. Conditions (4.2)–(4.4) say basically that if $i < j$, then j is the more risky state, and it is in fact easy to show that $\psi_i(u) \leq \psi_j(u)$ (this is used in the derivation of (4.9) below).

Theorem 4.1 *Assume that conditions* (4.2)–(4.4) *hold. Then* $\psi^* \prec_{\text{st}} \psi_{\boldsymbol{\pi}}$.

For the proof, we need two lemmas. The first is a standard result going back to Chebycheff and appearing in a more general form in Esary, Proschan & Walkup [357], the second follows from an extension of Theorem III.5.5 (cf. also Proposition 2.1) which with basically the same proof can be found in Asmussen & Schmidt [103].

Lemma 4.2 *If* $a_1 \leq \ldots \leq a_p$, $b_1 \leq \ldots \leq b_p$ *and* $\pi_i > 0$ $(i = 1, \ldots, p)$, $\sum_1^p \pi_i = 1$, *then*

$$\sum_{i=1}^{p} \pi_i a_i b_i \geq \sum_{i=1}^{p} \pi_i a_i \sum_{j=1}^{p} \pi_j b_j.$$

The equality holds if and only if $a_1 = \ldots = a_p$ *or* $b_1 = \ldots = b_p$.

Lemma 4.3 (a) $\mathbb{P}_\pi\big(J_{\tau(0)} = i, \tau(0) < \infty\big) = \rho\pi_i^{(+)}$, where $\pi_i^{(+)} = \beta_i \mu_{B_i} \pi_i / \rho$;
(b) $\mathbb{P}_\pi\big(S_{\tau(0)} \in dx \mid J_{\tau(0)} = i, \tau(0) < \infty\big) = \overline{B}_i(x)\, dx / \mu_{B_i}$.

Proof of Theorem 4.1. Conditioning upon the first ladder epoch, we obtain (cf. Proposition 2.1 for the first term in (4.7) and Lemma 4.3 for the second)

$$\psi^*(u) \;=\; \beta^* \overline{B^*}(u) + \beta^* \int_0^u \psi^*(u-x)\overline{B^*}(x)\, dx, \tag{4.6}$$

$$\psi_\pi(u) \;=\; \beta^* \overline{B^*}(u) + \rho \sum_{i=1}^p \pi_i^{(+)} \int_0^u \psi_i(u-x)\overline{B}_i(x)/\mu_{B_i}\, dx \tag{4.7}$$

$$=\; \beta^* \overline{B^*}(u) + \int_0^u \sum_{i=1}^p \pi_i \beta_i \overline{B}_i(x)\psi_i(u-x)\, dx \tag{4.8}$$

$$\geq\; \beta^* \overline{B^*}(u) + \int_0^u \sum_{i=1}^p \pi_i \beta_i \overline{B}_i(x) \cdot \sum_{i=1}^p \pi_i \psi_i(u-x)\, dx \tag{4.9}$$

$$=\; \beta^* \overline{B^*}(u) + \beta^* \int_0^u \overline{B}(x)\psi_\pi(u-x)\, dx. \tag{4.10}$$

Here (4.9) follows by considering the increasing functions $\beta_i \overline{B}_i(x)$ and $\psi_i(u-x)$ of i and using Lemma 4.2. Comparing (4.10) and (4.6), it follows by a standard argument from renewal theory that ψ_π dominates the solution ψ^* to the renewal equation (4.6). $\qquad\square$

Here is a counterexample showing that the inequality $\psi^*(u) \leq \psi_\pi(u)$ is not in general true:

Proposition 4.4 *Assume that* $\beta_i \mu_i < 1$ *for all* i, *that*

$$\sum_{i=1}^p \pi_i \beta_i^2 \mu_{B_i} \;<\; \sum_{i=1}^p \pi_i \beta_i \cdot \sum_{i=1}^p \pi_i \beta_i \mu_{B_i}, \tag{4.11}$$

and that Λ *has the form* $\epsilon \Lambda_0$ *for some fixed intensity matrix* Λ_0. *Then* $\psi^*(u) \leq \psi_\pi(u)$ *fails for all sufficiently small* $\epsilon > 0$.

Proof. Since $\psi_\pi(0) = \psi^*(0)$, it is sufficient to show that $\psi_\pi'(0) < \psi^{*'}(0)$ for ϵ small enough. Using (4.6), (4.8) we get

$$\psi^{*'}(0) \;=\; -\beta^* + \beta^* \psi^*(0) \;=\; \sum_{i=1}^p \pi_i \beta_i \cdot \sum_{i=1}^p \pi_i \beta_i \mu_{B_i} \;-\; \beta^*,$$

$$\psi_\pi'(0) \;=\; \sum_{i=1}^p \pi_i \beta_i \psi_i(0) \;-\; \beta^*.$$

But it is intuitively clear (see Theorem 3.2.1 of [370] for a formal proof) that $\psi_i(u)$ converges to the ruin probability for the compound Poisson model with parameters β_i, B_i as $\epsilon \downarrow 0$. For $u = 0$, this ruin probability is $\beta_i \mu_{B_i}$, and from this the claim follows.   □

To see that Proposition 4.4 is not vacuous, let

$$\boldsymbol{\pi} = \begin{pmatrix} 1/2 & 1/2 \end{pmatrix}, \quad \beta_1 = 10^{-3}, \ \beta_2 = 1, \ \mu_{B_1} = 10^2, \ \mu_{B_2} = 10^{-4}.$$

Then the l.h.s. of (4.11) is of order 10^{-4} and the r.h.s. of order 10^{-1}.

Notes and references The results are from Asmussen, Frey, Rolski & Schmidt [78]. As is seen, they are at present not quite complete. What is missing in relation to Theorem 4.1 and Proposition 4.4 is the understanding of whether the stochastic monotonicity condition (4.4) is essential (the present authors conjecture it is).

4b   Ordering of adjustment coefficients

Despite the fact that $\psi^*(u) \leq \psi_{\boldsymbol{\pi}}(u)$ may fail for some u, it will hold for all sufficiently large u, except possibly for a very special situation. Recall that the adjustment coefficient for the Markov-modulated model is defined as the solution $\gamma > 0$ of $\kappa(\gamma) = 0$ where $\kappa(\alpha)$ is the eigenvalue with maximal real part of the matrix $\boldsymbol{\Lambda} + (\kappa_i(\alpha))_{\mathrm{diag}}$ where $\kappa_i(\alpha) = \beta_i(\widehat{B}_i[\alpha] - 1) - \alpha$. The adjustment coefficient γ^* for the averaged compound Poisson model is the solution > 0 of $\kappa^*(\gamma^*) = 0$ where

$$\kappa^*(\alpha) = \beta^*(\widehat{B}^*[\alpha] - 1) - \alpha = \sum_{i \in E} \pi_i \kappa_i(\alpha). \tag{4.12}$$

Theorem 4.5 $\gamma \leq \gamma^*$, *with equality only when* $\kappa_i(\gamma^*)$ *does not depend on* $i \in E$.

Lemma 4.6 *Let* $(\delta_i)_{i \in E}$ *be a given set of constants satisfying* $\sum_{i \in E} \pi_i \delta_i = 0$ *and define* $\lambda(\alpha)$ *as the eigenvalue with maximal real part of the matrix* $\boldsymbol{\Lambda} + \alpha(\delta_i)_{\mathrm{diag}}$. *Then* $\lambda(\alpha) \geq 0$, *with strict inequality unless* $\alpha = 0$ *or* $\delta_i = 0$ *for all* $i \in E$.

Proof. Define

$$X_t = \int_0^t \delta_{J_s} \, ds.$$

Then $\{(J_t, X_t)\}$ is a Markov additive process (a so-called Markovian fluid model, cf. e.g. Asmussen [62]) as discussed in III.5, and by Proposition III.4.2 we have

$$\left(\mathbb{E}_i[e^{\alpha X_t}; J_t = j] \right)_{i,j \in E} = e^{\boldsymbol{\Lambda} + \alpha(\delta_i)_{\mathrm{diag}}}.$$

Further (see Corollary III.4.7) λ is convex with

$$\lambda'(0) \;=\; \lim_{t\to\infty} \frac{\mathbb{E}X_t}{t} \;=\; \sum_{i\in E} \pi_i \delta_i \;=\; 0, \tag{4.13}$$

$$\lambda''(0) \;=\; \lim_{t\to\infty} \frac{\mathbb{V}ar\, X_t}{t}. \tag{4.14}$$

By convexity, (4.13) implies $\lambda(\alpha) \geq 0$ for all α.

Now we can view $\{X_t\}$ as a cumulative process (see 1d) with generic cycle

$$\omega \;=\; \inf\bigl\{t > 0 :\, J_{t-} \neq k,\, J_t = k \,\big|\, J_0 = k\bigr\}$$

(the return time of k) where $k \in E$ is some arbitrary but fixed state. It is clear that the distribution of X_ω is non-degenerate except when δ_i does not depend on $i \in E$, which in view of $\sum_{i\in E} \pi_i \delta_i = 0$ is only possible if $\delta_i = 0$ for all $i \in E$. Hence if $\delta_i \neq 0$ for some $i \in E$, it follows by Proposition A1.4(b) that the limit in (4.14) is non-zero so that $\lambda''(0) > 0$. This implies that λ is strictly convex, in particular $\lambda(\alpha) > 0$ for all $\alpha \neq 0$. □

Proof of Theorem 4.5. Let $\delta_i = \kappa_i(\gamma^*)$, $\alpha = 1$ in Lemma 4.6. Then $\sum \pi_i \delta_i = 0$ because of (4.12) and $\kappa^*(\gamma^*) = 0$. Further $\lambda(1) = \kappa(\gamma^*)$ by definition of $\lambda(\cdot)$ and $\kappa(\cdot)$. Hence $\kappa(\gamma^*) \geq 0$. Since κ is convex with $\kappa'(0) < 0$, this implies that the solution $\gamma > 0$ of $\kappa(\gamma) = 0$ must satisfy $\gamma \leq \gamma^*$. If $\kappa_i(\gamma^*)$ is not a constant function of $i \in E$, we get $\kappa(\gamma^*) > 0$ which in a similar manner implies that $\gamma < \gamma^*$. □

Notes and references   Theorem 4.5 is from Asmussen & O'Cinneide [93], improving upon more incomplete results from Asmussen, Frey, Rolski & Schmidt [78].

4c   Sensitivity estimates for the adjustment coefficient

Now assume that the intensity matrix for the environment is $\boldsymbol{\Lambda}_\epsilon = \boldsymbol{\Lambda}_0/\epsilon$, whereas the β_i and B_i are fixed. The corresponding adjustment coefficient is denoted by $\gamma(\epsilon)$. Thus $\gamma(\epsilon) \to \gamma^*$ as $\epsilon \downarrow 0$, and our aim is to compute the sensitivity

$$\left.\frac{\partial\gamma}{\partial\epsilon}\right|_{\epsilon=0}.$$

A dual result deals with the limit $\epsilon \to \infty$. Here we put $a = 1/\epsilon$, note that $\gamma(a) \to \min_{i=1,\dots,p} \gamma_i$ and compute

$$\left.\frac{\partial\gamma}{\partial a}\right|_{a=0}.$$

In both cases, the basic equation is $\big(\mathbf{\Lambda} + (\kappa_i(\gamma))_{\text{diag}}\big)\boldsymbol{h} = 0$, where $\mathbf{\Lambda}, \gamma, \boldsymbol{h}$ depend on the parameter (ϵ or a).

In the case of ϵ, multiply the basic equation by ϵ to obtain

$$\mathbf{0} = \big(\mathbf{\Lambda}_0 + \epsilon(\kappa_i(\gamma))_{\text{diag}}\big)\boldsymbol{h},$$

$$\mathbf{0} = \big((\kappa_i(\gamma))_{\text{diag}} + \epsilon\gamma'(\kappa_i'(\gamma))_{\text{diag}}\big)\boldsymbol{h} + \big(\mathbf{\Lambda}_0 + \epsilon(\kappa_i(\gamma))_{\text{diag}}\big)\boldsymbol{h}'. \quad (4.15)$$

Normalizing $\boldsymbol{h}$ by $\boldsymbol{\pi}\boldsymbol{h} = 0$, we have $\boldsymbol{\pi}\boldsymbol{h}' = 0$, $\boldsymbol{h}(0) = \boldsymbol{e}$. Hence letting $\epsilon = 0$ in (4.15) yields

$$\mathbf{0} = \big(\kappa_i(\gamma^*)\big)_{\text{diag}}\boldsymbol{e} + \mathbf{\Lambda}_0\boldsymbol{h}'(0) = \big(\kappa_i(\gamma^*)\big)_{\text{diag}}\boldsymbol{e} + (\mathbf{\Lambda}_0 - \boldsymbol{e}\boldsymbol{\pi})\boldsymbol{h}'(0),$$

$$\boldsymbol{h}'(0) = -(\mathbf{\Lambda}_0 - \boldsymbol{e}\boldsymbol{\pi})^{-1}\big(\kappa_i(\gamma^*)\big)_{\text{diag}}\boldsymbol{e}. \quad (4.16)$$

Differentiating (4.15) once more and letting $\epsilon = 0$ we get

$$\mathbf{0} = 2\gamma'(0)\big(\kappa_i'(\gamma^*)\big)_{\text{diag}}\boldsymbol{e} + 2\big(\kappa_i(\gamma^*)r\big)_{\text{diag}}\boldsymbol{h}'(0) + \mathbf{\Lambda}_0\boldsymbol{h}''(0), \quad (4.17)$$

$$\mathbf{0} = 2\gamma'(0)\rho + 2\boldsymbol{\pi}\big(\kappa_i(\gamma^*)\big)_{\text{diag}}\boldsymbol{h}'(0), \quad (4.18)$$

multiplying (4.17) by $\boldsymbol{\pi}$ to the left to get (4.18). Inserting (4.16) yields

Proposition 4.7 $\left.\dfrac{\partial\gamma}{\partial\epsilon}\right|_{\epsilon=0} = \dfrac{1}{\rho}\boldsymbol{\pi}\big(\kappa_i'(\gamma^*)\big)_{\text{diag}}(\mathbf{\Lambda}_0 - \boldsymbol{e}\boldsymbol{\pi})^{-1}\big(\kappa_i(\gamma^*)\big)_{\text{diag}}\boldsymbol{e}$.

Now turn to the case of a. We assume that

$$0 < \gamma_1 < \gamma_i, \quad i = 2, \dots, p. \quad (4.19)$$

Then $\gamma \to \gamma_1$ as $a \downarrow 0$ and we may take $\boldsymbol{h}(0) = \boldsymbol{e}_1$ (the first unit vector). We get

$$\mathbf{0} = \big(a\mathbf{\Lambda}_0 + (\kappa_i(\gamma))_{\text{diag}}\big)\boldsymbol{h},$$

$$\mathbf{0} = \big(\mathbf{\Lambda}_0 + \gamma'(\kappa_i'(\gamma))_{\text{diag}}\big)\boldsymbol{h} + \big(a\mathbf{\Lambda}_0 + (\kappa_i(\gamma))_{\text{diag}}\big)\boldsymbol{h}'. \quad (4.20)$$

Letting $a = 0$ in (4.20) and multiplying by $\boldsymbol{e}_1$ to the left we get $0 = \lambda_{11} + \gamma'(0)\kappa_1'(0) + 0$ (here we used $\kappa_1(\gamma(0)) = 0$ to infer that the first component of $\boldsymbol{K}[\gamma(0)]\boldsymbol{h}'(0)$ is 0), and we have proved:

Proposition 4.8 *If* (4.19) *holds, then* $\left.\dfrac{\partial\gamma}{\partial a}\right|_{a=0} = -\dfrac{\lambda_{11}}{\kappa_1'(0)}$.

Notes and references The results are from Asmussen, Frey, Rolski & Schmidt [78]. The analogue of Proposition 4.8 when $\gamma_i < 0$ for some i is open.

5   The Markovian arrival process

We shall here briefly survey an extension of the model, which has recently received much attention in the queueing literature, and has some relevance in risk theory as well.

The additional feature of the model is the following:

- Certain transitions of $\{J_t\}$ from state i to state j are accompanied by a claim with distribution B_{ij}; the intensity for such a transition (referred to as *marked* in the following) is denoted by $\lambda_{ij}^{(2)}$ and the remaining intensity for a transition $i \to j$ by $\lambda_{ij}^{(1)}$ (thus $\lambda_{ij} = \lambda_{ij}^{(1)} + \lambda_{ij}^{(2)}$). For $i = j$, we use the convention that $\lambda_{ii}^{(2)} = \beta_i$ where β_i is the Poisson rate in state i, that $B_{ii} = B_i$, and that the $\lambda_{ii}^{(1)}$ are determined by $\boldsymbol{\Lambda} = \boldsymbol{\Lambda}^{(1)} + \boldsymbol{\Lambda}^{(2)}$ where $\boldsymbol{\Lambda}$ is the intensity matrix governing $\{J_t\}$.

Thus, the Markov-modulated compound Poisson model considered so far corresponds to $\boldsymbol{\Lambda}^{(2)} = (\beta_i)_{\mathrm{diag}}$, $\boldsymbol{\Lambda}^{(1)} = \boldsymbol{\Lambda} - (\beta_i)_{\mathrm{diag}}$, $B_{ii} = B_i$; the definition of B_{ij} is redundant for $i \neq j$.

Note that the case that $0 < q_{ij} < 1$, where q_{ij} is the probability that a transition $i \to j$ is accompanied by a claim, is covered by letting B_{ij} have an atom of size q_{ij} at 0.

Again, the claim surplus is a Markov additive process (cf. III.3). The extension of the model can also be motivated via Markov additive processes: if $\{N_t\}$ is the counting process of a point process, then $\{N_t\}$ is a Markov additive process if and only if it corresponds to an arrival mechanism of the type just considered.

Here are some main examples:

Example 5.1 (PHASE-TYPE RENEWAL ARRIVALS)   Consider a risk process where the claim sizes are i.i.d. with common distribution B, but the point process of arrivals is not Poisson but renewal with interclaim times having common distribution A of phase-type with representation $(\boldsymbol{\nu}, \boldsymbol{T})$. In the above setting, we may let $\{J_t\}$ represent the phase processes of the individual interarrival times glued together (see further IX.2 for details), and the marked transitions are then the ones corresponding to arrivals. This is the only way in which arrivals can occur, and thus

$$\beta_i = 0, \quad \boldsymbol{\Lambda}^{(1)} = \boldsymbol{T}, \quad \boldsymbol{\Lambda}^{(2)} = \boldsymbol{t}\boldsymbol{\nu}, \quad B_{ij} = B;$$

the definition of B_i is redundant because of $\beta_i = 0$. $\square$

Example 5.2 (SUPERPOSITIONS) A nice feature of the set-up is that it is closed under superposition of independent arrival streams. Indeed, let $\left\{J_t^{(1)}\right\}$, $\left\{J_t^{(2)}\right\}$ be two independent environmental processes and let $E^{(k)}$, $\mathbf{\Lambda}^{(1;k)}$, $\mathbf{\Lambda}^{(2;k)}$, $B_{ij}^{(k)}$ etc. refer to $\left\{J_t^{(k)}\right\}$. We then let (see the Appendix for the Kronecker notation)

$$E = E^{(1)} \times E^{(2)}, \quad J_t = \left(J_t^{(1)}, J_t^{(2)}\right),$$

$$\mathbf{\Lambda}^{(1)} = \mathbf{\Lambda}^{(1;1)} \oplus \mathbf{\Lambda}^{(1;2)}, \quad \mathbf{\Lambda}^{(2)} = \mathbf{\Lambda}^{(2;1)} \oplus \mathbf{\Lambda}^{(2;2)},$$

$$B_{ij,kj} = B_{ik}^{(1)}, \quad B_{ij,ik} = B_{jk}^{(2)}$$

(the definition of the remaining $B_{ij,k\ell}$ is redundant). In this way we can model, e.g., superpositions of renewal processes. □

Example 5.3 (AN INDIVIDUAL MODEL) In contrast to the collective assumptions (which underly most of the topics treated so far in this book and lead to Poisson arrivals), assume that there is a finite number N of policies. Assume further that the ith policy leads to a claim having distribution C_i after a time which is exponential, with rate α_i, say, and that the policy then expires. This means that the environmental states are of the form $i_1 i_2 \cdots i_N$ with $i_1, i_2, \ldots \in \{0, 1\}$, where $i_k = 0$ means that the kth policy has not yet expired and $i_k = 1$ that it has expired. Thus, claims occur only at state transitions for the environment so that

$$\lambda_{0i_2\cdots i_N, 1i_2\cdots i_N} = \alpha_1, \quad B_{0i_2\cdots i_N, 1i_2\cdots i_N} = C_1,$$

$$\lambda_{i_1 0\cdots i_N, i_1 1\cdots i_N} = \alpha_2, \quad B_{i_1 0\cdots i_N, i_1 1\cdots i_N} = C_2,$$

$$\vdots$$

All other off-diagonal elements of $\mathbf{\Lambda}$ are zero so that all other B_{ii} are redundant. Similarly, all $\beta_{i_1 i_2\cdots i_N}$ are zero and all B_i are redundant. Easy modifications apply to allow for

- the time until expiration of the kth policy is general phase-type rather than exponential;

- upon a claim, the kth policy enters a recovering state, possibly having a general phase-type sojourn time, after which it starts afresh.

□

Example 5.4 (A SINGLE LIFE INSURANCE POLICY) Consider the life insurance of a single policy holder which can be in one of several states, $E = \{$WORKING, RETIRED, MARRIED, DIVORCED, WIDOWED, INVALIDIZED, DEAD etc.$\}$. The individual pays at rate p_i when in state i and receives an amount having distribution B_{ij} when his/her state changes from i to j. □

Notes and references The point process of arrivals was studied in detail by Neuts [658] and is often referred to in the queueing literature as *Neuts' versatile point process*, or, more recently, as *the Markovian arrival process* (MAP). However, the idea of arrivals at transition epochs can be found in Hermann [460] and Rudemo [755].

The versatility of the set-up is even greater than for the Markov-modulated model. In fact, Hermann [460] and Asmussen & Koole [88] showed that in some appropriate sense any arrival stream to a risk process can be approximated by a model of the type studied in this section: any marked point process is the weak limit of a sequence of such models. For the Markov-modulated model, one limitation for approximation purposes is the inequality $\mathbb{V}ar\, N_t \geq \mathbb{E}N_t$ which needs not hold for all arrival streams.

Some main queueing references using the MAP are Ramaswami [722], Sengupta [794], Lucantoni [612], Lucantoni *et al.* [612], Neuts [662] and Asmussen & Perry [95]. For recent applications in risk theory, see e.g. Badescu, Drekic & Landriault [118] and Cheung & Landriault [239].

6 Risk theory in a periodic environment

6a The model

We assume as in the previous part of the chapter that the arrival mechanism has a certain time-inhomogeneity, but now exhibiting (deterministic) periodic fluctuations rather than (random) Markovian ones. Without loss of generality, let the period be 1; for $s \in E = [0,1)$, we talk of s as the 'time of the year'. The basic assumptions are as follows:

- The arrival intensity at time t of the year is $\beta(t)$ for a certain function $\beta(t)$, $0 \leq t < 1$;

- Claims arriving at time t of the year have distribution $B^{(t)}$;

- The premium rate at time t of the year is $p(t)$.

By periodic extension, we may assume that the functions $\beta(t)$, $p(t)$ and $B^{(t)}$ are defined also for $t \notin [0,1)$. Obviously, one needs to assume also (as a minimum) that they are measurable in t; from an application point of view, continuity would hold in presumably all reasonable examples.

We denote throughout the initial season by s and by $\mathbb{P}^{(s)}$ the corresponding governing probability measure for the risk process. Thus at time t the premium rate is $p(s+t)$, a claim arrives with rate $\beta(s+t)$ and is distributed according to $B^{(s+t)}$. Let

$$\beta^* = \int_0^1 \beta(t)\,\mathrm{d}t, \quad B^* = \int_0^1 B^{(t)}\frac{\beta(t)}{\beta^*}\,\mathrm{d}t, \quad p^* = \int_0^1 p(t)\,\mathrm{d}t. \qquad (6.1)$$

Then the average arrival rate is β^* and the safety loading η is $\eta = (p^* - \rho)/\rho$, where

$$\rho = \int_0^1 \beta(v)\,\mathrm{d}v \int_0^\infty x B^{(v)}(\mathrm{d}x) = \beta^* \mu_B^* .\qquad(6.2)$$

Note that ρ is the average net claim amount per unit time and $\mu^* = \rho/\beta^*$ the average mean claim size.

In a similar manner as in Proposition 1.8, one may think of the standard compound Poisson model with parameters β^*, B^*, p^* as an averaged version of the periodic model, or, equivalently, of the periodic model as arising from the compound Poisson model by adding some extra variability. Many of the results given below indicate that the averaged and the periodic model share a number of main features. In particular, it turns out that they have the same adjustment coefficient. In contrast, for Markov-modulated model typically the adjustment coefficient is larger than for the averaged model (cf. Section 4b), in agreement with the general principle of added variation increasing the risk (cf. the discussion in IV.9). The behavior of the periodic model does not need to be seen as a violation of this principle, since the added variation is deterministic, not random.

Example 6.1 As an example to be used for numerical illustration throughout this section, let $\beta(t) = 3\lambda(1 + \sin 2\pi t)$, $p(t) = \lambda$ and let $B^{(t)}$ be a mixture of two exponential distributions with intensities 3 and 7 and weights $w(t) = (1 + \cos 2\pi t)/2$ and $1 - w(t)$, respectively.

It is easily seen that $\beta^* = 3\lambda$, $p^* = \lambda$ whereas B^* is a mixture of exponential distributions with intensities 3 and 7 and weights $1/2$ for each ($1/2 = \int_0^1 w(t)\,\mathrm{d}t = \int_0^1(1 - w(t))\,\mathrm{d}t$). Thus, the average compound Poisson model is the same as in IV.(3.1) and Example 1.10, and we recall from there that the ruin probability is

$$\psi^*(u) = \frac{24}{35}\mathrm{e}^{-u} + \frac{1}{35}\mathrm{e}^{-6u}.\qquad(6.3)$$

Note that λ enters just as a scaling factor of the time axis, and thus the averaged standard compound Poisson models have the same risk for all λ. In contrast, we shall see that for the periodic model increasing λ increases the effect of the periodic fluctuations. □

Remark 6.2 Define

$$\theta(T) = \int_0^T p(t)\,\mathrm{d}t, \quad \widetilde{S}_t = S_{\theta^{-1}(t)}.$$

Then (by standard operational time arguments) $\{\widetilde{S}_t\}$ *is a periodic risk process with unit premium rate and the same infinite horizon ruin probabilities.* We assume in the rest of this section that $p(t) \equiv 1$. □

The arrival process $\{N_t\}_{t\geq 0}$ is a time-inhomogeneous Poisson process with intensity function $\{\beta(s+t)\}_{t\geq 0}$. The claim surplus process $\{S_t\}_{t\geq 0}$ is defined in the obvious way as $S_t = \sum_1^{N_t} U_i - t$. Thus, the conditional distribution of U_i given that the ith claim occurs at time t is $B^{(s+t)}$. As usual, $\tau(u) = \inf\{t > 0 : S_t > u\}$ is the time to ruin, and the ruin probabilities are

$$\psi^{(s)}(u) = \mathbb{P}^{(s)}\big(\tau(u) < \infty\big), \quad \psi^{(s)}(u,T) = \mathbb{P}^{(s)}\big(\tau(u) \leq T\big).$$

The claim surplus process $\{S_t\}$ may be seen as a Markov additive process, with the underlying Markov process $\{J_t\}$ being deterministic period motion on $E = [0,1)$, i.e.

$$J_t = (s+t) \bmod 1 \quad \mathbb{P}^{(s)}\text{-a.s.} \tag{6.4}$$

At a first sight this point of view may appear quite artificial, but it turns out to have obvious benefits in terms of guidelining the analysis of the model as a parallel of the analysis for the Markovian environment risk process.

Notes and references The model has been studied in risk theory by, e.g., Daykin *et.al.* [279], Dassios & Embrechts [273] and Asmussen & Rolski [97], [98] (the literature in the mathematical equivalent setting of queueing theory is somewhat more extensive, see the Notes to Section 7). The exposition of the present chapter is basically an extract from [98], with some variants in the proofs. Recently, Kötter & Bäuerle [558] addressed the stochastic optimization problem to minimize the ruin probability through investment in the framework of a periodic environment, see also Chapter XIV.

6b   Lundberg conjugation

Motivated by the discussion in Chapter III.4 (see in particular Remark III.4.8), we start by deriving formulas giving the m.g.f. of the claim surplus process. To this end, let

$$\kappa^*(\alpha) = \beta^*\big(\widehat{B}^*[\alpha] - 1\big) - \alpha = \int_s^{s+1} \beta(v)\big(\widehat{B}^{(v)}[\alpha] - 1\big)\,\mathrm{d}v - \alpha$$

be the c.g.f. of the averaged compound Poisson model (the last expression is independent of s by periodicity), and define

$$h(s;\alpha) = \exp\left\{-\int_0^s \Big[\beta(v)\big(\widehat{B}^{(v)}[\alpha] - 1\big) - \alpha - \kappa^*(\alpha)\Big]\,\mathrm{d}v\right\};$$

then $h(\cdot\,;\alpha)$ is periodic on $\mathbb{R}$.

Theorem 6.3 $\mathbb{E}^{(s)}e^{\alpha S_t} = \dfrac{h(s;\alpha)}{h(s+t;\alpha)}e^{t\kappa^*(\alpha)}.$

Proof. Conditioning upon whether a claim occurs in $[t, t+dt]$ or not, we obtain

$$\mathbb{E}^{(s)}\left[e^{\alpha S_{t+dt}}|\mathscr{F}_t\right]$$
$$= \left(1 - \beta(s+t)dt\right)e^{\alpha S_t - \alpha dt} + \beta(s+t)dt \cdot e^{\alpha S_t}\widehat{B}^{(s+t)}[\alpha]$$
$$= e^{\alpha S_t} \cdot \left(1 - \alpha dt + \beta(s+t)dt\left[\widehat{B}^{(s+t)}[\alpha] - 1\right]\right),$$

$$\mathbb{E}^{(s)}e^{\alpha S_{t+dt}} = \mathbb{E}^{(s)}e^{\alpha S_t}\left(1 - \alpha dt + \beta(s+t)dt\left[\widehat{B}^{(s+t)}[\alpha] - 1\right]\right),$$

$$\frac{d}{dt}\mathbb{E}^{(s)}e^{\alpha S_t} = \mathbb{E}^{(s)}e^{\alpha S_t}\left(-\alpha + \beta(s+t)\left[\widehat{B}^{(s+t)}[\alpha] - 1\right]\right),$$

$$\frac{d}{dt}\log\mathbb{E}^{(s)}e^{\alpha S_t} = -\alpha + \beta(s+t)\left[\widehat{B}^{(s+t)}[\alpha] - 1\right],$$

$$\log\mathbb{E}^{(s)}e^{\alpha S_t} = -\alpha t + \int_0^t \beta(s+v)\left(\widehat{B}^{(s+v)}[\alpha] - 1\right)dv$$
$$= \log\widetilde{h}(s+t;\alpha) - \log\widetilde{h}(s;\alpha),$$

where

$$\widetilde{h}(t;\alpha) = \exp\left\{\int_0^t \beta(v)\left(\widehat{B}^{(v)}[\alpha] - 1\right)dv - \alpha t\right\} = \frac{e^{t\kappa^*(\alpha)}}{h(t;\alpha)}.$$

Thus

$$\mathbb{E}^{(s)}e^{\alpha S_t} = \frac{\widetilde{h}(s+t;\alpha)}{\widetilde{h}(s;\alpha)} = \frac{h(s;\alpha)}{h(s+t;\alpha)}e^{t\kappa^*(\alpha)}. \tag{6.5}$$

□

Corollary 6.4 *For each θ such that the integrals in the definition of $h(t;\theta)$ exist and are finite,*

$$\{L_{\theta,t}\}_{t\geq 0} = \left\{\frac{h(s+t;\theta)}{h(s;\theta)}e^{\theta S_t - t\kappa^*(\theta)}\right\}_{t\geq 0}$$

is a $\mathbb{P}^{(s)}$-martingale with mean one.

Proof. In the Markov additive sense of (6.4), we can write

$$L_{\theta,t} = \frac{h(J_t;\theta)}{h(J_0;\theta)}e^{\theta S_t - t\kappa^*(\theta)}$$

$\mathbb{P}^{(s)}$-a.s. so that obviously $\{L_{\theta,t}\}$ is a multiplicative functional for the Markov process $\{(J_t, S_t)\}$. According to Remark III.1.8, it then suffices to note that $\mathbb{E}^{(s)}L_{\theta,t} = 1$ by Theorem 6.3. □

Remark 6.5 The formula for $h(s) = h(s; \alpha)$ as well as the fact that $\kappa = \kappa^*(\alpha)$ is the correct exponential growth rate of $\mathbb{E}e^{\alpha S_t}$ can be derived via Remark III.4.9 as follows. With $\mathscr{G}$ the generator of $\{X_t\} = \{(J_t, S_t)\}$ and $h_\alpha(s, y) = e^{\alpha y} h(s)$, the requirement is $\mathscr{G}h_\alpha(i, 0) = \kappa h(s)$. However, as above

$$\mathbb{E}^{(s)} h_\alpha(J_{\mathrm{dt}}, S_{\mathrm{dt}})$$
$$= h(s + \mathrm{dt})e^{-\alpha \mathrm{dt}}\big(1 - \beta(s)\mathrm{dt}\big) + \beta(s)\mathrm{dt} \cdot \widehat{B}^{(s)}[\alpha]h(s)$$
$$= h(s) + \mathrm{dt}\big\{-\alpha h(s) - \beta(s)h(s) + h'(s) + \beta(s)\widehat{B}^{(s)}[\alpha]h(s)\big\},$$
$$\mathscr{G}h_\alpha(s, 0) = -\alpha h(s) - \beta(s)h(s) + h'(s) + \beta(s)\widehat{B}^{(s)}[\alpha]h(s).$$

Equating this to $\kappa h(s)$ and dividing by $h(s)$ yields

$$\frac{h'(s)}{h(s)} = \alpha + \beta(s) - \beta(s)\widehat{B}^{(s)}[\alpha] + \kappa,$$
$$h(s) = \exp\left\{-\int_0^s \Big[\beta(v)\big(\widehat{B}^{(v)}[\alpha] - 1\big) - \alpha - \kappa\Big]\,\mathrm{d}v\right\}$$

(normalizing by $h(0) = 1$). That $\kappa = \kappa^*(\alpha)$ then follows by noting that $h(1) = h(0)$ by periodicity. □

For each θ satisfying the conditions of Corollary 6.4, it follows by Theorem III.1.7 that we can define a new Markov process $\{(J_t, S_t)\}$ with governing probability measures $\mathbb{P}_\theta^{(s)}$, say, such that for any s and $T < \infty$, the restrictions of $\mathbb{P}^{(s)}$ and $\mathbb{P}_\theta^{(s)}$ to $\mathscr{F}_t$ are equivalent with likelihood ratio $L_{\theta,T}$.

Proposition 6.6 *The $\mathbb{P}_\theta^{(s)}$, $0 \le s < 1$, correspond to a new periodic risk model with parameters*

$$\beta_\theta(t) = \beta(t)\widehat{B}^{(t)}[\theta], \qquad B_\theta^{(t)}(\mathrm{d}x) = \frac{e^{\theta x}}{\widehat{B}^{(t)}[\theta]}B^{(t)}(\mathrm{d}x).$$

Proof. (i) Check that m.g.f. of S_t is as for the asserted periodic risk model, cf. Proposition 6.3; (ii) use Markov-modulated approximations (Section 6c); (iii) use approximations with piecewiese constant $\beta(s), B^{(s)}$; (iv) finally, see [98] for a formal proof. □

Now define γ as the positive solution of the Lundberg equation for the averaged model. That is, γ solves $\kappa^*(\gamma) = 0$. When $\alpha = \gamma$, we put for short $h(s) = h(s; \gamma)$. A further important constant is the value γ_0 (located in $(0, \gamma)$) at which $\kappa^*(\alpha)$ attains its minimum. That is, γ_0 is determined by

$$0 = \kappa^{*'}(\gamma_0) = \beta\widehat{B}^{*'}[\gamma_0] - 1. \tag{6.6}$$

Lemma 6.7 *When $\alpha \geq \gamma_0$, $\mathbb{P}_\alpha^{(s)}\big(\tau(u) < \infty\big) = 1$ for all $u \geq 0$.*

Proof. According to (6.2), the mean number of claims per unit time is

$$
\rho_\alpha = \int_0^1 \beta(v)\,\mathrm{d}v \int_0^\infty x\mathrm{e}^{\alpha x} B^{(v)}(\mathrm{d}x)
$$
$$
= \beta^* \int_0^\infty x\mathrm{e}^{\alpha x} B^*(\mathrm{d}x) = \beta^* \widehat{B}^{*'}[\alpha] = \kappa^{*'}(\alpha) + 1,
$$

which is ≥ 1 by convexity. $\qquad\square$

The relevant likelihood ratio representation of the ruin probabilities now follows immediately from Corollary III.1.5. Here and in the following, $\xi(u) = S_{\tau(u)} - u$ is the overshoot and $\theta(u) = (\tau(u) + s) \bmod 1$ the season at the time of ruin.

Corollary 6.8 *The ruin probabilities can be computed as*

$$
\psi^{(s)}(u,T) = h(s;\alpha)\mathrm{e}^{-\alpha u}\mathbb{E}_\alpha^{(s)}\left[\frac{\mathrm{e}^{-\alpha\xi(u)+\tau(u)\kappa^*(\alpha)}}{h(\theta(u);\alpha)}; \tau(u) \leq T\right], \quad (6.7)
$$

$$
\psi^{(s)}(u) = h(s;\alpha)\mathrm{e}^{-\alpha u}\mathbb{E}_\alpha^{(s)}\frac{\mathrm{e}^{-\alpha\xi(u)+\tau(u)\kappa^*(\alpha)}}{h(\theta(u);\alpha)}, \quad \alpha \geq \gamma_0 \quad (6.8)
$$

$$
\psi^{(s)}(u) = h(s)\mathrm{e}^{-\gamma u}\mathbb{E}_\gamma^{(s)}\frac{\mathrm{e}^{-\gamma\xi(u)}}{h(\theta(u))}. \quad (6.9)
$$

To obtain the Cramér-Lundberg approximation from Corollary 3.1, we need the following auxiliary result. The proof involves machinery from the ergodic theory of Markov chains on a general state space, which is not used elsewhere in the book, and we refer to [98].

Lemma 6.9 *Assume that there exist open intervals $I \subseteq [0,1)$, $J \subseteq \mathbb{R}_+$ such that the $B^{(s)}$, $s \in I$, have components with densities $b^{(s)}(x)$ satisfying*

$$
\inf_{s\in I,\ x\in J} \beta(s)b^{(s)}(x) > 0. \quad (6.10)
$$

Then for each α, the Markov process $\big\{\big(\xi(u),\theta(u)\big)\big\}_{u\geq 0}$, considered with governing probability measures $\big\{\mathbb{P}_\alpha^{(s)}\big\}_{s\in[0,1)}$, has a unique stationary distribution, say the distribution of $\big(\xi(\infty),\theta(\infty)\big)$, and no matter what is the initial season s, $\big(\xi(u),\theta(u)\big) \xrightarrow{\mathscr{D}} \big(\xi(\infty),\theta(\infty)\big)$.

Letting $u \to \infty$ in (6.9) and noting that weak convergence entails convergence of $\mathbb{E}f\big(\xi(u),\theta(u)\big)$ for any bounded continuous function (e.g. $f(x,q) = \mathrm{e}^{-\gamma x}/h(q)$), we get:

Theorem 6.10 *Under the condition* (6.10) *of Lemma* 3.1,

$$\psi^{(s)}(u) \sim Ch(s)e^{-\gamma u}, \quad u \to \infty, \tag{6.11}$$

where $C = \mathbb{E}_\gamma \dfrac{e^{-\gamma \xi(\infty)}}{h\big(\theta(\infty)\big)}.$

Note that (6.11) gives an interpretation of $h(s)$ as a measure of how the risks of different initial seasons s vary. For our basic Example 6.1, elementary calculus yields

$$h(s) = \exp\left\{\lambda\left(\frac{1}{2\pi}\cos 2\pi s - \frac{1}{4\pi}\sin 2\pi s + \frac{1}{16\pi}\cos 4\pi s - \frac{9}{16\pi}\right)\right\}.$$

Plots of h for different values of λ are given in Fig. VII.4, illustrating that the effect of seasonality increases with λ.

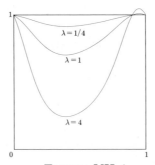

FIGURE VII.4

In contrast to h, it does not seem within the range of our methods to compute C explicitly, which may provide one among many motivations for the Markov-modulated approximation procedure to be considered in Section 6c. Among other things, this provides an algorithm for computing C as a limit. At this stage, Theorem 6.10 shows that certainly γ is the correct Lundberg exponent.

Noting that $\xi(u) \geq 0$ in (6.9), we obtain immediately the following version of Lundberg's inequality which is a direct parallel of the result given in Corollary 3.6 for the Markov-modulated model:

Theorem 6.11 $\psi^{(s)}(u) \leq C_+^{(0)}h(s)e^{-\gamma u},$ *where*

$$C_+^{(0)} = \frac{1}{\inf_{0 \leq t \leq 1} h(t)}.$$

Thus, e.g., in our basic example with $\lambda = 1$, we obtain $C_+^{(0)} = 1.42$ so that

$$\psi^{(s)}(u) \leq 1.42 \cdot \exp\left\{\frac{1}{2\pi}\cos 2\pi s - \frac{1}{4\pi}\sin 2\pi s + \frac{1}{16\pi}\cos 4\pi s - \frac{9}{16\pi}\right\}e^{-u}. \tag{6.12}$$

As for the Markovian environment model, Lundberg's inequality can be considerably sharpened and extended. We state the results below; the proofs are basically the same as in Section 3 and we refer to [98] for details.

Consider first the time-dependent version of Lundberg's inequality. Just as in V.4, we substitute $T = yu$ in $\psi(u, T)$ and replace the Lundberg exponent γ by $\gamma_y = \alpha_y - y\kappa(\alpha_y)$, where α_y is the unique solution of

$$\kappa'(\alpha_y) = \frac{1}{y}. \tag{6.13}$$

Elementary convexity arguments show that we always have $\gamma_y > \gamma$ and $\alpha_y > \gamma$, $\kappa(\alpha_y) > 0$ when $y < 1/\kappa'(\gamma)$, whereas $\alpha_y < \gamma$, $\kappa(\alpha_y) < 0$ when $y > 1/\kappa'(\gamma)$.

Theorem 6.12 *Let* $C_+^{(0)}(y) = \dfrac{1}{\inf_{0 \leq t \leq 1} h(t; \alpha_y)}$. *Then*

$$\psi^{(s)}(u, yu) \leq C_+^{(0)}(y)h(s)e^{-\gamma_y u}, \quad y < \frac{1}{\kappa'(\gamma)}, \tag{6.14}$$

$$\psi^{(s)}(u) - \psi^{(s)}(u, yu) \leq C_+^{(0)}(y)h(s)e^{-\gamma_y u}, \quad y > \frac{1}{\kappa'(\gamma)}. \tag{6.15}$$

The next result improves upon the constant $C_+^{(0)}$ in front of $e^{-\gamma u}$ in Theorem 6.11 as well as it supplements with a lower bound.

Theorem 6.13 *Let*

$$C_- = \inf_{0 \leq t \leq 1} \frac{1}{h(t)} \cdot \inf_{x \geq 0} \frac{\overline{B}^{(t)}(x)}{\int_x^\infty e^{\gamma(y-x)}B^{(t)}(\mathrm{d}y)},$$

$$C_+ = \sup_{0 \leq t \leq 1} \frac{1}{h(t)} \cdot \sup_{x \geq 0} \frac{\overline{B}^{(t)}(x)}{\int_x^\infty e^{\gamma(y-x)}B^{(t)}(\mathrm{d}y)}. \tag{6.16}$$

Then for all $s \in [0, 1)$ *and all* $u \geq 0$,

$$C_- h(s)e^{-\gamma u} \leq \psi^{(s)}(u) \leq C_+ h(s)e^{-\gamma u}. \tag{6.17}$$

In order to apply Theorem 6.13 to our basic example, we first note that the function

$$\frac{\int_u^\infty \{w \cdot 3e^{-3x} + (1-w) \cdot 7e^{-7x}\}\,\mathrm{d}x}{\int_u^\infty e^{x-u}\{w \cdot 3e^{-3x} + (1-w) \cdot 7e^{-7x}\}\,\mathrm{d}x} = \frac{6w + 6(1-w)e^{-4u}}{9w + 7(1-w)e^{-4u}}$$

attains its minimum 2/3 for $u = \infty$ and its maximum $6/(7 + 2w)$ for $u = 0$.
Thus

$$
\begin{aligned}
C_- &= \frac{2}{3} \inf_{0 \leq s \leq 1} \exp \left\{ -\lambda \left(\frac{1}{2\pi} \cos 2\pi s - \frac{1}{4\pi} \sin 2\pi s + \frac{1}{16\pi} \cos 4\pi s - \frac{9}{16\pi} \right) \right\} \\
&= \frac{2}{3} e^{-0.013\lambda},
\end{aligned}
$$

$$
C_+ = \sup_{0 \leq s \leq 1} \frac{6 \exp \left\{ -\lambda \left(\frac{1}{2\pi} \cos 2\pi s - \frac{1}{4\pi} \sin 2\pi s + \frac{1}{16\pi} \cos 4\pi s - \frac{9}{16\pi} \right) \right\}}{8 + \cos 2\pi s}.
$$

Thus e.g. for $\lambda = 1$ (where $\frac{2}{3} e^{-0.013\lambda} = 0.66$, $C_+ = 1.20$),

$$
\psi^{(s)}(u) \geq 0.66 \cdot \exp \left\{ \frac{1}{2\pi} \cos 2\pi s - \frac{1}{4\pi} \sin 2\pi s + \frac{1}{16\pi} \cos 4\pi s - \frac{9}{16\pi} \right\} e^{-u},
$$

$$
\psi^{(s)}(u) \leq 1.20 \cdot \exp \left\{ \frac{1}{2\pi} \cos 2\pi s - \frac{1}{4\pi} \sin 2\pi s + \frac{1}{16\pi} \cos 4\pi s - \frac{9}{16\pi} \right\} e^{-u}.
$$

Finally, we have the following result:

Theorem 6.14 *Let $C_+(\gamma_0)$ be as in (6.16) with γ replaced by γ_0 and $h(t)$ by $h(t; \gamma_0)$, and let $\delta = e^{\kappa^*(\gamma_0)}$. Then*

$$
0 \leq \psi^{(s)}(u) - \psi^{(s)}(u, T) \leq C_+(\gamma_0) h(s; \gamma_0) e^{-\gamma_0 u} \delta^T. \tag{6.18}
$$

Notes and references The material is from Asmussen & Rolski [98]. Some of the present proofs are more elementary by avoiding the general point process machinery of [98], but thereby also slightly longer.

6c   Markov-modulated approximations

A periodic risk model may be seen as a varying environment model, where the environment at time t is $(s + t)$ mod $1 \in [0, 1)$, with s the initial season. Of course, such a deterministic periodic environment may be seen as a special case of a Markovian one (allowing a continuous state space $E = [0, 1)$ for the environment), and in fact, much of the analysis of the preceding section is modelled after the techniques developed in the preceding sections for the case of a finite E. This observation motivates to look for a more formal connection between the periodic model and the one evolving in a finite Markovian environment.

The idea is basically to approximate the (deterministic) continuous clock by a discrete (random) Markovian one with n 'months'. Thus, the nth Markovian environmental process $\{J_t\}$ moves cyclically on $\{1, \ldots, n\}$, completing a cycle within one unit of time on the average, so that the intensity matrix is $\mathbf{\Lambda}^{(n)}$ given by

$$\mathbf{\Lambda}^{(n)} = \begin{pmatrix} -n & n & 0 & \cdots & 0 \\ 0 & -n & n & \cdots & 0 \\ \vdots & \vdots & \vdots & \ddots & \vdots \\ n & 0 & 0 & \cdots & -n \end{pmatrix}. \tag{6.19}$$

Arrivals occur at rate β_{ni} and their claim sizes are distributed according to B_{ni} if the governing Markov process is in state i. We want to choose the β_{ni} and B_{ni} in order to achieve good convergence to the periodic model. To this end, one simple choice is

$$\beta_{ni} = \beta\left(\frac{i-1}{n}\right) \quad \text{and} \quad B_{ni} = B^{((i-1)/n)}, \tag{6.20}$$

but others are also possible. We let $\left\{S_t^{(n)}\right\}_{t \geq 0}$ be the claim surplus process of the nth approximating Markov-modulated model, $M^{(n)} = \sup_{t \geq 0} S_t^{(n)}$, and the ruin probability corresponding to the initial state i of the environment is then

$$\psi_i^{(n)}(t) = \mathbb{P}_i(M^{(n)} > t), \tag{6.21}$$

which serves as an approximation to $\psi^{(s)}(u)$ whenever n is large and $i/n \approx s$.

Notes and references   See Rolski [745].

7   Dual queueing models

The essence of the results of the present section is that the ruin probabilities $\psi_i(u)$, $\psi_i(u, T)$ can be expressed in a simple way in terms of the waiting time probabilities of a queueing system with the input being the time-reversed input of the risk process. This queue is commonly denoted as the Markov-modulated M/G/1 queue and has received considerable attention in the last decades. Thus, since the settings are equivalent from a mathematical point of view, it is desirable to have formulas permitting freely to translate from one setting into the other.

Let β_i, B_i, $\mathbf{\Lambda}$ be the parameters defining the risk process in a random environment and consider a queueing system governed by a Markov process $\{J_t^*\}$ ('Markov-modulated') as follows:

- The intensity matrix for $\{J_t^*\}$ is the time-reversed intensity matrix $\mathbf{\Lambda}^* = (\lambda_{ij}^*)_{i,j \in E}$ of the risk process, $\lambda_{ij}^* = \lambda_{ji}\pi_j/\pi_i$.

- The arrival intensity is β_i when $J_t^* = i$;

- Customers arriving when $J_t^* = i$ have service time distribution B_i;

- The queueing discipline is FIFO.

The actual waiting time process $\{W_n\}_{n=1,2,\dots}$ and the virtual waiting time (workload) process $\{V_t\}_{t \geq 0}$ are defined exactly as for the renewal model in Chapter VI.

Proposition 7.1 *Assume* $V_0 = 0$. *Then*

$$\mathbb{P}_i\big(\tau(u) \leq T, J_T = j\big) \;=\; \frac{\pi_j}{\pi_i} \mathbb{P}_j\big(V_T > u, J_T^* = i\big). \tag{7.1}$$

In particular,

$$\psi_i(u, T) \;=\; \frac{1}{\pi_i} \mathbb{P}_{\boldsymbol{\pi}}\big(V_T > u, J_T^* = i\big) \;=\; \mathbb{P}_{\boldsymbol{\pi}}\big(V_T > u \,|\, J_T^* = i\big), \tag{7.2}$$

$$\psi_i(u) \;=\; \frac{1}{\pi_i} \mathbb{P}(V > u, J^* = i) \;=\; \mathbb{P}_{\boldsymbol{\pi}}(V > u \,|\, J^* = i), \tag{7.3}$$

where (V, J^*) *is the steady-state limit of* (V_t, J_t^*).

Proof. Consider stationary versions of $\{J_t\}_{0 \leq t \leq T}$, $\{J_t^*\}_{0 \leq t \leq T}$. Then we may assume that $J_t^* = J_{T-t}$, $0 \leq t \leq T$ and that the risk process $\{R_t\}_{0 \leq t \leq T}$ is coupled to the virtual waiting process $\{V_t\}_{0 \leq t \leq T}$ as in the basic duality lemma (Theorem III.2.1). The first conclusion of that result then states that the events $\{\tau(u) \leq T, J_0 = i, J_T = j\}$ and $\{V_T > u, J_0^* = j, J_T^* = i\}$ coincide. Taking probabilities and using the stationarity yields

$$\pi_i \mathbb{P}_i\big(\tau(u) \leq T, J_T = j\big) \;=\; \pi_j \mathbb{P}_j(V_T > u, J_T^* = i),$$

and (7.1) follows. For (7.2), just sum (7.1) over j, and for (7.3), let $T \to \infty$ in (7.2) and use that $\lim \mathbb{P}_j(V_T > u, J_T^* = i) = \mathbb{P}(V > u, J^* = i)$ for all j.   □

Now let I_n^* denote the environment when customer n arrives and I^* the steady-state limit.

Proposition 7.2 *The relation between the steady-state distributions of the actual and the virtual waiting time distribution is given by*

$$\mathbb{P}(W > u, I^* = i) \;=\; \frac{\beta_i}{\beta^*} \mathbb{P}(V > u, J^* = i), \tag{7.4}$$

where $\beta^* = \sum_{j \in E} \pi_j \beta_j$. *In particular,*

$$\psi_i(u) \;=\; \frac{\beta^*}{\pi_i \beta_i} \mathbb{P}(W > u, I^* = i). \tag{7.5}$$

Proof. Identifying the distribution of (W, I^*) with the time-average, we have

$$\frac{1}{N}\sum_{n=1}^{N} I(W_n > u, I_n^* = i) \overset{\text{a.s.}}{\to} \mathbb{P}(W > u, I^* = i), \quad N \to \infty.$$

However, if T is large, on average $\beta^* T$ customers arrive in $[0, T]$, and of these, on average $\beta_i T \mathbb{P}(V > u, J^* = i)$ see $W > u, I^* = i$. Taking the ratio yields (7.4), and (7.5) follows from (7.4) and (7.3). $\qquad\square$

Notes and references One of the earliest papers drawing attention to the Markov-modulated M/G/1 queue is Burman & Smith [210]. The first comprehensive solution of the waiting time problem is Regterschot & de Smit [729], a paper relying heavily on classical complex plane methods. A more probabilistic treatment was given by Asmussen [59], and further references (to which we add Prabhu & Zhu [714]) can be found therein.

Proposition 7.1 is from Asmussen [58], with (7.3) improving somewhat upon (2.7) of that paper. The relation (7.4) can be found in Regterschot & de Smit [729]; a general formalism allowing this type of conclusion is 'conditional PASTA', see Regterschot & van Doorn [327].

In the setting of the periodic model of Section 6, the dual queueing model is a periodic M/G/1 queue with arrival rate $\beta(-t)$ and service time distribution $B^{(-t)}$ at time t of the year (assuming w.l.o.g. that $\beta(t)$, $B^{(t)}$ have been periodically extended to negative t). With $\{V_t\}$ denoting the workload process of the periodic queue, $\rho < 1$ then ensures that $V^{(s)} = \lim_{N \to \infty} V_{N+s}$ exists in distribution, and one has

$$\mathbb{P}^{(s)}\big(\tau(u) \le T\big) \;=\; \mathbb{P}^{(-s-T)}(V_T > u), \tag{7.6}$$

$$\mathbb{P}^{(-s-T)}\big(\tau(u) \le T\big) \;=\; \mathbb{P}^{(s)}(V_T > u), \tag{7.7}$$

$$\mathbb{P}^{(1-s)}\big(\tau(u) < \infty\big) \;=\; \mathbb{P}^{(s)}(V^{(0)} > u). \tag{7.8}$$

For treatments of periodic M/G/1 queues, see in particular Harrison & Lemoine [450], Lemoine [579, 580], and Rolski [745].

Chapter VIII

Level-dependent risk processes

1 Introduction

We assume as in Chapter IV that the claim arrival process $\{N_t\}$ is Poisson with rate β, and that the claim sizes $U_1, U_2, \ldots$ are i.i.d. with common distribution B and independent of $\{N_t\}$. Thus, the aggregate claims in $[0, t]$ are

$$A_t = \sum_{i=1}^{N_t} U_i \tag{1.1}$$

(other terms are *accumulated claims* or *total claims*). However, the increase of the surplus process R_t in between the claim payments now does not have to be linear with constant slope, but can depend on the current surplus level. This can always be interpreted as a modified premium rate $p(r)$ charged at the current reserve $R_t = r$ (but note that the actual reason for the level dependence of the increase may be quite different, see the examples below). Thus, in between jumps, $\{R_t\}$ moves according to the differential equation $\dot{R} = p(R)$, and the evolution of the reserve may be described by the equation[1]

$$R_t = u - A_t + \int_0^t p(R_s) \, ds. \tag{1.2}$$

---

[1]Here it is assumed that $p(r)$ is a deterministic function. Stochastic $p(r)$ will be discussed in Sections 5 and 6.

As earlier,

$$\psi(u) \; = \; \mathbb{P}\Big(\inf_{t \geq 0} R_t < 0 \,\Big|\, R_0 = u\Big), \quad \psi(u, T) \; = \; \mathbb{P}\Big(\inf_{0 \leq t \leq T} R_t < 0 \,\Big|\, R_0 = u\Big)$$

denote the ruin probabilities with initial reserve u and infinite, resp. finite horizon, and $\tau(u) = \inf\{t > 0 : R_t < u\}$ is the time to ruin starting from $R_0 = u$ so that $\psi(u) = \mathbb{P}\big(\tau(u) < \infty\big)$, $\psi(u, T) = \mathbb{P}\big(\tau(u) \leq T\big)$.

The following examples provide some main motivation for studying the model:

Example 1.1 Assume that the company reduces the premium rate from p_1 to p_2 when the reserve comes above some critical value v. That is, $p_1 > p_2$ and

$$p(r) = \begin{cases} p_1 & r \leq v \\ p_2 & r > v. \end{cases} \tag{1.3}$$

One reason could be competition, where one would try to attract new customers as soon as the business has become reasonably safe. Another could be the payout of dividends: here the premium paid by the policy holders is the same for all r, but when the reserve comes above v, dividends are paid out at rate $p_1 - p_2$. Possibilities for more general level-dependent premium (dividend payment) schemes than the two-step rule above are obvious. □

Example 1.2 (INTEREST)   If the company charges a constant premium rate p but invests its money at interest rate i, we get $p(r) = p + ir$. □

Example 1.3 (ABSOLUTE RUIN)   Consider the same situation as in Example 1.2, but assume now that the company borrows the deficit in the bank when the reserve goes negative, say at interest rate i'. Thus at deficit $x > 0$ (meaning $R_t = -x$), the payout rate of interest is $i'x$ and *absolute ruin* occurs when this exceeds the premium inflow p, i.e. when $x > p/i'$, rather than when the reserve itself becomes negative. In this situation, we can put $\widetilde{R}_t = R_t + p/i'$,

$$\widetilde{p}(r) = \begin{cases} p + i(r - p/i') & r > p/i', \\ p - i'(p/i' - r) & 0 \leq r \leq p/i'. \end{cases}$$

Then the ruin problem for $\{\widetilde{R}_t\}$ is of the type defined above, and the probability of absolute ruin with initial reserve $u \in [-p/i', \infty)$ is given by $\widetilde{\psi}(u + p/i')$. □

Example 1.4 (TAX)   If the insurance company makes profit, it will have to pay tax. One way to model this is to assume that whenever the risk process R_t is in a running maximum, a certain proportion ϑ of the premium income is paid to the tax authority (such a model is related to the so-called *loss-carried-forward scheme*). The resulting premium rule is $p(r) = \vartheta\, p$ in the running maxima and $p(r) = p$ otherwise. Due to the non-Markovian character, the analysis for this model is somewhat different from the above examples, see Section 4. □

Now let us return to the general Markovian model.

Proposition 1.5 *Either $\psi(u) = 1$ for all u, or $\psi(u) < 1$ for all u.*

Proof. Obviously $\psi(u) \leq \psi(v)$ when $u \geq v$. Assume $\psi(u) < 1$ for some u. If $R_0 = v < u$, there is positive probability, say ϵ, that $\{R_t\}$ will reach level u before the first claim arrives. Hence in terms of survival probabilities, $1 - \psi(v) \geq \epsilon(1 - \psi(u)) > 0$ so that $\psi(v) < 1$. □

A basic question is thus which premium rules $p(r)$ ensure that $\psi(u) < 1$. No tractable necessary and sufficient condition is known in complete generality of the model. However, it seems reasonable to assume monotonicity ($p(r)$ is decreasing in Example 1.1 and increasing in Example 1.2) for r sufficiently large so that $p(\infty) = \lim_{r \to \infty} p(r)$ exists. This is basically covered by the following result (but note that the case $p(r) \downarrow \beta\mu_B$ requires a more detailed analysis and that $\mu_B < \infty$ is not always necessary for $\psi(u) < 1$ when $p(r) \to \infty$, cf. [APQ, pp. 388–389]):

Theorem 1.6 (a) *If $p(r) \leq \beta\mu_B$ for all sufficiently large r, then $\psi(u) = 1$ for all u;*
(b) *If $p(r) > \beta\mu_B + \epsilon$ for all sufficiently large r and some $\epsilon > 0$, then $\psi(u) < 1$ for all u, and $\mathbb{P}(R_t \to \infty) > 0$.*

Proof. This follows by a simple comparison with the compound Poisson model. Let $\psi_p(u)$ refer to the compound Poisson model with the same β, B and (constant) premium rate p.

In case (a), choose u_0 such that $p(r) \leq p = \beta\mu_B$ for $r \geq u_0$. Starting from $R_0 = u_0$, the probability that $R_t \leq u_0$ for some t is at least $\psi_p(0) = 1$ (cf. Proposition IV.1.2(d)), hence $R_t \leq u_0$ also for a whole sequence of t's converging to ∞. However, obviously $\inf_{u \leq u_0} \psi(u) > 0$, and hence by a geometric trials argument $\psi(u_0) = 1$ so that $\psi(u) = 1$ for all u by Proposition 1.5. In case (b), choose u_0 such that $p(r) \geq p = \beta\mu_B + \epsilon$ for $r \geq u_0$. Then if $u \geq u_0$, we have $\psi(u) \leq \psi_p(u - u_0)$ and, appealing to Proposition IV.1.2 once more, that $\psi_p(u - u_0) < 1$. Hence $\psi(u) < 1$ for all u by Proposition IV.1.2(d). □

We next recall the following result, which was proved in III.3. Here $\{V_t\}_{t \geq 0}$ is a storage process which has reflection at zero and initial condition $V_0 = 0$.

In between jumps, $\{V_t\}$ decreases at rate $p(v)$ when $V_t = v$ (i.e., $\dot{V} = -p(V)$). That is, instead of (1.2) we have

$$V_t = A_t - \int_0^t p(V_s)\,\mathrm{d}s, \tag{1.4}$$

and we use the convention $p(0) = 0$ to make zero a reflecting barrier (when hitting 0, $\{V_t\}$ remains at 0 until the next arrival).

Theorem 1.7 *For any $T < \infty$, one can couple the risk process and the storage process on $[0, T]$ in such a way that the events $\{\tau(u) \leq T\}$ and $\{V_T > u\}$ coincide. In particular,*

$$\psi(u, T) = \mathbb{P}(V_T > u), \tag{1.5}$$

and the process $\{V_t\}$ has a proper limit in distribution, say V, if and only if $\psi(u) < 1$ for all u. Then

$$\psi(u) = \mathbb{P}(V > u). \tag{1.6}$$

In order to make Theorem 1.7 applicable, we thus need to look more into the stationary distribution G, say, for the storage process $\{V_t\}$. It is intuitively obvious and not too hard to prove that G is a mixture of two components, one having an atom at 0 of size g_0, say, and the other being given by a density $g(x)$ on $(0, \infty)$. It follows in particular that

$$\psi(u) = \int_u^\infty g(y)\,\mathrm{d}y. \tag{1.7}$$

Proposition 1.8

$$p(x)g(x) = g_0\beta\overline{B}(x) + \beta\int_0^x \overline{B}(x-y)g(y)\,\mathrm{d}y. \tag{1.8}$$

Proof. In stationarity, the flow of mass from $[0, x]$ to (x, ∞) must be the same as the flow the other way. In view of the path structure of $\{V_t\}$, this means that the rate of upcrossings of level x must be the same as the rate of downcrossings. Now obviously, the l.h.s. of (1.8) is the rate of downcrossings (the event of an arrival in $[t, t + \mathrm{d}t]$ can be neglected so that a path of $\{V_t\}$ corresponds to a downcrossing in $[t, t + \mathrm{d}t]$ if and only if $V_t \in [x, x + p(x)\mathrm{d}t]$). An attempt of an upcrossing occurs as a result of an arrival, say when $\{V_t\}$ is in state y, and is successful if the jump size is larger than $x - y$. Considering the cases $y = 0$ and $0 < y \leq x$ separately, we arrive at the desired interpretation of the r.h.s. of (1.8) as the rate of upcrossings. $\square$

Define

$$w(x) = \int_0^x \frac{1}{p(t)} \, dt.$$

Then $w(x)$ is the time it takes for the reserve to reach level x provided it starts with $R_0 = 0$ and no claims arrive. Note that it may happen that $w(x) = \infty$ for all $x > 0$, say if $p(r)$ goes to 0 at rate r or faster as $r \downarrow 0$.

Corollary 1.9 *Assume that B is exponential with rate δ, $\overline{B}(x) = e^{-\delta x}$ and that $w(x) < \infty$ for all $x > 0$. Then the ruin probability is $\psi(u) = \int_u^\infty g(y) \, dy$, where*

$$g(x) = \frac{g_0 \beta}{p(x)} \exp\{\beta w(x) - \delta x\} \quad and \quad \frac{1}{g_0} = 1 + \int_0^\infty \frac{\beta}{p(x)} \exp\{\beta w(x) - \delta x\} dx \,.$$
$$(1.9)$$

Proof. We may rewrite (1.8) as

$$g(x) = \frac{1}{p(x)} \left\{ g_0 \beta e^{-\delta x} + \beta e^{-\delta x} \int_0^x e^{\delta y} g(y) \, dy \right\} = \frac{\beta}{p(x)} e^{-\delta x} \kappa(x)$$

where $\kappa(x) = g_0 + \int_0^x e^{\delta y} g(y) \, dy$ so that

$$\kappa'(x) = e^{\delta x} g(x) = \frac{\beta}{p(x)} \kappa(x).$$

Thus

$$\log \kappa(x) = \log \kappa(0) + \int_0^x \frac{\beta}{p(t)} dt = \log \kappa(0) + \beta w(x),$$
$$\kappa(x) = \kappa(0) e^{\beta w(x)} = g_0 e^{\beta w(x)},$$
$$g(x) = e^{-\delta x} \kappa'(x) = e^{-\delta x} g_0 \beta w'(x) e^{\beta w(x)}$$

which is the same as the expression in (1.9). That g_0 has the asserted value is a consequence of $1 = \|G\| = g_0 + \int_0^\infty g(y) dy$. $\qquad \square$

Remark 1.10 The exponential case in Corollary 1.9 is the only one in which explicit formulas are known (or almost so; see further the notes to Section 2), and thus it becomes important to develop algorithms for computing the ruin probabilities. We next outline one possible approach based upon the integral equation (1.8) (another one is based upon numerical solution of a system of differential equations which can be derived under phase-type assumptions, see further IX.7).

A *Volterra integral equation* has the general form

$$g(x) = h(x) + \int_0^x K(x,y)g(y)\,dy, \tag{1.10}$$

where $g(x)$ is an unknown function ($x \geq 0$), $h(x)$ is known and $K(x,y)$ is a suitable kernel. Dividing (1.8) by $p(x)$ and letting

$$K(x,y) = \frac{\beta \overline{B}(x-y)}{p(x)}, \quad h(x) = \frac{g_0 \beta \overline{B}(x)}{p(x)},$$

we see that for fixed g_0, the function $g(x)$ in (1.8) satisfies (1.10). For the purpose of explicit computation of $g(x)$ (and thereby $\psi(u)$), the general theory of Volterra equations does not seem to lead beyond the exponential case already treated in Corollary 1.9. However, one might try instead a numerical solution. We consider the simplest possible approach based upon the most basic numerical integration procedure, the trapezoidal rule

$$\int_{x_0}^{x_N} f(x)\,dx = \frac{h}{2}\left[f(x_0) + 2f(x_1) + 2f(x_2) + \cdots + 2f(x_{N-1}) + f(x_N)\right],$$

where $x_k = x_0 + kh$. Fixing $h > 0$, letting $x_0 = 0$ (i.e. $x_k = kh$) and writing $g_k = g(x_k)$, $K_{k,\ell} = K(x_k, x_\ell)$, this leads to

$$g_N = h_N + \frac{h}{2}\left\{K_{N,0}g_0 + K_{N,N}g_N\right\} + h\left\{K_{N,1}g_1 + \cdots + K_{N,N-1}g_{N-1}\right\},$$

i.e.

$$g_N = \frac{h_N + \frac{h}{2}K_{N,0}g_0 + h\left\{K_{N,1}g_1 + \cdots + K_{N,N-1}g_{N-1}\right\}}{1 - \frac{h}{2}K_{N,N}}. \tag{1.11}$$

In the case of (1.8), the unknown g_0 is involved. However, (1.11) is easily seen to be linear in g_0. One therefore first makes a trial solution $g^*(x)$ corresponding to $g_0 = 1$, i.e. $h(x) = h^*(x) = \beta \overline{B}(x)/p(x)$, and computes $\int_0^\infty g^*(x)dx$ numerically (by truncation and using the g_k^*). Then $g(x) = g_0 g^*(x)$, and $\|G\| = 1$ then yields

$$\frac{1}{g_0} = 1 + \int_0^\infty g^*(x)dx \tag{1.12}$$

from which g_0 and hence $g(x)$ and $\psi(u)$ can be computed.   □

Remark 1.11 Plugging (1.7) into (1.8), one obtains by partial integration and reordering

$$p(u)\psi'(u) - \beta\,\psi(u) + \beta \int_0^u \psi(u-y)\,dB(y) + \beta\,\overline{B}(u) = 0, \tag{1.13}$$

where in the last term it was used that $g_0 + \psi(0) = 1$. It is also possible to derive (1.13) directly (without reference to storage processes) in the following way. For $h > 0$, consider the time interval $(0, h)$ and condition on the time t and the amount y of the first claim in $(0, h)$. Since the probability that there is no claim in $(0, h)$ is $e^{-\beta h}$ and the probability that the first claim occurs between time t and $t + dt$ is $e^{-\beta t}\beta \, dt$, one obtains, using the Markov property of the process R_t,

$$\psi(u) = e^{-\beta h}\psi\Big(u + \int_0^h p(R_s)\,ds\Big) + \int_0^h e^{-\beta t}\beta\,dt\,\overline{B}\Big(u + \int_0^t p(R_s)\,ds\Big)$$

$$+ \int_0^h e^{-\beta t}\beta dt \int_0^{u+\int_0^t p(R_s)ds} \psi\Big(u + \int_0^t p(R_s)ds - y\Big)\,dB(y).$$

Since every other part of the above equation is differentiable w.r.t. h, also $\psi\big(u + \int_0^h p(R_s)\,ds\big)$ has to be (by symmetry this also establishes the differentiability w.r.t. u). Taking the derivative w.r.t. h and subsequently setting $h = 0$ then gives (1.13). The formal framework for this approach are of course generators (cf. Chapter II).

For the particular case $\overline{B}(x) = e^{-\delta x}$, one can multiply (1.13) by $e^{\delta u}$ and differentiate the resulting equation w.r.t. u. In this way one obtains the second-order differential equation

$$p(u)\,\psi''(u) + (p'(u) + \delta p(u) - \beta)\,\psi'(u) = 0$$

with the boundary conditions $p(0)\psi'(0) = \beta\big(\psi(0) - 1\big)$ and $\lim_{u\to\infty}\psi(u) = 0$. This leads to

$$\psi(u) = \frac{\beta \int_u^\infty \frac{1}{p(v)}\,e^{\beta\,\omega(v)-\delta\,v}dv}{1 + \beta \int_0^\infty \frac{1}{p(v)}\,e^{\beta\,\omega(v)-\delta\,v}dv}, \qquad (1.14)$$

which is again the result of Corollary 1.9.   □

1a   Two-step premium functions

We now assume the premium function to be constant in two levels as in Example 1.1,

$$p(r) = \begin{cases} p_1 & r \le v \\ p_2 & r > v. \end{cases} \qquad (1.15)$$

We may think of the risk reserve process R_t as pieced together from two risk reserve processes $R_t^{(1)}$ and $R_t^{(2)}$ with constant premiums p_1, p_2, such that R_t coincide with $R_t^{(1)}$ under level v and with $R_t^{(2)}$ above level v. If, as outlined in Example 1.1, the reduced income above v is due to dividend payments, this model is usually referred to as the *threshold dividend model.*[2] For an example of a sample path of such a refracted process, see Fig. VIII.1.

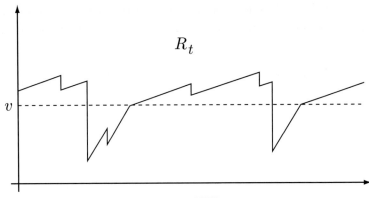

FIGURE VIII.1

Proposition 1.12 *Let $\psi^{(i)}(u)$ denote the ruin probability of $\{R_t^{(i)}\}$, define $\sigma = \inf\{t \geq 0: R_t < v\}$, let $\pi(u)$ be the probability of ruin between σ and the next upcrossing of v (including ruin possibly at σ), and let*

$$\phi_v(u) = \frac{1 - \psi^{(1)}(u)}{1 - \psi^{(1)}(v)}, \quad 0 \leq u \leq v. \tag{1.16}$$

Then

$$\psi(u) = \begin{cases} 1 - \phi_v(u) + \phi_v(u)\psi(v), & 0 \leq u \leq v, \\[2mm] \dfrac{\pi(v)}{1 + \pi(v) - \psi^{(2)}(0)}, & u = v, \\[2mm] \pi(u) + \left(\psi^{(2)}(u - v) - \pi(u)\right)\psi(v), & v \leq u < \infty. \end{cases}$$

Proof. Recall from Proposition II.2.6 that $\phi_v(u) = 1 - \psi_v(u)$ is the probability for $\{R_t^{(1)}\}$ (and hence also for $\{R_t\}$) of upcrossing level v before ruin given the process starts at $u \leq v$. Hence, for $u < v$ the probability of ruin for $\{R_t\}$ will be the sum of the probability of being ruined before upcrossing v, $1 - \phi_v(u)$, and the probability of ruin given we hit v first, $\phi_v(u)\psi(v)$.

---

[2]The corresponding dividend payout scheme, namely to pay nothing when $R_t < v$ and to pay out at rate $p_1 - p_2$ when $R_t \geq v$, turns out to maximize the expected discounted sum of dividend payments until ruin under certain assumptions, see e.g. Gerber & Shiu [412].

Similarly, if $u \geq v$ then the probability of ruin is the sum of being ruined between σ and the next upcrossing of v which is $\pi(u)$, and the probability of ruin given the process hits v before $(-\infty, 0)$ again after σ,

$$\left(\mathbb{P}_u(\sigma < \infty) - \pi(u)\right)\psi(v) = \left(\psi^{(2)}(u - v) - \pi(u)\right)\psi(v).$$

This yields the expression for $u \geq v$, and the one for $u = v$ then immediately follows. $\qquad\square$

Example 1.13 Assume that B is exponential, $\overline{B}(x) = \mathrm{e}^{-\delta x}$. Then

$$\psi^{(1)}(u) = \frac{\beta}{p_1\delta}\mathrm{e}^{-\gamma_1 u}, \quad \psi^{(2)}(u) = \frac{\beta}{p_2\delta}\mathrm{e}^{-\gamma_2 u},$$

where $\gamma_i = \delta - \beta/p_i$, so that

$$\phi_v(u) = \frac{1 - \dfrac{\beta}{p_1\delta}\mathrm{e}^{-\gamma_1 u}}{1 - \dfrac{\beta}{p_1\delta}\mathrm{e}^{-\gamma_1 v}}.$$

Furthermore, for $u \geq v$, $\mathbb{P}(\sigma < \infty) = \psi^{(2)}(u-v)$ and the conditional distribution of $v - R_\sigma$ given $\sigma < \infty$ is exponential with rate δ. If $v - R_\sigma < 0$, ruin occurs at time σ. If $v - R_\sigma = x \in [0, v]$, the probability of ruin before the next upcrossing of v is $1 - \phi_v(v - x)$. Hence

$$
\begin{aligned}
\pi(u) &= \psi^{(2)}(u - v)\left\{\mathrm{e}^{-\delta v} + \int_0^v \left(1 - \phi_v(v - x)\right)\delta\mathrm{e}^{-\delta x}\mathrm{d}x\right\} \\[2mm]
&= \frac{\beta}{p_2\delta}\mathrm{e}^{-\gamma_2(u-v)}\left\{1 - \int_0^v \frac{1 - \dfrac{\beta}{p_1\delta}\mathrm{e}^{-\gamma_1(v-x)}}{1 - \dfrac{\beta}{p_1\delta}\mathrm{e}^{-\gamma_1 v}}\delta\mathrm{e}^{-\delta x}\mathrm{d}x\right\} \\[2mm]
&= \frac{\beta}{p_2\delta}\mathrm{e}^{-\gamma_2(u-v)}\left\{1 - \frac{1 - \mathrm{e}^{-\delta v} - \dfrac{\beta}{\pi(\gamma_1 - \delta)}\mathrm{e}^{-\gamma_1 v}\left(\mathrm{e}^{(\gamma_1 - \delta)v} - 1\right)}{1 - \dfrac{\beta}{p_1\delta}\mathrm{e}^{-\gamma_1 v}}\right\} \\[2mm]
&= \frac{\beta}{p_2\delta}\mathrm{e}^{-\gamma_2(u-v)}\left\{1 - \frac{1 - \mathrm{e}^{-\gamma_1 v}}{1 - \dfrac{\beta}{p_1\delta}\mathrm{e}^{-\gamma_1 v}}\right\}.
\end{aligned}
$$

$\qquad\square$

Also for general phase-type distributions, all quantities in Proposition 1.12 can be found explicitly, see IX.7.

1b   Multi-step premium functions

In a similar manner one can investigate a more general model with a premium rule of the form

$$p(r) = \begin{cases} p_1 & 0 = v_0 \leq r < v_1, \\ p_2 & v_1 \leq r < v_2, \\ \vdots \\ p_k & r \geq v_{k-1}. \end{cases} \tag{1.17}$$

Assume that $p_k > \rho$. A similar approach as in Remark 1.11 then gives the piece-wise integro-differential equation

$$p_i \psi'(u) - \beta \psi(u) + \beta \int_0^u \psi(u - z) \mathrm{d}B(z) + \beta \overline{B}(u) = 0 \tag{1.18}$$

for $v_{i-1} \leq u < v_i$ and $i = 1, \ldots, k - 1$. For the solution $\psi(u)$ to be continuous, we require the boundary conditions (*contact conditions*)

$$\lim_{u \to v_i+} \psi(u) = \lim_{u \to v_i-} \psi(u) \tag{1.19}$$

and from $\eta > 0$ we have

$$\lim_{u \to \infty} \psi(u) = 0. \tag{1.20}$$

Note that $\psi(u)$ is not differentiable at the boundaries of the layers, as in view of (1.18) and continuity of ψ the conditions (1.19) can be rewritten in the form

$$p_{i+1} \frac{\partial^+}{\partial u} \psi(u) \bigg|_{u=v_i} = p_i \frac{\partial^-}{\partial u} \psi(u) \bigg|_{u=v_i}.$$

For exponential claim distribution $\overline{B}(x) = \mathrm{e}^{-\delta x}$, the system (1.18) can be solved explicitly in the following way. Akin to the procedure in Remark 1.11, first transform (1.18) into

$$p_i \psi''(u) + (p_i \delta - \beta) \psi'(u) = 0, \quad u \in [v_{i-1}, v_i). \tag{1.21}$$

Using the notation

$$\psi(u) = \sum_{i=1}^{k} I(b_{i-1} \leq u < b_i) \psi^{[i]}(u), \tag{1.22}$$

where the function $\psi^{[i]}(u)$ is the solution of (1.21) for $u \in [v_{i-1}, v_i)$ for each i, we obtain

$$\psi^{[i]}(u) = A^{(i)} + C^{(i)} \mathrm{e}^{-\gamma_i u},$$

where $\gamma_i = \delta - \beta/p_i$ is the (unique) positive solution of the Lundberg equation $\beta\left(\widehat{B}[\alpha] - 1\right) - p_i\,\alpha = 0$.

It remains to establish the constants in the above representation: from (1.20), we have $A^{(k)} = 0$. From the continuity conditions (1.19) we immediately get

$$A^{(i+1)} + C^{(i+1)}e^{-\gamma_{i+1}v_i} - A^{(i)} - C^{(i)}e^{-\gamma_i v_i} = 0, \quad (i = 1, 2, \ldots, k - 1). \quad (1.23)$$

Using (1.18) and comparing the coefficients of $e^{-\delta u}$, we further obtain after elementary algebra

$$-A^{(i+1)} - C^{(i+1)}\frac{\delta e^{-\gamma_{i+1}v_i}}{\delta - \gamma_{i+1}} + A^{(i)} + C^{(i)}\frac{\delta e^{-\gamma_i v_i}}{\delta - \gamma_i} \;=\; 0, \quad (i = 1, 2, \ldots, k - 1),$$
$$(1.24)$$

together with $A^{(1)} + C^{(1)}\frac{\delta}{\delta - \gamma_1} = 1$. Adding (1.23) and (1.24) leads to

$$
\begin{aligned}
C^{(i+1)} &= \frac{(\delta - \gamma_{i+1})\gamma_i}{(\delta - \gamma_i)\gamma_{i+1}}\,\exp\left\{(\gamma_{i+1} - \gamma_i)v_i\right\} C^{(i)} \\
&= \frac{\beta - \delta(p_i)}{\beta - \delta(p_{i+1})}\,\exp\left\{\beta\Big(\frac{1}{p_i} - \frac{1}{p_{i+1}}\Big)v_i\right\} C^{(i)}
\end{aligned}
$$

so that

$$
\begin{aligned}
C^{(i)} &= \frac{(\delta - \gamma_i)\gamma_1}{(\delta - \gamma_1)\gamma_i}\,\exp\left\{\sum_{j=1}^{i-1}(\gamma_{j+1} - \gamma_j)v_j\right\} C^{(1)} \\
&= \frac{\beta - \delta(p_1)}{\beta - \delta(p_i)}\,\exp\left\{\beta\sum_{j=1}^{i-1}(\frac{1}{p_j} - \frac{1}{p_{j+1}})v_j\right\} C^{(1)}
\end{aligned}
$$

for $i = 1, \ldots, k$. Define

$$L_i \;=\; \gamma_1 e^{-\gamma_1 v_1}\sum_{j=1}^{i-1}\left(\frac{1}{\gamma_j} - \frac{1}{\gamma_{j+1}}\right)\exp\left\{\sum_{\ell=2}^{j}(v_{\ell-1} - v_\ell)\gamma_\ell\right\}. \quad (1.25)$$

Then, again from (1.23), we have

$$A^{(i)} \;=\; A^{(1)} + C^{(1)}\frac{\delta}{\delta - \gamma_1}L_i = A^{(1)} + (1 - A^{(1)})L_i$$

and hence

$$A^{(1)} \;=\; -\frac{L_k}{1 - L_k} \quad \text{and} \quad C^{(1)} = \frac{\delta - \gamma_1}{\delta}\frac{1}{1 - L_k}.$$

Altogether we thus arrive at the explicit formula (1.22) with

$$
\psi^{[i]}(u) = \frac{1}{1-L_k}\left(L_i - L_k + \frac{(\delta-\gamma_i)\gamma_1}{\delta\,\gamma_i}\,\exp\left\{-\gamma_1 u + \sum_{j=1}^{i-1}(\gamma_{j+1}-\gamma_j)v_j\right\}\right)
$$

$$
= \frac{1}{1-L_k}\left(L_i - L_k + \frac{\gamma_1}{\gamma_i}\,\exp\left\{\sum_{j=1}^{i-1}(\gamma_{j+1}-\gamma_j)v_j\right\}\psi^{(i)}(u)\right)
$$

for $i = 1, \ldots, k$, where $\psi^{(i)}(u)$ again denotes the ruin probability in the classical model with constant premium intensity $p_i(u) \equiv p_i$. Note that for $k = 2$, the formula from Example 1.13 is retained.

Remark 1.14 The main tool in the above calculation was the reformulation of the integro-differential equations as ordinary differential equations, which allowed to find the fundamental solution for each layer locally and separately and subsequently to determine the coefficients through the continuity assumptions between the solutions in different layers (through a system of linear equations). This program can still be carried out for, say, Erlang(n) claim sizes, in which case the ODEs (with constant coefficients) that generalize (1.21) are of order $n + 1$. However, the solution of the resulting linear system of equations usually is highly involved and can only be evaluated numerically. □

Notes and references Some early references drawing attention to the model are Dawidson [277] and Segerdahl [790]. For the absolute ruin problem, see Gerber [396], Dassios & Embrechts [273] and, for a recent extension to finite-time horizons in a more general Lévy set-up, Loeffen & Patie [605].

Equation (1.6) was derived by Harrison & Resnick [450] by a different approach, whereas (1.5) is from Asmussen & Schock Petersen [104]; see further the notes to III.3.

The analytic derivation of (1.14) can be found in Tichy [851]. For some explicit solutions beyond Corollary 1.9, see the notes to Section 2.

Remark 1.10 is based upon Schock Petersen [695]; for complexity- and accuracy aspects, see the Notes to IX.7. Extensive discussion of the numerical solution of Volterra equations can be found in Baker [125]; see also Jagerman [499, 500]. An extension of Proposition 1.12, in which the switch from $R_t^{(2)}$ to $R_t^{(1)}$ only takes place if the risk process has gone below a threshold $w < v$ first, but the switch from $R_t^{(1)}$ to $R_t^{(2)}$ still takes place at the upcrossing of v is given in Bratiychuk & Derfla [196], see also Frostig [377].

The special case $p_2 = 0$ in the premium rule (1.15) refers to the situation when all original premium income is paid out as dividends to shareholders whenever the surplus level is above v. If in addition it is specified that for initial capital $u > v$, the difference $u - v$ is immediately paid out as a lump sum dividend payment, then the resulting strategy is known as the *horizontal dividend barrier strategy*. The ruin probability is

$\psi(u) = 1$ for the corresponding risk process (namely R_t reflected at v) and hence this case is not of interest for the main focus of this book. However, already back in 1957 de Finetti [283] suggested to consider the expected discounted accumulated dividend payments until ruin in a portfolio as an (economically motivated) alternative to the ruin probability for measuring the value of a portfolio, and the identification of the optimal dividend strategy to maximize this quantity leads to challenging sto chastic control problems. The horizontal barrier strategy often turns out to be optimal (see Gerber [394] for early results; the weakest currently known criteria on the risk process under which horizontal dividend barrier strategies are optimal among all admissible payout strategies are due to Loeffen [604], Kyprianou, Rivero & Song [568] and Loeffen & Renaud [606]). For duality considerations of the reflected risk processes with G/M/1 queues, see Löpker & Perry [609].

An analysis of exit problems for the threshold model (1.15) in a general Lévy set-up (with particular emphasis on the spectrally negative case) is given in Kyprianou & Loeffen [565].

The threshold and multi-step premium rules are a somewhat popular alternative to horizontal dividend strategies that are still to some extent analytically tractable and lead to ruin probabilities smaller than 1. The multi-step rule was first studied by Kerekesha [530], who looked at (1.18) for arbitrary $u > 0$ using correction terms for the different premium intensities outside the respective layer, which he expressed through truncated Fourier transforms. The explicit solution for exponential claims given above is due to Albrecher & Hartinger [25]; see also Zhou [919] and Lin & Sendova [594], where (1.18) is derived by a renewal approach that directly implies the continuity conditions (1.19). In [25, 594, 919] the analysis is also considerably extended to cover quantities like the time value of ruin, deficit at ruin and the surplus prior to ruin. To improve upon the problem described in Remark 1.14, an alternative recursive approach that iteratively calculates the full solution for the same model with one layer less is developed in [25]. For extensions of these results to the renewal model see Yang & Zhang [904].

In the literature also other surplus-dependent risk processes have been discussed in connection with dividend payout schemes that lead to a positive probability of survival. Among them are time-dependent barrier strategies, for which the barrier itself is an increasing function of time and if the risk process touches the barrier it stays at the barrier until the next claim occurs and the additional premium income is paid out as dividends. The ruin probability for the resulting risk process can then often be obtained as the solution of partial integro-differential equations, see Gerber [399], Siegl & Tichy [804, 805] and Albrecher & Tichy [41]. Albrecher, Hartinger & Thonhauser [26] analytically compare the performance of linear barrier strategies with the threshold strategies of (1.15). Albrecher & Kainhofer [30] and Albrecher, Kainhofer & Tichy [31] investigate risk processes with a non-linear time-dependent barrier structure including constant interest on the surplus. See also Garrido [390] in a diffusion setting. For general surveys on dividend models in risk theory, see Avanzi [109] and Albrecher & Thonhauser [40].

2 The model with constant interest

In this section, we assume that $p(x) = p + ix$. This example is of particular application relevance because of the interpretation of i as interest rate. However, it also turns out to have nice mathematical features.

A basic tool is a representation of the ruin probability in terms of a discounted stochastic integral

$$Z = -\int_0^\infty e^{-it} \mathrm{d}S_t \tag{2.1}$$

w.r.t. the claim surplus process $S_t = A_t - pt = \sum_1^{N_t} U_i - pt$ of the associated compound Poisson model without interest. Write $R_t^{(u)}$ when $R_0 = u$. We first note that:

Proposition 2.1 $R_t^{(u)} = e^{it} u + R_t^{(0)}$.

Proof. The result is obvious if one thinks in economic terms and represents the reserve at time t as the initial reserve u with added interest plus the gains/deficit from the claims and incoming premiums. For a more formal mathematical proof, note that

$$
\begin{aligned}
\mathrm{d}R_t^{(u)} &= p + iR_t^{(u)} - \mathrm{d}A_t, \\
\mathrm{d}\big[R_t^{(u)} - e^{it} u\big] &= p + i\big[R_t^{(u)} - e^{it} u\big] - \mathrm{d}A_t.
\end{aligned}
$$

Since $R_0^{(u)} - e^{i \cdot 0} u = 0$ for all u, $R_t^{(u)} - e^{it} u$ must therefore be independent of u which yields the result. $\qquad\square$

Let

$$Z_t = e^{-it} R_t^{(0)} = e^{-it}\left(\int_0^t \big(p + iR_s^{(0)}\big)\mathrm{d}s - A_t\right).$$

Then

$$
\begin{aligned}
\mathrm{d}Z_t &= e^{-it}\left(-i\,\mathrm{d}t \cdot \int_0^t (p + iR_s^{(0)})\mathrm{d}s + (p + iR_t^{(0)})\mathrm{d}t + i\,\mathrm{d}t \cdot A_t - \mathrm{d}A_t\right) \\
&= e^{-it}\,(p\,\mathrm{d}t - \mathrm{d}A_t) = -e^{-it}\mathrm{d}S_t.
\end{aligned}
$$

Thus

$$Z_v = -\int_0^v e^{-it}\mathrm{d}S_t,$$

where the last integral exists pathwise because $\{S_t\}$ is of locally bounded variation.

Proposition 2.2 *The r.v. Z in (2.1) is well-defined and finite, with distribution $H(z) = \mathbb{P}(Z \le z)$ given by the m.g.f.*

$$\widehat{H}[\alpha] = \mathbb{E}e^{\alpha Z} = \exp\left\{\int_0^\infty \kappa\left(-\alpha e^{-it}\right) dt\right\} = \exp\left\{\int_0^\alpha \frac{1}{iy}\kappa\left(-y\right) dy\right\},$$

where $\kappa(\alpha) = \beta\big(\widehat{B}[\alpha] - 1\big) - p\alpha$. Further $Z_t \overset{a.s.}{\to} Z$ as $t \to \infty$.

Proof. Let $M_t = A_t - t\beta\mu_B$. Then $S_t = M_t + t(\beta\mu_B - p)$ and $\{M_v\}$ is a martingale. From this it follows immediately that $\left\{\int_0^v e^{-it}dM_t\right\}$ is again a martingale. The mean is 0 and (since $\mathbb{V}ar(dM_t) = \beta\mu_B^{(2)}dt$)

$$\mathbb{V}ar\left(\int_0^v e^{-it}dM_t\right) = \int_0^v e^{-2it}\beta\mu_B^{(2)}dt = \frac{\beta\mu_B^{(2)}}{2i}(1 - e^{-2iv}).$$

Hence the limit as $v \to \infty$ exists by the convergence theorem for L_2-bounded martingales, and we have

$$
\begin{aligned}
Z_v &= -\int_0^v e^{-it}dS_t = -\int_0^v e^{-it}\big(dM_t + (\beta\mu_B - p)dt\big) \\
&\overset{a.s.}{\to} -\int_0^\infty e^{-it}\big(dM_t + (\beta\mu_B - p)dt\big) \\
&= -\int_0^\infty e^{-it}dS_t = Z.
\end{aligned}
$$

Now if $X_1, X_2, \ldots$ are i.i.d. with c.g.f. ϕ and $\rho < 1$, we obtain the c.g.f. of $\sum_1^\infty \rho^n X_n$ at α as

$$\log\mathbb{E}\prod_{n=1}^\infty e^{\alpha\rho^n X_n} = \log\prod_{n=1}^\infty e^{\phi(\alpha\rho^n)} = \sum_{n=1}^\infty \phi(\alpha\rho^n).$$

Letting $\rho = e^{-ih}$, $X_n = S_{nh} - S_{(n+1)h}$, we have $\phi(\alpha) = h\kappa(-\alpha)$, and obtain the c.g.f. of $Z = -\int_0^\infty e^{-it}dS_t$ as

$$\lim_{h\downarrow 0}\sum_{n=1}^\infty \phi(\alpha\rho^n) = \lim_{h\downarrow 0}h\sum_{n=1}^\infty \kappa(-\alpha e^{-inh}) = \int_0^\infty \kappa\left(-\alpha e^{-it}\right) dt\,;$$

the last expression for $\widehat{H}[\alpha]$ follows by the substitution $y = \alpha e^{-it}$. □

Theorem 2.3 $\psi(u) = \dfrac{H(-u)}{\mathbb{E}\left[H(-R_{\tau(u)})\,\middle|\,\tau(u) < \infty\right]}.$

Proof. Write $\tau = \tau(u)$ for brevity. On $\{\tau < \infty\}$, we have

$$
\begin{aligned}
u + Z &= (u + Z_\tau) + (Z - Z_\tau) = \mathrm{e}^{-i\tau}\left[\mathrm{e}^{i\tau}(u + Z_\tau) - \int_\tau^\infty \mathrm{e}^{-i(t-\tau)}\mathrm{d}S_t\right] \\
&= \mathrm{e}^{-i\tau}[R_\tau^{(u)} + Z^*],
\end{aligned}
$$

where $Z^* = -\int_\tau^\infty \mathrm{e}^{-i(t-\tau)}\mathrm{d}S_t$ is independent of $\mathscr{F}_\tau$ and distributed as Z. The last equality followed from $R_t^{(u)} = \mathrm{e}^{it}(Z_t + u)$, cf. Proposition 2.1, which also yields $\tau < \infty$ on $\{Z < -u\}$. Hence

$$
\begin{aligned}
H(-u) &= \mathbb{P}(u + Z < 0) = \mathbb{P}\big(R_\tau + Z^* < 0;\, \tau < \infty\big) \\
&= \psi(u)\mathbb{E}\big[\mathbb{P}(R_\tau + Z^* < 0 \,\big|\, \mathscr{F}_\tau, \tau < \infty)\big] \\
&= \psi(u)\mathbb{E}\big[H(-R_{\tau(u)}) \,\big|\, \tau(u) < \infty\big].
\end{aligned}
$$

$\square$

Corollary 2.4 *Assume that B is exponential, $\overline{B}(x) = \mathrm{e}^{-\delta x}$, and that $p(x) = p + ix$ with $p > 0$. Then*

$$
\psi(u) = \frac{\beta i^{\beta/i-1}\Gamma\left(\dfrac{\delta(p + iu)}{i}; \dfrac{\beta}{i}\right)}{\delta^{\beta/i}p^{\beta/i}\mathrm{e}^{-\delta p/i} + \beta i^{\beta/i-1}\Gamma\left(\dfrac{\delta p}{i}; \dfrac{\beta}{i}\right)}, \tag{2.2}
$$

where $\Gamma(x; \eta) = \int_x^\infty t^{\eta-1}\mathrm{e}^{-t}\mathrm{d}t$ is the incomplete Gamma function.

Proof 1. We use Corollary 1.9 and get

$$
\begin{aligned}
\omega(x) &= \int_0^x \frac{1}{p + it}\mathrm{d}t = \frac{1}{i}\log(p + ix) - \frac{1}{i}\log p, \\
g(x) &= \frac{\gamma_0\beta}{p + ix}\exp\left\{\frac{\beta}{i}\log(p + ix) - \frac{\beta}{i}\log p - \delta x\right\} \\
&= \frac{\gamma_0\beta}{p^{\beta/i}}(p + ix)^{\beta/i-1}\mathrm{e}^{-\delta x},
\end{aligned}
$$

$$
\begin{aligned}
\frac{1}{\gamma_0} &= 1 + \int_0^\infty \frac{\beta}{p(x)}\exp\{\beta\omega(x) - \delta x\}\,\mathrm{d}x \\
&= 1 + \int_0^\infty \frac{\beta}{p^{\beta/i}}(p + ix)^{\beta/i-1}\mathrm{e}^{-\delta x}\,\mathrm{d}x \\
&= 1 + \frac{\beta}{ip^{\beta/i}}\int_p^\infty y^{\beta/i-1}\mathrm{e}^{-\delta(y-p)/i}\,\mathrm{d}y \\
&= 1 + \frac{\beta i^{\beta/i-1}\mathrm{e}^{\delta p/i}}{\delta^{\beta/i}p^{\beta/i}}\Gamma\left(\frac{\delta p}{i}; \frac{\beta}{i}\right),
\end{aligned}
$$

$$\psi(u) = \gamma_0 \int_u^\infty \frac{\beta}{p(x)} \exp\{\beta\omega(x) - \delta x\} \, \mathrm{d}x$$

$$= \gamma_0 \frac{\beta i^{\beta/i-1} e^{\delta p/i}}{\delta^{\beta/i} p^{\beta/i}} \Gamma\left(\frac{\delta(p+iu)}{i}; \frac{\beta}{i}\right),$$

from which (2.2) follows by elementary algebra. $\qquad\square$

Proof 2. We use Theorem 2.3. From $\kappa(\alpha) = \beta\alpha/(\delta - \alpha) - p\alpha$, it follows that

$$\log \widehat{H}[\alpha] = \int_0^\alpha \frac{1}{iy} \kappa(-y) \, \mathrm{d}y = \frac{1}{i} \int_0^\alpha \left(p - \beta/(\delta + y)\right) \mathrm{d}y$$

$$= \frac{1}{i} \left[p\alpha + \beta \log \delta - \beta \log(\delta + \alpha)\right] = \log\left[e^{p\alpha/i} \left(\frac{\delta}{\delta + \alpha}\right)^{\beta/i}\right],$$

which shows that Z is distributed as $p/i - V$, where V is Gamma$(\delta, \beta/i)$, i.e. with density

$$f_V(x) = \frac{x^{\beta/i-1} \delta^{\beta/i}}{\Gamma(\beta/i)} e^{-\delta x}, \quad x > 0.$$

In particular,

$$H(-u) = \mathbb{P}(Z < -u) = \mathbb{P}(V > u + p/i) = \frac{\Gamma\left(\delta(p+iu)/i; \beta/i\right)}{\Gamma(\beta/i)}.$$

By the memoryless property of the exponential distribution, $-R_{\tau(u)}$ has an exponential distribution with rate (δ) and hence

$$\mathbb{E}\left[H(-R_{\tau(u)})\big|\, \tau(u) < \infty\right]$$

$$= \int_0^\infty \delta e^{-\delta x} \mathbb{P}(p/i - V \le x) \, \mathrm{d}x$$

$$= \left[-e^{-\delta x} \mathbb{P}(p/i - V \le x)\right]_0^\infty + \int_0^\infty e^{-\delta x} f_V(p/i - x) \, \mathrm{d}x$$

$$= \mathbb{P}(V \ge p/i) + \int_0^{p/i} \frac{(p/i - x)^{\beta/i-1} \delta^{\beta/i}}{\Gamma(\beta/i)} e^{-\delta p/i} \, \mathrm{d}x$$

$$= \frac{1}{\Gamma(\beta/i)} \left\{\Gamma(\delta p/i; \beta/i) + \frac{(p/i)^{\beta/i} \delta^{\beta/i} e^{-\delta p/i}}{\beta/i}\right\}.$$

From this (2.2) follows by elementary algebra. $\qquad\square$

Proof 3. Just insert $p(x) = p + i\,x$ into (1.14) and identify the resulting integral as the incomplete Gamma function. $\qquad\square$

Example 2.5 The analysis leading to Theorem 2.3 is also valid if $\{R_t\}$ is obtained by adding interest to a more general process with stationary independent increments. As an example, assume Brownian motion $\{B_t\}$ with drift μ and variance σ^2; then $\{R_t\}$ is the diffusion with drift function $\mu + ix$ and constant variance σ^2. The process $\{S_t\}$ corresponds to $\{-B_t\}$ so that $\kappa(\alpha) = \sigma^2\alpha^2/2 - \mu\alpha$, and the c.g.f. of Z is

$$
\begin{aligned}
\log \widehat{H}[\alpha] &= \int_0^\alpha \frac{1}{iy}\kappa(-y)\,\mathrm{d}y = \frac{1}{i}\int_0^\alpha\left(\frac{\sigma^2 y}{2} + \mu\right)\mathrm{d}y \\
&= \frac{\sigma^2\alpha^2}{4i} + \frac{\mu\alpha}{i}.
\end{aligned}
$$

I.e., Z is normal $\left(\mu/i, \sigma^2/2i\right)$, and since $R_\tau = 0$ by the continuity of Brownian motion, it follows that the ruin probability is

$$
\psi(u) = \frac{H(-u)}{H(0)} = \frac{\Phi\left(\dfrac{-u - \mu/i}{\sigma/\sqrt{2i}}\right)}{\Phi\left(\dfrac{-\mu/i}{\sigma/\sqrt{2i}}\right)}. \tag{2.3}
$$

$\square$

Notes and references Theorem 2.3 is from Harrison [449]; for a martingale proof, see e.g. Gerber [398, p. 134] (the time scale there is discrete but the argument is easily adapted to the continuous case). Corollary 2.4 is classical. Formula (2.3) was derived by Emanuel *et al.* [342] and Harrison [449]; it is also used as a basis for a diffusion approximation by these authors.

Paulsen & Gjessing [687] found some remarkable explicit formulas for $\psi(u)$ beyond the exponential case in Corollary 1.9. The solution is in terms of Bessel functions for an Erlang(2) B and in terms of confluent hypergeometric functions for a H_2 B (a mixture of two exponentials). It must be noted, however, that the analysis does not seem to carry over to general phase-type distributions, not even Erlang(3) or H_3, nor to non-linear premium rules $p(\cdot)$.

Explicit formulas for the finite-time ruin probabilities $\psi(u, T)$ for exponential claims in terms of finite gamma series were obtained in Albrecher, Teugels & Tichy [39] whenever $\beta = ki$ for some integer k. See also Knessl & Peters [548] for a detailed asymptotic study of $\psi(u, T)$ for $i > 0$ and exponential claims. Avram, Leonenko & Rabehasaina [110] extend the method of [39] to certain jump-diffusion models. A numerical algorithm for determining $\psi(u, T)$ based on discrete time Markov chains can be found in Cardoso & Waters [221, 222].

A r.v. of the form $\sum_1^\infty \rho^n X_n$ with the X_n i.i.d. as in the proof of Proposition 2.2 is a special case of a *perpetuity*; see e.g. Goldie & Grübel [422] and Section 5.

Further studies of the model with interest can be found in Boogaert & Crijns [180], Gerber [396], Delbaen & Haezendonck [288], Emanuel *et al.* [342], Paulsen [680, 681, 682], Paulsen & Gjessing [687], Sundt & Teugels [822, 823], Yang [899], Cai & Dickson [215] and Rullière & Loisel [758]. Some of these references also go into a stochastic interest rate.

3 The local adjustment coefficient. Logarithmic asymptotics

For the classical risk model with constant premium rule $p(x) \equiv p^*$, write γ^* for the solution of the Lundberg equation

$$\beta\big(\widehat{B}[\gamma^*] - 1\big) - \gamma^* p^* = 0,\qquad(3.1)$$

write $\psi^*(u)$ for the ruin probability etc., and recall Lundberg's inequality

$$\psi^*(u) \leq e^{-\gamma^* u}\qquad(3.2)$$

and the Cramér-Lundberg approximation

$$\psi^*(u) \sim C^* e^{-\gamma^* u}.\qquad(3.3)$$

When trying to extend these results to the model of this chapter where $p(x)$ depends on x, a first step is the following:

Theorem 3.1 *Assume that for some $0 < \delta_0 \leq \infty$, it holds that $\widehat{B}[s] \uparrow \infty$, $s \uparrow \delta_0$, and that $p(x) \to \infty$, $x \to \infty$. Then* $\limsup_{u\to\infty} \dfrac{\log \psi(u)}{u} \leq -\delta_0$. *If $\delta_0 < \infty$ and $e^{-\epsilon r}p(r) \to 0$, $e^{(\delta_0+\epsilon)x}\overline{B}(x) \to \infty$ for all $\epsilon > 0$, then* $\dfrac{\log \psi(u)}{u} \to -\delta_0$, $u \to \infty$.

In the proof as well as in the remaining part of the section, we will use the *local adjustment coefficient* $\gamma(x)$, which for a fixed x is defined as the adjustment coefficient of the classical risk model with $p^* = p(x)$, i.e. as the solution of the equation

$$\kappa\big(x, \gamma(x)\big) = 0 \quad \text{where} \quad \kappa(x, \alpha) = \beta(\widehat{B}[\alpha] - 1) - \alpha p(x).\qquad(3.4)$$

We assume existence of $\gamma(x)$ for all x, as will hold under the steepness assumption of Theorem 3.1, and (for simplicity) that

$$\inf_{x\geq 0} p(x) > \beta\mu_B,\qquad(3.5)$$

which implies $\inf_{x \geq 0} \gamma(x) > 0$. The intuitive idea behind introducing local adjustment coefficients is that the classical risk model with premium rate $p^* = p(x)$ serves as a 'local approximation' at level x for the general model when the reserve is close to x.[3]

Proof of Theorem 3.1. The steepness assumption and $p(x) \to \infty$ ensure $\gamma(x) \to \delta_0$. Let $\gamma^* < \delta_0$, let p^* be as in (3.1) and for a given $\epsilon > 0$, choose u_0 such that $p(x) \geq p^*$ when $x \geq u_0 \epsilon$. When $u \geq u_0$, obviously $\psi(u)$ can be bounded with the probability that the Cramér-Lundberg compound Poisson model with premium rate p^* downcrosses level $u\epsilon$ starting from u, which in turn by Lundberg's inequality can be bounded by $e^{-\gamma^*(1-\epsilon)u}$. Hence $\limsup_{u\to\infty} \log \psi(u)/u \leq -\gamma^*(1 - \epsilon)$. Letting first $\epsilon \to 0$ and next $\gamma^* \uparrow \delta_0$ yields the first statement of the theorem.

For the last assertion, choose $c_\epsilon^{(1)}, c_\epsilon^{(2)}$ such that $p(x) \leq c_\epsilon^{(1)} e^{\epsilon x}$, $\overline{B}(x) \geq c_\epsilon^{(2)} e^{-(\delta_0+\epsilon)x}$ for all x. Then we have the following lower bound for the time for the reserve to go from level u to level $u + v$ without a claim:

$$\omega(u + v) - \omega(u) \; = \; \int_0^v \frac{1}{p(u+t)} dt \; \geq \; c_\epsilon^{(3)} e^{-\epsilon u}$$

where $c_\epsilon^{(3)} = (1 - e^{-\epsilon v})/(\epsilon c_\epsilon^{(1)})$. Therefore the probability that a claim arrives before the reserve has reached level $u + v$ is at least $c_\epsilon^{(4)} e^{-\epsilon u}$. Given such an arrival, ruin will occur if the claim is at least $u + v$, and hence

$$\psi(u) \; \geq \; c_\epsilon^{(4)} e^{-\epsilon u} c^{(2)} e^{-(\delta_0+\epsilon)u} .$$

The truth of this for all $\epsilon > 0$ implies $\liminf \log \psi(u) \geq -\delta_0$.   □

Obviously, Theorem 3.1 only presents a first step, and in particular, the result is not very informative if $\delta_0 = \infty$. The rest of this section deals with tail estimates involving the local adjustment coefficient. The first main result in this direction is the following version of Lundberg's inequality:

Theorem 3.2 *Assume that $p(x)$ is a non-decreasing function of x and let $I(u) = \int_0^u \gamma(x) \, dx$. Then*

$$\psi(u) \; \leq \; e^{-I(u)}. \tag{3.6}$$

The second main result to be derived states that the bound in Theorem 3.2 is also an approximation under appropriate conditions. The form of the result is superficially similar to the Cramér-Lundberg approximation, noting that in many cases the constant C is close to 1. However, the limit is not $u \to \infty$ but

---

[3]Note that this was also the motivation behind the approach of Section 1b.

the *slow Markov walk* limit in large deviations theory (see e.g. Bucklew [207]). For $\epsilon > 0$, let $\psi_\epsilon(u)$ be evaluated for the process $\{R_t^{(\epsilon)}\}$ defined as in (1.2), only with β replaced by β/ϵ and U_i by ϵU_i.

Theorem 3.3 *Assume that either* (a) $p(r)$ *is a non-decreasing function of* r, *or* (b) *Condition* 3.13 *below holds. Then*

$$\lim_{\epsilon \downarrow 0} -\epsilon \log \psi_\epsilon(u) = I(u). \qquad (3.7)$$

Remarks:

1. Condition 3.13 is a technical condition on the claim size distribution B, which essentially says that an overshoot r.v. $U|U > x$ cannot have a much heavier tail than the claim U itself.

2. If $p(x) \equiv p$ is constant, then $R_t^{(\epsilon)} = \epsilon R_{t/\epsilon}$ for all t so that $\psi_\epsilon(u) = \psi(u/\epsilon)$, I.e., the asymptotics $u \to \infty$ and $\epsilon \to 0$ are the same.

3. The slow Markov walk limit is appropriate if $p(x)$ does not vary too much compared to the given mean interarrival time $1/\beta$ and the size U of the claims; one can then assume that $\epsilon = 1$ is small enough for Theorem 3.3 to be reasonably precise and use $e^{-I(u)}$ as approximation to $\psi(u)$.

4. As typical in large deviations theory, the logarithmic form of (3.7) only captures 'the main term in the exponent', but is not precise to describe the asymptotic form of $\psi(u)$ in terms of ratio limit theorems (the precise asymptotics could be $\log I(u) e^{-I(u)}$ or $I(u)^\alpha e^{-I(u)}$, say, rather than $e^{-I(u)}$).

3a Examples

Before giving the proofs of Theorems 3.2, 3.3, we consider some simple examples. First, we show how to rewrite the explicit solution for $\psi(u)$ in Corollary 1.9 in terms of $I(u)$ when the claims are exponential:

Example 3.4 Consider again the exponential case $\overline{B}(x) = e^{-\delta x}$ as in Corollary 1.9. Then $\gamma(x) = \delta - \beta/p(x)$, and

$$\int_0^u \gamma(x)\,dx = \delta u - \beta \int_0^u p(x)^{-1}dx = \delta u - \beta \omega(u).$$

Integrating by parts, we get

$$
\begin{aligned}
\frac{1}{\gamma_0} &= 1 + \int_0^\infty \frac{\beta}{p(x)} \exp\{\beta\omega(x) - \delta x\}\,\mathrm{d}x \\
&= 1 + \int_0^\infty \frac{\mathrm{d}\beta\omega(x)}{\mathrm{d}x} \exp\{\beta\omega(x) - \delta x\}\,\mathrm{d}x \\
&= 1 + \big[\exp\{\beta\omega(x) - \delta x\}\big]_0^\infty + \delta \int_0^\infty \exp\{\beta\omega(x) - \delta x\}\,\mathrm{d}x \\
&= 1 + 0 - 1 + \delta \int_0^\infty \mathrm{e}^{-I(x)}\,\mathrm{d}x,
\end{aligned}
$$

$$
\begin{aligned}
\frac{1}{\gamma_0} \int_u^\infty g(x)\,\mathrm{d}x &= \int_u^\infty \frac{\beta}{p(x)} \exp\{\beta\omega(x) - \delta x\}\,\mathrm{d}x \\
&= \big[\exp\{\beta\omega(x) - \delta x\}\big]_u^\infty + \delta \int_u^\infty \exp\{\beta\omega(x) - \delta x\}\,\mathrm{d}x \\
&= \delta \int_u^\infty \exp\{\beta\omega(x) - \delta x\}\,\mathrm{d}x - \exp\{\beta\omega(u) - \delta u\},
\end{aligned}
$$

and hence

$$
\psi(u) = \frac{\int_u^\infty \mathrm{e}^{-I(y)}\,\mathrm{d}y - \mathrm{e}^{-I(u)}/\delta}{\int_0^\infty \mathrm{e}^{-I(y)}\,\mathrm{d}y} = \mathrm{e}^{-I(u)} \frac{\int_0^\infty \mathrm{e}^{-\int_0^y \gamma(x+u)\,\mathrm{d}x}\,\mathrm{d}y - 1/\delta}{\int_0^\infty \mathrm{e}^{-\int_0^y \gamma(x)\mathrm{d}x}\,\mathrm{d}y}. \quad (3.8)
$$

$\square$

We next give direct derivations of Theorems 3.2, 3.3 in the particularly simple case of diffusions:

Example 3.5 Assume that $\{R_t\}$ is a diffusion on $[0, \infty)$ with drift $\mu(x)$ and variance $\sigma^2(x) > 0$ at x. The appropriate definition of the local adjustment coefficient $\gamma(x)$ is then as the one $2\mu(x)/\sigma^2(x)$ for the locally approximating Brownian motion. It is well known (see Theorem XIII.4.4 or Karlin & Taylor [522, pp. 191–195]) that

$$
\psi(u) = \frac{\int_u^\infty \mathrm{e}^{-I(y)}\mathrm{d}y}{\int_0^\infty \mathrm{e}^{-I(y)}\mathrm{d}y} = \mathrm{e}^{-I(u)} \frac{\int_0^\infty \mathrm{e}^{-\int_0^y \gamma(x+u)\mathrm{d}x}\,\mathrm{d}y}{\int_0^\infty \mathrm{e}^{-\int_0^y \gamma(x)\mathrm{d}x}\,\mathrm{d}y}. \quad (3.9)
$$

If $\gamma(x)$ is increasing, applying the inequality $\gamma(x+u) \geq \gamma(x)$ yields immediately the conclusion of Theorem 3.2. For Theorem 3.3, note first that the appropriate slow Markov walk assumption amounts to $\mu_\epsilon(x) = \mu(x)$, $\sigma_\epsilon^2(x) = \epsilon\sigma^2(x)$ so that $\gamma_\epsilon(x) = \gamma(x)/\epsilon$, $I_\epsilon(u) = I(u)/\epsilon$, and (3.9) yields

$$-\epsilon \log \psi_\epsilon(u) = I(u) + A_\epsilon - B_\epsilon, \qquad (3.10)$$

where

$$A_\epsilon = \epsilon \log\left(\int_0^\infty e^{-\int_0^y \gamma(x)dx/\epsilon}\, dy\right), \quad B_\epsilon = \epsilon \log\left(\int_0^\infty e^{-\int_0^y \gamma(x+u)dx/\epsilon}\, dy\right).$$

The analogue of (3.5) is $\inf_{x\geq 0}\gamma(x) > 0$ which implies that the integral in the definition of A_ϵ converges to 0. In particular, the integral is bounded by 1 eventually and hence $\limsup A_\epsilon \leq \limsup \epsilon \log 1 = 0$. Choosing $y_0, \gamma_0 > 0$ such that $\gamma(x) \leq \gamma_0$ for $y < y_0$, we get

$$\int_0^\infty e^{-\int_0^y \gamma(x)dx/\epsilon}\, dy \geq \int_0^{y_0} e^{-y\gamma_0/\epsilon}\, dy = \frac{\epsilon}{\gamma_0}(1 - e^{-y_0\gamma_0/\epsilon}) \sim \frac{\epsilon}{\gamma_0}.$$

This implies $\liminf A_\epsilon \geq \lim \epsilon \log \epsilon = 0$ and $A_\epsilon \to 0$. Similarly, $B_\epsilon \to 0$, and (3.7) follows. □

The analogue of Example 3.5 for risk processes with exponential claims is as follows:

Example 3.6 Assume that B is exponential with rate δ. Then the solution of the Lundberg equation is $\gamma^* = \delta - \beta/p^*$ so that

$$I(u) = \delta u - \beta \int_0^u \frac{1}{p(x)}\, dx.$$

Note that this expression shows up also in the explicit formula for $\psi(u)$ in the form given in Example 3.4. Ignoring $1/\delta$ in the formula there, this leads to (3.6) exactly as in Example 3.5. Further, the slow Markov walk assumption means $\delta_\epsilon = \delta/\epsilon$, $\beta_\epsilon = \beta/\epsilon$. Thus $\gamma_\epsilon(x) = \gamma(x)/\epsilon$ and (3.10) holds if we redefine A_ϵ as

$$A_\epsilon = \epsilon \log\left(\int_0^\infty e^{-\int_0^y \gamma(x)dx/\epsilon} dy - \epsilon/\delta\right)$$

and similarly for B_ϵ. As in Example 3.5,

$$\limsup_{\epsilon\to 0} A_\epsilon \leq \limsup_{\epsilon\to 0} \epsilon \log(1 - 0) = 0.$$

By (3.5) and $\gamma^* = \delta - \beta/p^*$, we have $\delta > \gamma_0$ and get

$$\liminf A_\epsilon \geq \lim \epsilon \log\left(\epsilon\left(\frac{1}{\gamma_0} - \frac{1}{\delta}\right)\right) \geq 0.$$

Now (3.7) follows just as in Example 3.5. □

We next investigate what the upper bound (or approximation) $e^{-I(u)}$ looks like in the case $p(x) = p + ix$ (interest) subject to various forms of the tail $\overline{B}(x)$ of B. Of course, $\gamma(x)$ is typically not explicit, so our approach is to determine standard functions $G_1(u), \ldots, G_q(u)$ representing the first few terms in the asymptotic expansion of $I(u)$ as $u \to \infty$. I.e.,

$$G_i(u) \to \infty, \quad \frac{G_{i+1}(u)}{G_i(u)} = o(1), \quad I(u) = G_1(u) + \cdots + G_q(u) + o\big(G_q(u)\big).$$

It should be noted, however, that the interchange of the slow Markov walk limit $\epsilon \to 0$ and the limit $u \to \infty$ is not formally justified and in fact, the slow Markov walk approximation deteriorates as x becomes large. Nevertheless, the results are suggestive in their form and much more explicit than anything else in the literature.

Example 3.7 Assume that

$$\overline{B}(x) \sim c_1 x^{\alpha-1} e^{-\delta x}, \quad x \to \infty \tag{3.11}$$

with $\alpha > 0$. This covers mixtures or convolutions of exponentials or, more generally, phase-type distributions (Example I.2.4) or gamma distributions; in the phase-type case, the typical case is $\alpha = 1$ which holds, e.g., if the phase generator is irreducible (Proposition IX.1.8). It follows from (3.11) that $\widehat{B}[s] \to \infty$ as $s \uparrow \delta$ and hence $\gamma^* \uparrow \delta$ as $p^* \to \infty$. More precisely,

$$\widehat{B}[s] = 1 + s \int_0^\infty e^{sx} \overline{B}(x) dx = 1 + \frac{c_1 \delta \Gamma(\alpha)}{(\delta - s)^\alpha} \big(1 + o(1)\big)$$

as $s \uparrow \delta$, and hence (3.1) leads to

$$(\delta - \gamma^*)^\alpha \approx \frac{\beta c_1 \Gamma(\alpha)}{p^*}, \quad \gamma^* \approx \delta - c_2 p^{*-1/\alpha}, \quad c_2 = \big(\beta c_1 \Gamma(\alpha)\big)^{1/\alpha},$$

$$I(u) \approx \delta u - c_2 \int_0^u \frac{1}{(p+ix)^{1/\alpha}} dx \approx \begin{cases} \delta u & \alpha < 1 \\ \delta u - c_3 \log u & \alpha = 1 \\ \delta u - c_4 u^{1-1/\alpha} & \alpha > 1 \end{cases},$$

where $c_3 = c_2/i$, $c_4 = c_2 i^{-1/\alpha}/(1 - 1/\alpha)$. $\qquad\square$

Example 3.8 Assume next that B has bounded support, say 1 is the upper limit and

$$\overline{B}(x) \sim c_5 (1-x)^{\eta-1}, \quad x \uparrow 1, \tag{3.12}$$

with $\eta \geq 1$. For example, $\eta = 1$ if B is degenerate at 1, $\eta = 2$ if B is uniform on $(0,1)$ and $\eta = k + 1$ if B is the convolution of k uniforms on $(0, 1/k)$. Here $\widehat{B}[s]$ is defined for all s and

$$
\begin{aligned}
\widehat{B}[s] - 1 &= s \int_0^1 e^{sx} \overline{B}(x) \, dx = e^s \int_0^s e^{-y} \overline{B}(1 - y/s) \, dy \\
&\approx \frac{c_5 e^s}{s^{\eta-1}} \int_0^\infty e^{-y} y^{\eta-1} dy = \frac{c_5 e^s \Gamma(\eta)}{s^{\eta-1}}
\end{aligned}
$$

as $s \uparrow \infty$. Hence (3.1) leads to $\beta c_5 e^{\gamma^*} \Gamma(\eta) \sim \gamma^{*\eta} p^*$,

$$
\gamma^* \approx \log p^* + \eta \log \log p^*, \quad I(u) \approx u(\log u + \eta \log \log u).
$$

$\square$

Example 3.9 As a case intermediate between (3.11) and (3.12), assume that

$$
\overline{B}(x) \sim c_6 e^{-x^2/2c_7}, \quad x \uparrow \infty. \tag{3.13}
$$

We get

$$
\begin{aligned}
\widehat{B}[s] - 1 &\approx c_6 s \int_0^\infty e^{sx} e^{-x^2/2c_7} \, dx = c_6 s e^{c_7 s^2/2} \int_0^\infty e^{-(x-c_7 s)^2/2c_7} \, dx \\
&= c_6 s \sqrt{2\pi c_7} e^{c_7 s^2/2} \Phi\left(\frac{s}{\sqrt{c_7}}\right) \sim c_6 s \sqrt{2\pi c_7} e^{c_7 s^2/2},
\end{aligned}
$$

$$
\frac{c_7}{2} \gamma^{*2} \sim \log p^*, \quad \gamma^* \sim c_8 \sqrt{\log p^*}, \quad I(u) \approx c_8 u \sqrt{\log u},
$$

where $c_8 = \sqrt{2/c_7}$. $\square$

3b Proof of Theorem 3.2

We first remark that the definition (3.4) of the local adjustment coefficient is not the only possible one: whereas the motivation for (3.4) is the formula

$$
\frac{1}{h} \log \mathbb{E}_u e^{s(R_h - u)} \sim \beta(\widehat{B}[s] - 1) - sp(u), \quad h \downarrow 0, \tag{3.14}
$$

for the m.g.f. of the increment in a small time interval $[0, h]$, one could also have considered the increment $r_u(T_1) - u - U_1$ up to the first claim (here $r_u(\cdot)$ denotes the solution of $\dot{r} = p(r)$ starting from $r_u(0) = u$). This leads to an alternative local adjustment coefficient $\gamma_0(u)$ defined as solution of

$$
1 = \mathbb{E} e^{\gamma_0(u)(U_1 + u - r_u(T_1))} = \widehat{B}[\gamma_0(u)] \cdot \int_0^\infty \beta e^{-\beta t} e^{\gamma_0(u)(u - r_u(t))} \, dt. \tag{3.15}
$$

Proposition 3.10 *Assume that $p(x)$ is a non-decreasing function of x. Then:*
(a) *$\gamma(x)$ and $\gamma_0(x)$ are also non-decreasing functions of x;*
(b) *$\gamma(x) \leq \gamma_0(x)$.*

Proof. That $\gamma(x)$ is non-decreasing follows easily by inspection of (3.4). The assumption implies that $r_u(t) - u$ is a non-decreasing function of u. Hence for $u < v$,

$$1 \;=\; \mathbb{E}e^{\gamma_0(u)(U_1 + u - r_u(T_1))} \geq \mathbb{E}e^{\gamma_0(u)(U_1 + v - r_v(T_1))}.$$

By convexity of the m.g.f. of $U_1 + v - r_v(T_1)$, this is only possible if $\gamma_0(v) \geq \gamma_0(u)$.

For (b), note that the assumption implies that $r_u(t) - u \geq tp(u)$. Hence

$$
\begin{aligned}
1 \;&=\; \mathbb{E}e^{\gamma_0(u)(U_1 + u - r_u(T_1))} \;\leq\; \mathbb{E}e^{\gamma_0(u)(U_1 - p(u)T_1)} \\
&=\; \widehat{B}\big[\gamma_0(u)\big] \frac{\beta}{\beta + \gamma_0(u)p(u)}, \\
0 \;&\leq\; \beta\big(\widehat{B}[\gamma_0(u)] - 1\big) - \gamma_0(u)p(u).
\end{aligned}
$$

Since (3.4) considered as function of γ is convex and equals 0 for $\gamma = 0$, this is only possible if $\gamma_0(u) \geq \gamma(u)$. $\qquad\square$

We prove Theorem 3.2 in terms of γ_0; the case of γ then follows immediately by Proposition 3.10(b):

Theorem 3.11 *Assume that $p(x)$ is a non-decreasing function of x. Then*

$$\psi(u) \;\leq\; e^{-\int_0^u \gamma_0(x)\,dx}. \tag{3.16}$$

Proof. Define $\psi^{(n)}(u) = \mathbb{P}\big(\tau(u) \leq \sigma_n\big)$ as the ruin probability after at most n claims ($\sigma_n = T_1 + \cdots + T_n$). We shall show by induction that

$$\psi^{(n)}(u) \;\leq\; e^{-\int_0^u \gamma_0(x)\,dx}, \tag{3.17}$$

from which the theorem follows by letting $n \to \infty$. The case $n = 0$ is clear since here $T_0 = 0$ so that $\psi^{(0)}(u) = 0$. Assume (3.17) shown for n and let $F_u(x) = \mathbb{P}(U_1 + u - r_u(T_1) \leq x)$. Separating after whether ruin occurs at the first claim or not, we obtain

$$
\begin{aligned}
\psi^{(n+1)}&(u) \\
&\leq\; 1 - F_u(u) + \int_{-\infty}^u \psi^{(n)}(u - x)F_u(dx) \\
&\leq\; \int_u^\infty F_u(dx) + \int_{-\infty}^u e^{-\int_0^{u-x}\gamma_0(y)dy}F_u(dx) \\
&=\; e^{-\int_0^u \gamma_0(x)dx}\left\{\int_u^\infty e^{\int_0^u \gamma_0(y)dy}F_u(dx) + \int_{-\infty}^u e^{\int_{u-x}^u \gamma_0(y)dy}F_u(dx)\right\}.
\end{aligned}
$$

Considering the cases $x \geq 0$ and $x < 0$ separately, it is easily seen that $\int_{u-x}^{u} \gamma_0(y)dy \leq x\gamma_0(u)$. Also, $\int_0^u \gamma_0(y)dy \leq u\gamma_0(u) \leq x\gamma_0(u)$ for $x \geq u$. Hence

$$
\begin{aligned}
\psi^{(n+1)}(u) &\leq e^{-\int_0^u \gamma_0(x)dx}\left\{\int_u^\infty e^{x\gamma_0(u)}F_u(dx) + \int_{-\infty}^u e^{x\gamma_0(u)}F_u(dx)\right\} \\
&= e^{-\int_0^u \gamma_0(x)dx}\widehat{F}_u\big[\gamma_0(u)\big] \\
&= e^{-\int_0^u \gamma_0(x)dx},
\end{aligned}
$$

where the last identity immediately follows from (3.15); we used also Proposition 3.10(a) for some of the inequalities. □

It follows from Proposition 3.10(b) that the bound provided by Theorem 3.11 is sharper than the one given by Theorem 3.2. However, $\gamma_0(u)$ appears more difficult to evaluate than $\gamma(u)$. Also, for either of Theorems 3.2, 3.11 be reasonably tight something like the slow Markov walk conditions in Theorem 3.3 is required, and here it is easily seen that $\gamma_0(u) \approx \gamma(u)$. For these reasons, we have chosen to work with $\gamma(u)$ as the fundamental local adjustment coefficient.

3c Proof of Theorem 3.3

The idea of the proof is to bound $\{R_t^{(\epsilon)}\}$ above and below in a small interval $[x - x/n, x + x/n]$ by two classical risk processes with a constant p and appeal to the classical results (3.2), (3.3). To this end, define

$$
u_{k,n} = \frac{k}{n}u, \quad \overline{p}_{k,n} = \sup_{u_{k-1,n} \leq x \leq u_{k+1,n}} p(x), \quad \underline{p}_{k,n} = \inf_{u_{k-1,n} \leq x \leq u_{k+1,n}} p(x),
$$

and, in accordance with the notation $\psi_\epsilon(u)$, $\psi_{p^*}^*(u)$, let $\psi_{p^*;\epsilon}^*(u)$ denote the ruin probability for the classical model with β replaced by β/ϵ and U_i by ϵU_i.

Lemma 3.12 $\limsup_{\epsilon \downarrow 0} -\epsilon \log \psi_\epsilon(u) \leq I(u)$.

Proof. For ruin to occur, $\{R_t^{(\epsilon)}\}$ (starting from $u = u_{n,n}$) must first downcross $u_{n-1,n}$. The probability of this is at least $\phi_{\overline{p}_{n,n};\epsilon}^*(u/n)$, the probability that ruin occurs in the Cramér-Lundberg model with $p^* = \overline{p}_{n,n}$ (starting from u/n) without that $2u/n$ is upcrossed before ruin. Further, given downcrossing occurs, the value of $\{R_t^{(\epsilon)}\}$ at the time of downcrossing is $< u_{n-1,n}$ so that

$$
\begin{aligned}
\psi_\epsilon(u) &\geq \phi_{\overline{p}_{n,n};\epsilon}^*(u/n)\psi_\epsilon(u_{n-1,n}) \\
&\geq \phi_{\overline{p}_{n,n};\epsilon}^*(u/n)\phi_{\overline{p}_{n-1,n};\epsilon}^*(u/n)\psi_\epsilon(u_{n-2,n}) \\
&\geq \cdots \geq \prod_{k=1}^n \phi_{\overline{p}_{k,n};\epsilon}^*(u/n).
\end{aligned}
$$

Now as $\epsilon \downarrow 0$,

$$\psi^*_{p^*;\epsilon}(u) = \psi^*_{p^*}(u/\epsilon) \sim C^* e^{-\gamma^* u/\epsilon},$$

where the first equality follows by an easy scaling argument and the approximation by (3.3). Let $C_{k,n}$, $\overline{\gamma}_{k,n}$ be C^*, resp. γ^* evaluated for $p^* = \overline{p}_{k,n}$; in particular, since γ^* is an increasing function of p^*, also

$$\overline{\gamma}_{k,n} = \sup_{u_{k-1,n} \leq x \leq u_{k,n}} \gamma(x).$$

Clearly,

$$\psi^*_{p^*;\epsilon}(u/n) - \phi^*_{p^*;\epsilon}(u/n) \leq \psi^*_{p^*;\epsilon}(2u/n),$$

$$\phi^*_{\overline{p}_{k,n};\epsilon}(u/n) \geq \psi^*_{\overline{p}_{k,n};\epsilon}(u/n) - \psi^*_{\overline{p}_{k,n};\epsilon}(2u/n)$$

$$\sim C_{k,n} e^{-\overline{\gamma}_{k,n} u/\epsilon n}\left(1 - e^{-\overline{\gamma}_{k,n} u/\epsilon n}\right)$$

$$= C_{k,n} e^{-\overline{\gamma}_{k,n} u/\epsilon n}\left(1 + o(1)\right),$$

where $o(1)$ refers to the limit $\epsilon \downarrow 0$ with n and u fixed. It follows that

$$-\log \psi_\epsilon(u) \leq -\sum_{k=1}^{n} \log \phi^*_{\overline{p}_{k,n};\epsilon}(u/n)$$

$$= \frac{u}{\epsilon n}\sum_{k=1}^{n}\overline{\gamma}_{k,n} - \sum_{k=1}^{n}\log C_{k,n} + o(1),$$

$$\limsup_{\epsilon \downarrow 0} -\epsilon \log \psi_\epsilon(u) \leq \frac{u}{n}\sum_{k=1}^{n}\overline{\gamma}_{k,n}.$$

Letting $n \to \infty$ and using a Riemann sum approximation completes the proof.
□

Theorem 3.3 now follows easily in case (a). Indeed, in obvious notation one has $\gamma_\epsilon(x) = \gamma(x)/\epsilon$, so that Theorem 3.2 gives

$$\psi_\epsilon(u) \leq e^{-I(u)/\epsilon} \quad \Rightarrow \quad \liminf_{\epsilon \downarrow 0} -\epsilon \log \psi_\epsilon(u) \geq I(u).$$

Combining with the upper bound of Lemma 3.12 completes the proof.

In case (b), we need the following condition:

Condition 3.13 *There exists a r.v. $V < \infty$ such that* (i) *for any $u < \infty$ there exist $C_u < \infty$ and $\delta(u) > \sup_{x \leq u} \gamma(x)$ such that*

$$\mathbb{P}(V > x) \leq C_u e^{-\delta(u)x}; \tag{3.18}$$

(ii) *the family of claim overshoot distributions is stochastically dominated by* V, *i.e. for all* $x, y > 0$ *it holds that*

$$\mathbb{P}\big(U > x + y \,|\, U > x\big) = \frac{\overline{B}(x+y)}{\overline{B}(x)} \leq \mathbb{P}(V > y). \tag{3.19}$$

To complete the proof, let $v \leq u$ and define

$$\tau^{(\epsilon)}(u, v) = \inf\big\{t > 0 : R_t^{(\epsilon)} < v \,|\, R_0^{(\epsilon)} = u\big\}, \quad \xi^{(\epsilon)}(u, v) = v - R_{\tau^{(\epsilon)}(u,v)}^{(\epsilon)}.$$

Then

$$
\begin{aligned}
\psi_\epsilon &(u) \\
&= \mathbb{E}\big[\psi_\epsilon\big(R_{\tau^{(\epsilon)}(u,u/n)}\big)\, ; \tau^{(\epsilon)}(u, u/n) < \infty\big] \\
&= \mathbb{E}\big[\psi_\epsilon\big(u/n - \xi^{(\epsilon)}(u, u/n)\big)\, ; \tau^{(\epsilon)}(u, u/n) < \infty\big] \\
&= \mathbb{E}\big[\psi_\epsilon\big(u/n - \xi^{(\epsilon)}(u, u/n)\big)\big|\tau^{(\epsilon)}(u, u/n) < \infty\big] \cdot \mathbb{P}\big(\tau^{(\epsilon)}(u, u/n) < \infty\big) \\
&\leq \mathbb{E}\psi_\epsilon\,(u/n - \epsilon V) \cdot \mathbb{P}\big(\tau^{(\epsilon)}(u, u/n) < \infty\big).
\end{aligned}
$$

Write $\mathbb{E}\psi_\epsilon\,(u/n - \epsilon V) = E_1 + E_2$, where E_1 is the contribution from the event that the process does not reach level $2u/n$ before ruin and E_2 is the rest. Then the standard Lundberg inequality yields

$$
\begin{aligned}
E_1 &\leq \mathbb{E}\psi^*_{\underline{p}_{1,n};\epsilon}\,(u/n - \epsilon V) = \mathbb{E}\psi^*_{\underline{p}_{1,n}}\,(u/\epsilon n - V) \\
&\leq e^{-\underline{\gamma}_{1,n}u/\epsilon n}\,\mathbb{E}\big[e^{\gamma_{1,n}V}; V \leq u/\epsilon n\big] + \mathbb{P}(V > u/\epsilon n) \\
&= e^{-\underline{\gamma}_{1,n}u/\epsilon n}O(1)
\end{aligned}
$$

(using (3.18) for the last equality). For E_2, we first note that the number of downcrossings of $2u/n$ starting from $R_0^{(\epsilon)} = 2u/n$ is bounded by a geometric r.v. N with

$$\mathbb{E}N \leq \frac{1}{1 - \psi^*_{\inf_{x \geq 2u/n} p(x);\epsilon}(0)} = \frac{\inf_{x \geq 2u/n} p(x)}{\inf_{x \geq 2u/n} p(x) - \beta\mathbb{E}U} = O(1),$$

cf. (3.5) and the standard formula for $\psi(0)$. The probability of ruin in between two downcrossings is bounded by

$$\mathbb{E}\psi^*_{\underline{p}_{1,n};\epsilon}\,(2u/n - \epsilon V) = e^{-2\underline{\gamma}_{1,n}u/\epsilon n}O(1)$$

so that

$$E_2 \leq e^{-2\underline{\gamma}_{1,n}u/\epsilon n}O(1), \quad E_1 + E_2 \leq e^{-\underline{\gamma}_{1,n}u/\epsilon n}O(1).$$

Hence

$$\liminf_{\epsilon \downarrow 0} -\epsilon \log \psi_\epsilon (u)$$

$$\geq \liminf_{\epsilon \downarrow 0} -\epsilon \big\{\log(E_1 + E_2) + \log \mathbb{P}\big(\tau^{(\epsilon)}(u, u/n) < \infty\big)\big\}$$

$$\geq \frac{u}{n}\gamma_{1,n} + \liminf_{\epsilon \downarrow 0} -\epsilon \log \mathbb{P}\big(\tau^{(\epsilon)}(u, u/n) < \infty\big)$$

$$\vdots$$

$$\geq \frac{u}{n}\sum_{i=1}^{n}\gamma_{i,n}.$$

Another Riemann sum approximation completes the proof. $\square$

Notes and references   With the exception of Theorem 3.1, the results are from Asmussen & Nielsen [92]; they also discuss simulation based upon 'local exponential change of measure' for which the likelihood ratio is

$$L_t = \exp\Big\{-\int_0^t \gamma(R_{s-})\, \mathrm{d}R_s\Big\} = \exp\Big\{-\int_0^t \gamma(R_s)p(R_s)\, \mathrm{d}s + \sum_{i=1}^{N_t} \gamma(R_{T_i-})U_i\Big\}.$$

An approximation similar to (3.7) for ruin probabilities in the presence of an upper barrier b appears in Cottrell *et al.* [263], where the key mathematical tool is the deep Wentzell-Freidlin theory of slow Markov walks (see e.g. Bucklew [207]). Djehiche [324] gives an approximation for $\psi(u, T) = \mathbb{P}_u(\inf_{0 \leq t \leq T} R_t < 0)$ via related large deviations techniques. Comparing these references with the present work shows that in the slow Markov walk set-up, the risk process itself is close to the solution of the differential equation

$$\dot{r}(x) = -\kappa'(x, 0) \quad (= p(x) - \beta \mathbb{E}U) \tag{3.20}$$

(with $\kappa(x, s)$ as in (3.4) and the prime meaning differentiation w.r.t. s), whereas the most probable path leading to ruin is the solution of

$$\dot{r}(x) = -\kappa'\big(x, \gamma(x)\big) \tag{3.21}$$

(the initial condition is $r(0) = u$ in both cases). Whereas the result of [324] is given in terms of an action integral which does not look very explicit, one can in fact arrive at the optimal path by showing that the approximation for $\psi(u, T)$ is maximized over T by taking T as the time for (3.21) to pass from u to 0; the approximation (3.7) then comes out (at least heuristically) by analytical manipulations with the action integral. Similarly, it might be possible to show that the limits $\epsilon \downarrow 0$ and $b \uparrow \infty$ are interchangeable in the setting of [263]. Typically, the rigorous implementation of these ideas via large deviations techniques would require slightly stronger smoothness conditions on $p(x)$ than ours and conditions somewhat different from Condition 3.13, the simplest being to require $\widehat{B}[s]$ to be defined for *all* $s > 0$ (thus excluding, e.g., the exponential distribution). We would like, however, to point out as a maybe much more important fact that the present approach is far more elementary and self-contained than that using large deviations theory. For different types of applications of large deviations to ruin probabilities, see XIII.3.

Asymptotic results for surplus-dependent premium under heavy-tailed claims are given in Section X.5.

4 The model with tax

Consider now a compound Poisson risk process $R_t^{(\vartheta)}$ with constant premium income intensity $(1 - \vartheta) p$ $(0 < \vartheta \leq 1)$ whenever the risk process is in its running maximum $M_t^{(\vartheta)} = \max\{R_s^{(\vartheta)}, 0 \leq s \leq t\}$ and constant premium income intensity p otherwise. So the dynamics of the risk process are given by

$$
\mathrm{d}R_t^{(\vartheta)} = \begin{cases} p\,\mathrm{d}t - \mathrm{d}A_t, & \text{if } R_t^{(\vartheta)} < M_t^{(\vartheta)}, \\ (1 - \vartheta)\,p\,\mathrm{d}t - \mathrm{d}A_t, & \text{if } R_t^{(\vartheta)} = M_t^{(\vartheta)}, \end{cases} \tag{4.1}
$$

where $A_t = \sum_{i=1}^{N_t} U_i$ are the aggregate claims up to time t. As outlined in Example 1.4, a natural interpretation for this model is that the insurance company needs to pay tax at rate ϑ whenever the risk process is at a new record height (which is considered to be profit) and does not need to pay tax if it is below the running maximum, as then the incoming premium is needed to amortize the previous claim payments until a new running maximum is reached. But one can also simply think of a profit-participation in terms of dividend payments to shareholders according to the above scheme. Figure VIII.2 depicts a sample path of the resulting risk process.

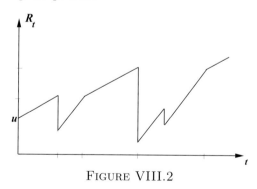

FIGURE VIII.2

The resulting ruin probability $\psi_\vartheta(u)$ has a strikingly simple relation to the ruin probability $\psi(u) = \psi_0(u)$ of the original risk process.

Theorem 4.1 *In case of positive safety loading $\eta > 0$ and $\vartheta < 1$, $\psi_\vartheta(u) < 1$ holds for all $u \geq 0$. In particular, in this case*

$$1 - \psi_\vartheta(u) = \left(1 - \psi_0(u)\right)^{1/(1-\vartheta)}. \tag{4.2}$$

Proof. Recall from Theorem II.2.3 that the survival probability $\phi(u) = 1 - \psi(u)$ in the classical compound Poisson risk model can be interpreted as the probability to have zero events during $[u, \infty)$ of an inhomogeneous Poisson process with rate $\beta(s) = \beta/p\, \mathbb{P}(V_{\max} > s)$, where $V_{\max}$ denotes the maximum workload of an M/G/1 queue.[4] But now one realizes that the survival probability with tax $\phi_\vartheta(u)$ can also be interpreted in a similar way, since when we cut out the excursions away from the running maximum that do not lead to ruin (which are identical to those witho ut tax), then a straight line with slope $p(1 - \vartheta)$ remains. After rescaling to slope 1, the probability to survive is to have no events during $[u, \infty)$ of the inhomogeneous Poisson process with rate $\beta(s) = \beta/\left(p(1-\vartheta)\right)\mathbb{P}(V_{\max} > s)$, where $V_{\max}$ is again the maximum workload of the original M/G/1 queue. In view of (2.5) from Chapter II, this leads to

$$\phi_\vartheta(u) = \exp\!\left(-\frac{\beta}{p(1 - \vartheta)} \int_u^\infty \mathbb{P}(V_{\max} > s)\, \mathrm{d}s\right) \;=\; \left(\phi_0(u)\right)^{1/(1-\vartheta)}.$$

$\square$

Remark 4.2 If one defines $\phi_\vartheta(u, v)$ for $u < v$ as the probability that, starting from level u at time 0, the process reaches level v before ruin occurs (clearly $\phi_\vartheta(u) = \phi_\vartheta(u, \infty)$), then — along the same line of arguments as above — we have

$$\phi_\vartheta(u, v) \;=\; \exp\!\left(-\frac{\beta}{p(1 - \vartheta)} \int_u^v \mathbb{P}(V_{\max} > s)\, \mathrm{d}s\right) \;=\; \left(\phi_0(u, v)\right)^{1/(1-\vartheta)}. \tag{4.3}$$

$\square$

It is now straightforward to extend identity (4.2) to a surplus-dependent tax rate.

Corollary 4.3 *If the tax rate $\vartheta(r)$ $(0 \leq \vartheta(r) < 1)$ depends on the current surplus level $R_t = r$, then the corresponding survival probability $\phi_\Gamma(u)$ is given by*

$$\phi_\Gamma(u) \;=\; \exp\!\left(-\int_u^\infty \left(\frac{\mathrm{d}}{\mathrm{d}s} \log \phi_0(s)\right) \frac{1}{1 - \vartheta(s)}\, \mathrm{d}s\right). \tag{4.4}$$

---

[4]In Theorem II.2.3 we had $p = 1$. Here we have a general p, so we first need to rescale time by the factor p, i.e. the Poisson process with intensity β/p and the new premium rate 1 leads to the same $\phi(u)$ as the original process, but now can be linked with the M/G/1 queue.

Proof. In this case $\phi_\Gamma(u)$ is the probability to have zero events during $[u, \infty)$ of an inhomogeneous Poisson process with rate $\beta(s) = \beta/(p(1 - \vartheta(s)))\, \mathbb{P}(V_{\max} > s)$, where $V_{\max}$ is again the maximum workload of a busy period in a classical M/G/1 queue. Now just again invoke Theorem II.2.3.   □

Remark 4.4 From (4.4) one sees that even in the case $\vartheta(r) \to 1$ as $r \to \infty$ there can be a positive probability of survival, as long as the convergence rate is sufficiently low.   □

Consider finally a further generalization of the risk process given by the surplus-dependent dynamics

$$
dR_t^g = \begin{cases} p_1(R_t^g)\, dt - dA_t, & \text{if } R_t^g < M_t^g, \\ p_2(R_t^g)\, dt - dA_t, & \text{if } R_t^g = M_t^g. \end{cases} \tag{4.5}
$$

Denote with $q(u)$ the probability that if a claim occurs in the running maximum u, then ruin occurs before R_t^g reaches level u again (note that $q(u)$ depends on $p_1(\cdot)$, but not on $p_2(\cdot)$). Then, the same reasoning as above (where now $q(u)$ takes the role of $\mathbb{P}(V_{\max} > u)$) shows that the survival probability is given by

$$
\phi^g(u) = \exp\left(-\beta \int_u^\infty \frac{q(s)}{p_2(s)}\, ds\right). \tag{4.6}
$$

It immediately follows that if $p_2(s) = (1 - \vartheta)p_1(s)$, then the tax identity (4.2) holds again, where $\psi_0(u)$ then refers to the ruin probability of the risk process with premium rule $p_1(x)$ for all $x \geq 0$. In particular, this shows that the tax identity also holds for the compound Poisson risk process with interest (where $p_1(x) = p + ix$) discussed in Section 2. As a by-product, one obtains with $p_1(x) = p_2(x) = p + ix$ that the ruin probability of the classical risk process with constant interest (without tax) can be expressed as

$$
\psi(u) = 1 - \exp\left(-\beta \int_u^\infty \frac{q(s)}{p + is}\, ds\right),
$$

which can be compared with (1.6). It then remains, however, to identify an explicit expression for the quantity $q(u)$.

Notes and references The tax model was introduced in Albrecher & Hipp [28], where the identity (4.2) was derived by a different approach. The simpler proof given here is based on Albrecher, Borst, Boxma & Resing [16], where also the treatment of the surplus-dependent tax rate can be found. Albrecher, Renaud & Zhou [35] used excursion theory to extend the identity (4.2) to arbitrary spectrally negative Lévy processes and also generalized a formula from [28] for the moments of the accumulated discounted tax payments until ruin in terms of scale functions. Kyprianou & Zhou [569] then further extended this approach to surplus-dependent tax rates and quantities like the time of ruin, the deficit at ruin and the surplus prior to ruin in the tax model; see also Renaud [736]. A more direct analysis of these latter quantities in the compound Poiss on setting with tax can be found in Ming & Wang [645] and absolute ruin under the tax payment scheme is studied in Ming, Wang & Xao [646].

A treatment of the tax problem for the model (4.5) can be found in Wei [878], where (4.6) is derived by a direct differential argument and in Wang *et al.* [868], where the accumulated discounted tax payments until ruin is considered. For an extension to a Markov-modulated model, see Wei, Yang & Wang [877]. A slightly different model is considered in Hao & Tang [448], where the authors give a fine asymptotic study of the ruin probabilities of a spectrally negative Lévy risk model that is subject to periodic taxation on its net gains during each period.

The effect of tax payments on a non-Markovian risk process is investigated in Albrecher, Badescu & Landriault [15] in the context of the dual risk model, in which the sign of the premium income and the aggregate claims are reverted. The simple tax identity (4.2) then does not hold anymore, but a similar relationship holds for arbitrary interarrival times and exponential jump sizes.

5   Discrete-time ruin problems with stochastic investment

We consider a discrete-time risk reserve process $R_0, R_1, \ldots$ given by $R_0 = u > 0$ and the recursion

$$R_n = A_n R_{n-1} - B_n \,, \tag{5.1}$$

where $\{A_n\}$, $\{B_n\}$ are independent sequences each consisting of i.i.d. r.v.'s and $A_n > 0$. The interpretation is that the reserve is invested in risky assets yielding a stochastic interest rate of $A_n - 1$ in period n, whereas B_n is the claim surplus, that is, the difference between claims and premiums received. The reserve may decrease if either $A_n < 1$ or $B_n > 0$, so that financial risk enters via the A_n and traditional insurance risk via the B_n. As usual, the ruin time $\tau(u)$ corresponding to $R_0 = u$ is the first n with $R_n < 0$, and the ruin probability $\psi(u)$ is the probability that $R_n < 0$ for some n. To avoid trivialities, we assume $\mathbb{P}(B_n > 0) > 0$ since otherwise $\psi(u) = 0$ for all $u > 0$, and also $\mathbb{P}(A_n < 1) > 0$ since otherwise there is no investment risk.

A first question is when $\psi(u) = 1$ for all u and when not. One could expect the first possibility to occur when $\mathbb{E}A_1 < 1$ but not when $\mathbb{E}A_1 > 1$ (then one would expect $R_n \to \infty$ with positive probability). However, the relevant criterion is in terms of $\mathbb{E} \log A_n$:

Proposition 5.1 *Assume* $\mathbb{E} \log |B_n| < \infty$. *If* $\mathbb{E} \log A_n < 0$, *then* $\psi(u) = 1$ *for all* $u > 0$. *If* $\mathbb{E} \log A_n > 0$, *then* $\psi(u) < 1$ *for all large* $u > 0$.

Proof for $\mathbb{E}\log A_n < 0$.

First note that for $R_{n-1} = x > 0$, we have

$$\log R_n^+ \;=\; \log A_n + \log x + \log(1 - B_n/x)^+ \;\leq\; \log A_n + \log x + \log\left(1 + |B_n|/x\right).$$

By assumption, there exists x_0 such that $\mathbb{E}\log A_n + \mathbb{E}\log\left(1 + |B_n|/x\right) \leq 0$ for all $x > x_0$. I.e.,

$$\mathbb{E}\left[\log R_n^+ - \log x^+ \mid R_{n-1} = x\right] \;\leq\; 0$$

for all $x > x_0$. By standard recurrence criteria for Markov processes ([APQ, p. 21), this implies that R_n cannot go to ∞ so that some interval of the form $[0, x_1]$ is visited i.o. by $\{R_n^+\}$. Since (unless for trivial cases) $\inf_{0 \leq u \leq x_1} \psi(u) > 0$, a geometric trial argument therefore gives $\psi(u) = 1$. $\qquad\square$

The proof for the case $\mathbb{E}\log A_n > 0$ will be given after Proposition 5.3.

In view of Proposition 5.1, we henceforth assume $\mathbb{E}\log|B_n| < \infty$ and $\mathbb{E}\log A_n > 0$. This implies in particular $\mathbb{P}(A_n > 1) > 0$ (recall that we also assumed $\mathbb{P}(A_n < 1) > 0$).

We next note some representations of R_n and the ruin time, which are similar to some from Section 2 in the constant interest rate case (with the $-Y$ chain taking the role of the Z process there).

Proposition 5.2 *Define* $D_n = A_1^{-1} \cdots A_n^{-1}$, $R_n^* = D_n R_n$ *and*

$$Y_n \;=\; D_1 B_1 + D_2 B_2 + \cdots + D_n B_n \;=\; \sum_{k=1}^{n} D_k B_k \,.$$

Then

$$R_n^* = u - Y_n, \quad \tau(u) = \inf\{n \geq 1 : Y_n > u\}\,. \tag{5.2}$$

Note that A_k^{-1} is simply the discounting factor for period k and D_n the one for the totality of periods $1, \ldots, n$. Thus R_n^* is simply the present value of the reserve, and the formula $R_n^* = u - Y_n$ tells that, as should be, this is the initial reserve minus the present value of claim surpluses from the different periods.

Proof. We can rewrite (5.1) (with n replaced by k) as $R_k^* - R_{k-1}^* = -D_k B_k$. Thus

$$R_n^* = u - \sum_{k=1}^{n}(R_k^* - R_{k-1}^*) \;=\; u - Y_n\,.$$

From this the claim on $\tau(u)$ follows by noting that $R_n < 0$ if and only if $R_n^* < 0$, because $A_k > 0$. $\qquad\square$

Proposition 5.3 *Assume* $A_1^{-1} > 0$, $\mu^* = \mathbb{E}\log A_1^{-1} < 0$ *and* $\mathbb{E}\log|B_1| < \infty$. *Then the r.v.* $Y = \sum_1^\infty D_n B_n$ *satisfies* $\mathbb{P}(-\infty < Y < \infty) = 1$, *and* $Y_n \overset{a.s.}{\to} Y$ *as* $n \to \infty$.

Proof. We have a.s. that $D_n = e^{\mu^* n (1 + o(1))}$. Furthermore, a standard application of the Borel-Cantelli lemma shows that $\mathbb{E} \log |B_n| < \infty$ ensures that for all c_1 one has $\log |B_n| > n c_1$, i.e. $|B_n| > e^{n c_1}$, only for finitely many n. Choosing $c_1 < -\mu^*$, it follows that the terms in the series in (5.3) decay at least geometrically fast, which implies the assertions.   □

As a corollary, we can give the
Proof of Proposition 5.1 *when* $\mathbb{E} \log A_n > 0$.
Note that

$$\psi(u) \;=\; \mathbb{P}(Y_n > u \text{ for some } n) \;\leq\; \mathbb{P}\Big(\sum_{i=1}^{\infty} D_i B_i > u\Big).$$

By the above, the sum is a finite r.v., and therefore the r.h.s. goes to 0 as $u \to \infty$.   □

The ruin probability can be represented in terms of Y as follows:

Theorem 5.4 *Define* $\overline{H}(u) = \mathbb{P}(Y > u)$. *Then* $\psi(u) = \overline{H}(u)/C_1(u)$ *where* $C_1(u) = \mathbb{E}\big[\overline{H}(R_{\tau(u)}) \,\big|\, \tau(u) < \infty\big]$.

Proof. For brevity, write $\tau = \tau(u)$. On $\{\tau < \infty\}$ we have $u - Y = u - Y_\tau - (Y - Y_\tau)$. Here

$$Y - Y_\tau \;=\; \sum_{i=\tau+1}^{\infty} D_i B_i \;=\; D_\tau \sum_{i=1}^{\infty} \frac{D_{\tau+i}}{D_\tau} B_{\tau+i} \;=\; D_\tau \widetilde{Y}$$

where

$$\widetilde{Y} \;=\; \sum_{i=1}^{\infty} \frac{D_{\tau+i}}{D_\tau} B_{i+\tau} \;=\; \sum_{i=1}^{\infty} A_{\tau+1}^{-1} \cdots A_{\tau+i}^{-1} B_{\tau+i}$$

is a copy of Y independent of τ and R_τ. Thus

$$u - Y \;=\; D_\tau (D_\tau^{-1}(u - Y_\tau) - \widetilde{Y}) \;=\; D_\tau(R_\tau - \widetilde{Y}).$$

Since $Y_n \to Y$, we also have $Y_n > u$ for some n (and hence $\tau < \infty$) when $Y > u$. Thus

$$\begin{aligned}
\overline{H}(u) &= \mathbb{P}(u - Y < 0) = \mathbb{P}\big(R_\tau - \widetilde{Y} < 0, \tau < \infty\big) \\
&= \psi(u) \mathbb{P}\big(R_\tau - \widetilde{Y} < 0 \,\big|\, \tau < \infty\big) = \psi(u) \mathbb{E}\big[\overline{H}(R_\tau) \,\big|\, \tau < \infty\big].
\end{aligned}$$
                                    □

Proposition 5.5 *The r.v.* Y *satisfies* $Y \overset{\mathscr{D}}{=} A_1^{-1}(B_1 + \check{Y})$ *where* $\check{Y}$ *is a copy of* Y *which is independent of* A_1, B_1.

Proof. Take

$$\check{Y} = \sum_{i=2}^{\infty} A_2^{-1} \dots A_i^{-1} B_i \,. \qquad \qquad \square$$

The representation in Theorem 5.4 is in general quite intractable, because usually $\overline{H}$ and $C_1(u)$ cannot be calculated. An obvious and important question is therefore to ask for tail asymptotics. The question that immediately comes up is how strongly $C_1(u)$ depends on u. The following result shows that under suitable conditions, this dependence is weak so that the important part of the tail asymptotics is $\overline{H}(u)$ itself.

Proposition 5.6 $0 < \liminf\limits_{u \to \infty} \dfrac{\psi(u)}{\overline{H}(u)} \leq \limsup\limits_{u \to \infty} \dfrac{\psi(u)}{\overline{H}(u)} \leq 1 \,.$

Proof. Since $-\infty < R_\tau \leq 0$, we have $\overline{H}(0) \leq C_1(u) \leq 1$, so all that remains to show is $\overline{H}(0) > 0$. Assume otherwise. Then the upper point a in the support of Y satisfies $-\infty < a \leq 0$. Assume first $a < 0$. Recall that by assumption, $p_1 = \mathbb{P}(A_1 \geq 1) > 0$ and/or that $B_1 \leq 0$ is excluded. Choose $\epsilon > 0$ with $p_2 = \mathbb{P}(B_1 > 2\epsilon) > 0$ and let $p_3 = \mathbb{P}(Y > a - \epsilon)$. Then $p_3 > 0$ and hence

$$\begin{aligned}
\mathbb{P}(Y > a + \epsilon) &= \mathbb{P}\big(A_1^{-1}(B_1 + \check{Y}) > a + \epsilon\big) \\
&\geq p_1 \mathbb{P}\big(B_1 + \check{Y} > a + \epsilon\big) \geq p_1 p_2 p_3 > 0 \,.
\end{aligned}$$

If instead $a = 0$, we have similarly that $\mathbb{P}(Y > a + \epsilon) \geq p_4 p_2 p_3$ where $p_4 = \mathbb{P}(A_1 \leq 1) > 0$. In both cases, we have reached a contradiction with the definition of a. $\qquad \square$

It thus remains to get some hold on $\overline{H}(u)$. We shall here involve a classical result due to Kesten [531] and Goldie [421] on *perpetuities*. By a perpetuity one understands a r.v. of the form

$$\widetilde{Y} = B_1 + D_1 B_2 + D_2 B_3 + \dots = \sum_{i=1}^{\infty} D_{i-1} B_i \qquad (5.3)$$

where $\{A_n\}$, $\{B_n\}$ are independent sequences each consisting of i.i.d. r.v.'s and $D_n = A_1^{-1} \cdots A_n^{-1}$ (the use of reciprocals is unusual but made to conform with the above risk theoretic setting). The result of [421, 531] states that under suitable conditions, $\overline{H}(u) = \mathbb{P}(Y > u)$ decays like an α-power of u. Before stating the result (the proof of which is outside the scope of this book), we give as a help for intuition some heuristic steps that give a heuristic motivation of the heavy tail and allow one to identify α.

Remark 5.7 Let T_n be the random walk $-\log A_1 - \cdots - \log A_n$ so that $Y = \sum_1^\infty \mathrm{e}^{T_n} B_n$. With light-tailed B_n, we expect that a single term $\mathrm{e}^{T_n} B_n$ can only be large if T_n is so. By assumption, T_n has negative drift and we will assume that the conditions of the Cramér Lundberg approximation are satisfied. Then the probability that $T_n > x$ for some x is approximately $C_T \mathrm{e}^{-\alpha x}$ where α solves

$$1 \;=\; \mathbb{E}\,\mathrm{e}^{-\alpha \log A} \;=\; \mathbb{E}\,\frac{1}{A^\alpha}\,. \tag{5.4}$$

Choose $b_0 > 0$ with $\overline{B}(b_0) = \mathbb{P}(B_n > b_0) > 0$. For an n with $T_n > x$ we then have $\mathbb{P}(\mathrm{e}^{T_n} B_n > \mathrm{e}^x b_0) > \overline{B}(b_0)$. Taking $x = \log u - \log b$, it follows that

$$\mathbb{P}(\mathrm{e}^{T_n} B_n > u \text{ for some } n) \;\geq\; C_T \mathrm{e}^{-\alpha x}\overline{B}(b_0) \;=\; \frac{C_T \overline{B}(b_0) b^\alpha}{u^\alpha}\,.$$

This motivates that Y is heavy-tailed and that the tail decays at least as a power.

The first trial solution for tail asymptotics is therefore a power tail, $\mathbb{P}(Y > u) \sim C/(1+u)^\beta$. Assume as before that B_1 is light-tailed. Then Proposition 5.5 implies that Y and $A_1^{-1}\breve{Y}$ have equivalent tails, i.e.

$$\begin{aligned}
\frac{C}{(1+u)^\beta} \;&\sim\; \mathbb{P}(Y > u) \;\sim\; \mathbb{P}(A_1^{-1}\breve{Y} > u) \\
&=\; \int_0^\infty \mathbb{P}(\breve{Y} > u/a)\,\mathbb{P}(A_1^{-1} \in \mathrm{d}a) \;\sim\; \int_0^\infty \frac{C}{(1+u/a)^\beta}\,\mathbb{P}(A_1^{-1} \in \mathrm{d}a) \\
&=\; \int_0^\infty \frac{Ca^\beta}{(a+u)^\beta}\,\mathbb{P}(A_1^{-1} \in \mathrm{d}a)\,.
\end{aligned}$$

Multiplying by $(1 + u)^\beta$ and going to the limit under the integral, the r.h.s. becomes $\mathbb{E}A^{-\beta}$. This suggests that $\beta = \alpha$, where α is as in (5.4). $\qquad\square$

Here is the result of Kesten [531] and Goldie [421]:

Theorem 5.8 *Assume that there exists $\alpha > 0$ such that $\mathbb{E}A_1^{-\alpha} = 1$ together with $\mathbb{E}A_1^{-\alpha}\log^- A_1 < \infty$ and $\mathbb{E}|B_1| < \infty$. Assume further that the distribution of A_1^{-1} is non-lattice. Then for some $C_2 > 0$,*

$$\mathbb{P}(Y > u) \;\sim\; \frac{C_2}{(1+u)^\alpha}, \quad u \to \infty\,. \tag{5.5}$$

The (untractable) expression for C_2 is given in [531, 421] (it is shown in Nyrhinen [669] that $C_2 > 0$). The proof of Theorem 5.8 is much too technical to be given here. However, up to the untractable constant we can now obtain the desired asymptotics of the ruin probability by combining with Theorem 5.4:

Corollary 5.9 *Under the conditions of Theorem 5.8,* $\psi(u) \sim \dfrac{C_2}{C_1(u)} \dfrac{1}{u^\alpha}.$ *In particular,*

$$0 < \liminf_{u\to\infty} u^\alpha \psi(u) \le \limsup_{u\to\infty} u^\alpha \psi(u) < \infty.$$

The result follows immediately by combining Theorems 5.8, 5.4 and Proposition 5.6 with the following lemma:

Lemma 5.10 *Under the conditions of Theorem 5.8,* $\mathbb{P}(\widetilde Y > u) \sim \mathbb{P}(Y > u) \sim C_2/u^\alpha$.

Proof. We have $Y = A_1^{-1}\widetilde Y^*$ where $\widetilde Y^* = \sum_1^\infty B_i \prod_2^n A_i^{-1}$ is a copy of $\widetilde Y$ independent of A_1. Rewrite Theorem 5.8 as $\mathbb{P}(\widetilde Y^* > y) \sim C_2/(1+y)^\alpha$. Then also $\mathbb{P}(\widetilde Y^* > y) \le C_3(1+y)^\alpha$ for all $y > 0$ since clearly such an inequality holds on any finite interval. Thus

$$
\begin{aligned}
u^\alpha \mathbb{P}(Y > u) &= u^\alpha \int_0^\infty \mathbb{P}(\widetilde Y^* > au)\,\mathbb{P}(A_1 \in \mathrm{d}a) \\
&= \int_0^\infty \mathbb{P}(\widetilde Y^* > au)(1+au)^\alpha \frac{u^\alpha}{(1+au)^\alpha}\,\mathbb{P}(A_1 \in \mathrm{d}a) \\
&\to \int_0^\infty C_2 \frac{1}{a^\alpha}\,\mathbb{P}(A_1 \in \mathrm{d}a) = C_2
\end{aligned}
$$

by dominated convergence. □

Remark 5.11 The reason that Corollary 5.9 gives a heavier tail asymptotics than in Section 2 is not so much that the interest is random, but that it is inherent in the set-up that negative returns are possible. Namely, if $A_1 \ge 1$ (and is not degenerate at 1), then $\mathbb{E}A_1^{-\alpha}$ is always < 1 so the conditions of Corollary 5.9 cannot hold. □

Remark 5.12 Nyrhinen [669] gives conditions under which $C_1(u)$ in Corollary 5.9 is not significant in terms of logarithmic asymptotics. □

Notes and references Rather than assuming that $\{A_n\}$, $\{B_n\}$ are independent sequences each consisting of i.i.d. r.v.'s, much of the literature relaxes this to the case that (A_n, B_n) being i.i.d. I.e. some dependence between A_n and B_n is allowed. See, for example, Nyrhinen [669, 668]. For models where the independence among the A_j themselves or among the B_j is relaxed see e.g. Cai [212], Cai & Dickson [214], Goovaerts et al. [424], Weng, Zhang & Tan [881], Shen, Lin & Zhang [796] and Collamore [252].

The recursion $R_n = A_n(R_{n-1} - B_n)$ also has some relevance as a model for the risk reserve in the presence of investments. The results are much as for (5.1), but will not be given here, see again Nyrhinen [669, 668].

Tail asymptotics for finite horizon ruin probabilities $\psi(u, n)$ in a setting where n goes to infinity with u and either A_1^{-1} or B_1 (or both) are heavy-tailed are given in Tang & Tsitsiashvili [830, 831]. For extensions to other ruin-related quantities see Yang & Zhang [903].

An important early paper is Paulsen [680]. Paulsen [682] surveys the literature up to 1998 and Paulsen [684] in the decade after that.

6 Continuous-time ruin problems with stochastic investment

Results for continuous-time models with stochastic investment can be obtained as a suitable limit of corresponding discrete-time set-ups (see the Notes). However, continuous-time models often also enable a direct analysis that can have a quite different flavor from its discrete counterpart. In this section this will be illustrated on a heuristic level for the case of a risk reserve process of compound Poisson type (with Poisson intensity β), where all the reserve is continuously invested in a financial market of Black-Scholes type, i.e. the risky asset is a geometric Brownian motion. More precisely, the resulting reserve is given by

$$R_t = u + t - \sum_{i=1}^{N_t} U_i + a \int_0^t R_{s^-} \, \mathrm{d}s + \sigma \int_0^t R_{s^-} \, \mathrm{d}B_s,$$

where $\{B_t\}$ is standard Brownian motion, σ is the volatility and a is the drift of a geometric Brownian motion. In view of the generators for the diffusion and the compound Poisson process derived in Examples II.4.1 and II.4.2 one observes that the generator of the resulting risk process is given by

$$\mathscr{A}f = au \, f'(u) + \frac{\sigma^2}{2} u^2 \, f''(u) + f'(u) + \beta \int_0^\infty \big(f(u - x) - f(u)\big) B(\mathrm{d}x).$$

Using Itô's Lemma, one can now show that if a twice continuously differentiable and bounded function $f(u)$ with $\lim_{u \to \infty} f(u) = 0$ satisfies $\mathscr{A}f(u) = 0$ for $u > 0$ and $f(u) = 1$ for $u < 0$, then $f(u)$ must be the ruin probability $\psi(u)$ of the process. Hence $\psi(u)$ satisfies

$$au \, \psi'(u) + \frac{\sigma^2 u^2}{2} \, \psi''(u) + \psi'(u) - \beta \, \psi(u) + \beta \int_0^u \psi(u - x) \, B(\mathrm{d}x) + \beta \overline{B}(u) = 0. \quad (6.1)$$

with $\lim_{u \to \infty} \psi(u) = 0$.[5] In this way the problem of studying the ruin probability with stochastic investment has been reduced to the purely analytical problem of solving an integro-differential equation.

---

[5] Note that for $\sigma = 0$ the investment is riskless and we get back to the risk model with constant interest rate (cf. (1.11) with $p(u) = p + a \, u$).

Example 6.1 Assume that the claim size distribution is exponential(ν). Then one can add the derivative of (6.1) to (6.1) multiplied by ν to get rid of the convolution term,[6] which leads to

$$\frac{\sigma^2 u^2}{2}\, \psi'''(u) + \left(au + \sigma^2 u + 1 + \nu\sigma^2 u^2/2\right)\psi''(u) + \left(a - \beta + ua\nu + \nu\right)\psi'(u) = 0$$

with additional boundary condition $\psi'(0) = \beta\psi(0)$. After substituting $\psi'(u)$ by another function, this is in fact a second-order ODE with polynomial coefficients which can be solved analytically in terms of special functions (of Heun type). Although the resulting formula is explicit, it is quite lengthy and we do not state it here. □

Looking at the drift of the geometric Brownian motion, one can show that for $\sigma^2 \geq 2a$, $\psi(u) = 1$ holds for all $u \geq 0$ (cf. e.g. Paulsen [681] or Pergamenshchikov & Zeitouny [691]), so it is enough to restrict to $\sigma^2 < 2a$. In general, it is impossible to obtain an explicit solution of the above IDE, but one can retrieve asymptotic results as $u \to \infty$ for a large class of claim size distributions.

Theorem 6.2 *Assume that the free reserve in the Cramér-Lundberg model is invested in a financial asset that is modeled by a geometric Brownian motion with drift $a > 0$ and volatility $\sigma > 0$ with $2a > \sigma^2$. If the claim size distribution is exponentially bounded, then*

$$\psi(u) \sim C\, u^{1 - 2a/\sigma^2}, \quad u \to \infty,$$

for some constant $C > 0$. If the claim size distribution is regularly varying $(\overline{B}(x) \sim L(x)x^{-\alpha}, \alpha > 0)$, then

$$\psi(u) \sim L_1(u)\, u^{\max\{1 - 2a/\sigma^2, -\alpha\}}, \quad u \to \infty, \tag{6.2}$$

where $L(u), L_1(u)$ are slowly varying functions.

Proof. For the full proof, we refer to Paulsen [683]. Here we only give a sketch of an analytical proof for $1 < 2a/\sigma^2 < 2$ to highlight the origin of the involved power terms. In view of the convolution term in (6.1), it is natural to take the Laplace transform of (6.1) and then try to apply Tauberian theorems to use the asymptotic behavior of $\widehat{\psi}[-s]$ for $s \to 0$ to infer information about $\psi(u)$ as $u \to \infty$. For simplicity of notation, call $g(s) = \widehat{\psi}[-s]$ the Laplace transform of the ruin probability. Since the Laplace transform of $u\psi'(u)$ is $-\left(sg(s)\right)'$ and the

---

[6]Cf. Section XII.3c for a more general procedure to eliminate convolution terms for a large class of claim size distributions.

Laplace transform of $u^2\psi''(u)$ is $\left(s^2 g(s)\right)''$, one obtains from (6.1) after some elementary calculations that

$$s^2 g''(s) + p_0 s\, g'(s) + (q_0 + q_1(s))\, g(s) \;=\; h(s)$$

with

$$p_0 = 4 - \frac{2a}{\sigma^2}, \quad q_0 = 2 - \frac{2a}{\sigma^2}, \quad q_1(s) = 2\big(s - \beta + \beta\widehat{B}[-s]\big)/\sigma^2,$$

$$h(s) \;=\; 2\psi(0)/\sigma^2 - 2\beta\mu_B\widehat{B_0}[-s]/\sigma^2,$$

where $\widehat{B_0}[-s] = \big(1 - \widehat{B}[-s]\big)/(\mu_B s)$ is the Laplace transform of the integrated tail distribution. It follows that $s = 0$ is a regular singular point of the homogeneous equation

$$s^2 g''(s) + p_0 s\, g'(s) + (q_0 + q_1(s))\, g(s) \;=\; 0 \tag{6.3}$$

which by the usual Frobenius method has a solution of the form

$$g(s) = s^r \sum_{k=0}^{\infty} c_k s^k. \tag{6.4}$$

Substituting this into (6.3) gives the condition $r(r-1) + p_0 r + q_0 = 0$ for r, i.e.

$$r_1 = -1 \quad \text{and} \quad r_2 = -2 + \frac{2a}{\sigma^2}. \tag{6.5}$$

For $1 < 2a/\sigma^2 < 2$, $r_1 - r_2$ is not an integer, so we obtain two independent solutions of the homogeneous ODE. The particular solution $g_p(s)$ can now be obtained by the classical method of variation of constants. With considerable but purely analytical effort (the details are omitted here) one can show that for exponentially bounded claim size distribution B, $g_p(s)$ tends to a constant for $s \to 0$. We now look at the asymptotics for $s \to 0$ of the full solution

$$g(s) \;=\; C_1 s^{-1}\eta_1(s) + C_2 s^{-2 + \frac{2a}{\sigma^2}}\eta_2(s) + g_p(s)$$

(due to (6.4), $\eta_1(s), \eta_2(s)$ also tend to a constant for $s \to 0$). A priori the first term would dominate, but from Theorem A6.1 this would translate into $\int_0^u \psi(x)\mathrm{d}x \sim C_1 u$ and by the Monotone Density Theorem $\psi(u) \to C_1$, which contradicts $\lim_{u\to\infty} \psi(u) = 0$. Hence we must have $C_1 = 0$ and the second term dominates the asymptotic behavior at $s \to 0$. Theorem A6.1 now gives $\int_0^u \psi(x)\mathrm{d}x \sim C_2 u^{2 - 2a/\sigma^2}/\Gamma(3 - 2a/\sigma^2)$ and the Monotone Density Theorem implies $\psi(u) \sim C \cdot u^{1-2a/\sigma^2}$ (one still needs to ensure that $C_2 \neq 0$ which is omitted here).

If, on the other hand, B is regularly varying with parameter α, then one can show that $g_p(s) \sim s^{\alpha-1}L(1/s)$, so that now the dominating term (which is either the particular solution or the second term above) is of order $s^{\min\{\alpha-1,-2+\frac{2a}{\sigma^2}\}}$. The same arguments as above then imply the claimed result. □

Remark 6.3 Theorem 6.2 states that (as in the discrete risk model of the previous section) full investment in geometric Brownian motion leads to Pareto-type asymptotic decay of the ruin probability even for light-tailed claim distributions. If the tail of the claim size distribution is heavy enough (and from (6.2) one sees that the tail needs to be very heavy), insurance risk can still dominate the financial risk. □

Remark 6.4 If only a constant fraction η ($0 < \eta < 1$) of the current wealth is invested in the risky asset, the analysis above is exactly the same, one just needs to replace a by $a\eta$ and σ by $\sigma\eta$. In Chapter XIV we will see how to improve the asymptotic behavior of the ruin probability by dynamically changing the investment fraction η as a function of the current risk reserve level. □

Notes and references Early results on ruin probabilities with investments can be found in Frolova, Kabanov & Pergamenshchikov [374] for a Cramér-Lundberg model with exponential claim sizes and investments into geometric Brownian motion (the explicit solution of Example 6.1 can also be found there) and bounds are derived in Kalashnikov & Norberg [520] in a more general set-up. For extensions to more general claim size distributions, see Constantinescu & Thomann [255]. Results for the continuous-time risk model can also be derived by using methods motivated by discrete-time models, see e.g. Nyrhinen [669] and for a fairly general account Paulsen [683]. Martingale techniques are exploited in Ma & Sun [618]. In order to assess $\psi(u)$ for moderate size u, Paulsen, Kasozi & Steigen [688] transform the IDE (6.1) into an ordinary Volterra integral equation of the second k ind and design an effective numerical solution procedure for the latter.

The proof technique of Theorem 6.2 outline here is made rigorous in Albrecher, Constantinescu & Thomann [22], and it is shown there that the method in principle also extends to renewal risk models with interclaim times of rational Laplace transform (in particular it turns out that the asymptotic result is insensitive to the choice of interclaim time distribution within that class). See also Wei [879] for a different method in the renewal set-up. Pergamenshchikov & Zeitouny [691] deal with more general premium rate functions and Cai & Xu [220] add perturbation to the original risk process. For absolute ruin probabilities in this context, see Gerber & Yang [414].

Klüppelberg & Kostadinova [540], Brokate *et al.* [205], Tang, Wang & Yuen [833] and Heyde & Wang [462] study the investment into exponential Lévy models in more detail. For further references in this active field of research we refer again to the recent survey by Paulsen [684].

Chapter IX

Matrix-analytic methods

1 Definition and basic properties of phase-type distributions

Phase-type distributions are the computational vehicle of much of modern applied probability. Typically, if a problem can be solved explicitly when the relevant distributions are exponentials, then the problem may admit an algorithmic solution involving a reasonable degree of computational effort if one allows for the more general assumption of phase-type structure, and not in other cases. A proper knowledge of phase-type distributions seems therefore a must for anyone working in an applied probability area like risk theory.

A distribution B on $(0, \infty)$ is said to be of *phase-type* if B is the distribution of the lifetime of a terminating Markov process $\{J_t\}_{t \geq 0}$ with finitely many states and time homogeneous transition rates. More precisely, a terminating Markov process $\{J_t\}$ with state space E and intensity matrix T is defined as the restriction to E of a Markov process $\{\bar{J}_t\}_{0 \leq t < \infty}$ on $E_\Delta = E \cup \{\Delta\}$ where Δ is some extra state which is *absorbing*, that is, $\mathbb{P}_i(\bar{J}_t = \Delta \text{ eventually}) = 1$ for all $i \in E$ [1] and where all states $i \in E$ are transient. This implies in particular that the intensity matrix for $\{\bar{J}_t\}$ can be written in block-partitioned form as

$$\left(\begin{array}{c|c} T & t \\ \hline 0 & 0 \end{array} \right). \tag{1.1}$$

We often write p for the number of elements of E. Note that since (1.1) is the intensity matrix of a non-terminating Markov process, the rows sum to zero

---

[1] Here as usual, $\mathbb{P}_i$ refers to the case $J_0 = i$; if $\boldsymbol{\nu} = (\nu_i)_{i \in E}$ is a probability distribution, we write $\mathbb{P}_{\boldsymbol{\nu}}$ for the case where J_0 has distribution $\boldsymbol{\nu}$ so that $\mathbb{P}_{\boldsymbol{\nu}} = \sum_{i \in E} \nu_i \mathbb{P}_i$.

which in matrix notation can be rewritten as $\boldsymbol{t} + \boldsymbol{T}\boldsymbol{e} = 0$ where $\boldsymbol{e}$ is the column E-vector with all components equal to one. In particular, $\boldsymbol{T}$ is a subintensity matrix,[2] and we have

$$\boldsymbol{t} = -\boldsymbol{T}\boldsymbol{e}. \tag{1.2}$$

The interpretation of the column vector $\boldsymbol{t}$ is as the *exit rate vector*, i.e. the ith component t_i gives the intensity in state i for leaving E and going to the absorbing state Δ.

We now say that B is of *phase-type with representation* $(E, \boldsymbol{\alpha}, \boldsymbol{T})$ (or sometimes just $(\boldsymbol{\alpha}, \boldsymbol{T})$) if B is the $\mathbb{P}_{\boldsymbol{\alpha}}$-distribution of $\zeta = \inf\{t > 0 : J_t = \Delta\}$ (the absorption time), i.e. $B = \mathbb{P}_{\boldsymbol{\alpha}}(\zeta \leq t)$. Equivalently, ζ is the lifetime $\sup\{t \geq 0 : J_t \in E\}$ of $\{J_t\}$. A convenient graphical representation is the *phase diagram* in terms of the entrance probabilities α_i, the exit rates t_i and the transition rates (intensities) t_{ij}:

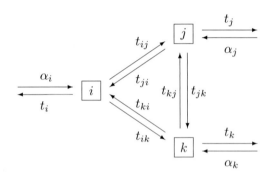

FIGURE IX.1: The phase diagram of a phase-type distribution with 3 phases, $E = \{i, j, k\}$.

The initial vector $\boldsymbol{\alpha}$ is written as a row vector.

Here are some important special cases:

Example 1.1 Suppose that $p = 1$ and write $\beta = -t_{11}$. Then $\alpha = \alpha_1 = 1$, $t_1 = \beta$, and the phase-type distribution is the lifetime of a particle with constant failure rate β, that is, an exponential distribution with rate parameter β. Thus *the phase-type distributions with $p = 1$ are exactly the class of exponential distributions*. □

---

[2]This means that $t_{ii} \leq 0$, $t_{ij} \geq 0$ for $i \neq j$ and $\sum_{j \in E} t_{ij} \leq 0$.

Example 1.2 The *Erlang distribution* E_p with p phases is defined as the Gamma distribution with integer parameter p and density

$$\delta^p \frac{x^{p-1}}{(p-1)!} e^{-\delta x}. \tag{1.3}$$

Since this corresponds to a convolution of p exponential densities with the same rate δ, the E_p distribution may be represented by the phase diagram ($p = 3$)

FIGURE IX.2

corresponding to $E = \{1, \ldots, p\}$, $\boldsymbol{\alpha} = (1\ 0\ 0 \ldots 0\ 0)$,

$$\boldsymbol{T} = \begin{pmatrix} -\delta & \delta & 0 & \cdots & 0 & 0 \\ 0 & -\delta & \delta & \cdots & 0 & 0 \\ \vdots & & & \ddots & & \vdots \\ 0 & 0 & 0 & \cdots & -\delta & \delta \\ 0 & 0 & 0 & \cdots & 0 & -\delta \end{pmatrix}, \quad \boldsymbol{t} = \begin{pmatrix} 0 \\ 0 \\ \vdots \\ 0 \\ \delta \end{pmatrix}.$$

□

Example 1.3 The *hyperexponential distribution* H_p with p parallel channels is defined as a mixture of p exponential distributions with rates $\delta_1, \ldots, \delta_p$ so that the density is

$$\sum_{i=1}^{p} \alpha_i \delta_i e^{-\delta_i x}. \tag{1.4}$$

Thus $E = \{1, \ldots, p\}$,

$$\boldsymbol{T} = \begin{pmatrix} -\delta_1 & 0 & 0 & \cdots & 0 & 0 \\ 0 & -\delta_2 & 0 & \cdots & 0 & 0 \\ \vdots & & & \ddots & & \vdots \\ 0 & 0 & 0 & \cdots & -\delta_{p-1} & 0 \\ 0 & 0 & 0 & \cdots & 0 & -\delta_p \end{pmatrix}, \quad \boldsymbol{t} = \begin{pmatrix} \delta_1 \\ \delta_2 \\ \vdots \\ \delta_{p-1} \\ \delta_p \end{pmatrix},$$

and the phase diagram is ($p = 2$)

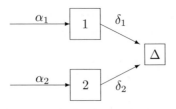

FIGURE IX.3

□

Example 1.4 (COXIAN DISTRIBUTIONS) This class of distributions is popular in much of the applied literature, and is defined as the class of phase-type distributions with a phase diagram of the following form:

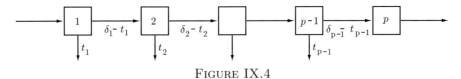

FIGURE IX.4

For example, the Erlang distribution is a special case of a Coxian distribution.

□

The basic analytical properties of phase-type distributions are given by the following result. Recall that the matrix-exponential $e^{\boldsymbol{K}}$ is defined by the standard series expansion $\sum_{n=0}^{\infty} \boldsymbol{K}^n / n!$. [3]

Theorem 1.5 *Let B be phase-type with representation $(E, \boldsymbol{\alpha}, \boldsymbol{T})$. Then:*
(a) *the c.d.f. is $B(x) = 1 - \boldsymbol{\alpha} e^{\boldsymbol{T} x} \boldsymbol{e}$;*
(b) *the density is $b(x) = B'(x) = \boldsymbol{\alpha} e^{\boldsymbol{T} x} \boldsymbol{t}$;*
(c) *the m.g.f. $\widehat{B}[r] = \int_0^\infty e^{rx} B(\mathrm{d}x)$ is $\boldsymbol{\alpha}(-r\boldsymbol{I} - \boldsymbol{T})^{-1}\boldsymbol{t}$;*
(d) *the nth moment $\int_0^\infty x^n B(\mathrm{d}x)$ is $(-1)^n n! \, \boldsymbol{\alpha} \boldsymbol{T}^{-n} \boldsymbol{e}$.*

Proof. Let $\overline{\boldsymbol{P}}^s = (\overline{p}_{ij}^s)$ be the s-step $E_\Delta \times E_\Delta$ transition matrix for $\{\overline{J}_t\}$ and $\boldsymbol{P}^s$ the s-step $E \times E$-transition matrix for $\{J_t\}$, i.e. the restriction of $\overline{\boldsymbol{P}}^s$ to E. Then for $i, j \in E$, the backwards equation for $\{\overline{J}_t\}$ (e.g. [APQ, p. 48]) yields

$$\frac{\mathrm{d}p_{ij}^s}{\mathrm{d}s} = \frac{\mathrm{d}\overline{p}_{ij}^s}{\mathrm{d}s} = t_i \overline{p}_{\delta j}^s + \sum_{k \in E} t_{ik} \overline{p}_{kj}^s = \sum_{k \in E} t_{ik} p_{kj}^s.$$

---

[3]For a number of additional important properties of matrix-exponentials and discussion of computational aspects, see Appendix A3.

That is, $\frac{d}{ds}\boldsymbol{P}^s = \boldsymbol{T}\boldsymbol{P}^s$, and since obviously $\boldsymbol{P}^0 = \boldsymbol{I}$, the solution is $\boldsymbol{P}^s = \mathrm{e}^{\boldsymbol{T}s}$. Since

$$1 - B(x) = \mathbb{P}_{\boldsymbol{\alpha}}(\zeta > x) = \mathbb{P}_{\boldsymbol{\alpha}}(J_x \in E) = \sum_{i,j \in E} \alpha_i p_{ij}^x = \boldsymbol{\alpha}\boldsymbol{P}^x\boldsymbol{e},$$

this proves (a), and (b) then follows from

$$B'(x) = -\boldsymbol{\alpha}\frac{d}{dx}\boldsymbol{P}^x\boldsymbol{e} = -\boldsymbol{\alpha}\mathrm{e}^{\boldsymbol{T}x}\boldsymbol{T}\boldsymbol{e} = \boldsymbol{\alpha}\mathrm{e}^{\boldsymbol{T}x}\boldsymbol{t}$$

(since $\boldsymbol{T}$ and $\mathrm{e}^{\boldsymbol{T}x}$ commute). For (c), the rule (A.12) for integrating matrix-exponentials yields

$$\widehat{B}[r] = \int_0^\infty \mathrm{e}^{rx}\boldsymbol{\alpha}\mathrm{e}^{\boldsymbol{T}x}\boldsymbol{t}\,dx = \boldsymbol{\alpha}\left(\int_0^\infty \mathrm{e}^{(r\boldsymbol{I}+\boldsymbol{T})x}\,dx\right)\boldsymbol{t} = \boldsymbol{\alpha}(-r\boldsymbol{I} - \boldsymbol{T})^{-1}\boldsymbol{t}.$$

Alternatively, define $h_i = \mathbb{E}_i\mathrm{e}^{r\zeta}$. Then

$$h_i = \frac{-t_{ii}}{-t_{ii} - r}\left\{\frac{t_i}{-t_{ii}} + \sum_{j \neq i}\frac{t_{ij}}{-t_{ii}}h_j\right\}. \tag{1.5}$$

Indeed, $-t_{ii}$ is the rate of the exponential holding time of state i and hence $(-t_{ii})/(-t_{ii} - r)$ is the m.g.f. of the initial sojourn in state i. After that, we either go to state $j \neq i$ w.p. $t_{ij}/-t_{ii}$ and have an additional time to absorption which has m.g.f. h_j, or w.p. $t_i/-t_{ii}$ we go to Δ, in which case the time to absorption is 0 with m.g.f. 1. Rewriting (1.5) as

$$h_i(t_{ii} + r) = -t_i - \sum_{j \neq i}t_{ij}h_j, \quad \sum_{j \in E}t_{ij}h_j + h_ir = -t_i,$$

this means in vector notation that $(\boldsymbol{T} + r\boldsymbol{I})\boldsymbol{h} = -\boldsymbol{t}$, i.e. $\boldsymbol{h} = -(\boldsymbol{T} + r\boldsymbol{I})^{-1}\boldsymbol{t}$, and since $\widehat{B}[r] = \boldsymbol{\alpha}\boldsymbol{h}$, we arrive once more at the stated expression for $\widehat{B}[r]$.

Part (d) follows by differentiating the m.g.f.,

$$\frac{d^n}{dr^n}\boldsymbol{\alpha}(-r\boldsymbol{I} - \boldsymbol{T})^{-1}\boldsymbol{t} = (-1)^{n+1}n!\,\boldsymbol{\alpha}(r\boldsymbol{I} + \boldsymbol{T})^{-n-1}\boldsymbol{t},$$

$$\widehat{B}^{(n)}[0] = (-1)^{n+1}n!\,\boldsymbol{\alpha}\boldsymbol{T}^{-n-1}\boldsymbol{t} = (-1)^n n!\boldsymbol{\alpha}\boldsymbol{T}^{-n-1}\boldsymbol{T}\boldsymbol{e}$$
$$= (-1)^n n!\,\boldsymbol{\alpha}\boldsymbol{T}^{-n}\boldsymbol{e}.$$

Alternatively, for $n = 1$ we may put $k_i = \mathbb{E}_i\zeta$ and get as in (1.5)

$$k_i = \frac{1}{-t_{ii}} + \sum_{j \neq i}\frac{t_{ij}}{-t_{ii}}k_j,$$

which is solved as above to get $\boldsymbol{k} = -\boldsymbol{\alpha}\boldsymbol{T}^{-1}\boldsymbol{e}$. $\qquad\square$

Example 1.6 Though typically the evaluation of matrix-exponentials is most conveniently carried out on a computer, there are some examples where it is appealing to write T on diagonal form, making the problem trivial. One obvious instance is the hyperexponential distribution, another the case $p = 2$ where explicit diagonalization formulas are always available, see the Appendix. Consider for example

$$\boldsymbol{\alpha} = (1/2 \ \ 1/2), \quad \boldsymbol{T} = \begin{pmatrix} -3/2 & 9/14 \\ 7/2 & -11/2 \end{pmatrix} \quad \text{so that} \quad \boldsymbol{t} = \begin{pmatrix} 6/7 \\ 2 \end{pmatrix}.$$

Then (cf. Example A3.7) the diagonal form of T is

$$\boldsymbol{T} \;=\; -\begin{pmatrix} 9/10 & 9/70 \\ 7/10 & 1/10 \end{pmatrix} - 6\begin{pmatrix} 1/10 & -9/70 \\ -7/10 & 9/10 \end{pmatrix},$$

where the two matrices on the r.h.s. are idempotent. This implies that we can compute the nth moment as

$$
\begin{aligned}
(-1)^n n!\, \boldsymbol{\alpha} \boldsymbol{T}^{-n} \boldsymbol{e} \;=\;& 1^n n!\,(1/2 \ \ 1/2) \begin{pmatrix} 9/10 & 9/70 \\ 7/10 & 1/10 \end{pmatrix}\begin{pmatrix} 1 \\ 1 \end{pmatrix} \\
& + 6^{-n} n!\,(1/2 \ \ 1/2)\begin{pmatrix} 1/10 & -9/70 \\ -7/10 & 9/10 \end{pmatrix}\begin{pmatrix} 1 \\ 1 \end{pmatrix} \\
\;=\;& n!\left(\frac{32}{35} + \frac{3}{35 \cdot 6^n} \right).
\end{aligned}
$$

Similarly, we get the density as

$$
\begin{aligned}
\boldsymbol{\alpha} e^{\boldsymbol{T}x} \boldsymbol{t} \;=\;& e^{-x} \left(\frac{1}{2} \ \ \frac{1}{2} \right)\begin{pmatrix} 9/10 & 9/70 \\ 7/10 & 1/10 \end{pmatrix}\begin{pmatrix} 6/7 \\ 2 \end{pmatrix} \\
& + e^{-6x}\left(\frac{1}{2} \ \ \frac{1}{2} \right)\begin{pmatrix} 1/10 & -9/70 \\ -7/10 & 9/10 \end{pmatrix}\begin{pmatrix} 6/7 \\ 2 \end{pmatrix} \\
\;=\;& \frac{32}{35} e^{-x} + \frac{18}{35} e^{-6x}.
\end{aligned}
$$

$\square$

The following result becomes basic in Sections 4, 5 and serves at this stage to introduce Kronecker notation and calculus (see Section 4b for definitions and basic rules):

Proposition 1.7 *If B is phase-type with representation $(\boldsymbol{\nu}, \boldsymbol{T})$, then the matrix m.g.f. $\widehat{B}[\boldsymbol{Q}]$ of B is*

$$\widehat{B}[\boldsymbol{Q}] = \int_0^\infty e^{\boldsymbol{Q}x} B(dx) = (\boldsymbol{\nu} \otimes \boldsymbol{I})(-\boldsymbol{T} \oplus \boldsymbol{Q})^{-1}(\boldsymbol{t} \otimes \boldsymbol{I}). \qquad (1.6)$$

Proof. According to (A.29) and Proposition A4.4,

$$\widehat{B}[\boldsymbol{Q}] = \int_0^\infty \boldsymbol{\nu} e^{\boldsymbol{T}x} \boldsymbol{t} e^{\boldsymbol{Q}x} \, dx = (\boldsymbol{\nu} \otimes \boldsymbol{I}) \left(\int_0^\infty e^{\boldsymbol{T}x} \otimes e^{\boldsymbol{Q}x} \, dx \right) (\boldsymbol{t} \otimes \boldsymbol{I})$$

$$= (\boldsymbol{\nu} \otimes \boldsymbol{I}) \left(\int_0^\infty e^{(\boldsymbol{T} \oplus \boldsymbol{Q})x} dx \right) (\boldsymbol{t} \otimes \boldsymbol{I}) = (\boldsymbol{\nu} \otimes \boldsymbol{I})(-\boldsymbol{T} \oplus \boldsymbol{Q})^{-1}(\boldsymbol{t} \otimes \boldsymbol{I}).$$

$\square$

Sometimes it is relevant also to consider phase-type distributions, where the initial vector $\boldsymbol{\alpha}$ is substochastic, $\|\boldsymbol{\alpha}\| = \sum_{i \in E} \alpha_i < 1$. There are two ways to interpret this:

- The phase-type distribution B is *defective*, i.e. $\|B\| = \|\boldsymbol{\alpha}\| < 1$; a random variable U having a defective phase-type distribution with representation $(\boldsymbol{\alpha}, \boldsymbol{T})$ is then defined to be ∞ on a set of probability $1 - \|\boldsymbol{\alpha}\|$, or one just lets U be undefined on this additional set.

- The phase-type distribution B is *zero-modified*, i.e a mixture of a phase-type distribution with representation $(\boldsymbol{\alpha}/\|\boldsymbol{\alpha}\|, \boldsymbol{T})$ with weight $\|\boldsymbol{\alpha}\|$ and an atom at zero with weight $1 - \|\boldsymbol{\alpha}\|$. This is the traditional choice in the literature, and in fact one also most often there allows $\boldsymbol{\alpha}$ to have a component α_Δ at Δ.

1a Asymptotic exponentiality

Writing $\boldsymbol{T}$ on the Jordan canonical form, it is easily seen that the asymptotic form of the tail of a general phase-type distribution has the form

$$\overline{B}(x) \sim C x^k e^{-\eta x},$$

where $C, \eta > 0$ and $k = 0, 1, 2 \ldots$ The Erlang distribution gives an example where $k > 0$ (in fact, then $k = p - 1$), but in many practical cases, one has $k = 0$. Here is a sufficient condition:

Proposition 1.8 *Let B be phase-type with representation $(\boldsymbol{\alpha}, \boldsymbol{T})$, assume that $\boldsymbol{T}$ is irreducible, let $-\eta$ be the eigenvalue of largest real part of $\boldsymbol{T}$, let $\boldsymbol{\nu}, \boldsymbol{h}$ be the corresponding left and right eigenvectors normalized by $\boldsymbol{\nu h} = 1$ and define $C = \boldsymbol{\alpha h} \cdot \boldsymbol{\nu e}$. Then the tail $\overline{B}(x)$ is asymptotically exponential,*

$$\overline{B}(x) \sim C e^{-\eta x}. \qquad (1.7)$$

Proof. By Perron-Frobenius theory (A.4c), η is real and positive, $\boldsymbol{\nu}, \boldsymbol{h}$ can be chosen with strictly positive component, and we have

$$e^{\boldsymbol{T}x} \sim \boldsymbol{h\nu}e^{-\eta x}, \quad x \to \infty.$$

Using $\overline{B}(x) = \boldsymbol{\alpha}e^{\boldsymbol{T}x}\boldsymbol{e}$, the result follows (with $C = (\boldsymbol{\alpha h})(\boldsymbol{\nu e})$). □

Of course, the conditions of Proposition 1.8 are far from necessary (a mixture of phase-type distributions with the respective $\boldsymbol{T}^{(i)}$ irreducible has obviously an asymptotically exponential tail, but the relevant $\boldsymbol{T}$ is not irreducible, cf. Example A5.8).

In Proposition A5.1 of the Appendix, we give a criterion for asymptotical exponentiality of a phase-type distribution B, not only in the tail but in the whole distribution.

Notes and references The idea behind using phase-type distributions goes back to Erlang, but today's interest in the topic was largely initiated by M.F. Neuts, see his book [660] (a historical important intermediate step is Jensen [505]). Other expositions of the basic theory of phase-type distributions can be found in [APQ], Lipsky [600], Rolski, Schmidli, Schmidt & Teugels [746] and Wolff [894]. All material of the present section is standard; the text is essentially identical to Section 2 of Asmussen [68].

In some of the literature and also in Section XII.3, the slightly larger class of distributions with a rational m.g.f. (or Laplace transform) is used which may seem less intuitive than phase-type distributions. See in particular the notes to Section 6. O'Cinneide [670] gave a necessary and sufficient criterion for a distribution B with a rational m.g.f. $\widehat{B}[s] = p(s)/q(s)$ to be phase-type: the density $b(x)$ should be strictly positive for $x > 0$ and the root of $q(s)$ with the smallest real part should be unique (not necessarily simple, cf. the Erlang case). No satisfying algorithm for finding a phase representation of a distribution B (which is known to be phase-type and for which the m.g.f. or the density is available) is, however, known. A related important unsolved problem deals with minimal representations: given a phase-type distribution, what is the smallest possible dimension of the phase space E?

2   Renewal theory

A summary of the renewal theory in general is given in A1 of the Appendix, but is in part repeated below. Let $U_1, U_2, \ldots$ be i.i.d. with common distribution B and define[4]

$$\begin{aligned}
\mathscr{U}(A) &= \mathbb{E}\# \{n = 0, 1, \ldots : U_1 + \cdots + U_n \in A\} \\
&= \mathbb{E}\sum_{n=0}^{\infty} I(U_1 + \cdots + U_n \in A).
\end{aligned}$$

---

[4]Here the empty sum $U_1 + \ldots + U_0$ is 0.

We may think of the U_i as the lifetimes of items (say electrical bulbs) which are replaced upon failure, and $\mathscr{U}(A)$ is then the expected number of replacements (renewals) in A. For this reason, we refer to $\mathscr{U}$ as the *renewal measure*; if $\mathscr{U}$ is absolutely continuous on $(0, \infty)$ w.r.t. Lebesgue measure, we denote the density by $u(x)$ and refer to u as the *renewal density*. If B is exponential with rate β, the renewals form a Poisson process and we have $u(x) = \beta$. The explicit calculation of the renewal density (or the renewal measure) is often thought of as infeasible for other distributions, but nevertheless, the problem has an algorithmically tractable solution if B is phase-type:

Theorem 2.1 *Consider a renewal process with interarrivals which are phase-type with representation $(\boldsymbol{\alpha}, \boldsymbol{T})$. Then the renewal density exists and is given by*

$$u(x) = \boldsymbol{\alpha} \mathrm{e}^{(\boldsymbol{T} + \boldsymbol{t}\boldsymbol{\alpha})x} \boldsymbol{t}. \tag{2.1}$$

Proof. Let $\left\{ J_t^{(k)} \right\}$ be the governing phase process for U_k and define $\left\{ \widetilde{J}_t \right\}$ by piecing the $\left\{ J_t^{(k)} \right\}$ together,

$$\widetilde{J}_t = J_t^{(1)}, \ 0 \le t < U_1, \quad \widetilde{J}_t = J_{t-U_1}^{(2)}, \ U_1 \le t < U_1 + U_2, \ \dots.$$

Then $\left\{ \widetilde{J}_t \right\}$ is Markov and has two types of jumps, the jumps of the $J_t^{(k)}$ and the jumps corresponding to a transition from one $J_t^{(k)}$ to the next $J_t^{(k+1)}$. A jump of the last type from i to j occurs at rate $t_i \alpha_j$, and the jumps of the first type are governed by $\boldsymbol{T}$. Hence the intensity matrix is $\boldsymbol{T} + \boldsymbol{t}\boldsymbol{\alpha}$, and the distribution of $\widetilde{J}_x$ is $\boldsymbol{\alpha} \mathrm{e}^{(\boldsymbol{T} + \boldsymbol{t}\boldsymbol{\alpha})x}$. The renewal density at x is now just the rate of jumps of the second type, which is t_i in state i. Hence (2.1) follows by the law of total probability. $\square$

The argument goes through without change if the renewal process is *terminating*, i.e. B is defective, and hence (2.1) remains valid for that case. However, the phase-type assumptions also yield the distribution of a further quantity of fundamental importance in later parts of this chapter, the *lifetime* of the renewal process. This is defined as $U_1 + \cdots + U_{\kappa-1}$ where κ is the first k with $U_k = \infty$, that is, as the time of the last renewal; since $U_k = \infty$ with probability $1 - \|B\|$ which is > 0 in the defective case, this is well-defined.

Corollary 2.2 *Consider a terminating renewal process with interarrivals which are defective phase-type with representation $(\boldsymbol{\alpha}, \boldsymbol{T})$, i.e. $\|\boldsymbol{\alpha}\| < 1$. Then the lifetime is zero-modified phase-type with representation $(\boldsymbol{\alpha}, \boldsymbol{T} + \boldsymbol{t}\boldsymbol{\alpha})$.*

Proof. Just note that $\left\{ \widetilde{J}_t \right\}$ is a governing phase process for the lifetime. $\square$

Returning to non-terminating renewal processes, define the *excess life* $\xi(t)$ at time t as the time until the next renewal following t, see Fig. IX.5.

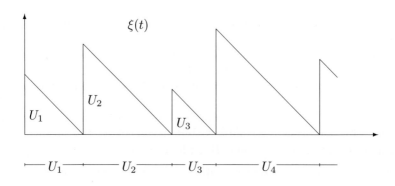

FIGURE IX.5

Corollary 2.3 *Consider a renewal process with interarrivals which are phase-type with representation* $(\boldsymbol{\alpha}, \boldsymbol{T})$, *and let* $\mu_B = -\boldsymbol{\alpha}\boldsymbol{T}^{-1}\boldsymbol{e}$ *be the mean of* B. *Then:*
(a) the excess life $\xi(t)$ *at time* t *is phase-type with representation* $(\boldsymbol{\nu}_t, \boldsymbol{T})$ *where*
$\boldsymbol{\nu}_t = \boldsymbol{\alpha}\mathrm{e}^{(\boldsymbol{T}+\boldsymbol{t}\boldsymbol{\alpha})t}$;
(b) $\xi(t)$ *has a limiting distribution as* $t \to \infty$, *which is phase-type with representation* $(\boldsymbol{\nu}, \boldsymbol{T})$ *where* $\boldsymbol{\nu} = -\boldsymbol{\alpha}\boldsymbol{T}^{-1}/\mu_B$. *Equivalently, the density is* $\boldsymbol{\nu}\mathrm{e}^{\boldsymbol{T}x}\boldsymbol{t} = \overline{B}(x)/\mu_B$.

Proof. Consider again the process $\{\widetilde{J}_t\}$ in the proof of Theorem 2.1. The time of the next renewal after t is the time of the next jump of the second type, hence $\xi(t)$ is phase-type with representation $(\boldsymbol{\nu}_t, \boldsymbol{T})$ where $\boldsymbol{\nu}_t$ is the distribution of $\widetilde{J}_t$ which is obviously given by the expression in (a). Hence in (b) it is immediate that $\boldsymbol{\nu}$ exists and is the stationary limiting distribution of $\widetilde{J}_t$, i.e. the unique positive solution of

$$\boldsymbol{\nu}\boldsymbol{e} = 1, \qquad \boldsymbol{\nu}(\boldsymbol{T} + \boldsymbol{t}\boldsymbol{\alpha}) = 0. \qquad (2.2)$$

Here are two different arguments that this yields the asserted expression:

(i) Just check that $-\boldsymbol{\alpha}\boldsymbol{T}^{-1}/\mu_B$ satisfies (2.2):

$$
\begin{aligned}
\frac{-\boldsymbol{\alpha}\boldsymbol{T}^{-1}}{\mu_B}\boldsymbol{e} &= \frac{\mu_B}{\mu_B} = 1, \\
\frac{-\boldsymbol{\alpha}\boldsymbol{T}^{-1}}{\mu_B}(\boldsymbol{T} + \boldsymbol{t}\boldsymbol{\alpha}) &= \frac{-\boldsymbol{\alpha} + \boldsymbol{\alpha}\boldsymbol{T}^{-1}\boldsymbol{T}\boldsymbol{e}\boldsymbol{\alpha}}{\mu_B} \\
&= \frac{-\boldsymbol{\alpha} + \boldsymbol{\alpha}\boldsymbol{e}\boldsymbol{\alpha}}{\mu_B} = \frac{-\boldsymbol{\alpha} + \boldsymbol{\alpha}}{\mu_B} = 0.
\end{aligned}
$$

(ii) First check the asserted identity for the density: since $\boldsymbol{T}$, $\boldsymbol{T}^{-1}$ and $\mathrm{e}^{\boldsymbol{T}x}$ commute, we get

$$\frac{\overline{B}(x)}{\mu_B} = \frac{\boldsymbol{\alpha}\mathrm{e}^{\boldsymbol{T}x}\boldsymbol{e}}{\mu_B} = \frac{\boldsymbol{\alpha}\boldsymbol{T}^{-1}\mathrm{e}^{\boldsymbol{T}x}\boldsymbol{T}\boldsymbol{e}}{\mu_B} = \boldsymbol{\nu}\mathrm{e}^{\boldsymbol{T}x}\boldsymbol{t}.$$

Next appeal to the standard fact from renewal theory that the limiting distribution of $\xi(x)$ has density $\overline{B}(x)/\mu_B$, cf. Section A.1e. □

Example 2.4 Consider a non-terminating renewal process with two phases. The formulas involve the matrix-exponential of the intensity matrix

$$\boldsymbol{Q} = \boldsymbol{T} + \boldsymbol{t}\boldsymbol{\alpha} = \begin{pmatrix} t_{11} + t_1\alpha_1 & t_{12} + t_1\alpha_2 \\ t_{12} + t_2\alpha_1 & t_{22} + t_2\alpha_2 \end{pmatrix} = \begin{pmatrix} -q_1 & q_1 \\ q_2 & -q_2 \end{pmatrix} \quad \text{(say)}.$$

According to Example A3.6, we first compute the stationary distribution of $\boldsymbol{Q}$,

$$\boldsymbol{\pi} = (\pi_1 \ \pi_2) = \left(\frac{q_2}{q_1 + q_2} \ \frac{q_1}{q_1 + q_2} \right),$$

and the non-zero eigenvalue $\lambda = -q_1 - q_2$. The renewal density is then

$$
\begin{aligned}
\boldsymbol{\alpha}\mathrm{e}^{\boldsymbol{Q}t}\boldsymbol{t} &= (\alpha_1 \ \alpha_2) \begin{pmatrix} \pi_1 & \pi_2 \\ \pi_1 & \pi_2 \end{pmatrix} \begin{pmatrix} t_1 \\ t_2 \end{pmatrix} \\
&\quad + \mathrm{e}^{\lambda t}(\alpha_1 \ \alpha_2) \begin{pmatrix} \pi_2 & -\pi_2 \\ -\pi_1 & \pi_1 \end{pmatrix} \begin{pmatrix} t_1 \\ t_2 \end{pmatrix} \\
&= (\pi_1 \ \pi_2) \begin{pmatrix} t_1 \\ t_2 \end{pmatrix} + \mathrm{e}^{\lambda t}(\alpha_1 \ \alpha_2) \begin{pmatrix} \pi_2(t_1 - t_2) \\ \pi_1(t_2 - t_1) \end{pmatrix} \\
&= \pi_1 t_1 + \pi_2 t_2 + \mathrm{e}^{\lambda t}(\alpha_1\pi_2 - \alpha_2\pi_1)(t_1 - t_2) \\
&= \frac{1}{\mu_B} + \mathrm{e}^{\lambda t}(\alpha_1\pi_2 - \alpha_2\pi_1)(t_1 - t_2).
\end{aligned}
$$

□

Example 2.5 Let B be Erlang(2). Then

$$\boldsymbol{Q} = \begin{pmatrix} -\delta & \delta \\ 0 & -\delta \end{pmatrix} + \begin{pmatrix} 0 \\ \delta \end{pmatrix}(1 \ 0) = \begin{pmatrix} -\delta & \delta \\ \delta & -\delta \end{pmatrix}.$$

Hence $\boldsymbol{\pi} = (1/2 \ 1/2)$, $\lambda = -2\delta$, and Example 2.4 yields the renewal density as

$$u(t) = \frac{\delta}{2}\left(1 - \mathrm{e}^{-2\delta t}\right).$$

□

Example 2.6 Let B be hyperexponential. Then

$$Q = \begin{pmatrix} -\delta_1 & 0 \\ 0 & -\delta_2 \end{pmatrix} + \begin{pmatrix} \delta_1 \\ \delta_2 \end{pmatrix} (\alpha_1 \; \alpha_2) = \begin{pmatrix} -\delta_1\alpha_2 & \delta_1\alpha_2 \\ \delta_2\alpha_1 & -\delta_2\alpha_1 \end{pmatrix}.$$

Hence

$$\pi = \left(\frac{\delta_2\alpha_1}{\delta_1\alpha_2 + \delta_2\alpha_1} \quad \frac{\delta_1\alpha_2}{\delta_1\alpha_2 + \delta_2\alpha_1} \right),$$

$\lambda = -\delta_1\alpha_2 - \delta_2\alpha_1$, and Example 2.4 yields the renewal density as

$$u(t) = \frac{\delta_1\delta_2}{\delta_1\alpha_2 + \delta_2\alpha_1} + \mathrm{e}^{-(\delta_1\alpha_2 + \delta_2\alpha_1)t} \frac{(\delta_1 - \delta_2)^2\alpha_1\alpha_2}{\delta_1\alpha_2 + \delta_2\alpha_1}.$$

$\square$

Notes and references Early expositions of renewal theory for phase-type distributions are Neuts [659] and Kao [521]. The present treatment, similar to that in [APQ], is somewhat more probabilistic.

3 The compound Poisson model

3a Phase-type claims

Consider the compound Poisson (Cramér-Lundberg) model in the notation of Chapter I, with β denoting the Poisson intensity, B the claim size distribution, $\tau(u)$ the time of ruin with initial reserve u, $\{S_t\}$ the claim surplus process, $G_+(\cdot) = \mathbb{P}\big(S_{\tau(0)} \in \cdot, \tau(0) < \infty\big)$ the ladder height distribution and $M = \sup_{t \geq 0} S_t$. We assume that B is phase-type with representation $(\boldsymbol{\alpha}, \boldsymbol{T})$.

Corollary 3.1 *Assume that the claim size distribution B is phase-type with representation $(\boldsymbol{\alpha}, \boldsymbol{T})$. Then:*
(a) *G_+ is defective phase-type with representation $(\boldsymbol{\alpha}_+, \boldsymbol{T})$ where $\boldsymbol{\alpha}_+$ is given by $\boldsymbol{\alpha}_+ = -\beta\boldsymbol{\alpha}\boldsymbol{T}^{-1}$, and M is zero-modified phase-type with representation $(\boldsymbol{\alpha}_+, \boldsymbol{T} + t\boldsymbol{\alpha}_+)$.*
(b) *$\psi(u) = \boldsymbol{\alpha}_+\mathrm{e}^{(\boldsymbol{T} + t\boldsymbol{\alpha}_+)u}\boldsymbol{e}$.*

Note in particular that $\rho = \|G_+\| = \boldsymbol{\alpha}_+\boldsymbol{e}$.

Proof. The result follows immediately by combining the Pollaczeck-Khinchine formula with general results on phase-type distributions: for (a), use the phase-type representation of B_0, cf. Corollary 2.3. For (b), represent the maximum M as the lifetime of a terminating renewal process and use Corollary 2.2.

Since the result is quite fundamental, we shall, however, add a more self-contained explanation of why the phase-type structure is preserved. The essence is contained in Fig. IX.6. Here we have taken the terminating Markov process underlying B with two states, marked by thin and thick lines on the figure. Then each claim (jump) corresponds to one (finite) sample path of the Markov process. The stars represent the ladder points $S_{\tau_+(k)}$. Considering the first, we see that the ladder height S_{τ_+} is just the residual lifetime of the Markov process corresponding to the claim causing upcrossing of level 0, i.e. itself phase-type with the same phase generator T and the initial vector α_+ being the distribution of the upcrossing Markov process at time $-S_{\tau_+-}$. Next, the Markov processes representing ladder steps can be pieced together to one $\{m_x\}$. Within ladder steps, the transitions are governed by T whereas termination of ladder steps may lead to some additional ones: a transition from i to j occurs if the ladder step terminates in state i, which occurs at rate t_i, and if there is a subsequent ladder step starting in j which occurs w.p. α_{+j}. Thus the total rate is $t_{ij}+t_i\alpha_{+j}$, and rewriting in matrix form yields the phase generator of $\{m_x\}$ as $T + t\alpha_+$. Now just observe that the initial vector of $\{m_x\}$ is α_+ and that the lifelength is M.

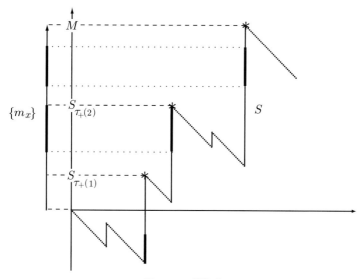

FIGURE IX.6

This derivation is a complete proof except for the identification of α_+ with $-\beta\alpha T^{-1}$. This is in fact a simple consequence of the form of the excess distribution B_0, see Corollary 2.3. □

Example 3.2 Assume that $\beta = 3$ and

$$b(x) = \frac{1}{2} \cdot 3e^{-3x} + \frac{1}{2} \cdot 7e^{-7x}.$$

Thus b is hyperexponential (a mixture of exponential distributions) with $\boldsymbol{\alpha} = (\frac{1}{2} \ \frac{1}{2})$, $\boldsymbol{T} = (-3 \ -7)_{\text{diag}}$ so that

$$\boldsymbol{\alpha}_+ = -\beta\boldsymbol{\alpha}\boldsymbol{T}^{-1} = -3(1/2 \ 1/2)\begin{pmatrix} -1/3 & 0 \\ 0 & -1/7 \end{pmatrix} = (1/2 \ 3/14),$$

$$\boldsymbol{T}+\boldsymbol{t\alpha}_+ = \begin{pmatrix} -3 & 0 \\ 0 & -7 \end{pmatrix} + \begin{pmatrix} 3 \\ 7 \end{pmatrix}\begin{pmatrix} 1 & 3 \\ 2 & 14 \end{pmatrix} = \begin{pmatrix} -3/2 & 9/14 \\ 7/2 & -11/2 \end{pmatrix}.$$

This is the same matrix as is Example 1.6, so that as there

$$e^{(\boldsymbol{T}+\boldsymbol{t\alpha}_+)u} = e^{-u}\begin{pmatrix} 9/10 & 9/70 \\ 7/10 & 1/10 \end{pmatrix} + e^{-6u}\begin{pmatrix} 1/10 & -9/70 \\ -7/10 & 9/10 \end{pmatrix}.$$

Thus

$$\psi(u) = \boldsymbol{\alpha}_+ e^{(\boldsymbol{T}+\boldsymbol{t\alpha}_+)u}\boldsymbol{e} = \frac{24}{35}e^{-u} + \frac{1}{35}e^{-6u}.$$

$\square$

Notes and references Corollary 3.1 can be found in Neuts [660] (in the setting of M/G/1 queues, cf. the duality result given in Corollary III.3.6), but that such a simple and general solution existed does not appear to have been well known to the risk theory community before a rather late stage. The result carries over to B being matrix-exponential, see Section 6. In the next sections, we encounter similar expressions for the ruin probabilities in the renewal- and Markov-modulated models, but there the vector $\boldsymbol{\alpha}_+$ is not explicit but needs to be calculated (typically by an iteration or a rootfinding).

The parameters of Example 3.2 are taken from Gerber [398]; his derivation of $\psi(u)$ is different.

For further more or less explicit computations of ruin probabilities, see Shiu [800].

It is notable that the phase-type assumption does not seem to simplify the computation of finite horizon ruin probabilities substantially (but see Section 8). For an attempt, see Stanford & Stroiński [817].

4 The renewal model

We consider the renewal model in the notation of Chapter VI, with A denoting the interarrival distribution and B the service time distribution. We assume

$\rho = \mu_B/\mu_A < 1$ and that B is phase-type with representation $(\boldsymbol{\alpha}, \boldsymbol{T})$. We shall derive phase-type representations of the ruin probabilities $\psi(u)$, $\psi^{(s)}(u)$ (recall that $\psi(u)$ refers to the zero-delayed case and $\psi^{(s)}(u)$ to the stationary case). For the compound Poisson model, this was obtained in Section 3, and the argument for the renewal case starts in just the same way (cf. the discussion around Fig. IX.6 which does not use that A is exponential) by noting that the distribution G_+ of the ascending ladder height S_{τ_+} is necessarily (defective) phase-type with representation $(\boldsymbol{\alpha}_+, \boldsymbol{T})$ for some vector $\boldsymbol{\alpha}_+ = (\alpha_{+;j})$. That is, if we define $\{m_x\}$ just as for the Poisson case (cf. Fig. IX.6):

Proposition 4.1 *In the zero-delayed case,*
(a) *G_+ is of phase-type with representation $(\boldsymbol{\alpha}_+, \boldsymbol{T})$, where $\boldsymbol{\alpha}_+$ is the (defective) distribution of m_0;*
(b) *The maximum claim surplus M is the lifetime of $\{m_x\}$;*
(c) *$\{m_x\}$ is a (terminating) Markov process on E, with intensity matrix $\boldsymbol{Q}$ given by $\boldsymbol{Q} = \boldsymbol{T} + \boldsymbol{t}\boldsymbol{\alpha}_+$.*

The key difference from the Poisson case is that it is more difficult to evaluate $\boldsymbol{\alpha}_+$. In fact, the form in which we derive $\boldsymbol{\alpha}_+$ for the renewal model is as the unique solution of a fixed point problem $\boldsymbol{\alpha}_+ = \varphi(\boldsymbol{\alpha}_+)$, which for numerical purposes can be solved by iteration. Nevertheless, the calculation of the first ladder height is simple in the stationary case:

Proposition 4.2 *The distribution $G_+^{(s)}$ of the first ladder height of the claim surplus process $\{S_t^{(s)}\}$ for the stationary case is phase-type with representation $(\boldsymbol{\alpha}^{(s)}, \boldsymbol{T})$, where $\boldsymbol{\alpha}^{(s)} = -\boldsymbol{\alpha}\boldsymbol{T}^{-1}/\mu_A$.*

Proof. Obviously, the Palm distribution of the claim size is just B. Hence by Theorem III.5.5, $G_+^{(s)} = \rho B_0$, where B_0 is the stationary excess life distribution corresponding to B. But by Corollary 2.3, B_0 is phase-type with representation $(-\boldsymbol{\alpha}\boldsymbol{T}^{-1}/\mu_B, \boldsymbol{T})$. $\qquad\square$

Proposition 4.3 *$\boldsymbol{\alpha}_+$ satisfies $\boldsymbol{\alpha}_+ = \varphi(\boldsymbol{\alpha}_+)$, where*

$$\varphi(\boldsymbol{\alpha}_+) \; = \; \boldsymbol{\alpha}\widehat{A}[\boldsymbol{T} + \boldsymbol{t}\boldsymbol{\alpha}_+] \; = \; \boldsymbol{\alpha}\int_0^\infty e^{(\boldsymbol{T}+\boldsymbol{t}\boldsymbol{\alpha}_+)y} A(\mathrm{d}y). \qquad (4.1)$$

Proof. We condition upon $T_1 = y$ and define $\{m_x^*\}$ from $\{S_{t+y} - S_{y-}\}$ in the same way as $\{m_x\}$ is defined from $\{S_t\}$, cf. Fig. IX.7. Then $\{m_x^*\}$ is Markov with the same transition intensities as $\{m_x\}$, but with initial distribution $\boldsymbol{\alpha}$ rather than $\boldsymbol{\alpha}_+$. Also, obviously $m_0 = m_y^*$. Since the conditional distribution of m_y^* given $T_1 = y$ is $\boldsymbol{\alpha}e^{\boldsymbol{Q}y}$, it follows by integrating y out that the distribution $\boldsymbol{\alpha}_+$ of m_0 is given by the final expression in (4.1). $\qquad\square$

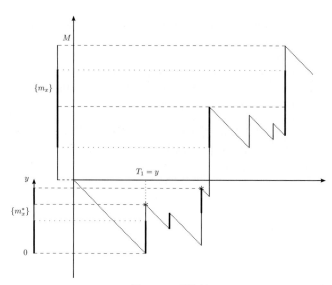

FIGURE IX.7

We have now almost collected all pieces of the main result of this section:

Theorem 4.4 *Consider the renewal model with interarrival distribution A and the claim size distribution B being of phase-type with representation $(\boldsymbol{\alpha}, \boldsymbol{T})$. Then*

$$\psi(u) \ = \ \boldsymbol{\alpha}_+ e^{(\boldsymbol{T}+\boldsymbol{t}\boldsymbol{\alpha}_+)x} \boldsymbol{e}, \quad \psi^{(s)}(u) \ = \ \boldsymbol{\alpha}^{(s)} e^{(\boldsymbol{T}+\boldsymbol{t}\boldsymbol{\alpha}_+)x} \boldsymbol{e}, \qquad (4.2)$$

where $\boldsymbol{\alpha}_+$ satisfies (4.1) and $\boldsymbol{\alpha}^{(s)} = -\boldsymbol{\alpha}\boldsymbol{T}^{-1}/\mu_A$. Furthermore, $\boldsymbol{\alpha}_+$ can be computed by iteration of (4.1), i.e. by

$$\boldsymbol{\alpha}_+ = \lim_{n\to\infty} \boldsymbol{\alpha}_+^{(n)} \quad \text{where} \quad \boldsymbol{\alpha}_+^{(0)} = 0, \ \boldsymbol{\alpha}_+^{(1)} = \varphi(\boldsymbol{\alpha}_+^{(0)}), \ \boldsymbol{\alpha}_+^{(2)} = \varphi(\boldsymbol{\alpha}_+^{(1)}), \ \ldots \qquad (4.3)$$

Proof. The first expression in (4.2) follows from Proposition 4.1 by noting that the distribution of m_0 is $\boldsymbol{\alpha}_+$. The second follows in a similar way by noting that only the first ladder step has a different distribution in the stationary case, and that this is given by Proposition 4.2; thus, the maximum claim surplus for the stationary case has a similar representation as in Proposition 4.1(b), only with initial distribution $\boldsymbol{\alpha}^{(s)}$ for m_0.

It remains to prove convergence of the iteration scheme (4.3). The term $\boldsymbol{t}\boldsymbol{\beta}$ in $\varphi(\boldsymbol{\beta})$ represents feedback with rate vector $\boldsymbol{t}$ and feedback probability vector $\boldsymbol{\beta}$. Hence $\varphi(\boldsymbol{\beta})$ (defined on the domain of subprobability vectors $\boldsymbol{\beta}$) is an increasing function of $\boldsymbol{\beta}$. In particular, $\boldsymbol{\alpha}_+^{(1)} \geq 0 = \boldsymbol{\alpha}_+^{(0)}$ implies

$$\boldsymbol{\alpha}_+^{(2)} \ = \ \varphi(\boldsymbol{\alpha}_+^{(1)}) \ \geq \ \varphi(\boldsymbol{\alpha}_+^{(0)}) \ = \ \boldsymbol{\alpha}_+^{(1)}$$

and (by induction) that $\left\{\boldsymbol{\alpha}_+^{(n)}\right\}$ is an increasing sequence such that $\lim_{n\to\infty} \boldsymbol{\alpha}_+^{(n)}$ exists. Similarly, $0 = \boldsymbol{\alpha}_+^{(0)} \le \boldsymbol{\alpha}_+$ yields

$$\boldsymbol{\alpha}_+^{(1)} = \varphi\big(\boldsymbol{\alpha}_+^{(0)}\big) \le \varphi\left(\boldsymbol{\alpha}_+\right) = \boldsymbol{\alpha}_+$$

and by induction that $\boldsymbol{\alpha}_+^{(n)} \le \boldsymbol{\alpha}_+$ for all n. Thus, $\lim_{n\to\infty} \boldsymbol{\alpha}_+^{(n)} \le \boldsymbol{\alpha}_+$.

To prove the converse inequality, we use an argument similar to the proof of Proposition VII.2.4. Let $F_n = \{T_1 + \cdots + T_{n+1} > \tau_+\}$ be the event that $\{m_x^*\}$ has at most n arrivals in $[T_1, \tau_+]$, and let $\widetilde{\alpha}_{+;i}^{(n)} = \mathbb{P}(m_{T_1}^* = i; F_n)$. Obviously, $\widetilde{\boldsymbol{\alpha}}_+^{(n)} \uparrow \boldsymbol{\alpha}_+$, so to complete the proof it suffices to show that $\widetilde{\boldsymbol{\alpha}}_+^{(n)} \le \boldsymbol{\alpha}_+^{(n)}$ for all n. For $n = 0$, both quantities are just 0. Assume the assertion shown for $n-1$. Then each subexcursion of $\{S_{t+T_1} - S_{T_1-}\}$ can contain at most $n-1$ arrivals (n arrivals are excluded because of the initial arrival at time T_1). It follows that on F_n the feedback to $\{m_x^*\}$ after each ladder step cannot exceed $\widetilde{\boldsymbol{\alpha}}_+^{(n-1)}$ so that

$$\widetilde{\boldsymbol{\alpha}}_+^{(n)} \le \boldsymbol{\alpha} \int_0^\infty \mathrm{e}^{(\boldsymbol{T}+t\widetilde{\boldsymbol{\alpha}}_+^{(n-1)})y} \boldsymbol{A}(\mathrm{d}y)$$

$$\le \boldsymbol{\alpha} \int_0^\infty \mathrm{e}^{(\boldsymbol{T}+t\boldsymbol{\alpha}_+^{(n-1)})y} \boldsymbol{A}(\mathrm{d}y) = \varphi\big(\boldsymbol{\alpha}_+^{(n-1)}\big) = \boldsymbol{\alpha}_+^{(n)}.$$

$\square$

We next give an alternative algorithm, which links together the phase-type setting and the classical complex plane approach of the renewal model (see further the Notes). To this end, let F be the distribution of $U_1 - T_1$. Then

$$\widehat{F}[r] = \boldsymbol{\alpha}(-r\boldsymbol{I} - \boldsymbol{T})^{-1}\boldsymbol{t} \cdot \widehat{A}[-r] \tag{4.4}$$

whenever $\mathbb{E}\mathrm{e}^{\Re(r)U} < \infty$. However, (4.4) makes sense and provides an analytic continuation of $\widehat{F}[\cdot]$ as long as $-r \notin \mathrm{sp}(\boldsymbol{T})$.

Theorem 4.5 *Let r be some complex number with $\Re(r) > 0$, $-r \notin \mathrm{sp}(\boldsymbol{T})$. Then $-r$ is an eigenvalue of $\boldsymbol{Q} = \boldsymbol{T} + t\boldsymbol{\alpha}_+$ if and only if $1 = \widehat{F}[r] = \widehat{A}[-r]\widehat{B}[r]$, with $\widehat{B}[r]$, $\widehat{F}[r]$ being interpreted in the sense of the analytical continuation of the m.g.f. In that case, the corresponding right eigenvector may be taken as $(-r\boldsymbol{I} - \boldsymbol{T})^{-1}\boldsymbol{t}$.*

Proof. Suppose first $\boldsymbol{Q}\boldsymbol{h} = -r\boldsymbol{h}$. Then $\mathrm{e}^{\boldsymbol{Q}x}\boldsymbol{h} = \mathrm{e}^{-rx}\boldsymbol{h}$ and hence

$$-r\boldsymbol{h} = \boldsymbol{Q}\boldsymbol{h} = \big(\boldsymbol{T} + t\boldsymbol{\alpha}\widehat{A}[\boldsymbol{Q}]\big)\boldsymbol{h} = \boldsymbol{T}\boldsymbol{h} + \widehat{A}[-r]t\boldsymbol{\alpha}\boldsymbol{h}. \tag{4.5}$$

Since $-r \notin \mathrm{sp}(T)$, this implies that $\alpha h \widehat{A}[-r] \neq 0$, and hence we may assume that h has been normalized such that $\alpha h \widehat{A}[-r] = 1$. Then (4.5) yields $h = (-rI - T)^{-1}t$. Thus by (4.4), the normalization is equivalent to $\widehat{F}[r] = 1$.

Suppose next $\widehat{F}[r] = 1$. Since $\Re(r) > 0$ and G_- is concentrated on $(-\infty, 0)$, we have $|\widehat{G}_-[r]| < 1$, and hence by the Wiener-Hopf factorization identity (A.9) we have $\widehat{G}_+[r] = 1$ which according to Theorem 1.5(c) means that $\alpha_+(-rI - T)^{-1}t = 1$. Hence with $h = (-rI - T)^{-1}t$ we get

$$Qh \;=\; (T + t\alpha_+)h \;=\; T(-rI - T)^{-1}t + t \;=\; -r(-rI - T)^{-1}t \;=\; -rh\,.$$

$\square$

Let d denote the number of phases.

Corollary 4.6 *Suppose $\mu < 0$, that the equation $\widehat{F}[r] = 1$ has d distinct roots $\rho_1, \ldots, \rho_d$ in the domain $\Re(r) > 0$, and define $h_i = (-\rho_i I - T)^{-1}t$, $Q = CD^{-1}$ where C is the matrix with columns $h_1, \ldots, h_d$, D that with columns $-\rho_1 h_1, \ldots, -\rho_d h_d$. Then G_+ is phase-type with representation (α_+, T) with $\alpha_+ = \alpha(Q - T)/\alpha t$. Further, letting ν_i be the left eigenvector of Q corresponding to $-\rho_i$ and normalized by $\nu_i h_i = 1$, Q has diagonal form*

$$Q \;=\; -\sum_{i=1}^{d} \rho_i\, \nu_i \otimes h_i \;=\; -\sum_{i=1}^{d} \rho_i\, h_i \nu_i\,. \tag{4.6}$$

Proof. Appealing to Theorem 4.5, the matrix Q in Theorem 2.1 has the d distinct eigenvalues $-\rho_1, \ldots, -\rho_d$ with corresponding eigenvectors $h_1, \ldots, h_d$. This immediately implies that Q has the form CD^{-1} and the last assertion on the diagonal form. Given T has been computed, we get

$$\frac{1}{\alpha t}\alpha(Q - T) \;=\; \frac{1}{\alpha t}\alpha t\alpha_+ \;=\; \alpha_+\,.$$

$\square$

Notes and references   Results like those of the present section have a long history, and the topic is classic both in risk theory and queueing theory (recall that we can identify $\psi(u)$ with the tail $\mathbb{P}(W > u)$ of the GI/PH/1 waiting time W; in turn, $W \overset{\mathscr{D}}{=} M^{(d)}$ in the notation of Chapter VI). In older literature, explicit expressions for the ruin/queueing probabilities are most often derived under the slightly more general assumption that $\widehat{B}$ is rational (say with degree d of the polynomial in the denominator) as discussed in Section 6. As in Corollary 4.6, the classical algorithm starts by looking for roots in the complex plane of the equation $\widehat{B}[\gamma]\widehat{A}[-\gamma] = 1$, $\Re(\gamma) > 0$. The roots are counted and located by Rouché's theorem (a classical result from complex analysis giving a criterion for two complex functions to have the same number of zeros within a defined region). This gives d roots $\gamma_1, \ldots, \gamma_d$ satisfying $\Re(\gamma_i) > 0$, and the solution is then in transform terms

$$1 + \alpha \int_0^\infty e^{\alpha u} \psi(u) \, du \; = \; \mathbb{E} e^{\alpha W} \; = \; \prod_{i=1}^d (-\gamma_i) \bigg/ \prod_{i=1}^d (\alpha - \gamma_i) \tag{4.7}$$

(see, e.g., Asmussen & O'Cinneide [94] for a short self-contained derivation). In risk theory, a pioneering paper in this direction is Täcklind [826], whereas the approach was introduced in queueing theory by Smith [814]; similar discussion appears in Kemperman [528] and much of the queueing literature like Cohen [249], see also Chapters XII and XIII.

This complex plane approach has been met with substantial criticism for a number of reasons like being lacking probabilistic interpretation and not giving the waiting time distribution / ruin probability itself but only the transform. In queueing theory, an alternative approach (the matrix-geometric method) has been developed largely by M.F. Neuts and his students, starting around in 1975. For surveys, see Neuts [660], [661] and Latouche & Ramaswami [574]. Here phase-type assumptions are basic, but the models solved are basically Markov chains and Markov processes with countably many states (for example queue length processes). The solutions are based upon iterations schemes like in Theorem 4.4; the fixed point problems look like

$$\boldsymbol{R} = \boldsymbol{A}_0 + \boldsymbol{R}\boldsymbol{A}_1 + \boldsymbol{R}^2 \boldsymbol{A}_2 + \cdots,$$

where $\boldsymbol{R}$ is an unknown matrix, and appears already in some early work by Wallace [870]. The distribution of W comes out from the approach but in a rather complicated form. The matrix-exponential form of the distribution was found by Sengupta [793] and the phase-type form by the first author [60].

The exposition here is based upon [60], which contains somewhat stronger results concerning the fixed point problem and the iteration scheme. Numerical examples appear in Asmussen & Rolski [97].

For further early explicit computations of ruin probabilities in the phase-type renewal case, see Dickson & Hipp [314, 315]; some recent extensions are discussed in Section XII.3. There is also much literature on the case where A is phase-type with a few phases.

5 Markov-modulated input

We consider a claim surplus process $\{S_t\}$ in a Markovian environment in the notation of Chapter VII. That is, the background Markov process with p states is $\{J_t\}$, the intensity matrix is $\boldsymbol{\Lambda}$ and the stationary row vector is $\boldsymbol{\pi}$. The arrival rate in background state i is λ_i and the distribution of an arrival claim is B_i. We assume that each B_i is phase-type, with representation say $(\boldsymbol{\alpha}^{(i)}, \boldsymbol{T}^{(i)}, E^{(i)})$. The number of elements of $E^{(i)}$ is denoted by q_i.

It turns out that subject to the phase-type assumption, the ruin probability can be found in matrix-exponential form just as for the renewal model, involving some parameters like the ones Q or α_+ for the renewal model which need to be determined by similar algorithms.

We start in Section 5a with an algorithm involving roots in a similar manner as Corollary 4.6. However, the analysis involves new features like an equivalence with first passage problems for Markovian fluids and the use of martingales (these ideas also apply to phase-type renewal models though we have not given the details). Section 5b then gives a representation along the lines of Theorem 4.4. The key unknown is the matrix K, for which the relevant fixed point problem and iteration scheme has already been studied in VII.2.

5a   Calculations via fluid models. Diagonalization

Consider a process $\left\{(I_t, V_t)\right\}_{t \geq 0}$ such that $\{I_t\}$ is a Markov process with a finite state space F and $\{V_t\}$ has piecewiese linear paths, say with slope $r(i)$ on intervals where $I_t = i$. The version of the process obtained by imposing reflection on the V component is denoted a *Markovian fluid* and is of considerable interest in telecommunications engineering as model for an ATM (Asynchronuous Transfer Mode) switch. The stationary distribution is obtained by finding the maximum of the V-component of the version of $\left\{(I_t, V_t)\right\}$ obtained by time reversing the I component. This calculation in a special case gives also the ruin probabilities for the Markov-modulated risk process with phase-type claims. The connection between the two models is a fluid representation of the Markov-modulated risk process given in Fig. IX.8.

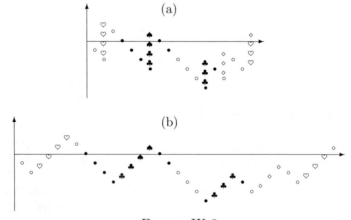

FIGURE IX.8

On Fig. IX.8, $p = q_1 = q_2 = 2$. The two environmental states are denoted $\circ, \bullet$, the phase space $E^{(\circ)}$ for $B_\circ$ has states $\diamond, \heartsuit$, and the one $E^{(\bullet)}$ for $B_\bullet$ states $\clubsuit, \spadesuit$. A claim in state i can then be represented by an $E^{(i)}$-valued Markov process as on Fig. IX.8(a). The fluid model $\{(I_t, V_t)\}$ on Fig. IX.8(b) is then obtained by changing the vertical jumps to segments with slope 1. Thus $F = \{\circ, \diamond, \heartsuit, \bullet, \clubsuit, \spadesuit\}$. In the general formulation, F is the disjoint union of E and the $E^{(i)}$,

$$F = E \cup \{(i, \alpha) : i \in E, \alpha \in E^{(i)}\}, \quad r(i) = -1, \; i \in E, \quad r(i, \alpha) = 1.$$

The intensity matrix for $\{I_t\}$ is (taking $p = 3$ for simplicity)

$$\mathbf{\Lambda}_I \;=\; \left(\begin{array}{ccc|ccc} & & & \beta_1 \boldsymbol{\alpha}^{(1)} & 0 & 0 \\ & \mathbf{\Lambda} - (\beta_i)_{\mathrm{diag}} & & 0 & \beta_2 \boldsymbol{\alpha}^{(2)} & 0 \\ & & & 0 & 0 & \beta_3 \boldsymbol{\alpha}^{(3)} \\ \hline \boldsymbol{t}^{(1)} & 0 & 0 & \boldsymbol{T}^{(1)} & 0 & 0 \\ 0 & \boldsymbol{t}^{(2)} & 0 & 0 & \boldsymbol{T}^{(2)} & 0 \\ 0 & 0 & \boldsymbol{t}^{(3)} & 0 & 0 & \boldsymbol{T}^{(3)} \end{array} \right).$$

The reasons for using the fluid representation are twofold. First, the probability in the Markov-modulated model of upcrossing level u in state i of $\{J_t\}$ and phase $\alpha \in E^{(i)}$ is the same as the probability that the fluid model upcrosses level u in state (i, α) of $\{I_t\}$. Second, in the fluid model $\mathbb{E}e^{sV_t} < \infty$ for all s, t, whereas $\mathbb{E}e^{sS_t} = \infty$ for all t and all $s \geq s_0$ where $s_0 < \infty$. This implies that in the fluid context, we have more martingales at our disposal.

Recall that in the phase-type case, $\widehat{B}_i[s] = -\boldsymbol{\alpha}^{(i)}(\boldsymbol{T}^{(i)} + s\boldsymbol{I})^{-1}\boldsymbol{t}^{(i)}$. Let $\mathbf{\Sigma}$ denote the matrix

$$\mathbf{\Delta}_r^{-1}\mathbf{\Lambda}_I \;=\; \left(\begin{array}{ccc|ccc} & & & -\beta_1 \boldsymbol{\alpha}^{(1)} & 0 & 0 \\ & (\beta_i)_{\mathrm{diag}} - \mathbf{\Lambda} & & 0 & -\beta_2 \boldsymbol{\alpha}^{(2)} & 0 \\ & & & 0 & 0 & -\beta_3 \boldsymbol{\alpha}^{(3)} \\ \hline \boldsymbol{t}^{(1)} & 0 & 0 & \boldsymbol{T}^{(1)} & 0 & 0 \\ 0 & \boldsymbol{t}^{(2)} & 0 & 0 & \boldsymbol{T}^{(2)} & 0 \\ 0 & 0 & \boldsymbol{t}^{(3)} & 0 & 0 & \boldsymbol{T}^{(3)} \end{array} \right),$$

with the four blocks denoted by $\mathbf{\Sigma}_{ij}$, $i, j = 1, 2$, corresponding to the partitioning of $\mathbf{\Sigma}$ into components indexed by E, resp. $E^{(1)} + \cdots + E^{(p)}$.

Proposition 5.1 *A complex number s satisfies*

$$\left| \mathbf{\Lambda} + (\beta_i(\widehat{B}_i[-s] - 1))_{\mathrm{diag}} + s\boldsymbol{I} \right| \;=\; 0 \tag{5.1}$$

if and only if s is an eigenvalue of Σ. If s is such a number, consider the vector $\boldsymbol{a}$ satisfying $\left(\boldsymbol{\Lambda} + \left(\beta_i(\widehat{B}_i[-s]-1)\right)_{\mathrm{diag}}\right)\boldsymbol{a} = -s\boldsymbol{a}$ and the eigenvector $\boldsymbol{b} = \begin{pmatrix} \boldsymbol{c} \\ \boldsymbol{d} \end{pmatrix}$ of $\boldsymbol{\Delta}_r^{-1}\boldsymbol{\Lambda}_I$, where $\boldsymbol{c}, \boldsymbol{d}$ correspond to the partitioning of $\boldsymbol{b}$ into components indexed by E, resp. $E^{(1)} + \cdots + E^{(p)}$. Then (up to a constant)

$$\boldsymbol{c} = \boldsymbol{a}, \quad \boldsymbol{d} = (s\boldsymbol{I} - \boldsymbol{\Sigma}_{22})^{-1}\boldsymbol{\Sigma}_{21}\boldsymbol{a} = \sum_{i\in E} a_i(s\boldsymbol{I} - \boldsymbol{T}^{(i)})^{-1}\boldsymbol{t}^{(i)} .$$

Proof. Using the well-known determinant identity

$$\begin{vmatrix} \boldsymbol{\Sigma}_{11} & \boldsymbol{\Sigma}_{12} \\ \boldsymbol{\Sigma}_{21} & \boldsymbol{\Sigma}_{22} \end{vmatrix} = \mid \boldsymbol{\Sigma}_{22} \mid \cdot \mid \boldsymbol{\Sigma}_{11} - \boldsymbol{\Sigma}_{12}\boldsymbol{\Sigma}_{22}^{-1}\boldsymbol{\Sigma}_{21} \mid ,$$

with $\boldsymbol{\Sigma}_{ii}$ replaced by $\boldsymbol{\Sigma}_{ii} - s\boldsymbol{I}$, it follows that if

$$\begin{vmatrix} & & & -\beta_1\boldsymbol{\alpha}^{(1)} & 0 & 0 \\ (\beta_i)_{\mathrm{diag}} - \boldsymbol{\Lambda} - s\boldsymbol{I} & & & 0 & -\beta_2\boldsymbol{\alpha}^{(2)} & 0 \\ & & & 0 & 0 & -\beta_1\boldsymbol{\alpha}^{(1)} \\ \hline \boldsymbol{t}^{(1)} & 0 & 0 & \boldsymbol{T}^{(1)} - s\boldsymbol{I} & 0 & 0 \\ 0 & \boldsymbol{t}^{(2)} & 0 & 0 & \boldsymbol{T}^{(2)} - s\boldsymbol{I} & 0 \\ 0 & 0 & \boldsymbol{t}^{(3)} & 0 & 0 & \boldsymbol{T}^{(3)} - s\boldsymbol{I} \end{vmatrix} = 0,$$

then also

$$\left| (\beta_i)_{\mathrm{diag}} - \boldsymbol{\Lambda} - s\boldsymbol{I} + \left(\beta_i\boldsymbol{\alpha}^{(i)}(\boldsymbol{T}^{(i)} - s\boldsymbol{I})^{-1}\boldsymbol{t}^{(i)}\right)_{\mathrm{diag}} \right| = 0$$

which is the same as (5.1).

For the assertions on the eigenvectors, assume that $\boldsymbol{a}$ is chosen as asserted which means

$$\left(\boldsymbol{\Sigma}_{11} - s\boldsymbol{I} + \boldsymbol{\Sigma}_{12}(s\boldsymbol{I} - \boldsymbol{\Sigma}_{22})^{-1}\boldsymbol{\Sigma}_{21}\right)\boldsymbol{a} = 0 ,$$

and let $\boldsymbol{d} = (s\boldsymbol{I} - \boldsymbol{\Sigma}_{22})^{-1}\boldsymbol{\Sigma}_{21}\boldsymbol{a}, \boldsymbol{c} = \boldsymbol{a}$. Then

$$\begin{aligned} \boldsymbol{\Sigma}_{21}\boldsymbol{c} + \boldsymbol{\Sigma}_{22}\boldsymbol{d} &= \boldsymbol{\Sigma}_{21}\boldsymbol{a} - (s\boldsymbol{I} - \boldsymbol{\Sigma}_{22} - s\boldsymbol{I})(s\boldsymbol{I} - \boldsymbol{\Sigma}_{22})^{-1}\boldsymbol{\Sigma}_{21}\boldsymbol{a} \\ &= \boldsymbol{\Sigma}_{21}\boldsymbol{a} - \boldsymbol{\Sigma}_{21}\boldsymbol{a} + s\boldsymbol{d} = s\boldsymbol{d}. \end{aligned}$$

Noting that $\boldsymbol{\Sigma}_{11}\boldsymbol{c} + \boldsymbol{\Sigma}_{12}\boldsymbol{d} = s\boldsymbol{c}$ by definition, it follows that

$$\begin{pmatrix} \boldsymbol{\Sigma}_{11} & \boldsymbol{\Sigma}_{12} \\ \boldsymbol{\Sigma}_{21} & \boldsymbol{\Sigma}_{22} \end{pmatrix}\begin{pmatrix} \boldsymbol{c} \\ \boldsymbol{d} \end{pmatrix} = s\begin{pmatrix} \boldsymbol{c} \\ \boldsymbol{d} \end{pmatrix} .$$

$\square$

Theorem 5.2 *Assume that* $\mathbf{\Sigma} = \mathbf{\Delta}_r^{-1}\mathbf{\Lambda}_I$ *has* $q = q_1 + \cdots + q_p$ *distinct eigenvalues* $s_1, \ldots, s_q$ *with* $\Re(s_\nu) < 0$ *and let* $\mathbf{b}^{(\nu)} = \begin{pmatrix} \mathbf{c}^{(\nu)} \\ \mathbf{d}^{(\nu)} \end{pmatrix}$ *be the right eigenvector corresponding to* s_ν, $\nu = 1, \ldots, q$. *Then*

$$\psi_i(u) = \mathbf{e}_i^{\mathsf{T}}\left(e^{s_1 u}\mathbf{c}^{(1)} \ldots e^{s_q u}\mathbf{c}^{(q)}\right)\left(\mathbf{d}^{(1)} \ldots \mathbf{d}^{(q)}\right)^{-1}\mathbf{e}.$$

Proof. Writing $\mathbf{\Delta}_r^{-1}\mathbf{\Lambda}_I\mathbf{b}^{(\nu)} = s_\nu\mathbf{b}^{(\nu)}$ as $(\mathbf{\Lambda}_I - \mathbf{\Delta}_r s_\nu)\mathbf{b}^{(\nu)} = 0$, it follows by Proposition III.4.4 that $\{e^{-s_\nu V_t}b_{I_t}^{(\nu)}\}$ is a martingale. For $u, v > 0$, define

$$\omega(u, v) = \inf\{t > 0 : V_t = u \text{ or } V_t = -v\}, \quad \omega(u) = \inf\{t > 0 : V_t = u\},$$
$$p_i(u, v; j, \alpha) = \mathbb{P}_i\big(V_{\omega(u,v)} = u, I_{\omega(u,v)} = (j, \alpha)\big),$$
$$p_i(u, v; j) = \mathbb{P}_i\big(V_{\omega(u,v)} = -v, I_{\omega(u,v)} = j\big),$$
$$p_i(u; j, \alpha) = \mathbb{P}_i\big(\omega(u) < \infty, I_{\omega(u,v)} = (j, \alpha)\big).$$

Optional stopping at time $\omega(u, v)$ yields

$$c_i^{(\nu)} = e^{-s_\nu u}\sum_{j,\alpha} p_i(u, v; j, \alpha)d_{j,\alpha}^{(\nu)} + e^{s_\nu v}\sum_j p_i(u, v; j)c_j^{(\nu)}.$$

Letting $v \to \infty$ and using $\Re(s_\nu) < 0$ yields

$$e^{s_\nu u}c_i^{(\nu)} = \sum_{j,\alpha} p_i(u; j, \alpha)d_{j,\alpha}^{(\nu)}.$$

Solving for the $p_i(u; j, \alpha)$ and noting that $\psi_i(u) = \sum_{j,\alpha} p_i(u; j, \alpha)$, the result follows. □

Example 5.3 Consider the Poisson model with exponential claims with rate δ. Here E has one state only. To determine $\psi(u)$, we first look for the negative eigenvalue s of $\mathbf{\Sigma} = \begin{pmatrix} \beta & -\beta \\ \delta & -\delta \end{pmatrix}$ which is $s = -\gamma$ with $\gamma = \delta - \beta$. We can take $a = c = 1$ and get $d = (s + \delta)^{-1}\delta = \delta/\beta = 1/\rho$. Thus $\psi(u) = e^{su}/d = \rho e^{-\gamma u}$ as should be. □

Example 5.4 Assume that E has two states and that B_1, B_2 are both exponential with rates δ_1, δ_2. Then we get $\psi_i(u)$ as the sum of two exponential terms where the rates s_1, s_2 are the negative eigenvalues of

$$\mathbf{\Sigma} = \left(\begin{array}{cc|cc} \lambda_1 + \beta_1 & -\lambda_1 & -\beta_1 & 0 \\ -\lambda_2 & \lambda_2 + \beta_2 & 0 & -\beta_2 \\ \hline \delta_1 & 0 & -\delta_1 & 0 \\ 0 & \delta_2 & 0 & -\delta_2 \end{array}\right).$$

□

5b   Computations via K

Recall the definition of the matrix K from VII.2. In terms of K, we get the following phase-type representation for the ladder heights (see the Appendix for the definition of the Kronecker product $\otimes$ and the Kronecker sum $\oplus$):

Proposition 5.5 $G_+(i,j;\cdot)$ *is phase-type with representation* $\left(E^{(j)}, \boldsymbol{\theta}_{j\cdot}^{(i)}, \boldsymbol{T}^{(j)}\right)$ *where*

$$\boldsymbol{\theta}_{j\cdot}^{(i)} = \beta_j (\boldsymbol{e}_i^{\mathsf{T}} \otimes \boldsymbol{\alpha}^{(j)}) (-\boldsymbol{K} \oplus \boldsymbol{T}^{(j)})^{-1} (\boldsymbol{e}_j \otimes \boldsymbol{I}).$$

Proof. We must show that

$$G_+\big(i,j;(y,\infty)\big) = \boldsymbol{\theta}_{j\cdot}^{(i)} \mathrm{e}^{\boldsymbol{T}^{(j)}y} \boldsymbol{e}. \tag{5.2}$$

However, according to VII.(2.2) the l.h.s. is

$$\begin{aligned}
\beta_j \int_{-\infty}^0 &R(i,j;\mathrm{d}x)\overline{B}_j(y-x) \\
&= \beta_j \int_{-\infty}^0 \boldsymbol{e}_i^{\mathsf{T}}\, \mathrm{e}^{-\boldsymbol{K}x}\boldsymbol{e}_j \cdot \boldsymbol{\alpha}^{(j)} \mathrm{e}^{\boldsymbol{T}^{(j)}(y-x)} \boldsymbol{e}\, \mathrm{d}x \\
&= \beta_j \int_0^\infty \boldsymbol{e}_i^{\mathsf{T}} \mathrm{e}^{\boldsymbol{K}x}\boldsymbol{e}_j \cdot \boldsymbol{\alpha}^{(j)} \mathrm{e}^{\boldsymbol{T}^{(j)}x}\, \mathrm{d}x\, \mathrm{e}^{\boldsymbol{T}^{(j)}y} \boldsymbol{e} \\
&= \beta_j (\boldsymbol{e}_i^{\mathsf{T}} \otimes \boldsymbol{\alpha}^{(j)}) \int_0^\infty \mathrm{e}^{\boldsymbol{K}x} \otimes \mathrm{e}^{\boldsymbol{T}^{(j)}x}\, \mathrm{d}x\, (\boldsymbol{e}_j \otimes \boldsymbol{I})\mathrm{e}^{\boldsymbol{T}^{(j)}y} \boldsymbol{e} \\
&= \beta_j (\boldsymbol{e}_i^{\mathsf{T}} \otimes \boldsymbol{\alpha}^{(j)}) \int_0^\infty \mathrm{e}^{\boldsymbol{K}x\oplus\boldsymbol{T}^{(j)}x}\, \mathrm{d}x\, (\boldsymbol{e}_j \otimes \boldsymbol{I})\mathrm{e}^{\boldsymbol{T}^{(j)}y} \boldsymbol{e} \\
&= \boldsymbol{\theta}_{j\cdot}^{(i)} \mathrm{e}^{\boldsymbol{T}^{(j)}y} \boldsymbol{e}.
\end{aligned}$$

$\square$

Theorem 5.6 *For* $i \in E$, *the* $\mathbb{P}_i$-*distribution of* M *is phase-type with representation* $\left(E^{(1)} + \cdots + E^{(p)}, \boldsymbol{\theta}^{(i)}, \boldsymbol{U}\right)$ *where*

$$u_{j\alpha,k\gamma} = \begin{cases} t_{\alpha\gamma}^{(j)} + t_\alpha^{(j)}\theta_{j\gamma}^{(j)} & j = k \\[2mm] t_\alpha^{(j)}\theta_{k\gamma}^{(j)} & j \neq k \end{cases}.$$

In particular,

$$\psi_i(u) = \mathbb{P}_i(M > u) = \boldsymbol{\theta}^{(i)}\mathrm{e}^{\boldsymbol{U}u}\boldsymbol{e}. \tag{5.3}$$

Proof. We decompose M in the familiar way as sum of ladder steps. Associated with each ladder step is a phase process, with phase space $E^{(j)}$ whenever the corresponding arrival occurs in environmental state j (the ladder step is of type j). Piecing together these phase processes yields a terminating Markov process with state space $\sum_{i \in E} E^{(i)}$, intensity matrix $\boldsymbol{U}$, say, and lifelength M, and it just remains to check that $\boldsymbol{U}$ has the asserted form. Starting from $J_0 = i$, the initial value of (i, α) is obviously chosen according to $\boldsymbol{\theta}^{(i)}$. For a transition from (j, α) to (k, γ) to occur when $j \neq k$, the current ladder step of type j must terminate, which occurs at rate $t_\alpha^{(j)}$, and a new ladder step of type k must start in phase γ, which occurs w.p. $\theta_{k\gamma}^{(j)}$. This yields the asserted form of $u_{j\alpha, k\gamma}$. For $j = k$, we have the additional possibility of a phase change from α to γ within the ladder step, which occurs at rate $t_{\alpha\gamma}^{(j)}$. □

Notes and references Section 5a is based upon Asmussen [63] and Section 5b upon Asmussen [59]. Numerical illustrations are given in Asmussen & Rolski [97].

The connection to fluid models is further exploited in a series of papers by Ahn & Ramaswami, e.g. [9, 10]. They also involve the connection to *quasi birth-death processes*, defined as birth-death processes in a Markovian environment and with some modification at the boundary 0. See also Badescu *et al.* [116]. First passage times for Markov additive processes with positive jumps of phase type are discussed in Breuer [201].

6 Matrix-exponential distributions

When deriving explicit or algorithmically tractable expressions for the ruin probability, we have so far concentrated on a claim size distribution B of phase-type. However, in many cases where such expressions are available there are classical results from the pre-phase-type-era which give alternative solutions under the slightly more general assumption that B has a Laplace-Stieltjes transform (or, equivalently, a m.g.f.) which is rational, i.e. the ratio between two polynomials (for the form of the density, see Example I.2.5). An alternative characterization is that such a distribution is *matrix-exponential*, i.e. that the density $b(x)$ can be written as $\boldsymbol{\alpha} e^{\boldsymbol{T} x} \boldsymbol{t}$ for some row vector $\boldsymbol{\alpha}$, some square matrix $\boldsymbol{T}$ and some column vector $\boldsymbol{t}$ (the triple $(\boldsymbol{\alpha}, \boldsymbol{T}, \boldsymbol{t})$ is the *representation* of the matrix-exponential distribution/density):

Proposition 6.1 *Let $b(x)$ be an integrable function on $[0, \infty)$ and $b^*[\theta] = \int_0^\infty e^{-\theta x} b(x) \, dx$ its Laplace transform. Then $b^*[\theta]$ is rational if and only if $b(x)$ is matrix-exponential. Furthermore, if*

$$b^*[\theta] = \frac{b_1 + b_2\theta + b_3\theta^2 + \ldots + b_n\theta^{n-1}}{\theta^n + a_1\theta^{n-1} + \ldots + a_{n-1}\theta + a_n}, \qquad (6.1)$$

then a matrix-exponential representation is given by $b(x) = \boldsymbol{\alpha}e^{\boldsymbol{T}x}\boldsymbol{t}$ where

$$\boldsymbol{\alpha} = (b_1 \; b_2 \; \ldots \; b_{n-1} \; b_n), \quad \boldsymbol{t} = (0 \; 0 \; \ldots \; 0 \; 1)^{\mathsf{T}}, \tag{6.2}$$

$$\boldsymbol{T} = \begin{pmatrix} 0 & 1 & 0 & 0 & 0 & \ldots & 0 & 0 \\ 0 & 0 & 1 & 0 & 0 & \ldots & 0 & 0 \\ .. & .. & .. & .. & .. & \ldots & .. & .. \\ 0 & 0 & 0 & 0 & 0 & \ldots & 0 & 1 \\ -a_n & -a_{n-1} & -a_{n-2} & -a_{n-3} & -a_{n-4} & \ldots & -a_2 & -a_1 \end{pmatrix}. \tag{6.3}$$

Proof. If $b(x) = \boldsymbol{\alpha}e^{\boldsymbol{T}x}\boldsymbol{t}$, then $b^*[\theta] = \boldsymbol{\alpha}(\theta\boldsymbol{I} - \boldsymbol{T})^{-1}\boldsymbol{t}$ which is rational since each element of $(\theta\boldsymbol{I} - \boldsymbol{T})^{-1}$ is so. Thus, matrix-exponentiality implies a rational transform. The converse follows from the last statement of the theorem. For a proof, see Asmussen & Bladt [74] (the representation (6.2), (6.3) was suggested by Colm O'Cinneide, personal communication). □

Remark 6.2 A remarkable feature of Proposition 6.1 is that it gives an explicit Laplace transform inversion which may appear more appealing than the first attempt to invert $b^*[\theta]$ one would do, namely to asssume the roots $\delta_1, \ldots, \delta_n$ of the denominator to be distinct and expand the r.h.s. of (6.1) as $\sum_{i=1}^n c_i/(\theta + \delta_i)$, giving $b(x) = \sum_{i=1}^n c_i e^{-\delta_i x}/\delta_i$. □

Example 6.3 A set of necessary and sufficient conditions for a distribution to be phase-type are given in O'Cinneide [670]. One of his elementary criteria, $b(x) > 0$ for $x > 0$, shows that the distribution B with density $b(x) = c(1 - \cos(2\pi x))e^{-x}$, where $c = 1 + 1/4\pi^2$, cannot be phase-type.
   Writing

$$b(x) = c(-e^{(2\pi i - 1)x}/2 - e^{(-2\pi i - 1)x}/2 + e^{-x}),$$

it follows that a matrix-exponential representation $(\boldsymbol{\beta}, \boldsymbol{S}, \boldsymbol{s})$ is given by

$$\boldsymbol{\beta} = (1\,1\,1), \quad \boldsymbol{S} = \begin{pmatrix} 2\pi i - 1 & 0 & 0 \\ 0 & -2\pi i - 1 & 0 \\ 0 & 0 & -1 \end{pmatrix}, \quad \boldsymbol{s} = \begin{pmatrix} -c/2 \\ -c/2 \\ c \end{pmatrix}. \tag{6.4}$$

This representation is complex, but as follows from Proposition 6.1, we can always obtain a real one $(\boldsymbol{\alpha}, \boldsymbol{T}, \boldsymbol{t})$. Namely, since

$$b^*[\theta] = \frac{1 + 4\pi^2}{\theta^3 + 3\theta^2 + (3 + 4\pi^2)\theta + 1 + 4\pi^2},$$

it follows by (6.2), (6.3) that we can take

$$\boldsymbol{\alpha} = (1 + 4\pi^2 \; 0 \; 0), \quad \boldsymbol{T} = \begin{pmatrix} 0 & 1 & 0 \\ 0 & 0 & 1 \\ -1 - 4\pi^2 & -3 - 4\pi^2 & -3 \end{pmatrix}, \quad \boldsymbol{t} = \begin{pmatrix} 0 \\ 0 \\ 1 \end{pmatrix}.$$

□

Example 6.4 This example shows why it is sometimes useful to work with matrix-exponential distributions instead of phase-type distributions: for dimensionality reasons. Consider the distribution with density

$$b(x) = \frac{15}{7 + 15\delta} e^{-x} \left((2e^{-2x} - 1)^2 + \delta \right).$$

Then it is known from O'Cinneide [670] that b is phase-type when $\delta > 0$, and that the minimal number of phases in a phase-type representation increases to ∞ as $\delta \downarrow 0$, leading to matrix calculus in high dimensions when δ is small. But since

$$b^*[\theta] = \frac{15(1 + \delta)\theta^2 + 120\delta\theta + 225\delta + 105}{(7 + 15\delta)\theta^3 + (135\delta + 63)\theta^2 + (161 + 345\delta)\theta + 225\delta + 105},$$

Proposition 6.1 shows that a matrix-exponential representation can always be obtained in dimension only 3 independent of δ. □

As for the role of matrix-exponential distributions in ruin probability calculations, we shall only consider the compound Poisson model with arrival rate β and a matrix-exponential claim size distribution B, and present two algorithms for calculating $\psi(u)$ in that setting.

For the first, we take as starting point a representation of $b^*[\theta]$ as $p(\theta)/q(\theta)$ where p, q are polynomials without common roots. Then (cf. Corollary IV.3.4) the Laplace transform of the ruin probability is

$$\widehat{\psi}[-\theta] = \int_0^\infty e^{-\theta u} \psi(u) \, du = \frac{\beta - \beta p(\theta)/q(\theta) - \rho\theta}{\theta \left(\beta - \theta - \beta p(\theta)/q(\theta) \right)}. \tag{6.5}$$

Thus, we have represented $\widehat{\psi}[-\theta]$ as a ratio of polynomials (note that θ must necessarily be a root of the numerator and cancels), and can use this to invert by the method of Proposition 6.1 to get $\psi(u) = \beta e^{Su} s$.

For the second algorithm, we use a representation (α, T, t) of $b(x)$. We recall (see Section 3; recall that $t = -Te$) that if B is phase-type and (α, T, t) a phase-type representation with α the initial vector, T the phase generator and $t = -Te$, then

$$\psi(u) = -\alpha_+ e^{(T + t\alpha_+)u} T^{-1} t \quad \text{where} \quad \alpha_+ = -\beta \alpha T^{-1}. \tag{6.6}$$

The remarkable fact is that, despite that the proof of (6.6) in Section 3 seems to use the probabilistic interpretation of phase-type in an essential way, then:

Proposition 6.5 (6.6) *holds true also in the matrix-exponential case.*

Proof. Write

$$b^* = \boldsymbol{\alpha}(\theta\boldsymbol{I} - \boldsymbol{T})^{-1}\boldsymbol{t}, \quad b_+^* = \boldsymbol{\alpha}_+(\theta\boldsymbol{I} - \boldsymbol{T})^{-1}\boldsymbol{t}, \quad b_+^{**} = \boldsymbol{\alpha}_+(\theta\boldsymbol{I} - \boldsymbol{T})^{-1}\boldsymbol{T}^{-1}\boldsymbol{t}.$$

Then in Laplace transform formulation, the assertion is equivalent to

$$-\boldsymbol{\alpha}_+(\theta\boldsymbol{I} - \boldsymbol{T} - \boldsymbol{t}\boldsymbol{\alpha}_+)^{-1}\boldsymbol{T}^{-1}\boldsymbol{t} = \frac{\beta - \beta b^* - \rho\theta}{\theta(\beta - \theta - \beta b^*)}, \qquad (6.7)$$

cf. (6.5), (6.6). Presumably, this can be verified by analytic continuation from the phase-type domain to the matrix-exponential domain, but we shall give an algebraic proof. From the general matrix identity ([789, p. 519])

$$(\boldsymbol{A} + \boldsymbol{U}\boldsymbol{B}\boldsymbol{V})^{-1} = \boldsymbol{A}^{-1} - \boldsymbol{A}^{-1}\boldsymbol{U}\boldsymbol{B}(\boldsymbol{B} + \boldsymbol{B}\boldsymbol{V}\boldsymbol{A}^{-1}\boldsymbol{U}\boldsymbol{B})^{-1}\boldsymbol{B}\boldsymbol{V}\boldsymbol{A}^{-1},$$

with $\boldsymbol{A} = \theta\boldsymbol{I} - \boldsymbol{T}$, $\boldsymbol{U} = -\boldsymbol{t}$, $\boldsymbol{B} = 1$ and $\boldsymbol{V} = \boldsymbol{\alpha}_+$, we get

$$(\theta\boldsymbol{I} - \boldsymbol{T} - \boldsymbol{t}\boldsymbol{\alpha}_+)^{-1}$$
$$= (\theta\boldsymbol{I} - \boldsymbol{T})^{-1} + (\theta\boldsymbol{I} - \boldsymbol{T})^{-1}\boldsymbol{t}\big(1 - \boldsymbol{\alpha}_+(\theta\boldsymbol{I} - \boldsymbol{T})^{-1}\boldsymbol{t}\big)^{-1}\boldsymbol{\alpha}_+(\theta\boldsymbol{I} - \boldsymbol{T})^{-1}$$
$$= (\theta\boldsymbol{I} - \boldsymbol{T})^{-1} + \frac{1}{1 - b_+^*}(\theta\boldsymbol{I} - \boldsymbol{T})^{-1}\boldsymbol{t}\boldsymbol{\alpha}_+(\theta\boldsymbol{I} - \boldsymbol{T})^{-1}$$

so that

$$-\boldsymbol{\alpha}_+(\theta\boldsymbol{I} - \boldsymbol{T} - \boldsymbol{t}\boldsymbol{\alpha}_+)^{-1}\boldsymbol{T}^{-1}\boldsymbol{t} = -b_+^{**} - \frac{b_+^* b_+^{**}}{1 - b_+^*} = \frac{b_+^{**}}{b_+^* - 1}.$$

Now, since

$$(\theta\boldsymbol{I} - \boldsymbol{T})^{-1}\boldsymbol{T}^{-1} = \frac{1}{\theta}\big(\boldsymbol{T}^{-1} + (\theta\boldsymbol{I} - \boldsymbol{T})^{-1}\big),$$
$$(\theta\boldsymbol{I} - \boldsymbol{T})^{-1}\boldsymbol{T}^{-2} = \frac{1}{\theta}\boldsymbol{T}^{-2} + \frac{1}{\theta^2}\boldsymbol{T}^{-1} + \frac{1}{\theta^2}(\theta\boldsymbol{I} - \boldsymbol{T})^{-1}$$

and

$$1 = \int_0^\infty b(x)\,\mathrm{d}x = -\boldsymbol{\alpha}\boldsymbol{T}^{-1}\boldsymbol{t},$$
$$\mu_B = \int_0^\infty x b(x)\,\mathrm{d}x = \boldsymbol{\alpha}\boldsymbol{T}^{-2}\boldsymbol{t},$$

we get

$$
\begin{aligned}
b_+^* &= -\beta \alpha T^{-1}(\theta I - T)^{-1} t = -\beta \alpha (\theta I - T)^{-1} T^{-1} t \\
&= -\frac{\beta}{\theta} \alpha (T^{-1} + (\theta I - T)^{-1}) t = \frac{\beta}{\theta}(1 - b^*), \\
b_+^{**} &= -\beta \alpha T^{-1}(\theta I - T)^{-1} T^{-1} t = -\beta \alpha (\theta I - T)^{-1} T^{-2} t \\
&= -\beta \alpha \left(\frac{1}{\theta} T^{-2} + \frac{1}{\theta^2} T^{-1} + \frac{1}{\theta^2} (\theta I - T)^{-1} \right) t \\
&= -\frac{\rho}{\theta} + \frac{\beta}{\theta^2} - \frac{\beta}{\theta^2} b^*.
\end{aligned}
$$

From this it is straightforward to check that $b_+^{**}/(b_+^* - 1)$ is the same as the r.h.s. of (6.7). □

Notes and references As noted in the references to Section 4, some key early references using distributions with a rational transform for applied probability calculations are Täcklind [826] (ruin probabilities) and Smith [814] (queueing theory). A key tool is identifying poles and zeros of transforms via Wiener-Hopf factorization. Much of the flavor of this classical approach and many examples are in Cohen [249]; see also Dufresne [332] and Kuznetsov [563] for a recent discussion.

For expositions on the general theory of matrix-exponential distributions, see Asmussen & Bladt [74], Lipsky [600] and Asmussen & O'Cinneide [94]; a key early paper is Cox [264] (from where the distribution in Example 6.3 is taken).

The proof of Proposition 6.5 is similar to arguments used in [74] for formulas in renewal theory.

Some relevant more recent references on matrix-exponential distributions are Bean, Fackrell & Taylor [150], Bladt & Neuts [173] and Fackrell [360].

7 Reserve-dependent premiums

We consider the model of Chapter VIII with Poisson arrivals at rate β, premium rate $p(r)$ at level r of the reserve $\{R_t\}$ and claim size distribution B which we assume to be of phase-type with representation (E, α, T).

In Corollary VIII.1.9, the ruin probability $\psi(u)$ was found in explicit form for the case of B being exponential (for some remarkable explicit formulas due to Paulsen & Gjessing [687], see the Notes to VIII.1, but the argument of [687] does not apply in any reasonable generality). We present here first a computational approach for the general phase-type case (Section 7a) and next (Section 7b) a set of formulas covering the case of a two-step premium rule, cf. VIII.1a.

7a   Computing $\psi(u)$ via differential equations

The representation we use is essentially the same as the one used in Sections 3 and 4, to piece together the phases at downcrossing times of $\{R_t\}$ (upcrossing times of $\{S_t\}$) to a Markov process $\{m_x\}$ with state space E. See Fig. IX.9, which is self-explanatory given Fig. IX.6.

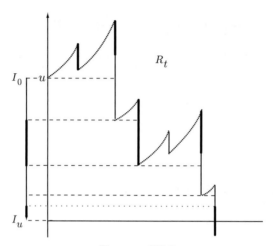

FIGURE IX.9

The difference from the case $p(r) \equiv p$ is that $\{m_x\}$, though still Markov, is no longer time-homogeneous. Let $\boldsymbol{P}(t_1, t_2)$ be the matrix with ijth element $\mathbb{P}\left(m_{t_2} = j \mid m_{t_1} = i\right)$, $0 \leq t_1 \leq t_2 \leq u$. Define further $\nu_i(u)$ as the probability that the risk process starting from $R_0 = u$ downcrosses level u for the first time in phase i. Note that in general $\sum_{i \in E} \nu_i(u) < 1$. In fact, $\sum_{i \in E} \nu_i(u)$ is the ruin probability for a risk process with initial reserve 0 and premium function $p(u + \cdot)$. Also, in contrast to Section 3, the definition of $\{m_x\}$ depends on the initial reserve $u = R_0$.

Since $\boldsymbol{\nu}(u) = \left(\nu_i(u)\right)_{i \in E}$ is the (defective) initial probability vector for $\{m_x\}$, we obtain

$$\psi(u) \;=\; \mathbb{P}(m_u \in E) \;=\; \boldsymbol{\nu}(u)\boldsymbol{P}(0, u)\boldsymbol{e} \;=\; \boldsymbol{\lambda}(u)\boldsymbol{e} \qquad (7.1)$$

where $\boldsymbol{\lambda}(t) = \boldsymbol{\nu}(u)\boldsymbol{P}(0, t)$ is the vector of state probabilities for m_t, i.e. $\lambda_i(t) = \mathbb{P}(m_t = i)$. Given the $\boldsymbol{\nu}(t)$ have been computed, the $\boldsymbol{\lambda}(t)$ and hence $\psi(u)$ is available by solving differential equations:

Proposition 7.1 $\boldsymbol{\lambda}(0) = \boldsymbol{\nu}(u)$ *and* $\boldsymbol{\lambda}'(t) \;=\; \boldsymbol{\lambda}(t)\big(\boldsymbol{T} + \boldsymbol{t}\boldsymbol{\nu}(u - t)\big)$, $0 \leq t \leq u$.

Proof. The first statement is clear by definition. By general results on time-inhomogeneous Markov processes,

$$\boldsymbol{P}(t_1, t_2) \;=\; \exp\left\{ \int_{t_1}^{t_2} \boldsymbol{Q}(v)\, \mathrm{d}v \right\} \tag{7.2}$$

where

$$\boldsymbol{Q}(t) \;=\; \frac{\mathrm{d}}{\mathrm{d}s}\left[\boldsymbol{P}(t, t+s) - \boldsymbol{I} \right]\Big|_{s=0}. \tag{7.3}$$

However, the interpretation of $\boldsymbol{Q}(t)$ as the intensity matrix of $\{m_x\}$ at time t shows that $\boldsymbol{Q}(t)$ is made up of two terms: obviously, $\{m_x\}$ has jumps of two types, those corresponding to state changes in the underlying phase process and those corresponding to the present jump of $\{R_t\}$ being terminated at level $u - t$ and being followed by a downcrossing. The intensity of a jump from i to j is t_{ij} for jumps of the first type and $t_i \nu_j(u - t)$ for the second. Hence $\boldsymbol{Q}(t) = \boldsymbol{T} + \boldsymbol{t}\boldsymbol{\nu}(u - t)$,

$$\boldsymbol{\lambda}'(t) \;=\; \boldsymbol{\lambda}(t)\boldsymbol{Q}(t) \;=\; \boldsymbol{\lambda}(t)\big(\boldsymbol{T} + \boldsymbol{t}\boldsymbol{\nu}(u - t) \big). \qquad \square$$

Thus, from a computational point of view the remaining problem is to evaluate the $\boldsymbol{\nu}(t)$, $0 \leq t \leq u$.

Proposition 7.2 *For $i \in E$,*

$$-\nu_i'(u)p(u) \;=\; \beta\alpha_i + \nu_i(u)\left\{ \sum_{j \in E} \nu_j(u)t_j p(u) - \beta \right\} + \sum_{j \in E} \nu_j(u)t_{ji}p(u). \tag{7.4}$$

Proof. Consider the event A that there are no arrivals in the interval $[0, \mathrm{d}t]$, the probability of which is $1 - \beta\mathrm{d}t$. Given A^c, the probability that level u is downcrossed for the first time in phase i is α_i. Given A, the probability that level $u + p(u)\mathrm{d}t$ is downcrossed for the first time in phase j is $\nu_j\big(u + p(u)\mathrm{d}t\big)$. Given this occurs, two things can happen: either the current jump continues from $u + p(u)\mathrm{d}t$ to u, or it stops between level $u + p(u)\mathrm{d}t$ and u. In the first case, the probability of downcrossing level u in phase i is

$$\delta_{ji}\big(1 + p(u)\mathrm{d}t \cdot t_{ii}\big) + (1 - \delta_{ji})p(u)\mathrm{d}t \cdot t_{ji} \;=\; \delta_{ji} + p(u)t_{ji}\,\mathrm{d}t,$$

whereas in the second case the probability is $p(u)\mathrm{d}t \cdot t_j \nu_i(u)$. Thus, given A, the probability of downcrossing level u in phase i for the first time is

$$\sum_{j \in E} \nu_j\big(u + p(u)\mathrm{d}t\big)\big[\delta_{ji} + p(u)\mathrm{d}t \cdot t_{ji} + p(u)\mathrm{d}t \cdot t_j \nu_i(u)\big]$$

$$=\; \nu_i(u) + \nu_i'(u)p(u)\,\mathrm{d}t \;+\; p(u)\,\mathrm{d}t \sum_{j \in E} \big\{ t_{ji} + t_j \nu_i(u) \big\}.$$

Collecting terms, we get

$$\nu_i(u) = \alpha_i \, dt + (1 - \beta dt)\nu_i(u) + \nu_i'(u)p(u) \, dt + p(u) \, dt \sum_{j \in E} \{t_{ji} + t_j \nu_i(u)\}.$$

Subtracting $\nu_i(u)$ on both sides and dividing by dt yields the asserted differential equation.   □

When solving the differential equation in Proposition 7.2, we face the difficulty that no boundary conditions is immediately available. To deal with this, consider a modification of the original process $\{R_t\}$ by linearizing the process with some rate ρ, say, after a certain level v, say. Let $p^v(t)$, R_t^v, $\mathbb{P}^v$ etc. refer to the modified process. Then

$$p^v(r) = \begin{cases} p(r) & r < v \\ \rho & r \geq v \end{cases},$$

and (no matter how ρ is chosen) we have:

Lemma 7.3 *For any fixed $u \geq 0$, $\nu_i(u) = \lim\limits_{v \to \infty} \nu_i^v(u)$.*

Proof. Let A be the event that the process downcrosses level u in phase i given that it starts at u and let B_v be the event

$$B_v = \left\{ \sigma < \infty, \ \sup_{t \leq \sigma} R_t > v \right\}$$

where σ denotes the time of downcrossing level u. Then $\mathbb{P}(B_v)$ is the tail of a (defective) random variable so that $\mathbb{P}(B_v) \to 0$ as $v \to \infty$, and similarly $\mathbb{P}^v(B_v) \to 0$.

Since the processes R_t and R_t^v coincide under level B_v, then $\mathbb{P}(A \cap B_v^c) = \mathbb{P}^v(A \cap B_v^c)$. Now since both $\mathbb{P}(A \cap B_v) \to 0$ and $\mathbb{P}^v(A \cap B_v) \to 0$ as $v \to \infty$ we have

$$
\begin{aligned}
\mathbb{P}(A) - \mathbb{P}^v(A) &= \mathbb{P}(A \cap B_v) + \mathbb{P}(A \cap B_v^c) - \mathbb{P}^v(A \cap B_v) - \mathbb{P}^v(A \cap B_v^c) \\
&= \mathbb{P}(A \cap B_v) - \mathbb{P}^v(A \cap B_v) \\
&\to 0
\end{aligned}
$$

as $v \to \infty$.   □

From Section 3, we have

$$p(r) \equiv \rho \ \Rightarrow \ \nu_i(u) \equiv -\frac{\beta}{\rho}\boldsymbol{\alpha T}\boldsymbol{e}_i, \tag{7.5}$$

which implies that $\nu_i^v(v)$ is given by the r.h.s. of (7.5). Thus, we can first for a given v solve (7.4) backwards for $\{\nu_i^v(t)\}_{v \geq t \geq 0}$, starting from $\nu^v(v) = -\beta \pi T^{-1}/\rho$. This yields $\nu_i^v(u)$ for any values of u and v such that $u \leq v$. Next consider a sequence of solutions obtained from a sequence of initial values $\{\nu_i^v(u)\}_v$ where, say, $v = u, 2u, 3u$ etc. Thus we obtain a convergent sequence of solutions that converges to $\{\nu_i(t)\}_{u \geq t \geq 0}$.

Notes and references The exposition is based upon Asmussen & Bladt [75] which also contains numerical illustrations.

The algorithm based upon the numerical solution of a Volterra integral equation (Remark VIII.1.10, numerically implemented in Schock Petersen [695]) and the present one based upon differential equations require both discretization along a discrete grid $0, 1/n, 2/n, \ldots$. However, typically the complexity in n is at best $O(n^2)$ for integral equations but $O(n)$ for differential equations. The actual precision depends on the particular numerical scheme being employed. The trapezoidal rule used in [695] gives a precision of $O(n^{-3})$, while the fourth-order Runge-Kutta method implemented in [75] gives $O(n^{-5})$.

7b Two-step premium rules

We now assume the premium function to be constant in two levels as in VIII.1a,

$$p(r) = \begin{cases} p_1 & r \leq v \\ p_2 & r > v. \end{cases} \qquad (7.6)$$

We may think of process R_t as pieced together of two standard risk processes R_t^1 and R_t^2 with constant premiums p_1, p_2, such that R_t coincide with R_t^1 under level v and with R_t^2 above level v. Let $\psi^i(u) = \alpha_+^{(i)} e^{(T + t\alpha_+^{(i)})u} e$ denote the ruin probability for R_t^i where $\alpha_+^i = \alpha_+^{(i)} = -\beta \alpha T^{-1}/p_i$, cf. Corollary 3.1. We recall from Proposition VIII.1.12 that in addition to the $\psi^{(i)}(\cdot)$, the evaluation of $\psi(u)$ requires $\phi_v(u) = 1 - \psi^{(1)}(u)/(1 - \psi^{(1)}(v))$, $0 \leq u \leq v$, which is available since the $\psi^{(i)}(\cdot)$ are so, as well $\pi(u)$, the probability of ruin between σ and the next upcrossing of v, where $\sigma = \inf\{t \geq 0 : R_t \leq v\}$.

To evaluate $\pi(u)$, let $\nu(u) = \alpha_+^{(2)} e^{(T + t\alpha_+^{(2)})(u-v)}$, assuming $u \geq v$ for the moment. Then $\nu(u)$ is the initial distribution of the undershoot when downcrossing level v given that the process starts at u, i.e. for $u \geq v$ the distribution of $v - R_\sigma$ (defined for $\sigma < \infty$ only) is defective phase-type with representation $(\nu(u), T)$. Recall that $\phi_v(w)$ is the probability of upcrossing level v before ruin given the process starts at $w \leq v$. Therefore

$$\pi(u) = \int_0^v \nu(u)e^{Tx} t\big(1 - \phi_v(v - x)\big)\, dx + \nu(u)e^{Tv} e \qquad (7.7)$$

(the integral is the contribution from $\{R_\sigma \geq 0\}$ and the last term the contribution from $\{R_\sigma < 0\}$). The integral in (7.7) equals

$$\int_0^v \boldsymbol{\nu}(u)\mathrm{e}^{\boldsymbol{T}x}\boldsymbol{t}\,\mathrm{d}x - \int_0^v \boldsymbol{\nu}(u)\mathrm{e}^{\boldsymbol{T}x}\boldsymbol{t}\frac{1 - \psi^{(1)}(v-x)}{1 - \psi^{(1)}(v)}\,\mathrm{d}x$$

$$= 1 - \boldsymbol{\nu}(u)\mathrm{e}^{\boldsymbol{T}v}\boldsymbol{e} - \frac{1}{1 - \psi^{(1)}(v)}\left\{1 - \boldsymbol{\nu}(u)\mathrm{e}^{\boldsymbol{T}v}\boldsymbol{e} - \int_0^v \boldsymbol{\nu}(u)\mathrm{e}^{\boldsymbol{T}x}\boldsymbol{t}\psi^{(1)}(v-x)\,\mathrm{d}x\right\}$$

from which we see that

$$\pi(u) = 1 + \frac{1}{1 - \psi^{(1)}(v)}\int_0^v \boldsymbol{\nu}(u)\mathrm{e}^{\boldsymbol{T}x}\boldsymbol{t}\psi^{(1)}(v-x)\,\mathrm{d}x - \frac{1}{1 - \psi^{(1)}(v)}\left(1 - \boldsymbol{\nu}(u)\mathrm{e}^{\boldsymbol{T}v}\boldsymbol{e}\right). \tag{7.8}$$

The integral in (7.8) equals

$$\int_0^v \boldsymbol{\nu}(u)\mathrm{e}^{\boldsymbol{T}x}\boldsymbol{t}\boldsymbol{\alpha}_+^{(2)}\mathrm{e}^{(\boldsymbol{T}+\boldsymbol{t}\boldsymbol{\alpha}_+^{(2)})(v-x)}\boldsymbol{e}\,\mathrm{d}x$$

which, using Kronecker calculus (see A.4), can be written as

$$\left(\boldsymbol{\nu}(u) \otimes \boldsymbol{\alpha}_+^{(2)}\mathrm{e}^{(\boldsymbol{T}+\boldsymbol{t}\boldsymbol{\alpha}_+^{(2)})v}\right)\left(\boldsymbol{T} \oplus (-\boldsymbol{T} - \boldsymbol{t}\boldsymbol{\alpha}_+^{(2)})\right)^{-1}\left\{\mathrm{e}^{\{\boldsymbol{T}\oplus(-\boldsymbol{T}-\boldsymbol{t}\boldsymbol{\alpha}_+^{(2)})\}v} - \boldsymbol{I}\right\}(\boldsymbol{t} \otimes \boldsymbol{e}).$$

Thus, all quantities involved in the computation of $\psi(u)$ have been found in matrix form.

Example 7.4 Let $\{R_t^1\}$ be as in Example 3.2. I.e., B is hyperexponential corresponding to

$$\boldsymbol{\alpha} = (\tfrac{1}{2}\ \tfrac{1}{2}), \quad \boldsymbol{T} = \begin{pmatrix} -3 & 0 \\ 0 & -7 \end{pmatrix}, \quad \boldsymbol{t} = \begin{pmatrix} 3 \\ 7 \end{pmatrix}.$$

The arrival rate is $\beta = 3$. Since $\mu_B = 5/21$, $p_2 \leq 3 \cdot \frac{5}{21} = \frac{5}{7}$ yields $\psi(u) = 1$, so we consider the non-trivial case example $p_2 = \frac{3}{4}$ and $p_1 = 1$.

From Example 3.2,

$$\psi^{(1)}(u) = \frac{24}{35}\mathrm{e}^{-u} + \frac{1}{35}\mathrm{e}^{-6u} \quad \Rightarrow \quad \phi_v(u) = \frac{35 - 24\mathrm{e}^{-u} - \mathrm{e}^{-6u}}{35 - 24\mathrm{e}^{-v} - \mathrm{e}^{-6v}}.$$

Let $\lambda_1 = -3 + 2\sqrt{2}$ and $\lambda_2 = -3 - 2\sqrt{2}$ be the eigenvalues of $\boldsymbol{T} + \boldsymbol{t}\boldsymbol{\alpha}_+^{(2)}$. Then one gets

$$\boldsymbol{\nu}(u) = \left(\frac{\sqrt{2}+1}{3}\mathrm{e}^{\lambda_1(u-v)} + \frac{1-\sqrt{2}}{3}\mathrm{e}^{\lambda_2(u-v)} \quad \frac{1}{7}\mathrm{e}^{\lambda_1(u-v)} + \frac{1}{7}\mathrm{e}^{\lambda_2(u-v)}\right),$$

$$\psi^{(2)}(u-v) = \left(\frac{10}{21} - \frac{1}{3}\sqrt{2}\right)\mathrm{e}^{\lambda_2(u-v)} + \left(\frac{10}{21} + \frac{1}{3}\sqrt{2}\right)\mathrm{e}^{\lambda_1(u-v)},$$

$$\psi^{(2)}(0) = \frac{20}{21}.$$

From (7.7) we see that we can write $\pi(u) = \boldsymbol{\nu}(u)\boldsymbol{V}_2$ where $\boldsymbol{V}_2$ depends only on v, and one gets

$$\boldsymbol{V}_2 = \begin{pmatrix} \dfrac{12e^{5v} - 2}{35e^{6v} - 24e^{5v} - 1} \\[2ex] \dfrac{4e^{5v} + 6}{35e^{6v} - 24e^{5v} - 1} \end{pmatrix}.$$

Thus, $\pi(u) = p_{12}(u)/p_{11}(u)$ where

$$
\begin{aligned}
p_{11}(u) &= 35e^{6v} - 24e^{5v} - 1, \\
p_{12}(u) &= \Big(\frac{32}{7} - 4\sqrt{2}\Big)e^{\lambda_2(u-v)}e^{5v} + \Big(\frac{2\sqrt{2}}{3} + \frac{4}{21}\Big)e^{\lambda_2(u-v)} \\
&\quad + \Big(\frac{32}{7} + 4\sqrt{2}\Big)e^{\lambda_1(u-v)}e^{5v} + \Big(\frac{4}{21} - \frac{2\sqrt{2}}{3}\Big)e^{\lambda_1(u-v)}.
\end{aligned}
$$

In particular,

$$
\begin{aligned}
\pi(v) &= \frac{192e^{5v} + 8}{21(35e^{6v} - 24e^{5v} - 1)}, \\
\psi(v) &= \frac{192e^{5v} + 8}{35e^{6v} + 168e^{5v} + 7}.
\end{aligned}
$$

Thus all terms involved in the formulae for the ruin probability have been explicitly derived. $\qquad\square$

Notes and references   The analysis and the example are from Asmussen & Bladt [75].

8   Erlangization for the finite horizon case

We consider the Cramér-Lundberg model with parameters β, B and recall from Corollary V.3.5 that an explicit formula for the Laplace transform w.r.t. u of $\mathbb{E}[e^{-\delta\tau(u)}; \tau(u) < \infty]$ can be found in terms of the root $-\rho_\delta < 0$ of

$$\kappa(r) = \beta\big(\widehat{B}[r] - 1\big) = \delta. \tag{8.1}$$

Thus the finite horizon ruin probability $\psi(u, T)$ can in principle be computed exactly via a double Laplace transform inversion. Now transform inversion is never entirely straightforward and even less so when it is higher-dimensional. We present in this section a numerical scheme that basically only requires a rootfinding and the computation of a matrix-exponential under the assumption that the claim size distribution is phase-type $(E, \boldsymbol{\alpha}, \boldsymbol{T})$. The basic idea is to replace the deterministic time horizon T by a r.v. H_k that has an Erlang distribution with k stages and mean T, that is, with density

$$\frac{\delta^k t^{k-1}}{(k-1)!} e^{-\delta t} \quad \text{where} \quad \delta = \delta_k = k/T .$$

That is, we compute

$$\psi_k(u) \;=\; \mathbb{E}\psi(u, H_k) \;=\; \int_0^\infty \psi(u, t) \frac{\delta^k t^{k-1}}{(k-1)!} e^{-\delta t} \, \mathrm{d}t . \qquad (8.2)$$

Since the s.c.v. of the Erlang distribution goes to 0 as $k \to \infty$, of course also $\psi_k(u) \to \psi(u, T)$. The case $k = 1$ of an exponential time horizon then comes out fairly easily, whereas a simple recursion scheme exists for going from k to $k+1$. Combining with an extrapolation idea yields a considerable improvement of the numerical scheme.

The approximation $\psi(u, T) \approx \psi_k(u)$ could be called *Erlang smoothing*. Namely, (8.2) means that we approximate $\psi(u, T)$ by the function $\psi(u, t)$ of t smoothed by the kernel which is the Erlang density with mean T. Cf. Fig. IX.10.

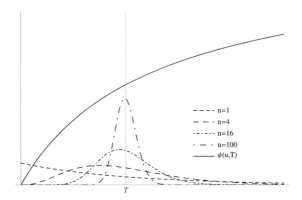

Figure IX.10: Erlang smoothing

Proceeding to the details, one may first note that the model is a special case of the Markovian environment model in Chapter VII. Namely, the state J_t of the environment at time t is the current exponential stage $1, \ldots, k$ of T_k. However, we have the difference that here the environmental process J is terminating, whereas Chapter VII concentrates on J being ergodic. Indeed, J has terminated by time t if $H_k < t$. Nevertheless, we may proceed along similar ideas as in VII.2, only do we now reverse the sign and not the time. More precisely, we define $Y_x \in E_k = \{1, \ldots, k\} \times E$ to have the value (i, j) if the upcrossing of $\{S_t\}$ of level x occurs in state j of the phase-type jump leading to the upcrossing and if H_k at that time is in stage i. Obviously, $\{Y_x\}$ is a Markov process, and it is terminating since $H_k < \infty$. Furthermore, a jump occurs in two ways: as consequence of a jump in the Markov process underlying the current claim. This changes only j, not i, so that the matrix of corresponding rates is $I \oplus T$. Or the current claim may terminate, in which case the new state will be (k, ℓ), with $k \geq i$ and $\ell \in E$ the phase at the next ladder point. Denoting by $\boldsymbol{\alpha}^{(k)}$ the row vector of state probabilities when the first ladder point of $\{S_t\}$ occurs in Erlang stage k, it follows that the intensity matrix U of Y is given by

$$U = I \oplus T + \begin{pmatrix} t\boldsymbol{\alpha}^{(1)} & t\boldsymbol{\alpha}^{(2)} & t\boldsymbol{\alpha}^{(3)} & \cdots & t\boldsymbol{\alpha}^{(k)} \\ 0 & t\boldsymbol{\alpha}^{(1)} & t\boldsymbol{\alpha}^{(2)} & \cdots & t\boldsymbol{\alpha}^{(k-1)} \\ 0 & 0 & t\boldsymbol{\alpha}^{(1)} & \cdots & t\boldsymbol{\alpha}^{(k-2)} \\ \vdots & & & \ddots & \vdots \\ 0 & 0 & 0 & \cdots & t\boldsymbol{\alpha}^{(1)} \end{pmatrix}$$

(recall that $t = -Te$ denotes the exit vector of the phase-type distribution). We further get $\psi_k(u) = \boldsymbol{\alpha}^* e^{Uu} e$ where $\boldsymbol{\alpha}^* = \left(\boldsymbol{\alpha}^{(1)} \ \boldsymbol{\alpha}^{(2)} \ \cdots \ \boldsymbol{\alpha}^{(k)} \right)$. Thus, it only remains to compute the $\boldsymbol{\alpha}^{(\ell)}$.

We first consider the exponential case $k = 1$.

Theorem 8.1 *Define* $\boldsymbol{\alpha}_\delta = \boldsymbol{\alpha}^{(1)}$. *Then* $\boldsymbol{\alpha}_\delta = \beta \boldsymbol{\alpha}(\rho_\delta I - T)^{-1}$ *where* $-\rho_\delta$ *is the negative root of* $\kappa(r) = \delta$, *i.e.* $\beta \left(\boldsymbol{\alpha}(-rI - T)^{-1} t - 1 \right) - r = \delta$.

Proof. We condition upon the time t of the first claim where $S_{t-} = -t$. The exponential time exceeds t w.p. $e^{-\delta t}$, and so, proceeding again along the lines of the second proof of Corollary 3.1, we conclude that

$$\boldsymbol{\alpha}_\delta = \int_0^\infty \beta e^{-\delta t} e^{-\delta t} \boldsymbol{\alpha} e^{(T + t\boldsymbol{\alpha}_\delta)t} \, dt = \beta \boldsymbol{\alpha} \left((\beta + \delta) I - T - t\boldsymbol{\alpha}_\delta \right)^{-1}.$$

Thus

$$(\beta + \delta)\boldsymbol{\alpha}_\delta - \boldsymbol{\alpha}_\delta T - \boldsymbol{\alpha}_\delta t\boldsymbol{\alpha}_\delta = \beta \boldsymbol{\alpha}. \tag{8.3}$$

For brevity, write $\boldsymbol{\nu} = \boldsymbol{\alpha}(\rho_\delta I - T)^{-1}$. We will show that $\boldsymbol{\alpha}_\delta = \beta \boldsymbol{\nu}$ satisfies (8.3). We first note that the definition of ρ_δ implies $\beta \boldsymbol{\nu} t = \beta + \delta - \rho_\delta$ and that

$$\boldsymbol{\nu} T = \boldsymbol{\alpha}(\rho_\delta I - T)^{-1}(-\rho_\delta I + T) + \boldsymbol{\alpha}(\rho_\delta I - T)^{-1}\rho_\delta I = -\boldsymbol{\alpha} + \rho_\delta \boldsymbol{\nu}.$$

Inserting $\boldsymbol{\alpha}_\delta = \beta\boldsymbol{\nu}$ in the l.h.s. of (8.2), we therefore obtain

$$(\beta^2 + \beta\delta)\boldsymbol{\nu} + \beta\boldsymbol{\alpha} - \beta\rho_\delta\boldsymbol{\nu} - \beta(\beta + \delta - \rho_\delta)\boldsymbol{\nu}$$

which equals $\beta\boldsymbol{\alpha}$, as should be. We omit the proof that $\boldsymbol{\alpha}(\rho_\delta\boldsymbol{I} - \boldsymbol{T})^{-1}$ is the correct one among the solutions of (8.2).   □

For a general k, we have the following recursion:

Theorem 8.2 $\boldsymbol{\alpha}_\delta^{(n+1)} = \left(\delta\boldsymbol{\alpha}_\delta^{(n)} + \sum\limits_{j=2}^{n} \boldsymbol{\alpha}_\delta^{n+2-j}t\boldsymbol{\alpha}_\delta^{(j)}\right)\left(\rho_\delta\boldsymbol{I} - \boldsymbol{T} - t\boldsymbol{\alpha}_\delta^{(1)}\right)^{-1}$.

The proof is more complicated, and we refer to Asmussen, Avram & Usabel [71].

The algorithm can be improved by *Richardson extrapolation*. This is a general method (see e.g. Press *et al.* [715]) for computing a number w accurately using a sequence $w_k \to w$ for which the convergence rate is known,

$$w - w_k = \frac{c}{k} + \frac{d}{k^{1+\epsilon}} + \cdots . \tag{8.4}$$

Here c is typically unknown but can be eliminated. Indeed, letting $w_k^* = (k + 1)w_{k+1} - kw_k$, it is clear that $w_k^* \to w$ and that one obtains an improved approximation of convergence rate $\mathrm{O}(k^{-1-\epsilon})$.

In the present setting, $w = \psi(u, T)$, $w_k = \psi_k(u)$ and (8.4) simply follows by the CLT for the underlying Erlang r.v. H_k:

$$
\begin{aligned}
\psi_k(u) &= \mathbb{E}\psi(u, H_k) \\
&= \mathbb{E}\left[\psi(u, T) + \psi_T(u, T)(H_k - T) + \psi_{TT}(u, T)(H_k - T)^2/2 + \cdots\right] \\
&= \psi(u, T) + 0 + \psi_{TT}(u, T)\mathbb{V}\mathrm{ar}(H_k) + \cdots \\
&= \psi(u, T) + \frac{c}{k} + \cdots
\end{aligned}
$$

where as usual ψ_T, ψ_{TT} are the first and second order partial derivatives of ψ w.r.t. T.

Example 8.3 For an illustration of the method, consider a highly skewed claim size distribution, namely a mixture of three exponential distributions with rates 0.015, 0.190, 5.51 and corresponding weights 0.004, 0.108 and 0.888. Choose the safety loading $\eta = 0.1$ and $T = 1$, $u = 0$. The exact value $\psi(u, T) = 2.28\%$ was calculated by transform inversion. Figure IX.11 shows the results of the Erlangization, with the circles corresponding to the simple method and the filled ones to the extrapolated values.

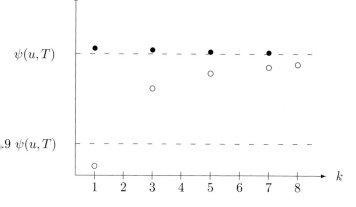

Figure IX.11: Erlangization and extrapolation

It is seen that the simple method produces good results even in the range $k = 5\text{–}7$. This is maybe somewhat surprising, since the Erlang(7) distribution is quite far from being degenerate. The precision of the extrapolation method is remarkable. Even the value for $k = 1$ would suffice for all practical purposes!

□

Notes and references The exposition is based upon Asmussen, Avram & Usabel [71] (who also consider general phase-type horizons). Note, however, that Theorem 8.1 appears already in Avram & Usabel [112]. See also Ramaswami, Woolford & Stanford [723].

Exponential/Erlangian time horizons have also been used in finance, where the idea is known as *Canadization*. An early classical reference for the exponential case is Carr [223]. Erlangian horizons occur, e.g., in Kyprianou & Pistorius [566].

Chapter X

Ruin probabilities in the presence of heavy tails

1 Subexponential distributions

We are concerned with distributions B with a heavy right tail $\overline{B}(x) = 1 - B(x)$.

A rough distinction between light and heavy tails is that the m.g.f. $\widehat{B}[r] = \int e^{rx} B(\mathrm{d}x)$ is finite for some $r > 0$ in the light-tailed case and infinite for all $r > 0$ in the heavy-tailed case. For example, the exponential change of measure techniques discussed in III.3, IV.4–6 and at numerous later occasions require a light tail. Some main cases where this light-tail criterion are violated are

(a) distributions with a regularly varying tail, $\overline{B}(x) = L(x)/x^{\alpha}$ where $\alpha > 0$ and $L(x)$ is slowly varying, $L(tx)/L(x) \to 1$, $x \to \infty$, for all $t > 0$;

(b) the lognormal distribution (the distribution of e^{U} where $U \sim N(\mu, \sigma^2)$) with density

$$\frac{1}{x\sqrt{2\pi\sigma^2}} e^{-(\log x - \mu)^2 / 2\sigma^2};$$

(c) the Weibull distribution with decreasing failure rate, $\overline{B}(x) = e^{-x^{\beta}}$ with $0 < \beta < 1$.

For further examples, see I.2b.

The definition $\widehat{B}[r] = \infty$ for all $r > 0$ of heavy tails is too general to allow for general non-trivial results on ruin probabilities, and instead we shall work within the class $\mathscr{S}$ of *subexponential* distributions. For the definition, we require that B is concentrated on $(0, \infty)$ and say then that B is subexponential ($B \in \mathscr{S}$) if

$$\frac{\overline{B^{*2}}(x)}{\overline{B}(x)} \rightarrow 2, \quad x \rightarrow \infty. \tag{1.1}$$

Here B^{*2} is the convolution square, that is, the distribution of the sum of independent r.v.'s X_1, X_2 with distribution B. In terms of r.v.'s, (1.1) then means $\mathbb{P}(X_1 + X_2 > x) \sim 2\mathbb{P}(X_1 > x)$.

To capture the intuition behind this definition, note first the following fact:

Proposition 1.1 *Let B be any distribution on* $(0, \infty)$. *Then:*
(a) $\mathbb{P}\big(\max(X_1, X_2) > x\big) \sim 2\overline{B}(x), \quad x \rightarrow \infty.$

(b) $\liminf\limits_{x \rightarrow \infty} \dfrac{\overline{B^{*2}}(x)}{\overline{B}(x)} \geq 2.$

Proof. By the inclusion-exclusion formula, $\mathbb{P}\big(\max(X_1, X_2) > x\big)$ is

$$\mathbb{P}(X_1 > x) + \mathbb{P}(X_2 > x) - \mathbb{P}(X_1 > x, X_2 > x) = 2\overline{B}(x) - \overline{B}(x)^2 \sim 2\overline{B}(x),$$

proving (a). Since B is concentrated on $(0, \infty)$, we have $\{\max(X_1, X_2) > x\} \subseteq \{X_1 + X_2 > x\}$, and thus the $\liminf$ in (b) is at least $\liminf \mathbb{P}(\max(X_1, X_2) > x)/\overline{B}(x) = 2.$ [1] □

The proof shows that the condition for $B \in \mathscr{S}$ is that the probability of the set $\{X_1 + X_2 > x\}$ is asymptotically the same as the probability of its subset $\{\max(X_1, X_2) > x\}$. That is, in the subexponential case *the only way $X_1 + X_2$ can get large is roughly by one of the X_i becoming large.* We later show:

Proposition 1.2 *If $B \in \mathscr{S}$, then*

$$\mathbb{P}\big(X_1 > x \,\big|\, X_1 + X_2 > x\big) \rightarrow \frac{1}{2}, \quad \mathbb{P}\big(X_1 \leq y \,\big|\, X_1 + X_2 > x\big) \rightarrow \frac{1}{2}B(y).$$

That is, given $X_1 + X_2 > x$, the r.v. X_1 is w.p. 1/2 'typical' (with distribution B) and w.p. 1/2 it has the distribution of $X_1 | X_1 > x$. In contrast, the behavior in the light-tailed case is illustrated in the following example:

Example 1.3 Consider the standard exponential distribution, $\overline{B}(x) = e^{-x}$. Then $X_1 + X_2$ has an Erlang(2) distribution with density ye^{-y} so that $\overline{B^{*2}}(x) \sim xe^{-x}$. Thus the $\liminf$ in Proposition 1.1(b) is ∞. In contrast to Proposition 1.2, one can check that

$$\left(\frac{X_1}{x}, \frac{X_2}{x}\right)\bigg|\, X_1 + X_2 > x \xrightarrow{\mathscr{D}} (U, 1 - U)$$

---

[1] Note that it can be shown that for any heavy-tailed distribution one has in fact the stronger result $\liminf_{x \rightarrow \infty} \overline{B^{*2}}(x)/\overline{B}(x) = 2$, see Foss & Korshunov [365].

where U is uniform on $(0,1)$. Thus, if $X_1 + X_2$ is large, then (with high probability) so are both of X_1, X_2 but none of them exceeds x. □

Here is the simplest example of subexponentiality:

Proposition 1.4 *Any B with a regularly varying tail is subexponential.*

Proof. Assume $\overline{B}(x) = L(x)/x^\alpha$ with L slowly varying and $\alpha > 0$. Let $0 < \delta < 1/2$. If $X_1 + X_2 > x$, then either one of the X_i exceeds $(1 - \delta)x$, or they both exceed δx. Hence

$$\limsup_{x\to\infty} \frac{\overline{B^{*2}}(x)}{\overline{B}(x)} \leq \limsup_{x\to\infty} \frac{2\overline{B}\big((1-\delta)x\big) + \overline{B}(\delta x)^2}{\overline{B}(x)}$$

$$= \limsup_{x\to\infty} \frac{2L\big((1-\delta)x\big)/\big((1-\delta)x\big)^\alpha}{L(x)/x^\alpha} + 0 = \frac{2}{(1-\delta)^\alpha}.$$

Letting $\delta \downarrow 0$, we get $\limsup \overline{B^{*2}}(x)/\overline{B}(x) \leq 2$, and combining with Proposition 1.1(b) we get $\overline{B^{*2}}(x)/\overline{B}(x) \to 2$. □

We now turn to the mathematical theory of subexponential distributions.

Proposition 1.5 *If $B \in \mathscr{S}$, then $\dfrac{\overline{B}(x-y)}{\overline{B}(x)} \to 1$ uniformly in $y \in [0, y_0]$ as $x \to \infty$.*

[In terms of r.v.'s: if $X \sim B \in \mathscr{S}$, then the overshoot $X - x | X > x$ converges in distribution to ∞. This follows since the probability of the overshoot to exceed y is $\overline{B}(x+y)/\overline{B}(x)$ which has limit 1.]

Proof. Consider first a fixed y. Using the identity

$$\frac{\overline{B^{*(n+1)}}(x)}{\overline{B}(x)} = 1 + \frac{B(x) - B^{*(n+1)}(x)}{\overline{B}(x)} = 1 + \int_0^x \frac{1 - B^{*n}(x-z)}{\overline{B}(x)} B(\mathrm{d}z) \quad (1.2)$$

with $n = 1$ and splitting the integral into two corresponding to the intervals $[0, y]$ and $(y, x]$, we get

$$\frac{\overline{B^{*2}}(x)}{\overline{B}(x)} \geq 1 + B(y) + \frac{\overline{B}(x-y)}{\overline{B}(x)}\big(B(x) - B(y)\big).$$

If $\limsup \overline{B}(x-y)/\overline{B}(x) > 1$, we therefore get $\limsup \overline{B^{*2}}(x)/\overline{B}(x) > 1 + B(y) + 1 - B(y) = 2$, a contradiction. Finally $\liminf \overline{B}(x-y)/\overline{B}(x) \geq 1$ since $y > 0$.

The uniformity now follows from what has been shown for $y = y_0$ and the obvious inequality

$$1 \leq \frac{\overline{B}(x-y)}{\overline{B}(x)} \leq \frac{\overline{B}(x-y_0)}{\overline{B}(x)}, \quad y \in [0, y_0].$$

$\square$

Corollary 1.6 *If $B \in \mathscr{S}$, then $e^{\epsilon x}\overline{B}(x) \to \infty$, $\widehat{B}[\epsilon] = \infty$ for all $\epsilon > 0$.*

Proof. For $0 < \delta < \epsilon$, we have by Proposition 1.5 that $\overline{B}(n) \geq e^{-\delta}\overline{B}(n-1)$ for all large n so that $\overline{B}(n) \geq c_1 e^{-\delta n}$ for all n. This implies $\overline{B}(x) \geq c_2 e^{-\delta x}$ for all x, and this immediately yields the desired conclusions. $\square$

Proof of Proposition 1.2.

$$\mathbb{P}\big(X_1 > x \,\big|\, X_1 + X_2 > x\big) = \frac{\mathbb{P}(X_1 > x)}{\mathbb{P}(X_1 + X_2 > x)} = \frac{\overline{B}(x)}{\overline{B^{*2}}(x)} \to \frac{1}{2},$$

$$\mathbb{P}\big(X_1 \leq y \,\big|\, X_1 + X_2 > x\big) \sim \frac{1}{2\overline{B}(x)} \int_0^y \overline{B}(x-z)\, B(dz)$$

$$\to \frac{1}{2} \int_0^y B(dz) = \frac{1}{2}B(y),$$

using Proposition 1.5 and dominated convergence. $\square$

The following result is extremely important and is often taken as definition of the class $\mathscr{S}$; its intuitive content is the same as discussed in the case $n = 2$ above.

Proposition 1.7 *If $B \in \mathscr{S}$, then for any n, $\overline{B^{*n}}(x)/\overline{B}(x) \to n$ as $x \to \infty$.*

Proof. We use induction. The case $n = 2$ is just the definition, so assume the proposition has been shown for n. Given $\epsilon > 0$, choose y such that $|\overline{B^{*n}}(x)/\overline{B}(x) - n| \leq \epsilon$ for $x \geq y$. Then by (1.2),

$$\frac{\overline{B^{*(n+1)}}(x)}{\overline{B}(x)} = 1 + \left(\int_0^{x-y} + \int_{x-y}^x \right) \frac{\overline{B^{*n}}(x-z)}{\overline{B}(x-z)} \frac{\overline{B}(x-z)}{\overline{B}(x)} B(dz).$$

Here the second integral can be bounded by

$$\sup_{v \geq 0} \frac{\overline{B^{*n}}(v)}{\overline{B}(v)} \frac{B(x) - B(x-y)}{\overline{B}(x)},$$

which converges to 0 by Proposition 1.5 and the induction hypothesis. The first integral is

$$(n + O(\epsilon)) \int_0^{x-y} \frac{\overline{B}(x-z)}{\overline{B}(x)} B(dz)$$

$$= (n + O(\epsilon)) \left\{ \frac{\overline{B}(x) - \overline{B^{*2}}(x)}{\overline{B}(x)} - \int_{x-y}^x \frac{\overline{B}(x-z)}{\overline{B}(x)} B(dz) \right\}.$$

Here the first term in $\{\cdot\}$ converges to 1 (by the definition of $B \in \mathscr{S}$) and the second to 0 since it is bounded by $(\overline{B}(x) - \overline{B}(x-y))/\overline{B}(x)$. Combining these estimates and letting $\epsilon \downarrow 0$ completes the proof. □

Lemma 1.8 *If $B \in \mathscr{S}$, $\epsilon > 0$, then there exists a constant $K = K_\epsilon$ such that $\overline{B^{*n}}(x) \le K(1+\epsilon)^n \overline{B}(x)$ for all n and x.*

Proof. Choose T such that $\left(\overline{B}(x) - \overline{B^{*2}}(x)\right)/\overline{B}(x) \le 1 + \epsilon$ for $x \ge T$ and let $A = 1/\overline{B}(T)$, $\alpha_n = \sup_{x \ge 0} \overline{B^{*n}}(x)/\overline{B}(x)$. Then by (1.2), for all n

$$\alpha_{n+1}$$

$$\le 1 + \sup_{x \le T} \int_0^x \frac{\overline{B^{*n}}(x-z)}{\overline{B}(x)} B(dz) + \sup_{x > T} \int_0^x \frac{\overline{B^{*n}}(x-z)}{\overline{B}(x-z)} \frac{\overline{B}(x-z)}{\overline{B}(x)} B(dz)$$

$$\le 1 + A + \alpha_n \sup_{x > T} \int_0^x \frac{\overline{B}(x-z)}{\overline{B}(x)} B(dz) \le 1 + A + \alpha_n (1 + \epsilon).$$

Iterating, we get with $\alpha_1 = 1$

$$\alpha_{n+1} \le (1+A)\frac{1 - (1+\epsilon)^n}{-\epsilon} + (1+\epsilon)^n.$$

Take $K = (1 + (1+A)/\epsilon)/(1+\epsilon)$. □

Proposition 1.9 *Let A_1, A_2 be distributions on $(0, \infty)$ such that $\overline{A}_i(x) \sim a_i \overline{B}(x)$ for some $B \in \mathscr{S}$ and some constants a_1, a_2 with $a_1 + a_2 > 0$. Then $\overline{A_1 * A_2}(x) \sim (a_1 + a_2)\overline{B}(x)$.*

Proof. Let X_1, X_2 be independent r.v.'s such that X_i has distribution A_i. Then by definition $\overline{A_1 * A_2}(x) = \mathbb{P}(X_1 + X_2 > x)$. For any fixed v, Proposition 1.5 easily yields

$$\mathbb{P}(X_1 + X_2 > x, X_i \le v) = \int_0^v \overline{A}_j(x-y)A_i(dy)$$

$$\sim a_j \overline{B}(x)A_i(v) = a_j \overline{B}(x)\big(1 + o_v(1)\big)$$

$(j = 3 - i)$. Since

$$\mathbb{P}(X_1 + X_2 > x, X_1 > x - v, X_2 > x - v) \leq \overline{A}_1(x - v)\overline{A}_2(x - v) \sim a_1 a_2 \overline{B}(x)^2$$

which can be neglected, it follows that it is necessary and sufficient for the assertion to be true that

$$\int_v^{x-v} \overline{A}_j(x - y)A_i(\mathrm{d}y) = \overline{B}(x)\mathrm{o}_v(1). \tag{1.3}$$

Using the necessity part in the case $A_1 = A_2 = B$ yields

$$\int_v^{x-v} \overline{B}(x - y)B(\mathrm{d}y) = \overline{B}(x)\mathrm{o}_v(1). \tag{1.4}$$

Now (1.3) follows if

$$\int_v^{x-v} \overline{B}(x - y)A_i(\mathrm{d}y) = \overline{B}(x)\mathrm{o}_v(1). \tag{1.5}$$

By a change of variables, the l.h.s. of (1.5) becomes

$$\overline{B}(x - v)\overline{A}_i(v) - \overline{A}_i(x - v)\overline{B}(v) + \int_v^{x-v} \overline{A}_i(x - y)B(\mathrm{d}y)\,.$$

Here approximately the last term is $\overline{B}(x)\mathrm{o}_v(1)$ by (1.4), whereas the two first yield $\overline{B}(x)(\overline{A}_i(v) - a_i\overline{B}(v)) = \overline{B}(x)\mathrm{o}_v(1)$. □

Corollary 1.10 *The class $\mathscr{S}$ is closed under tail-equivalence. That is, if $\overline{A}(x) \sim a\overline{B}(x)$ for some $B \in \mathscr{S}$ and some constant $a > 0$, then $A \in \mathscr{S}$.*

Proof. Taking $A_1 = A_2 = A$, $a_1 = a_2 = a$ yields $\overline{A^{*2}}(x) \sim 2a\overline{B}(x) \sim 2\overline{A}(x)$. □

Corollary 1.11 *Let $B \in \mathscr{S}$ and let A be any distribution with a lighter tail, $\overline{A}(x) = \mathrm{o}(\overline{B}(x))$. Then $A * B \in \mathscr{S}$ and $\overline{A * B}(x) \sim \overline{B}(x)$.*

Proof. Take $A_1 = A$, $A_2 = B$ so that $a_1 = 0$, $a_2 = 1$. □

It is tempting to conjecture that $\mathscr{S}$ is closed under convolution, i.e. $B_1 * B_2 \in \mathscr{S}$ and $\overline{B_1 * B_2}(x) \sim \overline{B}_1(x) + \overline{B}_2(x)$ when $B_1, B_2 \in \mathscr{S}$. However, $B_1 * B_2 \in \mathscr{S}$ does not hold in full generality (but once $B_1 * B_2 \in \mathscr{S}$ has been shown, $\overline{B_1 * B_2}(x) \sim \overline{B}_1(x) + \overline{B}_2(x)$ follows precisely as in the proof of Proposition 1.9). In the regularly varying case, it is easy to see that if L_1, L_2 are slowly varying, then so is $L = L_1 + L_2$. Hence:

Corollary 1.12 *Assume that $\overline{B}_i(x) = L_i(x)/x^\alpha$, $i = 1, 2$, with $\alpha > 0$ and L_1, L_2 slowly varying. Then $L = L_1 + L_2$ is slowly varying and $\overline{B_1 * B_2}(x) \sim L(x)/x^\alpha$.*

We next give a classical sufficient (and close to necessary) condition for subexponentiality due to Pitman [705]. Recall that the failure rate $\lambda(x)$ of a distribution B with density b is $\lambda(x) = b(x)/\overline{B}(x)$.

Proposition 1.13 *Let B have density b and failure rate $\lambda(x)$ such that $\lambda(x)$ is decreasing for $x \geq x_0$ with limit 0 at ∞. Then $B \in \mathscr{S}$ provided*

$$\int_0^\infty e^{x\lambda(x)} b(x) \, dx < \infty.$$

Proof. We may assume that $\lambda(x)$ is everywhere decreasing (otherwise, replace B by a tail equivalent distribution with a failure rate which is everywhere decreasing). Define $\Lambda(x) = \int_0^x \lambda(y) \, dy$. Then $\overline{B}(x) = e^{-\Lambda(x)}$. By (1.2),

$$\frac{\overline{B^{*2}}(x)}{\overline{B}(x)} - 1$$

$$= \int_0^x \frac{\overline{B}(x-y)}{\overline{B}(x)} b(y) \, dy = \int_0^x e^{\Lambda(x) - \Lambda(x-y) - \Lambda(y)} \lambda(y) \, dy$$

$$= \int_0^{x/2} e^{\Lambda(x) - \Lambda(x-y) - \Lambda(y)} \lambda(y) \, dy + \int_0^{x/2} e^{\Lambda(x) - \Lambda(x-y) - \Lambda(y)} \lambda(x - y) \, dy.$$

For $y < x/2$,

$$\Lambda(x) - \Lambda(x - y) \leq y\lambda(x - y) \leq y\lambda(y).$$

The rightmost bound shows that the integrand in the first integral is bounded by $e^{y\lambda(y) - \Lambda(y)} \lambda(y) = e^{y\lambda(y)} b(y)$, an integrable function by assumption. The middle bound shows that it converges to $b(y)$ for any fixed y since $\lambda(x - y) \to 0$. Thus by dominated convergence, the first integral has limit 1. Since $\lambda(x - y) \leq \lambda(y)$ for $y < x/2$, we can use the same domination for the second integral but now the integrand has limit 0. Thus $\overline{B^{*2}}(x)/\overline{B}(x) - 1$ has limit $1 + 0$, proving $B \in \mathscr{S}$. $\qquad\square$

Example 1.14 Consider the DFR Weibull case $\overline{B}(x) = e^{-x^\beta}$ with $0 < \beta < 1$. Then $b(x) = \beta x^{\beta-1} e^{-x^\beta}$, $\lambda(x) = \beta x^{\beta-1}$. Thus $\lambda(x)$ is everywhere decreasing, and $e^{x\lambda(x)} b(x) = \beta x^{\beta-1} e^{-(1-\beta)x^\beta}$ is integrable. Thus, the DFR Weibull distribution is subexponential. $\qquad\square$

Example 1.15 For the lognormal distribution,

$$\lambda(x) = \frac{e^{-(\log x - \mu)^2/2\sigma^2}/(x\sqrt{2\pi\sigma^2})}{\Phi(-(\log x - \mu)/\sigma)} \sim \frac{\log x}{\sigma^2 x}.$$

This yields easily that $e^{x\lambda(x)}b(x)$ is integrable. Further, elementary but tedious calculations (which we omit) show that $\lambda(x)$ is ultimately decreasing. Thus, the lognormal distribution is subexponential. □

In the regularly varying case, subexponentiality has already been proved in Corollary 1.12. To illustrate how Proposition 1.13 works in this setting, we first quote *Karamata's theorem* (Bingham, Goldie & Teugels [169]):

Proposition 1.16 *For $L(x)$ slowly varying and $\alpha > 1$,*

$$\int_x^\infty \frac{L(y)}{y^\alpha}\,dy \sim \frac{L(x)}{(\alpha-1)x^{\alpha-1}}.$$

From this we get

Proposition 1.17 *If B has a tail of the form $b(x) = \alpha L(x)/x^{\alpha+1}$ with $L(x)$ slowly varying and $\alpha > 1$, then $\overline{B}(x) \sim L(x)/x^\alpha$ and $\lambda(x) \sim \alpha/x$.*

Thus $e^{x\lambda(x)}b(x) \sim e^\alpha b(x)$ is integrable. However, the monotonicity condition in Proposition 1.13 may present a problem in some cases so that the direct proof in Proposition 1.4 is necessary in full generality.

We conclude with a property of subexponential distributions which is often extremely important: under some mild smoothness assumptions, the overshoot properly normalized has a limit which is Pareto if B is regularly varying and exponential for distributions like the lognormal or Weibull. More precisely, let $X^{(x)} = X - x | X > x$ and define the *mean excess function* $e(x) = \mathbb{E}X^{(x)}$ (in insurance mathematics and in particular in reinsurance, the term *stop-loss transform* for the unconditional expectation $\mathbb{E}(X - x)^+ = \int_x^\infty \overline{B}(y)\,dy$ is common). Then:

Proposition 1.18 (a) *If $\overline{B}(x) = L(x)/x^\alpha$ with $L(x)$ slowly varying and $\alpha > 1$, then $e(x) \sim x/(\alpha-1)$ and*

$$\mathbb{P}\big(X^{(x)}/e(x) > y\big) \rightarrow \frac{1}{\big(1 + y/(\alpha-1)\big)^\alpha}; \qquad (1.6)$$

(b) *Assume that for any y_0 the failure rate $\lambda(\cdot)$ satisfies*

$$\frac{\lambda\big(x + y/\lambda(x)\big)}{\lambda(x)} \rightarrow 1 \qquad (1.7)$$

uniformly for $y \in (0, y_0]$. Then $e(x) \sim 1/\lambda(x)$ and

$$\mathbb{P}\big(X^{(x)}/e(x) > y\big) \ \to \ e^{-y} \, ; \tag{1.8}$$

(c) *Under the assumptions of either* (a) *or* (b), $\int_x^\infty \overline{B}(y)\,\mathrm{d}y \sim e(x)\overline{B}(x)$.

Proof. (a): Using Karamata's theorem, we get

$$
\begin{aligned}
\mathbb{E}X^{(x)} \ &= \ \frac{\mathbb{E}(X - x)^+}{\mathbb{P}(X > x)} \ = \ \frac{1}{\mathbb{P}(X > x)} \int_x^\infty \mathbb{P}(X > y)\,\mathrm{d}y \\
&= \ \frac{1}{L(x)/x^\alpha} \int_x^\infty L(y)/y^\alpha\,\mathrm{d}y \ \sim \ \frac{L(x)/\big((\alpha - 1)x^{\alpha - 1}\big)}{L(x)/x^\alpha} \\
&= \ \frac{x}{\alpha - 1} \, .
\end{aligned}
$$

Further

$$
\begin{aligned}
\mathbb{P}\big((\alpha - 1)X^{(x)}/x > y\big) \ &= \ \mathbb{P}(X > x[1 + y/(\alpha - 1)] \mid X > x) \\
&= \ \frac{L\big(x[1 + y/(\alpha - 1)]\big)}{L(x)} \cdot \frac{x^\alpha}{\big(x[1 + y/(\alpha - 1)]\big)^\alpha} \\
&\sim \ 1 \cdot \frac{1}{\big(1 + y/(\alpha - 1)\big)^\alpha} \, .
\end{aligned}
$$

We omit the proof of (c) and that $\mathbb{E}X^{(x)} \sim 1/\lambda(x)$. The remaining statement (1.8) in (b) then follows from

$$
\begin{aligned}
&\mathbb{P}\big(\lambda(x)X^{(x)} > y\big) \\
&= \ \mathbb{P}\big(X > x + y/\lambda(x) \mid X > x\big) \ = \ \exp\big\{\Lambda(x) - \Lambda\big(x + y/\lambda(x)\big)\big\} \\
&= \ \exp\Big\{-\int_0^{y/\lambda(x)} \lambda(x + z)\,\mathrm{d}x\Big\} \ = \ \exp\Big\{-\int_0^y \frac{\lambda\big(x + u/\lambda(x)\big)}{\lambda(x)}\,\mathrm{d}u\Big\} \\
&= \ \exp\big\{-y\big(1 + o(1)\big)\big\}.
\end{aligned}
$$

$\square$

The property (1.7) is referred to as $1/\lambda(x)$ being *self-neglecting*. It is trivially verified to hold for the Weibull- and lognormal distributions, cf. Examples 1.14, 1.15.

The mean excess function will play a main role later in Section 4 in connection with finite-horizon ruin probabilities and in Section 6 in connection with tail estimation.

Notes and references Good general references for subexponential distributions are Embrechts, Klüppelberg & Mikosch [349] and Rolski *et al.* [746].

In the last decade, there has been a considerable literature on the theory of subexponential distributions. One direction is *local subexponentiality*, which in its simplest form has estimates for the density of the form $b^{*n}(x) \sim nb(x)$ and more generally gives conditions for $B^{*n}(x+y) - B^{*n}(x) \sim n\big(B(x+y) - B(x)\big)$ for any fixed y. See, e.g., Asmussen, Foss & Korshunov [77]. Another direction is variants of the definition. Some of these are slight generalizations like intermediate regular variation, following up on Cline [248], others are slightly less general classes designed typically for an ad hoc purpose of pursuing some specific line of applications. The perspective of such studies may be a matter of taste. We would like, however, to point out one specific class which has proved rather robust, the class $\mathscr{S}^* \subset \mathscr{S}$ originally introduced by Klüppelberg by the requirement

$$\int_0^x \overline{B}(x-y)\overline{B}(y)\,\mathrm{d}y \ \sim \ \mu_B\overline{B}(x)\,. \tag{1.9}$$

For the intuition, note that $\overline{B}(x-y)/\overline{B}(x) \to 1$ for any y so one expects the integral divided by $\overline{B}(x)$ to have a limit $\int_0^x \overline{B}(y)\,\mathrm{d}y = \mu_B$. However, there is nothing like a dominated convergence argument to justify this, and in fact (1.9) may fail in some exceptional cases.

2   The compound Poisson model

Consider the compound Poisson model with arrival intensity β and claim size distribution B. Let $S_t = \sum_{i=1}^{N_t} U_i - t$ be the claim surplus at time t and $M = \sup_{t \geq 0} S_t$, $\tau(u) = \inf\{t > 0;\, S_t > u\}$. We assume $\rho = \beta\mu_B < 1$ and are interested in the ruin probability $\psi(u) = \mathbb{P}(M > u) = \mathbb{P}(\tau(u) < \infty)$. Recall that B_0 denotes the stationary excess distribution, $B_0(x) = \int_0^x \overline{B}(y)\,\mathrm{d}y \,/\, \mu_B$.

Theorem 2.1 *If $B_0 \in \mathscr{S}$, then $\psi(u) \sim \dfrac{\rho}{1-\rho}\overline{B}_0(u)$.*

The proof is based upon the following lemma (stated slightly more generally than needed at present).

Lemma 2.2 *Let $Y_1, Y_2, \ldots$ be i.i.d. with common distribution $G \in \mathscr{S}$ and let K be an independent integer-valued r.v. with $\mathbb{E}z^K < \infty$ for some $z > 1$. Then $\mathbb{P}(Y_1 + \cdots + Y_K > u) \sim \mathbb{E}K\,\overline{G}(u)$.*

Proof. Recall from Section 1 that $\overline{G^{*n}}(u) \sim n\overline{G}(u)$, $u \to \infty$, and that for each $z > 1$ there is a $D < \infty$ such that $\overline{G^{*n}}(u) \leq \overline{G}(u)Dz^n$ for all u. We get

$$\frac{\mathbb{P}(Y_1 + \cdots + Y_K > u)}{\overline{G}(u)} \ = \ \sum_{n=0}^{\infty} \mathbb{P}(K = n)\frac{\overline{G^{*n}}(u)}{\overline{G}(u)} \ \to \ \sum_{n=0}^{\infty} \mathbb{P}(K = n)\cdot n \ = \ \mathbb{E}K,$$

using dominated convergence with $\sum \mathbb{P}(K = n) \, Dz^n$ as majorant. $\qquad \square$

Proof of Theorem 2.1. The Pollaczeck-Khinchine formula states that (in the set-up of Lemma 2.2) $M \overset{\mathscr{D}}{=} Y_1 + \cdots + Y_K$ where the Y_i have distribution B_0 and K is geometric with parameter ρ, $\mathbb{P}(K = k) = (1 - \rho)\rho^k$. Since $\mathbb{E}K = \rho/(1 - \rho)$ and $\mathbb{E}z^K < \infty$ whenever $\rho z < 1$, the result follows immediately from Lemma 2.2. $\quad \square$

The condition $B_0 \in \mathscr{S}$ is for all practical purposes equivalent to $B \in \mathscr{S}$. However, mathematically one must note that there exist (quite intricate) examples where $B \in \mathscr{S}$, $B_0 \notin \mathscr{S}$, as well as examples where $B \notin \mathscr{S}$, $B_0 \in \mathscr{S}$. The tail of B_0 is easily expressed in terms of the tail of B and the function $e(x)$ in Proposition 1.18,

$$\overline{B}_0(x) = \frac{1}{\mu_B} \int_x^\infty \overline{B}(y) \, dy = \frac{\overline{B}(x)\mathbb{E}X^{(x)}}{\mu_B} = \frac{e(x)\overline{B}(x)}{\mu_B}. \qquad (2.1)$$

In particular, in our three main examples (regular variation, lognormal, Weibull) one has

$$\overline{B}(x) \sim \frac{L(x)}{x^\alpha} \quad \Rightarrow \quad \overline{B}_0(x) \sim \frac{L(x)}{\mu_B(\alpha - 1)x^{\alpha-1}},$$

$$\overline{B}(x) = \overline{\Phi}\left(\frac{\log x - \mu}{\sigma}\right) \quad \Rightarrow \quad \mu_B = e^{\mu+\sigma^2/2}, \quad \overline{B}_0(x) \sim \frac{\sigma x e^{-(\log x - \mu)^2/2\sigma^2}}{e^{\mu+\sigma^2/2}(\log x)^2 \sqrt{2\pi}},$$

$$\overline{B}(x) = e^{-x^\beta} \quad \Rightarrow \quad \mu_B = \frac{\Gamma(1/\beta)}{\beta}, \quad \overline{B}_0(x) \sim \frac{1}{\Gamma(1/\beta)} x^{1-\beta} e^{-x^\beta}.$$

From this, $B_0 \in \mathscr{S}$ is immediate in the regularly varying case, and for the lognormal and Weibull cases it can be verified using Pitman's criterion (Proposition 1.13). In general it is known that $B \in \mathscr{S}^*$ is sufficient for $B_0 \in \mathscr{S}$.

Note that in these examples, B_0 is more heavy-tailed than B. In general:

Proposition 2.3 *If $B \in \mathscr{S}$, then $\overline{B}_0(x)/\overline{B}(x) \to \infty$, $x \to \infty$.*

Proof. Since $\overline{B}(x + y)/\overline{B}(x) \to 1$ uniformly in $y \in [0, a]$, we have

$$\liminf_{x\to\infty} \frac{\overline{B}_0(x)}{\overline{B}(x)} \geq \liminf_{x\to\infty} \frac{\int_x^{x+a} \overline{B}(y)dy}{\mu_B \overline{B}(x)} = \frac{a}{\mu_B}.$$

Let $a \to \infty$. $\qquad \square$

Remark 2.4 Note that for regularly varying claim size distributions, one can also use the Pollaczeck-Khinchine formula and the Tauberian Theorem A6.2

(given in the Appendix) to provide a somewhat alternative proof of Theorem 2.1. More precisely, combining IV.(3.4) and IV.(3.5) we have

$$\widehat{\psi}[-s] = \frac{\rho(\widehat{B}_0[-s] - 1)}{s(\rho\widehat{B}_0[-s] - 1)} = \rho\left(\frac{1 - \widehat{B}_0[-s]}{s}\right)\left(1 + \rho\widehat{B}_0[-s] + \rho^2\widehat{B}_0[-s]^2 + \cdots\right).$$

Assume that $\overline{B}(x) \sim L(x)/x^\alpha$ with $\alpha > 1$ and write $\alpha = n + \eta$ with $n = \lfloor\alpha\rfloor$ and $0 < \eta < 1$ (if $\eta = 0$ there are obvious amendments). Then from above

$$\overline{B}_0(x) \sim \frac{L(x)}{\mu_B(\alpha - 1)x^{\alpha-1}}, \quad x \to \infty$$

and with Theorem A6.2 this implies

$$\widehat{B}_0[-s] = 1 + \sum_{j=1}^{n-1}\frac{a_j(-s)^j}{j!} + \frac{(-1)^{n-1}}{\Gamma(-\alpha+1)}s^{\alpha-1}L(1/s)$$

for some constants a_j. Hence

$$\widehat{\psi}[-s] = \frac{\rho}{1-\rho}\left(\frac{1 - \widehat{B}_0[-s]}{s}\right)(1 + b_1 s + b_2 s^2 + \ldots)$$

for some constants b_j and after subtracting the resulting first $n - 2$ terms with powers s^k $(k = 1, \ldots, n - 2)$ the r.h.s. (and correspondingly also the l.h.s.) is regularly varying at $s = 0$ with index η. Since $\widehat{\psi}[-s]$ is the Laplace-Stieltjes transform of $\int_0^u \psi(y)\,dy$, another application of Theorem A6.2 shows that

$$\int_u^\infty \psi(y)\,dy \sim \frac{\rho}{1-\rho}\int_u^\infty \overline{B}_0(y)\,dy$$

and by the Monotone Density Theorem

$$\psi(u) \sim \frac{\rho}{1-\rho}\overline{B}_0(u).$$

$\square$

Notes and references   Theorem 2.1 was derived by several different authors and under varying assumptions. We mention here Teugels & Veraverbeke [842], von Bahr [122], Borovkov [182], Thorin & Wikstad [849], Pakes [677] and Embrechts & Veraverbeke [353].

The approximation in Theorem 2.1 is notoriously not very accurate. The problem is a usually very slow rate of convergence as $u \to \infty$. For some earlier numerical studies, see Abate, Choudhury & Whitt [1], Kalashnikov [517] and Asmussen & Binswanger [72]. E.g., in [517, p.195] there are numerical examples where $\psi(u)$ is of order 10^{-5} but Theorem 2.1 gives 10^{-10}. This shows that although the approximation is asymptotically correct in the tail, one may have to go out to values of $\psi(u)$ which are unrealistically small before the fit is reasonable. Second order terms were e.g. introduced in Abate *et al.* [1] and Baltrūnas [126, 127], but unfortunately the improvement is not very pronounced, see also Omey & Willekens [675, 676] for some related work. Based upon ideas of Hogan [473], Asmussen & Binswanger [72] suggested an approximation which is substantially better than Theorem 2.1 when u is small or moderately large. For US-Pareto and classical Pareto claim size distribution, the explicit Laplace transform in terms of an incomplete Gamma function can be used to obtain an integral representation (with non-oscillating integrand on the real line) for the ruin probability, which can be seen as an 'almost explicit' formula, see Ramsay [725, 726] and Albrecher & Kortschak [33].

In recent years, there has been a lot of research activity on higher-order asymptotic expansions of general compound distributions (under certain additional assumptions on the tail), see e.g. Geluk, Peng & de Vries [393], Borovkov & Borovkov [184], Barbe, McCormick & Zhang [131, 132], Mikosch & Nagaev [639], Kortschak & Albrecher [557] and Albrecher, Hipp & Kortschak [29]. In [29] it is also shown that a shift in the argument can substantially improve the accuracy of the first-order asymptotic approximation in Theorem 2.1, see also the Notes to Section XVI.2a. For higher-order approximations for absolute ruin probabilities, see Borovkov [183].

As any approximation valid as $u \to \infty$, the one in Theorem 2.1 can of course not be precise for small u. Olvera-Cravioto, Blanchet & Glynn [674] discuss the alternative of using the heavy traffic approximation for small and moderate u and identify the threshold where the subexponential approximation takes over.

Upper bounds for $\psi(u)$ in the heavy-tailed case can be found in Kalashnikov [517, 518] (see also Willmot & Lin [891, Sec.6.2]), but are in general quite complicated.

3 The renewal model

We consider the renewal model with claim size distribution B and interarrival distribution A as in Chapter VI. Let U_i be the ith claim, T_i the ith interarrival time and $X_i = U_i - T_i$,

$$S_n^{(d)} = X_1 + \cdots + X_n, \quad M = \sup_{\{n=0,1,\ldots\}} S_n^{(d)}, \quad \vartheta(u) = \inf\{n : S_n^{(d)} > u\}.$$

Then $\psi(u) = \mathbb{P}(M > u) = \mathbb{P}(\vartheta(u) < \infty)$. We assume positive safety loading, i.e. $\rho = \mu_B/\mu_A < 1$. The main result is:

Theorem 3.1 *Assume that* (a) *the stationary excess distribution B_0 of B is subexponential and that* (b) *B itself satisfies $\overline{B}(x - y)/\overline{B}(x) \to 1$ uniformly on* *compact y-intervals. Then*

$$\psi(u) \sim \frac{\rho}{1-\rho}\overline{B}_0(u), \quad u \to \infty. \tag{3.1}$$

[Note that (b) in particular holds if $B \in \mathscr{S}$.]

The proof is based upon the observation that also in the renewal setting, there is a representation of M similar to the Pollaczeck-Khinchine formula. To this end, let $\vartheta_+ = \vartheta(0)$ be the first ascending ladder epoch of $\{S_n^{(d)}\}$,

$$G_+(A) = \mathbb{P}\big(S_{\vartheta_+} \in A, \vartheta_+ < \infty\big) = \mathbb{P}\big(S_{\tau_+} \in A, \tau_+ < \infty\big)$$

where $\tau_+ = T_1 + \cdots + T_{\vartheta_+}$ as usual denotes the first ascending ladder epoch of the continuous time claim surplus process $\{S_t\}$. Thus G_+ is the ascending ladder height distribution (which is defective because of $\mu_B < \mu_A$). Define further $\theta = \|G_+\| = \mathbb{P}(\vartheta_+ < \infty)$. Then

$$M \stackrel{\mathscr{D}}{=} \sum_{i=1}^{K} Y_i \tag{3.2}$$

where K is geometric with parameter θ, $\mathbb{P}(K = k) = (1 - \theta)\theta^k$ and $Y_1, Y_2, \ldots$ are independent of K and i.i.d. with distribution G_+/θ (the distribution of S_{ϑ_+} given $\tau_+ < \infty$). As for the compound Poisson model, this representation will be our basic vehicle to derive tail asymptotics of M but we face the added difficulties that neither the constant θ nor the distribution of the Y_i are explicit.

Let F denote the distribution of the X_i and $\overline{F}_I$ the integrated tail, $\overline{F}_I(x) = \int_x^\infty \overline{F}(y)\,dy$, $x > 0$.

Lemma 3.2 $\overline{F}(x) \sim \overline{B}(x)$, $x \to \infty$, and hence $\overline{F}_I(x) \sim \mu_B \overline{B}_0(x)$.

Proof. By dominated convergence and (b),

$$\frac{\overline{F}(x)}{\overline{B}(x)} = \int_0^\infty \frac{\overline{B}(x+y)}{\overline{B}(x)}\,A(dy) \to \int_0^\infty 1 \cdot A(dy) = 1.$$

$\square$

The lemma implies that (3.1) is equivalent to

$$\mathbb{P}(M > u) \sim \frac{1}{|\mu_F|}\overline{F}_I(u), \quad u \to \infty, \tag{3.3}$$

and we will prove it in this form (in XIII.2, we will use the fact that the proof of (3.1) holds for a general random walk satisfying the analogues of (a), (b) and does not rely on the structure $X_i = U_i - T_i$).

Write $\overline{G}_+(x) = G_+(x, \infty) = \mathbb{P}(S_{\vartheta_+} > x, \vartheta_+ < \infty)$. Let further $\vartheta_- = \inf\{n > 0 : S_n^{(d)} \leq 0\}$ be the first descending ladder epoch, $G_-(A) = \mathbb{P}(S_{\vartheta_-} \in A)$ the descending ladder height distribution ($\|G_-\| = 1$ because of $\mu_B < \mu_A$) and let μ_{G_-} be the mean of G_-.

Lemma 3.3 $\overline{G}_+(x) \sim \overline{F}_I(x)/|\mu_{G_-}|$, $x \to \infty$.

Proof. Let $R_+(A) = \mathbb{E}\sum_0^{\vartheta_+ - 1} I(S_n^{(d)} \in A)$ denote the pre-ϑ_+ occupation measure and let $U_- = \sum_0^\infty G_-^{*n}$ be the renewal measure corresponding to G_-. Then

$$\overline{G}_+(x) = \int_{-\infty}^0 \overline{F}(x - y) R_+(dy) = \int_{-\infty}^0 \overline{F}(x - y) U_-(dy)$$

(the first identity is obvious and the second follows since an easy time reversion argument shows that $R_+ = U_-$, cf. A2). The heuristics is now that because of (b), the contribution from the interval $(-N, 0]$ to the integral is $O(\overline{F}(x)) = o(\overline{F}_I(x))$, whereas for large y, $U_-(dy)$ is close to Lebesgue measure on $(-\infty, 0]$ normalized by $|\mu_{G_-}|$ so that we should have

$$\overline{G}_+(x) \sim \frac{1}{|\mu_{G_-}|} \int_{-\infty}^0 \overline{F}(x - y)\, dy = \frac{1}{|\mu_{G_-}|}\overline{F}_I(x).$$

We now make this precise. If G_- is non-lattice, then by Blackwell's renewal theorem $U_-(-n - 1, -n] \to 1/|\mu_{G_-}|$. In the lattice case, we can assume that the span is 1 and then the same conclusion holds since then $U_-(-n - 1, -n]$ is just the probability of a renewal at n.

Given ϵ, choose N such that $\overline{F}(n - 1)/\overline{F}(n) \leq 1 + \epsilon$ for $n \geq N$ (this is possible by (b) and Lemma 3.2), and that $U_-(-n - 1, -n] \leq (1 + \epsilon)/|\mu_{G_-}|$ for $n \geq N$. We then get

$$\limsup_{x \to \infty} \frac{\overline{G}_+(x)}{\overline{F}_I(x)}$$

$$\leq \limsup_{x \to \infty} \int_{-N}^0 \frac{\overline{F}(x - y)}{\overline{F}_I(x)} U_-(dy) + \limsup_{x \to \infty} \int_{-\infty}^{-N} \frac{\overline{F}(x - y)}{\overline{F}_I(x)} U_-(dy)$$

$$\leq \limsup_{x \to \infty} \frac{\overline{F}(x)}{\overline{F}_I(x)} U_-(-N, 0]$$

$$+ \limsup_{x \to \infty} \frac{1}{\overline{F}_I(x)} \sum_{n=N}^\infty \overline{F}(x + n) U_-(-n - 1, -n]$$

$$\leq \quad 0 + \limsup_{x \to \infty} \frac{1}{\overline{F}_I(x)} \frac{1+\epsilon}{|\mu_{G_-}|} \sum_{n=N}^{\infty} \overline{F}(x+n)$$

$$\leq \quad \frac{(1+\epsilon)^2}{|\mu_{G_-}|} \limsup_{x \to \infty} \frac{1}{\overline{F}_I(x)} \int_N^{\infty} \overline{F}(x+y)\,dy$$

$$= \quad \frac{(1+\epsilon)^2}{|\mu_{G_-}|} \limsup_{x \to \infty} \frac{\overline{F}_I(x+N)}{\overline{F}_I(x)} \quad = \quad \frac{(1+\epsilon)^2}{|\mu_{G_-}|}.$$

Here in the third step we used that (b) implies $\overline{B}(x)/\overline{B}_0(x) \to 0$ and hence $\overline{F}(x)/\overline{F}_I(x) \to 0$, and in the last that $\overline{F}_I$ is asymptotically proportional to $B_0 \in \mathscr{S}$. Similarly,

$$\liminf_{x \to \infty} \frac{\overline{G}_+(x)}{\overline{F}_I(x)} \geq \frac{(1-\epsilon)^2}{|\mu_{G_-}|}.$$

Letting $\epsilon \downarrow 0$, the proof is complete.   □

Proof of Theorem 3.1. By Lemma 3.3, $\mathbb{P}(Y_i > x) \sim \overline{F}_I(x)/(\theta|\mu_{G_-}|)$. Hence using dominated convergence precisely as for the compound Poisson model, (3.2) yields

$$\mathbb{P}(M > u) \sim \sum_{k=1}^{\infty} (1-\theta)\theta^k\, k\, \frac{\overline{F}_I(u)}{\theta|\mu_{G_-}|} \quad = \quad \frac{\overline{F}_I(u)}{(1-\theta)|\mu_{G_-}|}.$$

Differentiating the Wiener-Hopf factorization identity (A.9)

$$1 - \widehat{F}[s] \quad = \quad \big(1 - \widehat{G}_-[s]\big)\big(1 - \widehat{G}_+[s]\big)$$

and letting $s = 0$ yields

$$-\mu_F \quad = \quad -(1-1)\widehat{G}'_+[0] - \big(1 - \|G_+\|\big)\mu_{G_-} \quad = \quad -(1-\theta)\mu_{G_-}.$$

Therefore by Lemma 3.2,

$$\frac{\overline{F}_I(u)}{(1-\theta)|\mu_{G_-}|} \quad \sim \quad \frac{\mu_B \overline{B}_0(u)}{\mu_A - \mu_B} \quad = \quad \frac{\rho \overline{B}_0(u)}{1-\rho}.$$

□

We conclude by a lemma needed in XIII.2:

Lemma 3.4 *For any* $a < \infty$, $\mathbb{P}(M > u, S_{\vartheta(u)} - S_{\vartheta(u)-1} \leq a) = o\big(\overline{F}_I(u)\big)$.

Proof. Let $\omega(u) = \inf\big\{n : S_n^{(d)} \in (u-a, u), M_n \leq u\big\}$. Then

$$\mathbb{P}\big(M \in (u-a, u)\big) \geq \mathbb{P}\big(\omega(u) < \infty\big)\big(1 - \psi(0)\big).$$

On the other hand, on the set $\{M > u,\ S_{\vartheta(u)} - S_{\vartheta(u)-1} \le a\}$ we have $\omega(u) < \infty$, and $\{S_{\omega(u)+n} - S_{\omega(u)}\}_{n=0,1,\dots}$ must attain a maximum > 0 so that

$$
\begin{aligned}
\mathbb{P}\big(M > u,\ S_{\vartheta(u)} - S_{\vartheta(u)-1} \le a\big) &\le \mathbb{P}\big(\omega(u) < \infty\big)\psi(0) \\
&\le \frac{\psi(0)}{1 - \psi(0)}\mathbb{P}\big(M \in (u-a, u)\big).
\end{aligned}
$$

But since $\mathbb{P}(M > u - a) \sim \mathbb{P}(M > u)$, we have

$$
\mathbb{P}\big(M \in (u-a, u)\big) \;=\; o\big(\mathbb{P}(M > u)\big) \;=\; o\big(\overline{F}_I(u)\big).
$$

$\square$

Notes and references Theorem 3.1 is due to Embrechts & Veraverbeke [353], with roots in von Bahr [122], Pakes [677] and Teugels & Veraverbeke [842]. Asymptotic results for maxima of random walks with heavy-tailed increments again carry over to corresponding statements for $\psi(u, T)$ in the renewal model, cf. for instance Veraverbeke & Teugels [865] and more recently Baltrūnas [128]; see also Baltrūnas & Klüppelberg [129]. Further results on tails of the discounted aggregate claims in this model are given in Hao & Tang [447]. Wei & Yang [880] extend the integral representation for the ruin probability for US Pareto claims to an Erlang renewal model.

A recent reference containing much relevant information on heavy-tailed asymptotics for random walks is Borovkov & Borovkov [185].

4 Finite-horizon ruin probabilities

We consider the compound Poisson model with $\rho = \beta \mu_B < 1$ and the stationary excess distribution B_0 subexponential. Then $\psi(u) \sim \rho/(1-\rho)\overline{B}_0(u)$, cf. Theorem 2.1. The asymptotic behavior of the finite-horizon ruin probability $\psi(u, T)$ for fixed T is trivially given by

$$
\psi(u, T) \sim \beta T \overline{B}(u), \tag{4.1}
$$

since the coarse inequality $\mathbb{P}(A_T - T > u) \le \psi(u, T) \le \mathbb{P}(A_T > u)$ and Proposition 1.5 in this case already suffice to see that $\psi(u, T) \sim \mathbb{P}(A_T > u)$ from which (4.1) follows from Lemma 2.2. It is therefore clear that one can expect deeper results only for the case when the time horizon itself scales with u, and this will be the subject of this section.

As usual, $\tau(u)$ is the time of ruin and as in V.7, we let $\mathbb{P}^{(u)} = \mathbb{P}(\cdot \mid \tau(u) < \infty)$. The main result of this section, Theorem 4.4, states that under mild additional conditions, there exist constants $e(u)$ such that the $\mathbb{P}^{(u)}$-distribution of $\tau(u)/e(u)$ has a limit which is either Pareto (when B is regularly varying) or exponential

(for B's such as the lognormal or DFR Weibull); this should be compared with
the normal limit for the light-tailed case, cf. V.4. Combined with the approximation for $\psi(u)$, this then easily yields approximations for the finite horizon
ruin probabilities $\psi(u, e(u))$ (Corollary 4.7).

We start by reviewing some general facts which are fundamental for the
analysis. Essentially, the discussion provides an alternative point of view to
some results in Chapter V, in particular Proposition V.2.3.

4a   Excursion theory for Markov processes

Let until further notice $\{S_t\}$ be an arbitrary Markov process with state space
E (we write $\mathbb{P}_x$ when $S_0 = x$) and m a stationary measure, i.e. m is a (σ-finite)
measure on E such that

$$\int_E m(\mathrm{d}x)\mathbb{P}_x(S_t \in A) \;=\; m(A) \tag{4.2}$$

for all measurable $A \subseteq E$ and all $t > 0$. Then there is a Markov process $\{R_t\}$
on E such that

$$\int_E m(\mathrm{d}x)h(x)\mathbb{E}_x k(R_t) \;=\; \int_E m(\mathrm{d}y)k(y)\mathbb{E}_y h(S_t) \tag{4.3}$$

for all bounded measurable functions h, k on E; in the terminology of general
Markov process theory, $\{S_t\}$ and $\{R_t\}$ are in *classical duality w.r.t. m.*

The simplest example is a discrete time discrete state space chain, where we
can take h, k as indicator functions, for states i, j, say, and (4.3) with $t = 1$
means $m_i r_{ij} = m_j s_{ji}$ where r_{ij}, s_{ij} are the transition probabilities for $\{S_t\}$,
resp. $\{R_t\}$. Thus, a familiar case is time reversion (here m is the stationary
distribution); but the example of relevance for us is the following:

Proposition 4.1 *A compound Poisson risk process $\{R_t\}$ and its associated
claim surplus process $\{S_t\}$ are in classical duality w.r.t. Lebesgue measure.*

Proof. Starting from $R_0 = x$, R_t is distributed as $x + t - \sum_1^{N_t} U_i$, and starting
from $S_0 = y$, S_t is distributed as $y - t + \sum_1^{N_t} U_i$ (note that we allow x, y to vary
in the whole of $\mathbb{R}$ and not as usual impose the restrictions $x \geq 0$, $y = 0$). Let G
denote the distribution of $\sum_1^{N_t} U_i - t$. Then (4.3) means

$$\iint h(x)k(x - z)\,\mathrm{d}x\,G(\mathrm{d}z) \;=\; \iint h(y + z)\,k(y)\,\mathrm{d}y\,G(\mathrm{d}z)\,.$$

The equality of the l.h.s. to the r.h.s. follows by the substitution $y = x - z$.   □

For $F \subset E$, an *excursion in F starting from* $x \in F$ is the (typically finite) piece of sample path[2]

$$\{S_t\}_{0 \leq t < \omega(F^c)} \,\big|\, S_0 = x \quad \text{where} \;\; \omega(F^c) = \inf\{t > 0 : S_t \notin F\}\,.$$

We let Q_x^S be the corresponding distribution and

$$Q_{x,y}^S \;=\; Q_x^S\big(\,\cdot\,\big|\, S_{\omega(F^c)-} = y,\, \omega(F^c) < \infty\big)\,,$$

$y \in F$ (in discrete time, $S_{\omega(F^c)-}$ should be interpreted as $S_{\omega(F^c)-1}$). Thus, $Q_{x,y}^S$ is the distribution of an excursion of $\{S_t\}$ conditioned to start in $x \in F$ and terminate in $y \in F$. Q_x^R and $Q_{x,y}^R$ are defined similarly, and we let $\widetilde{Q}_{x,y}^S$ refer to the time reversed excursion. That is,

$$\widetilde{Q}_{x,y}^S(\cdot) \;=\; \mathbb{P}\left(\{S_{\omega(F^c)-t-}\}_{0 \leq t < \omega(F^c)} \in \cdot \,\Big|\, S_0 = x,\, S_{\omega(F^c)-} = y\right)\,.$$

Theorem 4.2 $\widetilde{Q}_{x,y}^S \;=\; Q_{y,x}^R$.

The theorem is illustrated in Fig. X.1 for the case $F = (-\infty, 0]$, $x = 0$.

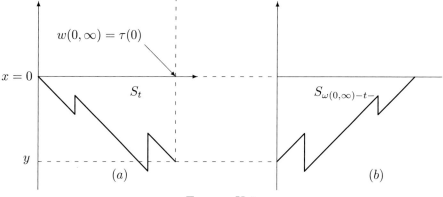

FIGURE X.1

The sample path in (a) is the excursion of $\{S_t\}$ conditioned to start in $x = 0$ and to end in $y > 0$, the one in (b) is the time reversed path. The theorem states that the path in (b) has the same distribution as an excursion of $\{R_t\}$ conditioned to start in $y < 0$ and to end in $x = 0$. But in the risk theory example (corresponding to which the sample paths are drawn), this simply means *the distribution of the path of $\{R_t\}$ starting from y and stopped when 0 is hit*. In particular:

---

[2]In general Markov process theory, a main difficulty is to make sense to such excursions also when $\mathbb{P}_x(\omega(F^c) = 0) = 1$. Say $\{S_t\}$ is reflected Brownian motion on $[0, \infty)$, $x = 0+$ and $F = (0, \infty)$. For the present purposes it suffices, however, to consider only the case $\mathbb{P}_x(\omega(F^c) = 0) = 0$.

Corollary 4.3 *The distribution of $\tau(0)$ given $\tau(0) < \infty$, $S_{\tau(0)-} = y < 0$ is the same as the distribution of $\omega(-y)$ where $\omega(z) = \inf\{t > 0 : R_t = z\}$, $z > 0$.*

[Note that $\omega(z) < \infty$ a.s. when $\rho = \beta\mu_B < 1$.]

Proof of Theorem 4.2. We consider the discrete time discrete state space case only (well-behaved cases such as the risk process example can then easily be handled by discrete approximations). We can then view $Q^S_{x,y}$ as a measure on all strings of the form $i_0 i_1 \ldots i_n$ with $i_0, i_1, \ldots, i_n \in F$, $i_0 = x$, $i_n = y$,

$$Q^S_{x,y}(i_0 i_1 \ldots i_n) = \frac{\mathbb{P}_x\big(S_1 = i_1, \ldots, S_n = i_n = y; S_{n+1} \in F^c\big)}{\mathbb{P}_x\big(\omega(F^c) < \infty, S_{\omega(F^c)-1} = y\big)};$$

note that

$$\mathbb{P}_x\big(\omega(F^c) < \infty, S_{\omega(F^c)-1} = y\big)$$
$$= \sum_{n=1}^{\infty} \sum_{i_1, \ldots, i_{n-1} \in F} \mathbb{P}_x\big(S_1 = i_1, \ldots, S_n = i_n = y; S_{n+1} \in F^c\big).$$

Similarly, $\widetilde{Q}^S_{x,y}$ and $Q^R_{y,x}$ are measures on all strings of the form $i_0 i_1 \ldots i_n$ with $i_0, i_1, \ldots, i_n \in F$, $i_0 = y$, $i_n = x$,

$$Q^R_{y,x}(i_0 i_1 \ldots i_n) = \frac{\mathbb{P}_y\big(R_1 = i_1, \ldots, R_n = i_n = x; R_{n+1} \in F^c\big)}{\mathbb{P}_y\big(\omega(F^c) < \infty, R_{\omega(F^c)-1} = y\big)}$$

and $\widetilde{Q}^S_{x,y}(i_0 i_1 \ldots i_n) = Q^S_{x,y}(i_n i_{n-1} \ldots i_0)$.

To show $Q^R_{y,x}(i_0 i_1 \ldots i_n) = Q^S_{x,y}(i_n i_{n-1} \ldots i_0)$ when $i_0, i_1, \ldots, i_n \in F$, $i_0 = y$, $i_n = x$, note first that

$$\mathbb{P}_y\big(R_1 = i_1, \ldots, R_n = i_n = x; R_{n+1} \in F^c\big)$$
$$= r_{i_0 i_1} r_{i_1 i_2} \ldots r_{i_{n-1} i_n} \sum_{j \in F^c} r_{xj}$$
$$= \frac{m_{i_1} s_{i_1 i_0}}{m_{i_0}} \cdot \frac{m_{i_2} s_{i_2 i_1}}{m_{i_1}} \cdots \frac{m_{i_n} s_{i_n i_{n-1}}}{m_{i_{n-1}}} \sum_{j \in F^c} \frac{m_j s_{jx}}{m_x}$$
$$= \frac{1}{m_y} s_{i_n i_{n-1}} \cdots s_{i_1 i_0} \sum_{j \in F^c} m_j s_{jx}.$$

Thus

$$Q^R_{y,x}(i_0 i_1 \ldots i_n) = \frac{s_{x i_{n-1}} \cdots s_{i_1 y}}{\displaystyle\sum_{k=1}^{\infty} \sum_{i_1, \ldots, i_{k-1} \in F} s_{x i_{k-1}} \cdots s_{i_1 y}}.$$

Similarly but easier

$$
\begin{aligned}
Q^S_{x,y}(i_n i_{n-1} \ldots i_0) &= \frac{s_{xi_{n-1}} \cdots s_{i_1 y} \sum\limits_{j \in F^c} s_{yj}}{\sum\limits_{k=1}^{\infty} \sum\limits_{i_1,\ldots,i_{k-1} \in F} s_{xi_{k-1}} \cdots s_{i_1 y} \sum\limits_{j \in F^c} s_{yj}} \\
&= \frac{s_{xi_{n-1}} \cdots s_{i_1 y}}{\sum\limits_{k=1}^{\infty} \sum\limits_{i_1,\ldots,i_{k-1} \in F} s_{xi_{k-1}} \cdots s_{i_1 y}}.
\end{aligned}
$$

□

4b The time to ruin

Our approach to the study of the asymptotic distribution of the ruin time is to decompose the path of $\{S_t\}$ in ladder segments. To clarify the ideas we first consider the case where ruin occurs already in the first ladder segment, that is, the case $\tau(0) < \infty, S_{\tau(0)} > u$.

Let $Y = Y_1 = S_{\tau_+(1)}$ be the value of the claim surplus process just after the first ladder epoch, $Z = Z_1 = S_{\tau_+(1)-}$ the value just before the first ladder epoch (these r.v.'s are defined w.p. 1 w.r.t. $\mathbb{P}^{(0)}$), see Fig. X.2.

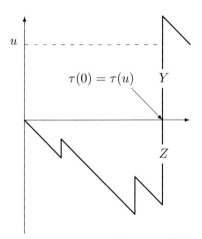

FIGURE X.2

The distribution of (Y, Z) is described in Theorem IV.2.2. The formulation relevant for the present purposes states that Y has distribution B_0 and that conditionally upon $Y = y$, Z follows the excess distribution $B^{(y)}$ given by $\overline{B^{(y)}}(x) = \overline{B}(y + x)/\overline{B}(y)$.

We are interested in the conditional distribution of $\tau(u) = \tau(0)$ given

$$\{\tau(0) < \infty, S_{\tau(0)} > y\} \;=\; \{\tau(0) < \infty, Y > y\},$$

that is, the distribution w.r.t. $\mathbb{P}^{(u,1)} = \mathbb{P}(\cdot \mid \tau(0) < \infty,\, Y > u)$. Now the $\mathbb{P}^{(u,1)}$-distribution of $Y - u$ is $B_0^{(u)}$. That is, the $\mathbb{P}^{(u,1)}$-density of Y is $\overline{B}(y)/[\mu_B \overline{B}_0(u)]$, $y > u$. $B_0^{(u)}$ is also the $\mathbb{P}^{(u,1)}$-distribution of Z since

$$
\begin{aligned}
\mathbb{P}(Z > a \mid Y > u) &= \int_u^\infty \frac{\overline{B}(y)}{\mu_B \overline{B}_0(u)} \cdot \frac{\overline{B}(y + a)}{\overline{B}(y)}\, dy \\
&= \int_{u+a}^\infty \frac{\overline{B}(z)}{\mu_B \overline{B}_0(u)}\, dy \;=\; \overline{B_0^{(u)}}(a)\,.
\end{aligned}
$$

Let $\{\omega(z)\}_{z \geq 0}$ be defined by $\omega(z) = \inf\{t > 0 : R_t = z\}$ where $\{R_t\}$ is independent of $\{S_t\}$, in particular of Z. Then Corollary 4.3 implies that the $\mathbb{P}^{(u,1)}$-distribution of $\tau(u) = \tau(0)$ is that of $\omega(Z)$. Now $B_0 \in \mathscr{S}$ implies that the $\overline{B_0^{(u)}}(a) \to 0$ for any fixed a, i.e. $\mathbb{P}(Z \leq a \mid Y > u) \to 0$. Since $\omega(z)/z \overset{\text{a.s.}}{\to} 1/(1 - \rho)$, $z \to \infty$, it therefore follows that $\tau(u)/Z$ converges in $\mathbb{P}^{(u,1)}$-probability to $1/(1 - \rho)$.

Since the conditional distribution of Z is known (viz. $B_0^{(u)}$), this in principle determines the asymptotic behavior of $\tau(u)$. However, a slight rewriting may be more appealing. Recall the definition of the auxiliary function $e(x)$ in Section 1. It is straightforward that under the conditions of Proposition 1.18(c)

$$\overline{B_0^{(u)}}\big(ye(u)\big) \;\to\; \mathbb{P}(W > y) \tag{4.4}$$

where the distribution of W is Pareto with mean one in case (a) and exponential with mean one in case (b). That is, $Z/e(u) \to W$ in $\mathbb{P}^{(u,1)}$-distribution. $\tau(u)/Z \to 1/(1 - \rho)$ then yields the final result $\tau(u)/e(u) \to W/(1 - \rho)$ in $\mathbb{P}^{(u,1)}$-distribution.

We now turn to the general case and will see that this conclusion is also true in $\mathbb{P}^{(u)}$-distribution:

Theorem 4.4 *Assume that $B_0 \in \mathscr{S}$ and that (4.4) holds. Then $\tau(u)/e(u) \to W/(1 - \rho)$ in $\mathbb{P}^{(u)}$-distribution.*

In the proof, let $\tau_+(1) = \tau(0), \tau_+(2), \ldots$ denote the ladder epochs and let Y_k, Z_k be defined similarly as $Y = Y_1, Z = Z_1$ but relative to the kth ladder segment, cf. Fig. X.3. Then, conditionally upon $\tau_+(n) < \infty$, the random vectors $(Y_1, Z_1), \ldots, (Y_n, Z_n)$ are i.i.d. and distributed as (Y, Z).

We let $K(u) = \inf \{n = 1, 2, \ldots : \tau_+(n) < \infty, Y_1 + \cdots + Y_n > u\}$ denote the number of ladder steps leading to ruin and $\mathbb{P}^{(u,n)} = \mathbb{P}(\cdot \mid \tau(u) < \infty, K(u) = n)$. The idea is now to observe that if $K(u) = n$, then by the subexponential property Y_n must be large, i.e. $> u$ with high probability, and $Y_1, \ldots, Y_{n-1}$ 'typical'. Hence Z_n must be large and $Z_1, \ldots, Z_{n-1}$ 'typical' which implies that the first $n-1$ ladder segment must be short and the last long; more precisely, the duration $\tau_+(n) - \tau_+(n-1)$ of the last ladder segment can be estimated by the same approach as we used above when $n = 1$, and since it dominates the first $n-1$, we get the same asymptotics as when $n = 1$.

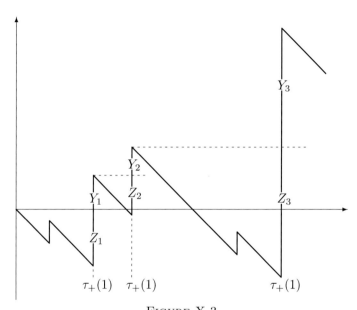

FIGURE X.3

In the following, $\|\cdot\|$ denotes the total variation norm between probability measures and $\otimes$ product measure.

Lemma 4.5 $\left\| \mathbb{P}^{(u,n)}\big((Y_1, \ldots, Y_{n-1}, Y_n - u) \in \cdot \big) - B_0^{\otimes(n-1)} \otimes B_0^{(u)} \right\| \to 0.$

Proof. We shall use the easily proved fact that if $A'(u), A''(u)$ are events such that $\mathbb{P}\big(A'(u) \Delta A''(u)\big) = o\big(\mathbb{P}(A'(u))\big)$ (Δ = symmetrical difference of events), then

$$\left\| \mathbb{P}\big(\cdot \mid A'(u)\big) - \mathbb{P}\big(\cdot \mid A''(u)\big) \right\| \to 0.$$

Taking $A'(u) = \{Y_n > u\}$,

$$A''(u) \;=\; \{K(u) = n\} \;=\; \{Y_1 + \cdots + Y_{n-1} \le u, \; Y_1 + \cdots + Y_n > u\},$$

the condition on $A'(u) \,\Delta\, A''(u)$ follows from B_0 being subexponential (Proposition 1.2, suitably adapted). Further, $\mathbb{P}(\cdot \mid A'(u)) \;=\; \mathbb{P}^{(u,n)}$,

$$\mathbb{P}(Y_1, \ldots, Y_{n-1}, Y_n - u) \in \cdot \mid A'(u)) \;=\; B_0^{\otimes(n-1)} \otimes B_0^{(u)}.$$

$\square$

Lemma 4.6 $\left\| \mathbb{P}^{(u,n)}\big((Z_1, \ldots, Z_n) \in \cdot\big) - B_0^{\otimes(n-1)} \otimes B^{(u)} \right\| \;\to\; 0.$

Proof. Let $(Y_1', Z_1'), \ldots, (Y_n', Z_n')$ be independent random vectors such that the conditional distribution of Z_k' given $Y_k' = y$ is $B^{(y)}$, $k = 1, \ldots, n$, and that Y_k' has marginal distribution B_0 for $k = 1, \ldots, n-1$ and $Y_n' - u$ has distribution $B_0^{(u)}$. That is, the density of Y_n' is $\overline{B}(y)/[\mu_B \overline{B}_0(u)]$, $y > u$. The same calculation as given above when $n = 1$ shows then that the marginal distribution of Z_n' is $B_0^{(u)}$. Similarly (replace u by 0), the marginal distribution of Z_k' is B_0 for $k < n$, and clearly $Z_1', \ldots, Z_n'$ are independent. Now use that if the conditional distribution of $\boldsymbol{Z}'$ given $\boldsymbol{Y}'$ is the same as the conditional distribution of $\boldsymbol{Z}$ given $\boldsymbol{Y}$ and $\|\mathbb{P}(\boldsymbol{Y} \in \cdot) - \mathbb{P}(\boldsymbol{Y}' \in \cdot)\| \to 0$, then $\|\mathbb{P}(\boldsymbol{Z} \in \cdot) - \mathbb{P}(\boldsymbol{Z}' \in \cdot)\| \to 0$ (here $\boldsymbol{Y}, \boldsymbol{Y}', \boldsymbol{Z}, \boldsymbol{Z}'$ are arbitrary random vectors, in our example $\boldsymbol{Y} = (Y_1, \ldots, Y_n)$ etc.). $\square$

Proof of Theorem 4.4. The first step is to observe that $K(u)$ has a proper limit distribution w.r.t. $\mathbb{P}^{(u)}$ since by Theorem 2.1,

$$\mathbb{P}^{(u)}(K(u) = n) \;=\; \frac{1}{\psi(u)} \rho^n \mathbb{P}\big(Y_1 + \cdots + Y_{n-1} \le u, \; Y_1 + \cdots + Y_n > u\big)$$

$$\sim \;\; \frac{1}{\rho/(1-\rho)\,\overline{B}_0(u)} \rho^n \mathbb{P}(Y_n > u) \;=\; (1 - \rho)\rho^{n-1}$$

for $n = 1, 2, \ldots$. It therefore suffices to show that the $\mathbb{P}^{(u,n)}$-distribution of $\tau(u)$ has the asserted limit. Let $\{\omega_1(z)\}, \ldots, \{\omega_n(z)\}$ be i.i.d. copies of $\{\omega(z)\}$. Then according to Section 4a, the $\mathbb{P}^{(u,n)}$-distribution of $\tau(u)$ is the same as the $\mathbb{P}^{(u,n)}$-distribution of $\omega_1(Z_1) + \cdots + \omega_n(Z_n)$. By Lemma 4.6, $\omega_k(Z_k)$ has a proper limit distribution as $u \to \infty$ for $k < n$, whereas $\omega_n(Z_n)$ has the same limit behavior as when $n = 1$ (cf. the discussion just before the statement of Theorem 4.4). Thus

$$\mathbb{P}^{(u,n)}\big(\tau(u)/e(u) > y\big) \;=\; \mathbb{P}^{(u,n)}\big([\omega_1(Z_1) + \cdots + \omega_n(Z_n)]/e(u) > y\big)$$

$$\sim \;\; \mathbb{P}^{(u,n)}\big(\omega_n(Z_n)/e(u) > y\big) \;\to\; \mathbb{P}(W/(1-\rho) > y).$$

$\square$

Corollary 4.7 $\psi(u, e(u)T) \sim \dfrac{\rho}{1-\rho} \overline{B}_0(u) \cdot \mathbb{P}\big(W/(1-\rho) \leq T\big).$

For a growth rate $d(u)T$ of the time horizon with $d(u) = o\big(e(u)\big)$, Corollary 4.7 implies $\psi\big(u, d(u)T\big) = o\big(\psi(u)\big)$. The following theorem gives some more explicit information for this case and identifies the bridge between the asymptotic behavior of $\psi(u, T)$ of the order of $\overline{B}(u)$ (cf. (4.1)) and the one of $\psi\big(u, e(u)T\big)$ of the order of $\overline{B}_0(u)$ (which is already the order of $\psi(u)$).

Theorem 4.8 *If* $B \in \mathcal{S}$ *and* $B_0 \in \mathcal{S}$, *then for* $d(u) \uparrow \infty$ *with* $d(u) = o\big(e(u)\big)$,

$$\psi\big(u, d(u)T\big) \sim \beta \, \overline{B}(u) \, d(u) \, T.$$

Proof. Consider the random walk $A_n = \sum_{i=1}^{n} \xi_i$ with $\mathbb{E}(\xi_i) = 0$ and distribution function $F(x) = \mathbb{P}(\xi_i \leq x) \in \mathcal{S}$ and also its stationary excess distribution $F_0 \in \mathcal{S}$. Let $M_\sigma^c := \max_{n \leq \sigma}(A_n - c\,n)$ for some stopping time σ of the random walk and some constant $c \geq 0$. According to a result of Foss, Palmowski & Zachary [366], one then has

$$\mathbb{P}(M_\sigma^c > u) \sim \sum_{n \geq 1} \mathbb{P}(\sigma \geq n)\overline{F}(u + c\,n) \tag{4.5}$$

as $u \to \infty$, uniformly over all stopping times σ. Let h be fixed and choose $\xi_i = \sum_{j=1}^{N_h} U_j - \beta\mu_B h$ (implying $F \in \mathcal{S}$ and $F_0 \in \mathcal{S}$) and furthermore $\sigma = d(u)\,T$. Denote the ruin probability of the discrete-time process $R_n^{(h)}$ ($n \in \mathbb{N}$) (i.e. the Cramér-Lundberg process viewed at time points $n\,h$ only) by $\psi^{(h)}$. Relation (4.5) then translates into

$$\psi^{(h)}\big(u, d(u)\,T\big) \sim \sum_{n \geq 1} \mathbb{P}\big(d(u)\,T \geq nh\big)\overline{F}\big(u + (1 - \beta\mu_B)nh\big)$$

$$= \sum_{1 \leq n \leq d(u)\,T/h} \overline{F}\big(u + (1 - \beta\mu_B)nh\big).$$

It follows that an asymptotic upper bound for $\psi^{(h)}\big(u, d(u)\,T\big)$ is

$$\frac{d(u)\,T}{h} \overline{F}\big(u + (1 - \beta\mu_B)h\big) = \frac{d(u)\,T}{h} \mathbb{P}\Big(\sum_{j=1}^{N(h)} U_j > u + h\Big)$$

$$\sim \beta\,d(u)\,T\,\overline{B}(u + h) \sim \beta\,d(u)\,T\,\overline{B}(u).$$

and similarly an asymptotic lower bound is

$$\frac{d(u)\,T}{h}\,\overline{F}\bigl(u + (1 - \beta\mu_B)\,h\,d(u)T\bigr)$$

$$= \frac{d(u)\,T}{h}\,\mathbb{P}\Bigl(\sum_{j=1}^{N(h)} U_j > u + (1 - \beta\mu_B)h\,d(u)\,T + \beta\mu_B\,h\Bigr)$$

$$\sim \beta\,d(u)\,T\,\overline{B}\bigl(u + (1 - \beta\mu_B)\,h\,d(u)\,T + \beta\mu_B\,h\bigr)$$

$$\sim \beta\,d(u)\,T\,\overline{B}(u),$$

where the last asymptotic relation uses $\overline{B}\bigl(u + d(u)\bigr) \sim \overline{B}(u)$, which holds for $d(u) = \mathrm{o}\bigl(e(u)\bigr)$ under the stated assumptions on B (this property in fact characterizes the special role of the mean excess function $e(u)$ in this context). Thus $\psi^{(h)}\bigl(u, d(u)\,T\bigr) \sim \beta\,d(u)\,T\,\overline{B}(u)$. From

$$\max_{t \le \lfloor d(u)T/h\rfloor\,h}\,(A_t - t) \ge \max_{n \le d(u)T/h}\,(A_{nh} - n\,h) \ge \max_{t \le \lfloor d(u)T/h\rfloor\,h}\,(A_t - t) - h$$

one finally observes that for $h \to 0$, $\psi^{(h)}\bigl(u, d(u)\,T\bigr)$ can be replaced by $\psi(u, d(u)T)$.
$\square$

Notes and references   Excursion theory for general Markov processes is a fairly abstract and advanced topic. For Theorem 4.2, see Fitzsimmons [363], in particular his Proposition 2.1.

Most of the results of Section 5b are from Asmussen & Klüppelberg [86] who also treated the renewal model and gave a sharp total variation limit result. Extensions to the Markov-modulated model of Chapter VII are in Asmussen & Højgaard [80] and to Lévy processes in Klüppelberg, Kyprianou & Maller [543]. Theorem 4.8 can be found in Albrecher & Asmussen [12].

For extensions of (4.1) that hold uniformly for t in renewal models, see Tang [828] and Leipus & Siaulys [578].

Asmussen & Teugels [107] studied approximations of $\psi(u, T)$ when $T \to \infty$ with u fixed; the results only cover the regularly varying case.

5   Reserve-dependent premiums

We consider the model of Chapter VIII with Poisson arrivals at rate β, claim size distribution B, and premium rate $p(x)$ at level x of the reserve.

Theorem 5.1 *Assume that B is subexponential and that $p(x) \to \infty$, $x \to \infty$. Then*

$$\psi(u) \;\sim\; \beta \int_u^\infty \frac{\overline{B}(y)}{p(y)}\,\mathrm{d}y\,. \tag{5.1}$$

The key step in the proof is the following lemma on the cycle maximum of the associated storage process $\{V_t\}$, cf. Corollary III.2.2. Assume for simplicity that $\{V_t\}$ regenerates in state 0, i.e. that $\int_0^\epsilon p(x)^{-1}\,dx < \infty$, and define the cycle as

$$\sigma = \inf\left\{t > 0 : V_t = 0, \max_{0 \le s \le t} V_s > 0 \,\middle|\, V_0 = 0\right\}.$$

Lemma 5.2 *Define* $M_\sigma = \sup_{0 \le t < \sigma} V_t$. *Then* $\mathbb{P}(M_\sigma > u) \sim \beta\,\mathbb{E}\sigma \cdot \overline{B}(u)$.

The heuristic motivation is the usual in the heavy-tailed area, that M_σ becomes large as consequence of one big jump. The form of the result then follows by noting that the process has mean time $\mathbb{E}\sigma$ to make this big jump and that it then occurs with intensity $\beta\overline{B}(u)$. More precisely, one expects the level y from which the big jump occurs to be $O(1)$; the probability that this exceeds u is then $\overline{B}(u - y) \sim \overline{B}(u)$. The rigorous proof is, however, non-trivial and we refer to Asmussen [64] (with a gap of that paper being filled in Asmussen *et al.* [83]).

Proof of Theorem 5.1. We will show that the stationary density $f(x)$ of $\{V_t\}$ satisfies

$$f(x) \sim \frac{\beta\overline{B}(x)}{p(x)}. \tag{5.2}$$

We then get

$$\psi(u) = \mathbb{P}(V > u) = \int_u^\infty f(y)\,dy \sim \beta \int_u^\infty \frac{\overline{B}(y)}{p(y)}\,dy,$$

and the result follows.

Define $D(u)$ as the steady-state rate of downcrossings of $\{V_t\}$ of level u and $D_\sigma(u)$ as the expected number of downcrossings of level u during a cycle. Then $D(u) = f(u)p(u)$ and, by regenerative process theory, $D(u) = D_\sigma(u)/\mu$. Further the conditional distribution of the number of downcrossings of u during a cycle given $M_\sigma > u$ is geometric with parameter $q(u) = \mathbb{P}(M_\sigma > u \mid V_0 = u)$. Hence

$$f(u)p(u) = D(u) = \frac{D_\sigma(u)}{\mu} = \frac{\mathbb{P}(M_\sigma > u)}{\mu(1 - q(u))} \sim \frac{\beta\overline{B}(u)}{1 - q(u)}.$$

Now just use that $p(x) \to \infty$ implies $q(x) \to 0$. □

Notes and references The results are from Asmussen [64], where also the (easier) case of $p(x)$ having a finite limit is treated. It is also shown in that paper that typically, there exist constants $c(u) \to 0$ such that the limiting distribution of $\tau(u)/c(u)$, given $\tau(u) < \infty$, is exponential. An early reference for linear $p(x)$ is Klüppelberg & Stadtmüller [546]. Note that for linear $p(x)$ and regularly varying claim size distribution, the result is consistent with the limit $\sigma \to 0$ of X.(6.2). For extensions and variants see Foss, Palmowski & Zachary [366] and Robert [741]. There are a number of further papers of Tang and co-authors dealing with aspects of ruin with subexponential claims under interest force, e.g. [551, 829]. See also Kalashnikov & Konstantinides [519].

6 Tail estimation

The fact that the order of ruin probabilities usually depends crucially on the tail and that the asymptotics are very different in light- and heavy-tailed regimes poses the problem of which distributional tail $\overline{F}$ to employ. Of course, this is a general statistical problem but definitely something that needs to be taken seriously. We give here only a brief introduction and refer in the Notes to standard textbooks for more detailed and broader expositions.

When computing ruin probabilities, assumptions on the tail do only partially suffice — more precise estimates like the Cramér-Lundberg approximations with light tails or the subexponential approximation require the whole distribution, but we will not discuss here how to combine tail fitting with fitting in the whole support.

We will consider the problem of fitting (with particular emphasis on the tail) a distribution F to a set of data $X_1, \ldots, X_n \geq 0$ assumed to be i.i.d. with common distribution F. As usual, $X_{(1)}, \ldots, X_{(n)}$ denote the order statistics.

Inference on $\overline{F}(x)$ beyond $x = X_{(n)}$ is of course extrapolation of the data, and in a given situation, it will far from always be obvious that this makes sense. However, some extrapolation seems inevitable: because the empirical distribution F_n^* has the finite upper bound $X_{(n)}$, most methods are likely to underestimate the tail and in particular often postulate that it is light.

6a The mean excess plot

A first question is to decide whether to use a light- or a heavy-tailed model. The approach most widely used is based on the *mean excess function*

$$e(x) \; = \; \mathbb{E}\big[X - x \mid X > x\big] \; = \; \frac{1}{\overline{F}(x)} \int_x^\infty \overline{F}(y)\,\mathrm{d}y$$

introduced in Section 1.

The reason that the mean excess function $e(x)$ is useful is that it typically asymptotically behaves quite differently for light and heavy tails. Namely, for a subexponential heavy-tailed distribution one has $e(x) \to \infty$, whereas with light tails it will typically hold that $\limsup e(x) < \infty$; say a sufficient condition is

$$\overline{F}(x) \; \sim \; \ell(x)\mathrm{e}^{-\alpha x} \tag{6.1}$$

for some $\alpha > 0$ and some $\ell(x)$ such that $\ell(\log x)$ is slowly varying (e.g., $\ell(x) = x^\gamma$ with $-\infty < \gamma < \infty$).

The mean excess test proceeds by plotting the empirical version

$$e_n(x) = \frac{1}{\#j : X_j > x} \sum_{j : X_j > x} (X_j - x)$$

of $e(x)$, usually only at the (say) K largest X_j. That is, the plot consists of the pairs formed by $X_{(n-k)}$ and

$$\frac{1}{k} \sum_{l=n-k+1}^{n} \left(X_{(l)} - X_{(n-k)} \right),$$

where $k = 1, \ldots, K$. If the plot shows a clear increase to ∞ except possibly at very small k, one takes this as an indication that F is heavy-tailed, otherwise one settles for a light-tailed model.

Example 6.1 Figure X.4 contains the mean excesses of simulated data with $n = 1,000$ from six different distributions. Each row is generated from i.i.d. r.v.'s $Y_1, Y_2, \ldots$ such that $X = Y_1$ in the left column and $X = Y_1 + Y_2 + Y_3$ in the right. In row 1, Y is Pareto with $\alpha = 3/2$; in row 2, Y is Weibull with $\beta = 1/2$; and in row 3, Y is exponential; the scale is chosen such that $\mathbb{E}Y = 3$ in all cases.

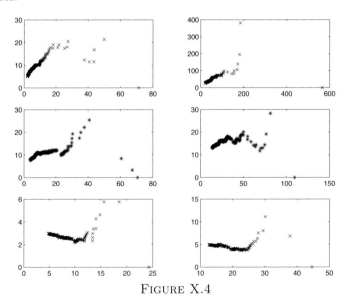

FIGURE X.4

To our mind, the story told by these pictures is not all that clear, so one conclusion is certainly that using the mean excess plot is not entirely straightforward.   □

6b   Extreme values and POT

Another method for making qualitative statements about the tail of the underlying distribution F of the data is based on extreme value theory, more precisely the results from that area that describe the asymptotics of the maximum $M_n = X_{(n)}$ and large order statistics like $X_{(n-1)}, \ldots, X_{(n-k)}$.

To describe the method, we first recall the Fisher-Tippett theorem that states that when M_n can be scaled and centered such that $(M_n - d_n)/c_n$ has a non-degenerate limit in distribution H for suitable constants c_n, d_n, then H must be of one of three types, the three classical extreme value distributions Fréchet, Weibull or Gumbel. Here a Fréchet limit typically occurs for very heavy-tailed distributions (in fact, if and only if F is regularly varying), whereas a Gumbel limit occurs for light-tailed distributions and moderately heavy-tailed distributions like the lognormal and Weibull tails e^{-ax^b} with $b < 1$; Weibull limits[3] need not concern us here since they only occur for distributions with a bounded support.

The Fréchet c.d.f. is $H(x) = e^{-x^{-\alpha}}$, $x > 0$ and $\alpha > 0$, and the Gumbel c.d.f. is $H(x) = e^{-e^{-x}}$, $x \in \mathbb{R}$. The qualifier 'type' above refers to the fact that obviously H can only be given up to scaling and location constants. It is customary to work in the class of *generalized extreme value distributions* defined by

$$H_\xi(x) = \begin{cases} \exp\{-(1+\xi x)^{-1/\xi}\} & \xi \neq 0,\ 1+\xi x > 0 \\ \exp\{-e^{-x}\} & \xi = 0,\ -\infty < x < \infty \end{cases}.$$

The particular reason for the normalization of the Fréchet c.d.f. ($\xi > 0$) is to ensure continuity at $\xi = 0$, i.e. $H_\xi(x) \to H_0(x)$ for all x as $\xi \to 0$. The class of all possible limits is obtained by adding a location parameter μ and a scale parameter $\sigma > 0$, i.e. by considering the $H_{\xi,\mu,\sigma}(x) = H_\xi\big((x - \mu)/\sigma\big)$.

A distribution F such that $(M_n - d_n)/c_n$ has limit H is said to be in the *maximum domain of attraction* of H. For applications, one can safely assume a given distribution F with infinite support to be in the maximum domain of attraction of either the Fréchet or the Gumbel (but there are exceptions, in particular discrete distributions like the geometric sometimes disturb the

---

[3]Note that the Weibull extreme value distribution is the negative analogue of the classical Weibull distribution.

picture). This means that the distribution of M_k with k large is likely to be close to some $H_{\xi,\mu,\sigma}$.

The statistical procedure based on this observation is to obtain m approximately i.i.d. replicates $M_{k,1}, \ldots, M_{k,m}$ of M_k and use these as data for maximum likelihood estimation of ξ, μ, σ. Writing $n = km$, the $M_{k,i}$ can be obtained by splitting the n observations into m blocks of size k and letting $M_{k,i}$ be the maximum over block i. The density $h_{\xi,\mu,\sigma}(x)$ of $H_{\xi,\mu,\sigma}$ is non-zero only when $1 + \xi(x - \mu)/\sigma > 0$ and is given by

$$h_{\xi,\mu,\sigma}(x) = \frac{1}{\sigma(1 + \xi(x-\mu)/\sigma)^{1/\xi+1}} \exp\left\{-\left(1 + \xi(x-\mu)/\sigma\right)^{-1/\xi}\right\}$$

(taking $\xi > 0$ for simplicity). The log likelihood is therefore

$$-m\log\sigma - (1/\xi + 1)\sum_{i=1}^{m}\log\left(1 + \xi(M_{k,i} - \mu)/\sigma\right) - \sum_{i=1}^{m}\left(1 + \xi(M_{k,i} - \mu)/\sigma\right)^{-1/\xi}$$

and has to be maximized over the region

$$\xi \geq 0, \quad \sigma > 0, \quad \left\{\mu : 0 < \min_{i=1,..,m}\left(1 + \xi(M_{k,i} - \mu)/\sigma\right)\right\}.$$

Obviously, the maximization has to be done numerically.

The most interesting parameter to estimate is ξ. Namely, an estimate $\widehat{\xi}$ that is significantly larger than 0 indicates regular variation of F, one that is close to 0 indicates that most likely the tail is lighter than for regular variation.

In practice the uncertainty on the estimates is usually high. One reason is that the block size k needs to be taken large in order that the $M_{k,i}$ have a distribution reasonably close to the asymptotics predicted by extreme value theory. Thus, the sample size m for fitting the parameters will be orders of magnitude smaller than the actual number n of observations. Consequently, the resulting estimates should be interpreted with great care.

Due to the difficulty with the waste of data by blocking, another method is more popular in practice. It is based on the *generalized Pareto distribution* $G_{\xi,\beta}$ with tail

$$\overline{G}_{\xi,\beta}(x) = \begin{cases} \dfrac{1}{(1 + \xi x/\beta)^{1/\xi}} & \xi \neq 0, \\ e^{-x/\beta} & \xi = 0. \end{cases}$$

Thus for $\xi > 0$, $G_{\xi,\beta}$ is the distribution of $\beta X/\xi$ where X has the standard Pareto tail $(1 + x)^{-\alpha}$ with $\alpha = 1/\xi$, and $G_{0,\beta}$ is the exponential$(1/\beta)$ limit as $\xi \downarrow 0$. One has:

Lemma 6.2 *If X has distribution $G_{\xi,\beta}$, then the distribution F_x of the overshoot $X - x \mid X > x$ is $G_{\xi,\beta(x)}$ where $\beta(x) = \beta + \xi x$.*

The proof is elementary and omitted. Furthermore:

Theorem 6.3 *A distribution F is in the maximum domain of a generalized extreme value distribution H_ξ with $\xi \geq 0$ if and only if there exist constants $\beta(x)$ such that*

$$\lim_{x \to \infty} \sup_{y \geq 0} \left| F_x(y) - G_{\xi, \beta(x)}(y) \right| = 0.$$

The proof is also omitted, but is not elementary!

As noted above, one can almost always safely assume a given distribution F with infinite support to satisfy the assumptions of Theorem 6.3. This motivates that for tail estimation, one assumes that F_x has a $G_{\xi, \beta}$ distribution for all large x (where β depends on x), selects some large but fixed threshold x and estimates ξ, β from the observations exceeding x. Then, letting $N(x) = \#\{i : X_i > x\}$, the final estimate of the tail of F is

$$\widehat{\overline{F}}(y) = \frac{N(x)}{n} \overline{G}_{\widehat{\xi}, \widehat{\beta}}(y - x), \quad y > x, \tag{6.2}$$

where $\widehat{\xi}, \widehat{\beta}$ are the estimates of $\xi, \beta = \beta(x)$. To obtain $\widehat{\xi}, \widehat{\beta}$, one lets $\widetilde{Y}_i = X_{j_i} - x$, $i = 1, \ldots, N(x)$ where $j_0 = 0$, $j_i = \inf\{j > j_{i-1} : X_j > x\}$ and maximizes the log likelihood

$$\sum_{i=1}^{N(x)} \log g_{\xi, \beta}(Y_i) = -N(x) \log \beta - (1/\xi + 1) \sum_{i=1}^{N(x)} \log(1 + \xi Y_i / \beta),$$

where $g_{\xi, \beta}$ is the density of $G_{\xi, \beta}$.

The Y_i represent peaks over the threshold x, and for this reason, the method and its extensions go under the name POT.

6c   The Hill estimator

We now assume that F is either regularly varying, $\overline{F}(x) = L(x)/x^\alpha$, or light-tailed satisfying (6.1). The problem is to estimate α.

Even with L or ℓ completely specified, the maximum likelihood estimator (MLE) is not adequate in this connection, because maximum likelihood will try to adjust α so that the fit is good in the center of the distribution, without caring too much about the tail, where there are fewer observations. The Hill estimator is the most commonly used (though not the only) estimator designed specifically to take this into account.

To explain the idea, consider first the setting of (6.1). If we ignore fluctuations in $\ell(x)$ by replacing $\ell(x)$ by a constant, the $X_j - x$ with $X_j > x$ are i.i.d. exponential(α). Since the standard MLE of α in the (unshifted) exponential distribution is $n/(X_1 + \cdots + X_n)$, the MLE α based on these selected X_j alone is

$$\frac{\#j : X_j > x}{\sum_{j: X_j > x}(X_j - x)}.$$

The *Hill plot* is this quantity plotted as function of x or the number $\#j : X_j > x$ of observations used. As for the mean excess plot, one usually plots only at the (say) k largest j or the k largest X_j. That is, one plots

$$\frac{k}{\sum_{\ell=n-k+1}^{n}\left(X_{(\ell)} - X_{(n-k)}\right)} \tag{6.3}$$

as function of either k or $X_{(n-k)}$. The *Hill estimator* $\alpha_{n,k}^H$ is (6.3) evaluated at some specified k. However, most often one checks graphically whether the Hill plot looks reasonably constant in a suitable range and takes a typical value from there as the estimate of α.

The regularly varying case can be treated by entirely the same method, or one may remark that it is 1-to-1 correspondence with (6.1) because X has tail $L(x)/x^\alpha$ if and only if $\log X$ has tail (6.1). Therefore, the Hill estimator in the regularly varying case is

$$\frac{k}{\sum_{\ell=n-k+1}^{n}\left(\log X_{(\ell)} - \log X_{(n-k)}\right)}. \tag{6.4}$$

It can be proved that if $k = k(n) \to \infty$ but $k/n \to 0$, then weak consistency $\alpha_{n,k}^H \xrightarrow{\mathbb{P}} \alpha$ holds. No conditions on L are needed for this. One might think that the next step would be the estimation of the slowly varying function L, but this is in general considered impossible among statisticians. In fact, we will see below that there are already difficulties enough with $\alpha_{n,k}^H$ itself.

Example 6.4 Figure X.5 contains the Hill plot (6.4) of simulated data (now with $n = 10\,000$) from the same six distributions as in Example 6.1 and the number $k = 10, \ldots, 2\,000$ of order statistics used on the horizontal axis.

Of course, only the first row, Pareto(3/2), is meaningful, since the distributions in the remaining rows are not regularly varying. Nevertheless, the appearance of the second row of plots, Weibull, is so close to the first that it is hard to assert from this alone that the distribution is not regularly varying (the evidence from the mean excess plot in Figure X.4 is not that conclusive either). The same holds, though maybe in a somewhat weaker form, for the exponential case in the third row. The first row also clearly demonstrates the difficulty in choosing k. Maybe one would settle for a value between 50 and 500 in the left panel, giving an estimate of α between 1.6 and 1.4.

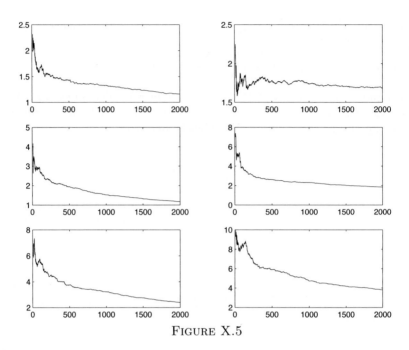

FIGURE X.5

The high statistical uncertainty on an estimate $\widehat{\alpha}$ of α is of course reflected in a high uncertainty on the estimates of the tail we obtain by plugging in $\widehat{\alpha}$ instead of α in the parametric expression for the tail. To quantify this point, consider again the Pareto(3/2) example, where it was not easy to assess from our simulation studies whether one would use $\widehat{\alpha} = 1.4$, 1.5 or 1.6 or values even further from 1.5. In the following table, we use these three α-values and compute the tail probabilities $\overline{F}_\alpha(x)$ for the 4 x-values in the first row (chosen as the 99%, 99.9%, 99.99% and 99.999% quantiles of $F_{1.5}$).

	20.5	99.0	463	2153
1.6	0.007	0.0006	0.00005	0.000005
1.5	0.010	0.0010	0.00010	0.000010
1.4	0.014	0.0016	0.00018	0.000022

There is also a CLT $k^{1/2}\bigl(\alpha^{H}_{n,k} - \alpha\bigr) \to N(0, \alpha^2)$. For this, however, stronger conditions on $k = k(n)$ and L are needed. In particular, the correct choice of k requires delicate estimates of L. That L can present a major complication has also been observed in the many "Hill horror plots" in the literature.   □

Notes and references Some standard textbooks on the topic are Embrechts, Klüppelberg, & Mikosch [349], McNeil *et al.* [633], Resnick [738], Beirlant *et al.* [154]

and de Haan & Ferreira [284]. Another simple technique to estimate $\alpha \in (0, 2)$ from i.i.d. observations is given in Albrecher & Teugels [37].

One should note that the area is rapidly expanding and that much literature recently also deals with dependence contexts which are not considered here.

Chapter XI

Ruin probabilities for Lévy processes

1 Preliminaries

An important family of stochastic processes arising in many areas of applied probability is the class of Lévy processes. A process $X = \{X_t\}_{t \geq 0}$ is said to be a *Lévy process* if it has D-paths and stationary and independent increments. Often one requires also $X_0 = 0$, but at some instances we will also allow for starting values $X_0 = u \neq 0$ and then write $\mathbb{P}_u$ for the governing probability measure (if $u = 0$, we simply write $\mathbb{P}$). For the purposes of ruin theory we will usually think of X_t as claim surplus S_t at time t, in which case indeed $S_0 = 0$ and as usual, the ruin time is then $\tau(u) = \inf \{t \geq 0 : X_t > u\}$ and the infinite horizon ruin probability is $\psi(u) = \mathbb{P}_0\big(\tau(u) < \infty\big)$. At some points it will alternatively be convenient to think of X_t as the reserve process R_t with starting value $R_0 = u \geq 0$. One easily checks that under the D-path assumption the strong Markov property holds for Lévy processes (see e.g. [APQ, p. 35]).

Standard Brownian motion B is a Lévy process, and so is a Brownian motion $\{\mu t + \sigma B_t\}$ with general drift and variance parameters. A further fundamental example is the counting process N_β of a Poisson process, where β is the rate. In fact, one of the central results in the foundations of Lévy processes is that any Lévy process can be represented as an independent sum of a Brownian motion and a 'compound Poisson'-like process. In particular, any Lévy process exhibiting finitely many jumps per unit time can be represented as

$$X_t \;=\; \mu t + \sigma B_t \;+\; \sum_{i=1}^{N_\beta(t)} Y_i$$

for $t \geq 0$, where the Y_i are i.i.d. and independent of B, N_β. This covers in particular the compound Poisson claim surplus process, where $\sigma^2 = 0$, $\mu = -1$ and the Y_i are positive. However, there are Lévy processes for which the non-Brownian jump component $J = \{J(t)\}_{t \geq 0}$ exhibits infinitely many jumps per unit time. Dealing with such processes is the main topic of this chapter.

The jump process J is characterized by its Lévy measure $\nu(\mathrm{d}x)$, which can be any non-negative measure on $\mathbb{R}$ satisfying $\nu(\{0\}) = 0$ and

$$\int_{-\infty}^{\infty} (y^2 \wedge 1)\, \nu(\mathrm{d}y) \;<\; \infty. \tag{1.1}$$

Equivalently, $\int_{|y|>\epsilon} \nu(\mathrm{d}y)$ and $\int_{-\epsilon}^{\epsilon} y^2\, \nu(\mathrm{d}y)$ are finite for some (and then all) $\epsilon > 0$.

A rough description of J is that jumps of size y occur at intensity $\nu(\mathrm{d}y)$. In particular, if ν has finite mass $\lambda = \int_{-\infty}^{\infty} \nu(\mathrm{d}y) < \infty$, then J is a compound Poisson process with intensity λ and jump size distribution $\nu(\mathrm{d}y)/\lambda$. In general, for any bounded interval K separated from 0, the sum of the jumps of size $\in K$ in the time-interval $[s, s + t)$ is a compound Poisson r.v. with intensity $t\lambda_K = t \int_K \nu(\mathrm{d}y)$ and jump-size distribution $\nu(\mathrm{d}y)I(y \in K)/\lambda_K$. Jumps in disjoint intervals are independent, and so we can describe the totality of jumps by the points in a planar Poisson process $N(\mathrm{d}y, \mathrm{d}t)$ with intensity measure $\nu(\mathrm{d}y) \otimes \mathrm{d}t$. A point of N at (Y_i, T_i) then corresponds to a jump of size Y_i at time T_i for J. If in addition to (1.1) one has

$$\int_{-\infty}^{\infty} (|y| \wedge 1)\, \nu(\mathrm{d}y) \;<\; \infty \tag{1.2}$$

(this is equivalent to the paths of J being of finite variation), one can simply write

$$J_t \;=\; \int_{\mathbb{R} \times [0,t]} y\, N(\mathrm{d}y, \mathrm{d}s). \tag{1.3}$$

If (1.2) fails, this Poisson integral does not converge absolutely, and J has to be defined by a *compensation* (centering) procedure. For example, letting

$$Y_0(t) = \int_{\{y:\, |y|>1\} \times [0,t]} y\, N(\mathrm{d}y, \mathrm{d}s), \quad Y_n(t) = \int_{|y| \in (y_{n+1}, y_n]} y\, N(\mathrm{d}y, \mathrm{d}s),$$

one can let

$$J(t) = Y_0(t) + \sum_{n=1}^{\infty} \left\{ Y_n(t) - \mathbb{E}Y_n(t) \right\}, \qquad (1.4)$$

where $1 = y_1 > y_2 > \cdots \downarrow 0$ and

$$\mathbb{E}Y_n(t) = t \int_{|y| \in (y_{n+1}, y_n]} y\, \nu(\mathrm{d}y).$$

The series converges a.s. since

$$\sum_{n=1}^{\infty} \mathbb{V}ar\big(Y_n(t)\big) = \sum_{n=1}^{\infty} t \int_{|y| \in (y_{n+1}, y_n]} y^2 \nu(\mathrm{d}y) = t \int_{-1}^{1} y^2 \nu(\mathrm{d}y) < \infty,$$

and the sum is easily seen to be independent of the particular partitioning $\{y_n\}$. But note that since the role of the interval $[-1, 1]$ is arbitrary, a compensated Lévy jump process is given canonically only up to a drift term.

If J_t has non-decreasing paths, then J is called a *subordinator*. The Lévy measure for a subordinator necessarily satisfies (1.2), and any Lévy jump process satisfying (1.2) can be written as the independent difference between two subordinators, defined in terms of the restriction of ν to $(0, \infty)$ and, respectively, the restriction of ν to $(-\infty, 0)$ reflected to $(0, \infty)$ (possibly a positive drift term has to be added).

The property of stationary independent increments implies that $\log \mathbb{E}e^{rX_t}$ has the form $t\kappa(r)$. Here $\kappa(r)$ is called the *Lévy exponent* (also often referred to as *Laplace exponent*); its domain includes the imaginary axis $\Re(r) = 0$ and frequently larger sets depending on properties of ν, say $\{r : \Re(r) \leq 0\}$ in the case of a subordinator. Thus, $\kappa(r)$ is the cumulant g.f. of an infinitely divisible distribution, having the *Lévy-Khinchine representation*

$$\kappa(r) = cr + \frac{\sigma^2 r^2}{2} + \int_{-\infty}^{\infty} \left(e^{ry} - 1 - ryI(|y| \leq 1) \right) \nu(\mathrm{d}y), \qquad (1.5)$$

where one refers to $\big(c, \sigma^2, \nu\big)$ as the *characteristic triplet*.

In the finite variation case (1.2), the Lévy-Khinchine representation (1.5) is often written

$$\kappa(r) = c_1 r + \frac{\sigma^2 r^2}{2} + \int_{-\infty}^{\infty} \left(e^{ry} - 1 \right) \nu(\mathrm{d}y), \qquad (1.6)$$

where $c_1 = c - \int_{-1}^{1} y\, \nu(\mathrm{d}y).$[1]

---

[1]Note that in much of the literature on Lévy processes c or c_1 is referred to as the drift, whereas in the sequel we will refer to $\mathbb{E}[X_1]$ as the drift.

The Lévy exponent's derivatives at 0 give the cumulants of X_1. In particular, $\mathbb{E}X_t = t\kappa'(0)$, $\mathbb{V}arX_t = t\kappa''(0)$.

Notes and references Bertoin [157] and Sato [763] are the classical references for Lévy processes, but there are also some good recent texts such as Applebaum [49] and Kyprianou [564]. A good impression of the many directions into which the topic has been developed and applied can be obtained from the volume edited by Barndorff-Nielsen *et al.* [137].

An early appearance of using Lévy processes (beyond the compound Poisson model and diffusion) for risk reserve modeling is Dufresne, Gerber & Shiu [335], for a more recent discussion see Morales & Schoutens [650]. Apart from mathematical elegance and generality, one often used argument to justify the use of these more general Lévy processes for risk modeling is that they lead to explicit knowledge of the distribution of aggregate claims (by construction via the infinitely divisible generating distribution), so instead of modeling the individual claims and compounding them, here the approach is rather to model (and eventually calibrate) the aggregate claim and then break down this information to infer consequences about the behavior on the individual claim level. Models with infinitely many jumps in finite time intervals are obviously not directly useful in the claims modelling. At the same time, the idea to calibrate the aggregate effects directly and to check the suitability of the resulting model through robustness and recalibration techniques is very popular in quantitative finance and may have a certain degree of attractiveness in the insurance context as well.

1a Special Lévy processes

In the examples we treat, the Lévy measure will have a density w.r.t. Lebesgue measure, which we denote by $n = \mathrm{d}\nu/\mathrm{d}x$. The density of X_t is denoted by $f_t(x)$ throughout.

Example 1.1 For $1 < \alpha < 2$, $\alpha \neq 1$, the *α-stable $S_\alpha(\sigma, \beta, \mu)$ distribution* is defined as the distribution with c.g.f. of the form

$$\kappa(r) = -\sigma^\alpha |r|^\alpha \left(1 - \beta \operatorname{sign}(r/\mathrm{i}) \tan \frac{\pi\alpha}{2}\right) + r\mu, \quad \Re r = 0,$$

for some $\sigma > 0$, $\beta \in [-1, 1]$, and $\mu \in \mathbb{R}$. There is a similar but somewhat different expression, which we omit, when $\alpha = 1$. The reader should note that the theory is somewhat different according to whether $0 < \alpha < 1$, $\alpha = 1$, or $1 < \alpha < 2$.

If the r.v. Y has an $S_\alpha(\sigma, \beta, \mu)$ distribution, then $Y + a$ has an $S_\alpha(\sigma, \beta, \mu + a)$ distribution and aY an $S_\alpha(\sigma |a|, \operatorname{sign}(a)\beta, a\mu)$ distribution. Thus, μ is a translation parameter and σ a scale parameter. The interpretation of β is as a skewness parameter, as will be clear from the discussion of stable processes to follow.

A *stable process* is defined as a Lévy jump process in which X_1 has an α-stable $S_\alpha(\sigma, \beta, 0)$ distribution. This can be obtained by choosing the Lévy density as

$$n(y) = \begin{cases} C_+/y^{\alpha+1} & y > 0, \\ C_-/|y|^{\alpha+1} & y < 0, \end{cases} \tag{1.7}$$

with

$$C_\pm = \frac{1-\alpha}{\Gamma(2-\alpha)\cos(\pi\alpha/2)} \sigma^\alpha \frac{1\pm\beta}{2}.$$

One can reconstruct β from the Lévy measure as $\beta = (C_+ - C_-)/(C_+ + C_-)$. If $0 < \alpha < 1$, then (1.2) holds and the process can be defined by (1.3). If $1 \le \alpha < 2$, compensation is needed and care must be taken to choose the drift term to get $\mu = 0$. Stable processes have a scaling property (self-similarity) similar to Brownian motion, $\{T^{-1/\alpha}X_{tT}\}_{t\ge0} \stackrel{\mathcal{D}}{=} \{X_t\}_{t\ge0}$ ($\mu = 0$ is crucial for this!).

Stable processes and some of their modifications are treated in depth in Samorodnitsky & Taqqu [764]. □

Example 1.2 An important property of stable processes is that the Lévy density and hence the marginals have heavy tails. A modification with light tails corresponds to the Lévy density

$$n(x) = \begin{cases} C_+ e^{-Mx}/x^{1+Y} & x > 0, \\ C_- e^{Gx}/|x|^{1+Y} & x < 0, \end{cases}$$

where $C_+ \ge, C_- \ge 0, C_+ + C_- > 0, G, M > 0, 0 \le Y < 2$. Such a Lévy process is called a *tempered stable process*. For $Y > 0$ and $C_+ = C_- = C$, the corresponding Lévy process is called the *CGMY process*[2]; for $Y = 0$ and $C_+ = C_- = C$, the process is called the *Variance Gamma process*. The Lévy exponent is

$$\kappa(r) = C_+\Gamma(-Y)\left[(M-r)^Y - M^Y\right] + C_-\left[(G+r)^Y - G^Y\right]. \quad \square$$

Example 1.3 Since the Gamma distribution with a density proportional to $x^{\alpha-1}e^{-\lambda x}$ is infinitely divisible, there is a Lévy process with this distribution of X_1. For obvious reasons, it is called the *Gamma process*. The Lévy measure can be shown to have density $n(x) = \alpha e^{-\lambda x}/x$ for $x > 0$; note that $n(x) \sim x^{-1}$, $x \downarrow 0$, so the Lévy measure is infinite but at the borderline of being so. Hence

---

[2]CGMY = Carr-Geman-Madan-Yor; cf. the notation for the parameters!

small jumps play a relatively small role for the Gamma process. By standard properties of the Gamma distribution,

$$\kappa(r) = \alpha \log \frac{\lambda}{\lambda - r}, \quad f_t(x) = \frac{\lambda^{\alpha t}}{\Gamma(\alpha t)} x^{\alpha t - 1} e^{-\lambda x}. \qquad \square$$

The Variance Gamma process in Example 1.2 is the difference between two independent Gamma processes.

Example 1.4 The *Normal Inverse Gaussian* (NIG) *Lévy process* has four parameters $\alpha, \delta > 0$, $\beta \in (-\alpha, \alpha)$, $\mu \in \mathbb{R}$, and

$$\kappa(r) = \mu r - \delta\left(\sqrt{\alpha^2 - (\beta + r)^2} - \sqrt{\alpha^2 - \beta^2}\right).$$

The Lévy measure has density

$$\frac{\alpha \delta}{\pi |x|} K_1(\alpha |x|) e^{\beta x}, \quad x \in \mathbb{R}, \tag{1.8}$$

(here as usual K_1 denotes the modified Bessel function of the third kind with index 1), and the density of X_1 is

$$f_1(x) = \frac{\alpha \delta}{\pi} \exp\left\{\delta\sqrt{\alpha^2 - \beta^2} - \beta\mu\right\} \frac{K_1\left(\alpha\sqrt{\delta^2 + (x - \mu)^2}\right)}{\sqrt{\delta^2 + (x - \mu)^2}} e^{\beta x},$$

which is called the $\mathrm{NIG}(\alpha, \beta, \mu, \delta)$ density; clearly the density $f_t(x)$ of X_t is $\mathrm{NIG}(\alpha, \beta, t\mu, t\delta)$. $\qquad \square$

Example 1.5 Let X be any Lévy process with nonnegative drift. Then $T(x) = \inf\{t : X(t) > x\}$ is finite a.s., and clearly $\{T(x)\}_{x \geq 0}$ has stationary independent increments, so it is a Lévy process (in fact a subordinator, since the sample paths are nondecreasing).

The most notable example is the *Inverse Gaussian Lévy process*, which corresponds to X being Brownian motion with drift $\gamma > 0$ and variance 1. Here

$$f_x(t) = \frac{x}{t^{3/2}\sqrt{2\pi}} \exp\left\{\gamma x - \frac{1}{2}\left(\frac{x^2}{t} + \gamma^2 t\right)\right\}$$

(cf. also Corollary III.1.6) and the Lévy measure has density

$$n(x) = \frac{1}{\sqrt{2\pi}\, x^{3/2}} e^{-x\gamma^2/2}, \quad x > 0. \qquad \square$$

1b Exponential change of measure

As for the compound Poisson model and random walks, exponential change of measure also plays a main role for Lévy processes. It is also clear from the analogy with these classical models what should be the appropriate definition of an exponential θ-tilting for a θ satisfying $\kappa(\theta) < \infty$: change the Lévy exponent $\kappa(r)$ to $\kappa_\theta(r) = \kappa(r+\theta) - \kappa(\theta)$.

Proposition 1.6 *Assume that X has characteristic triplet (c, σ^2, ν). Then $\kappa_\theta(r)$ is the Lévy exponent of the Lévy process with characteristic triplet $(c_\theta, \sigma_\theta^2, \nu_\theta)$, where $\sigma_\theta^2 = \sigma^2$, $\nu_\theta(\mathrm{d}x) = \mathrm{e}^{\theta x}\nu(\mathrm{d}x)$, and*

$$c_\theta = c + \sigma^2\theta + \int_{-1}^{1} (\mathrm{e}^{\theta x} - 1)x\,\nu(\mathrm{d}x)\,.$$

Proof. In view of (1.5) we have

$$
\begin{aligned}
\kappa_\theta(r) &= \kappa(r+\theta) - \kappa(\theta) \\
&= (c + \sigma^2\theta)r + \sigma^2 r^2/2 + \int_{-\infty}^{\infty} \left(\mathrm{e}^{(\theta+r)x} - \mathrm{e}^{\theta x} - rxI(|x| \le 1)\right)\nu(\mathrm{d}x) \\
&= \left(c + \sigma^2\theta + \int_{-1}^{1} (\mathrm{e}^{\theta x} - 1)x\,\nu(\mathrm{d}x)\right)r + \sigma^2 r^2/2 \\
&\quad + \int_{-\infty}^{\infty} \left(\mathrm{e}^{rx} - 1 - rxI(|x| \le 1)\right)\mathrm{e}^{\theta x}\,\nu(\mathrm{d}x)\,.
\end{aligned}
$$

$\square$

Letting $\mathbb{P}$ be the governing probability measure for X and $\mathbb{P}^\theta$ the one for the exponentially tilted Lévy process, the likelihood ratio on $[0, T]$ takes the form

$$\frac{\mathrm{d}\mathbb{P}}{\mathrm{d}\mathbb{P}^\theta}\bigg|_{\mathscr{F}_T} = \mathrm{e}^{-\theta X(T)+T\kappa(\theta)}\,,$$

as may be seen, for example, by discrete-random-walk approximations.

Often exponential change of measure and other calculations involve roots r of an equation of the form $\kappa(r) = \delta$ where κ is the Lévy exponent and $\delta \ge 0$. By convexity and $\kappa(0) = 0$, $\kappa(r) = \delta$ has (depending on the domain of existence of κ) either zero, one or two real roots and we denote by $-\rho_\delta$ the smallest one.

Notes and references Many of the examples mentioned above are frequently used for asset price modeling in finance. The NIG Lévy process and the Variance Gamma process are particular examples of generalized hyperbolic Lévy processes. The generalized hyperbolic distribution Y was introduced by Barndorff-Nielsen [135] as a normal variance-mean mixture (e.g. both the mean and variance of the normal distribution are distributed according to an (appropriately scaled) mixing distribution W) with a generalized inverse Gaussian mixing distribution W. For the NIG distribution W is the inverse Gaussian distribution, for the Variance Gamma distribution W is a Gamma distribution, giving the corresponding processes their names. Another popular approach is to interpret the NIG and VG Lévy process as Brownian motion subordinated by an inverse Gaussian and a Gamma process, respectively. For details see e.g. Bibby & Sørensen [164] or Schoutens [783].

2   One-sided ruin theory

In this section, we give the results (both asymptotic and exact) for the infinite horizon ruin probability $\psi(u)$ that can be derived with reasonable effort. We assume throughout that the claim surplus process $\{S_t\}$ is a Lévy process with negative drift, i.e. $\kappa'(0) < 0$ and, to avoid trivialities, that it is not the negative of a subordinator (in which case $\psi(u) = 0$ for all $u > 0$).

Going beyond the compound Poisson model to general Lévy processes, heavy tails are a remarkably simple case, and we have the following analogue of results from Chapter X:

Theorem 2.1 *Assume that $\{S_t\}$ is a Lévy process with $\mathbb{E}S_1 = \kappa'(0) < 0$ and the Lévy measure ν satisfying $\overline{\nu}(x) = \int_x^\infty \nu(\mathrm{d}y) \sim \overline{B}(x)$ as $x \to \infty$ for some distribution B such that the integrated tail B_0 of B is subexponential. Then*

$$\psi(u) \;\sim\; \frac{1}{|\mathbb{E}S_1|} \int_u^\infty \overline{\nu}(y)\,\mathrm{d}y \,. \tag{2.1}$$

Lemma 2.2 $\mathbb{P}(S_1 > x) \sim \overline{\nu}(x)\,.$

Proof. Write $S = S' + S'' + S'''$ where S', S'', S''' have characteristic triplets (c, σ^2, ν'), $(0, 0, \nu'')$ and $(0, 0, \nu''')$, resp., with ν', ν'', ν''' being the restrictions of ν to $[-1, 1]$, $(-\infty, -1)$ and $(1, \infty)$, respectively.

With $\beta''' = \overline{\nu}(1)$, the r.v. S_1''' is a compound Poisson sum of r.v.'s, with Poisson parameter β''' and distribution ν'''/β'''. Thus by X.2.1, we have

$$\mathbb{P}(S_1''' > x) \;\sim\; \beta''' \frac{\overline{\nu}'''(x)}{\beta'''} \;=\; \overline{\nu}(x), \quad x > 1 \,.$$

The independence of S_1'' and $S_1''' > 0$ therefore implies

$$\mathbb{P}(S_1'' + S_1''' > x) \;\sim\; \overline{\nu}(x),$$

cf. the proof of X.3.2. From (1.5) it is immediate that $\kappa'(r) < \infty$ for all r. In particular, S_1' is light-tailed, and the desired estimate for $S_1 = S_1' + S_1'' + S_1'''$ then follows by X.1.11. $\qquad\square$

Proof of Theorem 2.1. Define

$$M^d = \sup_{n=0,1,2,\dots} S_n, \quad M = \sup_{0 \le t < \infty} S_t.$$

Then

$$\mathbb{P}(M^d > u) \sim \frac{1}{|\mathbb{E}S_1|} \int_u^\infty \bar{\nu}(y)\,\mathrm{d}y \tag{2.2}$$

by the general random walk result X.(3.3) and Lemma 2.2. Also clearly $\mathbb{P}(M^d > u) \le \mathbb{P}(M > u) = \psi(u)$. Given $\epsilon > 0$, choose $a > 0$ with $\mathbb{P}\big(\inf_{0 \le t \le 1} S_t > -a\big) \ge 1 - \epsilon$. Then $\mathbb{P}(M^d > u - a) \ge (1-\epsilon)\mathbb{P}(M > u) = (1-\epsilon)\psi(u)$. But by subexponentiality, $\mathbb{P}(M^d > u - a) \sim \mathbb{P}(M^d > u)$. Putting these estimates together completes the proof. $\qquad\square$

Let us now move to general tail behavior of the Lévy measure, but restrict to one-sided jumps only. A Lévy process $\{S\}$ is called *spectrally negative* if there are no positive jumps, i.e. $\nu(0, \infty) = 0$. Equivalently, the paths are skipfree upwards. In this case, $\psi(u)$ is in fact of exact exponential form:

Theorem 2.3 *Assume that the claim surplus process $\{S_t\}$ is spectrally negative with $\mathbb{E}S_1 < 0$ (i.e. ruin can only be caused by diffusion). Then $\kappa(r)$ as defined in (1.5) has a positive zero $\gamma > 0$ and $\psi(u) = \mathrm{e}^{-\gamma u}$, $u > 0$.*

Proof. Spectral negativity implies $\kappa(r) < \infty$ for all $r > 0$, and since $\kappa(r) \to \infty$ as $r \to \infty$ and $\kappa'(0) < 0$, the desired root γ exists by continuity. Under the change of measure with tilting factor γ, $\mathbb{E}_\gamma S_1 = \kappa'(\gamma) > 0$ so that $\mathbb{P}_\gamma(\tau(u) < \infty) = 1$. Due to the absence of up-jumps we have $S_{\tau(u)} = u$ and correspondingly

$$\psi(u) = \mathbb{P}\big(\tau(u) < \infty\big) = \mathbb{E}_\gamma\big[\mathrm{e}^{-\gamma S_{\tau(u)}}; \tau(u) < \infty\big] = \mathrm{e}^{-\gamma u}.$$

$\qquad\square$

If $\{S_t\}$ is *spectrally positive* (i.e. $\nu(-\infty, 0) = 0$), $\psi(u)$ is not explicit but the Laplace transform can be found in closed form:

Theorem 2.4 *Assume that the claim surplus process $\{S_t\}$ is spectrally positive with $\mathbb{E}S_1 = \mu < 0$. Then (at least) for all s with $\Re(s) > 0$*

$$\widehat{\psi}[-s] = \int_0^\infty \mathrm{e}^{-su}\psi(u)\,\mathrm{d}u = \frac{1}{s} + \frac{\mu}{\kappa(-s)}. \tag{2.3}$$

Proof. Define $M = \max_{t \geq 0} S_t$ and recall that the distribution of M is also the stationary distribution of the Lévy process reflected at 0,

$$V_t = V_0 + S_t + L_t\,, \tag{2.4}$$

where $L_t = \left(-\inf_{s \leq t} S_s - V_0\right)^+$ (cf. [APQ, p. 250]). Taking $V_0 = M^*$ where M^* is an independent copy of M, $\{V_t\}$ becomes stationary. In particular, $\mathbb{E}V_1 = \mathbb{E}V_0$ and therefore (2.4) yields $\mathbb{E}L_1 = -\mu$. Further, by spectral positivity $\{L_t\}$ is continuous so that the Kella-Whitt martingale becomes

$$\kappa(r) \int_0^t e^{rV_u}\,\mathrm{d}u \;+\; e^{rV_0} \;-\; e^{rV_t} \;+\; rL_t\,.$$

Optional stopping at $t = 1$ gives

$$0 \;=\; \kappa(r)\mathbb{E}e^{rM} \;+\; \mathbb{E}e^{rM} \;-\; \mathbb{E}e^{rM} \;+\; r\mathbb{E}L_1 \;=\; \kappa(r)\mathbb{E}e^{rM} - r\mu\,.$$

Hence

$$\int_0^\infty e^{-su}\psi(u)\,\mathrm{d}u \;=\; \int_0^\infty e^{-su}\mathbb{P}(M > u)\,\mathrm{d}u \;=\; \frac{1 - \mathbb{E}e^{-sM}}{s} \;=\; \frac{1 + s\mu/\kappa(-s)}{s}\,.$$

$\square$

Remark 2.5 Note that for the Cramér-Lundberg model $\kappa(r) = \beta(\widehat{B}[r] - 1) - r$ and $\mu = \beta\mu_B - 1$, so that (2.3) then simplifies to

$$\widehat{\psi}[-s] = \frac{1}{s} - \frac{1 - \beta\mu_B}{\beta\widehat{B}[-s] - \beta + s}\,,$$

which indeed coincides with IV.(3.4). $\square$

Next consider light tails for the up-jumps, meaning $\int_1^\infty e^{rx}\nu(\mathrm{d}x) < \infty$ for some $r > 0$ (which is the same as $\kappa(r) < \infty$). If $\mathbb{E}S_1 < 0$ and the adjustment coefficient $\gamma > 0$ (i.e. the positive root of $\kappa(r) = 0$) exists, one expects from the Cramér-Lundberg theory that $\psi(u)$ decays asymptotically exponentially at rate γ. Indeed, in the general case:

Theorem 2.6 *Consider a general Lévy process $\{S_t\}$ with $\mathbb{E}S_1 < 0$. Assume that $\gamma > 0$ exists and satisfies $\kappa'(\gamma) < \infty$, and further that $\{S_t\}$ is not a compound Poisson process with lattice support of ν. Then $\psi(u) \sim Ce^{-\gamma u}$, $u \to \infty$, for some constant $0 < C < \infty$.*

We will see that the proof is straightforward, given some small technicalities on the non-lattice property. However, the really difficult step in the Cramér-Lundberg theory for Lévy processes is identifying C (for this one needs the Wiener-Hopf factorization briefly discussed in Section 4). Here we note only the following special case, which comprises the compound Poisson case and where the expression for C is entirely analogous to the one there, cf. IV.5.5:

Corollary 2.7 *If in the spectrally positive case $\gamma > 0$ exists and $\kappa'(\gamma) < \infty$, then $C = -\mu/\kappa'(\gamma) = -\kappa'(0)/\kappa'(\gamma)$.*

Proof. Because of Theorem 2.6 and Remark IV.5.6 it suffices to calculate the constant $C = \lim_{u\to\infty} e^{\gamma u}\psi(u) = \lim_{s\to 0} s\,\widehat{\psi}[-s+\gamma]$. In view of Theorem 2.4 the result hence follows by a simple application of L'Hôpital's rule. □

Proof of Theorem 2.6. The spectrally negative case is covered by Theorem 2.3. In the (additional) presence of positive jumps, let $\xi(x) = S_{\tau(x)} - x$ denote the overshoot and

$$Y = Y_1 = \inf\big\{x > 1 : \xi(x-) = 0\big\}, \quad Y_n = \inf\big\{x > 1 + Y_{n-1} : \xi(x-) = 0\big\}.$$

Then the Y_n are finite $\mathbb{P}_\gamma$-a.s. since then $\mathbb{E}_\gamma S_1 > 0$ and hence $\tau(x) < \infty$ for all x, and $\mathbb{E}_\gamma S_1 > 0$ implies that there exists an infinity of x with $\xi(x) = 0$ [note that we cannot use $x > 0$ in the definition instead of $x > 1$ since it may then happen that $Y_1 = 0$ a.s.]. Thus $\big\{\xi(x)\big\}_{x\geq 0}$ is a regenerative process with regeneration points $Y_1, Y_2, \ldots$. The assumption that S is not a compound Poisson process with lattice support of ν is easily seen to imply that the distribution of Y_1 is non-lattice (see Kyprianou [564] for details). Hence $\xi(x) \overset{\mathscr{D}}{\to} \xi(\infty)$ for some $\xi(\infty) < \infty$, and using exponential change of measure, we get

$$\psi(u) = \mathbb{E}_\gamma\big[e^{-\gamma S_{\tau(u)}}; \tau(u) < \infty\big] = \mathbb{E}_\gamma e^{-\gamma S_{\tau(u)}} = e^{-\gamma u}\mathbb{E}_\gamma e^{-\gamma\xi(u)} \sim C e^{-\gamma u}$$

where $C = \mathbb{E}_\gamma e^{-\gamma\xi(\infty)}$. □

Notes and references Asymptotic results on ruin probabilities for Lévy insurance risk processes can be found in Klüppelberg, Kyprianou & Maller [543]. For Theorem 2.1, see also Klüppelberg & Kyprianou [542]. Corollary 2.7 goes back to Doney [325], where it is given as a consequence of a more involved argument. Huzak et al. [490] derive a ladder-height decomposition of the ruin probability and in that way generalize the Pollaczeck-Khinchine formula to certain Lévy set-ups, see also Schmidli [776]. At the same time, this implies that one can formulate a defective renewal equation for the ruin probability, see also Section XII.4. Bernyk, Dalang & Peskir [156] use fractional derivatives to derive accurate information on finite-time ruin probabilities for α-stable

Lévy processes ($1 < \alpha < 2$) with only down-sided jumps. For extensions to more general Lévy processes with two-sided jumps, see for instance Bertoin & Doney [158] and Lewis & Mordecki [582].

For asymptotic results on finite-time ruin probabilities, see Palmowski & Pistorius [679].

3 The scale function and two-sided ruin problems

The concept of a scale function as discussed in II.2 for diffusions and giving two-sided exit probabilities generalizes to Lévy processes. One even can go one step further and include information on the exit time as well. So, let $\{X_t\}_{t \geq 0}$ be a Lévy process with Lévy exponent $\kappa(r)$, and for $0 < u < a$, define

$$\tau_0^- \;=\; \inf\{t > 0 : X_t \leq 0 \mid X_0 = u\}, \qquad \tau_a^+ \;=\; \inf\{t > 0 : X_t \geq a \mid X_0 = u\}.$$

To avoid trivialities, we assume that $\{X\}$ is not a subordinator or the negative of a subordinator. We will also need the assumption of spectral negativity (this ensures $\kappa(s) < \infty$ for all $s > 0$); so in this Section $\{X_t\}$ refers to the reserve process $\{R_t\}$. Of course, results for the spectrally positive case (and thus for the claim surplus process $\{S_t\}$) follow immediately by sign reversion.

For $\delta > 0$, the equation $\kappa(s) = \delta$ has a unique positive solution which we denote by $\rho_\delta > 0$. If $\delta = 0$ and $\mathbb{E}(X_1) = \kappa'(0) > 0$, then $\rho_\delta = \rho_0 = 0$. Note that since we now deal with R_t (rather than S_t), the sign of the argument of κ is reversed,[3] so the positive solution is now not the adjustment coefficient!

Lemma 3.1 $\mathbb{E}_u\big[\mathrm{e}^{-\delta\tau_a^+} ; \tau_a^+ < \infty\big] \;=\; \mathrm{e}^{-\rho_\delta(a-u)}$.

Proof. As a simple adaptation of Lemma X.3.1, just note that $\big\{\mathrm{e}^{\rho_\delta X_t - \delta t}\big\}$ is a martingale, apply optional stopping at $\tau_a^+ \wedge T$ and let $T \to \infty$ with dominated convergence.   □

Theorem 3.2 (a) *For each $\delta \geq 0$, there exists a function $W^{(\delta)}(u)$ (the scale function) such that*

$$\mathbb{E}_u\big[\mathrm{e}^{-\delta\tau_a^+} ; \tau_a^+ < \tau_0^-\big] \;=\; \frac{W^{(\delta)}(u)}{W^{(\delta)}(a)}. \tag{3.1}$$

(b) *$W^{(\delta)}(u)$ is unique up to a multiplicative constant, which may be chosen such that $W^{(\delta)}(u)$ is given via its Laplace transform in u by*

$$\int_0^\infty \mathrm{e}^{-su} W^{(\delta)}(u)\,\mathrm{d}u \;=\; \frac{1}{\kappa(s) - \delta} \quad \text{for } s > \rho_\delta. \tag{3.2}$$

---

[3]which we emphasize by using the argument s instead of r.

Note that taking $\delta = 0$ we obtain the probability of exiting the interval $(0, a)$ to the right as $W^{(0)}(u)/W^{(0)}(a)$. Note also that the one-sided survival probability can be computed by taking the limit $a \to \infty$. In particular, if $\kappa'(0) > 0$, then $\lim_{a \to \infty} W^{(0)}(a) = \lim_{s \to 0} s/\kappa(s) = 1/\kappa'(0)$, so that

$$\psi(u) = 1 - \kappa'(0)\,W^{(0)}(u)$$

(if $\kappa'(0) \leq 0$, $\psi(u) = 1$).

A further fundamental function for two-sided ruin problems is

$$Z^{(\delta)}(u) \;=\; 1 + \delta \int_0^u W^{(\delta)}(y)\,dy. \tag{3.3}$$

In fact:

Theorem 3.3 (a)

$$\mathbb{E}_u\!\left[e^{-\delta\tau_0^-}\,;\,\tau_0^- < \infty\right] \;=\; Z^{(\delta)}(u) - \frac{\delta}{\rho_\delta}W^{(\delta)}(u). \tag{3.4}$$

(b)

$$\mathbb{E}_u\!\left[e^{-\delta\tau_0^-}\,;\,\tau_0^- < \tau_a^+\right] \;=\; Z^{(\delta)}(u) - Z^{(\delta)}(a)\,\frac{W^{(\delta)}(u)}{W^{(\delta)}(a)}. \tag{3.5}$$

In the proofs, we will need the running minimum and maximum,

$$\underline{X}_t \;=\; \min_{0 \leq s \leq t} X_s\,, \quad \overline{X}_t \;=\; \max_{0 \leq s \leq t} X_s\,.$$

Further e_δ will denote an exponential r.v. with rate δ and which is independent of the Lévy process $\{X\}$. e_δ will become useful via the following lemma:

Lemma 3.4 $\mathbb{E}e^{-s\overline{X}_{e_\delta}} \;=\; \dfrac{\rho_\delta}{\rho_\delta + s}$, $\quad \mathbb{E}e^{s\underline{X}_{e_\delta}} \;=\; \dfrac{\delta}{\rho_\delta}\dfrac{\rho_\delta - s}{\delta - \kappa(s)}$.

Proof. Using exponential change of measure, we get

$$\mathbb{P}\!\left(\overline{X}_{e_\delta} > a\right) \;=\; \mathbb{P}\!\left(\tau_a^+ < e_\delta\right)$$

$$=\; \mathbb{E}_{\rho_\delta}\!\left[e^{-\rho_\delta X_{\tau_a^+} + \tau_a^+ \kappa(\rho_\delta)}\,;\,\tau_a^+ < e_\delta\right]$$

$$=\; \mathbb{E}_{\rho_\delta}\!\left[e^{-\rho_\delta a + \tau_a^+ \delta}e^{-\delta\tau_a^+}\right] \;=\; e^{-\rho_\delta a}.$$

I.e., $\overline{X}_{e_\delta}$ is exponentially distributed with parameter ρ_δ which is equivalent to the first statement of the lemma.

For the second, we use the Kella-Whitt martingale M_t (say) with exponential parameter $-s$ on $Z_t = -X_t + L_t$ with $L_t = -\inf_{0 \le s \le t} -X_s = \sup_{0 \le s \le t} X_s$. I.e., Z is $-X$ reflected at 0 which by the continuous-time analogue of III.(3.2) implies that

$$Z_t \overset{\mathscr{D}}{=} \max_{0 \le v \le t} -X_t = -\underline{X}_t .$$

Note that spectral negativity implies that L can only increase when Z is at 0 and that L has no jumps. Therefore

$$M_t = \kappa(s) \int_0^t \mathrm{e}^{-sZ_v} \, \mathrm{d}v + 1 - \mathrm{e}^{-sZ_t} - s \int_0^t L(\mathrm{d}v) .$$

Since optional stopping at an independent random time is permissible for any martingale, we have

$$0 = M_0 = \mathbb{E}M_{e_\delta} . \tag{3.6}$$

Here

$$\mathbb{E} \int_0^{e_\delta} \mathrm{e}^{-sZ_v} \, \mathrm{d}v = \int_0^\infty \mathrm{e}^{-v\delta} \mathrm{e}^{-sZ_v} \, \mathrm{d}v = \frac{1}{\delta} \mathbb{E}\mathrm{e}^{s\underline{X}_{e_\delta}} .$$

Using the just established fact that $\overline{X}_{e_\delta}$ has an exponential(ρ_δ) distribution, (3.6) therefore becomes

$$0 = \frac{\kappa(s)}{\delta} \mathbb{E}\mathrm{e}^{s\underline{X}_{e_\delta}} + 1 - \mathbb{E}\mathrm{e}^{s\underline{X}_{e_\delta}} - \frac{s}{\rho_\delta} ,$$

which gives the desired conclusion concerning $\underline{X}_{e_\delta}$.   □

Proof of Theorem 3.2(a) when $\kappa'(0) > 0$, $\delta = 0$.
The assumption $\kappa'(0) > 0$ ensures that $-\underline{X}_\infty$ is finite a.s., and we can define $W^{(0)}(u) = \mathbb{P}_u(\underline{X}_\infty \ge 0)$. Sample path arguments beyond the scope of this book show that either at τ_0^- or immediately after, X will attain strictly negative values (one needs to consider the case of a Brownian component or none separately; see [564, pp. 216, 177–179]). Therefore, under the event $\tau_0^- < \tau_a^+, \underline{X}_\infty \ge 0$ is impossible so that by the strong Markov property

$$\mathbb{P}_u(\underline{X}_\infty \ge 0) = \mathbb{E}_u[\mathbb{P}_a(\underline{X}_\infty \ge 0); \tau_a^+ < \tau_0^-] = \mathbb{P}_a(\underline{X}_\infty \ge 0)\,\mathbb{P}_u(\tau_a^+ < \tau_0^-),$$

which gives the desired conclusion in the form

$$\mathbb{P}_u(\tau_a^+ < \tau_0^-) = \frac{W^{(0)}(u)}{W^{(0)}(a)} . \tag{3.7}$$

□

Proof of Theorem 3.2(a) when $\delta > 0$ or $\kappa'(0) < 0$, $\delta = 0$.
In this case $\rho_\delta > 0$ and we can use exponential tilting with factor ρ_δ and define

$$W_{\rho_\delta}^{(0)}(u) = \mathbb{P}_{u,\rho_\delta}(\underline{X}_\infty \geq 0), \quad W^{(\delta)}(u) = e^{\rho_\delta u}W_{\rho_\delta}^{(0)}(u). \tag{3.8}$$

Easy convexity arguments show that the $\mathbb{P}_{\rho_\delta}$-drift is positive, so by (3.7),

$$\mathbb{P}_{u,\rho_\delta}(\tau_a^+ < \tau_0^-) = \frac{W_{\rho_\delta}^{(0)}(u)}{W_{\rho_\delta}^{(0)}(a)}. \tag{3.9}$$

But using the exponential change of measure, the l.h.s. can also be written as

$$\mathbb{E}_u\left[\exp\{\rho_\delta(X_{\tau_a^+} - u) - \delta\tau_a^+\}; \tau_a^+ < \tau_0^-\right] = e^{\rho_\delta(a-u)}\mathbb{E}_u\left[e^{-\delta\tau_a^+}; \tau_a^+ < \tau_0^-\right].$$

Combining these two expressions gives (3.1).   □

Proof of Theorem 3.2(a) when $\delta = 0$ and $\kappa'(0) = 0$.
An easy continuity argument, letting $\delta \downarrow 0$. We omit the details.   □

Proof of Theorem 3.2(b).
Consider again first the case $\kappa'(0) > 0$. It is obvious that $W^{(\delta)}$ may be modified by a multiplicative constant, so we redefine $W^{(0)}(u)$ as $W^{(0)}(u) = \mathbb{P}_u(\underline{X}_\infty \geq 0)/\kappa'(0)$.
    Using integration by parts and noting that $\mathbb{P}(\underline{X}_\infty = 0) = 0$, we have

$$\begin{aligned}
\mathbb{E}e^{s\underline{X}_\infty} &= \mathbb{E}e^{-s(-\underline{X}_\infty)} = 1 - \int_0^\infty se^{-su}\mathbb{P}(-\underline{X}_\infty > u)\,du \\
&= \int_0^\infty se^{-su}\mathbb{P}(-\underline{X}_\infty \leq u)\,du = \int_0^\infty se^{-su}\mathbb{P}_u(\underline{X}_\infty \geq 0)\,du.
\end{aligned}$$

On the other hand, $\rho_\delta \to 0$ as $\delta \downarrow 0$, more precisely $\rho_\delta \sim \delta/\kappa'(0)$. Letting $\delta \downarrow 0$ in the second part of Lemma 3.4 therefore yields $\mathbb{E}e^{s\underline{X}_\infty} = \kappa'(0)s/\kappa(s)$. Comparing these two expressions gives

$$\int_0^\infty e^{-su}W^{(0)}(u)\,du = \frac{1}{\kappa(s)},$$

which is the desired conclusion for the case $\kappa'(0) > 0$. The proofs for the remaining cases are then easy by involving the connections between W, W_{ρ_δ} and $W^{(\delta)}$.   □

Remark 3.5 There is a simple, but slightly heuristic way to see Theorem 3.2 for arbitrary drift and $\delta \geq 0$: Define $C(u, a) = \mathbb{E}_u\big[e^{-\delta \tau_a^+}; \tau_a^+ < \tau_0^-\big]$. By the absence of upward jumps, X with $X_0 = u$ can only reach an arbitrary level b (with $b > a > u$) without ruin in between, if level a is passed before that. Consequently, by the strong Markov property of X one has $C(u, b) = C(u, a)C(a, b)$, so that $C(u, a) = C(u, b)/C(a, b) = h(u)/h(a)$ and one may identify h with the scale function. This argument shows that Theorem 3.2 is in fact valid beyond Lévy processes, namely for stationary Markov processes without upward jumps. □

We next turn to the proof of Theorem 3.3. The first step is to note:

Lemma 3.6 *There exists a measure $W^{(\delta)}(\mathrm{d}u)$ on $[0, \infty)$ such that $W^{(\delta)}[0, u] = W^{(\delta)}(u)$. This measure has Laplace transform*

$$\int_0^\infty e^{-su} W^{(\delta)}(\mathrm{d}u) = \frac{s}{\kappa(s) - \delta}. \tag{3.10}$$

Proof. The first statement is clear since $W^{(\delta)}(u)$ is strictly increasing.[4] The l.h.s. of (3.10) then comes out as

$$\int_0^\infty W^{(\delta)}(\mathrm{d}u) \int_u^\infty s e^{-sy}\,\mathrm{d}y = \int_0^\infty s e^{-sy}\,\mathrm{d}y \int_0^y W^{(\delta)}(\mathrm{d}u)$$

$$= s \int_0^\infty e^{-sy} W^{(\delta)}(y)\,\mathrm{d}y$$

which is the same as the r.h.s. □

Proof of Theorem 3.3(a).
For $\delta > 0$, we have from Lemma 3.4 that

$$\int_0^\infty e^{-su}\left[\frac{\delta}{\rho_\delta} W^{(\delta)}(\mathrm{d}u) - \delta W^{(\delta)}(u)\,\mathrm{d}u\right] = \frac{\delta}{\rho_\delta}\frac{s}{\kappa(s) - \delta} - \frac{\delta}{\kappa(s) - \delta} = \mathbb{E}e^{s\underline{X}_{e_\delta}}.$$

I.e.

$$\mathbb{P}(-\underline{X}_{e_\delta} \in \mathrm{d}u) = \frac{\delta}{\rho_\delta} W^{(\delta)}(\mathrm{d}u) - \delta W^{(\delta)}(u)\,\mathrm{d}u.$$

It follows that

$$\mathbb{E}_u\big[e^{-\delta \tau_0^-}; \tau_0^- < \infty\big] = \mathbb{P}_u(e_\delta > \tau_0^-) = \mathbb{P}_u(\underline{X}_{e_\delta} < 0)$$

$$= 1 - \mathbb{P}(-\underline{X}_{e_\delta} \leq u) = 1 + \delta \int_0^u W^{(\delta)}(y)\,\mathrm{d}y - \frac{\delta}{\rho_\delta} W^{(\delta)}(u)$$

$$= Z^{(\delta)}(u) - \frac{\delta}{\rho_\delta} W^{(\delta)}(u).$$

---

[4]Strictly speaking, we would need a right-continuous version. However, this turns out to be inessential for the following, and in fact $W^{(\delta)}(u)$ can be shown to be continuous.

The case $\delta = 0$ is again easy by taking limits. □

Proof of Theorem 3.3(b).

We can write

$$\mathbb{E}_u\big[e^{-\delta\tau_0^-} ; \tau_0^- < \tau_a^+\big] = \mathbb{E}_u\big[e^{-\delta\tau_0^-} ; \tau_0^- < \infty\big] - \mathbb{E}_u\big[e^{-\delta\tau_0^-} ; \tau_a^+ < \tau_0^-\big].$$

Since $X_{\tau_a^+} = a$, the second $\mathbb{E}_u$ is

$$\mathbb{E}_u\big[e^{-\delta\tau_a^+} ; \tau_a^+ < \tau_0^-\big] \cdot \mathbb{E}_a\big[e^{-\delta\tau_0^-} ; \tau_0^- < \infty\big].$$

Thus $\mathbb{E}_u\big[e^{-\delta\tau_0^-} ; \tau_0^- < \tau_a^+\big]$ becomes

$$Z^{(\delta)}(u) - \frac{\delta}{\rho_\delta}W^{(\delta)}(u) - \frac{W^{(\delta)}(u)}{W^{(\delta)}(a)}\Big[Z^{(\delta)}(a) - \frac{\delta}{\rho_\delta}W^{(\delta)}(a)\Big],$$

which is the asserted expression. □

Notes and references Scale functions are a classical tool for spectrally one-sided Lévy processes with roots in Zolotarev [922], Takacs [827] and Korolyuk [554]. Parts of the exposition above are close to Kyprianou [564]. For a recent survey of available explicit forms of scale functions and methods how to construct them see Hubalek & Kyprianou [482] and also Kyprianou & Rivero [567]. The argument of Remark 3.5 can be found in Gerber, Lin & Yang [407].

As for ruin probabilities themselves, one can naturally also use the generator approach to identify the scale function, leading to an integro-differential equation and subsequently to a Volterra integral equation. The connection between this approach and the more standard one pursued here is highlighted in Biffis & Kyprianou [165].

For an extension of the above analysis to one- and two-sided exit problems with non-constant boundaries see Bertoin, Doney & Maller [159]. Extensions to Lévy processes that are reflected at the supremum or infimum are worked out by Zhou [920]. Loeffen & Patie [605] give a fine analysis of one- and two-sided exit problems with interest rates and absolute ruin for the case when the aggregate claim process is a subordinator.

Lemma 3.4 can be exploited to design efficient numerical procedures for determining finite-time ruin probabilities $\psi(u,t)$; for an application in credit risk see e.g. Madan & Schoutens [622].

4 Further topics

This section gives a brief overview of some topics which are basic in fluctuation theory for Lévy processes but more advanced than what we have looked at so far. The treatment should basically be seen as a heuristical introduction (to be followed up by the interested reader by more detailed and rigorous treatments such as Kyprianou [564]). Thus the 'proofs' we present should mainly be considered as heuristical motivations that the results are true (in fact, the theory is so advanced that even [564] has to skip certain steps).

The topics under consideration are certainly relevant for ruin theory. However, one problem is that explicit results beyond what we have already presented are rarely available.

4a Local time at the maximum

In the following, denote by $\overline{X}_t = \sup_{0 \le s \le t} X_t$ the running maximum. A nondecreasing process $\{L_t\}$ with D-paths and $L_0 = 0$ is called a (version of) *the local time at the maximum* if

(i) The support of the measure dL_t is the closure of the set $\{\overline{X}_t = X_t\}$;

(ii) For every stopping time τ such that $\overline{X}_\tau = X_\tau$ on $\{\tau < \infty\}$, the shifted trivariate process

$$\left\{X_{\tau+t} - X_\tau, \overline{X}_{\tau+t} - X_{\tau+t}, L_{\tau+t} - L_\tau\right\}_{t \ge 0}$$

is independent of $\mathscr{F}_\tau$ on $\{\tau < \infty\}$ and has the same distribution as

$$\left\{X_t, \overline{X}_t - X_t, L_t\right\}_{t \ge 0}.$$

Note that this definition identifies L only up to a multiplicative constant, and that existence is not a priori obvious. Note also that the term 'local time' occurs in various different, though often related, meanings in the probability literature.

For some Lévy processes an obvious candidate for L easily suggests itself and the verification that it indeed is a local time is straightforward. In particular:

(a) If X is spectrally negative, one can take $L_t = \overline{X}_t$.

(b) For a compound Poisson process with positive drift and negative jumps, the set $\{\overline{X}_t = X_t\}$ is a union of disjoint intervals, and one may take

$$L_t = a \int_0^t I\left(\overline{X}_s = X_s\right) ds$$

with $a > 0$ arbitrary. In particular, this covers the reserve process in the Cramér-Lundberg model.

(c) If the set of times of maxima of X is discrete (as for the claim surplus process of the Cramér-Lundberg model), one may take $L_t = M_t$ where M_t is the number of maxima before t.

The intuition behind the definition of local time at the maximum is to give an indication of how much time X has spent at its running maximum before

t. For this reason, none of the definitions in (a), (b), (c) are applicable in the whole class of Lévy processes. More precisely, the definition $L_t = \overline{X}_t$ would not have the required intuitive properties for (say) a compound Poisson process without drift, and the definitions in (b), (c) would not be appropriate for (say) Brownian motion because $\{\overline{X}_t = X_t\}$ is a Lebesgue null set, excluding (b), and not discrete, excluding (c).

The general definition of L requires the notion of *regularity*. We say that B (say $B = (0,\infty)$ or $B = [0,\infty)$) is regular if $\mathbb{P}(\tau_B = 0) = 1$ where $\tau_B = \inf\{t > 0 : X_t \in B\}$ (note that by Blumenthal's 0–1 law, $\mathbb{P}(\tau_B = 0)$ is either 0 or 1). For example, $(0,\infty)$ is regular for Brownian motion, but $(0,\infty)$ and $[0,\infty)$ are both irregular for the Cramér-Lundberg claim surplus process. There are then the following three cases:

1) X has bounded variation and $[0,\infty)$ is irregular. Then the set of maxima is discrete and we can take $L_t = \sum_{i=1}^{M_t} E_{\lambda,i}$ where M_t is as in (c) and the $E_{\lambda,i}$ are i.i.d. exponential(λ).[5]
2) X has bounded variation and $(-\infty,0)$ is irregular. One may define L as in (b) above. If X is spectrally negative, then L is proportional to $\overline{X}$.
3) X has unbounded variation (this can be shown to imply that $[0,\infty)$ is regular). A local time exists, but there is no simple known expression in terms of the path of X. Again, if X is spectrally negative, then L is proportional to $\overline{X}$.

4b   The ladder height process

First note that $L_\infty = \lim_{t \to \infty} L_t$ may be finite (the main case is negative drift). Then define

$$L_t^{-1} = \begin{cases} \inf\{s > 0 : L_s > t\} & t < L_\infty \\ \infty & t \geq L_\infty \end{cases} .$$

Further let the ladder height process be

$$M_t = \begin{cases} X_{L_t^{-1}} & t < L_\infty \\ \infty & t \geq L_\infty \end{cases} .$$

Theorem 4.1 *The process* $\mathbf{Y} = \left\{\left(L_t^{-1}, M_t\right)\right\}_{t \geq 0}$ *is a bivariate Lévy process, possibly terminating if* $L_\infty < \infty$.

Proof. The definition of $\left(L_t^{-1}, M_t\right)$ immediately implies that X is at a maximum at time L_t^{-1}. Therefore

$$\mathbf{Y}_{t+s} - \mathbf{Y}_t = \left(L_{t+s}^{-1} - L_t^{-1}, M_{t+s} - M_t\right)$$

---

[5]The variation from (c) is motivated from the desire of L_t^{-1} to have certain properties, cf. the following Section 4b.

has the same distribution as $Y_s = (L_s^{-1}, M_s)$ and is independent of $\{Y_v\}_{v \le t}$. This implies that Y has stationary independent increments and the assertion.
□

It follows that for some suitable function $\phi(\cdot, \cdot)$ we can write

$$\log \mathbb{E} \exp\{-aL_t^{-1} - bM_t\} = -\phi(a, b)t. \qquad (4.1)$$

4c Excursions

By an *excursion from the maximum* we understand a segment $\{X_t\}_{u \le t \le v}$ of X such that $\overline{X}_u = X_u \le X_v$ and $X_t < X_u$ for $u < t < v$.

The fundamental fact about excursions is that they roughly occur according to a Poisson process in the time scale given by L^{-1}. However, for example for X a Brownian motion and s a time where X is at a maximum, sample path properties of X imply that each interval $[s, s+\epsilon]$ contains infinitely many excursions. Of course, the sum of their lengths has to be finite, so for each $\delta > 0$ there must be finitely many excursions of length $> \delta$ and infinitely many of length $\le \delta$. The same phenomenon typically occurs for general Lévy processes, so a careful formulation of the Poisson property is needed, for example the following one:

Theorem 4.2 *Let $\delta > 0$ and let $\eta_1 < \eta_2 < \dots$ be the times > 0 where an excursion of length $> \delta$ starts.[6] Then the points $L_{\eta_1}, L_{\eta_2}, \dots$ form a homogeneous Poisson process.*

Proof. We only treat the case of L_t being continuous in t. For brevity, denote the excursions of length $> \delta$ as δ-excursions. The counting process N of δ-excursions on the L^{-1}-scale is given by $N_t = \max\{i : \eta_i \le s\}$ where $L_s = t$. Let $t_1 < t_2 < t_3 < \cdots$. Then in the ordinary time scale for X, $L_{t_1}^{-1}, L_{t_2}^{-1}, L_{t_3}^{-1}$ correspond to times $s_1 < s_2 < s_3$ with $L_{s_i} = t_i$ and X is at a maximum at each s_i. It is clear that the number $N_{t_2} - N_{t_1}$ of $\eta_i \in [s_1, s_2]$ is independent of the number $N_{t_3} - N_{t_2}$ of $\eta_i \in [s_2, s_3]$, and similarly for further intervals of the same type. Further, it is not difficult to see by considering a sample path that the distribution of $N_{t_2} - N_{t_1}$ only depends on $t_2 - t_1$. Also, if a δ-excursion starts and ends at say u, v, then X is at a maximum at v so the local time has to increase at v which implies that the local time at the time w (say) where the next δ-excursion starts satisfies $L_w > L_v = L_u$. This implies that N cannot have multiple points, which together with the already noted fact of stationary and independent increments implies the Poisson property.
□

---

[6]The characteristic 'length $> \delta$' could be replaced by many others, for example that the maximal deviation from the maximum during the excursion exceeds δ.

The intensity parameter of N or similar point processes of excursions is in general not available. A notable exception is Itô's excursion law for Brownian motion (e.g. Rogers & Williams [744, Sec. VI.8]), where a complete description of the probability mechanism governing Brownian excursions is possible. Another one is the reserve process R_t of the Cramér-Lundberg model, where excursions occur at intensity β and have the distribution of the busy period in the dual M/G/1 queue (see also Theorem III.2.3).

4d   The Wiener-Hopf factorization

The Wiener-Hopf factorization occurs in many alternative forms in the literature, but its currently most used version is the following one. Recall that e_δ is an independent exponential(δ) time. Further, define

$$\overline{G}_t = \sup\{s < t : X_s = \overline{X}_s\}, \quad \underline{G}_t = \sup\{s < t : X_s = \underline{X}_s\}$$

(the times of the last maximum, resp. minimum, before t).

Theorem 4.3 (i) *The pairs* $\left(\overline{G}_{e_\delta}, \overline{X}_{e_\delta}\right)$ *and* $\left(e_\delta - \overline{G}_{e_\delta}, \overline{X}_{e_\delta} - X_{e_\delta}\right)$ *are independent. Therefore*

$$\frac{\delta}{\delta - a - \kappa(b)} = \Psi^+(a,b)\Psi^-(a,b) \tag{4.2}$$

where

$$\Psi^+(a,b) = \mathbb{E}e^{a\overline{G}_{e_\delta} + b\overline{X}_{e_\delta}}, \quad \Psi^-(a,b) = \mathbb{E}e^{a\underline{G}_{e_\delta} + b\underline{X}_{e_\delta}}. \tag{4.3}$$

(ii) *The functions* Ψ^+, Ψ^- *in* (4.3) *can be identified* (involving analytic continuation, if needed) *via the function* ϕ *in* (4.1) *and the corresponding one* $\check{\phi}$ *for the descending ladder process by means of*

$$\mathbb{E}e^{a\overline{G}_{e_\delta} + b\overline{X}_{e_\delta}} = \frac{\phi(\delta, 0)}{\phi(\delta - a, -b)}, \quad \mathbb{E}e^{a\underline{G}_{e_\delta} + b\underline{X}_{e_\delta}} = \frac{\check{\phi}(\delta, 0)}{\check{\phi}(\delta - a, -b)}. \tag{4.4}$$

The functions Ψ^+, Ψ^- are called the Wiener-Hopf factors of X. They are obviously at best given up to a multiplicative constant, but can in fact be shown to be unique modulo this.

Proof. Given $e_\delta > t$, the distribution of $e_\delta - t$ is again exponential(δ). If t is the time of the last maximum before e_δ, this changes the distribution to that of an exponential e'_δ (which is an independent copy of e_δ) given that an excursion away from the maximum occurs at time 0 and lasts at least e'_δ. However, independence pertains and obviously, $\overline{X}_{e_\delta} - X_{e_\delta}$ is conditionally independent of (G_t, X_t) and independent of t, which implies the claimed independence.

For (4.2), first note that

$$\frac{\delta}{\delta - a - \kappa(b)} = \int_0^\infty \delta e^{-\delta t} e^{at} e^{\kappa(b)t}\, dt = \mathbb{E} e^{ae_\delta + bX_{e_\delta}}.$$

Using the independence just established, this becomes

$$\mathbb{E} e^{a\overline{G}_{e_\delta} + b\overline{X}_{e_\delta}} \mathbb{E} e^{a(e_\delta - \overline{G}_{e_\delta}) + b(X_{e_\delta} - \overline{X}_{e_\delta})} = \Psi^+(a,b)\mathbb{E} e^{a(e_\delta - \overline{G}_{e_\delta}) + b(X_{e_\delta} - \overline{X}_{e_\delta})}.$$

$$(4.5)$$

However, a sign reversion argument easily gives that

$$\left(e_\delta - \overline{G}_{e_\delta},\, X_{e_\delta} - \overline{X}_{e_\delta}\right) \overset{\mathscr{D}}{=} \left(e_\delta, \underline{X}_{e_\delta}\right).$$

Hence the final expectation in (4.5) reduces to $\Psi^-(a,b)$, completing the proof of (i).

We do not give the proof of (ii), see for instance Kyprianou [564]. □

Example 4.4 For the spectrally negative case with positive drift, $L_t^{-1} = \tau_t^+$ and $M_t = t$, so that from (4.1) and Lemma 3.1 it follows that $\phi(a,b) = \rho_a + b$ ($\tau_t^+ < \infty$ a.s. for positive drift). Hence

$$\Psi^+(a,b) = \rho_\delta/(\rho_{\delta-a} - b),$$

which for the special case $a = 0$ was already established in Lemma 3.4. From (4.2) we then easily identify

$$\Psi^-(a,b) = \frac{\delta}{\rho_\delta} \frac{\rho_{\delta-a} - b}{\delta - a - \kappa(b)}.$$

From this one can, in view of (4.3), read off $\breve{\phi}(a,b) = \left(a - \kappa(b)\right)/(\rho_a - b)$. □

4e A quintuple identity

Consider the quintuple $(V_1, V_2, V_3, V_4, V_5)$ given by the r.v.'s

$$\overline{G}_{\tau^+(x)-},\ \overline{X}_{\tau^+(x)-},\ \tau^+(x) - \overline{G}_{\tau^+(x)-},\ X_{\tau^+(x)-}, X_{\tau^+(x)}$$

(the time of the last maximum before first passage, the value of that maximum, the time from that maximum to first passage, the value just before first passage, and the value just after, see Figure XI.1).

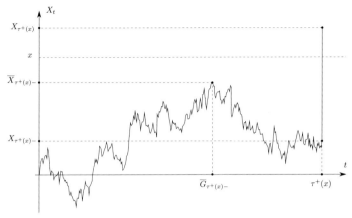

FIGURE XI.1

Further define the measures $\mathcal{U}, \breve{\mathcal{U}}$ (often called potential measures) by

$$\mathcal{U}(\mathrm{d}s, \mathrm{d}x) = \int_0^\infty \mathbb{P}(L_t^{-1} \in \mathrm{d}s, M_t \in \mathrm{d}x)\,\mathrm{d}t\,,$$

$$\breve{\mathcal{U}}(\mathrm{d}s, \mathrm{d}x) = \int_0^\infty \mathbb{P}(\breve{L}_t^{-1} \in \mathrm{d}s, \breve{M}_t \in \mathrm{d}x)\,\mathrm{d}t\,,$$

where as before $\breve{L}_t^{-1}$ and $\breve{M}_t$ refer to the corresponding quantities of the descending ladder height process. By Fubini's Theorem and (4.1), the bivariate Laplace transform of $\mathcal{U}$ has the simple form

$$\int_{[0,\infty)^2} \mathrm{e}^{-as-bx}\mathcal{U}(\mathrm{d}s, \mathrm{d}x) = \int_0^\infty \mathrm{d}t \cdot \mathbb{E}(\mathrm{e}^{-aL_t^{-1}-bM_t}) = \frac{1}{\phi(a,b)}. \tag{4.6}$$

Remark 4.5 From (4.6), obviously $\int_{[0,\infty)^2} \mathcal{U}(\mathrm{d}s, \mathrm{d}x) = 1/\phi(0,0)$. With the definition $U(\mathrm{d}x) = \int_{s=0}^\infty \mathcal{U}(\mathrm{d}s, \mathrm{d}x)$ one then sees by normalization that

$$\psi(u) = \mathbb{P}\big(\tau(u) < \infty\big) = \phi(0,0)\,U(u,\infty). \tag{4.7}$$

This representation of the ruin probability can be interpreted as the continuous-time extension of the Pollaczeck-Khinchine formula of Theorem IV.2.1, because M_t is the (ascending) ladder height process. $\qquad\square$

Theorem 4.6 *The conditional distribution of V_3, V_4 given V_1, V_2 depends only on V_2, and the conditional distribution of V_5 given V_1, V_2, V_3, V_4 depends only on V_4. Further, there exists a normalization of the local time such that the density of $\big(V_1, x - V_2, V_3, x - V_4, V_5 - x\big)$ at (s, y, t, v, z) can be written as*

$$\mathcal{U}(\mathrm{d}s, x - \mathrm{d}y)\,\breve{\mathcal{U}}(\mathrm{d}t, \mathrm{d}v - y)\,\nu(\mathrm{d}z + v)\,.$$

Proof. The claims on the conditional distributions are clear from Figure XI.1 and the strong Markov property. This gives a factorization of the density of $(V_1, x - V_2, V_3, x - V_4, V_5 - x)$ as $h_1(\mathrm{d}s, x - \mathrm{d}y)h_2(\mathrm{d}t, \mathrm{d}v - y)h_3(\mathrm{d}z + v)$. Here it is clear that $h_3(\mathrm{d}z + v) = \nu(\mathrm{d}z + v)$. The claim on h_1, h_2 will not be shown here (see e.g. again [564]). $\qquad\square$

The quintuple law usually does not lead to explicit formulas. For spectrally positive Lévy processes $\{X_t\}$, one can however obtain the following simpler expression for the joint density of the last maximum before first passage V_2, the value just before passage V_4 and the value after passage V_5:

Corollary 4.7 *If $\{X_t\}$ is a spectrally positive Lévy process drifting to $-\infty$ with scale function $W(u) = W^{(0)}(u)$, then the density of $\left(x - V_2, x - V_4, V_5 - x\right)$ at (y, v, z) is given by*

$$W'(x - y)\,\nu(\mathrm{d}z + v)\,\mathrm{d}v\,\mathrm{d}y, \qquad 0 \le y \le \min(x, v),\ z > 0. \qquad (4.8)$$

Proof. We can use the expressions obtained in Example 4.4, but have to reverse the role of the ascending and descending ladder process, because now we have a spectrally positive process. From (4.6) with $a = 0$ we see that

$$\int_0^\infty \mathrm{e}^{-bx} U(\mathrm{d}x) = \frac{b}{\kappa(-b)}. \qquad (4.9)$$

At the same time $\check{\phi}(0, b) = b$, so we can choose $\check{U}(\mathrm{d}x) = \mathrm{d}x$. But in view of Theorem 4.6 this implies that the density in (4.8) is

$$U(x - \mathrm{d}y)\,\nu(\mathrm{d}z + v)\,.$$

However, comparing with the definition (3.2) of the scale function $W^{(0)}$, it is clear that U can be identified with $W^{(0)}$, because the latter is only unique up to a constant (which can be controlled by the normalization of the local time of X at its supremum). $\qquad\square$

Remark 4.8 If in addition $\{X_t\}$ has bounded variation, then from (1.6)

$$\kappa(-b) \;=\; -c_1 b + \int_0^\infty (\mathrm{e}^{-by} - 1)\nu(\mathrm{d}y)$$

with $\mathbb{E}(X_1) < 0$ and one can infer from (4.9) by expanding the resulting geometric series that

$$U(\mathrm{d}x) \;=\; \frac{1}{c_1} \sum_{n=0}^\infty \chi^{*n}(\mathrm{d}x)$$

where $\chi(\mathrm{d}x) = \nu(x, \infty)\,\mathrm{d}x/c_1$. Correspondingly, (4.8) can in this case also be written as

$$\frac{1}{c_1} \sum_{n=0}^{\infty} \chi^{*n}(x - \mathrm{d}y)\,\nu(\mathrm{d}z + v)\,\mathrm{d}v, \qquad 0 \leq y \leq x,\ y \leq v,\ z > 0.$$

$\square$

In terms of the risk reserve process R_t, the above result gives a formula for the joint density of the surplus prior to ruin, the deficit at ruin and the size of the last minimum before ruin in terms of the scale function and the Lévy measure; see Chapter XII for a further discussion.

Notes and references A complete proof of Theorem 4.6 is given in Doney & Kyprianou [326], where also asymptotics of the quintuple law for $x \to \infty$ are given. Kuznetsov [563] recently gave quite general criteria under which the Wiener-Hopf factors are of semi-explicit form and identified a set of tractable special cases.

5 The scale function for two-sided phase-type jumps

In this section, we assume that $\{X_t\}_{t \geq 0}$ is the superposition of a Brownian motion with drift μ and variance constant $\sigma^2 > 0$ and two compound Poisson processes, one having upward jumps at rate λ^+ and being phase-type with representation $(E^+, \boldsymbol{\alpha}^+, \boldsymbol{T}^+)$ and the other having downward jumps at rate λ^- and being phase-type with representation $(E^-, \boldsymbol{\alpha}^-, \boldsymbol{T}^-)$ (the cardinalities of E^+, E^- are denoted by p^+, resp. p^-). That is, the Lévy exponent $\kappa(s)$ as defined by $\log \mathbb{E}e^{sX_t}/t$ equals

$$s\mu + s^2\sigma^2/2 + \lambda^+\left(\boldsymbol{\alpha}^+(-s\boldsymbol{I} - \boldsymbol{T}^+)^{-1}\boldsymbol{t}^+ - 1\right) + \lambda^-\left(\boldsymbol{\alpha}^-(s\boldsymbol{I} - \boldsymbol{T}^-)^{-1}\boldsymbol{t}^- - 1\right). \quad (5.1)$$

It is well-defined in the strip $\mathscr{D} = \{s \in \mathbb{C} : \rho^- < \Re s < \rho^+\}$ where ρ^+ is the eigenvalue with largest real part of $-\boldsymbol{T}^+$ and ρ^- is the eigenvalue with smallest real part of $\boldsymbol{T}^-$.

Theorem 5.1 *Assume that there exist $p = p^+ + p^- + 2$ distinct complex numbers s_k such that $\kappa(s_k) = \delta$. Define $c_0^+(s) = c_0^-(s) = 1$ and $c_i^+(s) = \boldsymbol{e}_i^\top(-s\boldsymbol{I} - \boldsymbol{T}^+)^{-1}\boldsymbol{t}^+$, $c_i^-(s) = \boldsymbol{e}_i^\top(s\boldsymbol{I} - \boldsymbol{T}^-)^{-1}\boldsymbol{t}^-$, and denote by $b_1^+, \ldots, b_{p^+}^+$, $b_1^-, \ldots, b_{p^+}^-$, b_0^+, b_0^- the solutions to the p linear equations*

$$\mathrm{e}^{s_k u} = -\sum_{i=0}^{p^+} c_i^+(s_k)\mathrm{e}^{sa}b_i^+ - \sum_{i=0}^{p^-} c_i^-(s_k)b_i^-. \quad (5.2)$$

Then

$$\mathbb{E}_u\left[e^{-\delta \tau_a^+} \, ; \, \tau_a^+ < \tau_0^-\right] \;\; = \;\; \sum_{i=0}^{p^+} b_i^+ \, , \tag{5.3}$$

$$\mathbb{E}_u\left[e^{-\delta \tau_0^-} \, ; \, \tau_0^- < \tau_a^+\right] \;\; = \;\; \sum_{i=0}^{p^-} b_i^- \, . \tag{5.4}$$

Remark 5.2 The r.h.s. of (5.1) is well-defined not just for $\Re(s) \in (\rho^-, \rho^+)$, but for any $s \in \mathbb{C}$ that is not an eigenvalue of $-\boldsymbol{T}^+$ or $\boldsymbol{T}^-$, and is analytic in this domain. The roots s_k should be looked for in this entire domain, not just $\mathscr{D}$.

Note that we can write

$$\boldsymbol{\alpha}^+(-s\boldsymbol{I} - \boldsymbol{T}^+)^{-1}\boldsymbol{t}^+ \;\; = \;\; n^+(s)/d^+(s), \quad \boldsymbol{\alpha}^-(s\boldsymbol{I} - \boldsymbol{T}^-)^{-1}\boldsymbol{t}^- \;\; = \;\; n^-(s)/d^-(s) \, ,$$

where (assuming minimal PH representations) $n^+(s), d^+(s), n^-(s), d^-(s)$ are polynomials of degree $p^+ - 1, p^+, p^- - 1$ and p^-, respectively. Thus, the defining equation for the s_k can be written

$$\begin{aligned}
\delta d^+(s)d^-(s) \;\; = \;\; & sd^+(s)d^-(s)\mu + s^2 d^+(s)d^-(s)\sigma^2/2 \\
& + \lambda^+\left(n^+(s) - d^+(s)d^-(s)\right) + \lambda^-\left(n^-(s) - d^+(s)d^-(s)\right) .
\end{aligned}$$

This is a polynomial equation of degree $p = p^+ + p^- + 2$, so that indeed p roots exist. For the question of the roots being distinct, see the Notes and References.

If $\sigma^2 = 0, \mu \neq 0$, one only has to look for $p^+ + p^- + 1$ roots, and if $\sigma^2 = 0, \mu \neq 0$, only for $p^+ + p^-$ roots. The modifications are obvious and will not be spelled out. $\qquad \square$

Proof of Theorem 5.1. Let $0 \leq s \leq \rho^+$ and $Z_t = X_t - \delta t/s$. The Kella-Whitt martingale then becomes $M_t = M_t(s)$ where

$$M_t(s) \;\; = \;\; \kappa(s)\int_0^t e^{sZ_v}\,dv \; + \; e^{su} \; - e^{sZ_t} \; - \; \delta\int_0^t e^{sZ_v}\,dv \, .$$

Let $\tau = \tau_a^+ \vee \tau_0^-$. Then

$$|M_t| \;\; \leq \;\; |\kappa(s) - \delta|e^{sa}\tau + e^{su} + e^{s(a+V^+)}$$

for $0 \leq t \leq \tau$ where we can think of V^+ as the possible overshoot over a. Since V^+ is phase-type with phase generator $\boldsymbol{T}^+$, we have $\mathbb{E}e^{s(a+V^+)} < \infty$. Also $\mathbb{E}\tau < \infty$, and it follows that $\sup_{t \leq \tau}|M_t|$ is integrable so that optional stopping at τ is permissible.

Let B_0^+ be the event that a is upcrossed before 0 is downcrossed and that the upcrossing results from the Brownian motion and not a jump, and let B_i^+ be the event that a is upcrossed before 0 is downcrossed and that this results from a jump being in phase i at the upcrossing. Similarly, let B_0^- be the event that 0 is downcrossed before a is upcrossed and that the downcrossing results from the Brownian motion and not a jump, and let B_i^- be the event that 0 is downcrossed before a is upcrossed and that this results from a jump being in phase i at the upcrossing. Write

$$b_i^+ = \mathbb{E}_u[e^{-\delta \tau_0^-} \,;\, \tau_a^+ < \tau_0^-, B_i^+], \quad b_i^- = \mathbb{E}_u[e^{-\delta \tau_0^-} \,;\, \tau_a^+ > \tau_0^-, B_i^-].$$

Optional stopping at τ now gives

$$M_0 = e^{su} = \mathbb{E}_u M_\tau = \big(\kappa(s) - \delta\big) \int_0^\tau e^{sZ_v}\,dv - \mathbb{E}_u e^{-\delta\tau + sX_\tau}. \tag{5.5}$$

Given B_i^+, the overshoot over a equals 0 for $i = 0$ and is phase-type with representation $(E^+, e_i^\mathsf{T}, T^+)$ for $i > 0$. A similar argument applies to B_i^-, and so expanding the r.h.s. of (5.5), we obtain

$$e^{su} = \big(\kappa(s) - \delta\big) \int_0^\tau e^{sZ_v}\,dv - \sum_{i=0}^{q^+} c_i^+(s) e^{sa} b_i^+ - \sum_{i=0}^{q^-} c_i^-(s) b_i^-. \tag{5.6}$$

Now note that

$$0 \le \mathbb{E}_u \int_0^\tau e^{sZ_v}\,dv \le \mathbb{E}_u \tau e^{sa} \tag{5.7}$$

for any $s \in \mathbb{C}$. This readily implies that $\mathbb{E}_u \int\{\cdot\}$ is an analytic function defined for all $s \in \mathbb{C}$. It therefore follows by analytic continuation that

$$e^{su} = \big(\kappa(s) - \delta\big) \int_0^\tau e^{sZ_v}\,dv - \sum_{i=0}^{q^+} c_i^+(s) e^{sa} b_i^+ - \sum_{i=0}^{q^-} c_i^-(s) b_i^-$$

for all $s \notin \mathscr{D}$. In particular, taking $s = s_k$ this becomes (5.2). Formulas (5.3) and (5.4) are then clear. □

Notes and references Theorem 3.3 occurs in Asmussen, Avram & Pistorius [70], though with a somewhat different proof. Further references on fluctuation theory for Lévy processes under phase-type assumptions include Pistorius [704], Dieker [323] and a series of papers by Mordecki and co-authors, e.g. Lewis & Mordecki [582]. A phase-type approximation for CGMY Lévy processes with applications for the pricing of equity default swaps is given in Asmussen, Madan & Pistorius [90].

In practice, the roots s_k will more or less always be found to be distinct. In the rare cases with one or more roots having multiplicity > 1, modifications are needed. For an example of how this can be done, see D'Auria et al. [276].

Chapter XII

Gerber-Shiu functions

1 Introduction

At some places in previous chapters we have seen results on the time of ruin $\tau(u)$, the deficit at ruin $\xi(u) = |R_{\tau(u)}|$ and the surplus prior to ruin $R_{\tau(u)-}$. In this chapter we will study a combination of these quantities simultaneously, which leads to a tractable and elegant treatment. This combination is of the form of an expected discounted penalty at ruin

$$
m(u) = \mathbb{E}\left[e^{-\delta\tau(u)}\, w\big(R_{\tau(u)-}, \xi(u)\big); \tau(u) < \infty\right], \tag{1.1}
$$

where the penalty w is a non-negative function of the surplus prior to ruin and the deficit at ruin. The expression $m(u)$ is usually referred to as the *Gerber-Shiu function*. Clearly, for $w \equiv 1$ and $\delta = 0$, (1.1) reduces to the ruin probability $\psi(u)$, and for $w \equiv 1$ and $\delta > 0$ one arrives (with a slight abuse of terminology) at the Laplace transform of the time to ruin $\tau(u)$. Alternatively, if $\delta = 0$ and w is the bivariate Dirac-delta function, (1.1) represents the joint density of the surplus prior to ruin and the deficit at ruin. The parameter $\delta \geq 0$ can be interpreted both as a discount rate and the Laplace transform argument. The refinement to include time-dependence in the analysis is a natural step towards a better understanding of the behavior of the risk process.

If $f(x, y, t|u)$ denotes the (defective) joint density of surplus prior to ruin, deficit at ruin and time of ruin given that $R_0 = u$, then the finite-time ruin probability can be expressed as

$$
\psi(u, t) = \int_0^t \left(\int_0^\infty \int_0^\infty f(x, y, s|u)\, \mathrm{d}x\, \mathrm{d}y\right) \mathrm{d}s,
$$

from which an integration by parts and the representation

$$m(u) \;=\; \int_0^\infty \int_0^\infty \int_0^\infty \mathrm{e}^{-\delta t}\, w(x,y)\, f(x,y,t|u)\, \mathrm{d}t\, \mathrm{d}x\, \mathrm{d}y$$

yields that for $w \equiv 1$

$$\mathbb{E}\big[\mathrm{e}^{-\delta\tau(u)};\, \tau(u) < \infty\big] \;=\; \int_0^\infty \mathrm{e}^{-\delta t}\frac{\partial}{\partial t}\psi(u,t)\, \mathrm{d}t \;=\; \delta \int_0^\infty \mathrm{e}^{-\delta t}\psi(u,t)\, \mathrm{d}t. \quad (1.2)$$

Thus the Gerber-Shiu function also contains as a special case the Laplace transform (w.r.t. time) of the finite-time ruin probability (or, equivalently, the ruin probability up to a random exponential time horizon with parameter δ). Other choices of the penalty w lead to interpretations of $m(u)$ as the expected present value of deferred continuous annuities during the first negative excursion of R_t or the price of a perpetual American put option on an asset with dynamics given by $\{R_t\}$ as well as the price of reset guarantees for mutual funds (see e.g. Gerber & Shiu [410]).

Define the discounted (defective) joint density of surplus prior to ruin and deficit at ruin as

$$f(x,y|u) = \int_0^\infty \mathrm{e}^{-\delta t} f(x,y,t|u)\, \mathrm{d}t$$

and the discounted density of the surplus prior to ruin as

$$f(x|u) = \int_0^\infty f(x,y|u)\, \mathrm{d}y.$$

Let us start with some general considerations for renewal risk models. Unless stated otherwise, we will always assume a positive safety loading $\eta > 0$. Although not always necessary, we usually assume that the claim size distribution B has a density b. With the notation $\omega(x) = \int_x^\infty w(x, y - x)\, B(\mathrm{d}y)$, an alternative representation of the Gerber-Shiu function is

$$m(u) \;=\; \int_0^\infty \int_0^\infty w(x,y) f(x|u)\frac{b(x+y)}{1 - B(x)}\, \mathrm{d}x\, \mathrm{d}y \;=\; \int_0^\infty \omega(x)\frac{f(x|u)}{1 - B(x)}\, \mathrm{d}x.$$

It is now easy to derive a defective renewal equation for $m(u)$ that holds for general (zero-delayed) renewal models. Conditioning on the first time that the surplus falls below the initial level u and the size of this jump, one has

$$
\begin{aligned}
m(u) &= \int_0^u \int_0^\infty \int_0^\infty e^{-\delta t} m(u-y) f(x,y,t|0) \, dt \, dx \, dy \\
&\quad + \int_u^\infty \int_0^\infty \int_0^\infty e^{-\delta t} w(x+u, y-u) f(x,y,t|0) \, dt \, dx \, dy \\
&= \int_0^u \int_0^\infty m(u-y) f(x,y|0) \, dx \, dy \\
&\quad + \int_u^\infty \int_0^\infty w(x+u, y-u) f(x,y|0) \, dx \, dy. \quad (1.3)
\end{aligned}
$$

Denoting with

$$
g_\delta(y) = \int_0^\infty f(x,y|0) \, dx \quad (1.4)
$$

the defective discounted density of deficit at ruin when $u = 0$, the defective renewal equation can be written as $m = m * g_\delta + h$ for the function h specified in (1.3). This equation will be useful at a number of places later on. A first consequence is

$$
m(0) = \int_0^\infty \int_0^\infty w(x,y) f(x,y|0) \, dx \, dy. \quad (1.5)
$$

Throughout the chapter, we tacitly assume

$$
\int_0^\infty \int_0^\infty w(x,y) b(x+y) \, dx \, dy < \infty, \quad (1.6)
$$

which is a natural condition to ensure that $m(u)$ is finite for all $u \geq 0$. In view of $\eta > 0$, we will also assume the natural boundary condition

$$
\lim_{u \to \infty} m(u) = 0 \quad (1.7)
$$

(which in many cases is automatically fulfilled under additional assumptions on the interplay between the penalty w and the claim size distribution B).

Notes and references The investigation of extensions of ruin probabilities has a long history, see e.g. Segerdahl [791], Siegmund [806], Gerber, Goovaerts & Kaas [405] and Dickson [305]. The definition of $m(u)$ and the derivation of many of its properties goes back to Gerber & Shiu [408, 409]. Since then, this topic has experienced an enormous interest and activity. In a diffusion set-up, another function involving the time value of ruin was discussed in Powers [710] under the name expected discounted cost of insolvency.

In the literature, often additional conditions on the penalty function w like boundedness and continuity are imposed, which for instance ensure absolute continuity of m.

However, with some effort (and sometimes at the expense of regularity properties of m) these assumptions can usually be relaxed to condition (1.6). To avoid a too technical exposition, we will therefore not always be precise on the conditions for w, with the implicit understanding that for the respective proof method w is chosen appropriately and subsequently this choice can be relaxed (for a detailed discussion see e.g. Schmidli [780]).

2   The compound Poisson model

If $\{R_t\}$ is the classical Cramér-Lundberg process, one can derive an IDE for m, for instance via the following direct argument:

Let $h > 0$. By conditioning on the time and amount of the first jump before time h (if there is such a jump), we get

$$
m(u) = \int_0^h \beta\, e^{-(\beta+\delta)t} \mathrm{d}t \left(\int_0^{u+t} m(u+t-x)B(\mathrm{d}x) \right.
$$
$$
\left. + \int_{u+t}^\infty w(u+t, x-u-t)B(\mathrm{d}x) \right) + e^{-(\beta+\delta)h} m(u+h).
$$

We differentiate this equation with respect to h and set $h = 0$ in the resulting equation.[1] This yields

$$
\beta \int_0^u m(u-x)B(\mathrm{d}x) + \beta \int_u^\infty w(u, x-u)B(\mathrm{d}x) - (\beta+\delta)m(u) + m'(u) = 0. \quad (2.1)
$$

Under the boundary condition $\lim_{u\to\infty} m(u) = 0$, equation (2.1) has a unique solution.

The equation discussed in the following lemma will turn out to be of crucial importance throughout the whole section.

Lemma 2.1 *If $\widehat{B}[r]$ exists for an $r > 0$ and is steep (cf. p. 91) and $\delta > 0$, then, within the domain of $\widehat{B}[r]$, the Lundberg fundamental equation*

$$
\kappa(r) = \beta\big(\widehat{B}[r] - 1\big) - r = \delta \quad (2.2)
$$

has one positive root $\gamma_\delta > 0$ and one negative root $-\rho_\delta < 0$.

Proof. The only difference to equation IV.(5.2) is that now $\widehat{B}[r] = 1+\delta/\beta+r/\beta$, so the result is obvious by the convexity of $\widehat{B}[r]$ (see Figure XII.1). Note that $\gamma_\delta > \gamma_0 = \gamma$.   □

---

[1] Note that the differentiability is again guaranteed by the same argument as in Remark VIII.1.11. Other ways to derive equation (2.1) include the (essentially equivalent) generator approach (cf. Chapter II) and the method given in Section 3c.

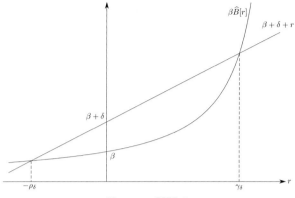

FIGURE XII.1

2a A Laplace transform approach

In view of the convolution term in the integro-differential equation (2.1), the analysis becomes particularly transparent with Laplace transforms. Let

$$\widehat{m}[-s] = \int_0^\infty e^{-su} m(u) \, du$$

and $\widehat{\omega}[-s] = \int_{u=0}^\infty e^{-su} \int_u^\infty w(u, x - u) \, B(dx) \, du$. Then taking Laplace transforms in (2.1) leads to

$$\widehat{m}[-s] = \frac{m(0) - \beta\widehat{\omega}[-s]}{\kappa(-s) - \delta}.$$

By (1.7), $m(u)$ is bounded in u, so its Laplace transform $\widehat{m}[-s]$ must be an analytic function for (at least) $\Re(s) > 0$ and hence the positive zero $s = \rho_\delta$ of the denominator must also be a zero of the numerator. In this way, one obtains by purely analytic arguments the identity

$$m(0) = \beta\widehat{\omega}[-\rho_\delta] = \beta \int_{x=0}^\infty e^{-\rho_\delta x} \int_0^\infty w(x, y) \, b(x + y) \, dy \, dx. \qquad (2.3)$$

From this we arrive at

$$\widehat{m}[-s] = \frac{\beta\big(\widehat{\omega}[-\rho_\delta] - \widehat{\omega}[-s]\big)}{s - \delta - \beta + \beta\widehat{B}[-s]}. \qquad (2.4)$$

Remark 2.2 Equation (2.3) contains surprisingly explicit information: if one chooses for $w(x, y)$ the Dirac-delta function for the second argument, one obtains the discounted probability density function of the deficit at ruin for initial surplus zero

$$g_\delta(y) \;=\; \beta \int_0^\infty \mathrm{e}^{-\rho_\delta x} b(x + y)\, \mathrm{d}x, \qquad y > 0, \qquad (2.5)$$

which provides an alternative proof of Lemma V.(3.2).

On the other hand, the choice $w(x) \equiv 1$ (and using $\widehat{\omega}[-\rho_\delta] = (1 - \widehat{B}[-\rho_\delta])/\rho_\delta$ and $\kappa(-\rho_\delta) = \delta$) leads to

$$\mathbb{E}\big[\mathrm{e}^{-\delta\tau(0)}; \tau(0) < \infty\big] \;=\; 1 - \frac{\delta}{\rho_\delta}, \qquad (2.6)$$

which already appeared in Corollary V.3.4 with a related, but somewhat different proof.

In view of (1.2), (2.6) implies that the Laplace transform of the finite-time survival probability $\phi(0, t) = 1 - \psi(0, t)$ w.r.t. t is simply given by

$$\int_0^\infty \mathrm{e}^{-\delta t} \phi(0, t)\, \mathrm{d}t \;=\; 1/\rho_\delta.$$

$\square$

Remark 2.3 Note that for $\delta \to 0$, under the net profit condition $\eta > 0$, we have $\rho_\delta \to 0$. An application of L'Hôpital's rule in (2.2) then gives $\delta/\rho_\delta \to 1 - \beta\mu_B$, and so the formulas in Remark 2.2 correspondingly simplify to $g_0(y) = \beta\overline{B}(y)$ (which is the ladder height density of the compound Poisson process as derived in IV.2) and (2.6) reduces to $\psi(0) = \beta\mu_B$.

Similarly, for $\delta = 0$ equation (2.4) simplifies to

$$\widehat{m}[-s] \;=\; \frac{\beta\big(\widehat{\omega}[0] - \widehat{\omega}[-s]\big)}{\kappa(-s)}. \qquad (2.7)$$

If further $w \equiv 1$, this gives

$$\widehat{\psi}[-s] \;=\; \frac{\beta\big(\mu_B - (1 - \widehat{B}[-s])/s\big)}{\kappa(-s)} \;=\; \frac{1}{s} - \frac{1 - \beta\mu_B}{\kappa(-s)}, \qquad (2.8)$$

which is another way to write the Pollaczeck-Khinchine formula IV.(2.2). $\square$

This link to the classical model can be used to obtain some further nice identities in a simple way:

Proposition 2.4 *The defective (non-discounted) density of the surplus prior to ruin in the compound Poisson risk model with initial capital u is given by*

$$f_0(x|u) = \frac{\beta}{1 - \beta\,\mu_B}\left(\overline{B}(x)\big(1 - \psi(u)\big) - I(x < u)\overline{B}(x)\big(1 - \psi(u - x)\big)\right) \quad (2.9)$$

and the defective (non-discounted) density of the deficit at ruin is

$$f_0(y|u) = \frac{\beta}{1 - \beta\,\mu_B}\left(\overline{B}(y)\big(1 - \psi(u)\big) - \int_0^u \big(1 - \psi(u - x)\big)\,b(x + y)\,\mathrm{d}x\right). \quad (2.10)$$

Proof. Replacing the denominator in (2.7) by (2.8) gives, after inverting the Laplace transform,

$$m(u) = \frac{\beta}{1 - \beta\,\mu_B}\left(\widehat{\omega}(0)\big(1 - \psi(u)\big) - \int_0^u \big(1 - \psi(u - x)\big)\int_x^\infty w(x, y - x)\,B(\mathrm{d}y)\,\mathrm{d}x\right).$$

If we now choose $w(x, y) = \mathrm{e}^{-ax}$, i.e. $\widehat{\omega}[-s] = \big(1 - \widehat{B}[-a - s]\big)/(a + s)$, then this leads to

$$\mathbb{E}\big[\mathrm{e}^{-a\,R_{\tau(u)}^-};\,\tau(u) < \infty\big]$$
$$= \frac{\beta}{1 - \beta\,\mu_B}\left(\frac{1 - \widehat{B}(-a)}{a}\big(1 - \psi(u)\big) - \int_0^u \big(1 - \psi(u - x)\big)\mathrm{e}^{-a\,x}\overline{B}(x)\,\mathrm{d}x\right),$$

and the inverse Laplace transform w.r.t. a is (2.9). On the other hand, the choice $w(x, y) = \mathrm{e}^{-ay}$, i.e. $\widehat{\omega}[-s] = \big(\widehat{B}[-a] - \widehat{B}[-s]\big)/(s - a)$, gives

$$\mathbb{E}\big[\mathrm{e}^{-a\,\xi(u)};\,\tau(u) < \infty\big] =$$
$$\frac{\beta}{1 - \beta\,\mu_B}\left(\frac{1 - \widehat{B}(-a)}{a}\big(1 - \psi(u)\big) - \int_0^u \big(1 - \psi(u - x)\big)\int_x^\infty \mathrm{e}^{-a(y - x)}B(\mathrm{d}y)\,\mathrm{d}x\right)$$

and its inverse Laplace transform is (2.10).[2] $\qquad\square$

Taking derivatives at $a = 0$ of the above Laplace transforms now leads to the following identities:

Corollary 2.5 *The moments $\mathbb{E}\big[(R_{\tau(u)-})^n \,\big|\, \tau(u) < \infty\big]$ of the surplus prior to ruin are*

$$\frac{\beta}{(1 - \beta\,\mu_B)\psi(u)}\left(\frac{\mu_B^{(n+1)}(1 - \psi(u))}{n + 1} - \int_0^u x^n\big(1 - \psi(u - x)\big)\overline{B}(x)\,\mathrm{d}x\right),$$

---

[2]Note that from these expressions one can again observe the somewhat curious fact that in the compound Poisson model for $u = 0$ the distributions of the surplus prior to ruin and of the deficit at ruin coincide, see also Theorem IV.2.2.

and the moments $\mathbb{E}\left[\xi(u)^n \,|\, \tau(u) < \infty\right]$ *of the deficit at ruin are given by*

$$\frac{\beta}{(1 - \beta\,\mu_B)\psi(u)} \left(\frac{\mu_B^{(n+1)}\left(1 - \psi(u)\right)}{n + 1} - \int_0^u \left(1 - \psi(u - x)\right) \int_x^\infty (y - x)^n B(\mathrm{d}y)\,\mathrm{d}x \right).$$

Proposition 2.6 *Let* $\mu_B^{(2)} < \infty$ *and define* $\psi_n(u) = \mathbb{E}\left[\tau(u)^n;\, \tau(u) < \infty\right]$ *for* $n \in \mathbb{N}_0$. *Then for* $n \geq 1$, $\psi_n(u)$ *is given by*

$$\frac{n}{1 - \beta\,\mu_B} \left(\int_0^u \psi(u - y)\psi_{n-1}(y)\,\mathrm{d}y + \int_u^\infty \psi_{n-1}(y)\,\mathrm{d}y - \psi(u) \int_0^\infty \psi_{n-1}(y)\,\mathrm{d}y \right).$$

In particular,

$$\mathbb{E}\left[\tau(u)|\,\tau(u) < \infty\right] = \frac{\int_0^u \psi(u - y)\psi(y)\,\mathrm{d}y + \int_u^\infty \psi(y)\,\mathrm{d}y - \frac{\beta\mu_B^{(2)}}{2(1 - \beta\mu_B)}\psi(u)}{(1 - \beta\mu_B)\,\psi(u)}.$$

Proof. For $w = 1$, $v_n(s) = \left. \frac{\partial^n \widehat{m}_\delta[-s]}{\partial \delta^n} \right|_{\delta = 0}$ is the Laplace transform of $(-1)^n \psi_n(u)$ w.r.t. u. Consequently, differentiation w.r.t. δ of

$$\left(\kappa(-s) - \delta\right) \widehat{m}_\delta[-s] = m(0) - \beta\,\widehat{\omega}[-s]$$

and choosing $\delta = 0$ gives

$$\kappa(-s)\, v_n(s) = \left. \frac{\partial^n m(0)}{\partial \delta^n} \right|_{\delta = 0} + n\,v_{n-1}(s).$$

Since $v_n(s)$ is an analytic function for $\Re(s) > 0$ and $s = 0$ is the only zero of $\kappa(-s)$ in the right halfplane, it follows that

$$\left. \frac{\partial^n m(0)}{\partial \delta^n} \right|_{\delta = 0} = -n \lim_{s \to 0} v_{n-1}(s) = (-1)^n\, n \int_0^\infty \psi_{n-1}(y)\,\mathrm{d}y.$$

Hence, together with (2.8), we obtain

$$v_n(s) = n \left((-1)^n \int_0^\infty \psi_{n-1}(y)\,\mathrm{d}y + v_{n-1}(s) \right) \frac{1/s - \widehat{\psi}[-s]}{1 - \beta\mu_B},$$

from which the desired formula for $\psi_n(u)$ follows immediately. Finally, the formula for $n = 1$ follows from $\psi_0(x) = \psi(x)$ and the identity $\int_0^\infty \psi(x)\,\mathrm{d}x = \beta\,\mu_B^{(2)}/\left[2(1 - \beta\mu_B)\right]$ (which is itself a direct consequence of (2.8) for $s \to 0$; see also IV.(3.7)). $\square$

Let us now use equation (2.4) to derive another representation of the defective renewal equation for $m(u)$:

Proposition 2.7 *The Gerber-Shiu function $m(u)$ in the compound Poisson model satisfies the defective renewal equation*

$$m(u) = (1 - \delta/\rho_\delta) \int_0^u m(u - y)\, g_p(y)\, \mathrm{d}y + h(u), \qquad (2.11)$$

where the (proper) density $g_p(y)$ is given by

$$g_p(y) = \frac{\beta}{1 - \delta/\rho_\delta} \int_y^\infty \mathrm{e}^{-\rho_\delta(x-y)} B(\mathrm{d}x), \quad y \ge 0 \qquad (2.12)$$

and

$$h(u) = \beta \int_u^\infty \mathrm{e}^{-\rho_\delta(x-u)} \int_x^\infty w(x, y - x)\, B(\mathrm{d}y)\, \mathrm{d}x. \qquad (2.13)$$

Proof. Replacing $\delta + \beta$ in the denominator of (2.4) by $\rho_\delta + \beta \widehat{B}[-\rho_\delta]$ (which holds because $\kappa(-\rho_\delta) = \delta$), one gets

$$\widehat{m}[-s] = \frac{\beta(\widehat{\omega}[-\rho_\delta] - \widehat{\omega}[-s])}{s - \rho_\delta - \beta\,\widehat{B}[-\rho_\delta] + \beta\widehat{B}[-s]} = \frac{\beta(\widehat{\omega}[-\rho_\delta] - \widehat{\omega}[-s])/(s - \rho_\delta)}{1 - \beta\,\frac{\widehat{B}[-\rho_\delta] - \widehat{B}[-s]}{s - \rho_\delta}}, \tag{2.14}$$

which is of the form (2.11) for

$$\widehat{h}[-s] = \beta \frac{\widehat{\omega}[-\rho_\delta] - \widehat{\omega}[-s]}{s - \rho_\delta}$$

and

$$\widehat{g}_p[-s] = \frac{\beta}{1 - \delta/\rho_\delta} \frac{\widehat{B}[-\rho_\delta] - \widehat{B}[-s]}{s - \rho_\delta}.$$

Taking the inverse Laplace transform of the latter two quantities then gives the assertion. $\qquad \square$

As a by-product, in view of (1.3) and (1.4) this again leads to

$$g_\delta(y) = (1 - \delta/\rho_\delta)\, g_p(y) = \beta \int_0^\infty \mathrm{e}^{-\rho_\delta x}\, b(x + y)\, \mathrm{d}x, \quad y \ge 0, \qquad (2.15)$$

which is (2.5).

At the same time, with $w \equiv 1$ it follows from Proposition 2.7 (or directly from $\widehat{\omega}[-s] = (1 - \widehat{B}[-s])/s$ in (2.4)) that

$$\int_0^\infty \mathrm{e}^{-su} \mathbb{E}\big[\mathrm{e}^{-\delta\tau(u)}; \tau(u) < \infty\big]\, \mathrm{d}u = \frac{\kappa(-s)/s - \delta/\rho_\delta}{\kappa(-s) - \delta},$$

which already appeared in Corollary V.3.5.

A related consequence of Proposition 2.7 is the asymptotic behavior of $m(u)$ for subexponential claim sizes:

Theorem 2.8 *Assume that $w \equiv 1$ and $B \in \mathscr{S}$ with finite mean. Then, for $\delta > 0$*

$$m(u) \;\sim\; \frac{\beta}{\delta}\,\overline{B}(u) \quad as \; u \to \infty.$$

Proof. For $w \equiv 1$ we have $\widehat{\omega}[-s] = \big(1 - \widehat{B}[-s]\big)/s$, and hence (again exploiting $\kappa(-\rho_\delta) = \delta$) the expression $\widehat{h}[-s]$ in the proof of Proposition 2.7 simplifies to

$$\widehat{h}[-s] \;=\; (1 - \delta/\rho_\delta)\,\frac{1 - \widehat{g_p}[-s]}{s}.$$

But this implies that for all $s \geq 0$

$$\widehat{m}[-s] \;=\; \frac{(1 - \delta/\rho_\delta)\,\frac{1-\widehat{g_p}[-s]}{s}}{1 - (1 - \delta/\rho_\delta)\,\widehat{g_p}[-s]} \;=\; \frac{1 - \delta/\rho_\delta}{s} - \frac{\delta}{s\,\rho_\delta}\sum_{n=1}^{\infty}(1 - \delta/\rho_\delta)^n\,\widehat{g_p}[-s]^n,$$

where the geometric series converges since both $m(0) = 1 - \delta/\rho_\delta < 1$ and $\widehat{g_p}[-s] < 1$ (g_p is a probability density). Taking the inverse Laplace transform now gives a representation of $m(u)$ as the geometric compound tail

$$m(u) \;=\; \frac{\delta}{\rho_\delta}\sum_{n=1}^{\infty}\left(1 - \frac{\delta}{\rho_\delta}\right)^n\,\overline{G_p^{*n}}(u), \tag{2.16}$$

where G_p is the c.d.f. of the density g_p in (2.12). Its tail is

$$\begin{aligned}
\overline{G_p}(y) &= \frac{\beta}{1 - \delta/\rho_\delta}\int_0^\infty \mathrm{e}^{-\rho_\delta x}\,\overline{B}(x+y)\,\mathrm{d}x \\
&= \frac{\beta}{\rho_\delta - \delta}\left(\overline{B}(y) + \int_0^\infty \mathrm{e}^{-\rho_\delta x}\,\overline{B}(\mathrm{d}x + y)\right) \\
&= \frac{\beta}{\rho_\delta - \delta}\,\overline{B}(y)\,\big(1 + \mathrm{o}(1)\big), \tag{2.17}
\end{aligned}$$

where the last step follows from $B \in \mathscr{S}$ and Proposition X.1.5. Corollary X.1.10 now implies $G_p \in \mathscr{S}$. As in Lemma X.2.2, one obtains from (2.16) by dominated convergence that

$$\lim_{u\to\infty}\frac{m(u)}{\overline{G_p}(u)} \;=\; \frac{1 - \delta/\rho_\delta}{\delta/\rho_\delta}$$

and the result finally follows from (2.17). $\qquad\qquad\square$

Remark 2.9 Since for $\delta = 0$, $m(u) = \psi(u)$ for $w \equiv 1$, a comparison of the above result with Theorem X.2.1 shows that for subexponential claim sizes the introduction of the discount rate $\delta > 0$ moves the asymptotic behavior of $m(u)$ away from the magnitude of the integrated tail $\overline{B}_0$ to the one of the tail $\overline{B}$ (for $\delta \to 0$ one has $1 - \delta/\rho_\delta \to \beta\mu_B$ and the density $g_0(y)$ in the defective renewal equation is correspondingly replaced by $\overline{B}(y)/\mu_B$, cf. IV.(3.2)).

Since for general penalty functions the representation of $m(u)$ as a compound geometric tail is usually not available, one needs slightly different methods to establish corresponding asymptotic results. We shall not pursue this further; the interested reader is referred to Tang & Wei [834] for details. $\quad\square$

2b   Change of measure

As in Section IV.1, consider the Wald martingale $L_t = \exp\{rS_t - \kappa(r)t\}$ as the likelihood ratio process. Then we have by a change of measure that

$$
m(u) = \mathbb{E}_r\left[e^{-rS_{\tau(u)} + \kappa(r)\tau(u)}\,e^{-\delta\tau(u)}\,w\big(R_{\tau(u)-},\xi(u)\big); \tau(u) < \infty\right].
$$

If the Lundberg coefficient $\gamma_\delta > 0$ exists, then $\mathbb{P}_{\gamma_\delta}\big(\tau(u) < \infty\big) = 1$ and hence

$$
m(u) = \mathbb{E}_{\gamma_\delta}\left[e^{-\gamma_\delta S_{\tau(u)}}\,w\big(R_{\tau(u)-},\xi(u)\big)\right] = \mathbb{E}_{\gamma_\delta}\left[e^{-\gamma_\delta \xi(u)}\,w\big(R_{\tau(u)-},\xi(u)\big)\right]e^{-\gamma_\delta u}.
$$

Note that under the new measure $\mathbb{P}_{\gamma_\delta}$, not only the event of ruin is certain, but also the time-dependence of the penalty function has disappeared (or, rather, hides in the value of γ_δ). If the penalty w is bounded, then a Lundberg-type inequality of the form $m(u) \leq \sup_{x,y} w(x,y)e^{-\gamma_\delta u}$ immediately follows.

The next result gives the asymptotic behavior for general continuous penalty functions:

Proposition 2.10 *Assume that the penalty function w is continuous. If $\gamma_\delta > 0$ exists, then*

$$
\lim_{u\to\infty} m(u)e^{\gamma_\delta u} = C_\delta = \frac{\beta \int_0^\infty \int_z^\infty w(z, x-z)B(\mathrm{d}x)\,(e^{\gamma_\delta z} - e^{-\rho_\delta z})\,\mathrm{d}z}{\beta\widehat{B}'[\gamma_\delta] - 1}.
$$

Proof. Define $\widetilde{m}(u) = m(u)e^{\gamma_\delta u} = \mathbb{E}_{\gamma_\delta}\left[e^{-\gamma_\delta \xi(u)}\,w\big(R_{\tau(u)-},\xi(u)\big)\right]$ and denote by $H(x,y) = \mathbb{P}_{\gamma_\delta}[R_{\tau(0)-} \leq x, \xi(0) \leq y]$ the joint distribution of the surplus prior to ruin and the deficit at ruin under the tilted measure given that the risk process starts in $u = 0$. Then, just as in (1.3) but now under the new measure, it immediately follows that

$$
\widetilde{m}(u) = \int_0^u \widetilde{m}(u-y)H(\infty,\mathrm{d}y) + \int_{y=u}^\infty \int_{x=0}^\infty w(x+u, y-u)e^{-\gamma_\delta(y-u)}H(\mathrm{d}x,\mathrm{d}y).
$$

This is a (now proper) renewal equation for $\widetilde{m}(u)$ and according to Proposition A1.1 we need to show that the second summand above is directly Riemann integrable. Since w is continuous, it is enough to show that there is a directly Riemann integrable upper bound. Since $\gamma_\delta > 0$ exists, all moments of the claim size distribution (and consequently of the surplus prior to ruin and the deficit at ruin) exist and in view of (1.6) it is then enough to show that $1 - H(\infty, y)$ is directly Riemann integrable, but the latter follows from the existence of all moments of the claim size distribution B. Applying now Proposition A1.1, it just remains to calculate the limiting constant

$$C_\delta = \frac{\int_0^\infty z(u)\,\mathrm{d}u}{\mu_F} = \frac{\int_{u=0}^\infty \int_{y=u}^\infty \int_{x=0}^\infty w(x+u, y-u)\mathrm{e}^{-\gamma_\delta(y-u)} H(\mathrm{d}x, \mathrm{d}y)\,\mathrm{d}u}{\int_0^\infty (1 - H(\infty, y))\,\mathrm{d}y}.$$

(2.18)

Following the idea of Remark IV.5.6, the simplest way to identify its value is from $C_\delta = \lim_{s\to 0} s\,\widehat{m}[-s + \gamma_\delta]$ and expression (2.4). However, we shall here directly evaluate (2.18): Recall from Theorem IV.2.2 that under the original measure

$$\mathbb{P}\big[R_{\tau(0)^-} \le x, \xi(0) \le y, \tau(0) < \infty\big] = \beta \int_{z=0}^x \int_{v=z}^{z+y} B(\mathrm{d}v)\,\mathrm{d}z.$$

(2.19)

Under $\mathbb{P}_{\gamma_\delta}$, the risk process is again compound Poisson with $\beta_{\gamma_\delta} = \beta\,\widehat{B}[\gamma_\delta]$ and $B_{\gamma_\delta}(\mathrm{d}x) = \dfrac{\mathrm{e}^{\gamma_\delta x}}{\widehat{B}[\gamma_\delta]} B(\mathrm{d}x)$, so the safety loading is negative. Hence we need a further exponential tilting by the factor $-(\rho_\delta + \gamma_\delta)$ to obtain a classical compound Poisson process R_t^* with positive safety loading, claim distribution B^* and Poisson parameter β^*, for which we can apply (2.19). This leads to

$$
\begin{aligned}
H(x, y) &= \mathbb{E}\big[\mathrm{e}^{(\gamma_\delta + \rho_\delta)\xi^*(0)}; R_{\tau^*(0)^-}^* \le x, \xi^*(0) \le y, \tau^*(0) < \infty\big] \\
&= \beta^* \int_{v=0}^y \mathrm{e}^{(\gamma_\delta + \rho_\delta)v} \int_{z=0}^x B^*(z + \mathrm{d}v)\mathrm{d}z \\
&= \beta^* \int_0^x \mathrm{e}^{-(\gamma_\delta + \rho_\delta)z} \int_z^{z+y} \mathrm{e}^{(\gamma_\delta + \rho_\delta)\,v} B^*(\mathrm{d}v)\,\mathrm{d}z.
\end{aligned}
$$

Since β^* and B^* are related to β and B through exponential tilting by $-\rho_\delta$, we finally arrive at

$$H(x, y) = \beta \int_0^x \mathrm{e}^{-(\gamma_\delta + \rho_\delta)z} \int_z^{z+y} \mathrm{e}^{\gamma_\delta v} B(\mathrm{d}v)\,\mathrm{d}z.$$

The denominator of (2.18) then is, by changing the order of integration,

$$\int_0^\infty \left(1 - H(\infty, y)\right) dy = \beta \int_0^\infty \int_0^\infty \int_{y+z}^\infty e^{\gamma_\delta v} \, dB(v) e^{-(\gamma_\delta + \rho_\delta)z} \, dz \, dy$$

$$= \frac{\beta}{\gamma_\delta + \rho_\delta} \left(\widehat{B}'[\gamma_\delta] - \frac{\widehat{B}[\gamma_\delta] - \widehat{B}[-\rho_\delta]}{\gamma_\delta + \rho_\delta}\right) = \frac{\beta \widehat{B}'[\gamma_\delta] - 1}{\gamma_\delta + \rho_\delta}.$$

Similarly, the numerator of (2.18) simplifies to

$$\beta \int_{u=0}^\infty \int_{y=u}^\infty \int_{x=0}^\infty w(x+u, y-u) e^{-\rho_\delta x + \gamma_\delta u} B(x+dy) \, dx \, du$$

which finally leads to the assertion. $\qquad \square$

Remark 2.11 Note that for $w \equiv 1$ the constant simplifies to

$$C_\delta = \frac{\delta}{\kappa'(\gamma_\delta)} \left(\frac{1}{\gamma_\delta} + \frac{1}{\rho_\delta}\right). \qquad (2.20)$$

$\qquad \square$

2c Martingales

It is easy to see that the stochastic process $\left\{e^{-\delta t - r R_t}\right\}_{t \geq 0}$ is a martingale w.r.t. its natural filtration $\mathscr{F}$ if and only if $r = \gamma_\delta > 0$ or $r = -\rho_\delta \leq 0$. This can be exploited in various ways.

Proposition 2.12 *If $\gamma_\delta > 0$ exists, then*

$$\mathbb{E}\left[e^{-\delta \tau(u) + \gamma_\delta \xi(u)}; \tau(u) < \infty\right] = e^{-\gamma_\delta u}, \qquad \delta \geq 0, u \geq 0.$$

Proof. The martingale $\left\{e^{-\delta t - \gamma_\delta R_t}\right\}_{t \geq 0}$ is bounded by 1 for $0 \leq t < \tau(u)$. Hence we can apply the optional sampling theorem for the stopping time $\tau(u)$ to obtain

$$\mathbb{E}\left[e^{-\delta \tau(u) - \gamma_\delta R_{\tau(u)}}\right] = e^{-\gamma_\delta u}.$$

Due to $\lim_{u \to \infty} R_t = \infty$ a.s., one has $\mathbb{E}\left[e^{-\delta \tau(u) - \gamma_\delta R_{\tau(u)}}; \tau(u) = \infty\right] = 0$ for $\delta \geq 0$ and the result follows. $\qquad \square$

Note that for $\delta = 0$, $\mathbb{E}\left[e^{\gamma \xi(u)}; \tau(u) < \infty\right] = e^{-\gamma u}$, in line with II.(3.1).

Exploiting the martingale for $r = -\rho_\delta$ leads to the following result (which is Lemma V.3.1, but we state it here again for completeness).

Proposition 2.13 *Let* $\tau_a^+ = \min\{t > 0 : R_t \geq a \mid R_0 = u\}$ *for* $a > u$. *Then*

$$\mathbb{E}[e^{-\delta \tau_a^+}] = e^{-\rho_\delta(a-u)}, \qquad \delta > 0, a > u. \qquad (2.21)$$

Proof. For fixed $a > u$, the martingale $\{e^{-\delta t + \rho_\delta R_t}\}_{t \geq 0}$ is bounded by $e^{\rho_\delta a}$ for $0 \leq t \leq \tau_a^+$. Hence we can apply the optional sampling theorem for the stopping time τ_a^+ to obtain $\mathbb{E}[e^{-\delta \tau_a^+ + \rho_\delta a}] = e^{\rho_\delta u}$. $\qquad \square$

Define $\psi_\delta(u) = \mathbb{E}[e^{-\delta \tau(u) + \rho_\delta R_\tau(u)}; \tau(u) < \infty]$. Let $T_0 = \min\{t > \tau(u)|R_t = 0\}$ be the time of recovery after ruin. Since (2.21) holds for arbitrary $a, u \in \mathbb{R}$, one has for $a < b$ that $\mathbb{E}[e^{-\delta(\tau_b^+ - \tau_a^+)}|\tau_a^+ < \tau_b^+] = e^{-\rho_\delta(b-a)}$ and consequently

$$\mathbb{E}[e^{-\delta(T_0 - \tau(u))} \,\big|\, \tau(u) < \infty, \mathscr{F}_{\tau(u)}] = e^{-\rho_\delta \xi(u)}.$$

This leads to

$$\mathbb{E}[e^{-\delta T_0}; \tau(u) < \infty] = \mathbb{E}[e^{-\delta(T_0 - \tau(u))}e^{-\delta \tau(u)}; \tau(u) < \infty] = \psi_\delta(u),$$

which gives $\psi_\delta(u)$ the interpretation as the expected present value of a payment of 1 made at the time of recovery, if ruin occurs.

Proposition 2.14 *The discounted density of the surplus prior to ruin satisfies*

$$f(x|u) = f(x|0) \frac{e^{\rho_\delta u} \, I(x > u) + e^{\rho_\delta x} \psi_\delta(u - x) \, I(x \leq u) - \psi_\delta(u)}{1 - \psi_\delta(0)}, \qquad x > 0.$$

At $x = u$, $f(x|u)$ *has a discontinuity of size* $f(x|0) \, e^{\rho_\delta u} = \beta \overline{B}(u)$.

Proof. We will use Laplace transforms. Since $\psi_\delta(u)$ is the Gerber-Shiu function with $w(x, y) = e^{-\rho_\delta y}$ (and correspondingly $\widehat{\omega}[-s] = (\widehat{B}[-\rho_\delta] - \widehat{B}[-s])/(s - \rho_\delta) = \widehat{g}_\delta[-s]/\beta$), it follows from (2.14) and $\beta \widehat{\omega}[-\rho_\delta] = m(0)$ that

$$\widehat{\psi}_\delta[-s] = \frac{\psi_\delta(0) - \widehat{g}_\delta[-s]}{(s - \rho_\delta)(1 - \widehat{g}_\delta[-s])}. \qquad (2.22)$$

Similarly to (1.3), by conditioning whether or not ruin occurs at the first time when the surplus falls below the initial value u, one can write down the renewal equation

$$f(x, y|u) = \int_0^u f(x, y|u - z)g_\delta(z)\, \mathrm{d}z + f(x - u, y + u|0) \, I(x > u). \qquad (2.23)$$

By (2.15)

$$f(x - u, y + u|0) = \beta e^{-\rho_\delta(x-u)} \, b(x + y) = f(x, y|0)e^{\rho_\delta u}.$$

Hence integrating the renewal equation w.r.t. y, we have

$$f(x|u) = \int_0^u f(x|u-z)g_\delta(z)\,\mathrm{d}z + f(x|0)\,\mathrm{e}^{\rho_\delta u}\,I(x>u).$$

The function $\zeta(u)$ defined through $f(x|u) = f(x|0)\,\zeta(u)$ then fulfills the renewal equation

$$\zeta(u) = \int_0^u \zeta(u-z)g_\delta(z)\,\mathrm{d}z + \mathrm{e}^{\rho_\delta u}\,I(x>u),$$

so that its Laplace transform is given by

$$\widehat{\zeta}[-s] = \frac{1}{1-\widehat{g}_\delta[-s]}\,\frac{\mathrm{e}^{(\rho_\delta-s)x}-1}{\rho_\delta-s}.$$

The statement of the proposition is that this expression is also the Laplace transform (w.r.t. u) of the function

$$\zeta_2(u) = \frac{\mathrm{e}^{\rho_\delta u}\,I(x>u) + \mathrm{e}^{\rho_\delta x}\psi_\delta(u-x)\,I(x\le u) - \psi_\delta(u)}{1-\psi_\delta(0)}.$$

Standard calculations show that

$$\widehat{\zeta_2}[-s] = \frac{\frac{\mathrm{e}^{(\rho_\delta-s)x}-1}{\rho_\delta-s} + \mathrm{e}^{(\rho_\delta-s)x}\widehat{\psi}_\delta[-s] - \widehat{\psi}_\delta[-s]}{1-\psi_\delta(0)} = \frac{\left(\mathrm{e}^{(\rho_\delta-s)x}-1\right)\left(\frac{1}{\rho_\delta-s)}+\widehat{\psi}_\delta[-s]\right)}{1-\psi_\delta(0)}.$$

Substituting (2.22) into the latter equation gives

$$\widehat{\zeta_2}[-s] = \frac{\mathrm{e}^{(\rho_\delta-s)x}-1}{\rho_\delta-s}\,\frac{1}{1-\widehat{g}_\delta[-s]},$$

which indeed coincides with $\widehat{\zeta}[-s]$. □

Since $\rho_\delta = 0$ for $\delta = 0$, $\psi_0(u)$ is the usual ruin probability $\psi(u)$, and we obtain the (defective) non-discounted density of the surplus prior to ruin

$$f_0(x|u) = f_0(x|0)\,\frac{I(x>u) + \psi(u-x)\,I(x\le u) - \psi(u)}{1-\psi(0)}, \qquad x>0,$$

which is another way of writing (2.9).

2d   Further ruin-related quantities

Somewhat surprisingly, it just turned out that the time of recovery T_0 plays a crucial role for the surplus prior to ruin. A related natural question is to consider the maximum severity of ruin prior to recovery, i.e. the r.v. $M(u) = \sup\{|R_t| \,|\, \tau(u) \le t \le T_0\}$. Its distribution function (given that ruin occurs) turns out to have a strikingly simple form in terms of the ruin probability $\psi(u)$:

Proposition 2.15 *For positive safety loading $\eta > 0$,*

$$\mathbb{P}\big(M(u) \le z \,|\, \tau(u) < \infty\big) \;=\; \frac{\psi(u) - \psi(u+z)}{\psi(u)\big(1 - \psi(z)\big)}.$$

Proof. Given ruin occurs, the event $M(u) \le z$ happens if ruin occurs with some deficit $y \le z$ and if the reserve process does not fall below level $-z$ from there on before it is positive again. The latter is equivalent to the event that a risk reserve process starting in $z - y$ attains level z before ruin, which happens with probability $\big(1 - \psi(z - y)\big)/\big(1 - \psi(z)\big)$. This gives

$$\mathbb{P}\big(M(u) \le z | \tau(u) < \infty\big) \;=\; \int_0^z \frac{f_0(y|u)}{\psi(u)} \frac{1 - \psi(z - y)}{1 - \psi(z)} \,\mathrm{d}y\,.$$

On the other hand,

$$\psi(u+z) \;=\; \int_z^\infty f_0(y|u)\,\mathrm{d}y + \int_0^z f_0(y|u)\psi(z - y)\,\mathrm{d}y, \qquad (2.24)$$

because for a risk process starting at level $u + z$, ruin can only occur if the reserve falls below level z and the first integral gives the probability that ruin occurs directly then, whereas the second integral gives the probability that ruin occurs later. Equation (2.24) can be rewritten as

$$\int_0^z f_0(y|u)\big(1 - \psi(z - y)\big)\,\mathrm{d}y \;=\; \int_0^\infty f_0(y|u)\,\mathrm{d}y - \psi(u+z) \;=\; \psi(u) - \psi(u+z)$$

from which the result follows. $\square$

If the reserve process recovers after ruin, it may again first become negative before it reaches the previous maximum of the process, possibly leading to a larger ruin severity before reaching again this running maximum. The maximum severity of the ruin excursion out of the running maximum can be studied by very simple means thanks to the duality with queueing models, leading to another simple formula in terms of the survival probability $\phi(u) = 1 - \psi(u)$:

Proposition 2.16 *Define $M_r(u)$ as the (absolute value of the) maximum severity during the excursion out of the running maximum of R_t that causes ruin. Then, for positive safety loading,*

$$\mathbb{P}\big(M_r(u) > z \,|\, \tau(u) < \infty\big) = \int_u^\infty \frac{\phi'(w+z)}{\phi(w+z)} \frac{\phi(u)}{\phi(w)} \, \mathrm{d}w.$$

Proof. Denote by $F_k(u, z)$ the probability that for a reserve process starting in u, ruin occurs at the kth excursion out of the running maximum and the maximum severity is below level $-z$. Recall from Theorem III.2.3 the close connection between the maximum workload $V_{\max}$ of an M/G/1 queue and the survival probability $\phi(u)$ of the compound Poisson risk process. If $G(u) = \mathbb{P}(V_{\max} < u)$, then

$$F_k(u, z) = \int_{t=0}^\infty \beta^k \frac{t^{k-1}}{(k-1)!} \mathrm{e}^{-\beta t} \left(\int_{v=0}^t G(u+v) \frac{\mathrm{d}v}{t} \right)^{k-1} \overline{G}(u+t+z) \, \mathrm{d}t,$$

because each excursion out of the running maximum occurs at an exponential(β) distributed time and whenever the excursion does not lead to ruin it can be excised from the process. Accordingly, the kth excursion occurs after an Erlang(k) distributed time and the previous ones are uniformly distributed over this interval and must not lead to ruin. Now the assertion follows by noting that $\mathbb{P}\big(M_r(u) > z \,|\, \tau(u) < \infty\big) = \sum_{k=1}^\infty F_k(u, z)$, some simple algebra and application of Theorem III.2.3. $\square$

Notes and references There are several ways to derive the results of this section. The seminal paper of Gerber & Shiu [409] is a rich source of calculations in this context and much of the material presented in this section can be found there. The transparency of Laplace transforms in the analysis of Gerber-Shiu functions is apparent, see also Dufresne & Gerber [333], Gerber & Shiu [408] and Dickson [305, 308]. Parts of the exposition of Section 2a follow from Albrecher & Boxma [19]; see also Willmot & Lin [892] and Schmidli [773]. In Albrecher, Gerber & Yang [23] one can find a transparent approach to some of the derived formulas by just using rational functions. Starting with the defective renewal equation for m, Lin & Willmot [596] derive some of the above and further results via compound geometric tails. Computational aspects of the calculation of ruin time moments are discussed in Drekic *et al.* [329, 330], see also Dermitzakis *et al.* [296]. The duration of negative surplus $T_0 - \tau(u)$ was studied by other techniques in Dickson & Egidio dos Reis [311]. For the time and area spent at negative surplus levels up to a fixed time T, see Loisel [607]. Borovkov, Palmowski & Boxma [190] give a detailed related analysis of such quantities in a queueing context. Pitts & Politis [708] use a functional approach to approximate the Gerber-Shiu function with the one from a 'near' claim distribution for which more explicit results exist. An algorithmic procedure to obtain moments of the ruin time for discrete claim sizes in terms of generalized Appell polynomials was developed in Picard & Lefèvre [702].

The result of Theorem 2.8 is from Siaulys & Asanaviciute [803]. Generalizing a number of earlier results, Landriault & Willmot [573] use the Lagrange implicit function theorem to determine an explicit expression for the trivariate distribution of the time to ruin, the deficit at ruin and the surplus prior to ruin. This expression contains an infinite series of the integral of convolutions of the claim size density. Extending an idea of Frey & Schmidt [373], Usabel [860] develops a recursive computational technique to approximate the above trivariate distribution using its Taylor expansion in terms of the Poisson parameter β around $\beta = 0$. Tail bounds for this distribution obtained from the integral equation can be found in Psarrakos & Politis [720].

The Gerber-Shiu function in a compound Poisson model with interest is considered in Cai & Dickson [214], see also Yang & Zhang [901] and Wu, Wang & Zhang [897]. Cai [213] and Yuen & Wang [911] deal with stochastic interest, whereas Badescu, Drekic & Landriault [118] study these ruin-related quantities with a multi-step premium rule under a MAP arrival process. For absolute ruin and the inclusion of tax payments see Ming, Wang & Xao [646]. Albrecher, Hartinger & Tichy [27] study the Gerber-Shiu function under a time-dependent threshold model for the premium income.

Using the renewal measure of the defective renewal sequence of the zero points of R_t, calculations involving the maximum of the surplus process up to ruin, the last time the surplus process passes zero before going to infinity ultimately and the minimum of the surplus process up to that time are provided in Wu, Wang & Wei [896]. Proposition 2.15 is from Picard [700], where also the quantity $\int_{\tau(u)}^{T_0} |R_t| \mathrm{d}t$ (i.e. the area below zero until the time of recovery) is studied. For Proposition 2.16 and further related formulas see Albrecher, Borst, Boxma & Resing [17]. Baigger [124] studies general criteria under which ruin occurs only finitely often.

The Kella-Whitt martingale and the martingale introduced in [84] are used in Frostig [376] to derive results about the time of ruin in the presence of a reflecting barrier. Extensions are of course possible in many directions, Cheung et al. [242] include for instance information on the surplus after the second-last claim before ruin.

Note that all techniques presented for the renewal model in the next section are by definition directly applicable for the compound Poisson model as well.

3   The renewal model

In the following we consider the zero-delayed renewal model with interarrival time distribution $A(t)$ and density $a(t)$. If T_1 is the epoch of the first claim, the standard renewal argument gives $m(u) = \mathbb{E}\left(e^{-\delta T_1} m(u + T_1 - U_1)\right)$. I.e. $m(u)$ is given by

$$\int_0^\infty e^{-\delta t} a(t) \left(\int_0^{u+t} m(u+t-y)\, B(\mathrm{d}y) + \int_{u+t}^\infty w(u+t, y-u-t)\, B(\mathrm{d}y) \right) \mathrm{d}t. \quad (3.1)$$

3a Change of measure

Recall from Section VI.3a the imbedded random walk structure of the renewal model. I.e. if we only consider the discrete time points at which a claim just occurred the resulting discrete time process is a random walk and in particular Markovian. However, since in this chapter we want to keep information on the time to ruin and the surplus prior to ruin as well (which is both lost when only in the imbedded random walk view), this type of markovization of the process is too crude for the present purpose. An alternative and elegant way to markovize the process is to consider the random variable $V_t = T_{N_t+1} - t$ as an additional state variable, which is the time remaining until the next claim.[3] Define $\kappa(r)$ as the solution of

$$\widehat{B}[r]\widehat{A}\big[-r - \kappa(r)\big] = 1, \tag{3.2}$$

for every r with $\widehat{B}[r] < \infty$. It is easy to show by properties of moment-generating functions that for $r \geq 0$ this solution $\kappa(r)$ exists, is unique, and that $\kappa(r)$ is a strictly convex function on the set where it exists. Also, $\kappa(0) = 0$ and $\kappa'(0) < 0$ under the net profit condition. With some further effort, one can then show that

$$L_t = \widehat{B}[r]e^{-(r+\kappa(r))V_t}\, e^{rS_t}e^{-\kappa(r)t}$$

is a martingale with respect to the filtration generated by $\{(R_t, V_t)\}$ (see e.g. [746, Th.11.5.2]). L_t can now be used as a likelihood ratio process. Under the measure $\mathbb{P}_r[\cdot] = \mathbb{E}[L_t; \cdot]$, the risk process R_t remains a Sparre Andersen risk process with claim distribution $B_r(dy) = \widehat{A}[-r - \kappa(r)]e^{ry}B(dy)$ and the interclaim time distribution changed to $A_r(dt) = \widehat{B}[r]e^{-(r+\kappa(r))t}A(dt)$. If $r > \operatorname{argmin}\kappa(r)$, then the drift under the new measure is negative $(-\kappa'(r) < 0)$ and consequently $\mathbb{P}_r\big(\tau(u) < \infty\big) = 1$.

Under the measure $\mathbb{P}_r$, the Gerber-Shiu function $m(u)$ can be expressed as

$$\widehat{A}[-r - \kappa(r)] \cdot \mathbb{E}_r\left[e^{(r+\kappa(r))V_{\tau(u)} - rS_{\tau(u)} + (\kappa(r)-\delta)\tau(u)}\, w\big(R_{\tau(u)-}, \xi(u)\big); \tau(u) < \infty\right].$$

Since $V_{\tau(u)}$ is the time to the next claim after ruin and hence independent of $\mathscr{F}_{\tau(u)-}$ and $R_{\tau(u)}$, this identity simplifies to

$$m(u) = \mathbb{E}_r\left[e^{-r\xi(u)}e^{(\kappa(r)-\delta)\tau(u)}\, w\big(R_{\tau(u)-}, \xi(u)\big); \tau(u) < \infty\right]e^{-ru}.$$

As in the compound Poisson case, the time dependence disappears if $\kappa(r) = \delta$.

---

[3]This is called *forward markovization* and leads to some subtleties concerning the interpretation of the appropriate filtration. Alternatively, one could also use the time since the last claim (*backward markovization*) with a more intuitive appropriate filtration, but then the resulting equations are usually more cumbersome, see the References at the end of the section.

Proposition 3.1 *Assume that the equation $\kappa(r) - \delta = 0$ with $\kappa(r)$ defined in (3.2) has a positive solution $\gamma_\delta > \operatorname{argmin} \kappa(r)$. Then*

$$m(u) = \mathbb{E}_{\gamma_\delta}\left[e^{-\gamma_\delta \xi(u)} w\left(R_{\tau(u)-}, \xi(u)\right)\right] e^{-\gamma_\delta u} \qquad (3.3)$$

and for a continuous penalty function w

$$\lim_{u \to \infty} e^{\gamma_\delta u} m(u) = C_\delta$$

for some constant $C_\delta > 0$.

Proof. Expression (3.3) follows from $r = \gamma_\delta$ and $\mathbb{P}_{\gamma_\delta}(\tau(u) < \infty) = 1$. Now the same procedure as in the proof of Proposition 2.10 gives a renewal equation for $e^{\gamma_\delta u} m(u)$ and establishes the asymptotic result. □

The constant C_δ is now more difficult to evaluate. We will give its form for a large class of interclaim time distributions A in Corollary 3.9.

Formula (3.3) can be helpful in a number of situations. We give one particular example:

Corollary 3.2 *In the renewal model, for exponential(ν) claim sizes the Laplace transform of the time to ruin is given by*

$$\mathbb{E}\left[e^{-\delta \tau(u)}; \tau(u) < \infty\right] = \frac{\nu - \gamma_\delta}{\nu} e^{-\gamma_\delta u}.$$

Proof. In this case $\gamma_\delta > 0$ clearly exists and is the solution of $\widehat{A}[-\gamma_\delta - \delta] = 1 - \gamma_\delta/\nu$. The lack-of-memory property implies $\mathbb{P}\left(\xi(u) > x \mid \tau(u) < \infty\right) = e^{-\nu x}$, and correspondingly $\mathbb{P}_{\gamma_\delta}\left(\xi(u) > x\right) = e^{-(\nu - \gamma_\delta)x}$. The result then follows from (3.3) with $w \equiv 1$. □

Remark 3.3 Note that Corollary 3.2 extends Theorem VI.2.2 with a quite different proof. □

3b A modified random walk

Another way to remove the discounting is to interpret $a_\delta(t) = e^{-\delta t} a(t)$ in (3.1) as a new (now defective) interclaim time density for a non-discounted risk process. This leads to a modified imbedded random walk $S_{\delta,k} = \sum_{i=1}^{k}(U_i - T_{\delta,i})$, $k \geq 1$, where $T_{\delta,i}$ has defective density $a_\delta(t)$ and a point mass of size $1 - \int_0^\infty a_\delta(t)\mathrm{d}t = 1 - \widehat{A}[-\delta]$ at infinity. Consequently, $\sup_k S_{\delta,k}$ is finite with probability 1. By definition, the ruin probability of this modified random walk is the Laplace transform of the ruin time of the original risk process R_t,

$$\mathbb{E}\big[e^{-\delta\tau(u)}; \tau(u) < \infty\big] = \mathbb{P}\big(\sup_k S_{\delta,k} > u\big),$$

and the discounted distribution of the deficit at ruin of R_t is

$$\mathbb{E}\big[e^{-\delta\tau(u)}; \xi(u) \leq y, \tau(u) < \infty\big] = \mathbb{P}\big(N_{\delta,\tau} < \infty, S_{\delta,N_{\delta,\tau}} \leq u + y\big),$$

where $N_{\delta,\tau} = \inf\{k : S_{\delta,k} > u\}$. Hence for these penalty functions the calculations are reduced to random walk techniques with defective increment distribution (for which e.g. the Wiener-Hopf factorization can still be done in the same way). If the claim sizes are phase-type, this leads to the following generalization of Corollary 3.2 and also of Theorem IX.4.4:

Theorem 3.4 *Consider the renewal model with arbitrary interarrival distribution A and phase-type claim size distribution B with representation $(\boldsymbol{\alpha}, \boldsymbol{T})$. Denote with $\boldsymbol{\alpha}_{\delta+}$ the minimal non-negative solution of*

$$\boldsymbol{\alpha}_{\delta+} = \boldsymbol{\alpha} \int_0^\infty e^{(\boldsymbol{T} + \boldsymbol{t}\boldsymbol{\alpha}_{\delta+})t} a_\delta(t)\mathrm{d}t.$$

Then the Laplace transform of the ruin time is

$$\mathbb{E}\big[e^{-\delta\tau(u)}; \tau(u) < \infty\big] = \boldsymbol{\alpha}_{\delta+}\, e^{(\boldsymbol{T} + \boldsymbol{t}\boldsymbol{\alpha}_{\delta+})u}\, \boldsymbol{e},$$

and the discounted distribution of the deficit at ruin is given by

$$\mathbb{E}\big[e^{-\delta\tau(u)}; \xi(u) \geq y, \tau(u) < \infty\big] = \boldsymbol{\alpha}_{\delta+}\, e^{(\boldsymbol{T} + \boldsymbol{t}\boldsymbol{\alpha}_{\delta+})u}\, e^{\boldsymbol{T}y}\boldsymbol{e}.$$

Proof. The proof is a straightforward extension of the one of Theorem IX.4.4; see also Ren [735]. □

3c Integro-differential equations

If the interarrival time density $a(t)$ has rational Laplace transform, the integral equation (3.1) can be transformed into an integro-differential equation (IDE). For that purpose assume that $a(t)$ satisfies an nth order linear differential equation with constant coefficients, written in operator notation as

$$p_A\Big(\frac{\mathrm{d}}{\mathrm{d}t}\Big)a(t) = 0, \tag{3.4}$$

with the polynomial

$$p_A(x) = x^n + c_{n-1}x^{n-1} + \cdots + c_0, \qquad c_j \in \mathbb{R}, c_0 \neq 0.$$

The first initial condition is determined by the fact that $a(t)$ is a density. For ease of exposition, assume that the remaining $n-1$ initial conditions of this ordinary differential equation (ODE) are homogeneous,[4] i.e.

$$a^{(k)}(0) = 0 \quad (k = 0, \ldots, n-2). \tag{3.5}$$

Integrating (3.4) from 0 to ∞ and using $\int_0^\infty a(t)\mathrm{d}t = 1$, with (3.5) the first initial condition then is

$$a^{(n-1)}(0) = c_0. \tag{3.6}$$

Proposition 3.5 *Let the interarrival density $a(t)$ fulfill (3.4) with initial conditions (3.5) and (3.6) and let $m(u)$ be sufficiently smooth. Then $m(u)$ is the solution of the IDE*

$$p_A\left(\delta - \frac{\mathrm{d}}{\mathrm{d}u}\right) m(u) = c_0 \int_0^u m(u-y)\, B(\mathrm{d}y) + c_0\, \omega(u), \tag{3.7}$$

with boundary condition

$$\lim_{u\to\infty} m(u) = 0. \tag{3.8}$$

Proof. Rewrite (3.1) as

$$m(u) = \int_0^\infty e^{-\delta t} a(t) g(u+t)\, \mathrm{d}t$$

with

$$g(u) = \int_0^u m(u-y)\, B(\mathrm{d}y) + \int_u^\infty w(u, y-u)\, B(\mathrm{d}y).$$

By dominated convergence and partial integration one has

$$
\begin{aligned}
\left(\delta - \frac{\mathrm{d}}{\mathrm{d}u}\right) m(u) &= \int_0^\infty \left(\delta - \frac{\mathrm{d}}{\mathrm{d}u}\right) \left(e^{-\delta t} a(t) g(u+t)\right)\, \mathrm{d}t \\
&= \int_0^\infty e^{-\delta t} a(t) \left[\left(\delta - \frac{\mathrm{d}}{\mathrm{d}u}\right) g(u+t)\right]\, \mathrm{d}t \\
&= \int_0^\infty e^{-\delta t} a(t) \left[(\delta - \frac{\mathrm{d}}{\mathrm{d}t}) g(u+t)\right]\, \mathrm{d}t \\
&= g(u) a(0) + \int_0^\infty e^{-\delta t} g(u+t)\, \frac{\mathrm{d}}{\mathrm{d}t} a(t)\, \mathrm{d}t.
\end{aligned}
$$

---

[4]Inhomogeneous initial conditions can be dealt with analogously, one just gets additional terms in the calculations. Recall from Chapter I that any $a(t)$ with rational Laplace transform can be represented as the solution of (3.4) and general initial conditions. But already the subclass with homogeneous conditions (3.5) is relevant. For instance, any density which is a convolution of n exponential densities with parameters β_i can be expressed through (3.4) and (3.5) with $p_A(x) = \prod_{i=1}^n (x + \beta_i)$.

Analogously we have under (3.5)

$$\left(\delta - \frac{\mathrm{d}}{\mathrm{d}u}\right)^k m(u) \;=\; g(u)a^{(k-1)}(0) + \int_0^\infty e^{-\delta t} g(u+t)\,\frac{\mathrm{d}^k}{\mathrm{d}t^k}a(t)\,\mathrm{d}t, \qquad k = 1,\ldots,n,$$

and combining these identities in such a way that (3.4) appears inside the integral on the right-hand side, we obtain (3.7) in view of the initial conditions. $\square$

Ordinary differential equations

Assume now that the claim size density $b(y)$ is also the solution of an ODE of the form

$$p_B\left(\frac{\mathrm{d}}{\mathrm{d}y}\right)b(y) = 0, \tag{3.9}$$

with the polynomial

$$p_B(x) = x^\ell + d_{\ell-1}x^{\ell-1} + \cdots + d_0, \qquad d_j \in \mathbb{R}, d_0 \neq 0$$

and some initial conditions $b^{(k)}(0)$ $(k = 0,\ldots,\ell-1)$ (where one initial condition is again determined by the fact that $b(x)$ is a density). Then the IDE for $m(u)$ can further be reduced to a linear ODE:

Proposition 3.6 *Assume that the claim size density $b(y)$ satisfies (3.9). Then, under the assumptions of Proposition 3.5, $m(u)$ satisfies the ODE*

$$\left(p_B\left(\frac{\mathrm{d}}{\mathrm{d}u}\right)p_A\left(\delta - \frac{\mathrm{d}}{\mathrm{d}u}\right) - p_I\left(\frac{\mathrm{d}}{\mathrm{d}u}\right)\right)m(u) \;=\; c_0\,p_B\left(\frac{\mathrm{d}}{\mathrm{d}u}\right)\omega(u), \tag{3.10}$$

with the polynomial

$$p_I(x) = c_0 \sum_{j=0}^{\ell-1} \sum_{k=j+1}^{\ell} d_k\, b^{(k-j-1)}(0)\, x^j$$

and $d_\ell = 1$. One boundary condition is (3.8) and ℓ more boundary conditions need to be specified.

Proof. For $k = 1,\ldots,\ell$

$$\frac{\mathrm{d}^k}{\mathrm{d}u^k}\left(\int_0^u m(y)\,b(u-y)\,\mathrm{d}y\right) = \sum_{i=0}^{k-1} m^{(k-i-1)}(u)\,b^{(i)}(0) + \int_0^u m(y)\,\frac{\mathrm{d}^k}{\mathrm{d}u^k}b(u-y)\mathrm{d}y.$$

The appropriate linear combination of derivatives of (3.7) according to (3.9) cancels the integral term on the r.h.s. of (3.7) and leaves instead

$$c_0 \sum_{k=1}^{\ell} d_k \sum_{i=0}^{k-1} b^{(i)}(0) \frac{\mathrm{d}^{k-i-1}}{\mathrm{d}u^{k-i-1}} m(u) = c_0 \sum_{j=0}^{\ell-1} \left(\sum_{k=j+1}^{\ell} d_k b^{(k-j-1)}(0) \right) \frac{\mathrm{d}^j}{\mathrm{d}u^j} m(u)$$

with $d_\ell = 1$. $\qquad\square$

It immediately follows from the representation (3.10) that the Lundberg fundamental equation for this model is given by the polynomial equation

$$p_B(s) p_A(\delta - s) - p_I(s) = 0. \qquad (3.11)$$

From the definition of p_B and p_I, it becomes clear that $p_I(s)/p_B(s) = c_0 \widehat{B}[-s]$, so that (3.11) can also be written as

$$p_A(\delta - s) - c_0 \widehat{B}[-s] = 0. \qquad (3.12)$$

Lemma 3.7 *For $\delta > 0$, the Lundberg fundamental equation (3.12) has exactly n roots with positive and ℓ roots with negative real part.*

Proof. From (3.11) it is clear that the Lundberg fundamental equation is a polynomial of degree $n + \ell$, so that it has $n + \ell$ complex roots. The location of the roots then follows by an application of Rouché's theorem. $\qquad\square$

Example 3.8 Assume that the initial conditions of (3.9) are given by

$$b^{(k)}(0) = 0 \quad (k = 0, \ldots, \ell - 2). \qquad (3.13)$$

$b(y)$ is a density, so it automatically follows that $b^{(\ell-1)}(0) = d_0$ and the inhomogeneity polynomial in (3.10) simplifies to $p_I(x) = c_0 d_0$. A particular case is when the interclaim time is Erlang(n, β) distributed and the claim size is Erlang(ℓ, ν) distributed, in which case we have $p_A(x) = (x + \beta)^n$ with (3.5) and $c_0 = \beta^n$ as well as $p_B(x) = (x + \nu)^\ell$ with (3.13) and $d_0 = \nu^\ell$. Consequently, the ODE (3.10) then simplifies to

$$\left(\frac{\mathrm{d}}{\mathrm{d}u} + \nu\right)^\ell \left(-\frac{\mathrm{d}}{\mathrm{d}u} + \delta + \beta\right)^n m(u) - \beta^n \nu^\ell m(u) = \beta^n \left(\frac{\mathrm{d}}{\mathrm{d}u} + \nu\right)^\ell \omega(u),$$

which is a popular choice in the literature. $\qquad\square$

In order to use the ODE (3.10) for concrete calculations, one needs to determine the remaining $\ell - 1$ boundary conditions, usually using the negative solutions of the Lundberg fundamental equation, which can be a quite cumbersome task in general, but is possible in particular cases (for instance by the method of so-called integrating factors, see the Notes). An alternative is to determine the fundamental solution of (3.10) and substitute that back in the original IDE (3.7), or to use Laplace transforms.

Laplace transforms

As in Section 2, the convolution term in the IDE (3.7) suggests Laplace transforms as a natural tool in this context. Since $\int_0^\infty e^{-su} m^{(k)}(u)\,du = s^k \widehat{m}[-s] - s^{k-1} m(0) - s^{k-2} m'(0) - \cdots - m^{(k-1)}(0)$, one gets

$$p_A(\delta - s)\widehat{m}[-s] + q(s) = c_0\,\widehat{m}[-s]\widehat{B}[-s] + c_0\,\widehat{\omega}[-s]$$

for some polynomial $q(s)$ of degree $n-1$ and subsequently

$$\widehat{m}[-s] = \frac{c_0\,\widehat{\omega}[-s] - q(s)}{p_A(\delta - s) - c_0\,\widehat{B}[-s]}, \quad s \ge 0. \tag{3.14}$$

Note that the denominator in this expression is again the Lundberg fundamental equation, which from Lemma 3.7 for $\delta > 0$ is known to have n roots $s = \rho_1, \ldots, \rho_n$ with positive real part. Since the Laplace transform is an analytic function for $s \ge 0$, these n roots must also be zeros of the numerator, which determines the n coefficients of $q(s)$. Assuming for simplicity that the roots are distinct, the usual Lagrange interpolation formula gives

$$q(s) = c_0 \sum_{j=1}^n \widehat{\omega}[-\rho_j] \prod_{k=1, k\neq j}^n \frac{s - \rho_k}{\rho_j - \rho_k}. \tag{3.15}$$

Using $m(0) = \lim_{s\to\infty} s\,\widehat{m}[-s]$, it then follows from (3.14) that

$$m(0) = \frac{-c_0 \sum_{j=1}^n \widehat{\omega}[-\rho_j] \prod_{k=1,k\neq j}^n \frac{1}{\rho_j - \rho_k}}{(-1)^n} = c_0 \sum_{j=1}^n \widehat{\omega}[-\rho_j] \prod_{k=1,k\neq j}^n \frac{1}{\rho_k - \rho_j}. \tag{3.16}$$

Since $\widehat{\omega}[-s] = \int_0^\infty \int_0^\infty e^{-sx} w(x,y)\, b(x+y)\, dx\, dy$, a comparison with (1.5) now yields the pleasant formula

$$f(x,y|0) = c_0\, b(x+y) \sum_{j=1}^n e^{-\rho_j x} \prod_{k=1,k\neq j}^n \frac{1}{\rho_k - \rho_j} \tag{3.17}$$

for the discounted joint density of surplus prior to and at ruin (given zero initial capital), expressed in terms of the zeros of the Lundberg fundamental equation. For the compound Poisson case ($n = 1$ and $c_0 = \beta$), (3.17) simplifies to (2.15).

Since (2.23) holds in the present renewal setting as well, one gets the representation

$$\int_0^u f(x,y|u - z) g_\delta(z)\, dz + c_0\, b(x+y) \sum_{j=1}^n e^{-\rho_j(x-u)} \prod_{k=1,k\neq j}^n \frac{1}{\rho_k - \rho_j}\, I(x > u)$$

of $f(x, y|u)$. Integration with respect to y gives the expression

$$\int_0^u f(x|u - z)g_\delta(z)\,\mathrm{d}z + c_0\,\overline{B}(x)\sum_{j=1}^{n} \mathrm{e}^{-\rho_j(x-u)}\prod_{k=1,k\neq j}^{n}\frac{1}{\rho_k - \rho_j}\,I(x > u)$$

for $f(x|u)$. Correspondingly, as a function of x, at $x = u$ the discounted density of the surplus prior to ruin has a discontinuity of size

$$c_0\,\overline{B}(u)\sum_{j=1}^{n}\prod_{k=1,k\neq j}^{n}\frac{1}{\rho_k - \rho_j}.$$

But for $n \geq 2$ this sum equals zero, so that the discontinuity disappears! [5]

The explicit form of the Laplace transform also allows to sharpen Proposition 3.1:

Corollary 3.9 *If the interarrival density $a(t)$ fulfills (3.4) with initial conditions (3.5), then under the assumptions of Proposition 3.1 with a simple positive zero $\gamma_\delta > 0$ of $\kappa(r) = \delta$ and distinct roots $-\rho_1, \ldots, -\rho_n$ with negative real part, one has*

$$\lim_{u\to\infty}\mathrm{e}^{\gamma_\delta u}m(u) = \frac{\widehat{\omega}[\gamma_\delta] - \sum_{j=1}^{n}\widehat{\omega}[-\rho_j]\prod_{k=1,k\neq j}^{n}\dfrac{-\gamma_\delta - \rho_k}{\rho_j - \rho_k}}{-p_A'(\delta + \gamma_\delta)/c_0 + \widehat{B}'[\gamma_\delta]}.$$

Proof. In view of Proposition 3.1, it suffices to determine the constant C_δ. With the formula $C_\delta = \lim_{s\to 0} s\,\widehat{m}[-s + \gamma_\delta]$, we obtain the result through an application of L'Hôpital's rule in (3.14), using (3.15) and the fact that the solution γ_δ of the Lundberg fundamental equation has multiplicity 1. □

Finally, we illustrate how the simple formula (2.6) for the Laplace transform of the time to ruin with zero initial capital can be generalized to certain renewal models:

Example 3.10 Assume that the interarrival time is a generalized Erlang r.v. (that is, an independent sum of not necessarily identically distributed exponential r.v.'s) with $p_A(x) = \prod_{i=1}^{n}(x + \beta_i)$ and correspondingly $c_0 = \prod_{i=1}^{n}\beta_i$. For $w \equiv 1$ one has $\widehat{\omega}[-s] = (1 - \widehat{B}[-s])/s$, and formula (3.16) together with

---

[5]Note that here an underlying assumption was the homogeneity condition (3.5). The discontinuity does not necessarily disappear if the boundary conditions for $a(t)$ are inhomogeneous, see Ren [734].

$c_0 \widehat{B}[-\rho_j] = p_A(\delta - \rho_j)$ for $j = 1, \ldots, n$ (note that ρ_j are solutions of the Lundberg fundamental equation) then implies

$$\mathbb{E}\left[e^{-\delta\tau(0)}; \ \tau(0) < \infty\right] = \sum_{j=1}^{n} \left(\beta_1 \cdots \beta_n - p_A(\delta - \rho_j)\right) \prod_{k=1, k \neq j}^{n} \frac{1}{\rho_j(\rho_k - \rho_j)}.$$

But by an induction argument this expression can be simplified to

$$\mathbb{E}\left[e^{-\delta\tau(0)}; \ \tau(0) < \infty\right] = 1 - \frac{\prod_{i=1}^{n}(\delta + \beta_i) - \beta_1 \cdots \beta_n}{\rho_1 \cdots \rho_n}.$$

$\square$

Notes and references Early studies of penalty-related quantities in renewal models include Dickson & Hipp [316], Cheng & Tang [237], Tsai & Sun [857], Sun & Yang [819] and Drekic et al. [328]. Extensions to general discounted penalty functions in renewal models with Erlang interclaim times go back to Li & Garrido [589] and Gerber & Shiu [411]. For this model, Li [584] and Li & Garrido [589] give an alternative representation of the renewal equation (1.3) in terms of certain integral transforms T_r that can be interpreted as pseudo-resolvents of the differentiation operator (evaluated at $r = \rho_1, \ldots, \rho_n$). These transforms turn out to be helpful in related models as well (originally studied by Redheffer [728]; they are nowadays usually referred to as *Dickson-Hipp operators*), see [411] for a detailed comparison of methods. Since then there have been numerous further papers on the subject, and the following list is by no means exhaustive. More general interclaim times are treated in Li & Garrido [590], Schmidli [778] and Song et al. [815]. Ren [734] extends Proposition 2.14 and Li [585] extends Proposition 2.13 to phase-type interclaim times. In Li [586], the latter result is used to study the time to recovery T_0 and the maximum severity of ruin for phase-type interclaim times. Biard et al. [163] study the asymptotic behavior of the expected time-integrated negative part of the risk process. Li & Dickson [587] investigate the maximum surplus before ruin in general Sparre Andersen models. Willmot, Cai & Lin [888] derive general bounds for solutions of ren ewal equations and apply them to the present set-up.

The derivation of the integro-differential equation with operators is from Constantinescu [254], where the formulation is in terms of adjoint operators. Albrecher et al. [21] start from (3.10) to factorize the differential operator and subsequently lift this factorization to the equation level, which leads to an iterative solution of first-order boundary value problems and allows to obtain a number of explicit expressions for $m(u)$ using Gröbner bases. Landriault & Willmot [572] give explicit expressions for the Laplace transform $\widehat{m}[-s]$ for arbitrary interclaim times and Coxian claim sizes. However, its explicit inversion is in general difficult. Section 3a is based on Schmidli [780], who uses the same technique to work out Lundberg-type approximations also in more general models including certain Cox models. The trick to use forward markovization (instead of backward markovization) needs some care w.r.t. the appropriate filtration, but can be quite powerful. For a detailed discussion see e.g. Rolski *et al.* [746].

The idea to get rid of the discounting by modifying the interclaim distribution can be found in Avram & Usabel [113] and Ren [735].

Using duality relations to a compound Poisson model with arbitrary claim size distribution, a closed-form formula for the density of the time to ruin for arbitrary interclaim times and exponential claim sizes (in terms of an infinite series of convolutions of A) is derived in Borovkov & Dickson [189], see also Dickson & Li [317].

Necessary amendments of the above results for stationary renewal models are for instance discussed in Willmot *et al.* [889, 890] and Ng [664], for other delayed renewal models see Willmot [887]. For bounds on the distribution of the deficit, see Chadji-constantinidis & Politis [228] and Psarrakos [719]. The asymptotic behavior of $m(u)$ for large u in the presence of heavy and semi-heavy tails depends in a subtle way on the shape of the penalty function w, see Tang & Wei [834] for a fine and complete analysis. For an extension to a model with constant interest rate, see Wu, Lu & Fang [895].

4   Lévy risk models

As already discussed in Chapter XI, there may be certain reasons to consider more general Lévy processes in the risk reserve modeling procedure, let alone the appeal of generality on the mathematical level. Recall from the quintuple identity of Section XI.4e that the joint distribution of the surplus prior to ruin and the deficit at ruin of a general Lévy process can be expressed through potential measures. The resulting expression, however, usually does not lead to explicit expressions unless one adds further restrictions on the model. We will consider in the sequel two cases that admit a rather explicit treatment, namely the case with one-sided jumps and the compound Poisson process with two-sided jumps.

4a   Spectrally negative Lévy processes

If the risk reserve process is a Lévy process that can only have downward jumps, then it is possible to find an integral representation of the Gerber-Shiu function through the corresponding scale function. As in Section XI.3, ρ_δ denotes the positive solution of the Lundberg equation $\kappa(s) = \delta$.

Theorem 4.1 *Suppose that $\{R_t\}$ is a spectrally negative Lévy process with positive drift. Then for a bounded measurable penalty function $w(x, y)$ with*

$w(\cdot, 0) = 0$,

$$m(u) = \int_0^\infty \int_0^\infty w(x, y) \left(e^{-\rho_\delta x} W^{(\delta)}(u) - W^{(\delta)}(u - x) \right) \nu(dy + x) \, dx.$$

Remark 4.2 Note that ruin can occur either by a jump or through diffusion and the assumption $w(\cdot, 0) = 0$ simply restricts the discounted penalty function to the case when ruin happens through jumps. If ruin is caused by diffusion, then the problem is somewhat degenerate with $R_{\tau(u)-} = R_{\tau(u)} = 0$. In that case one knows from Pistorius [703] that

$$\mathbb{E}\left[e^{-\delta \tau(u)}; R_{\tau(u)} = 0 \right] = \frac{\sigma^2}{2} \left(W^{(\delta)'}(u) - \rho_\delta W^{(\delta)}(u) \right).$$

If $\{R_t\}$ has bounded variation, the assumption $w(\cdot, 0) = 0$ is not needed. Also, if $\{R_t\}$ has unbounded variation and $\sigma = 0$, the assumption is not needed for $u > 0$. □

Proof of Theorem 4.1. Our model for R_t is equivalent to a spectrally positive Lévy process with negative drift and $R_0 = 0$, where the ruin event then refers to an overshoot of level u. Hence we can directly use Corollary XI.4.7 to write down the (defective) joint density of the surplus prior to ruin and the deficit at ruin. In particular, integrating XI.(4.8) w.r.t. y and translating into the present notation we obtain

$$\mathbb{P}\left(R_{\tau(u)-} \in dx, |R_{\tau(u)}| \in dy \right) = \left(W^{(0)}(u) - W^{(0)}(u - x) \right) \nu(x + dy) \, dx.$$

We can now use the fact that exponential tilting by ρ_δ leaves the drift of the process positive and by XI.(3.8) the zero-scale function under the tilted measure is given by $W^{(0)}_{\rho_\delta}(u) = e^{-\rho_\delta u} W^{(\delta)}(u)$ and the Lévy measure changes to $\nu_{\rho_\delta}(dx) = e^{-\rho_\delta x} \nu(dx)$. Hence we can write

$$\begin{aligned}
\mathbb{E}&\left[e^{-\delta \tau(u)}; R_{\tau(u)-} \in dx, |R_{\tau(u)}| \in dy \right] \\
&= e^{\rho_\delta(u+y)} \mathbb{P}_{\rho_\delta}\left(R_{\tau(u)-} \in dx, |R_{\tau(u)|} \in dy \right) \\
&= e^{\rho_\delta(u+y)} \left(W^{(0)}_{\rho_\delta}(u) - W^{(0)}_{\rho_\delta}(u - x) \right) \nu_{\rho_\delta}(x + dy) \, dx \\
&= \left(e^{-\rho_\delta x} W^{(\delta)}(u) - W^{(\delta)}(u - x) \right) \nu(x + dy) \, dx.
\end{aligned}$$

From the latter the assertion follows. □

Remark 4.3 In the absence of a diffusion component (i.e. $\sigma = 0$), the jumps larger than a fixed $\epsilon > 0$ form a compound Poisson process. As $\epsilon \to 0$, this compound Poisson process converges weakly to the original spectrally negative Lévy process.[6] One can now use this fact to observe that a number of results derived for the compound Poisson process still hold for more general pure-jump Lévy processes. The recipe is to just replace $\beta(1 - B(x))$ by $\overline{\nu}(x)$. For instance, from (2.13) and (2.15) it follows that for a spectrally negative Lévy process with triplet $(1, 0, \nu)$ and some restrictions on the penalty function, $m(u)$ satisfies the defective renewal equation $m = m * g + h$ with

$$g(y) \; = \; \int_y^\infty e^{-\rho_\delta(x-y)}\, \nu(\mathrm{d}x)$$

and

$$h(u) \; = \; \int_u^\infty \int_0^\infty e^{-\rho_\delta(x-u)} w(x, y)\, \nu(\mathrm{d}y + x)\, \mathrm{d}x.$$

□

The compound Poisson risk model with perturbation

Consider the risk reserve process

$$R_t \; = \; u + t - \sum_{i=1}^{N_t} U_i + \sigma W_t, \quad t \geq 0, \tag{4.1}$$

where N_t is again a homogeneous Poisson process and $\{W_t\}$ is independent standard Brownian motion. The interpretation is that the diffusion part accounts for small perturbations of the risk process that can come from various sources (inaccuracies in the estimation or measurement, local deviations in the premium income or claim payouts etc.). As usual, the justification of using a diffusion to model such effects is that it can be thought of as the sum of many small independent effects and in the absence of further knowledge it is natural to assume that its drift is zero. Mathematically, (4.1) is clearly a special case of a spectrally negative Lévy process and so the above fluctuation theory and its results apply, but due to its simplicity this model can also be treated by other self-contained techniques which can give additional insight.

In the following we give an illustration of this. Recall that in the presence of the Brownian component, ruin can occur in two ways, either by a claim (which results in a non-zero deficit at ruin) or by oscillation (which is also often referred to as *creeping*). Assume that the penalty for the second is given by the constant

---

[6]In fact, it converges even almost surely uniformly on bounded time intervals, see Bertoin [157].

$w(0) = w_0$. With the generator technique of Chapter II it is clear that the discounted penalty function now satisfies the integro-differential equation

$$\beta \int_0^u m(u-x)B(\mathrm{d}x) + \beta \int_u^\infty w(u, x-u)B(\mathrm{d}x) - (\beta+\delta)m(u) + m'(u) + \frac{\sigma^2}{2} m''(u) = 0.$$
(4.2)

One can now again proceed with Laplace transform techniques as in Section 2a. Then

$$\widehat{m}[-s] = \frac{\frac{\sigma^2}{2}\left(s\,m(0) + m'(0)\right) - \beta\widehat{\omega}[-s]}{\kappa(-s) - \delta},$$

where now $\kappa(r) = \beta\widehat{B}[r] - r - \beta + \sigma^2 r^2/2$. Starting with zero initial capital leads to ruin immediately (due to the oscillation), so that $m(0) = w_0$. Since $\kappa(-s) - \delta$ has exactly one positive zero $-\rho_\delta > 0$, this must be a zero of the denominator as well (again $\widehat{m}[-s]$ is analytic for $\Re(s) > 0$). Hence we arrive at

$$\widehat{m}[-s] = \frac{\frac{\sigma^2}{2} w_0(s - \rho_\delta) + \beta\,\widehat{\omega}[-\rho_\delta] - \beta\,\widehat{\omega}[-s]}{\kappa(-s) - \delta},$$
(4.3)

which extends formula (2.4).

By adapting the techniques of Section 2b one can show that for general penalty functions $w(x, y)$ (under mild assumptions, see Sarkar & Sen [762] for details) the Cramér-Lundberg approximation

$$\lim_{u \to \infty} m(u)\mathrm{e}^{\gamma_\delta u} = C_\delta$$

holds for the perturbed model (4.1). The constant is again given by $C_\delta = \lim_{s \to 0} s\,\widehat{m}[-s + \gamma_\delta]$, i.e. with L'Hôpital's rule we obtain from (4.3)

$$C_\delta = \frac{1}{\kappa'(\gamma_\delta)} \left(\beta \int_0^\infty w(x, y) \int_0^\infty (\mathrm{e}^{\gamma_\delta\,x} - \mathrm{e}^{-\rho_\delta x})b(x+y)\,\mathrm{d}x\,\mathrm{d}y + \frac{\sigma^2 w_0(\gamma_\delta + \rho_\delta)}{2} \right).$$
(4.4)

For $w(x) = w_0 = 1$, this further simplifies to

$$C_\delta = \frac{\delta}{\kappa'(\gamma_\delta)} \left(\frac{1}{\gamma_\delta} + \frac{1}{\rho_\delta} \right),$$

which formally coincides with (2.20), but note that the underlying $\kappa(r)$ is different. If furthermore $\delta = 0$ (i.e. $m(u) = \psi(u)$), then $C_0 = C = (1 - \beta\mu_B)/\kappa'(\gamma)$, in accordance with Corollary XI.2.7.

With $\delta = 0$ in (4.4) and the previously established fact that $\delta/\rho_\delta \to 1 - \beta\mu_B$, the choice $w(x, y) = 0$ and $w_0 = 1$ finally gives the asymptotic probability of ruin caused by oscillation to be

$$\psi_d(u) \sim \frac{\sigma^2 \gamma}{2\kappa'(\gamma)} \, e^{-\gamma u} \quad \text{as } u \to \infty,$$

and the probability of ruin caused by a claim ($w(x,y) \equiv 1$ and $w_0 = 0$) behaves as

$$\psi_s(u) \sim \frac{1 - \beta \mu_B - \sigma^2 \gamma / 2}{\kappa'(\gamma)} \, e^{-\gamma u} \quad \text{as } u \to \infty.$$

Notes and references In Biffis & Kyprianou [165], Theorem 4.1 is given in a more general form, where the Gerber-Shiu function also includes the size of the last minimum before ruin; see also Biffis & Morales [166] for a convolution-type approach. Chiu & Yin [245] give expressions for the duration of ruin and the time of the last visit of the ruin boundary for spectrally negative Lévy processes. The idea of Remark 4.3 goes back to Dufresne, Gerber & Shiu [335]; see also Garrido & Morales [391]. Klüppelberg, Kyprianou & Maller [543] derive explicit asymptotic results for ruin-related quantities like the deficit at ruin and the surplus prior to ruin for $u \to \infty$; see also Doney & Kyprianou [326].

For a detailed study of the Gerber-Shiu function in a compound Poisson model with Brownian perturbation including defective renewal equations and asymptotics, we refer to Gerber & Landry [406] and Tsai & Willmot [858]. Lin & Wang [595] apply these results to the pricing of perpetual American catastrophe put options. The same argument as in Remark 4.3 applies to extend the resulting formulas to general spectrally negative Lévy processes (see Morales [649] for details). Some explicit calculations for this model under phase-type claims can be found in Ren [733]. For inclusion of interest rates see Wang & Wu [872] and a recent extension with more general investment is given in Avram & Usabel [114] and Wang, Xu & Yao [867]. Gerber-Shiu functions under Brownian perturbation in renewal models are for instance studied by Li & Garrido [591] and in Markov-modulated compound Poisson models by Lu & Tsai [611]. There have also been studies with more general perturbations than Brownian motion. Among them, Furrer [381] deals with α-stable motion for the perturbation and Chi, Jaimungal & Lin [244] use singular perturbation theory to deal with the Gerber-Shiu function under perturbation with stochastic volatility of Ornstein-Uhlenbeck type.

4b The compound Poisson model with two-sided jumps

Another case that admits a direct treatment is a compound Poisson model where jumps can be both upward and downward. Complementing the first passage expressions given in Section XI.5, let us briefly revisit risk models of the type

$$R_t = u + \sum_{i=1}^{N_t^u} P_i - \sum_{i=1}^{N_t} U_i. \tag{4.5}$$

Here the linear drift t from the Cramér-Lundberg process is replaced by a compound Poisson process with i.i.d. positive up-jumps P_i (with density $p(x)$) that occur according to a homogeneous Poisson process $\{N_t^u\}$ with intensity β^u, independent of the claims process. For transparency of the exposition, we neither include an additional drift term nor a further Brownian perturbation component, although each is easily possible, see the Notes.

Let $h > 0$. By conditioning on the time and amount of the first jump before time h, one has

$$
\begin{aligned}
m(u) &= \beta \int_0^h e^{-(\beta+\beta^u+\delta)t} \int_0^u m(u-x)b(x)\,\mathrm{d}x\,\mathrm{d}t \\
&\quad + \beta \int_0^h e^{-(\beta+\beta^u+\delta)t} \int_u^\infty w(u,x-u)b(x)\,\mathrm{d}x\,\mathrm{d}t \\
&\quad + \beta^u \int_0^h e^{-(\beta+\beta^u+\delta)t} \int_0^\infty m(u+x)p(x)\,\mathrm{d}x\,\mathrm{d}t + e^{-(\beta+\beta^u+\delta)h}m(u),
\end{aligned}
$$

which after differentiation with respect to h and setting $h = 0$ gives[7]

$$
\begin{aligned}
\beta \int_0^u m(u-x)b(x)\mathrm{d}x + \beta \int_u^\infty w(u,x-u)b(x)\mathrm{d}x \\
+ \beta^u \int_0^\infty m(u+x)p(x)\mathrm{d}x - (\beta+\beta^u+\delta)m(u) = 0. \tag{4.6}
\end{aligned}
$$

The function $m(u)$ is the unique solution of (4.6), since the mapping

$$
\begin{aligned}
m(u) \to \frac{\beta}{\beta+\beta^u+\delta} \int_0^u m(u-x)b(x)\mathrm{d}x + \frac{\beta}{\beta+\beta^u+\delta} \int_u^\infty w(u,x-u)b(x)\mathrm{d}x \\
+ \frac{\beta^u}{\beta+\beta^u+\delta} \int_0^\infty m(u+x)p(x)\mathrm{d}x
\end{aligned}
$$

is a contraction and has a unique fixed point. Let us again impose the boundary condition $\lim_{u\to\infty} m(u) = 0$.

Instead of using Laplace transforms, we shall here proceed in a related, but slightly heuristic way and restrict the analysis to a penalty function that only

---

[7]As before, the formal background for this type of reasoning is the generator approach of Section II.4a.

depends on the deficit, i.e. $w(x, y) \equiv w(y)$. Assume first that the claim size distribution is a combination of n exponentials, i.e.

$$b(x) = \sum_{i=1}^{n} A_i \alpha_i e^{-\alpha_i x}, \qquad x > 0, \tag{4.7}$$

where $0 < \alpha_1 < \alpha_2 < \ldots < \alpha_n$ and $A_1 + \ldots + A_n = 1$. Some of the A_i's may be negative as long as $b(x) \geq 0$. Then the discounted penalty function is of the form

$$m(u) = \sum_{k=1}^{n} C_k e^{-r_k u}, \qquad u \geq 0. \tag{4.8}$$

To see this, one substitutes (4.8) into (4.6) and $r_1, \ldots, r_n$ turn out to be the n solutions with positive real part of the generalized Lundberg equation

$$\beta \sum_{i=1}^{n} A_i \frac{\alpha_i}{\alpha_i - r} + \beta^u \int_0^\infty e^{-rx} p(x) \, dx - (\beta + \beta^u + \delta) = 0 \tag{4.9}$$

(potential negative solutions would violate $\lim_{u \to \infty} m(u) = 0$). This equation has indeed exactly n solutions with positive real part (by the usual Rouché argument). The one with the smallest real part is real and is the adjustment coefficient $\gamma_\delta < \alpha_1$. The coefficients $C_1, \ldots, C_n$ are the solutions of

$$\sum_{k=1}^{n} \frac{C_k}{\alpha_i - r_k} = \frac{\Pi_i}{\alpha_i}, \qquad i = 1, \ldots, n, \tag{4.10}$$

with the notation

$$\Pi_i = \alpha_i \int_0^\infty w(y) e^{-\alpha_i y} \, dy.$$

One way to solve this system of n linear equations for $C_1, \ldots, C_n$ goes as follows. Define a rational function

$$Q(r) = \sum_{k=1}^{n} \frac{C_k}{r - r_k}$$

(note that $\widehat{m}[-s] = -Q(-s)$). Obviously,

$$C_h = \lim_{r \to r_h} (r - r_h) Q(r), \qquad h = 1, \ldots, n. \tag{4.11}$$

One can now find more tractable expressions for $Q(r)$ and apply (4.11) to these expressions. Note that $Q(r)$ is completely determined by the following three properties:

- It is a rational function of the type polynomial of degree at most $n-1$ divided by polynomial of degree n.

- Its poles are $r_1, ..., r_n$.

- $Q(\alpha_i) = \frac{\Pi_i}{\alpha_i}$, $i = 1, ..., n$, according to (4.10).

The rational function

$$Q_1(r) = \frac{\displaystyle\sum_{j=1}^{n} \frac{\Pi_j}{\alpha_j} \prod_{k=1}^{n} (\alpha_j - r_k) \prod_{i=1, i \neq j}^{n} \frac{r - \alpha_i}{\alpha_j - \alpha_i}}{\displaystyle\prod_{k=1}^{n} (r - r_k)}$$

also fulfills these properties, and from this together with (4.11) gives a full specification of (4.8).

If we now restrict to $p(x) = \eta e^{-\eta x}$ (i.e. exponential up-jumps), then the Lundberg equation (4.9) specializes to

$$\beta \sum_{i=1}^{n} A_i \frac{\alpha_i}{\alpha_i - r} + \beta^u \frac{\eta}{\eta + r} - (\beta + \beta^u + \delta) = 0.$$

In addition to $r_1, \ldots, r_n$, this equation has one negative solution $-\rho_\delta$. We can hence represent $Q(r)$ also as

$$Q_2(r) = \frac{\beta}{\eta + r} \frac{(\eta + r) \displaystyle\sum_{i=1}^{n} A_i \frac{\Pi_i}{\alpha_i - r} - (\eta - \rho_\delta) \sum_{i=1}^{n} A_i \frac{\Pi_i}{\alpha_i + \rho_\delta}}{\beta \displaystyle\sum_{i=1}^{n} A_i \frac{\alpha_i}{\alpha_i - r} + \beta^u \frac{\eta}{\eta + r} - (\beta + \beta^u + \delta)},$$

which immediately leads to

$$C_h = \frac{\beta}{\eta + r_h} \frac{(\eta + r_h) \displaystyle\sum_{i=1}^{n} A_i \frac{\Pi_i}{\alpha_i - r_h} - (\eta - \rho_\delta) \sum_{i=1}^{n} A_i \frac{\Pi_i}{\alpha_i + \rho_\delta}}{\beta \displaystyle\sum_{i=1}^{n} A_i \frac{\alpha_i}{(\alpha_i - r_h)^2} - \beta^u \frac{\eta}{(\eta + r_h)^2}}. \tag{4.12}$$

Furthermore, with $m(0) = \lim_{r \to \pm\infty} r Q_2(r)$ we get

$$m(0) = \frac{\beta}{\beta + \beta^u + \delta} \left[\sum_{i=1}^{n} A_i \Pi_i \left(1 + \frac{\eta - \rho_\delta}{\alpha_i + \rho_\delta} \right) \right] = \frac{\beta}{\beta + \beta^u + \delta} \sum_{i=1}^{n} A_i \Pi_i \frac{\alpha_i + \eta}{\alpha_i + \rho_\delta}.$$

The class of distributions (4.7) is dense in the class of all positive distributions, so (heuristic, but intuitive!) one can deduce from the above that $m(0)$ is given by

$$\frac{\beta}{\beta + \beta^u + \delta} \left[\int_0^\infty w(y)b(y)\mathrm{d}y + (\eta - \rho_\delta) \int_0^\infty w(y) \int_0^\infty \mathrm{e}^{-\rho_\delta x} b(x+y)\,\mathrm{d}x\,\mathrm{d}y \right].$$

In this general case, $-\rho_\delta$ is the negative solution of the equation

$$\kappa(r) = \beta\widehat{B}[r] + \beta^u \frac{\eta}{\eta+r} - (\beta + \beta^u) = \delta, \qquad (4.13)$$

which is defined for all $r > -\eta$ with $\widehat{B}[r] < \infty$ (see Figure XII.2, which shows that with up-jumps the straight line from Figure XII.1 for the classical model is replaced by a more general curve).

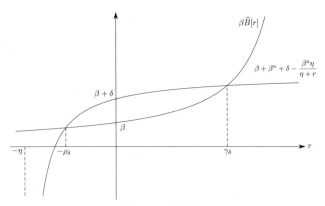

FIGURE XII.2

Note that $\beta/(\beta + \beta^u + \delta)$ is the discounted probability that the surplus process has a downward jump before the first upward jump (ruin occurs at that time), which explains the first summand in (4b). Let again $g_\delta(x)$ denote the discounted probability density function of the deficit at ruin for initial surplus zero (i.e. the discounted descending ladder height density). Because (4b) holds for arbitrary $w(y)$, by choosing the Dirac-delta function we get

$$g_\delta(y) = \frac{\beta}{\beta + \beta^u + \delta} \left[b(y) + (\eta - \rho_\delta) \int_0^\infty \mathrm{e}^{-\rho_\delta x} b(x+y)\,\mathrm{d}x \right], \qquad x > 0. \,(4.14)$$

If the Lundberg approximation

$$m_\delta(u) \sim C_\delta \mathrm{e}^{-\gamma_\delta u} \qquad \text{as } u \to \infty$$

holds,[8] then the corresponding extension of (4.12) with $r_h = \gamma_\delta$ immediately gives

$$C_\delta = \frac{\beta}{\eta + \gamma_\delta} \frac{\int_0^\infty w(y) \int_0^\infty [(\eta + \gamma_\delta)e^{\gamma_\delta x} - (\eta - \rho_\delta)e^{-\rho_\delta x}] \, b(x+y) \, \mathrm{d}x \, \mathrm{d}y}{\kappa'(\gamma_\delta)}. \quad (4.15)$$

Corollary 4.4 *The Laplace transform of the time to ruin with zero initial capital in the two-sided compound Poisson model with exponential(η) up-jumps is given by*

$$\mathbb{E}\left[e^{-\delta\tau(0)}; \tau(0) < \infty\right] = 1 - \frac{\eta\,\delta}{\rho_\delta(\beta + \beta^u + \delta)}.$$

With positive safety loading, the ruin probability with zero initial capital is

$$\psi(0) = \frac{\beta}{\beta + \beta^u} (1 + \eta\,\mu_B). \quad (4.16)$$

Furthermore, the constant C in the Cramér-Lundberg approximation $\psi(u) \sim Ce^{-\gamma_0 u}$ is given by

$$C = \frac{\beta^u - \eta\,\beta\,\mu_B}{(\eta + \gamma_0)\kappa'(\gamma_0)}. \quad (4.17)$$

Proof. Just choose $w(x) \equiv 1$ in (4.12) and use the identity

$$\int_0^\infty \int_0^\infty e^{ry} b(x+y) \, \mathrm{d}y \, \mathrm{d}x = \frac{1}{r}\left(\widehat{B}[r] - 1\right) \quad (4.18)$$

for $r = -\rho_\delta$ together with $\kappa(-\rho_\delta) = \delta$.

For the ruin probability, $\rho_\delta \to 0$ as $\delta \to 0$, so $\psi(0)$ follows from the limit $\delta/\rho_\delta \to \beta^u/\eta - \beta\mu_B$ as $\delta \to 0$. Alternatively, take $w \equiv 1$ and $\delta \to 0$ directly in (4.12).

Finally, the constant C follows from (4.15) by another application of (4.18), this time for $r = \gamma_0$ with $\kappa(\gamma_0) = \delta$. $\qquad\square$

Remark 4.5 Formula (4.16) can be reformulated as

$$\psi(0) = \frac{\beta}{\beta + \beta^u} + \frac{\beta^u}{\beta + \beta^u} \frac{\mathbb{E}[\sum_{i=1}^{N_1^u} P_i]}{\mathbb{E}[\sum_{i=1}^{N_1} U_i]},$$

---

[8]Conditions under which a defective renewal equation for m can be derived (and hence a Lundberg approximation for light-tailed B exists by the key renewal theorem) can e.g. be found in Labbé & Sendova [570].

which has the following interpretation: The first term is the probability a down-jump occurs before an up-jump, in which case ruin occurs at that time. If not (the probability of which is $\beta^u/(\beta + \beta^u)$), the conditional probability of ruin is the ratio of expected claim payments per time unit over expected income per time unit, which is a natural extension of Corollary IV.3.1 from the one-sided jumps model.   □

Remark 4.6 Choosing $\delta = 0$ in (4.14) gives the (non-discounted defective descending) ladder height density

$$g_0(y) \;=\; \frac{\beta}{\beta + \beta^u} \left[b(y) + \eta \, \overline{B}(y) \right], \qquad x > 0,$$

which is the extension of $g_0(y) = \beta \overline{B}(y)$ for the compound Poisson process with one-sided jumps derived in Theorem III.5.1.   □

If we take the limit $\beta^u \to \infty$, $\eta \to \infty$ such that $\beta^u/\eta = 1$, then $\sum_{i=1}^{N_t^u} P_i \to t$, so the up-jumps converge to a linear drift with slope 1 and we arrive at the classical Cramér-Lundberg model. Correspondingly, one can retrieve from each of the above results the corresponding Cramér-Lundberg analogues in Section 2 as a limit case.

Remark 4.7 If the up-jump distribution is extended to a combination of exponentials, in principle analogous formulas hold, but the Lundberg equation $\kappa(r) = \delta$ then has additional zeros in the negative halfplane, so that the exposition gets more cumbersome and is omitted here (see the Notes below).

□

Notes and references   In an obvious way, one can derive the analogous expressions for an added Brownian perturbation in (4.5), where the polynomials in $Q(r)$ will then have its degree increased by 1. Also, the inclusion of a drift term is just a notational issue (and was left out here deliberately for transparency, and also in order to identify the Cramér-Lundberg model as a simple limit). A Pollaczeck-Khinchine -type formula for the ruin probability in the two-sided pure-jump model was derived by Boucherie, Boxma & Sigman [191] in a queueing context, exploiting the observation that up-jumps can be equivalently described as an increase of inter-arrival times in a renewal model with constant premium intensity as long as the desired quantities are invariant to scaling of time; for early time-dependent considerations we refer to Perry, Stadje & Zacks [692], Kou & Wang [559] and Jacobsen [498] . Since then many papers appeared on the subject. The slightly heuristical procedure given above is from Albrecher, Gerber & Yang [23] and can be formally backed up by the usual IDE and renewal techniques.

In more finance-oriented contexts, a compound Poisson model with two-sided jumps and perturbation is usually referred to as a *jump-diffusion* (see e.g. Kou & Wang [560]). This model is for instance investigated by IDE methods in Chen, Lee & Sheu [233] and by renewal techniques in Zhang, Yang & Li [915] under weaker assumptions on the up-jumps. For methodological links between ruin theory and credit ratings assessments under this model assumption, see e.g. Chen & Panjer [232].

Related considerations of the Wiener-Hopf factorization for more general up-jumps with rational Laplace transform are given in Section XI.5, see also Lewis & Mordecki [582], Pistorius [704], Dieker [323], Levendorskii [581] with finance applications and Chi [243] for formulations in terms of Gerber-Shiu functions. Roynette, Vallois & Volpi [750] identify the limit distribution as $u \to \infty$ of the surplus prior to ruin and the deficit at ruin in such a model. For a Sparre Andersen model with two-sided jumps, see Zhang & Yang [914].

For most of the above results, the zeros of the generalized Lundberg function play a crucial role. Accordingly, for possible extensions of the results to more general Lévy models with two-sided jumps, a fine study of these zeros is essential, see e.g. Kuznetsov [563] for more general scenarios that are still somewhat tractable.

The Gerber-Shiu function has also been extensively studied for risk processes that are reflected at horizontal barrier b (in which case we denote it by $m(u; b)$), with one possible interpretation that above the barrier, all premium income is paid out to shareholders as dividends. A closely related question is then to determine the expected present value $V(u; b)$ of the corresponding aggregate dividend payouts until ruin, a quantity that in certain economic approaches is interpreted as the 'value' of the insurance portfolio (see the Notes of Section VIII.1). If the discount factor is the same δ as the one for $m(u)$, it is clear that the dynamics of the process for $m(u; b)$ and $V(u; b)$ between 0 and b are identical, differences occur only upon exit of this interval. Under particular model assumptions, this translates into integro-differential equations only differing in inhomogeneity terms and boundary conditions. For the compound Poisson model Lin, Willmot & Drekic [597] identified in this way the so-called *dividends-penalty identity*

$$m(u) = m(u; b) - m'(b)V(u; b), \quad 0 \le u \le b.$$

See also Yuen, Wang & Li [912] and Cai, Feng & Willmot [216] for inclusion of interest rates. Gerber, Lin & Yang [407] established this identity for arbitrary stationary Markov risk processes with only downward jumps by a strikingly simple probabilistic argument. An even more direct argument can be used for the model (4.5) with exponential up-jumps, since then the dividends are paid out discretely and the compound Poisson process identity is then again obtained as a limit, see Gerber & Yang [415]. Extensions for Markov-modulated processes (in which case one obtains a matrix identity) are investigated in Li & Lu [592]; see also Cheung & Landriault [240] for a Markovian arrival process set-up. As discussed in the Notes to Section VIII.1, the literature on dividend processes is huge and is not treated in this book. A general formalism under which the Gerber-Shiu function, the expected discounted dividends, but also more general utilities of paths of the risk process can be accommodated is proposed in Cai, Feng & Willmot [217].

Further results on discounted penalty functions for dependent risk models will be treated in Chapter XVI.

Chapter XIII

Further models with dependence

Many classical results in ruin theory rest on the independence assumption of claim sizes, of interclaim times and the independence between claims and interclaim times. However, examples of risk processes with a certain degree of dependence appeared already at several places in this book (in particular, the Markov-modulated and the general Markov additive processes discussed in Chapters VII and IX). In this chapter, a number of further risk models with dependence will be discussed, some of which allow a quite analytic treatment.

Naturally, for more involved model assumptions, the possible calculations will be less explicit and there is a trade-off between considering a flexible dependence model (that can be calibrated to practical portfolio situations) and its tractability. In any case, it is crucial to understand how (possibly neglected) dependence may influence the actual values of ruin probabilities and related measures of riskiness in the portfolio.

We will start with some general considerations on large deviations, which are of independent interest, but also provide a powerful tool to generalize asymptotic ruin results for light-tailed claim sizes to certain dependent scenarios. It will turn out that for weak forms of dependence the asymptotic behavior of the ruin probability is still exponential, but often with a modified adjustment coefficient (dependence situations where the adjustment coefficient remains unchanged include certain types of delay in claim settlement). For stronger (long-range) dependence it can happen that the ruin probability becomes heavy-tailed although the claim sizes are light-tailed. On the other hand, for heavy-tailed claim size distributions we will see in Section 2 that the asymptotic ruin probability is relatively insensitive to weak forms of dependence (which is consistent with the 'one large claim' heuristic). Sections 3-7 then deal with some more specific dependence models and the chapter finishes with some results on ordering and multivariate risk processes.

1   Large deviations

The area of large deviations is a set of asymptotic results on rare event probabilities and a set of methods to derive such results. The last decades have seen a boom in the area and a considerable body of applications in queueing theory and insurance risk.

The classical result in the area is *Cramér's theorem*. Cramér considered a random walk $S_n = X_1 + \cdots + X_n$ such that the cumulant generating function $\kappa(\theta) = \log \mathbb{E}e^{\theta X_1}$ is defined for sufficiently many θ, and gave sharp asymptotics for probabilities of the form $\mathbb{P}(S_n/n \in I)$ for intervals $I \subset \mathbb{R}$. For example, if $x > \mathbb{E}X_1$, then

$$\mathbb{P}\left(\frac{S_n}{n} > x\right) \sim e^{-\eta n} \frac{1}{\theta \sigma \sqrt{2\pi n}} \qquad (1.1)$$

where we return to the values of θ, η, σ^2 later.

The limit result (1.1) is an example of sharp asymptotics: $\sim$ means (as at other places in the book) that the ratio is one in the limit (here $n \to \infty$). However, large deviations results have usually a weaker form, logarithmic asymptotics, which in the setting of (1.1) amounts to the weaker statement

$$\lim_{n \to \infty} \frac{1}{n} \log \mathbb{P}\left(\frac{S_n}{n} > x\right) = -\eta. \qquad (1.2)$$

Note in particular that (1.2) does not capture the $\sqrt{n}$ in (1.1) but only the dominant exponential term — the correct sharp asymptotics might as well have been, e.g., $c_1 e^{-\eta n}$ or $c_2 e^{-\eta n + c_3 n^\alpha}$ with $\alpha < 1$. Thus, *large deviations results typically only give the dominant term in an asymptotic expression*. Accordingly, logarithmic asymptotics is usually much easier to derive than sharp asymptotics but also less informative. The advantage of the large deviations approach is, however, its generality, in being capable of treating many models beyond simple random walks which are not easily treated by other models, and that a considerable body of theory has been developed.

For sequences f_n, g_n with $f_n \to 0$, $g_n \to 0$, we will write $f_n \overset{\log}{\sim} g_n$ if

$$\lim_{n \to \infty} \frac{\log f_n}{\log g_n} = 1$$

(later in this section, the parameter will be u rather than n). Thus, (1.2) can be rewritten as $\mathbb{P}\left(S_n/n > x\right) \overset{\log}{\sim} \mathrm{e}^{-\eta n}$.

Example 1.1 We will go into some more detail concerning (1.1), (1.2).
Define κ^* as the convex conjugate of κ,

$$\kappa^*(x) = \sup_\theta \left(\theta x - \kappa(\theta)\right)$$

(other names are the *entropy*, the *Legendre-Fenchel transform* or just the *Legendre transform*, or the *large deviations rate function*). Most often, the sup in the definition of κ^* can be evaluated by differentiation: $\kappa^*(x) = \theta x - \kappa(\theta)$ where $\theta = \theta(x)$ is the solution of $x = \kappa'(\theta)$, which is a *saddlepoint equation* — the mean $\kappa'(\theta)$ of the distribution of X_1 exponentially tilted with θ, i.e. of

$$\widetilde{\mathbb{P}}(X_1 \in \mathrm{d}x) = \mathbb{E}\left[\mathrm{e}^{\theta X_1 - \kappa(\theta)}; X_1 \in \mathrm{d}x\right], \tag{1.3}$$

is put equal to x. In fact, exponential change of measure is a key tool in large deviations methods.
Define $\eta = \kappa^*(x)$. Since

$$\mathbb{P}\left(\frac{S_n}{n} > x\right) = \widetilde{\mathbb{E}}\left[\mathrm{e}^{-\theta S_n + n\kappa(\theta)}; \frac{S_n}{n} > x\right],$$

replacing S_n by nx in the exponent and ignoring the indicator yields the *Chernoff bound*

$$\mathbb{P}\left(\frac{S_n}{n} > x\right) \leq \mathrm{e}^{-\eta n}. \tag{1.4}$$

Next, since S_n is asymptotically normal w.r.t. $\widetilde{\mathbb{P}}$ with mean nx and variance $n\sigma^2$ where $\sigma^2 = \sigma^2(x) = \kappa''(\theta)$, we have

$$\widetilde{\mathbb{P}}\left(nx < S_n < nx + 1.96\sigma\sqrt{n}\right) \to 0.475,$$

and hence for large n

$$\begin{aligned}
\mathbb{P}(S_n/n > x) &\geq \widetilde{\mathbb{E}}\left[\mathrm{e}^{-\theta S_n + n\kappa(\theta)}; nx < S_n < nx + 1.96\sigma\sqrt{n}\right] \\
&\geq 0.4\,\mathrm{e}^{-\eta n + 1.96\theta\sigma\sqrt{n}},
\end{aligned}$$

which in conjunction with (1.4) immediately yields (1.2).
More precisely, if we replace S_n by $nx + \sigma\sqrt{n}V$ where V is $N(0,1)$, we get

$$\begin{aligned}
\mathbb{P}(S_n/n > x) &\approx \widetilde{\mathbb{E}}\left[\mathrm{e}^{-\theta nx + n\kappa(\theta) - \theta\sigma\sqrt{n}V}; V > 0\right] \\
&= \mathrm{e}^{-\eta n} \int_0^\infty \mathrm{e}^{-\theta\sigma\sqrt{n}y} \frac{1}{\sqrt{2\pi}} \mathrm{e}^{-y^2/2}\,\mathrm{d}y \\
&= \mathrm{e}^{-\eta n} \frac{1}{\theta\sigma\sqrt{2\pi n}} \tag{1.5}
\end{aligned}$$

which is the same as (1.1), commonly denoted as the *saddlepoint approximation*. The substitution by V needs, however, to be made rigorous; see Jensen [506] or [APQ, pp. 355–356] for details.   □

Further main results in large deviations theory are the *Gärtner-Ellis theorem*, which is a version of Cramér's theorem where independence is weakened to the existence of $\kappa(\theta) = \lim_{n\to\infty} \log \mathbb{E}e^{\theta S_n}/n$, *Sanov's theorem* which give rare events asymptotics for empirical distributions, *Mogul'skiĭ's theorem* which gives path asymptotics, that is, asymptotics for probabilities of the form

$$\mathbb{P}\big(\{S_{\lfloor nt\rfloor}/n\}_{0\le t\le 1} \in \Gamma\big)$$

for a suitable set Γ of functions on $[0,1]$, and the *Wentzell-Freidlin theory* of slow Markov walks, which is of similar spirit as the dicussion in VIII.3.

In the application of large deviations to ruin probabilities, we shall concentrate on a result which give asymptotics under conditions similar to the Gärtner-Ellis theorem:

Theorem 1.2 (GLYNN & WHITT [419]) *Let $X_1, X_2, \ldots$ be a sequence of r.v.'s, and write $S_n = X_1 + \cdots + X_n$, $\tau(u) = \inf\{n : S_n > u\}$ and $\psi(u) = \mathbb{P}\big(\tau(u) < \infty\big)$. Assume that there exist $\gamma, \epsilon > 0$ such that*
(i) *$\kappa_n(\theta) = \log \mathbb{E}e^{\theta S_n}$ is well-defined and finite for $\gamma - \epsilon < \theta < \gamma + \epsilon$;*
(ii) *$\limsup_{n\to\infty} \mathbb{E}e^{\theta X_n} < \infty$ for $-\epsilon < \theta < \epsilon$;*
(iii) *$\kappa(\theta) = \lim_{n\to\infty} \dfrac{1}{n}\kappa_n(\theta)$ exists and is finite for $\gamma - \epsilon < \theta < \gamma + \epsilon$;*
(iv) *$\kappa(\gamma) = 0$ and κ is differentiable at γ with $0 < \kappa'(\gamma) < \infty$.*
Then $\psi(u) \overset{\log}{\sim} e^{-\gamma u}$ as $u \to \infty$.

For the proof, we introduce a change of measure for $X_1, \ldots, X_n$ given by

$$\widetilde{F}_n(dx_1, \ldots, dx_n) = e^{\gamma s_n - \kappa_n(\gamma)} F_n(dx_1, \ldots, dx_n)$$

where F_n is the distribution of $(X_1, \ldots, X_n)$ and $s_n = x_1 + \cdots + x_n$ (note that the r.h.s. integrates to 1 by the definition of κ_n). We further write $\widetilde{\mu} = \kappa'(\gamma)$. We shall need:

Lemma 1.3 *For each $\eta > 0$, there exists $z \in (0,1)$ and n_0 such that*

$$\widetilde{\mathbb{P}}_n\left(\left|\frac{S_n}{n} - \widetilde{\mu}\right| > \eta\right) \le z^n, \quad \widetilde{\mathbb{P}}_n\left(\left|\frac{S_{n-1}}{n} - \widetilde{\mu}\right| > \eta\right) \le z^n$$

for $n \ge n_0$.

Proof. Let $0 < \theta < \epsilon$ where ϵ is as in Theorem 1.2. Clearly,

$$\widetilde{\mathbb{P}}_n(S_n/n > \widetilde{\mu} + \eta) \;\leq\; e^{-n\theta(\widetilde{\mu}+\eta)}\widetilde{\mathbb{E}}_n e^{\theta S_n} \;=\; e^{-n\theta(\widetilde{\mu}+\eta)}e^{\kappa_n(\theta+\gamma)-\kappa_n(\gamma)}.$$

Hence by (iii) and (iv),

$$\limsup_{n\to\infty} \frac{1}{n}\log\widetilde{\mathbb{P}}_n(S_n/n > \widetilde{\mu} + \eta) \;\leq\; \kappa(\theta+\gamma) - \theta\widetilde{\mu} - \theta\eta$$

and by Taylor expansion and (iv), the r.h.s. is of order $-\theta\eta + o(\theta)$ as $\theta \downarrow 0$, in particular the r.h.s. can be chosen strictly negative by taking θ small enough. This proves the existence of $z < 1$ and n_0 such that $\widetilde{\mathbb{P}}_n(S_n/n > \widetilde{\mu} + \eta) \leq z^n$ for $n \geq n_0$. The corresponding claim for $\widetilde{\mathbb{P}}_n(S_n/n < \widetilde{\mu} - \eta)$ follows by symmetry (note that the argument did not use $\widetilde{\mu} > 0$). This establishes the first claim of the lemma, for S_n.

For S_{n-1}, we have

$$\begin{aligned}
\widetilde{\mathbb{P}}_n(S_{n-1}/n > \widetilde{\mu} + \eta) &\leq e^{-n\theta(\widetilde{\mu}+\eta)}\widetilde{\mathbb{E}}_n e^{\theta S_{n-1}} = e^{-n\theta(\widetilde{\mu}+\eta)}\widetilde{\mathbb{E}}_n e^{\theta S_n - \theta X_n}\\
&= e^{-n\theta(\widetilde{\mu}+\eta)}\mathbb{E}e^{(\theta+\gamma)S_n - \theta X_n - \kappa_n(\gamma)}\\
&\leq e^{-n\theta(\widetilde{\mu}+\eta)-\kappa_n(\gamma)}\big[\mathbb{E}e^{p(\theta+\gamma)S_n}\big]^{1/p}\big[\mathbb{E}e^{-q\theta X_n}\big]^{1/q}\\
&= e^{-n\theta(\widetilde{\mu}+\eta)-\kappa_n(\gamma)}e^{\kappa_n(p(\theta+\gamma))/p}\big[\mathbb{E}e^{-q\theta X_n}\big]^{1/q}
\end{aligned}$$

where we used Hölder's inequality with $1/p + 1/q = 1$ and p chosen so close to 1 and θ so close to 0 that $|p(\theta+\gamma) - \gamma| < \epsilon$ and $|q\theta| < \epsilon$. Since $\big[\mathbb{E}e^{-q\theta X_n}\big]^{1/q}$ is bounded for large n by (ii), we get

$$\limsup_{n\to\infty} \frac{1}{n}\log\widetilde{\mathbb{P}}_n(S_{n-1}/n > \widetilde{\mu} + \eta) \;\leq\; -\theta(\widetilde{\mu}+\eta) + \kappa\big(p(\theta+\gamma)\big)/p$$

and by Taylor expansion, it is easy to see that the r.h.s. can be chosen strictly negative by taking p close enough to 1 and θ close enough to 0. The rest of the argument is as before. □

Proof of Theorem 1.2. We first show that $\liminf_{u\to\infty} \log\psi(u)/u \geq -\gamma$. Let

$\eta > 0$ be given and let $m = m(\eta) = \lfloor u(1+\eta)/\widetilde{\mu} \rfloor + 1$. Then

$$
\begin{aligned}
\psi(u) \;&\geq\; \mathbb{P}(S_m > u) \;=\; \widetilde{\mathbb{E}}_m\left[e^{-\gamma S_m + \kappa_m(\gamma)}; \, S_m > u \right] \\
&\geq\; \widetilde{\mathbb{E}}_m\left[e^{-\gamma S_m + \kappa_m(\gamma)}; \, S_m > \frac{m\widetilde{\mu}}{1+\eta} \right] \\
&=\; \widetilde{\mathbb{E}}_m\left[e^{-\gamma S_m + \kappa_m(\gamma)}; \, \frac{S_m}{m} - \widetilde{\mu} > -\frac{\widetilde{\mu}\eta}{1+\eta} \right] \\
&\geq\; \widetilde{\mathbb{E}}_m\left[e^{-\gamma S_m + \kappa_m(\gamma)}; \, \left|\frac{S_m}{m} - \widetilde{\mu}\right| < \frac{\widetilde{\mu}\eta}{1+\eta} \right] \\
&\geq\; \exp\left\{ -\gamma\widetilde{\mu}\frac{1+2\eta}{1+\eta}m + \kappa_m(\gamma) \right\} \widetilde{\mathbb{P}}_m\left(\left|\frac{S_m}{m} - \widetilde{\mu}\right| < \frac{\widetilde{\mu}\eta}{1+\eta} \right).
\end{aligned}
$$

Here $\widetilde{\mathbb{P}}_m(\cdot)$ goes to 1 by Lemma 1.3, and since $\kappa_m(\gamma)/u \to 0$ and $m/u \to (1+\eta)/\widetilde{\mu}$, we get

$$
\liminf_{u \to \infty} \psi(u) \;\geq\; -\gamma\frac{1+2\eta}{1+\eta}.
$$

Letting $\eta \downarrow 0$ yields $\liminf_{u \to \infty} \log \psi(u)/u \geq -\gamma$.

For $\limsup_{u \to \infty} \log \psi(u)/u \leq -\gamma$, we write

$$
\psi(u) \;=\; \sum_{n=1}^{\infty} \mathbb{P}(\tau(u) = n) \;=\; I_1 + I_2 + I_3 + I_4
$$

where

$$
I_1 \;=\; \sum_{n=1}^{n(\delta)} \mathbb{P}(\tau(u) = n), \qquad I_2 \;=\; \sum_{n=n(\delta)+1}^{\lfloor u(1-\delta)/\widetilde{\mu} \rfloor} \mathbb{P}(\tau(u) = n),
$$

$$
I_3 \;=\; \sum_{\lfloor u(1-\delta)/\widetilde{\mu} \rfloor + 1}^{\lfloor u(1+\delta)/\widetilde{\mu} \rfloor} \mathbb{P}(\tau(u) = n), \qquad I_4 \;=\; \sum_{\lfloor u(1+\delta)/\widetilde{\mu} \rfloor + 1}^{\infty} \mathbb{P}(\tau(u) = n)
$$

and $n(\delta)$ is chosen such that $\kappa_n(\gamma)/n < \min\left\{ \delta, (-\log z)/2 \right\}$ and

$$
\widetilde{\mathbb{P}}_n\left(\left|\frac{S_n}{n} - \widetilde{\mu}\right| > \frac{\delta\widetilde{\mu}}{1+\delta} \right) \leq z^n, \qquad \widetilde{\mathbb{P}}_n\left(\left|\frac{S_{n-1}}{n} - \widetilde{\mu}\right| > \frac{\delta\widetilde{\mu}}{1+\delta} \right) \leq z^n \qquad (1.6)
$$

for some $z < 1$ and all $n \geq n(\delta)$; this is possible by (iii), (iv) and Lemma 1.3.

Obviously,

$$
\begin{aligned}
\mathbb{P}(\tau(u) = n) \;&\leq\; \mathbb{P}(S_n > u) \;=\; \widetilde{\mathbb{E}}_n\left[e^{-\gamma S_n + \kappa_n(\gamma)}; \, S_n > u \right] \\
&\leq\; e^{-\gamma u + \kappa_n(\gamma)} \widetilde{\mathbb{P}}_n(S_n > u) \qquad (1.7)
\end{aligned}
$$

so that

$$I_1 \leq e^{-\gamma u} \sum_{n=1}^{n(\delta)} e^{\kappa_n(\gamma)}, \tag{1.8}$$

$$I_2 \leq e^{-\gamma u} \sum_{n=n(\delta)+1}^{\lfloor u(1-\delta)/\widetilde{\mu} \rfloor} e^{\kappa_n(\gamma)} \widetilde{\mathbb{P}}(S_n > u)$$

$$\leq e^{-\gamma u} \sum_{n=n(\delta)+1}^{\lfloor u(1-\delta)/\widetilde{\mu} \rfloor} e^{-n \log z/2} \widetilde{\mathbb{P}}_n \left(\left| \frac{S_n}{n} - \widetilde{\mu} \right| > \frac{\delta \widetilde{\mu}}{1+\delta} \right)$$

$$\leq e^{-\gamma u} \sum_{n=n(\delta)+1}^{\lfloor u(1-\delta)/\widetilde{\mu} \rfloor} \frac{1}{z^{n/2}} z^n \leq e^{-\gamma u} \sum_{n=0}^{\infty} z^{n/2} = \frac{e^{-\gamma u}}{1 - z^{1/2}}, \tag{1.9}$$

$$I_3 \leq e^{-\gamma u} \sum_{\lfloor u(1-\delta)/\widetilde{\mu} \rfloor+1}^{\lfloor u(1+\delta)/\widetilde{\mu} \rfloor} e^{\kappa_n(\gamma)} \leq e^{-\gamma u} \sum_{\lfloor u(1-\delta)/\widetilde{\mu} \rfloor+1}^{\lfloor u(1+\delta)/\widetilde{\mu} \rfloor} e^{n\delta}$$

$$\leq e^{-\gamma u} \left(\frac{2\delta u}{\widetilde{\mu}} + 1 \right) e^{\delta u(1+\delta)/\widetilde{\mu}}. \tag{1.10}$$

Finally,

$$I_4 \leq \sum_{\lfloor u(1+\delta)/\widetilde{\mu} \rfloor+1}^{\infty} \mathbb{P}(S_{n-1} \leq u, \, S_n > u)$$

$$= \sum_{\lfloor u(1+\delta)/\widetilde{\mu} \rfloor+1}^{\infty} \widetilde{\mathbb{E}}_n \left[e^{-\gamma S_n + \kappa_n(\gamma)}; \, S_{n-1} \leq u, \, S_n > u \right]$$

$$\leq e^{-\gamma u} \sum_{\lfloor u(1+\delta)/\widetilde{\mu} \rfloor+1}^{\infty} e^{\kappa_n(\gamma)} \widetilde{\mathbb{P}}_n \left(\left| \frac{S_{n-1}}{n} - \widetilde{\mu} \right| > \frac{\delta \widetilde{\mu}}{1+\delta} \right)$$

$$\leq e^{-\gamma u} \sum_{\lfloor u(1+\delta)/\widetilde{\mu} \rfloor+1}^{\infty} \frac{1}{z^{n/2}} z^n \leq \frac{e^{-\gamma u}}{1 - z^{1/2}}. \tag{1.11}$$

Thus an upper bound for $\psi(u)$ is

$$e^{-\gamma u} \left\{ \sum_{n=1}^{n(\delta)} e^{\kappa_n(\gamma)} + \frac{2}{1 - z^{1/2}} + \left(\frac{2\delta u}{\widetilde{\mu}} + 1 \right) e^{\delta u(1+\delta)/\widetilde{\mu}} \right\},$$

and using (i), we get

$$\limsup_{u \to \infty} \frac{\log \psi(u)}{u} \leq -\gamma + \frac{\delta(1+\delta)}{\widetilde{\mu}}.$$

Let $\delta \downarrow 0$.   □

The following corollary shows that given that ruin occurs, the typical time is $u/\kappa'(\gamma)$ just as for the compound Poisson model, cf. V.4.

Corollary 1.4 *Under the assumptions of Theorem 1.2, it holds for each $\delta > 0$ that*

$$\psi(u) \overset{\log}{\sim} \mathbb{P}\big(\tau(u) \in \big(u(1-\delta)/\kappa'(\gamma), u(1+\delta)/\kappa'(\gamma)\big)\big).$$

Proof. Since

$$\psi(u) = I_1 + I_2 + I_3 + I_4 \overset{\log}{\sim} e^{-\gamma(u)}, \quad I_3 = \mathbb{P}\big(\tau(u) \in \big(u(1-\delta)/\kappa'(\gamma), u(1+\delta)/\kappa'(\gamma)\big)\big),$$

it suffices to show that for $j = 1, 2, 4$ there is an $\alpha_j > 0$ and a $c_j < \infty$ such that $I_j \le c_j e^{-\gamma u} e^{-\alpha_j u}$. For I_4, this is straightforward since the last inequality in (1.11) can be sharpened to

$$I_4 \le e^{-\gamma u} \frac{z^{\lfloor u(1+\delta)/\widetilde{\mu} \rfloor / 2}}{1 - z^{1/2}}.$$

For I_1, I_2, we need to redefine $n(\delta)$ as $\lfloor \beta u \rfloor$ where β is so small that $\omega = 1 - 4\beta\kappa'(\gamma) > 0$. For I_2, the last steps of (1.9) can then be sharpened to

$$I_2 \le e^{-\gamma u} \frac{z^{\lfloor \beta u \rfloor / 2}}{1 - z^{1/2}}$$

to give the desired conclusion.

For I_1, we replace the bound $\widetilde{\mathbb{P}}(S_n > u) \le 1$ used in (1.8) by

$$\widetilde{\mathbb{P}}(S_n > u) \le e^{-\alpha u} \widetilde{\mathbb{E}} e^{\alpha S_n} = e^{-\alpha u} e^{\kappa_n(\alpha+\gamma) - \kappa_n(\gamma)}$$

where $0 < \alpha < \epsilon$ and α is so small that $\kappa(\gamma + \alpha) \le 2\alpha\kappa'(\gamma)$. Then for n large, say $n \ge n_1$, we have

$$\kappa_n(\alpha + \gamma) \le 2n\kappa(\gamma + \alpha) \le 4n\alpha\kappa'(\gamma).$$

Letting $c_{11} = \max_{n \le n_1} e^{\kappa_n(\alpha+\gamma)}$, we get

$$I_1 \le \sum_{n=1}^{\lfloor \beta u \rfloor} \exp\{-(\gamma + \alpha)u + \kappa_n(\alpha + \gamma)\}$$

$$\le \exp\{-(\gamma + \alpha)u\}\Big\{c_{11}n_1 + \sum_{n=1}^{\lfloor \beta u \rfloor} \exp\{4n\alpha\kappa'(\gamma)\}\Big\}$$

$$\le \exp\{-(\gamma + \alpha)u\}c_1 \exp\{4\beta u\alpha\kappa'(\gamma)\} = c_1 e^{-\gamma u} e^{-\alpha_1 u},$$

where $\alpha_1 = \alpha\omega$. ☐

The criteria given in Theorem 1.2 are the somewhat natural extension of those for the renewal model discussed in Chapter VI, as due to the independence of the increments $X_i = U_i - T_i$ condition (iv) in that case simplifies to $\kappa(\gamma) = \frac{1}{n}\log\mathbb{E}(e^{\gamma\sum_{i=1}^{n}X_i}) = \log\mathbb{E}(e^{\gamma(U_i - T_i)})$ (cf. VI.(3.1)). In the renewal set-up, of course, also the stronger result of the Cramér-Lundberg approximation holds (cf. Theorem VI.3.2).

Example 1.5 Assume the X_n form a stationary Gaussian sequence with mean $\mu < 0$. It is then well-known and easy to prove that S_n has a normal distribution with mean $n\mu$ and a variance ω_n^2 satisfying

$$\lim_{n\to\infty}\frac{1}{n}\omega_n^2 = \omega^2 = \mathbb{V}ar(X_1) + 2\sum_{k=1}^{\infty}\mathbb{C}ov(X_1, X_{k+1})$$

provided the sum converges absolutely. Hence

$$\frac{1}{n}\kappa_n(\theta) = \frac{1}{n}\left(n\theta\mu + \frac{\theta^2\omega_n^2}{2}\right) \to \kappa(\theta) = \theta\mu + \frac{\theta^2\omega^2}{2}$$

for all $\theta \in \mathbb{R}$, and we conclude that Theorem 1.2 is in force with $\gamma = -2\mu/\omega^2$. ☐

Inspection of the proof of Theorem 1.2 shows that the discrete time structure is used in an essential way. Obviously many of the most interesting examples have a continuous time scale. If $\{S_t\}_{t\geq 0}$ is the claims surplus process, the key condition similar to (iii), (iv) becomes existence of a limit $\kappa(\theta)$ of $\kappa_t(\theta) = \log\mathbb{E}e^{\theta S_t}/t$ and a $\gamma > 0$ with $\kappa(\gamma) = 0$, $\kappa'(\gamma) > 0$. Assuming that the further regularity conditions can be verified, Theorem 1.2 then immediately yields the estimate

$$\mathbb{P}\left(\sup_{k=0,1,\dots} S_{kh} > u\right) \stackrel{\log}{\sim} e^{-\gamma u} \tag{1.12}$$

for the ruin probability $\psi_h(u)$ of any discrete skeleton $\{S_{kh}\}_{k=0,1,\dots}$. The problem is whether this is also the correct logarithmic asymptotics for the (larger) ruin probability $\psi(u)$ of the whole process, i.e. whether

$$\mathbb{P}\left(\sup_{0\leq t<\infty} S_t > u\right) \stackrel{\log}{\sim} e^{-\gamma u}. \tag{1.13}$$

One would expect this to hold in considerable generality, and in fact, criteria are given in Duffield & O'Connell [331]. To verify these in concrete examples may well present considerable difficulties, but nevertheless, we shall give two continuous time examples and tacitly assume that this can be done. The reader not satisfied by this gap in the argument can easily construct a discrete time version of the models!

The following formula (1.14) is needed in both examples. Let $\{N_t\}_{t\geq 0}$ be a possibly inhomogeneous Poisson process with arrival rate $\beta(s)$ at time s. An event occurring at time s is rewarded by a r.v. $V(s)$ with m.g.f. $\phi_s(\theta)$. Thus the total reward in the interval $[0, t]$ is

$$A_t \;=\; \sum_{n:\,\sigma_n \leq t} V(\sigma_n)$$

where the σ_n are the event times.[1] Then

$$\log \mathbb{E} e^{\theta A_t} \;=\; \int_0^t \beta(s)\big(\phi_s(\theta) - 1\big)\,\mathrm{d}s \tag{1.14}$$

(to see this, derive, e.g., a differential equation in t).

Example 1.6 We assume that claims arrive according to a homogeneous Poisson process with intensity β, but that a claim is not settled immediately. More precisely, if the nth claim arrives at time σ_n, then the corresponding payment from the company in $[\sigma_n, \sigma_n + s]$ is a r.v. $U_n(s)$. Thus, assuming a continuous premium inflow at unit rate, we have

$$S_t \;=\; \sum_{n:\,\sigma_n \leq t} U_n(t - \sigma_n) \;-\; t,$$

which is a *shot-noise process*. We further assume that the processes $\{U_n(s)\}_{s\geq 0}$ are i.i.d., non-decreasing and with finite limits $U_n(\infty)$ as $s \uparrow \infty$ (thus, $U_n(\infty)$ represents the total payment for the nth claim). We let $\kappa(\theta) = \beta\big(\mathbb{E} e^{\theta U_n(\infty)} - 1\big) - \theta$ and assume there are $\gamma, \epsilon > 0$ such that $\kappa(\gamma) = 0$ and that $\kappa(\theta) < \infty$ for $\theta < \gamma + \epsilon$.

If the nth claim arrives at time $\sigma_n = s$, it contributes to S_t by the amount $U_n(t - s)$. Thus by (1.14),

$$\kappa_t(\theta) \;=\; \beta \int_0^t \big(\mathbb{E} e^{\theta U_n(t-s)} - 1\big)\,\mathrm{d}s \;-\; \theta t \;=\; \beta \int_0^t \big(\mathbb{E} e^{\theta U_n(s)} - 1\big)\,\mathrm{d}s \;-\; \theta t,$$

and since $\mathbb{E} e^{\theta U_n(s)} \to \mathbb{E} e^{\theta U_n(\infty)}$ as $s \to \infty$, we have $\kappa_t(\theta)/t \to \kappa(\theta)$. Since the remaining conditions of Theorem 1.2 are trivial to verify, we conclude that $\psi(u) \overset{\log}{\sim} e^{-\gamma u}$ (cf. the above discussion of discrete skeletons).

---

[1]Another interpretation is to consider $V(s)$ as a claim whose distribution depends on the time s of its occurrence and A_t as the aggregate sum of such claims.

It is interesting to note and intuitively reasonable that the adjustment coefficient γ for the shot-noise model is the same as the one for the Cramér-Lundberg model where a claim is immediately settled by the amount $U_n(\infty)$. Of course, the Cramér-Lundberg model has the larger ruin probability. □

Example 1.7 Given the safety loading η, the Cramér-Lundberg model implicitly assumes that the Poisson intensity β and the claim size distribution B (or at least its mean μ_B) are known. Of course, this is often not realistic. An apparent solution to this problem is to calculate the premium rate $p = p(t)$ at time t based upon claims statistics. Most obviously, the best estimator of $\beta\mu_B$ based upon $\mathscr{F}_{t-}$, where $\mathscr{F}_t = \sigma(A_s : 0 \le s \le t)$, $A_t = \sum_{i=1}^{N_t} U_i$, is A_{t-}/t. Thus, one would take $p(t) = (1+\eta)A_{t-}/t$, leading to

$$S_t = A_t - (1+\eta)\int_0^t \frac{A_s}{s}\,ds. \tag{1.15}$$

With the σ_i being the arrival times, we have

$$S_t = \sum_{i=1}^{N_t} U_i - (1+\eta)\int_0^t \frac{\sum_{i=1}^{N_s} U_i}{s}\,ds = \sum_{i=1}^{N_t} U_i\Big(1 - (1+\eta)\log\frac{t}{\sigma_i}\Big). \tag{1.16}$$

Let $\kappa_t(\alpha) = \log \mathbb{E}e^{\alpha S_t}$. It then follows from (1.14) that

$$\kappa_t(\alpha) = \beta\int_0^t \phi\Big(\alpha\Big[1 - (1+\eta)\log\frac{t}{s}\Big]\Big)\,ds - \beta t = t\kappa(\alpha) \tag{1.17}$$

where

$$\kappa(\alpha) = \beta\int_0^1 \phi\big(\alpha[1 + (1+\eta)\log u]\big)\,du - \beta. \tag{1.18}$$

Thus (iii) of Theorem 1.2 holds, and since the remaining conditions are trivial to verify, we conclude that $\psi(u) \overset{\log}{\sim} e^{-\gamma u}$ (cf. again the above discussion of discrete skeletons) where γ solves $\kappa(\gamma) = 0$.

It is interesting to compare the adjustment coefficient γ with the one γ^* of the Cramér-Lundberg model, i.e. the solution of

$$\beta(\mathbb{E}e^{\gamma U} - 1) - (1+\eta)\beta\mu_B = 0. \tag{1.19}$$

Indeed, one has

$$\gamma \ge \gamma^* \tag{1.20}$$

with equality if and only if U is degenerate. Thus, typically the adaptive premium rule leads to a ruin probability which is asymptotically smaller than for the Cramér-Lundberg model. To see this, rewrite first κ as

$$\kappa(\alpha) = \beta\mathbb{E}\left[\frac{e^{\alpha U}}{1 + (1+\eta)\alpha U}\right] - \beta. \tag{1.21}$$

This follows from the probabilistic interpretation $S_1 \overset{\mathscr{D}}{=} \sum_{i=1}^{N_1} Y_i$ where

$$Y_i = U_i(1 + (1+\eta)\log\Theta_i) = U_i(1 - (1+\eta)V_i)$$

where the Θ_i are i.i.d. uniform$(0,1)$ or, equivalently, the $V_i = -\log\Theta_i$ are i.i.d. standard exponential, which yields

$$\mathbb{E}e^{\alpha Y} = \mathbb{E}\left[\Theta^{(1+\eta)\alpha U}e^{\alpha U}\right] = \mathbb{E}\left[e^{\alpha U}\int_0^1 t^{(1+\eta)\alpha U}\,dt\right] = \mathbb{E}\left[\frac{e^{\alpha U}}{1 + (1+\eta)\alpha U}\right].$$

Next, the function $k(x) = e^{\gamma^* x} - 1 - (1+\eta)\gamma^* x$ is convex with $k(\infty) = \infty$, $k(0) = 0$, $k'(0) < 0$, so there exists a unique zero $x_0 = x_0(\eta) > 0$ such that $k(x) > 0$, $x > x_0$, and $k(x) < 0$, $0 < x < x_0$. Therefore

$$\mathbb{E}\left[\frac{e^{\gamma^* U}}{1 + (1+\eta)\gamma^* U}\right] - 1 = \mathbb{E}\left[\frac{k(U)}{1 + (1+\eta)\gamma^* U}\right]$$

$$= \int_0^{x_0}\frac{k(y)}{1+(1+\eta)\gamma^* y}\,B(dy) + \int_{x_0}^\infty\frac{k(y)}{1+(1+\eta)\gamma^* y}\,B(dy)$$

$$\leq \frac{1}{1+(1+\eta)\gamma^* x_0}\left\{\int_0^{x_0}k(y)\,B(dy) + \int_{x_0}^\infty k(y)\,B(dy)\right\} = 0,$$

using that $\mathbb{E}k(U) = 0$ because of (1.19). This implies $\kappa(\gamma^*) \leq 0$, and since $\kappa(s)$, $\kappa^*(s)$ are convex with $\kappa'(0) < 0$, $\kappa^{*'}(0) < 0$, this in turn yields $\gamma \geq \gamma^*$. Further, $\gamma = \gamma^*$ can only occur if $U \equiv x_0$. □

Condition (iii) of Theorem 1.2 reflects that the ruin probability still decays exponentially if the involved dependence is weak enough such that the logarithmic average of the moment generating functions of the (light-tailed) increment distributions converges. As Example 1.5 indicates, this will usually only be the case for short-range dependence in the risk process.

If $\kappa_n(\theta)/v_n$ does not converge for $v_n = n$, but converges for another rate function v_n, it is also sometimes possible to derive the limiting behavior of $\psi(u)$ by large deviations techniques. The needed technical assumptions are then more involved and we just mention the type of result that one can expect in such situations (formulated for a continuous-time risk process): if

$$\kappa(\theta) = \lim_{t \to \infty} \frac{1}{v(t)} \log \mathbb{E} e^{\theta \, v(t) \, S_t / a(t)} \tag{1.22}$$

exists for some scaling functions $a(t) : \mathbb{R}^+ \to \mathbb{R}^+$ and $v(t) : \mathbb{R}^+ \to \mathbb{R}^+$ with $a(t), v(t) \uparrow \infty$, and there exists another increasing scaling function $h(t)$ such that $g(d) = \lim_{t \to \infty} v\big(a^{-1}(t/d)\big)/h(t)$ exists for all $d > 0$, then under some additional technical assumptions,

$$\lim_{u \to \infty} \frac{1}{h(u)} \log \psi(u) = -\inf_{d>0} \Big[g(d) \sup_{\theta \in \mathbb{R}} \big(\theta \, d - \kappa(\theta) \big) \Big] = -\gamma. \tag{1.23}$$

In particular, $\psi(u) \overset{\log}{\sim} e^{-\gamma \, v(a^{-1}(u))}$.

Example 1.8 Consider a continuous-time stationary zero-mean Gaussian process $\{Z_t\}$ with arbitrary covariance function $\mathbb{C}ov(s,t) = \mathbb{E}(X_s X_t)$ and let $S_t = Z_t - \mu t$ for some $\mu > 0$. Then (1.22) holds with $a(t) = t$ and $v(t) = t^2/\sigma_t^2$, where $\sigma_t^2 = \mathbb{E}(Z_t^2)$. For the choice $h(t) = v(t)$, the expression $g(d) = \lim_{t \to \infty} \sigma_t^2 / \big(d^2 \, \sigma_{t/d}^2\big)$ has to be finite for all $d > 0$, which introduces a condition on σ_t. It turns out that one can in fact verify all the technical assumptions underlying the above result and obtains

$$\lim_{u \to \infty} \frac{\sigma_u^2}{u^2} \log \psi(u) = -\inf_{d>0} \Big[g(d)(d+\mu)^2/2 \Big].$$

If $\sigma_t^2/t \to \sigma^2 > 0$, then this formula simplifies to $\lim_{u \to \infty} \frac{1}{u} \log \psi(u) = -2\mu/\sigma^2$, which is the continuous-time version of Example 1.5 (and corresponds to short-range dependence of S_t).

On the other hand, for $\mathbb{E}(X_s X_t) = \frac{1}{2}(s^{2H} + t^{2H} - |s-t|^{2H})$ we arrive at the case of *Fractional Brownian Motion* with Hurst exponent $H \in (0,1)$ (which is long-range dependent). From $\sigma_t^2 = t^{2H}$, one can easily derive that in this case

$$\lim_{u \to \infty} \frac{1}{u^{2-2H}} \log \psi(u) = \frac{\big[\mu(1-1/H) \big]^{2H}}{2(1-H)^2}, \tag{1.24}$$

which shows that the ruin probability has a Weibull-type tail. For $H > 0.5$ (positive dependence of the increments) this is an instance where, despite light-tailed increments, the involved dependence leads to a heavy-tailed ruin probability.

□

Notes and references Some standard textbooks on large deviations are Bucklew [207], Dembo & Zeitouni [290] and Shwartz & Weiss [799].

In addition to Glynn & Whitt [419], see also Nyrhinen [667] for Theorem 1.2. Müller & Pflug [652] give an elementary proof of this result in terms of exponential inequalities. Variants of the claims delay model of Example 1.6 can be found in Klüppelberg & Mikosch [545], Gao & Yan [388] and Ganesh, Macci & Torrisi [387]. For Example 1.7, see Nyrhinen [667] and Asmussen [67]; the proof of (1.20) is due to Tatyana Turova. A more general adaptive premium rule was considered in Müller & Pflug [652]. Result (1.23) is from Duffield & O'Connell [331], where details on the derivation can be found; see also Chang, Yao & Zajic [231]. Fractional Brownian Motion will be discussed in more detail in the framework of Gaussian processes in Section 7.

Further applications of large deviations ideas in risk theory occur e.g. in Djehiche [324], Lehtonen & Nyrhinen [576, 577], Martin-Löf [630, 631] and Nyrhinen [667].

2   Heavy-tailed risk models with dependent input

In the previous section we saw the effect of dependence on the adjustment coefficient in the case of light-tailed claims. We now turn to heavy-tailed claim size distributions. In view of the 'one large claim' heuristics from independent increments it seems reasonable to expect a certain insensitivity of the asymptotic behavior w.r.t. dependence, as long as the dependence is not too strong. Various criteria (on dependence types of interclaim times, but also for possible dependence between the arrival process and the claim sizes) for this to be true were given by Asmussen, Schmidli & Schmidt [101]. We give here one of them, Theorem 2.1 based upon a regenerative assumption, and apply it to the Markov-modulated model of Chapter VII. For further approaches, examples and counterexamples, see [101].

Assume that the claim surplus process $\{S_t\}_{t \geq 0}$ has a regenerative structure in the sense that there exists a renewal process $\chi_0 = 0 \leq \chi_1 \leq \chi_2 < \dots$ such that

$$\{S_{\chi_0 + t} - S_{\chi_0}\}_{0 \leq t < \chi_1 - \chi_0}, \quad \{S_{\chi_1 + t} - S_{\chi_1}\}_{0 \leq t < \chi_2 - \chi_1}, \quad \dots$$

(viewed as random elements of the space of D-functions with finite lifelengths) are independent and the distribution of $\{S_{\chi_k + t} - S_{\chi_k}\}_{0 \leq t < \chi_{k+1} - \chi_k}$ is the same for all $k = 1, 2, \dots$. The zero-delayed case corresponds to $\chi_0 = \chi_1 = 0$ and we write then $\mathbb{P}_0$, $\mathbb{E}_0$, $\psi_0(u)$ etc. We let F^* denote the $\mathbb{P}_0$-distribution of S_1^*, assume $\mu_{F^*} < 0$ and $\mathbb{E}_0 \chi < \infty$ where $\chi = \chi_2 - \chi_1$ is the generic cycle. See Fig. XIII.1 where the filled circles symbolize a regeneration in the path.

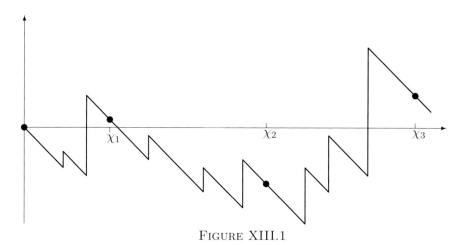

FIGURE XIII.1

Note that no specific sample path structure of $\{S_t\}$ (like in Fig. XIII.1) is assumed. We return to this point in Example 2.4 below.

Define

$$S_n^* = S_{\chi_{n+1}}, \quad M_n^* = \max_{k=0,\ldots,n} S_k^*, \quad M^* = \max_{n=0,1,\ldots} S_n^*, \quad M = \sup_{t\geq 0} S_t.$$

The idea is now to observe that in the zero-delayed case, $\{S_n^*\}_{n=0,1,\ldots}$ (corresponding to the filled circles on Fig. IV.1 except for the first one) is a random walk. Thus the assumption

$$\overline{F}^*(x) \;=\; \mathbb{P}_0(S_1^* > x) \;\sim\; \overline{G}(x) \tag{2.1}$$

for some G such that both $G \in \mathscr{S}$ and $G_0 \in \mathscr{S}$ makes X.(3.3) applicable so that

$$\mathbb{P}(M^* > u) \;\sim\; \frac{1}{|\mu_{F^*}|}\overline{F}_I^*(u)\,, \quad u \to \infty. \tag{2.2}$$

Imposing suitable conditions on the behavior of $\{S_t\}$ within a cycle will then ensure that M and M^* are sufficiently close to be tail equivalent. The one we focus on is

$$\mathbb{P}_0\big(M_1^{(\chi)} > x\big) \;\sim\; \mathbb{P}_0(S_1^* > x)\,, \tag{2.3}$$

where

$$M_n^{(\chi)} \;=\; \sup_{0\leq t < \chi_{n+1}-\chi_n} S_{\chi_n+t} - S_{\chi_n} \;=\; \sup_{0\leq t < \chi_{n+1}-\chi_n} S_{\chi_n+t} - S_{n-1}^*\,.$$

Since clearly $M_1^{(\chi)} \geq S_1^*$, the assumption means that $M_1^{(\chi)}$ and S_1^* are not too far away. See Fig. XIII.2.

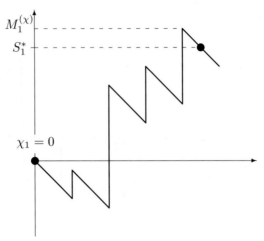

FIGURE XIII.2

Theorem 2.1 *Assume that (2.1) and (2.3) hold. Then*

$$\psi_0(u) \;=\; \mathbb{P}_0(M > u) \;\sim\; \frac{1}{|\mu_{F^*}|}\overline{F}_I^*(u).$$

Proof. Since $M \geq M^*$, it suffices by (2.2) to show

$$\liminf_{u \to \infty} \frac{\mathbb{P}(M^* > u)}{\mathbb{P}(M > u)} \;\geq\; 1. \tag{2.4}$$

Define

$$\vartheta^*(u) \;=\; \inf\{n = 1, 2, \ldots : S_n^* > u\},$$
$$\beta(u) \;=\; \inf\{n = 1, 2, \ldots : S_n^* + M_{n+1}^{(x)} > u\}$$

(note that $\{M > u\} = \{\beta(u) < \infty\}$). Let $a > 0$ be fixed. We shall use the estimate

$$\mathbb{P}_0\big(M > u,\, M_{\beta(u)+1}^{(x)} \leq a\big) \;=\; o\big(\mathbb{P}_0(M > u)\big) \tag{2.5}$$

which follows since

$$\mathbb{P}_0\big(M > u,\, M_{\beta(u)+1}^{(x)} \leq a\big)$$
$$\leq\; \mathbb{P}_0\Big(\bigcup_{n=1}^{\infty}\{M_n^* \in (u - a, u)\}\Big)$$
$$\leq\; \mathbb{P}\big(M^* \in (u - a, u)\big)/\mathbb{P}(M^* = 0) \;=\; o\big(\mathbb{P}_0(M^* > u)\big).$$

Given $\epsilon > 0$, choose a such that $\mathbb{P}_0(S_1^* > x) \geq (1 - \epsilon)\mathbb{P}_0\big(M_1^{(\chi)} > x\big)$, $x \geq a$. Then by Lemma 3.4,

$$
\begin{aligned}
\mathbb{P}_0(M^* > u) \;\sim\;& \mathbb{P}_0\big(M^* > u,\, S_{\vartheta^*(u)}^* - S_{\vartheta^*(u)-1}^* > a\big) \\
=\;& \sum_{n=1}^{\infty} \mathbb{P}_0\big(M_n^* \leq u,\, S_{n+1}^* - S_n^* > a \vee (u - S_n^*)\big) \\
\geq\;& (1 - \varepsilon) \sum_{n=1}^{\infty} \mathbb{P}_0\big(M_n^* \leq u,\, M_{n+1}^{(\chi)} > a \vee (u - S_n^*)\big) \\
\geq\;& (1 - \varepsilon) \sum_{n=1}^{\infty} \mathbb{P}_0\Big(\max_{0 < t \leq \chi_{n+1}} S_t \leq u,\, M_{n+1}^{(\chi)} > a \vee (u - S_n^*)\Big) \\
=\;& (1 - \varepsilon)\mathbb{P}_0\big(M > u,\, M_{\beta(u)+1}^{(\chi)} > a\big) \;\sim\; (1 - \varepsilon)\mathbb{P}_0\,(M > u)\,.
\end{aligned}
$$

Letting first $u \to \infty$ and next $\epsilon \downarrow 0$ yields (2.4). $\qquad\square$

Under suitable conditions, Theorem 2.1 can be rewritten as

$$
\psi_0(u) \;\sim\; \frac{\rho}{1 - \rho}\,\overline{B}_0(u) \tag{2.6}
$$

where B is the Palm distribution of claims and $\rho - 1 = \lim_{t\to\infty} S_t/t$. To this end, assume the path structure

$$
S_t \;=\; \sum_{i=1}^{N_t} U_i - t + Z_t \tag{2.7}
$$

with $\{Z_t\}$ continuous, independent of $\big\{\sum_{i=1}^{N_t} U_i\big\}$ and satisfying $Z_t/t \overset{\text{a.s.}}{\to} 0$. Then the Palm distribution of claims is

$$
B(x) \;=\; \frac{1}{\mathbb{E}_0 N_\chi}\mathbb{E}_0 \sum_{i=1}^{N_\chi} I(U_i \leq x)\,. \tag{2.8}
$$

Write $\beta = \mathbb{E}_0 N_\chi / \mathbb{E}_0 \chi$.

Corollary 2.2 *Assume that $\{S_t\}$ is regenerative and satisfies (2.7). Assume further that*
(i) *both B and B_0 are subexponential;*
(ii) *$\mathbb{E}_0 z^{N_\chi} < \infty$ for some $z > 1$;*
(iii) *For some σ-field $\mathscr{F}$, χ and N_χ are $\mathscr{F}$-measurable and*

$$
\mathbb{P}_0\Big(\sum_{i=1}^{N_\chi} U_i > x \,\Big|\, \mathscr{F}\Big) \;\sim\; N_\chi \cdot \overline{B}(x)
$$

(iv) $\mathbb{P}_0\left(\sup_{0 \le t < \chi} Z_t > x\right) = o\left(\overline{B}(x)\right)$.

Then (2.6) holds with $\rho = \beta \mu_B$.

Proof. It is easily seen that the r.v.'s $\sum_{i=1}^{N_\chi} U_i$ and $\sum_{i=1}^{N_\chi} U_i - \chi$ both have tails of order $\mathbb{E}_0 N_\chi \cdot \overline{B}(x)$, cf. the proof of Lemma 2.6 below. The same is true for S_1^*, since the tail of Z_χ is lighter than $\overline{B}(x)$ by (iv), and also for $M_1^{(\chi)}$ since

$$M_1^{(\chi)} \le \sum_{i=1}^{N_\chi} U_i + \sup_{0 \le t < \chi} Z_t.$$

Thus Theorem 2.1 is in force, and the rest is just rewriting of constants: since

$$\rho = 1 + \lim_{t \to \infty} \frac{S_t}{t} = 1 + \frac{\mu_{F^*}}{\mathbb{E}_0 \chi}$$

(see Proposition A1.4), we get

$$\begin{aligned}
\psi_0(u) &\sim \frac{1}{|\mu_{F^*}|} \int_u^\infty \mathbb{P}_0(S_1^* > x)\,\mathrm{d}x \\
&= \frac{1}{\mathbb{E}_0\chi(1-\rho)} \int_u^\infty \mathbb{E}_0 N_\chi \overline{B}(x)\,\mathrm{d}x \\
&= \frac{\rho}{1-\rho} \overline{B}_0(u).
\end{aligned}$$

$\square$

Example 2.3 As a first quick application, consider the periodic model of VII.6 with arrival rate $\beta(t)$ at time t (periodic with period 1) and claims with distribution B (independent of the time at which they arrive). Assume that $B \in \mathcal{S}$, $B_0 \in \mathcal{S}$, i.e. (i) holds. The regenerative assumption is satisfied if we take $\chi_0 = \chi_1 = 0, \chi_2 = 1, \chi_3 = 2, \ldots, Z_t \equiv 0$ (thus (iv) is trivial). The number N_χ of claims arriving in $[0,1)$ is Poisson with rate $\beta = \int_0^1 \beta(s)\,\mathrm{d}s$ so that (ii) holds, and taking $\mathcal{F} = \sigma(N_\chi)$, (iii) is obvious. Thus we conclude that (2.6) holds. $\square$

Example 2.4 Assume that $S_t = W_t - t + \sum_{i=1}^{N_t} U_i$ where $\{\sum_{i=1}^{N_t} U_i - t\}$ is standard compound Poisson and $\{W_t\}$ an independent Brownian motion with mean zero and variance constant σ^2. Again, we assume that $B \in \mathcal{S}$, $B_0 \in \mathcal{S}$; then (iv) holds since the distribution of $\sup_{0 \le t \le 1} W(t)$ is the same as that of $|W_1|$, in particular light-tailed. Taking again $\chi_0 = \chi_1 = 0, \chi_2 = 1, \chi_3 = 2, \ldots$, we conclude just as in Example 2.3 that (2.6) holds. In particular, note that *the asymptotics of $\psi_0(u)$ is the same irrespective of whether the Brownian term W_t in S_t is present or not.* $\square$

We now return to the Markov-modulated risk model of Chapter VII with background Markov process $\{J_t\}$ with $p < \infty$ states and stationary distribution $\boldsymbol{\pi}$. The arrival rate is β_i and the claim size distribution B_i when $J_t = i$. We consider the case where one or more of the claim size distributions B_i are heavy-tailed. More precisely, we will assume that

$$\lim_{x \to \infty} \frac{\overline{B}_i(x)}{\overline{G}(x)} = c_i \qquad (2.9)$$

for some distribution G such that both G and the integrated tail $\int_x^\infty \overline{G}(y)\,dy$ are subexponential, and for some constants $c_i < \infty$ such that $c_1 + \cdots + c_p > 0$. The average arrival rate β and the Palm distribution B of the claim sizes are given by

$$\beta = \sum_{i=1}^p \pi_i \beta_i, \quad B = \frac{1}{\beta} \sum_{i=1}^p \pi_i \beta_i B_i$$

and we assume $\rho = \beta \mu_B = \sum_{i=1}^p \pi_i \beta_i \mu_{B_i} < 1$.

Theorem 2.5 *Consider the Markov-modulated risk model with claim size distributions satisfying (2.9). Then (2.6) holds.*

The key step of the proof is the following lemma.

Lemma 2.6 *Let $(N_1, \ldots, N_p)$ be a random vector in $\{0, 1, 2, \ldots\}^p$, $\chi \geq 0$ a r.v. and $\mathscr{F}$ a σ-algebra such that $(N_1, \ldots, N_p)$ and χ are $\mathscr{F}$-measurable. Let $\{F_i\}_{i=1,\ldots,p}$ be a family of distributions on $[0, \infty)$ and define*

$$Y_\chi = \sum_{i=1}^p \sum_{j=1}^{N_i} X_{ij} - \chi$$

where conditionally upon $\mathscr{F}$ the X_{ij} are independent with distribution F_i for X_{ij}. Assume $\mathbb{E}z^{N_1 + \cdots + N_p} < \infty$ for some $z > 1$ and all i, and that for some distribution G on $[0, \infty)$ such that $G \in \mathscr{S}$ and some $c_1, \ldots, c_p$ with $c_1 + \cdots + c_p > 0$ it holds that $\overline{F}_i(x) \sim c_i \overline{G}(x)$. Then

$$\mathbb{P}(Y_\chi > x) \sim c\,\overline{G}(x) \quad \text{where } c = \sum_{i=1}^p c_i \mathbb{E}N_i \ .$$

Proof. Consider first the case $\chi = 0$. It follows by a slight extension of Section X.1 that

$$\mathbb{P}(Y_0 > x \mid \mathscr{F}) \sim \overline{G}(x) \sum_{i=1}^p c_i N_i, \quad \mathbb{P}(Y_0 > x \mid \mathscr{F}) \leq C\overline{G}(x) z^{N_1 + \cdots + N_p}$$

for some $C = C(z) < \infty$. Thus dominated convergence yields

$$\frac{\mathbb{P}(Y_0 > x)}{\overline{G}(x)} = \mathbb{E}\left(\frac{\mathbb{P}(Y_0 > x \mid \mathscr{F})}{\overline{G}(x)}\right) \to \mathbb{E}\left(\sum_{i=1}^{p} c_i N_i\right) = c.$$

In the general case, as $x \to \infty$,

$$\mathbb{P}(Y_\chi > x \mid \mathscr{F}) = \mathbb{P}(Y_0 > \chi + x \mid \mathscr{F}) \sim \overline{G}(\chi + x)\sum_{i=1}^{p} c_i N_i \sim \overline{G}(x)\sum_{i=1}^{p} c_i N_i,$$

and

$$\mathbb{P}(Y_\chi > x \mid \mathscr{F}) \le \mathbb{P}(Y_0 > x \mid \mathscr{F}) \le C\overline{G}(x)z^{N_1 + \cdots + N_p}.$$

The same dominated convergence argument completes the proof.   □

Proof of Theorem 2.5. If $J_0 = i$, we can define the regenerations points as the times of returns to i, and the rest of the argument is then just as the proof of Corollary 2.2. An easy conditioning argument then yields the result when J_0 is random.   □

For light-tailed distributions, Markov-modulation typically decreases the adjustment coefficient γ and thereby changes the order of magnitude of the ruin probabilities for large u, cf. VII.4. It follows from Theorem 2.5 that the effect of Markov-modulation is in some sense less dramatical for heavy-tailed distributions: the order of magnitude of the ruin probabilities remains $\int_u^\infty \overline{B}(x)\,\mathrm{d}x$.

Within the class of risk processes in a Markovian environment, Theorem 2.5 shows that basically only the tail dominant claim size distributions (those with $c_i > 0$) matter for determining the order of magnitude of the ruin probabilities in the heavy-tailed case. In contrast, for light-tailed distributions the value of the adjustment coefficient γ is given by a delicate interaction between *all* B_i.

Notes and references Theorem 2.5 was first proved by Asmussen, Fløe Henriksen & Klüppelberg [76] by a lengthy argument which did not provide the constant in front of $\overline{B}_0(u)$ in final form. An improvement was given in Asmussen & Højgaard [80], and the final reduction by Jelenkovic & Lazar [504]. The present approach via Theorem 2.1 is from Asmussen, Schmidli & Schmidt [101]. That paper also contains further criteria for regenerative input (in particular also a treatment of the delayed case which we have omitted here), as well as a condition for (2.6) to hold in a situation where the inter-claim times $(T_1, T_2, \ldots)$ form a general stationary sequence and the U_i i.i.d. and independent of $(T_1, T_2, \ldots)$; this is applied for example to risk processes with Poisson cluster arrivals. See also Araman & Glynn [52]. For further studies of perturbations like in Corollary 2.2 and Example 2.4, see Schlegel [766] and Zwart, Borst & Dębicki [923]. In the latter reference, situations are identified under which perturbations by general Gaussian processes do change the asymptotic behavior of $\psi(u)$.

3 Linear models

Let us consider a discrete-time risk model, where $R_n = u + Z_1 + \cdots + Z_n$ denotes the surplus of the portfolio at the end of year n and Z_n correspondingly the gain incurred in year n (of course any other time unit may be considered). Assume the auto-regressive moving average (ARMA) structure

$$Z_n = a_1 Z_{n-1} + \cdots + a_m Z_{n-m} + X_n + b_1 X_{n-1} + \cdots + b_k X_{n-k}, \qquad (3.1)$$

where $X_1, X_2, \ldots$ are i.i.d. r.v.'s with $\mathbb{E}[X_1] > 0$ and $a_1, \ldots a_m, b_1, \ldots b_k$ are constants. In compact notation one can write

$$p(\Delta)Z_n = q(\Delta)X_n$$

with the polynomials $p(x) = 1 - a_1 x - \cdots - a_m x^m$ and $q(x) = 1 + b_1 x + \cdots + b_k x^k$ and Δ the backward shift operator. Assume that $q(1) > 0$, $p(x)$ and $q(x)$ do not have any common factor and all zeros of $p(x)$ lie outside the unit disk of the complex plane (hence $p(1) > 0$).

Proposition 3.1 *Assume that $\{R_n\}$ follows an ARMA structure of the form (3.1) with the above assumptions and with given initial values $z_0, \ldots, z_{-m+1}$, $x_0, \ldots, x_{-k+1}$. Assume further that a positive solution $r = \gamma$ of the adjustment equation $\mathbb{E}[e^{-rX_1}] = 1$ exists. Then*

$$\mathbb{P}\big(\tau(u) < \infty \,\big|\, z_0, \ldots, z_{-m+1}, x_0, \ldots, x_{-k+1}\big)$$
$$= \frac{\exp\big\{-\gamma\big(u + \sum_{\ell=0}^{\infty}(\sum_{i=\ell+1}^{\infty} b_i')x_{-\ell}\big)/b'\big\}}{\mathbb{E}\big[\exp\big\{-\gamma\big(R_{\tau(u)} + \sum_{\ell=0}^{\infty}(\sum_{i=\ell+1}^{\infty} b_i')X_{\tau(u)-\ell}\big)/b'\big\}\,\big|\,\tau(u) < \infty\big]}, \qquad (3.2)$$

where b_ℓ and b' are defined by (3.4) and (3.6) and $x_{-k,\ldots,-m-k+1}$ are determined by (3.5).

Proof. One can equivalently express the ARMA model through a moving average (MA) model

$$Z_n = X_n + \sum_{\ell=1}^{\infty} b_\ell' X_{n-\ell}, \qquad (3.3)$$

where (for instance) $X_n = 0$ for $n \leq -m - k$ and b_ℓ' is determined by

$$q(x)/p(x) = 1 + \sum_{\ell=1}^{\infty} b_\ell' x^\ell. \qquad (3.4)$$

The needed m additional starting values $x_{-k}, \ldots, x_{-m-k+1}$ can then be determined in such a way that

$$z_n = x_n + \sum_{\ell=1}^{m+k+n-1} b'_\ell x_{n-\ell} \quad \text{for} \quad n = 0, \ldots, -m+1 \tag{3.5}$$

(which is a linear system of equations with a unique solution). From the location of the zeros of $p(x)$, it follows that b'_ℓ tend to zero exponentially fast and consequently $\sum_{\ell=1}^\infty \ell|b'_\ell| < \infty$. Define

$$b' = 1 + \sum_{\ell=1}^\infty b'_\ell = \frac{q(1)}{p(1)} > 0. \tag{3.6}$$

It is not difficult to check that $\exp\left\{ -\gamma \left(R_n + \sum_{\ell=0}^\infty (\sum_{i=\ell+1}^\infty b'_i) X_{n-\ell}\right)/b' \right\}$ is a martingale and the assertion then follows as usual by optimal stopping for $\tau(u) \wedge T$ and $T \to \infty$ (for bounded r.v. X_i, this limit operation can be justified by dominated convergence; for the unbounded case more work is needed, see Promislow [716]). □

Remark 3.2 The appearance of the factor b' in the above result is natural since in view of (3.1) the overall contribution of X_n to the surplus over time is $b'X_n$. Hence a 'fair' comparison of the ARMA model (3.1) (or equivalently (3.3)) with a classical risk model with independent increments would be to consider for the latter $\tilde{R}_n = \tilde{u} + \tilde{Z}_1 + \cdots + \tilde{Z}_n$, where $\tilde{Z}_n = b'X_n$ and $\tilde{u} = u + \sum_{\ell=0}^\infty (\sum_{i=\ell+1}^\infty b'_i) x_{-\ell}$ is the sum of contributions of all the deterministic starting values. The adjustment coefficient in this independence model is then $\tilde{\gamma} = \gamma/b'$. Hence the adjustment coefficient of the ARMA model and the one of its independence counterpart are equivalent, i.e. the introduced dependence is weak enough to leave the adjustment coefficient unchanged. □

Notes and references Proposition 3.1 is (for bounded r.v.'s) given in Gerber [400, 401] who also showed that for a finite-order MA model the denominator in (3.2) converges to a constant for $u \to \infty$, which establishes a Cramér-Lundberg approximation. Promislow [716] extended the proof to unbounded r.v.'s and slightly weaker conditions on the coefficients of the resulting MA model. Chan & Yang [229] include a force of interest and consider separate time series for the premium income and the annual claim payments. Particular cases of the ARMA model have immediate interpretations for a credibility model (see [401]) as well as for models including underwriting cycles effects on premiums and certain IBNR models for delay of claim payments (see e.g. Trufin, Albrecher & Denuit [853, 855]). Linear processes of the above type can also be addressed by large deviation techniques, which leads to logarithmic asymptotics only, but asymptotic information about the time of ruin can then be achieved as well, see e.g. Nyrhinen [666]. Other types of short-range dependence structures are e.g. discussed in Albrecher & Kantor [32] and Afonso, Egidio dos Reis & Waters [7], where the size of an annually changing premium may depend on previous loss experience. An extension of Cramér-type estimates to certain non-Gaussian long-range dependent processes of fractional auto-regressive integrated moving average (FARIMA) type is given in Barbe & McCormick [133]. Two-sided infinite-order MA processes with regularly varying tails were investigated by Mikosch & Samorodnitsky [640] and it was shown that under a tail-balance condition and some conditions on the coefficients (that imply short-range dependence!) the asymptotic ruin probability has the same asymptotic order as the case with independent increments given in Theorem X.3.1 (namely, the tail of the stationary excess distribution of the increments), but the constant in front changes (a similar conclusion is found for a first-order AR process with random coefficients in Konstantinides & Mikosch [550]; see also Hult & Samorodnitsky [486] for a recent extension to general two-sided linear processes). Barbe & McCormick [134] show that for non-stationary and long-range dependent FARIMA processes with regularly varying innovations this insensitivity no longer holds and the asymptotic order changes. Mikosch & Samorodnitsky [641] study the ruin probability of stationary ergodic symmetric α-stable processes for $\alpha \in (1,2)$ and show that its asymptotic decay can become significantly slower than the one for independent increments; further refinements of these results are given in Alparslan & Samorodnitsky [44].

4   Risk processes with shot-noise Cox intensities

Consider the surplus process $R_t = u + t - \sum_{i=1}^{N_t} U_i$ with i.i.d. claim sizes U_i that are independent of N_t, but now the claim number process N_t is a doubly stochastic Poisson process (Cox process) with a Poisson shot noise intensity process of the form

$$\beta_t = \beta + \sum_{n \in \mathbb{N}} h(t - \sigma_n, Y_n), \qquad (4.1)$$

where $\{\sigma_n\}_{n \in \mathbb{N}}$ is the sequence of arrival epochs of a homogeneous Poisson process of rate ζ, $\{Y_n\}_{n \in \mathbb{N}}$ is an i.i.d. sequence of positive r.v. (with distribution function F_Y) independent of the Poisson process, and the function $h(t, x)$ is non-negative with $h(t, x) = 0$ for $t < 0$ (here $\beta > 0$ is assumed to be constant). An interpretation of model (4.1) is as follows: In addition to the occurrence of 'normal' claims described by a homogeneous Poisson process with constant rate β, there are also claims triggered by external events (such as natural catastrophes). These events occur at times $\{\sigma_n\}_{n \in \mathbb{N}}$ (according to a homogeneous Poisson process with rate ζ). Due to reporting lags of the claims that originate from a given external event, the resulting increase in intensity will develop according to the function $h(t - \sigma_n, Y_n)$. This model captures the effect that such events can lead to a dramatic increase of the number of claims, whereas the individual claim sizes still follow the same distribution B. Figure XIII.3 shows a sample path of the intensity β_t for $h(t, x) = x\,\mathrm{e}^{-t}$ $(t > 0)$ (with Y_n being exponential(1), $\beta = 0.5$ and $\zeta = 0.7$).

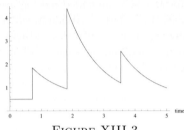

$$\text{FIGURE XIII.3}$$

Let $H(t, y) = \int_0^t h(s, y)\,\mathrm{d}s$ and

$$\Lambda_t \;=\; \int_0^t \beta_s\,\mathrm{d}s \;=\; \beta\,t + \sum_{n \in \mathbb{N}} H(t - \sigma_n, Y_n)\,.$$

The limiting average claim amount arriving per unit time turns out to be $\mu = \big(\beta + \zeta\,\mathbb{E}H(\infty, Y_1)\big)\mu_B$ and the safety loading condition here is $\mu < 1$. Similarly to (1.14), one can derive by a differential equation in t that

$$\log \mathbb{E}\mathrm{e}^{\theta A_t} \;=\; \beta\,t\,\big(\widehat{B}(\theta) - 1\big) + \zeta \int_0^t \Big(\mathbb{E}_Y\big(\mathrm{e}^{\int_s^t h(w-s,Y)(\widehat{B}(\theta)-1)\,\mathrm{d}w}\big) - 1\Big)\mathrm{d}s$$

$$\;=\; \beta\,t\,\big(\widehat{B}(\theta) - 1\big) + \zeta \int_0^t \Big(\mathbb{E}_Y\big[\mathrm{e}^{(\widehat{B}(\theta)-1)H(s,Y)}\big] - 1\Big)\mathrm{d}s, \qquad (4.2)$$

where $A_t = \sum_{i=1}^{N_t} X_i$ again denotes the aggregate claim size at time t. For $S_t = A_t - t$ and $\kappa_t(\theta) = \log \mathbb{E}\mathrm{e}^{\theta S_t}$ we then have $\kappa_t(\theta)/t \to \kappa(\theta)$ with

$$\kappa(\theta) \;=\; \beta\big(\widehat{B}(\theta) - 1\big) - \theta + \zeta\,\Big(\mathbb{E}_Y(\mathrm{e}^{(\widehat{B}(\theta)-1)H(\infty,Y)}) - 1\Big)\,. \qquad (4.3)$$

Theorem 4.1 *Let both the m.g.f. $\widehat{B}(\theta)$ and $\mathbb{E}\exp\big(\theta\,H(\infty, Y)\big)$ exist for all θ in a neighborhood of the origin and be steep (cf. p. 91). Then, the risk process with claim occurrence according to the shot-noise intensity process (4.1) satisfies*

$$\lim_{u \to \infty} \frac{1}{u} \log \psi(u) \;=\; -\gamma\,,$$

where γ is the positive solution of $\kappa(\gamma) = 0$ and κ is given by (4.3).

Proof. Considering a discrete skeleton $\{A_{nh}\}_{n\in\mathbb{N}}$, (4.3) implies that $\kappa_{nh}(\theta)/n$ has a limit of the form

$$\kappa^{(h)}(\theta) \;=\; h\,\kappa(\theta).$$

Since an easy calculation shows that $\kappa^{(h)}(\theta)'' > 0$ for every $\theta \geq 0$, $\kappa^{(h)}(0) = 0$ and $\kappa^{(h)'}(0) = h\,(\mu-1) < 0$ by the net profit condition, it follows that $\kappa^{(h)'}(\gamma) > 0$. Here the required steepness implies that $\kappa(\theta)$ is unbounded in a neighborhood of its abscissa of convergence and hence guarantees the existence of the solution $\gamma > 0$. Consequently, Theorem 1.2 applies and $\mathbb{P}(\max_n S_{nh} > u) \overset{\log}{\sim} e^{-\gamma u}$. Finally, since

$$\max_t S_t \;\geq\; \max_n S_{nh} \;\geq\; \max_t S_t - h,$$

the maximum over nh can be replaced by the continuous time maximum over t and the theorem follows from $\psi(u) = \mathbb{P}(\max_t S_t > u)$. □

We now intend to refine Theorem 4.1. For that purpose, consider the compound Poisson batch process $\widetilde{R}_t$, which is obtained by moving all arrivals of claims that are caused by a catastrophic event at σ_n to σ_n. This risk process has intensity $\widetilde{\beta} = \beta + \zeta$ for arrivals of claims and a claim size distribution $\widetilde{B}$ which is a mixture of B and the distribution of the random sum $Z = \sum_{i=1}^{N(Y)} U_i$, where $N(Y)$ is Poisson with parameter $H(\infty, Y)$ given Y and independent of the U_i; the weights are $\beta/\widetilde{\beta}$, resp. $\zeta/\widetilde{\beta}$, and the premium rate is 1 (Z can be interpreted as the total claim amount caused by the $N(Y)$ claims triggered by a specific event). Let L be the time from the event until the last of the $N(Y)$ claims occurs and $\widetilde{\psi}(u)$ the ruin probability of this compound Poisson batch process. Obviously, $\psi(u) \leq \widetilde{\psi}(u)$.

Theorem 4.2 *For some constant $C_- > 0$, $C_- e^{-\gamma u} \leq \psi(u) \leq e^{-\gamma u}$ for all u.*

Proof. The upper inequality is clear from Lundberg's inequality for $\widetilde{\psi}(u)$. For the lower, it is well known that $\widetilde{R}_{\widetilde{\tau}(u)-}$ has a limit distribution given $\widetilde{\tau}(u) < \infty$ as $u \to \infty$ (see Proposition V.7.4). Hence there exists an A such that

$$\mathbb{P}\big(\widetilde{\tau}(u) < \infty,\ \widetilde{R}_{\widetilde{\tau}(u)-} \leq A\big) \;\geq\; (1-\epsilon)\widetilde{\psi}(u) \tag{4.4}$$

for all large u. Define the pre-$\widetilde{\tau}(u)$ occupation measure $Q^{(u)}$ by

$$Q^{(u)}(G) \;=\; \mathbb{E}\int_0^{\widetilde{\tau}(u)} I(\widetilde{S}_t \in G)\,\mathrm{d}t\,, \quad G \subseteq (-\infty, u)\,.$$

Then the l.h.s. of (4.4) is

$$\int_{u-A}^{u} \widetilde{\beta}\left(1 - \widetilde{B}(u - x)\right) Q^{(u)}(\mathrm{d}x)$$

which is bounded above by $\widetilde{\beta} Q^{(u)}(u - A, u)$. Clearly, we can choose ℓ_1 with $\mathbb{P}(Z > A, L \leq \ell_1) > 0$. Every ruin event for $\widetilde{R}_t$ will also cause ruin for R_t, if the initial surplus u is lowered by ℓ_1, given that the variable L corresponding to the batch claim causing ruin does not exceed ℓ_1. Moreover, considering the situation only where the surplus prior to ruin is bounded above by A, we obtain a lower bound for the ruin probability of R_t, getting

$$
\begin{aligned}
\psi(u - \ell_1) &\geq \int_{u-A}^{u} \widetilde{\beta}\, \mathbb{P}(Z > u - x,\, L \leq \ell_1)\, Q^{(u)}(\mathrm{d}x) \\
&\geq \widetilde{\beta}\, Q^{(u)}(u - A, u)\, \mathbb{P}(Z > A,\, L \leq \ell_1) \\
&\geq \mathbb{P}(Z > A,\, L \leq \ell_1)\, (1 - \epsilon)\widetilde{\psi}(u)\,.
\end{aligned}
$$

Appealing to the Cramér-Lundberg asymptotics for $\widetilde{\psi}(u)$, the proof is complete.
□

Let us now turn to heavy-tailed claim size distributions B.

Theorem 4.3 *Assume both $B \in \mathscr{S}$ and $B_0 \in \mathscr{S}$, and $\mathbb{E}e^{\theta H(\infty, Y)} < \infty$ for some $\theta > 0$. Then*

$$
\psi(u) \sim \frac{\mu}{1 - \mu}\, \overline{B}_0(u)\,. \tag{4.5}
$$

In the proof, we shall employ coupling with the batch process $\widetilde{R}_t$ defined above. Clearly $\widetilde{S}_t \geq S_t$ in the sense of sample paths, and so it is trivial that $\psi(u) \leq \widetilde{\psi}(u)$. The next lemma shows that $\widetilde{\psi}(u)$ has the claimed asymptotics, establishing the asymptotic upper bound in (4.5).

Lemma 4.4 *Under the assumptions of Theorem 4.3, $\widetilde{\psi}(u) \sim \dfrac{\mu}{1 - \mu}\, \overline{B}_0(u)$.*

Proof. Conditioning upon Y, we get

$$
\mathbb{E}z^{N(Y)} = \mathbb{E}\exp\{H(\infty, Y)(z - 1)\}
$$

which, under the assumptions of Theorem 4.3, is finite for some $z > 1$ (implying that $\mathbb{P}(N(Y) = n)$ decreases geometrically fast in n). Hence Lemma X.2.2 implies

$$
\mathbb{P}(Z > x) \sim \mathbb{E}H(\infty, Y)\overline{B}_0(x) \tag{4.6}
$$

and subsequently

$$
1 - \widetilde{B}(x) \sim \frac{\beta + \zeta\, \mathbb{E}H(\infty, Y)}{\widetilde{\beta}}\overline{B}(x),
$$

$$
1 - \widetilde{B}_0(x) \sim \frac{(\beta + \zeta\, \mathbb{E}H(\infty, Y))}{\widetilde{\beta}\, \mu_{\widetilde{B}}} \int_x^\infty \overline{B}(z)\, \mathrm{d}z = \frac{\mu}{\widetilde{\beta}\, \mu_{\widetilde{B}}}\, \overline{B}_0(x) = \overline{B}_0(x)\,.
$$

Finally, we have by Theorem X.2.1

$$\widetilde{\psi}(u) \sim \frac{\mu}{1-\mu} \, \overline{B}_0(u). \qquad \qquad \square$$

Proof of Theorem 4.3. Consider the aggregate claim process $\breve{A}_t$ obtained from A_t by moving all claims triggered by a catastrophic event and occurring at most ℓ_0 time units later to all occurring precisely ℓ_0 time units after the catastrophic event, whereas claims occurring more than ℓ_0 time units later are deleted. Then $\psi(u) \geq \breve{\psi}(u)$ for all u. Standard results on translation of Poisson processes imply that the restriction of $\breve{A}_t - t$ to $t \in [\ell_0, \infty)$ is an ordinary Cramér-Lundberg risk process, and by reasoning as in the proof of Lemma 4.4, we obtain

$$\mathbb{P}\left(\sup_{t \in [\ell_0, \infty)} \left(\breve{A}_t - \breve{A}_{\ell_0} - (t - \ell_0) \right) > u \right) \sim \frac{\mu(\ell_0)}{1 - \mu(\ell_0)} \, \overline{B}_0(u) \qquad (4.7)$$

where $\mu(\ell_0) = \mu_B \big(\beta + \zeta \, \mathbb{E} H(\ell_0, Y) \big)$. Now

$$\sup_{t \in [0, \infty)} \left(\breve{A}_t - t \right) \geq \left(\breve{A}_{\ell_0} - \ell_0 \right) + \sup_{t \in [\ell_0, \infty)} \left(\breve{A}_t - \breve{A}_{\ell_0} - (t - \ell_0) \right). \qquad (4.8)$$

Here the two terms are independent. Since $\breve{A}_{\ell_0}$ is the sum of a Poisson$(\beta \ell_0)$ number of claims, $\mathbb{P}(\breve{A}_{\ell_0} - \ell_0 > u) \sim \beta \ell_0 \, \overline{B}(u)$, which is dominated by (4.7). Hence the tail of $\sup_{t \in [0, \infty)}(\breve{A}_t - t)$ is asymptotically given by (4.7), and we get

$$\liminf_{u \to \infty} \frac{\psi(u)}{\overline{B}_0(u)} \geq \liminf_{u \to \infty} \frac{\breve{\psi}(u)}{\overline{B}_0(u)} = \frac{\mu(\ell_0)}{1 - \mu(\ell_0)}.$$

Letting $\ell_0 \to \infty$ and using $\mu(\ell_0) \uparrow \mu$, we obtain

$$\liminf_{u \to \infty} \frac{\psi(u)}{\overline{B}_0(u)} \geq \frac{\mu}{1 - \mu}.$$

Combining this with the bound $\psi(u) \leq \widetilde{\psi}(u)$ and Lemma 4.4 completes the proof. $\qquad \square$

Remark 4.5 Note that for both light- and heavy-tailed claims, the asymptotic behavior of the ruin probability is the same as for the compound Poisson batch process, which is the process where all claims triggered from a particular event occur directly at that time as one 'batch claim'. In other words, on an asymptotic scale, the ruin probability turns out to be insensitive to the introduced dependence of claim arrivals (delay of claim arrivals, respectively) in this model. $\qquad \square$

Notes and references The risk model with a Poisson shot-noise intensity was first proposed in Dassios & Jang [274] for the specific form $h(t, x) = xe^{-t}$ which makes R_t a piecewise deterministic Markov process and then in principle enables an analysis with tools developed in Embrechts, Grandell & Schmidli [345]. Palmowski [678] uses a generator approach to derive an upper bound for the ruin probability for a general class of Cox processes generated by a diffusion process. For the estimation of the intensity from claim data, see Dassios & Jang [275]. The results given above are from Albrecher & Asmussen [12], where some further results on the corresponding aggregate claim sizes, finite horizon ruin probabilities and the inclusion of adaptive premium rules can be found. It is also possible to add a further stochastic process ν_t in (4.1) that represents some transient behavior. The particular choice $\nu_t = \sum_{n \in \mathbb{Z}^-} h(t - \sigma_n, Y_n)$ then makes the resulting process stationary in time, in which case R_t is a Poisson cluster process and Theorem 4.3 is covered by Theorem 3.1 of Asmussen, Schmidli & Schmidt [101]. Albrecher & Macci [34] provide sample path large deviations for the ruin probability of such a model in a Bayesian framework, where there is some uncertainty about involved parameters (see also Macci & Petrella [619]). Another approach to model ruin probabilities in the presence of catastrophes can be found in Cossette, Duchesne & Marceau [258].

5   Causal dependency models

Most of the models discussed so far contain dependence between claim sizes and/or their occurrence times through some common environment conditions. However, sometimes a causal dependence model may be needed in practice, where for instance the size of a claim determines the distribution of the next interclaim times (think e.g. of insurance of earthquake damages, where a large claim coming from an earthquake event may be followed by one of an afterquake etc.). It turns out that an example of a dependency model of that kind, where each interclaim time depends on the size of the previous claim, can conveniently be embedded in a semi-Markovian framework and in that way even allows explicit formulas for the ruin probability and related quantities. To see this, consider the surplus process $R_t = u + t - \sum_{i=1}^{N_t} U_i$ with i.i.d. claims U_i (and generic claim size distribution U), and assume that the time T_{i+1} between the ith claim U_i and the $(i + 1)$th claim U_{i+1} is exponentially distributed with parameter β_j if $U_i \in F_j$, where $(F_j)_{j=1,\ldots,M}$ is a (possibly random) partition of the positive halfline.

Let on the other hand $\{Z_n\}_{n \geq 0}$ be an irreducible discrete-time Markov chain with state space $\{1, \ldots, M\}$ and transition matrix $\boldsymbol{P} = (p_{ij})_{1 \leq i,j \leq M}$ and consider the semi-Markovian model

$$\mathbb{P}\big(T_{n+1} \le x, U_{n+1} \le y, Z_{n+1} = j \mid Z_n = i, (T_r, U_r, Z_r), 0 \le r \le n\big)$$
$$= (1 - e^{-\beta_i x}) p_{ij} B_j(y) .$$

Then the choices $p_{ij} = \mathbb{P}(U \in F_j)$ and $B_j \sim U|U \in F_j$ exactly correspond to the above causal dependency model. The net profit condition in this model is $\sum_{j=1}^{M} \pi_i \mu_i < \sum_{j=1}^{M} \pi_i \beta_i^{-1}$, where $\boldsymbol{\pi} = (\pi_1, \dots, \pi_M)$ is the stationary distribution of $\{Z_n\}$ and μ_i is the mean of the distribution B_i. Let $m_i(u)$ denote the Gerber-Shiu function (cf. XII.(1.1)) given that $Z_0 = i$. By the usual conditioning technique on the time interval $(0, dt)$, or (more formally) using the generator approach, one obtains the system of IDEs $(i = 1, \dots, M)$

$$m_i'(u) - (\beta_i + \delta) m_i(u) + \beta_i \sum_{j=1}^{M} p_{ij} \int_0^u m_j(u-y) B_j(\mathrm{d}y) + \beta_i \sum_{j=1}^{M} p_{ij} \int_u^\infty w(u, y-u) B_j(\mathrm{d}y) = 0 ,$$

and via Laplace transforms we arrive at the matrix equation

$$\big((s - \delta)\boldsymbol{I} - \boldsymbol{\Lambda} + \boldsymbol{\Lambda} \, \boldsymbol{P} \, \widehat{\boldsymbol{B}}[-s]\big) \, \widehat{\boldsymbol{m}}[-s] = \boldsymbol{m}(0) - \boldsymbol{\Lambda} \, \boldsymbol{P} \, \widehat{\boldsymbol{\omega}}[-s], \qquad (5.1)$$

where $\boldsymbol{m}(u) = \big(m_1(u), \dots, m_M(u)\big)$, $\widehat{\boldsymbol{m}}[-s] = \big(\widehat{m}_1[-s], \dots, \widehat{m}_M[-s]\big)$, $\widehat{\boldsymbol{\omega}}[-s] = \big(\widehat{\omega}_1[-s], \dots, \widehat{\omega}_M[-s]\big)$ with $\widehat{\omega}_i[-s] = \int_{x=0}^\infty e^{-sx} \int_x^\infty w(x, y-x) B_i(\mathrm{d}y) \, \mathrm{d}x$ and $\boldsymbol{\Lambda} = \mathrm{diag}(\beta_1, \dots, \beta_M)$, $\widehat{\boldsymbol{B}}(-s) = \mathrm{diag}\big(\widehat{B}_1[-s], \dots, \widehat{B}_M[-s]\big)$. As usual, we assume the boundary condition $\lim_{u \to \infty} m_i(u) = 0$ $(i = 1, \dots, M)$.

First, the quantities $m_i(0)$ have to be determined. For that purpose, denote $\boldsymbol{A}_\delta(s) = (s - \delta)\boldsymbol{I} - \boldsymbol{\Lambda} + \boldsymbol{\Lambda} \, \boldsymbol{P} \, \widehat{\boldsymbol{B}}[-s]$. The equation

$$\det\big(\boldsymbol{A}_\delta(s)\big) = 0 \qquad (5.2)$$

now generalizes the Lundberg fundamental equation XII.(2.2). By a combination of complex analysis and linear algebra, one can show that (5.2) has M zeros $\rho_1, \dots, \rho_M$ with $\Re(\rho_i) > 0$ for $\delta > 0$ and $\det\big(\boldsymbol{A}_0(s)\big) = 0$ has one zero $\rho_1 = 0$ and $M - 1$ zeros $\rho_2, \dots, \rho_M$ with $\Re(\rho_i) > 0$ (see [5, 19] for details).

The $m_i(u)$ are bounded functions due to the boundary conditions, so $\widehat{m}_i[-s]$ are analytic functions for $\Re(s) > 0$ (for $s = 0$ we further need integrability of $m_i(u)$), and for each of the M zeros $\rho_1, \dots, \rho_M$ we can now proceed in the following way: determine a non-trivial solution $\boldsymbol{k}_i$ of

$$\boldsymbol{A}_\delta^\mathsf{T}(\rho_i) \boldsymbol{k}_i = \boldsymbol{0}$$

for each $i = 1, \dots, M$. Since we then have

$$0 = \widehat{\boldsymbol{m}}(\rho_i)^\mathsf{T} \boldsymbol{A}_\delta^T(\rho_i) \boldsymbol{k}_i = \big(\boldsymbol{m}(0) - \boldsymbol{\Lambda} \, \boldsymbol{P} \, \widehat{\boldsymbol{\omega}}[-\rho_i]\big)^\mathsf{T} \boldsymbol{k}_i,$$

this gives M linear equations for $m_1(0), \dots, m_M(0)$.

Remark 5.1 For $\delta = 0$, the zeros $\rho_1, \ldots, \rho_M$ can always be obtained numerically. Moreover, if the involved claim size distributions have a rational Laplace transform, then $m(u)$ can be obtained explicitly by inversion of the Laplace transform of the solution of (5.1). $\square$

Example 5.2 To see how this can be put into practice, consider a causal dependency model, where the $(n+1)$th interclaim time T_{n+1} is exponential(β_1) if $U_n > \Theta_n$ for some random threshold Θ_n and T_{n+1} is exponential(β_2) if $U_n \leq \Theta_n$. This corresponds to $M = 2$ and

$$dB_1(y) = \frac{1}{\mathbb{P}(\Theta < U)} \, \mathbb{P}(\Theta \leq y) \, dB(y) \quad \text{and} \quad dB_2(y) = \frac{1}{\mathbb{P}(\Theta > U)} \, \mathbb{P}(\Theta > y) \, dB(y)$$

and $p_{i1} = \mathbb{P}(U > \Theta)$ and $p_{i2} = \mathbb{P}(U \leq \Theta)$ for $i = 1, 2$. Let Θ be exponential(2) and B exponential(1), $\beta_1 = 1.5, \beta_2 = 0.5$. Then

$$\boldsymbol{P} = \begin{pmatrix} 2/3 & 1/3 \\ 2/3 & 1/3 \end{pmatrix}, \quad \boldsymbol{\Lambda} = \begin{pmatrix} 1.5 & 0 \\ 0 & 0.5 \end{pmatrix},$$

$$\widehat{B}_1[-s] = \frac{3}{2} \left(\frac{1}{1+s} - \frac{1}{3+s} \right), \quad \widehat{B}_2[-s] = \frac{3}{3+s}.$$

For $\delta = 0$, we obtain the determinant

$$\det A_0(s) = 3 - 8s + 4s^2 + \frac{6s - 3}{1+s} - \frac{4s}{3+s},$$

which has one zero $\rho_1 = 0$ and one positive zero $\rho_2 = 1.226$, the two remaining zeros $s = -0.065$ and $s = -3.161$ are negative. E.g., for the ruin probabilities one obtains

$$\psi_1(u) = 0.007 \, e^{-3.161 \, u} + 0.938 \, e^{-0.065 \, u}, \quad \psi_2(u) = 0.003 \, e^{-3.161 \, u} + 0.867 \, e^{-0.065 \, u}.$$

$\square$

Notes and references The explicit treatment of causal dependency models of the above kind can be found in Albrecher & Boxma [18, 19]; see also Adan & Kulkarni [5] for related dependency models in a queueing context. An extension to MAP is given in Cheung & Landriault [240]. Note that for time-independent quantities ($\delta = 0$) the change of the Poisson intensity can also be reinterpreted as a change of the premium intensity for constant Poisson intensity and so an equivalent interpretation of the above model is to have dependence of the premium intensity between two claims on the size of the previous claim. Extensions of the model to include diffusion perturbation are studied in Zhou & Cai [917]; for an investigation of the Gerber-Shiu function for more general Markovian arrival processes via a fluid flow approach that avoids determining the roots of the Lundberg fundamental equation, see Ahn & Badescu [8] and the recent survey Badescu & Landriault [119]. Yang [900] studies ruin-related quantities for a risk process that is itself a Markov chain, which also has relevance in credit risk applications.

Finite-time ruin probabilities for regularly varying claim sizes and dependence that varies according to a Markovian environment process are studied in Biard, Lefèvre & Loisel [162].

Portfolios of life-insurance contracts contain certain dependencies that are different from the ones of non-life portfolios. For the calculation of ruin probabilities in such a situation we refer to Frostig & Denuit [378].

6   Dependent Sparre Andersen models

As discussed in Section VI.3a, in the Sparre Andersen model the representation

$$R_n = u + \sum_{k=1}^{n}(T_k - U_k), \quad n \geq 0,$$

reveals an imbedded random walk structure of the risk process with independent increments $T_k - U_k$ (which is the difference of the inter-occurrence time and the claim size). This random walk description enables the application of a number of classical random walk techniques to the sudy of ruin probabilities and related quantities. If one now assumes that T_k and U_k are not independent, but have some joint distribution, then the random walk structure is still preserved as long as (T_k, U_k), $k \geq 1$ is an i.i.d. sequence of bivariate random variables. In other words, one can allow the inter-occurrence time and the following claim to be dependent (which will change the increment distribution of $T_k - U_k$) and still use the random walk framework. Recall that A and B are the distribution functions of the r.v. T_k and U_k, respectively. Let $\kappa(s)$ denote the c.g.f. of the increment r.v. $T_k - U_k$, i.e. $e^{\kappa(s)} = \mathbb{E}e^{s(T_k - U_k)}$. If the dependence between T_k and U_k is described by a copula function $C(a, b)$, then a simple calculation gives that $\kappa(s)$ (in its domain of convergence) is given by

$$\widehat{B}[-s]\widehat{A}[s] - s^2 \int_0^1 e^{-sB^{-1}(a)} \int_0^1 e^{sA^{-1}(b)} \left(C(a,b) - ab\right) dA^{-1}(b) dB^{-1}(a). \quad (6.1)$$

This formula shows quite explicitly how the dependence structure (expressed through the copula) and the marginal distributions A and B influence the shape of $\kappa(s)$. In particular, for independent inter-occurrence times and claims we have $C(a, b) = ab$, so the second term in (6.1) represents the correction for the introduced dependence. Since a number of asymptotic random walk properties can be read off from the shape of $\kappa(s)$, one can now study the effect of dependence by investigating the resulting $\kappa(s)$. For instance, it is clear from (6.1) that *positive quadrant dependence* between T and U (i.e. $C(a,b) \geq ab$ for all $0 \leq a,b \leq 1$) implies that $\kappa(s)$ is for all s smaller than the one for independence. In case an adjustment coefficient γ exists, it will be the solution of $\kappa(s) = 0$ and so γ will be larger for this kind of positive dependence. More generally, whenever there is *concordance ordering* for two copulas (i.e. $C_1(a,b) \geq C_2(a,b)$ for all $0 \leq a,b \leq 1$), then $\gamma_1 \geq \gamma_2$. Also, the minimum of $\kappa(s)$ (which is modified through the dependence) reveals convergence rates of finite-time ruin probabilities (see the related Theorem V.4.5 and Veraverbeke & Teugels [864]). For particular choices of the copula and the marginal distributions, explicit expressions are possible.

Notes and references   The model discussed in this section was introduced in Albrecher & Teugels [36], where asymptotics of finite- and infinite time ruin probabilities and their orderings were investigated. Boudriault, Landriault & Marceau [192], Cossette, Marceau & Marri [262], Badescu, Cheung & Landriault [117] and Ambagaspitiya [47] establish explicit formulas for the ruin probability and Gerber-Shiu function for specific dependence structures within this model. An approach based on defective renewal equations is given in Cheung, Landriault, Willmot & Woo [241]. For a survey on dependence concepts and copulas in general, see e.g. Joe [508], Nelsen [656] and McNeil, Frey & Embrechts [633]. Models in which dependence is introduced through the aggregation of several lines of business are discussed in Section 9.

7   Gaussian models.   Fractional Brownian motion

When modeling the reserve process R or the claim surplus process $S = u - R$ of an insurance company, individual claims may be more or less important to take into account compared to aggregation. That is, one may either choose to incorporate jumps such as in the Cramèr-Lundberg model or Lévy processes, or to use a continuous approximation.

As examples of continuous approximations, we have already seen Brownian motion and more general diffusions. However, in the overall class of stochastic processes the most obvious other model choice that comes to mind is Gaussian processes. In fact, this alternative has within the last decade become popular within the area of queueing theory, and in many cases the problems studied there have as their main ingredient a ruin problem, as explained in more detail below.

A process $\{X_t\}$ (with $t \geq 0$ or $-\infty < t < \infty$) is *Gaussian* if for all $t_1 < t_2 <$

$\ldots < t_p$, the joint distribution of $X_{t_1}, X_{t_2}, \ldots, X_{t_p}$ is p-dimensional normal. By properties of the multivariate normal distribution, all that is needed to specify the distribution of the process, is the mean function $\mathbb{E}X_t$ and the covariance function $\mathbb{C}ov(X_{t_1}, X_{t_2})$ (as usual, we also assume D-paths).

The process is *stationary* if the distribution remains the same if $t_1 < t_2 < \ldots < t_p$ above are replaced by $s + t_1, s + t_2 < \ldots < s + t_p$ for arbitrary s. In the specification, one then only needs $\mu = \mathbb{E}X_t$ (that is independent of t) and the covariance function. The process has *stationary increments* if the distribution of $X(t + s) - X(t)$ only depends on s. The mean of $X(t + s) - X(t)$ must then be of the form μs, and in the specification of the process, it suffices to know in addition to μ only the variance function $v(s) = \mathbb{V}ar(X_{t+s} - X_t)$. Covariances do not need to be specified since for a process with stationary increments

$$\mathbb{C}ov(X_{t_1}, X_{t_2}) = \frac{1}{2}\big(v(t_1) + v(t_2) - v(t_2 - t_1)\big). \tag{7.1}$$

A main motivation for using Gaussian processes in queueing is the fact that many of the standard queueing processes after appropriate scaling and centering converge to a Gaussian process with stationary increments. In risk theory, we have also already seen such an example in connection with the diffusion approximation for the Cramér-Lundberg process in V.5 where the limit is Brownian motion (BM). Another main limit process in queueing is *fractional Brownian motion* (fBm), a process B^H with stationary increments, drift $\mu = 0$ and $v(s) = s^{2H}$ for some $H \in (0, 1)$ (the *Hurst parameter*; obviously, BM corresponds to $H = 1/2$). It arises in connection with so-called ON-OFF models where we have n i.i.d. sources. A source is either in an ON or an OFF state and feeds work into the system at rate 1 in ON periods. The time of an on-period (off-period) follows a distribution F_{on} and F_{off}, respectively, and all period lengths are assumed independent. Under stationarity, the total rate at which work is fed into the system is then $n\mu_P$ where $\mu_P = \mu_{\text{on}}/(\mu_{\text{on}} + \mu_{\text{off}})$ and $\mu_{\text{on}}, \mu_{\text{off}}$ are the means of $F_{\text{on}}, F_{\text{off}}$. Denote by $X_{t,n}$ the total work feed into the system before t. Because of the CLT, one intuitively expects that as $n \to \infty$, after appropriate scaling and centering, $X_{t,n}$ converges to a Gaussian process. Indeed this is true, but the limit will in general depend on $F_{\text{on}}, F_{\text{off}}$ as well as on the ordering in which the scaling and the $n \to \infty$ limits are taken. However, fBm arises as follows (Taqqu *et al.* [835], Whitt [883]):

Theorem 7.1 *Assume in the* ON-OFF *model that* $F_{\text{on}}, F_{\text{off}}$ *are both regularly varying with indices* $\alpha_{\text{on}}, \alpha_{\text{off}} \in (1, 2)$ *and that* $\alpha_{\text{on}} \neq \alpha_{\text{off}}$. *Then*

$$\lim_{K \to \infty} \lim_{n \to \infty} \left\{ \frac{K}{\sigma(K)} \frac{X_{Kt,n} - n\mu_P K t}{\sqrt{n}} \right\}_{t \geq 0} = \left\{ B_t^H \right\}_{t \geq 0}, \tag{7.2}$$

where $H = (3 - \alpha_{\mathrm{on}} \vee \alpha_{\mathrm{off}})/2$ and the convergence is in $D[0, \infty)$.[2]

fBm is not itself particularly useful as a model for the netput process S (work fed in minus the work done by the server(s)) because it has mean zero and therefore cannot lead to stationarity. Instead, one typically assumes that $S_t = X_t - \mu t$ where X_t is Gaussian with stationary increments (fBm or some other process). The stationary distribution of the workload (the netput process reflected at 0) is then the same as that of $\sup_{t \geq 0} S_t$; no time reversion is needed because stationary Gaussian processes are automatically time reversible. In particular, the probability that the stationary workload exceeds u is $\psi(u) = \mathbb{P}(\tau(u) < \infty)$ where as usual $\tau(u) = \inf\{t : S_t > u\}$. Thus, we are back to a ruin problem.

Ruin problems for Gaussian processes (or equivalently to say something on their maxima over infinite or finite time horizons) are notoriously difficult. We shall here concentrate on one approximation method, that of the *largest term*, which consists in approximating $\psi(u)$ by the tail

$$\int_u^\infty \frac{1}{\sqrt{2\pi v(t)}} \exp\left\{ -(u + \mu t)^2 / 2v(t) \right\} \, \mathrm{d}t$$

of S_t at u for that $t = t^*$ for which the density is maximal. One uses Mill's ratio to approximate the above tail by

$$\frac{1}{\sqrt{2\pi v(t)} u} \exp\left\{ -(u + \mu t)^2 / 2v(t) \right\}.$$

As a final approximation, one ignores the prefactor to the exponential, so that the largest term approximation becomes

$$
\begin{aligned}
\psi(u) \;\approx\; & \max_{t \geq 0} \exp\left\{ -(u + \mu t)^2 / 2v(t) \right\} \;=\; \exp\left\{ -\min_{t \geq 0}(u + \mu t)^2 / 2v(t) \right\} \\
=\; & \exp\left\{ -(u + \mu t^*)^2 / 2v(t^*) \right\}.
\end{aligned}
\tag{7.3}
$$

Example 7.2 Assume that S is standard Brownian motion, so that $v(t) = t$. The minimization problem is equivalent to minimizing $2\log(u + \mu t) - \log t$, which by differentation gives

$$0 = \frac{2\mu}{u + \mu t^*} - \frac{1}{t^*}, \quad \text{i.e. } t^* = \frac{u}{\mu}.$$

Insertion in (7.3) gives $\psi(u) \approx \mathrm{e}^{-2\mu u}$ which we recognize as the exact value (cf. II.(2.5) with $\sigma = 1$). □

---

[2] $\sigma(K)$ is a constant that does not need to concern us here.

Example 7.3 Assume, more generally, that $S = B_H$ is fBm, so that $v(t) = t^{2H}$. Proceeding in the same way, we get

$$0 = \frac{2\mu}{u + \mu t^*} - \frac{2H}{t^*} , \quad \text{i.e. } t^* = \frac{u}{\mu} \frac{H}{1 - H} .$$

Insertion in (7.3) gives the approximation

$$\psi(u) \approx \exp\left\{-\frac{1}{2}\left(\frac{u}{1 - H}\right)^{2-2H} \left(\frac{\mu}{H}\right)^{2H}\right\}. \tag{7.4}$$

$\square$

Remark 7.4 Approximation (7.4) shows that in the fBm case, the largest term approximation for $\psi(u)$ has a 'Weibull-like' decay with exponent $r = 2 - 2H$ (for BM, we of course refind the exponential form). That is, the decay is slower the smaller r is (see also Section 1). This phenomenon can be explained from covariance properties of fBm. Indeed, the covariances between increments can be shown to be negative when $H < 1/2$ and positive when $H > 1/2$. Thus a period of increase is typically followed by one of decrease when $H < 1/2$. In other words, the increments compete to keep S low, whereas they collaborate when $H > 1/2$. A similar phenomenon exhibits itself in the path properties: fBm has smoother paths the smaller H is (cf. Figure XIII.4, which contains sample paths of fBm with $H = 0.25, 0.5, 0.75$ and 0.95).

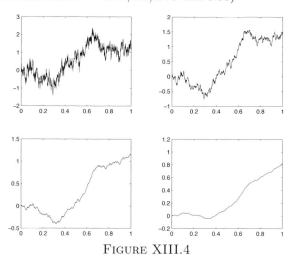

FIGURE XIII.4

A third example is integrability properties of the covariance function: the sum

$$\sum_{n=1}^{\infty} \mathbb{E}\left[S_1(S_{n+1} - S_n)\right]$$

converges for $H < 1/2$ but diverges for $H > 1/2$. Divergence of such sums or integrals is often referred to as *long-range dependence*, and the reason to focus on precisely this property is that certain CLT's hold if and only if there is convergence.   □

The exact asymptotics of $\psi(u)$ is in fact known for fBm (Piterbarg & Hüsler [487], Narayan [655]):

$$\psi(u) \ \sim \ C_H u^{2H-3+1/H} \exp\left\{ -\frac{1}{2}\left(\frac{u}{1-H}\right)^{2-2H}\left(\frac{\mu}{H}\right)^{2H} \right\} \qquad (7.5)$$

for some constant C_H (that is, there is a power-prefactor to (7.4)). The constant C_H can, however, hardly said to be explicit since it involves the so-called *Pickands constant*, a quantity that shows up also in other aspects of Gaussian process theory and is basically unknown.

Example 7.5 As a final example, we consider another popular model from queueing studies, an *integrated Ornstein-Uhlenbeck process* of the form $X_t = \int_0^t Y_v \, dv$ where Y is a stationary version of the Ornstein-Uhlenbeck process defined as the solution to $dY_t = -Y_t \, dt + dB_t$. Here one can check that $v(t) = t + 1 - e^{-t}$. The optimizer t^* is not readily computed, but one can use large-u asymptotics: if u is large, then so of course is t^*, and one has $v(t) \sim t$ as $t \to \infty$. Thus, for large u we expect t^* to have the same asymptotic form as in Example 7.2 with Brownian motion, and we get the same approximation $e^{-2\mu u}$ as for that case.   □

A main justification for the largest term approximation is that it is simple to compute, as seen from the examples. Another one is that it provides the correct logarithmic asymptotics:

$$\psi(u) \ \overset{\log}{\sim} \ \exp\left\{ -(u + \mu t^*)^2 / 2v(t^*) \right\}, \qquad (7.6)$$

see Dębicki [280].

Finally it should be mentioned that the largest term approach also suggests that $\tau(u)$ is of order t^*, thereby giving some information on the time horizon where ruin is most likely. Such information could be valuable, e.g., in a simulation study where attacking an infinite horizon is in general unfeasible and one could choose to simulate only up to time kt^* for some suitably chosen $k > 1$.

Notes and references The literature on ruin problems for Gaussian processes is huge. An accessible recent survey is in Mandjes [629]. Other main textbook references on aspects of Gaussian processes are Adler [6] and Rue & Held [756]. For convergence of stochastic processes, see Whitt [883] (motivating in many cases Gaussian models).

Apart from the largest term approach studied here, some trends in the literature on ruin problems for Gaussian processes are many-sources asymptotics (generalizing $n \to \infty$ in the ON-OFF model) and the so-called double sum approach, see again [629]. Ruin-type problems for fBm and other self-similar processes were also investigated in Michna [637], Dębicki, Michna & Rolski [282], Hüsler & Piterbarg [488] and Frangos, Vrontos & Yannacopoulos [369]. For asymptotic results on the time of ruin see e.g. Hüsler & Piterbarg [489].

8 Ordering of ruin probabilities

We have already seen some ordering results on ruin probabilities in Section IV.8 and Section VII.4. Such ordering results can be very helpful, especially in situations where quantitative results for ruin probabilities under dependence are difficult to obtain. In this section, we collect a few further ordering results in connection with discrete-time models with dependent increments.

Consider the discrete-time risk process $R_n = u + \sum_{i=1}^{n} X_i$, where X_i is the net income of year n. Assume that the r.v.'s X_i are dependent and light-tailed and that the assumptions of Theorem 1.2 are fulfilled. Then we still have an exponential decay of the ruin probability with adjustment coefficient γ defined by $\kappa(\gamma) = 0$. In this set-up one can now compare streams of net incomes w.r.t. their resulting adjustment coefficient (recall the notion of convex ordering of Section IV.8).

Proposition 8.1 *Assume that $X_1, X_2, \ldots$ and $\widetilde{X}_1, \widetilde{X}_2, \ldots$ both fulfill the assumptions of Theorem 1.2. If $\sum_{i=1}^{n} X_i \prec_{\mathrm{cx}} \sum_{i=1}^{n} \widetilde{X}_i$ for all $n \in \mathbb{N}$, then $\gamma \geq \widetilde{\gamma}$.*

Proof. The exponential function is convex, so we have $\mathbb{E}e^{\theta \sum_{i=1}^{n} X_i} \leq \mathbb{E}e^{\theta \sum_{i=1}^{n} \widetilde{X}_i}$ and subsequently $\kappa(\theta) \leq \widetilde{\kappa}(\theta)$ for all $\theta \in \mathbb{R}$, from which the assertion follows. $\square$

If we now want to compare streams of net incomes with the same marginal distributions, but different dependence structure, the so-called supermodular order is a helpful concept. A function $f : \mathbb{R}^n \to \mathbb{R}$ is a supermodular function, if for any $\mathbf{x}, \mathbf{y} \in \mathbb{R}^n$

$$f(\mathbf{x}) + f(\mathbf{y}) \leq f(\mathbf{x} \wedge \mathbf{y}) + f(\mathbf{x} \vee \mathbf{y}),$$

where the operators $\wedge$ and $\vee$ denote the component-wise minimum and maximum, respectively (if f is twice differentiable, then supermodularity means that $\partial^2 f/(\partial x_i \partial x_j) \geq 0$ for all $1 \leq i < j \leq n$). Two random vectors $\mathbf{X}, \widetilde{\mathbf{X}}$ are in *supermodular order* ($\mathbf{X} \prec_{\mathrm{sm}} \widetilde{\mathbf{X}}$) if $\mathbb{E}f(\mathbf{X}) \leq \mathbb{E}f(\widetilde{\mathbf{X}})$ for all supermodular functions $f : \mathbb{R}^n \to \mathbb{R}$. Since both functions $\mathbf{y} \mapsto I(\mathbf{y} > \mathbf{x})$ and $\mathbf{y} \mapsto I(\mathbf{y} \leq \mathbf{x})$

are supermodular for each fixed $\mathbf{x}$, it is clear that if $\mathbf{X} \prec_{\text{sm}} \widetilde{\mathbf{X}}$, the marginal distributions of $\mathbf{X}$ and $\widetilde{\mathbf{X}}$ have to coincide.

Proposition 8.2 *If* $\mathbf{X} \prec_{\text{sm}} \widetilde{\mathbf{X}}$, *then* $\sum_{i=1}^{n} X_i \prec_{\text{cx}} \sum_{i=1}^{n} \widetilde{X}_i$.

Proof. Simply note that $\mathbb{E}f(\mathbf{X}) \leq \mathbb{E}f(\widetilde{\mathbf{X}})$ for all supermodular functions $f :$ $\mathbb{R}^n \to \mathbb{R}$ and in particular also for those supermodular functions that are non-decreasing and convex in each component. Let $\phi(\mathbf{x}) = x_1 + \cdots + x_n$. Then for every non-decreasing convex function h clearly $g(\mathbf{x}) = h\big(\phi(\mathbf{x})\big)$ is non-decreasing and convex. But the supermodularity implies $\mathbb{E}\phi(\mathbf{X}) = \mathbb{E}\phi(\widetilde{\mathbf{X}})$ so that we obtain $\phi(\mathbf{X}) \prec_{\text{cx}} \phi(\widetilde{\mathbf{X}})$. $\qquad\square$

From Proposition 8.1 we thus get the following criterion:

Corollary 8.3 *Assume that* $X_1, X_2, \ldots$ *and* $\widetilde{X}_1, \widetilde{X}_2, \ldots$ *both fulfill the assumptions of Theorem 1.2. If* $(X_1, \ldots, X_n) \prec_{\text{sm}} (\widetilde{X}_1, \ldots, \widetilde{X}_n)$ *for all* $n \in \mathbb{N}$, *then* $\gamma \geq \widetilde{\gamma}$.

A random vector $\mathbf{X} = (X_1, \ldots, X_n)$ is called *associated* if $\mathbb{C}ov\big(f(\mathbf{X}), g(\mathbf{X})\big) \geq 0$ for all non-decreasing functions $f, g : \mathbb{R}^n \to \mathbb{R}$. The following result is another indication that positive dependence among the risks in the insurance portfolio is dangerous.

Proposition 8.4 *Assume that* $(X_1, \ldots, X_n)$ *is associated for all* $n \in \mathbb{N}$ *and that* $\widetilde{X}_1, \ldots, \widetilde{X}_n$ *is a sequence of independent random variables with the same marginals. Then* $\gamma \leq \widetilde{\gamma}$.

Proof. By Proposition 8.2 it suffices to show that $\sum_{i=1}^{n} X_i \prec_{\text{cx}} \sum_{i=1}^{n} \widetilde{X}_i$, and since $\mathbb{E}\big(\sum_{i=1}^{n} X_i\big) = \mathbb{E}\big(\sum_{i=1}^{n} \widetilde{X}_i\big)$, it even suffices to show that

$$\sum_{i=1}^{n} X_i \prec_{\text{icx}} \sum_{i=1}^{n} \widetilde{X}_i. \tag{8.1}$$

We proceed inductively. For $n = 1$, this statement is clearly fulfilled. Assume now that (8.1) holds. Since the $\prec_{\text{icx}}$-order is closed under convolution, we then have $\sum_{j=1}^{n+1} X_j \prec_{\text{icx}} \sum_{j=1}^{n} \widetilde{X}_j + X_{n+1}$. Choosing appropriate indicator functions in the definition of association, it is clear that

$$\mathbb{P}\Big(\sum_{i=1}^{n} \widetilde{X}_i \leq x_1\Big)\mathbb{P}(X_{n+1} \leq x_2) = \mathbb{P}\Big(\sum_{i=1}^{n} \widetilde{X}_i \leq x_1, X_{n+1} \leq x_2\Big)$$

$$\leq \mathbb{P}\Big(\sum_{i=1}^{n} \widetilde{X}_i \leq x_1, \widetilde{X}_{n+1} \leq x_2\Big)$$

for all $x_1, x_2 \geq 0$. But in view of the stop-loss order interpretation IV.(8.1) of the $\prec_{\text{icx}}$-order, it then follows from the general representation

$$\mathbb{E}(Z_1 + Z_2 - d)^+ = \mathbb{E}(Z_1) + \mathbb{E}(Z_2) - d + \int_0^d \mathbb{P}(Z_1 \leq x, Z_2 \leq d - x)\,\mathrm{d}x$$

that $\sum_{i=1}^n \widetilde{X}_i + X_{n+1} \prec_{\text{icx}} \sum_{i=1}^{n+1} \widetilde{X}_i$. The assertion now follows from the transitivity of the $\prec_{\text{icx}}$-order. □

This result is often useful, as in many situations association can be shown by a combination of the following properties:

- If $(X_1, \ldots, X_n)$ are independent, then the vector $\mathbf{X} = (X_1, \ldots, X_n)$ is associated.

- If $\mathbf{X}$ is associated and $f_1, \ldots, f_n : \mathbb{R} \to \mathbb{R}$ are non-decreasing functions, then $\big(f_1(X_1), \ldots, f_n(X_n)\big)$ is associated.

- If $\mathbf{X}$ is associated and $f_1, \ldots, f_k : \mathbb{R}^n \to \mathbb{R}$ are non-decreasing functions, then $\big(f_1(\mathbf{X}), \ldots, f_k(\mathbf{X})\big)$ is also associated.

Notes and references A general reference for stochastic orderings in the context of actuarial science is Denuit *et al.* [292]. Some ordering results of the adjustment coefficient under dependence can be found in Müller & Pflug [652]; see also Frostig [375]. Stochastic orderings for random sums have (in view of the Pollaczeck-Khinchine formula) also implications on ruin probabilities. Such results for a given ordering of the involved claim number r.v. are given by Denuit, Genest & Marceau [294]; for dependence between the number of claims and their individual distribution, see Belzunce *et al.* [155].

9 Multi-dimensional risk processes

Assume now that we have n possibly dependent portfolios (or lines of business) described through the vector $\boldsymbol{R}_t = (R_t^1, \ldots, R_t^n)$ of risk reserve processes with initial capital vector $\boldsymbol{u} = (u_1, \ldots, u_n)$ and one Poisson process N_t with intensity β that generates a claim in each of the components represented through the claim vector $\boldsymbol{U}_i = (U_{i1}, \ldots, U_{in})$. With a premium intensity vector $\boldsymbol{p}$, the multivariate risk reserve process is given by

$$\boldsymbol{R}_t = \boldsymbol{u} + t\,\boldsymbol{p} - \sum_{i=1}^{N_t} \boldsymbol{U}_i, \quad t \geq 0. \tag{9.1}$$

Here $U_1, U_2, \ldots$ is a sequence of i.i.d. random vectors with joint distribution function $B(x_1, \ldots, x_n)$, joint m.g.f. $\widehat{B}[r_1, \ldots, r_n] = \mathbb{E}\big[\exp(r_1 U_{11} + \cdots + r_n U_{1n})\big]$ and marginal distributions $B_1(x_1), \ldots, B_n(x_n)$ (so in general the components of the claim vector U_i may be dependent). It is easy to think of a number of situations where such a model applies, namely that one event or accident causes a claim in several lines of business or several portfolios. For such a risk process, there are now several ways to define the event of ruin and it will depend on the situation which one is appropriate. Let $\tau_{\max}$ be the first time when all of the components are negative, i.e.

$$\tau_{\max}(u) \;=\; \inf\{t > 0 \,|\, R_t < 0\} \;=\; \inf\{t > 0 \,|\, \max\{R_t^1, \ldots, R_t^n\} < 0\}\,,$$

where inequalities for vectors are meant component-wise. The corresponding finite-time ruin probability is

$$\psi_{\max}(u, T) \;=\; \mathbb{P}\big(\tau_{\max}(u) \le T\big)$$

and the infinite-time ruin probability is

$$\psi_{\max}(u) \;=\; \mathbb{P}\big(\tau_{\max}(u) < \infty\big)\,.$$

Other types of ruin times are

$$\tau_{\min}(u) \;=\; \inf\big\{t > 0 \,|\, \min(R_t^1, \ldots, R_t^n) < 0\big\}$$

and

$$\tau_{\mathrm{sum}}(u) \;=\; \inf\{t > 0 \,|\, R_t^1 + \ldots + R_t^n < 0\}.$$

Remark 9.1 Obviously the ruin probability $\psi_{\mathrm{sum}}(u) = \mathbb{P}(\tau_{\mathrm{sum}}(u) < \infty)$ reduces the problem again to a univariate problem with $u = u_1 + \cdots + u_n$ and i.i.d. claims $U_i = U_{i1} + \cdots + U_{in}$ (so each U_i is a sum of n dependent r.v.). In this case the multivariate framework is then just the model set-up to specify the dependence that determines the distribution of U_i and through that the ruin probability.[3] In particular, one can now ask how dependence influences $\psi_{\mathrm{sum}}(u)$ either by quantifying the dependence structure or by studying stochastic ordering. In the Notes some references to corresponding work in the literature are given (in particular for more general multivariate point processes). In the remainder of this section we focus however on ruin definitions that leave the problem in a 'truly' multivariate setting. □

The following martingale is an extension of the Wald martingale of the univariate case.

---

[3]For the distribution of dependent sums, see Section XVI.2d.

Lemma 9.2 *Let $r_1, \ldots, r_n \in \mathbb{R}$ be such that $\widehat{B}[r_1, \ldots, r_n] < \infty$. Define*

$$\kappa(r_1, \ldots, r_n) = \beta \widehat{B}[r_1, \ldots, r_n] - \beta - p_1 r_1 - \ldots - p_n r_n.$$

Then

$$M_t = \exp\{-r_1 R_t^1 - \ldots - r_n R_t^n - t\kappa(r_1, \ldots, r_n)\}, \quad t \geq 0$$

is a martingale w.r.t. the natural filtration $\mathscr{F}$.

Proof. Since N_t is a homogeneous Poisson process, we get for all $t, h \geq 0$

$$\mathbb{E}\left[\exp\left\{-\sum_{i=1}^{n} r_i(R_{t+h}^i - R_t^i)\right\}\right]$$

$$= \exp\left\{-\sum_{i=1}^{n} r_i p_i h + \beta h(\widehat{B}[r_1, \ldots, r_n] - 1)\right\} = e^{h\kappa(r_1, \ldots, r_n)}.$$

From this it follows that

$$\mathbb{E}(M_{t+h}|\mathscr{F}_t)$$
$$= \mathbb{E}\left[\exp\left\{-r_1 R_{t+h}^1 - \ldots - r_n R_{t+h}^n - (t+h)\kappa(r_1, \ldots, r_n)\right\} \middle| \mathscr{F}_t\right]$$
$$= \exp\left\{-r_1 R_t^1 - \ldots - r_n R_t^n - t\kappa(r_1, \ldots, r_n)\right\}.$$

$\square$

Let us consider the situation of light-tailed marginal claim size distributions and define $r_i^0 = \sup\{r_i \,|\, \widehat{B}[0, \ldots, 0, r_i, 0, \ldots, 0] < \infty\}$ as the abscissa of convergence of the m.g.f. of the marginal r.v. U_{1i}. Define further the sets $G = \{(r_1, \ldots, r_n) \in \mathbb{R}^n \,|\, \widehat{B}[r_1, \ldots, r_n] < \infty\}$, $G^0 = G \cap (0, \infty)^n$ and $\Delta = \{(r_1, \ldots, r_n) \in G \,|\, \kappa(r_1, \ldots, r_n) = 0\}$, $\Delta^0 = \Delta \cap (0, \infty)^n$.

Let $\boldsymbol{\mu}$ be the vector containing the expected values of the marginal claim size distributions.

Proposition 9.3 *Assume that the component-wise net profit condition $\beta\boldsymbol{\mu} < \boldsymbol{p}$ holds. If $r_i^0 > 0$ for all $i = 1, \ldots, n$ and $\sup_{(r_1, \ldots, r_n) \in G^0} \kappa(r_1, \ldots, r_n) > 0$, then*

$$\psi_{\max}(\boldsymbol{u}) \leq \inf_{(r_1, \ldots, r_n) \in \Delta^0} e^{-r_1 u_1 - \ldots - r_n u_n}.$$

An example of the shape of the set Δ for dimension $n = 2$ is illustrated in Fig. XIII.5 (the arrows are unimportant for the moment but will show up below in Remark 9.5).

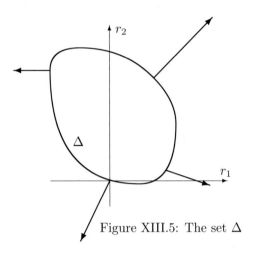

Figure XIII.5: The set Δ

Proof of Proposition 9.3. Due to Lemma 9.2, M_t is a martingale and $\tau_{\max}(u)$ is a stopping time. For every $(r_1, \ldots, r_n)$ with $\widehat{B}[r_1, \ldots, r_n] < \infty$ we know from Lemma 9.2 that

$$
\begin{aligned}
e^{-\sum_{i=1}^{n} r_i u_i} &= \mathbb{E}[M_t] \geq \mathbb{E}[M_t; \tau_{\max}(u) \leq t] \\
&= \mathbb{E}\big[M_{\tau_{\max}(u)} \mid \tau_{\max}(u) \leq t\big] \, \mathbb{P}\big(\tau_{\max}(u) \leq t\big).
\end{aligned}
$$

For all $(r_1, \ldots, r_n) \in G^0$

$$
e^{-\sum_{i=1}^{n} r_i R^i_{\tau_{\max}(u)}} \geq 1,
$$

which thus leads to

$$
\mathbb{P}\big(\tau_{\max}(u) \leq t\big) \leq e^{-\sum_{i=1}^{n} r_i u_i} \sup_{0 \leq h \leq t} e^{h \, \kappa(r_1, \ldots, r_n)}.
$$

It is easy to check by taking partial derivatives that along every ray from $\mathbf{0}$ into $(0, \infty)^n$, $\kappa(r_1, \ldots, r_n)$ is a continuous and convex function that (with the positive safety loading for each component) has negative derivative in $\mathbf{0}$ and by $\kappa(0, \ldots, 0) = 0$ and continuity will hence satisfy $\kappa(r_1, \ldots, r_n) = 0$ for at least one $(r_1, \ldots, r_n) \in G^0$, i.e. Δ^0 is not empty. Hence we can write

$$
\mathbb{P}\big(\tau_{\max}(u) \leq t\big) \leq \inf_{(r_1, \ldots, r_n) \in \Delta^0} e^{-\sum_{i=1}^{n} r_i u_i}.
$$

Letting $t \to \infty$ then gives the result. $\square$

Example 9.4 A particular boundary case of this multivariate risk model with practical relevance of its own is the two-dimensional model

$$\begin{pmatrix} R_t^1 \\ R_t^2 \end{pmatrix} = \begin{pmatrix} u_1 \\ u_2 \end{pmatrix} + t \begin{pmatrix} p_1 \\ p_2 \end{pmatrix} - \sum_{i=1}^{N_t} \begin{pmatrix} a \\ 1-a \end{pmatrix} U_i, \quad t \geq 0.$$

Here the dependence between the two claim components is the strongest possible, namely comonotonic dependence. The interpretation is that the (univariate) claims U_i are proportionally shared by two portfolios who may have different premium intensities p_1, p_2 (reflecting different safety loadings) and the question is for instance how to allocate initial capital u_1 and u_2 in a sensible way so as to minimize the ruin probability $\psi_{\min}$. One immediately observes that $\tau_{\min}(\boldsymbol{u})$ can also be represented as $\tau_{\min}(\boldsymbol{u}) = \inf\{t > 0 \mid \sum_{i=1}^{N_t} U_i > q(t)\}$ with $q(t) = \min\{(u_1 + p_1 t)/a, (u_2 + p_2 t)/(1-a)\}$. So in fact this two-dimensional model can be treated as a one-dimensional crossing problem of a compound Poisson process over a piecewise linear barrier. If $p_1/a > p_2/(1-a)$ and $u_1/a > u_2/(1-a)$, then the barrier is linear and one bounces back to the classical risk model. Extensions of this capital allocation problem to higher dimensions and more general claim arrival processes are obvious. $\square$

Remark 9.5 In Collamore [250, 251] a related multi-dimensional ruin problem is considered, namely to estimate the probability that a random walk $\{\boldsymbol{S}_n\}$ in $\mathbb{R}^d$ hits a rare set. More precisely, we will assume that the rare set has the form $xA = \{xa : a \in A\}$, where A is convex and x is a large parameter, and that the random walk in itself would typically avoid xA. For this, the drift vector $\boldsymbol{\mu} = \mathbb{E}\boldsymbol{S}_1$ should as a minimum satisfy $t\boldsymbol{\mu} \notin xA$ for all t and x (technically, the existence of a separating hyperplane is sufficient). For simplicity, we take $d = 2$. Define $\tau(x) = \inf\{n : \boldsymbol{S}_n \in xA\}$ and $z(x) = \mathbb{P}(\tau(x) < \infty)$.

The situation is as in Figure XIII.6. Here $x\boldsymbol{\xi}(\boldsymbol{k})$ is the point at which the line with direction $\boldsymbol{k}$ hits xA. The mean drift vector $\boldsymbol{\mu}$ points away from A, so an obvious possibility is to use an exponential change of measure changing the drift to some $\boldsymbol{k}$ pointing towards A.

Exponential change of measure is defined along similar lines as for the multidimensional continuous-time risk process. We let

$$\kappa(\boldsymbol{\theta}) = \kappa(\theta_1, \theta_2) = \log \mathbb{E} e^{\theta_1 X_1 + \theta_2 X_2}$$

where $(X_1, X_2) = \boldsymbol{S}_1$. The exponentially tilted measure is a random walk with increment distribution satisfying

$$\mathbb{E}_{\theta_1, \theta_2} h(X_1, X_2) = \mathbb{E}\big[h(X_1, X_2) \exp\{\theta_1 X_1 + \theta_2 X_2 - \kappa(\theta_1, \theta_2)\}\big]. \tag{9.2}$$

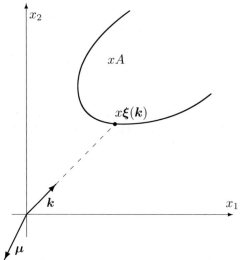

Figure XIII.6: The ruin set xA

It easily follows that the changed drift under $\mathbb{P}_{\theta_1,\theta_2}$ is given by

$$\boldsymbol{\mu}_{\theta_1,\theta_2} \;=\; \mathbb{E}_{\theta_1,\theta_2}(X_1,X_2) \;=\; \big(\kappa_1,\kappa_2\big) \;=\; \nabla\kappa(\theta_1,\theta_2), \qquad (9.3)$$

where ∇ denotes the gradient. In Fig. XIII.5, the arrows pointing outward from Δ are the gradients. The gradient is orthogonal to Δ and at any given point, its length is twice the radius of curvature at the given point of Δ.

Thus, we face the problem of which (row-vector) $\boldsymbol{\gamma} \in \Delta$ to work with. A lower bound for $z(x)$ is given by the probability of the path following the $\mathbb{P}_{\boldsymbol{\gamma}}$-description, i.e. by

$$z(x) \;=\; \mathbb{P}\big(\tau(x)<\infty\big) \;=\; \mathbb{E}_{\boldsymbol{\gamma}}\mathrm{e}^{-\boldsymbol{\gamma}\cdot\boldsymbol{S}_{\tau(x)}} \;\approx\; \mathrm{e}^{-x\boldsymbol{\gamma}\cdot\boldsymbol{\xi}(\boldsymbol{\mu}_{\boldsymbol{\gamma}})}. \qquad (9.4)$$

This suggests to take $\boldsymbol{\gamma} = \boldsymbol{\gamma}^*$ where

$$\boldsymbol{\gamma}^* \;=\; \operatorname{argmin}\; \boldsymbol{\gamma}\cdot\boldsymbol{\xi}(\boldsymbol{\mu}_{\boldsymbol{\gamma}}).$$

It can be shown that (under appropriate conditions) indeed the correct logarithmic asymptotics corresponds to taking $\boldsymbol{\gamma} = \boldsymbol{\gamma}^*$ in (9.4) and that the corresponding exponential change of measure with $\boldsymbol{\theta} = \boldsymbol{\gamma}^*$ leads to logarithmic efficiency; see Collamore [251].   □

If the claim components are heavy-tailed, the general picture is less complete. Here is a simple result on asymptotic finite-time ruin probabilities for a risk process of the type (9.1) with independent components, but with a possibly more general claim number process:

Proposition 9.6 *Assume that* $B(x_1, \ldots, x_n) = \prod_{i=1}^{n} B_i(x_i)$ *and* $B_i \in \mathscr{S}$. *Assume further that* $\mathbb{E}z^{N_T} < \infty$ *for some* $z > 1$, *where* N_T *is the number of claims up to time* T. *Then for fixed* $T > 0$

$$\psi_{\max}(\boldsymbol{u}, T) \ \sim \ \mathbb{E}\big[(N_T)^n\big] \prod_{i=1}^{n} \overline{B}_i(u_i), \quad \boldsymbol{u} \to \infty. \tag{9.5}$$

Proof. Since

$$\psi_{\max}(\boldsymbol{u}, T) = \mathbb{P}\Big(\sum_{i=1}^{N_t} \boldsymbol{U}_i - t\,\boldsymbol{p} > \boldsymbol{u} \ \text{ for some } 0 < t \leq T\Big),$$

we have the simple upper bound

$$\psi_{\max}(\boldsymbol{u}, T) \ \leq \ \mathbb{P}\Big(\sum_{i=1}^{N_T} \boldsymbol{U}_i > \boldsymbol{u}\Big) \ = \ \sum_{m=0}^{\infty} \mathbb{P}(N_T = m) \prod_{j=1}^{n} \mathbb{P}\Big(\sum_{i=1}^{m} U_{ij} > u_j\Big).$$

By the subexponential property of the marginals, Lemma X.1.8, Lemma X.2.2 and dominated convergence, the latter is asymptotically equal to

$$\sum_{m=0}^{\infty} \mathbb{P}(N_T = m)\, m^n \prod_{i=1}^{n} \overline{B}_i(u_i) \ = \ \mathbb{E}\big[(N_T)^n\big] \prod_{i=1}^{n} \overline{B}_i(u_i) \tag{9.6}$$

so that we have the upper bound

$$\psi_{\max}(\boldsymbol{u}, T) \leq \big(1 + o(1)\big)\, \mathbb{E}\big[(N_T)^n\big] \prod_{i=1}^{n} \overline{B}_i(u_i).$$

Similarly, a lower bound for the finite-time ruin probability is

$$\psi_{\max}(\boldsymbol{u}, T) \ \geq \ \mathbb{P}\Big(\sum_{i=1}^{N_T} \boldsymbol{U}_i - T\boldsymbol{p} > \boldsymbol{u}\Big) \ = \ \sum_{m=0}^{\infty} \mathbb{P}(N_T = m) \prod_{j=1}^{n} \mathbb{P}\Big(\sum_{i=1}^{m} U_{ij} > u_j + p_j T\Big).$$

By the long-tailed property of the subexponential distribution, this is asymptotically also equal to (9.6) so that

$$\psi_{\max}(\boldsymbol{u}, T) \ \geq \ \big(1 + o(1)\big)\, \mathbb{E}\big[(N_T)^n\big] \prod_{i=1}^{n} \overline{B}_i(u_i)$$

and we have asymptotic equivalence. $\qquad\square$

Remark 9.7 Note that by Jensen's inequality $\mathbb{E}\big[(N_T)^n\big] \geq \mathbb{E}[N_T]^n$, so that $\psi_{\max}(\boldsymbol{u}, T)$ is asymptotically larger than the product of the n marginal one-dimensional finite-time ruin probabilities, which can be explained by the common claim number process that governs the n components.

Notes and references Although multivariate ruin theory is a very natural extension of classical ruin theory with a lot of potential applications also in fields outside of insurance (such as credit risk or barrier option pricing), this research field is not yet very far developed. As in Remark 9.5, the event of ruin can in general be defined as the first passage of $\boldsymbol{R}_t$ into an n-dimensional open set A that does not contain $\boldsymbol{0}$. An early paper in such a framework is Dembo, Karlin & Zeitouni [289] for multivariate Lévy processes. In the framework of Remark 9.5, Collamore [250, 251] also derived asymptotic results for the time of ruin; see in addition Borovkov & Mogul'skiĭ [186, 187, 188]. Huh & Kolkiewicz [483] deal with ruin probabilities for multivariate diffusions and applications to the pricing of credit risk products.

The particular risk model (9.1) was investigated for two dimensions in Chan, Yang & Zhang [230]. The martingale approach was also implemented in Li, Liu & Tang [583] who worked out some concrete examples of dependence structures for two dimensions and illustrated that it is easily possible to extend Proposition 9.3 for the situation where a Brownian perturbation is added in each component of (9.1) with a joint correlation matrix (the form of κ is then correspondingly modified). Finite-time ruin probability approximations via a bivariate compound binomial model for two dimensions as well as some ordering results are given in Yuen, Guo & Wu [910] and Cai & Li [219]. In the latter paper also an explicit solution for $\psi_{\text{sum}}(\boldsymbol{u})$ for multivariate phase-type claims is derived, see also Eisele [341] for a Panjer-type recursion and Sundt & Vernic [824] for a more general treatment.

Explicit results for $\psi_{\max}(\boldsymbol{u})$ and $\psi_{\min}(\boldsymbol{u})$ are usually out of reach (except for very simple situations, see e.g. Dang *et al.* [272]); however, it was shown in Cai & Li [219] that if the claim vectors are associated, then $\prod_{i=1}^{n} \psi_i(u_i) \leq \psi_{max}(\boldsymbol{u}) \leq \min_{1 \leq i \leq n} \psi_i(u_i)$, where $\psi_i(u_i)$ is the marginal ruin probability of the ith component. From this it is not hard to establish in two dimensions the bound

$$\max\{\psi_1(u_1), \psi_2(u_2)\} \leq \psi_{min}(u_1, u_2) \leq \psi_1(u_1) + \psi_2(u_2) - \psi_1(u_1)\psi_2(u_2).$$

The model of Example 9.4 with N_t being a renewal process is studied by Avram, Palmowski & Pistorius [111]. In particular, asymptotic results for light-tailed claim size distributions are derived for two dimensions. Related two-sided barrier crossing problems for compound Poisson processes and analogies to queueing problems are studied in Perry, Stadje & Zaks [693].

If $\boldsymbol{U}_i$ in (9.1) is multivariate regularly varying, quite explicit and intuitive asymptotic results can be obtained for renewal claim number processes, see Hult & Lindskog [485] with a slightly different definition of ruin.

As mentioned in Remark 9.1, if the claim sizes in each business line and between business lines (i.e. components) are i.i.d. and the aim is to assess ψ_{sum}, then it is often possible to transform more complicated multivariate point processes into simpler ones. One example is that N_t is a superposition of counting processes each of which causes claims only in a selection of components, in which case one can usually identify a one-dimensional reformulation of the model with a modified (mixed) claim distribution (see e.g. Yuen, Guo & Wu [909], Ambagaspitiya [46] and for stochastic orderings Frostig [375] and Lindskog & McNeil [599]). Pfeifer & Nešlehová [696] and Bäuerle & Grübel [143] use copulas and random time shifts to generate multivariate claim counting distributions with Poisson marginals; for another general flexible multivariate counting process, see Bäuerle & Grübel [144].

Ruin probabilities of the type ψ_{sum} and Lundberg bounds in a discrete multivariate autoregressive model are investigated in Zhang, Yuen & Li [916]. Since $\mathrm{e}^{r\sum_{i=1}^n x_i}$ is a supermodular function, it is immediately clear that if $\mathbf{U} \prec_{\mathrm{sm}} \widetilde{\mathbf{U}}$ in (9.1), then the respective adjustment coefficients for ψ_{sum} (if they exist) fulfill $\widetilde{\gamma} \leq \gamma$. Extensions of this result to Cox models and finite-time Lundberg inequalities are given in Juri [513], see also Macci, Stabile & Torrisi [620].

Bregman & Klüppelberg [198] show that if two compound Poisson processes are coupled by a Clayton Lévy copula, one can obtain quite explicit asymptotic results for ψ_{sum}. For stochastic ordering results of component-wise ruin times and ψ_{sum} in a general multivariate set-up with Lévy copulas see Bäuerle, Blatter & Müller [142].

Another multivariate aspect is competing claim processes. That is, the reserve process is $R_t = R_t^1 + \cdots + R_t^n$, and if ruin occurs, one may ask which of the k components caused it. In particular, if all of the R^i can only go downwards by a jump, it is a well-defined question which R^i actually performed the jump to make the reserve go negative. However, especially in the light-tailed case this does not tell the whole story: some other R^j may have taken R close to zero before ruin. For work in this direction, see Huzak *et al.* [491].

Chapter XIV

Stochastic control

1 Introduction

The purpose of stochastic control is to find strategies that are optimal in the sense of maximizing a suitably defined reward function. In the setting of this book, consider the risk reserve process. Time may be discrete or continuous, and the time horizon finite (deterministic or a stopping time) or infinite, and we will denote by T its upper limit. Assume given a set of possible actions. At each time t the controller then chooses one particular action u_t, and the function $U = (u_t)_{t \leq T}$ is denoted a *strategy*, the set of admissible strategies over which to maximize is $\mathscr{U}$, and the reserve process governed by a particular strategy is $\{R_t^U\}$ (for notational convenience $\{R_t\}$ when it is clear what U is). We further assume that the reserve has a given initial value $x = x_0$.[1]

In discrete time, the reward to be maximized over U will have the form

$$V_T^U(x) = \mathbb{E}_x \sum_{t=0}^{T} r(R_t^U, u_t, t). \tag{1.1}$$

Thus $r(x, u, t)$ is the gain of using strategy u at reserve level x at time t. The function r may be negative, which corresponds to a loss, not a gain. It is common to assume that t only enters via a discounting factor δ, i.e. (with a slight abuse of notation) that $r(x, u, t) = \mathrm{e}^{-\delta t} r(x, u)$, but we will not always make this assumption. It is, however, convenient in continuous time problems, and in infinite-horizon problems one certainly needs $r(x, u, t)$ to somehow decrease with t since otherwise the total $V_T^U(x)$ may well be infinite.

---

[1] Note that u is used in most of the rest of the book. We use a different symbol here to avoid confusion with the control.

The value $V_T(x)$ of holding initial reserve x is obtained by maximizing over the set $\mathscr{U}$ of strategies under consideration, i.e.

$$V_T(x) = \sup_{U \in \mathscr{U}} V_T^U(x). \qquad (1.2)$$

The supremum may or may not be attained. When it is attained, the maximizer is denoted by U^*. The function $x \to V_T(x)$ is denoted the *value function*.

In continuous time, the reward will have the form

$$V_T^U(x) = r_T(R_T^U) + \int_0^T r(R_t^U, u_t, t)\, dt. \qquad (1.3)$$

The added term $r_T(R_T^U)$ corresponds to a terminal reward (or punishment). The value function is again defined by (1.2).

Example 1.1 Consider a risk process in discrete time, such that the amount of premiums received at each time instant $t = 1, 2, \ldots$ is 1 and the claim amont for the time period $(t-1, t]$ is a r.v. Y_t, such that the Y_t are i.i.d. and satisfy $\mathbb{E}Y_t < 1$. Thus, without any form of control we have

$$R_0 = x, \quad R_1 = x + 1 - Y_1, \quad R_2 = R_1 + 1 - Y_2 \ldots \qquad (1.4)$$

and the time horizon T is the time τ of ruin when $\tau < \infty$, ∞ otherwise. As an example of a control problem, consider dynamic proportional reinsurance and minimization of the ruin probability. That is, at time t the company chooses to reinsure a proportion $u_t \in [0,1]$ of its portfolio so that Y_t is replaced by $(1-u_t)Y_t$. If reinsurance is cheap, i.e. the premium income changes in the same proportion (replace 1 by $1-u_t$ in (1.4)), this reduces variability and so potentially the ruin probability. However, in practice reinsurance is not cheap: the premium $b(u_t)$ to pay for reinsurance will typically satisfy $b(u) > u$ so that the drift is reduced which potentially increases the ruin probability and there is a trade-off. Note that $\mathscr{U} = [0,1]^{\mathbb{N}}$.

The problem can be put into the framework (1.1) by taking $T = \infty$, $[0, \infty) \cup \{\Delta\}$ as state space, modifying R by letting $R_t = \Delta$ when $\tau < t < \infty$ (τ is the time of ruin) and taking the reward function as $r(x, u, t) = -1$ for $x < 0$ and 0 otherwise. Then the sum in (1.1) is -1 when $\tau < \infty$ and 0 otherwise, and so $V_T^U(x)$ is minus the probability of ruin under strategy U. $\square$

In this chapter we will only work with *feedback strategies*. This means that we assume R^U to be (possibly time-inhomogeneous) Markov with transition mechanism at time t depending on u_t, and that u_t is chosen as a function of R_t^U and t only.

Notes and references In the Markovian setting it may certainly seem counter-intuitive that a u_t depending on some further characteristics of $\left\{R_s^U\right\}_{s \leq t}$ could be optimal. However, the problem is more complicated than it may look. For some discussion in discrete time, see Blackwell [171] and Bertsekas & Shreve [160]. In continuous time, one typically first finds a feedback strategy that is a candidate for being optimal and subsequently gives a proof of the optimality by a so-called verification theorem.

2   Stochastic dynamic programming

Stochastic dynamic programming is a method for solving the optimization problem (1.2) in discrete time. The model then means that $R^U = (R_0^U, \ldots, R_T^U)$ is a Markov chain whose transition probabilities $p(t, x, y, u)$ from t to $t+1$ and from state x to state y depend on $u = u_t$.

The idea is most readily explained by first assuming R^U to have a finite state space E, the set of all controls u to be finite, and $T < \infty$ to be deterministic. We then define the value function $V(t, x)$ at time $t \leq T$ and in state x as

$$V(t, x) = \max_{U^t \in \mathscr{U}^t} \mathbb{E}_x \sum_{s=t}^{T} r\left(R_t^{U^t}, u_t, t\right) \tag{2.1}$$

where $\mathscr{U}^t$ is the set of all admissible $U^t = (u_t, \ldots, u_T)$, and we denote by $u^*(t, x)$ a control which is optimal (it does not necessarily have to be unique). Clearly, the strategy U^* given by $u_0 = u_0^*(0, x_0), u_1 = u^*(1, R_1), \ldots, u_T = u^*(T, R_T)$ is then optimal.

To compute $u^*(t, x)$, one proceeds backward in time. At $t = T$, it is obvious that $u^*(T, x) = \operatorname{argmax}_u r(x, u, T)$ and that $V(T, x) = r(x, u^*(x), T)$.

Assume the values $V(t + 1, y)$ have been computed for all $y \in E$. Then clearly

$$V(t, x) = \max_u \sum_{y \in E} p(t, x, y, u) V(t + 1, y), \tag{2.2}$$

and $u^*(t, x)$ is a maximizer.

Note that in this setting where everything is finite, we in principle have a finite maximization problem:

$$V(0, x) = \max_{u_0, \ldots, u_T} \mathbb{E}_x \sum_{t=0}^{T} r(R_t^U, u_t, t), \tag{2.3}$$

where $U = (u_0, \ldots, u_T)$. The advantage of stochastic dynamic programming is to reduce complexity. Say we have p Markov states and q possible controls.

Using (2.3), the expectation is then a sum over all p^T possible Markov chain paths, where each sum contains $T + 1$ terms, and this has to be evaluated over all q^{T+1} possible control combinations. Thus the total complexity is $(T + 1)p^T q^{T+1}$. In contrast, each sum in (2.2) has p terms and has to be evaluated for q controls u and p Markov states x. Thus the complexity of each backward step is $p^2 q$ and the total complexity of the stochastic dynamic programming algorithm is $Tp^2 q + pq$ (the second term comes from the initial $t = T$ step). This is typically an enormous saving over $(T + 1)p^T q^{T+1}$, not least in situations that are a discretization of a problem containing continuous components so that p and/or q and/or T may be huge.

Beyond the finite setting just considered, one can in principle set up a rather similar scheme, but a considerable amount of difficulties will typically arise. If E and/or the set of controls is countable or continuous, it may be difficult just to compute $\mathbb{E}[R_{t+1} \mid R_t = x, u_t = u]$ and also to find the maximizer in closed form (these two steps are the analogues of (2.2)). Even more serious problems arise when there is no upper bound on T because the initial step is then unfeasible. Finally, the supremum in (1.2) may not be attained.

A warning should be issued that the obvious idea of discretizing and truncating requires continuity properties that (maybe as a surprise!) need not always hold even in non-artificial settings. For example, if $T = \infty$ or T is a stopping time, replacing the time horizon by $T \wedge n$, using the above approach going back from $T \wedge n$ and finally letting $n \to \infty$ to get the $T = \infty$ optimal strategy as limit of the $T \wedge n$ optimal strategies does not always work, see Schmidli [779, pp.6–8].

Notes and references For a list of standard texts in stochastic dynamic programming, see Schmidli [779, p. 8]. A ruin probability problem treated by stochastic dynamic programming is in Schäl [765].

3 The Hamilton-Jacobi-Bellman equation

We now turn to the continuous time setting. The approach will be similar to the dynamic programming one, but in continuous time we go back to t from $t + \mathrm{d}t$ rather than from $t + 1$.

We will state two simplifying assumptions which, however, will be sufficient to cover the ruin probability applications we have in mind. One is $r(x, u, t) = \mathrm{e}^{-\delta t} r(x, u)$, the other that T is an exit time (for example, the time of ruin). This ensures that the value function $V(t, x)$ as defined by

$$V(t, x) = \sup_{U^t \in \mathscr{U}^t} \mathbb{E}_{x,t}\left[\mathrm{e}^{-\delta(T-t)} r_T(R_T^U) + \int_t^T \mathrm{e}^{-\delta(s-t)} r(R_s^U, u_s)\,\mathrm{d}s\right] \quad (3.1)$$

(where $\mathbb{E}_{x,t}$ denotes $\mathbb{E}[\,\cdot\,|\,R_t = x,\ t < T]$) only depends on $T - t$, not on t itself. One is then eventually interested in $V(x) = V(0, x)$.

Generators (cf. Chapter II) turn out to be an essential tool. Denote by $\mathscr{A}^u$ the generator of the (time-homogeneous) Markov process according to which R^U evolves when control $u_t \equiv u$ is used. Then:

Theorem 3.1 *Under certain assumptions, the value function $V(\cdot)$ is the solution of*

$$0 = \sup_u \big[\mathscr{A}^u V(x) - \delta V(x) + r(x, u)\big]. \tag{3.2}$$

Remark 3.2 Equation (3.2) goes under the name *Hamilton-Jacobi-Bellman* (HJB) *equation*. Its derivation typically requires some assumptions that are difficult to verify directly. For instance, suitable regularity conditions are needed, in particular that $V(\cdot)$ is in the domain of $\mathscr{A}^u$ for all u (we do not specify the remaining ones and the reader should not take the proof below for more than a heuristic justification). The argument suggests that the maximizer u^* (when it exists) is the optimal control when $R_t = x$. However, to establish that the solution of (3.2) indeed solves the control problem, more work is needed (in that sense, the custom to use V in the formulation of the HJB equation is a slight abuse of notation).

Another complication is that it is not a priori clear whether the HJB equation has a unique solution. If it has, then one usually needs to prove separately (in a so-called *verification step*) that the obtained solution is indeed the value function of the optimal control problem. This can be done by either justifying all steps of the derivation of the HJB derivation rigorously, or by proving that the solution of the HJB equation dominates all other possible value functions (such a procedure often involves martingale arguments and (extensions of) Itô's formula). The second possibility is usually the more feasible one.

If the solution of the HJB equation is not unique (which may for instance happen if the initial condition cannot be specified), then the stochastic control problem can become very difficult. This effect can for instance occur if the value function is not as regular as the HJB equation would ask for. In that case one can often still work with either *weak solutions* or so-called *viscosity solutions*, see the references at the end of the chapter.

Proof of Theorem 3.1. Let u be an arbitrary control and assume that u is used as control in $[t, t + h)$ and the optimal control is used in $[t + h, T)$. Then $V(x)$ has two parts, the contribution from $[t, t + h)$ and the one from $[t + h, T)$. This gives

$$\begin{aligned}
V(x) &\geq r(x, u)h + o(h) + e^{-\delta h}\mathbb{E}_{x,t}V(R^u_{t+h}) \\
&= r(x, u)h - \delta\,h\,V(x) + \mathscr{A}^u V(x)h + V(x) + o(h),
\end{aligned}$$

which shows that (3.2) holds with $=$ replaced by $\geq$. To see that the sup is actually 0, choose a control u such that the above scheme gives an expected reward of at least $V(x) - \epsilon$. The same calculation then gives

$$V(x) - \epsilon \;\leq\; \mathscr{A}^u V(x) + r(x, u) + V(x) - \delta V(x)\,.$$

Let $\epsilon \downarrow 0$.   □

Remark 3.3 When the value function is determined, the next step is to identify the corresponding control strategy that realizes this value function (this is not always simple and it may even happen that such a strategy does not exist!). In any case, by definition at least ϵ-optimal strategies always exist, i.e. for each $\epsilon > 0$ there is a strategy that leads to a value of $V(x) - \epsilon$.   □

As may become clear from the above remarks, giving a rigorous and systematic treatment of stochastic control theory in insurance is outside the scope of this book. In the sequel we shall rather consider a few particular examples to get the flavor of the topic.

Example 3.4 (OPTIMAL INVESTMENT FOR A DIFFUSION) As a first example, we consider the investment-ruin problem of Browne [206]. The model is given by two stochastic differential equations

$$\mathrm{d}R^0_t \;=\; a_1 \, \mathrm{d}t + a_2 \, \mathrm{d}B^0_t\,, \quad \mathrm{d}M_t \;=\; b_1 M_t \, \mathrm{d}t + b_2 M_t \, \mathrm{d}B^1_t\,,$$

where B^0, B^1 are independent standard Brownian motions and $R^0_0 = x$. Here R^0 describes the evolution of the reserve of the company without investment and M the price process of a risky asset. Thus, R^0 is Brownian motion with drift and M is geometric Brownian motion. It is now assumed that the company is free to invest an amount u_t in the risky asset[2] at any time t so that in the presence of investment, the reserve evolves according to

$$\mathrm{d}R_t \;=\; a_1 \, \mathrm{d}t + a_2 \, \mathrm{d}B^0_t + u_t b_1 \, \mathrm{d}t + u_t b_2 \, \mathrm{d}B^1_t\,.$$

The purpose is to minimize the infinite horizon ruin probability or, equivalently, to maximize the survival probability $\phi(u) = 1 - \psi(u)$. Thus in the general formulation we may take T as the ruin time (which is a stopping time), $r \equiv 0$, $\delta = 0$ and $r_T \equiv -1$. Adding the constant 1 to the value function, V indeed

---

[2]Here and in the sequel u_t can exceed the present surplus level x, i.e. it is possible to use additional sources (or borrow money) for the investment. See the Notes for references which deal with constraints on u_t such as $u_t \leq x$.

corresponds to the survival probability of the controlled process. Consequently the HJB equation is simply $0 = \sup_{u \geq 0} \mathscr{A}^u V(x)$, i.e.

$$0 = \sup_{u \geq 0} \left[(a_1 + ub_1)V'(x) + \frac{1}{2}(a_2^2 + u^2 b_2^2)V''(x) \right]. \tag{3.3}$$

For the solution, we first note that (since V is increasing and hence $V'(x) > 0$), a maximizer u^* can only exist when $V'' < 0$. We then simply compute u^* by differentiating $\mathscr{A}^u V(x)$ w.r.t. u to get $b_1 V'(x) + u^* b_2^2 V''(x) = 0$, i.e.

$$u^* = -\frac{b_1}{b_2^2}\frac{V'(x)}{V''(x)}.$$

Substituting back in the HJB equation gives the ODE

$$0 = a_1 V'(x) - \frac{b_1^2}{b_2^2}\frac{V'(x)}{V''(x)}V'(x) + \frac{1}{2}a_2^2 V''(x) + \frac{1}{2}b_2^2 \frac{b_1^2}{b_2^4}\frac{V'(x)^2}{V''(x)^2}V''(x).$$

Dividing by $V'(x)$ and collecting terms shows that $z(x) = V'(x)/V''(x)$ must be the solution of

$$0 = a_1 + \frac{1}{2}a_2^2\frac{1}{z(x)} - \frac{1}{2}\frac{b_1^2}{b_2^2}z(x).$$

Multiplying by $z(x)$ gives a quadratic equation, and since we assumed $V'' < 0$, $z(x)$ must be the negative solution, say k. In particular $z(x)$ and hence u^* does not depend on x, and we get our final solution $u^* = -kb_1/b_2^2$.

The (somewhat surprising) conclusion is that, no matter how large the current capital x, it is optimal to always invest the constant amount $-kb_1^2/b_2^2$ of money into the risky asset for minimizing the probability of ruin. The resulting minimal ruin probability can be calculated by substituting u^* into (3.3) and using the boundary conditions $V(0) = 0$ (with diffusion, starting in zero leads to immediate ruin) and $V(\infty) = 1$. This results in

$$\psi_I(x) = 1 - V(x) = e^{-\gamma_I x} \quad \text{with } \gamma_I = \frac{2(a_1 + b_1 u^*)}{a_2^2 + b_2^2 (u^*)^2} = \frac{2(a_1 - b_1^2 k/b_2^2)}{a_2^2 + b_1^2 k^2/b_2^2}.$$

$\square$

The example shows a common feature of control problems in diffusion models, that the HJB equation often takes a form of a non-standard ODE. Here it was rather easily solvable, but in general much ingenuity may be required. It should also be stressed that the calculations do not provide a rigorous proof that the candidate for the optimal strategy that was found indeed is optimal. For this one needs a verification step, that for diffusion models is often done by checking that the solution to the HJB equation that was found is twice differentiable (as here). Optimality then follows by a simple application of Itô's formula.

Example 3.5 (OPTIMAL PROPORTIONAL REINSURANCE FOR A DIFFUSION)
Similar techniques as in Example 3.4 can be used to treat optimal proportional
reinsurance for a diffusion that evolves according to

$$dR_t^0 = a_1 \, dt + a_2 \, dB_t^0 \, .$$

I.e., at each time t there is the possibility to pass on a fraction $1 - u_t \in [0, 1]$
of the risk (which in the diffusion approximation is represented by the second
part above), at the expense of a reduced drift (due to the subtraction of the
reinsurance premium drift a_θ). Correspondingly, in the presence of proportional
reinsurance, the process follows the dynamics

$$
\begin{aligned}
dR_t^U &= (u_t \, a_1 + (1 - u_t)(a_1 - a_\theta)) \, dt + a_2 \, u_t dB_t^0 \\
&= (u_t a_\theta + (a_1 - a_\theta)) \, dt + a_2 \, u_t dB_t^0 \, .
\end{aligned}
$$

We can restrict to the case $a_\theta > a_1$ (otherwise the ruin probability will trivially
be minimized (namely be equal to 0) by $u_t = 0$ for all t). We want to maximize
the probability of survival $\phi(u)$, so as in Example 3.4 we choose T as the ruin
time, $r \equiv 0$, $\delta = 0$ and $r_T \equiv -1$ (and add the constant 1 to the value function)
to arrive at the HJB equation $0 = \sup_{u \geq 0} \mathscr{A}^u V(x)$, i.e.

$$0 = \sup_{u \in [0,1]} \left[(a_1 - a_\theta + u \, a_\theta) \, V'(x) + \frac{u^2 a_2^2}{2} V''(x) \right] .$$

It can be solved in much the same way as in Example 3.4 (one just has to
additionally take care of the bound $u \in [0, 1]$) and one obtains that the optimal
strategy is to have a constant fraction of proportional reinsurance given by
$u^* = \min\{2(1 - a_1/a_\theta), 1\}$. If this value is now substituted in the HJB equation,
its solution (for the boundary conditions $V(0) = 0$ and $V(\infty) = 1$) gives the
resulting minimal ruin probability

$$\psi_R(x) = 1 - V(x) = 1 - e^{\gamma_R x}$$

with

$$
\gamma_R = \begin{cases}
a_\theta^2 / (2a_2^2(a_\theta - a_1)) & \text{if } a_1 < a_\theta < 2a_1, \\
2a_1 / a_2^2 & \text{if } a_\theta \geq 2a_1.
\end{cases}
$$

One finally needs a verification theorem showing that the obtained form of $V(x)$
indeed dominates all other admissible strategies which again can be done by applying Itô's formula. □

Stochastic control problems for jump processes turn out to be more subtle:

Example 3.6 (Optimal investment for the Cramér-Lundberg model)
Let us now consider a Cramér-Lundberg risk reserve process $R_t^0 = x + t - A_t$
(where $\{A_t\}$ is a compound Poisson process with rate β and individual claim
distribution function B) and the possibility to dynamically invest an amount of
u_t into a financial asset that is modeled by geometric Brownian motion M_t with

$$\mathrm{d}M_t = b_1 M_t \, \mathrm{d}t + b_2 M_t \, \mathrm{d}B_t.$$

The controlled process then satisfies

$$\mathrm{d}R_t^U = (1 + u_t b_1)\mathrm{d}t + u_t b_2 \, \mathrm{d}B_t - dA_t.$$

The goal is again to minimize the ruin probability of $\{R_t^U\}$ over all admissible
strategies u_t which are assumed to be predictable (in particular, the value of an
admissible strategy at time t may depend on the history of the process up to t,
but not on the size of a claim that may occur at t). In the present model, the
HJB equation $0 = \sup_{u \geq 0} \mathscr{A}^u V(x)$ translates into

$$0 = \sup_{u \geq 0} \left[(1 + b_1 u) \, V'(x) + \frac{u^2 b_2^2}{2} \, V''(x) + \beta \left(\int_0^x V(x - y)B(\mathrm{d}y) - V(x) \right) \right].$$
(3.4)

Note that $u^*(x) \to 0$ as $x \to 0$ (otherwise the investment will lead to $\phi(0) = V(0) = 0$ which cannot be optimal), so we obtain a boundary condition $V'(0) = \beta V(0)$. The second boundary condition is again $\lim_{x \to \infty} V(x) = 1$.

Since there is no solution to this equation for $V''(x) \geq 0$, we assume $V''(x) < 0$. Then the supremum is attained for

$$u^*(x) = -\frac{b_1 V'(x)}{b_2^2 V''(x)}$$

and plugging this into the HJB equation one gets

$$V'(x) - \frac{b_1^2}{2b_2^2} \frac{V'(x)^2}{V''(x)} + \beta \left(\int_0^x V(x - y)B(\mathrm{d}y) - V(x) \right) = 0.$$
(3.5)

It is now considerably more difficult than in the diffusion case to solve this
equation and retrieve further information about the optimal strategy. In a first
step one can show in a verification theorem using Itô's lemma that if (3.4) has an
increasing twice continuously differentiable solution, then the feedback strategy
u^* is indeed optimal among all admissible investment strategies. Under further
assumptions on B (like the existence of a bounded density $b(x)$), one can then
show with considerable effort that indeed a unique increasing twice continuously
differentiable solution of (3.4) exists; however its form can only be determined

numerically. Remarkably, one can still retrieve substantial information about the asymptotic behavior of both $u^*(x)$ and $\psi_I(x) = 1 - V(x)$ as $x \to \infty$: For light-tailed claim size distribution B, if the adjustment coefficient γ_I exists as the positive solution of

$$\beta(\widehat{B}[r] - 1) - r = \frac{b_1^2}{2b_2^2}, \qquad (3.6)$$

then $\psi_I(x) \le e^{-\gamma_I x}$ and (under a mild additional condition) the Cramér-Lundberg approximation $\lim_{x \to \infty} e^{\gamma_I x} \psi_I(x) = C$ holds for some constant C. Without investment, the r.h.s. of (3.6) is zero, so clearly $\gamma_I > \gamma$, and hence optimal investment can substantially decrease the probability of ruin.[3] Furthermore $\lim_{x \to \infty} u^*(x) = b_1/(b_2^2 \gamma_I)$, so asymptotically the optimal strategy is to invest a constant amount into the risky asset, which is somewhat surprising at first sight.

On the other hand, for heavy-tailed B (i.e. $\widehat{B}[r] = \infty$ for all $r > 0$), one can show that $u^*(x)$ is unbounded. If the failure rate of B tends to zero, then the optimal strategy converges and $\lim_{x \to \infty} u^*(x) = \infty$. A quite pleasant result (whose proof is beyond the scope of this book) is that for $B, B_0 \in \mathscr{S}$ the optimal investment strategy leads to

$$\psi_I(x) \sim \frac{2\beta b_2^2}{b_1^2} \int_x^\infty \frac{1}{\int_0^y \frac{1}{1-B(z)}dz} \, dy, \qquad x \to \infty. \qquad (3.7)$$

This can be compared with Theorem VIII.2.1 to see the reduction of $\psi(u)$ through investment. If further the failure rate of B tends to zero, then the rate at which $u^*(x)$ goes to infinity can be identified to be

$$u^*(x) \sim \frac{b_1}{b_2^2} \int_0^x \frac{1 - B(x)}{1 - B(z)} \, dz, \qquad x \to \infty.$$

In particular, if $\overline{B}(x)$ is regularly varying with index $-\alpha$, then simple applications of Karamata's theorem translate (3.7) into

$$\psi_I(x) \sim \frac{2\beta b_2^2(\alpha + 1)}{b_1^2 \alpha} (1 - B(x)), \qquad x \to \infty,$$

and

$$u^*(x) \sim \frac{b_1}{b_2^2(\alpha + 1)} x, \qquad x \to \infty.$$

---

[3]This is in contrast to investment of a constant fraction of the reserve, which led to a Pareto-type tail even for light-tailed B, cf. Theorem VIII.6.2.

Accordingly, for $x \to \infty$ it is then optimal to invest the constant fraction $b_1/(b_2^2(\alpha + 1))$ of the surplus into the risky asset.[4] For a derivation and detailed discussion of the above results the reader is referred to Schmidli [779, Ch. IV]. □

Example 3.7 (OPTIMAL REINSURANCE FOR THE CRAMÉR-LUNDBERG MODEL) Assume again the Cramér-Lundberg risk reserve process $R_t^0 = x + t - \sum_{n=1}^{N_t} U_i$ and the purchase of reinsurance on individual claims according to some reinsurance form u_t under which the cedent reduces a possible claim payment U_i at time t to $u_t(U_i)$ (with the implicit understanding that $u_t(y)$ is a continuous function satisfying $0 \le u_t(y) \le y$). The goal is to minimize the ruin probability through dynamically adapting the reinsurance form u_t (to avoid trivialities, the admissible strategies u_t are again assumed to be predictable, cf. Example 3.6). The premium intensity for such a reinsurance contract is $p_R(u_t)$ for a continuous function p_R with the understanding that more reinsurance is more expensive and that full reinsurance is more expensive than first insurance (i.e. if $u_t(U_i) = U_i$, then $p_R(u_t) > 1$), as otherwise it would be optimal to reinsure the entire insurance risk, leading to a ruin probability of zero. The controlled surplus process is now given by

$$R_t^U = x + \int_0^t (1 - p_R(u_s)) \mathrm{d}s - \sum_{n=1}^{N_t} u_t(U_i).$$

The HJB equation for this optimization problem then reads

$$\sup_{u \in \mathscr{U}} \left\{ (1 - p_R(u)) V'(x) + \beta \left(\int_0^\infty V\big(x - u(y)\big) B(\mathrm{d}y) - V(x) \right) \right\} = 0, \quad (3.8)$$

where at this stage u is a function (rather than a constant) representing the reinsurance form and $\mathscr{U}$ is the (compact) set of all admissible reinsurance forms. Here V again corresponds to the survival probability of the controlled process. Since we are interested in strictly increasing solutions of (3.8), we can restrict $\mathscr{U}$ to those admissible strategies for which $p_R(u) < 1$. If one specifies a boundary value, then (with quite some effort) this equation can be shown to have a unique, strictly increasing and continuously differentiable solution and that this solution (after appropriate scaling) indeed minimizes the ruin probability of the controlled process.

---

[4]Theorem VIII.6.2 together with Remark VIII.6.4 show that adopting this constant fraction strategy for all x also leads to the same asymptotic behavior of $\psi_I(x)$, but for finite x the performance can be quite different!

More explicit results can be obtained, if the set $\mathcal{U}$ is restricted to particular reinsurance forms. For instance, if one tries to find the optimal dynamic proportional reinsurance $u(y) = u\,y$ with $0 \le u \le 1$, then (3.8) simplifies to

$$\sup_{u \in [0,1]} \left\{ (1 - p_R(u))\, V'(x) + \beta \left(\int_0^{x/u} V(x - uy)\, B(\mathrm{d}y) - V(x) \right) \right\} = 0. \quad (3.9)$$

One can then show that if $\inf_{u \uparrow 1} (1 - p_R(u))/(1 - u) > 0$ (i.e. the reinsurer charges more than the net premium for u close to 1), then it is optimal to purchase no reinsurance for any initial capital x below some positive level.

Equation (3.9) cannot be solved explicitly, so one has to approximate the solution numerically in practical examples. Asymptotic results for $x \to \infty$ can however be obtained. In particular, if a strictly positive solution γ_R to the adjustment equation

$$\inf_{u \in [0,1]} \left\{ \beta \widehat{B}[ur] - \beta - (1 - p_R(u))r \right\} = 0 \quad (3.10)$$

exists, then under some mild additional assumptions the Cramér-Lundberg approximation

$$\lim_{x \to \infty} e^{\gamma_R x} \psi_R(x) = C \quad (3.11)$$

holds for some constant $C > 0$. If moreover the value u^* for which the infimum in (3.10) is attained is unique, then one can show that $\lim_{x \to \infty} u(x) = u^*$, i.e. for increasing initial capital x the optimal strategy converges to a constant reinsurance fraction.

If on the other hand $\overline{B}$ is regularly varying with index $-\alpha$, then

$$\psi_R(x) \sim \left(\inf_{u \in [0,1]} \frac{\beta u^\alpha}{(1 - p_R(u) - \beta \mu_B u)^+} \right) \overline{B}_0(x)$$

and if the infimum is attained for a unique value u^*, then again $\lim_{x \to \infty} u(x) = u^*$. For subexponential, but lighter tails $\overline{B}$, available results are not as explicit, but in that case the optimal strategy can be shown to satisfy $\limsup_{x \to \infty} u(x) = \inf\{u : 1 - p_R(u) > \beta \mu_B u\}$. I.e. for large x one tries to reinsure as much as still possible without resulting negative drift. The intuitive reason is that (in contrast to regularly varying tails) proportional reinsurance makes the tail of the distributions smaller and so one tries to purchase as much reinsurance as possible.

As another example one can consider the restricted class of excess-of-loss strategies $u(y) = \min(y, u)$, so in this case u is the retention. This restriction leads to the HJB equation

$$\sup_{u \in [0,\infty]} \left\{ (1 - p_R(u))\, V'(x) + \beta \int_0^{a(x,u)} V(x - \min(y, u))\, B(\mathrm{d}y) - \beta V(x) \right\} = 0,$$

where $a(x, u)$ is x if $u \geq x$ and infinity otherwise. Under mild additional assumptions one can then show that if the adjustment coefficient γ_R, this time defined as the positive solution of

$$\inf_{u \in [0, \infty]} \left\{ \beta r \int_0^u \left(1 - B(z) \right) e^{rz} \, dz - \left(1 - p_R(u) \right) r \right\} = 0,$$

exists, then the Cramér-Lundberg approximation (3.11) again holds and if this infimum is attained for a unique value u^*, then $\lim_{x \to \infty} u(x) = u^*$. Note that γ_R now also exists for heavy-tailed claim size distributions and both the Cramér-Lundberg approximation and the limiting strategy result still apply. □

Notes and references Schmidli [779] is a recent and rich source where one finds rigorous treatments of the above examples and numerous further stochastic control problems in insurance. A short survey of the topic is in Hipp [466]. Browne [206] also extends Example 3.4 in several further directions, including dependent Brownian motions, minimizing expected penalty at ruin and maximizing exponential utility in finite time. Optimal investment problems for the Cramér-Lundberg model were first studied by Hipp & Plum [469] and Gaier, Grandits & Schachermayer [386] and in a more general framework in [470]; since then many further results have been added. Among them, Gaier & Grandits [385] and Grandits [434] extend the optimal investment problem for regularly varying claims to the case when in addition a riskless asset is available, see also Liu & Yang [601] and Yang & Zhang [902]. For a periodic risk model, Kötter & Bäuerle [558] investigate the control problem to maximize the adjustment coefficient through investment.

The classical references for optimal reinsurance programs are Højgaard & Taksar [480], Schmidli [775] and Hipp & Vogt [472]. In a similar fashion, investment and reinsurance can be controlled simultaneously, usually without substantial additional complexity, see e.g. Schmidli [777] and Luo & Taksar [617] for minimizing the probability of absolute ruin.

Whereas in most considered investment problems it is allowed to borrow money for the purchase of the risky asset if necessary, Promislow & Young [717] and Azcue & Muler [115] deal with the effects of borrowing constraints, in which case one has to work with weak solutions to the HJB equation. Luo [616] and Bai & Guo [123] deal with several available risky assets.

The continuous-time risk model leads to elegant and often very explicit solutions for the control problems. However, similar to the dynamic hedging problem in finance, it will be practically impossible to continuously adjust the investment/reinsurance fraction. It is still a challenge for future research to incorporate frictions such as transaction costs and/or limited possibilities for portfolio adjustment into the model; for a step in this direction, see Højgaard & Taksar [481]. Optimal control strategies of reinsurance and investment to minimize the ruin probability in a multivariate discrete-time risk model can be found in Bäuerle & Blatter [141].

Other types of control problems with the objective to minimize the ruin probability include the possibility to accumulate new business (see Hipp & Taksar [471]) and to choose between proportional insurance and the issuing of catastrophe bonds which are correlated with the insurer's losses (see Bäuerle [149]). The related topic of maintaining solvency for pension plans is considered in Olivieri & Pitacco [673]. Optimal investment and reinsurance when instead of minimizing the ruin probability the objective is to maximize the utility of terminal wealth is investigated in Irgens & Paulsen [495]. See also Korn & Wiese [553] for the case where the size of the resulting ruin probability is a constraint, Zhang & Siu [913] for a game-theoretic approach that involves model uncertainty and Xia & Zhang [874] where a martingale approach is employed to identify mean-variance efficient inv estment strategies. For a general model set-up, see Liu & Ma [603]. Another methodological bridge between problems of the above kind and more finance-oriented control problems can be found in Bayraktar & Young [145, 146, 147], who consider an individual who consumes at a certain (possibly surplus-dependent or random) rate and can investment in a risk-less and a risky asset in such a way that the probability of ruin before a random time horizon is minimized; for the objective to maximize the expected utility of consumption and the size of the ruin probability being a constraint, see [148].

Another classical stochastic control problem in insurance (originally raised by de Finetti [283]) is how to pay out dividends from the risk reserves to shareholders in such a way that the expected (utility of the) discounted sum of dividend payments until ruin is maximized. The resulting value function is a profitability type measure for the value of an insurance portfolio and as such may be interpreted as an alternative to the ruin probability which rather measures the safety. There also have been studies of optimal strategies that balance between profitability and safety, expressed through a penalty term on early ruin, see e.g. Thonhauser & Albrecher [843]. The corresponding control problems lead to intricate mathematical challenges and have developed into an active field of research, which cannot be covered in this book (see e.g. Albrecher & Thonhauser [40] for a recent survey and Schmidli [779] for a detailed treatment. Cf. also the Notes of VIII.1).

In addition to the analytic approach outlined in this chapter, sometimes a more probabilistic approach also works in which the control problem is solved within a restricted smaller class of admissible strategies and then by comparison one can show that the optimal strategy is also optimal within the whole class of admissible strategies (see e.g. Loeffen [604]).

In finance applications, a popular and workable alternative to the dynamic programming principle and the HJB equation is the so-called dual method. For insurance applications, however, it seems that due to the intervention of the control into the underlying surplus process, the resulting set of possible trajectories is too restricted to make the dual method work here. Concerning general methods, some standard textbooks on continuous time stochastic control are Davis [278], Fleming & Soner [364], Øksendal & Sulem [672], Pham [698] and, for numerics, Kushner & Dupuis [541].

Chapter XV

Simulation methodology

1 Generalities

This section gives a summary of some basic issues in simulation and Monte Carlo methods. We shall be brief concerning general aspects and refer to standard textbooks like Asmussen & Glynn [79], Bratley, Fox & Schrage [197], Ripley [740] or Rubinstein & Kroese [751] for more detail (a treatment with a special view towards insurance is Korn, Korn & Kroisandt [552]); topics of direct relevance for the study of ruin probabilities are treated in more depth.

1a The crude Monte Carlo method

Let Z be some random variable and assume that we want to evaluate $z = \mathbb{E}Z$ in a situation where z is not available analytically but Z can be simulated. The crude Monte Carlo (CMC) method then amounts to simulating i.i.d. replicates $Z_1, \ldots, Z_N$, estimating z by the empirical mean $\bar{z} = (Z_1 + \cdots + Z_N)/N$ and the variance of Z by the empirical variance

$$s^2 \;=\; \frac{1}{N-1}\sum_{i=1}^{N}(Z_i - \bar{z})^2 \;=\; \frac{1}{N-1}\Big(\sum_{i=1}^{N}Z_i^2 - N\bar{z}^2\Big). \tag{1.1}$$

According to standard central limit theory, $\sqrt{N}(\bar{z} - z) \xrightarrow{\mathscr{D}} \mathrm{N}(0, \sigma_Z^2)$, where $\sigma_Z^2 = \mathbb{V}ar(Z)$. Hence

$$\bar{z} \pm \frac{1.96\, s}{\sqrt{N}} \tag{1.2}$$

is an asymptotic 95% confidence interval, and this is the form in which the result of the simulation experiment is commonly reported.

461

In the setting of ruin probabilities, it is straightforward to use the CMC method to simulate the finite horizon ruin probability $z = \psi(u, T)$: just simulate the risk process $\{R_t\}$ up to time T (or $T \wedge \tau(u)$) and let Z be the indicator that ruin has occurred,

$$Z = I\left(\inf_{0 \leq t \leq T} R_t < 0\right) = I\big(\tau(u) \leq T\big).$$

The situation is more intricate for the infinite horizon ruin probability $\psi(u)$. The difficulty in the naive choice $Z = I(\tau(u) < \infty)$ is that Z cannot be simulated in finite time: no finite segment of $\{S_t\}$ can tell whether ruin will ultimately occur or not. Sections 2–5 deal with alternative representations of $\psi(u)$ allowing to overcome this difficulty.

1b   Variance reduction techniques

The purpose of the techniques we study is to reduce the variance on a CMC estimator Z of z, typically by modifying Z to an alternative estimator Z' with $\mathbb{E}Z' = \mathbb{E}Z = z$ and (hopefully) $\mathbb{V}ar(Z') < \mathbb{V}ar(Z)$. This is a classical area of the simulation literature, and many sophisticated ideas have been developed. Typically variance reduction involves both some theoretical idea (in some cases also a mathematical calculation), an added programming effort, and a longer CPU time to produce one replication. Therefore, one can argue that unless $\mathbb{V}ar(Z')$ is considerable smaller than $\mathbb{V}ar(Z)$, variance reduction is hardly worthwhile. Consider for instance $\mathbb{V}ar(Z') = \mathbb{V}ar(Z)/2$. Then replacing the number of replications N by $2N$ will give the same precision for the CMC method as when simulating $N' = N$ replications of Z', and in most cases this modest increase of N is totally unproblematic.

We survey two methods which will be used below to study ruin probabilities, conditional Monte Carlo and importance sampling. However, there are others which are widely used in other areas and potentially useful also for ruin probabilities. We mention in particular (regression adjusted) control variates, stratification and common random numbers.

Conditional Monte Carlo

Let Z be a CMC estimator and Y some other r.v. generated at the same time as Z. Letting $Z' = \mathbb{E}[Z \,|\, Y]$, we then have $\mathbb{E}Z' = \mathbb{E}Z = z$, so that Z' is a candidate for a Monte Carlo estimator of z. Further, writing

$$\mathbb{V}ar(Z) = \mathbb{V}ar\big(\mathbb{E}[Z \,|\, Y]\big) + \mathbb{E}\big(\mathbb{V}ar[Z \,|\, Y]\big) = \mathbb{V}ar(Z') + \mathbb{E}\big(\mathbb{V}ar[Z \,|\, Y]\big)$$

and ignoring the last term shows that $\mathbb{V}ar(Z') \leq \mathbb{V}ar(Z)$ so that *conditional Monte Carlo always leads to variance reduction.*

Importance sampling

The idea is to compute $z = \mathbb{E}Z$ by simulating from a probability measure $\widetilde{\mathbb{P}}$ different from the given probability measure $\mathbb{P}$ and having the property that there exists a r.v. L such that

$$z = \mathbb{E}Z = \widetilde{\mathbb{E}}[LZ]. \tag{1.3}$$

Thus, using the CMC method one generates $(Z_1, L_1), \ldots, (Z_N, L_N)$ from $\widetilde{\mathbb{P}}$ and uses the estimator

$$\overline{z}_{IS} = \frac{1}{N} \sum_{i=1}^{N} L_i Z_i$$

and the confidence interval

$$\overline{z}_{IS} \pm \frac{1.96\, s_{IS}}{\sqrt{N}} \text{ where } s_{IS}^2 = \frac{1}{N-1} \sum_{i=1}^{N} (L_i Z_i - \overline{z}_{IS})^2 = \frac{1}{N-1} \Big(\sum_{i=1}^{N} L_i^2 Z_i^2 - N \overline{z}_{IS}^2 \Big).$$

In order to achieve (1.3), the obvious possibility is to take $\mathbb{P}$ and $\widetilde{\mathbb{P}}$ mutually equivalent and $L = \mathrm{d}\mathbb{P}/\mathrm{d}\widetilde{\mathbb{P}}$ as the likelihood ratio.

Variance reduction may or may not be obtained: it depends on the choice of the alternative measure $\widetilde{\mathbb{P}}$, and the problem is to make an efficient choice.

To this end, a crucial observation is that there is an optimal choice of $\widetilde{\mathbb{P}}$: define $\widetilde{\mathbb{P}}$ by $\mathrm{d}\widetilde{\mathbb{P}}/\mathrm{d}\mathbb{P} = Z/\mathbb{E}Z = Z/z$, i.e. $L = z/Z$ (the event $\{Z = 0\}$ is not a concern because $\widetilde{\mathbb{P}}(Z = 0) = 0$). Then

$$\widetilde{\mathrm{Var}}(LZ) = \widetilde{\mathbb{E}}(LZ)^2 - \big[\widetilde{\mathbb{E}}(LZ)\big]^2 = \widetilde{\mathbb{E}}\Big[\frac{z^2}{Z^2} Z^2\Big] - \mathbb{E}\Big[\frac{z}{Z} Z\Big]^2 = z^2 - z^2 = 0.$$

Thus, it appears that we have produced an estimator with variance zero. However, the argument cheats because we are simulating since z is not avaliable analytically. Thus we cannot compute $L = Z/z$ (further, it may often be impossible to describe $\widetilde{\mathbb{P}}$ in such a way that it is straightforward to simulate from $\widetilde{\mathbb{P}}$).

Nevertheless, even if the optimal change of measure is not practical, it gives a guidance: choose $\widetilde{\mathbb{P}}$ such that $\mathrm{d}\widetilde{\mathbb{P}}/\mathrm{d}\mathbb{P}$ is as proportional to Z as possible. This may also be difficult to assess, but tentatively, one would try to choose $\widetilde{\mathbb{P}}$ to make large values of Z more likely.

1c Rare events simulation

The problem is to estimate $z = \mathbb{P}(A)$ when z is small, say of the order 10^{-3} or less. I.e., $Z = I(A)$ and A is a *rare event*. In ruin probability theory, $A =$

$\{\tau(u) \leq T\}$ or $A = \{\tau(u) < \infty\}$ and the rare events assumption amounts to u being large, as is the case of typical interest.

The CMC method leads to a variance of $\sigma_Z^2 = z(1 - z)$ which tends to zero as $z \downarrow 0$. However, the issue is not so much that the precision is good as that relative precision is bad:

$$\frac{\sigma_Z}{z} = \frac{\sqrt{z(1 - z)}}{z} \sim \frac{1}{\sqrt{z}} \to \infty.$$

In other words, a confidence interval of width 10^{-4} may look small, but if the point estimate $\bar{z}$ is of the order 10^{-5}, it does not help telling whether z is of the magnitude 10^{-4}, 10^{-5} or even much smaller. Another way to illustrate the problem is in terms of the sample size N needed to acquire a given relative precision, say 10%, in terms of the half-width of the confidence interval. This leads to the equation $1.96\,\sigma_Z/(z\sqrt{N}) = 0.1$, i.e.

$$N = \frac{100 \cdot 1.96^2 z(1 - z)}{z^2} \sim \frac{100 \cdot 1.96^2}{z}$$

increases like z^{-1} as $z \downarrow 0$. Thus, if z is small, large sample sizes are required.

We shall focus on importance sampling as a potential (though not the only) way to overcome this problem. The optimal change of measure (as discussed above) is given by

$$\widetilde{\mathbb{P}}(B) = \mathbb{E}\left[\frac{Z}{z}; B\right] = \frac{1}{z}\mathbb{P}(AB) = \mathbb{P}(B|A).$$

I.e., *the optimal $\widetilde{\mathbb{P}}$ is the conditional distribution given A.* However, just the same problem as for importance sampling in general comes up: we do not know z which is needed to compute the likelihood ratio and thereby the importance sampling estimator, and further it is usually not practicable to simulate from $\mathbb{P}(\cdot|A)$. Again, we may try to make $\widetilde{\mathbb{P}}$ look as much like $\mathbb{P}(\cdot|A)$ as possible. An example where this works out nicely is given in Section 3.

Two established efficiency criteria in rare events simulation are *bounded relative error* and *logarithmic efficiency*. To introduce these, assume that the rare event $A = A(u)$ depends on a parameter u (say $A = \{\tau(u) < \infty\}$). For each u, let $z(u) = \mathbb{P}(A(u))$, assume that the $A(u)$ are rare in the sense that $z(u) \to 0$, $u \to \infty$, and let $Z(u)$ be a Monte Carlo estimator of $z(u)$. We then say that $\{Z(u)\}$ has bounded relative error if $\mathbb{V}ar(Z(u))/z(u)^2$ remains bounded as $u \to \infty$. According to the above discussion, this means that the sample size $N = N_\epsilon(u)$ required to obtain a given fixed relative precision (say $\epsilon = 10\%$) remains bounded. Logarithmic efficiency is defined by the slightly weaker requirement that one can get as close to the power 2 as desired: $\mathbb{V}ar(Z(u))$ should go to 0 as least as fast as $z(u)^{2-\epsilon}$, i.e.

$$\limsup_{u\to\infty} \frac{\mathbb{V}ar\big(Z(u)\big)}{z(u)^{2-\epsilon}} < \infty \tag{1.4}$$

for any $\epsilon > 0$. This allows $\mathbb{V}ar\big(Z(u)\big)$ to decrease slightly slower than $z(u)^2$, so that $N_\epsilon(u)$ may go to infinity. However, the mathematical definition puts certain restrictions on this growth rate, and in practice, logarithmic efficiency is almost as good as bounded relative error. The term *logarithmic* comes from the equivalent form

$$\liminf_{u\to\infty} \frac{-\log \mathbb{V}ar\big(Z(u)\big)}{-\log z(u)} \geq 2 \tag{1.5}$$

of (1.4).

Notes and references A survey on rare events simulation is in Asmussen & Glynn [79, Ch. VI]. See also Juneja & Shahabuddin [512].

For details on random variate generation to implement CMC methods and its refinements we refer to the textbooks mentioned at the beginning of the section. The traditional approach is pseudo-random numbers generated by some recursion. In finance applications, quasi-random numbers ([79, IX.3]) have recently become popular and often lead to a substantial improvement of precision. However, it is folklore that quasi-random numbers perform less well when the time horizon is random (say a stopping time like the ruin time) rather than fixed. For an illustration, see [79, p. 274]. If however, an algorithm can be designed which, instead of the risk process, needs the simulation of some other quantities with fixed dimension, quasi-random numbers can be competitive, see e.g. Coulibaly & Lefèvre [253].

2 Simulation via the Pollaczeck-Khinchine formula

Consider the compound Poisson model, let $X_1, X_2, \ldots$ be i.i.d. with common density $b_0(x) = \overline{B}(x)/\mu_B$, let $S_n = X_1 + \cdots + X_n$ and let K be independent and geometric with parameter ρ, $\mathbb{P}(K = k) = (1 - \rho)\rho^k$. The Pollaczeck-Khinchine formula IV.(2.2) may be written as $\psi(u) = \mathbb{P}(M > u)$, where $M = S_K$. Thus $\psi(u) = z = z(u) = \mathbb{E}Z$, where $Z = I(M > u)$ may be generated as follows:

1. Generate K as geometric, $\mathbb{P}(K = k) = (1 - \rho)\rho^k$.

2. Generate $X_1, \ldots, X_K$ from the density $b_0(x)$. Let $M \leftarrow S_K$.

3. If $M > u$, let $Z \leftarrow 1$. Otherwise, let $Z \leftarrow 0$.

The algorithm gives a solution to the infinite horizon problem, but as a CMC method, it is not efficient for large u. Therefore, it is appealing to combine it with some variance reduction method.

2a Light tails: importance sampling

With light tails, there is a standard way to perform importance sampling for geometric sums. In the present ruin context (assuming the conditions of the Cramér-Lundberg approximation), it amounts to the following. As set-up, note that an easy argument using integration by parts shows that the Lundberg equation for the adjustment coefficent γ can alternatively be written as

$$1 \;=\; \rho \int_0^\infty \mathrm{e}^{\gamma y}\, B_0(\mathrm{d}y)\,. \tag{2.1}$$

Let B_0^* be the distribution defined by $\mathrm{d}B_0^*/\mathrm{d}B_0(x) = \rho \mathrm{e}^{\gamma x}$, and to generate one replication of the estimator, generate $X_1^*, X_2^*, \dots$ from B_0^*. Let $S_n^* = X_1^* + \cdots + X_n^*$. Stop the simulation at $\tau^*(u) = \inf\{n : S_n^* > u\}$ and return the estimator $Z^*(u) = \mathrm{e}^{-\gamma S_{\tau^*(u)}}$.

To understand the algorithm, note first that $z(u) = \mathbb{P}\big(\tau(u) \le N\big)$. Next let $\mathbb{P}^*$ be the probability measure where the X_i^* are i.i.d. with distribution B_0^* and N remains independent and geometric(ρ). Then by the definition of B_0^*,

$$\mathbb{P}^*\big(X_1^* \in \mathrm{d}u\big) \;=\; \frac{1}{\rho}\,\mathbb{E}^*\big[\mathrm{e}^{-\gamma X_1^*};\, X_1^* \in \mathrm{d}u\big]\,.$$

By a standard extension to stopping times (see, e.g., [79, pp. 131–132]), this implies

$$z(u) \;=\; \mathbb{E}^*\left[\frac{1}{\rho^{\tau^*(u)}}\mathrm{e}^{-\gamma^* S_{\tau^*(u)}};\, \tau^*(u) \le N\right] \;=\; \mathbb{E}^*\mathrm{e}^{-\gamma^* S_{\tau(u)}^*}\,,$$

where we used that N remains geometric and independent of the X_i^* under $\mathbb{P}^*$. I.e., the estimator $Z^*(u)$ is unbiased.

Further

$$\mathbb{E}^* Z^*(u)^2 \;=\; \mathbb{E}^*\mathrm{e}^{-2\gamma S_{\tau(u)}} \;\le\; \mathrm{e}^{-2\gamma u} \;=\; O\big(z(u)^2\big)\,,$$

where the last step used the standard Cramér-Lundberg asymptotics $z(x) \sim C\mathrm{e}^{-\gamma u}$. This shows:

Theorem 2.1 *The estimator $Z^*(u)$ has bounded relative error.*

2b Heavy tails: conditional Monte Carlo

With heavy tails, the first efficient algorithm seems to be that of Asmussen & Binswanger [72], which gives a logarithmically efficient estimator when the claim size distribution B (and hence B_0) has a regularly varying tail. So, assume in the following that $\overline{B}_0(x) \sim L(x)/x^\alpha$ with $\alpha > 0$ and $L(x)$ slowly varying. Then (cf. Theorem X.2.1) $\psi(u) \sim \rho/(1 - \rho)\overline{B}_0(x)$, and the problem is to produce an estimator $Z(u)$ with a variance going to zero not slower (in the logarithmic sense) than $\overline{B}_0(u)^2$.

A first obvious idea when using conditional Monte Carlo is to write

$$
\begin{aligned}
\psi(u) &= \mathbb{P}(X_1 + \cdots + X_K > u) \\
&= \mathbb{E}\mathbb{P}\big[X_1 + \cdots + X_K > u \,\big|\, X_1, \ldots, X_{K-1}\big] \\
&= \mathbb{E}\overline{B}_0(u - X_1 - \cdots - X_{K-1}).
\end{aligned}
$$

Thus, we generate only $X_1, \ldots, X_{K-1}$, compute $Y = u - X_1 - \cdots - X_{K-1}$ and let $Z^{(1)}(u) = \overline{B}_0(Y)$ (if $K = 0$, $Z^{(1)}(u)$ is defined as 0). As a conditional Monte Carlo estimator, $Z^{(1)}(u)$ has a smaller variance than $Z_1(u)$. However, asymptotically it presents no improvement: the variance is of the same order of magnitude $\overline{F}(x)$. To see this, just note that

$$
\begin{aligned}
\mathbb{E}Z^{(1)}(u)^2 &\geq \mathbb{E}[\overline{B}_0(u - X_1 - \cdots - X_{K-1})^2; \, X_1 > u, K \geq 2] \\
&= \rho^2 \mathbb{P}(X_1 > u) = \rho^2 \overline{B}_0(u)
\end{aligned}
$$

(here we used that by positivity of the X_i, $X_1 + \cdots + X_{K-1} > u$ when $X_1 > u$, and that $\overline{B}_0(y) = 1$, $y < 0$).

This calculation shows that the reason that this algorithm does not work well is that the probability of one single X_i to become large is too big. The idea of [72] is to avoid this problem by discarding the largest X_i and considering only the remaining ones. For the simulation, we thus generate K and $X_1, \ldots, X_K$, form the order statistics

$$
X_{(1)} < X_{(2)} < \cdots < X_{(K)} \,,
$$

throw away the largest one $X_{(K)}$, and let

$$
\begin{aligned}
Z^{(2)}(u) &= \mathbb{P}\big(S_K > u \mid X_{(1)}, X_{(2)}, \ldots, X_{(K-1)}\big) \\
&= \frac{\overline{B}_0\big((u - S_{(K-1)}) \vee X_{(K-1)}\big)}{\overline{B}_0(X_{(K-1)})} \,,
\end{aligned}
$$

where $S_{(K-1)} = X_{(1)} + X_{(2)} + \cdots + X_{(K-1)}$. To check the formula for the conditional probability, note first that

$$\mathbb{P}\big(X_{(n)} > x \mid X_{(1)}, X_{(2)}, \ldots, X_{(n-1)}\big) = \frac{\overline{B}_0(X_{(n-1)} \vee x)}{\overline{B}_0(X_{(n-1)})}.$$

We then get

$$\begin{aligned}
&\mathbb{P}\big(S_n > x \mid X_{(1)}, X_{(2)}, \ldots, X_{(n-1)}\big) \\
&= \mathbb{P}\big(X_{(n)} + S_{(n-1)} > x \mid X_{(1)}, X_{(2)}, \ldots, X_{(n-1)}\big) \\
&= \mathbb{P}\big(X_{(n)} > x - S_{(n-1)} \mid X_{(1)}, X_{(2)}, \ldots, X_{(n-1)}\big) \\
&= \frac{\overline{B}_0\big((x - S_{(n-1)}) \vee X_{(n-1)}\big)}{\overline{B}_0(X_{(n-1)})}.
\end{aligned}$$

Theorem 2.2 *Assume that $\overline{B}_0(x) = L(x)/x^{\alpha}$ with $L(x)$ slowly varying. Then the algorithm given by $\big\{Z^{(2)}(u)\big\}$ is logarithmically efficient.*

The proof of Theorem 2.2 is elementary but lengthy. We will omit it, since another equally simple conditional Monte Carlo estimator developed later by Asmussen & Kroese [89] performs better. The idea there is to partition according to which X_i is the largest, i.e., for which i one has $M_n = X_{(n)} = X_i$, and condition on the X_j with $j \neq i$. Since clearly by symmetry $\mathbb{P}(S_n > u) = n\mathbb{P}(S_n > u, M_n = X_n)$, this gives the estimator

$$\begin{aligned}
Z^{(3')}(u) &= n\,\mathbb{P}\big(S_n > u, M_n = X_n \mid n, X_1, \ldots, X_{N-1}\big) \\
&= n\,\overline{B}_0\big(M_{n-1} \vee (u - S_{n-1})\big) \tag{2.2}
\end{aligned}$$

when $N = n$ is deterministic (note that for $M_n = X_n$ we need $X_n > M_{n-1}$ and for $S_n > u$, we need $X_n > u - S_{n-1}$), and

$$\begin{aligned}
Z^{(3'')}(u) &= N\,\mathbb{P}\big(S_N > u, M_N = X_N \mid N, X_1, \ldots, X_{N-1}\big) \\
&= N\,\overline{B}_0\big(M_{N-1} \vee (u - S_{N-1})\big) \tag{2.3}
\end{aligned}$$

when N is random.

Theorem 2.3 *The estimator $Z^{(3')}(u)$ has bounded relative error in the regularly varying case, and is logarithmically efficient in the Weibull case provided $\beta < \overline{\beta} = \log(3/2)/\log 2 = 0.585$. The same holds for $Z^{(3'')}(u)$ in the regularly varying case provided $L(\cdot)$ satisfies*

$$\limsup_{u \to \infty} \frac{1}{L(u)^2} \mathbb{E}\big[L(u/N)^2 N^{2\alpha+2}\big] < \infty.$$

Proof. We consider only the regularly varying case and the case of a deterministic $N = n$. If $M_{n-1} \leq u/n$, then $S_{n-1} \leq (n-1)u/n$ and therefore always $M_{n-1} \vee (u - S_{n-1}) \geq u/n$. Therefore

$$
\begin{aligned}
\frac{\mathbb{E}Z^{(3')}(u)^2}{\overline{B}_0(u)^2} &\leq n^2 \frac{\overline{B}_0(u/n)^2}{\overline{B}_0(u)^2} = n^2 \frac{L(u/n)^2/(u/n)^{2\alpha}}{L(u)^2/u^{2\alpha}} \\
&= n^{2+2\alpha} \frac{L(u/n)^2}{L(u)^2} \sim n^{2+2\alpha}.
\end{aligned}
$$

Noting that $z(u) \sim n\overline{F}(u)$ by subexponentiality completes the proof. $\square$

Notes and references In the case $Z^{(3'')}(u)$ of a random N, it is suggested in [89] that either N be used as a control variate or that N be stratified, and a substantial variance reduction was obtained. A theoretical support for the control-variate approach was provided by Hartinger & Kortschak [453], who showed that in fact, in this setting the relative error goes to 0 as $u \to \infty$.

2c Heavy tails: importance sampling

Asmussen, Binswanger & Højgaard [73] suggested an importance distribution $\widetilde{B}_0$ that is much heavier than B_0. They showed for example that for the regularly varying case and the tail of $\widetilde{B}_0$ being of order $1/\log x$, this gives bounded relative error. The practical experience with the algorithm is, however, discouraging, and a much better importance distribution was suggested by Juneja & Shahabuddin [511]. They suggested that the tail of B_0 be changed to $c_1\overline{B}_0(x)^{\theta(u)}$ on $[x_0, \infty)$ and that the density $c_2 b_0(x)$ be used on $(0, x_0)$, where $\theta(u) \to 0$ and c_1, c_2 have to be chosen in a certain way. We will not give the details but only present a simplified version of the algorithm in the Pareto case $b_0 = (\alpha - 1)/(1+x)^\alpha$, where we again choose $\widetilde{b}_0(x) = (\widetilde{\alpha} - 1)/(1+x)^{\widetilde{\alpha}}$ as Pareto with $\widetilde{\alpha} = \widetilde{\alpha}(u) = \alpha\theta(u) \to 0$ (the regularly varying case is an easy extension). Thus the estimator is

$$
Z^{(4)}(u) = I(S_N > u) \prod_{i=1}^{N} \frac{b_0(X_i)}{\widetilde{b}_0(X_i)} \tag{2.4}
$$

with the r.v.'s simulated as independent under the measure $\widetilde{\mathbb{P}}$ where N is geometric(ρ) and $X_1, \ldots X_N$ have density $\widetilde{b}_0(x)$.

Theorem 2.4 *The estimator $Z^{(4)}(x)$ is logarithmically efficient in the Pareto case provided $\log \widetilde{\alpha}/\log u \to 0$.*

Proof. Let $Z_n^{(4)}(u)$ denote (2.4) with N replaced by some fixed n. Then

$$
\begin{aligned}
\widetilde{\mathbb{E}} Z_n^{(4)}(u)^2 &= \int \cdots \int_{x_1 + \cdots + x_n > x} \frac{b_0(x_1)^2}{\widetilde{b}_0(x_1)^2} \cdots \frac{b_0(x_n)^2}{\widetilde{b}_0(x_n)^2} \widetilde{b}_0(x_1) \cdots \widetilde{b}_0(x_n) \, \mathrm{d}x_1 \cdots \mathrm{d}x_n \\
&= c_\#^{-n} \int \cdots \int_{x_1 + \cdots + x_n > x} b_0^\#(x_1) \cdots b_0^\#(x_n) \, \mathrm{d}x_1 \cdots \mathrm{d}x_n \\
&= c_\#^{-n} \mathbb{P}^\#(S_n > x) \,,
\end{aligned}
$$

where $(c_\#)^{-1} = \int b_0^2 / \widetilde{b}_0$ and $b_0^\# = c_\# b_0^2 / \widetilde{b}_0$. Now

$$
c_\#^{-1} = \int_0^\infty \frac{(\alpha - 1)^2 / (1 + x)^{2\alpha}}{(\widetilde{\alpha} - 1) / (1 + x)^{\widetilde{\alpha}}} \, \mathrm{d}x = \frac{(\alpha - 1)^2}{\widetilde{\alpha}(2\alpha - \widetilde{\alpha})} \sim \frac{\alpha}{2\widetilde{\alpha}} \,. \tag{2.5}
$$

Bounding $\mathbb{P}_{2\alpha - \widetilde{\alpha}}(S_n > x)$ above and below by

$$
\mathbb{P}_{2\alpha - \epsilon}(S_n > x) \sim \frac{n}{x^{2\alpha - \epsilon}}, \quad \text{respectively} \quad \mathbb{P}_{2\alpha + \epsilon}(S_n > x) \sim \frac{n}{x^{2\alpha + \epsilon}} \,,
$$

letting $\epsilon \downarrow 0$ and using $\log \widetilde{\alpha} / \log u \to 0$ gives easily that $Z_n^{(4)}(u)$ is logarithmically efficient for $\mathbb{P}(S_n > u)$. We omit the details that are needed to deal with a geometric N.   □

Notes and references Asmussen, Binswanger and Højgaard [73] give a general survey of rare events simulation for heavy-tailed distributions. In many aspects the findings of [73] are quite negative: the large deviations ideas which are the main approach to rare events simulation in the light-tailed case do not seem to work for heavy tails. It must be noted that a main restriction of all algorithms considered in this section is that they are so intimately tied up with the compound Poisson model because the explicit form of the Pollaczeck-Khinchine formula is crucial (say, in the renewal or Markov-modulated model $\mathbb{P}(\tau_+ < \infty)$ and G_+ are not explicit).

A further interesting and useful idea applicable in the Pollaczeck-Khinchine framework with heavy tails was given by Juneja [510]. He noted that

$$
\mathbb{P}\big(S_N > u, M_N > u \,\big|\, N\big) = \mathbb{P}\big(M_N > u \,\big|\, N\big) = 1 - B_0(u)^N
$$

is explicit, so that only $\mathbb{P}(S_N > u, M_N \le u)$ needs to be simulated.

3   Static importance sampling via Lundberg conjugation

We consider again the compound Poisson model and assume the conditions of the Cramér-Lundberg approximation so that $z(u) = \psi(u) \sim C e^{-\gamma u}$, use the representation $\psi(u) = \mathbb{E}e^{-\gamma S_{\tau(u)}} = e^{-\gamma u}\mathbb{E}_L e^{-\gamma \xi(u)}$ where $\xi(u) = S_{\tau(u)} - u$ is the overshoot (cf. IV.5), and simulate from $\mathbb{P}_L$, that is, using β_L, B_L instead of β, B, for the purpose of recording $Z(u) = e^{-\gamma S_{\tau(u)}}$.

For practical purposes, the continuous-time process $\{S_t\}$ is simulated by considering it at the discrete epochs $\{\sigma_k\}$ corresponding to claim arrivals. Thus, the algorithm for generating $Z = Z(u)$ is:

1. Compute $\gamma > 0$ as solution of the Lundberg equation

$$0 = \kappa(\gamma) = \beta(\widehat{B}[\gamma] - 1) - \gamma,$$

and define β_L, B_L by $\beta_L = \beta\widehat{B}[\gamma]$, $B_L(\mathrm{d}x) = e^{\gamma x}B(\mathrm{d}x)/\widehat{B}[\gamma]$.

2. Let $S \leftarrow 0$.

3. Generate T as being exponential with parameter $\widetilde{\beta}$ and U from $\widetilde{B}$. Let $S \leftarrow S + U - T$.

4. If $S > u$, let $Z \leftarrow e^{-\gamma S}$. Otherwise, return to 3.

There are various intuitive reasons that this should be a good algorithm. It resolves the infinite horizon problem since $\mathbb{P}_L(\tau(u) < \infty) = 1$. We may expect a small variance since we have used our knowledge of the form of $\psi(u)$ to isolate what is really unknown, namely $\mathbb{E}_L e^{-\gamma \xi(u)}$, and avoid simulating the known part $e^{-\gamma u}$. More precisely, the results of V.7 tell that $\mathbb{P}(\cdot \mid \tau(u) < \infty)$ and $\mathbb{P}_L$ (both measures restricted to $\mathscr{F}_{\tau(u)}$) asymptotically coincide on $\{\tau(u) < \infty\}$, so that changing the measure to $\mathbb{P}_L$ is close to the optimal scheme for importance sampling, cf. the discussion at the end of Section 1b. In fact:

Theorem 3.1 *The estimator $Z(u) = e^{-\gamma S_{\tau(u)}}$ (simulated from $\mathbb{P}_L$) has bounded relative error.*

Proof. Just note that $\mathbb{E}Z(u)^2 \leq e^{-2\gamma u} \sim z(u)^2/C^2$. $\qquad\qquad\square$

It is tempting to ask whether choosing importance sampling parameters $\widetilde{\beta}, \widetilde{B}$ different from β_L, B_L could improve the variance of the estimator. The answer is no. In detail, to deal with the infinite horizon problem, one must restrict attention to the case $\widetilde{\beta}\mu_{\widetilde{B}} \geq 1$. The estimator is then

$$Z(u) = \prod_{i=1}^{M(u)} \frac{\beta e^{-\beta T_i}}{\widetilde{\beta}e^{-\widetilde{\beta}T_i}} \frac{\mathrm{d}B}{\mathrm{d}\widetilde{B}}(U_i) \tag{3.1}$$

where $M(u)$ is the number of claims leading to ruin, and we have:

Theorem 3.2 *The estimator* (3.1) *(simulated with parameters $\widetilde{\beta}, \widetilde{B}$) is not logarithmically efficient when* $(\widetilde{\beta}, \widetilde{B}) \neq (\beta_L, B_L)$.

The proof is given below as a corollary to Theorem 3.3.

The algorithm generalizes easily to the renewal model. We formulate this in a slightly more general random walk setting.[1] Let $X_1, X_2, \ldots$ be i.i.d. with distribution F, let $S_n = X_1 + \cdots + X_n$, $M(u) = \inf\{n : S_n > u\}$, and assume that $\mu_F < 0$ and that $\widehat{F}[\gamma] = 1$, $\widehat{F}'[\gamma] < \infty$ for some $\gamma > 0$. Let $F_L(\mathrm{d}x) = \mathrm{e}^{\gamma x} F(\mathrm{d}x)$. The importance sampling estimator is then $Z(u) = \mathrm{e}^{-\gamma S_{M(u)}}$. More generally, let $\widetilde{F}$ be an importance sampling distribution equivalent to F and

$$Z(u) \;=\; \prod_{i=1}^{M(u)} \frac{\mathrm{d}F}{\mathrm{d}\widetilde{F}}(X_i)\,. \qquad (3.2)$$

Theorem 3.3 *The estimator* (3.2) *(simulated with distribution $\widetilde{F}$ of the X_i) has bounded relative error when* $\widetilde{F} = F_L$. *When* $\widetilde{F} \neq F_L$, *it is not logarithmically efficient.*

Proof. The first statement is proved exactly as Theorem 3.1. For the second, write

$$W(F \mid \widetilde{F}) \;=\; \frac{\mathrm{d}F}{\mathrm{d}\widetilde{F}}(X_1) \cdots \frac{\mathrm{d}F}{\mathrm{d}\widetilde{F}}(X_{M(u)}).$$

By the chain rule for Radon-Nikodym derivatives,

$$\begin{aligned}
\mathbb{E}_{\widetilde{F}} Z(u)^2 \;&=\; \mathbb{E}_{\widetilde{F}} W^2(F|\widetilde{F}) \;=\; \mathbb{E}_{\widetilde{F}} \left[W^2(F|F_L) W^2(F_L|\widetilde{F}) \right] \\
&=\; \mathbb{E}_L \left[W^2(F|F_L) W(F_L|\widetilde{F}) \right] \;=\; \mathbb{E}_L \exp\{K_1 + \cdots + K_{M(u)}\} \,,
\end{aligned}$$

where

$$K_i \;=\; \log\left(\frac{\mathrm{d}F_L}{\mathrm{d}\widetilde{F}}(X_i) \left(\frac{\mathrm{d}F}{\mathrm{d}F_L}(X_i) \right)^2 \right) \;=\; -\log \frac{\mathrm{d}\widetilde{F}}{\mathrm{d}F_L}(X_i) - 2\gamma X_i.$$

Here $\mathbb{E}_L K_i \;=\; \epsilon' - 2\gamma \mathbb{E}_L X_i$, where

$$\epsilon' \;=\; -\mathbb{E}_L \log \frac{\mathrm{d}\widetilde{F}}{\mathrm{d}F_L}(X_i) \;>\; 0$$

---

[1] For the renewal model, $X_i = U_i - T_i$, and the change of measure $F \to F_L$ corresponds to $B \to B_L$, $A \to A_L$ as in Chapter VI.

by the information inequality. Since $K_1, K_2, \ldots$ are i.i.d., Jensen's inequality and Wald's identity yield

$$
\begin{aligned}
\mathbb{E}_{\widetilde{F}} Z(u)^2 & \geq \exp\big\{\mathbb{E}_L(K_1 + \cdots + K_{M(u)})\big\} \\
& = \exp\big\{\mathbb{E}_L M(u)(\epsilon' - 2\gamma \mathbb{E}_L X_i)\big\}.
\end{aligned}
$$

Since $\mathbb{E}_L M(u)/u \to 1/\mathbb{E}_L X_i$, it thus follows that for $0 < \epsilon < \epsilon'/\mathbb{E}_L X_i$,

$$
\begin{aligned}
\limsup_{u \to \infty} \frac{\mathbb{E}_{\widetilde{F}} Z(u)^2}{z(u)^2 \mathrm{e}^{\epsilon u}} & = \limsup_{u \to \infty} \frac{\mathbb{E}_{\widetilde{F}} Z(u)^2}{C^2 \mathrm{e}^{-2\gamma u + \epsilon' u}} \\
& \geq \limsup_{u \to \infty} \frac{\mathrm{e}^{-2\gamma u}}{C^2 \mathrm{e}^{-2\gamma u}} = \frac{1}{C^2} > 0,
\end{aligned}
$$

which completes the proof. □

Proof of Theorem 3.2. Consider compound Poisson risk process with intensities β', β'', generic interarrival times T', T'', claim size distributions B', B'' and generic claim sizes U', U''. Then according to Theorem 3.3, all that needs to be shown is that if $U' - T' \overset{\mathscr{D}}{=} U'' - T''$, then $\beta' = \beta''$, $B' = B''$. First by the memoryless property of the exponential distribution, $U' - T'$ has a left exponential tail with rate β' and $U'' - T''$ has a left exponential tail with rate β'. This immediately yields $\beta' = \beta''$. Next, from

$$
\begin{aligned}
& \mathbb{P}(U' - T' > x) \\
& = \int_0^\infty \beta' \mathrm{e}^{-\beta' y} \overline{B}'(x + y)\, \mathrm{d}y = \beta' \mathrm{e}^{\beta' x} \int_x^\infty \mathrm{e}^{-\beta' z} \overline{B}'(z)\, \mathrm{d}z\,, \\
& \mathbb{P}(U'' - T'' > x) \\
& = \int_0^\infty \beta'' \mathrm{e}^{-\beta'' y} \overline{B}''(x + y)\, \mathrm{d}y = \beta'' \mathrm{e}^{\beta'' x} \int_x^\infty \mathrm{e}^{-\beta'' z} \overline{B}''(z)\, \mathrm{d}z
\end{aligned}
$$

$(x > 0)$ and $\beta' = \beta''$, $U' - T' \overset{\mathscr{D}}{=} U'' - T''$, we conclude by differentiation that $\overline{B}'(x) = \overline{B}''(x)$ for all $x > 0$, i.e. $B' = B''$. □

Notes and references The importance sampling method was suggested by Siegmund [807] for discrete time random walks and further studied by Asmussen [56] in the setting of compound Poisson risk models. The optimality result Theorem 3.1 is from Lehtonen & Nyrhinen [576], with the present (shorter and more elementary) proof taken from Asmussen & Rubinstein [99]. In [56], optimality is discussed in a heavy traffic limit $\eta \downarrow 0$ rather than when $u \to \infty$.

The extension to the Markovian environment model is straightforward and was suggested in Asmussen [58]. Further discussion is in Lehtonen & Nyrhinen [577].

The queueing literature on related algorithms is extensive, see e.g. the references in Asmussen & Rubinstein [99], Heidelberger [455] and Juneja & Shahabuddin [512].

4   Static importance sampling for the finite horizon case

The problem is to produce efficient simulation estimators for $\psi(u, T)$ with $T < \infty$. As in V.4, we write $T = yu$. The results of V.4 indicate that we can expect a major difference according to whether $y < 1/\kappa'(\gamma)$ or $y > 1/\kappa'(\gamma)$. The easy case is $y > 1/\kappa'(\gamma)$ where $\psi(u, yu)$ is close to $\psi(u)$, so that one would expect the change of measure $\mathbb{P} \to \mathbb{P}_L$ to produce close to optimal results. In fact:

Proposition 4.1 *If $y > 1/\kappa'(\gamma)$, then the estimator $Z(u) = e^{-\gamma S_{\tau(u)}} I(\tau(u) \leq yu)$ (simulated with parameters β_L, B_L) has bounded relative error.*

Proof. The assumption $y > 1/\kappa'(\gamma)$ ensures that $\psi(u, yu)/\psi(u) \to 1$ (Theorem V.4.1) so that $z(u) = \psi(u, yu)$ is of order of magnitude $e^{-\gamma u}$. Bounding $\mathbb{E}_L Z(u)^2$ above by $e^{-\gamma u}$, the result follows as in the proof of Theorem 3.1. $\square$

We next consider the case $y < 1/\kappa'(\gamma)$. We recall that α_y is defined as the solution of $\kappa'(\alpha) = 1/y$, that $\gamma_y = \alpha_y - y\kappa(\alpha_y)$ determines the order of magnitude of $\psi(u, yu)$ in the sense that

$$\frac{-\log \psi(u)}{u} \to \gamma_y \tag{4.1}$$

(Theorem V.4.9), and that $\gamma_y > \gamma$. Further

$$\psi(u, yu) = e^{-\alpha_y u} \mathbb{E}_{\alpha_y} \left[e^{-\alpha_y \xi(u) + \tau(u)\kappa(\alpha_y)}; \tau(u) \leq yu \right]. \tag{4.2}$$

Since the definition of α_y is equivalent to $\mathbb{E}_{\alpha_y} \tau(u) \sim yu$, one would expect that the change of measure $\mathbb{P} \to \mathbb{P}_{\alpha_y}$ is in some sense optimal. The corresponding estimator is

$$Z(u) = e^{-\alpha_y S_{\tau(u)} + \tau(u)\kappa(\alpha_y)} I\big(\tau(u) \leq yu\big), \tag{4.3}$$

and we have:

Theorem 4.2 *The estimator (4.3) (simulated with parameters $\beta_{\alpha_y}, B_{\alpha_y}$) is logarithmically efficient.*

Proof. Since $\gamma_y > \gamma$, we have $\kappa(\alpha_y) > 0$ and get

$$\begin{aligned}
\mathbb{E}_{\alpha_y} Z(u)^2 &= \mathbb{E}_{\alpha_y} \left[e^{-2\alpha_y S_{\tau(u)} + 2\tau(u)\kappa(\alpha_y)}; \tau(u) \leq yu \right] \\
&\leq e^{-2\gamma_y u} \mathbb{E}_{\alpha_y} \left[e^{-2\alpha_y \xi(u)}; \tau(u) \leq yu \right] \\
&\leq e^{-2\gamma_y u}.
\end{aligned}$$

Hence by (4.1),

$$\liminf_{u\to\infty} \frac{-\log \mathbb{V}ar(Z(u))}{-\log z(u)} = \liminf_{u\to\infty} \frac{-\log \mathbb{V}ar(Z(u))}{\gamma_y u} \geq 2$$

so that (1.5) follows. □

Remark 4.3 Theorem V.4.9 has a stronger conclusion than (4.1), and in fact, (4.1) (which is all that is needed here) can be shown more easily. Let $\sigma_y^2 = \lim_{u\to\infty} \mathbb{V}ar_{\alpha_y}(\tau(u))/u$ so that $(\tau(u)-yu)/(\sigma_y u^{1/2}) \overset{\mathscr{D}}{\to} N(0,1)$ (see Proposition V.4.2). Then

$$
\begin{aligned}
z(u) \\
&= \mathbb{E}_{\alpha_y} Z(u) \geq \mathbb{E}_{\alpha_y}\left[e^{-\alpha_y S_{\tau(u)}+\tau(u)\kappa(\alpha_y)}; \ yu - \sigma_y u^{1/2} < \tau(u) \leq yu\right] \\
&= e^{-\alpha_y u + yu\kappa(\alpha_y)}\mathbb{E}_{\alpha_y}\left[e^{-\alpha_y \xi(u)+(\tau(u)-yu)\kappa(\alpha_y)}; \ yu - \sigma_y u^{1/2} < \tau(u) \leq yu\right] \\
&\geq e^{-\gamma_y u + \sigma_y u^{1/2}\kappa(\alpha_y)}\mathbb{E}_{\alpha_y}\left[e^{-\alpha_y \xi(u)}; \ yu - \sigma_y u^{1/2} < \tau(u) \leq yu\right] \\
&\sim e^{-\gamma_y u + \sigma_y u^{1/2}\kappa(\alpha_y)}\mathbb{E}_{\alpha_y}e^{-\alpha_y \xi(\infty)}\big(\Phi(1) - 1/2\big)
\end{aligned}
$$

where the last step follows by Stam's lemma (Proposition V.4.4). Hence

$$\liminf_{u\to\infty} \frac{\log z(u)}{u} \geq \liminf_{u\to\infty} \frac{-\gamma_y u + \sigma_y u^{1/2}\kappa(\alpha_y)}{u} = -\gamma_y.$$

That $\limsup \leq$ follows similarly (but more easily) as when estimating $\mathbb{E}_{\alpha_y} Z(u)^2$ above. □

Notes and references The algorithms in the present section are the obvious ones, but seem to have been discussed for the first time in the first edition of this book. See also Nyrhinen [667]. In Asmussen [56], related discussion is given in a heavy traffic limit $\eta \downarrow 0$ rather than when $u \to \infty$.

5 Dynamic importance sampling

The terms *dynamic* importance samplingor *adaptive* importance sampling are used in at least two different meanings. One meaning is algorithms that, during the execution, change the importance distribution or seek for a good one; a good example is the cross-entropy algorithm, Rubinstein & Kroese [751]. The sense in which we will understand these terms is in describing algorithms that are level- and time-dependent: the importance distribution for (say) the Poisson rate and the claim size distribution at time t in a compound Poisson claim surplus process $\{S_t\}$ depends on the current value S_t as well as on t. Algorithms of this type have received considerable attention in recent years in areas such as queueing theory and have managed to provide efficient algorithms in situations where traditional (static) importance sampling got into difficulties. The basic idea in most of the papers in the area is to implement the principle of looking for a description of the conditional distribution given the rare event. We will exemplify this in two settings. Most steps in the variance calculations leading to asymptotic efficiency results are omitted since they are always very lengthy and technical in the dynamical setting.

5a   An algorithm by Dupuis, Leder and Wang

We follow Dupuis, Leder & Wang [336]. The setting is again that of estimating $\mathbb{P}(S_n > u)$ where $S_n = X_1 + \cdots + X_n$ with $X_1, \ldots, X_n$ non-negative and i.i.d. with common subexponential distribution F with density f (F can in particular be B_0 as discussed earlier).

In dynamic importance sampling, the importance distribution $\widetilde{\mathbb{P}}$ will generate X_k from a density $\widetilde{f}_{u,k,x}$ depending both on u, k and $S_{k-1} = x$. Thus, the estimator is

$$Z(u) \;=\; \prod_{k=1}^{n} \frac{f(X_k)}{\widetilde{f}_{u,k,S_{k-1}}(X_k)} \, I(S_n > u). \tag{5.1}$$

If $x > u$, obviously no importance sampling is needed. If $x \le u$, x will typically be much smaller than u. Basically, the event $S_n > u$ then occurs by one of the X_ℓ, $\ell = k, \ldots, n$, exceeding $u - x$, and the probability that $k = \ell$ is $1/(n - k + 1)$; otherwise X_k is 'typical'. This suggests taking

$$\widetilde{f}_{u,k,x}(y) \;=\; \frac{n-k}{n-k+1} f(y) \;+\; \frac{1}{n-k+1} \frac{f(y)}{\overline{F}(u-x)} I(y > u - x) \tag{5.2}$$

(note that $I(y > u - x) \, f(y)/\overline{F}(u - x)$ is the conditional density of X_k given $X_k > u - x$).

Unfortunately, this idea is too naive to produce efficient estimators, see Remark 5.2 below. One needs to replace the conditioning $X_k > u - x$ by $X_k > a(u - x)$ for some $a < 1$. As a generalization, [336] also allows weights different from the ones in (5.2). Thus, instead of (5.2) one has

$$\widetilde{f}_{u,k,x}(y) \;=\; p_k f(y) \;+\; q_k \frac{f(y)}{\overline{F}(a(u-x))} I\big(y > a(u - x)\big) \tag{5.3}$$

where $p_k + q_k = 1$.

Theorem 5.1 *Assume that F is regularly varying with index α and that the importance distribution is given by (5.3). Then for any fixed n, the estimator (5.1) has bounded relative error. More precisely,*

$$\frac{\mathbb{E}Z(u)^2}{\overline{F}(u)} \to \prod_{k=1}^{n-1} \frac{1}{p_k} + \frac{1}{a^\alpha} \sum_{\ell=1}^{n-1} \frac{1}{q_\ell} \prod_{k=1}^{\ell-1} \frac{1}{p_k}, \quad u \to \infty. \tag{5.4}$$

As said above, the proof is too lengthy to be given here.

Remark 5.2 Relation (5.4) shows that the closer a is to 1, the more asymptotically efficient is the estimator (5.3). It is therefore tempting to take $a = 1$. However, it turns out that there is a discontinuity at $a = 1$, and for $a = 1$, there is in fact not even logarithmic efficiency.

The problem is that the first-order heavy-tailed asymptotics are more imprecise than with light tails: realizations with $\max X_k < u$ but $S_n > u$ are asymptotically unimportant, but cannot be neglected for a finite u. This phenomenon is somewhat related to the slow rate of convergence of heavy-tailed approximations. □

Remark 5.3 An obvious question is to find the minimizers $p_1^*, \ldots, p_n^*$ of the r.h.s. of (5.4). They are in fact not given by $p_k = (n-k)/(n-k+1)$ but by

$$p_k^* = \frac{(n-k-1)/a^{\alpha/2} + 1}{(n-k)/a^{\alpha/2} + 1}$$

(of course, these two expressions coincide as $a \uparrow 1$). □

Notes and references Further relevant papers in the same direction are Dupuis & Wang [337] and Hult & Svensson [484].

5b An algorithm by Blanchet and Glynn

For a more general discussion of the distribution of a stochastic process given a rare event (e.g. ruin), consider a discrete state Markov chain $\{X_n\}$ with transition probabilities $p(x, y)$, Let the state space be E, let $G \subset E$ and $\tau_G = \inf \{n : X_n \in G\}$, $h(x) = \mathbb{P}_x(\tau_G < \infty)$. Then for any initial value $x_0 \notin G$, the conditional distribution of $\{X_n\}$ given $\tau_G < \infty$ is a Markov chain with transition probabilities

$$p^*(x, y) = p(x, y) \frac{h(y)}{h(x)}. \tag{5.5}$$

See [79, VI.7]; the transition function in (5.5) is referred to as an *h-transform*.

In a ruin context, one is interested in evaluating $h(x_0)$ or several of the $h(x)$. Simulating using the $p^*(x, y)$ is of course not practicable since one is simulating precisely because the function h is unknown. However, one may try to plug in an approximation and adapt this as an importance sampling scheme.

Example 5.4 Assume that $X_n = S_n$ is a random walk with negative drift, $G = (u, \infty)$ and $x_0 = 0$ (of course, the compound Poisson case or the renewal model can be handled in this way by looking at the risk process at claim arrival instants only). Then $h(0) = \psi(u)$. In (5.5), write $y = x + z$ and assume for simplicity that the increment distribution has a density f. With light tails, we then have the Cramér-Lundberg approximation $h(v) \approx Ce^{-\gamma(u-v)}$ for $v < u$, so that (5.5) suggests that the transition density from x to $x + z < u$ be taken roughly as

$$f^*(x + z|x) = f(z)\frac{Ce^{-\gamma(u-x-z)}}{Ce^{-\gamma(u-x)}} = f(z)e^{\gamma z}.$$

That is, we are back to the Siegmund algorithm discussed in Section 3 and that was shown there to give bounded relative error.

Obviously, this is a promising start for implementing the h-transform ideas. However, light tails are the easy case! We will see below that much more care is needed for heavy-tailed increments. Here the suggestion from the standard subexponential approximations in Chapter X is that the transition density $\widetilde{p}(x, x + z)$ from x to $x + z < 0$ be taken roughly as

$$f^*(x + z|x) = D(x) f(z)\frac{\overline{F}_I(u - x - z)}{\overline{F}_I(u - x)},$$

where $D(x)$ is a normalizing constant. However, there are at least two difficulties in this choice. First, $f^*(x + z|x)$ is not a standard density even in simple cases as the Pareto where

$$f^*(x + z|z) \approx \frac{\alpha(1 + u - x)^{\alpha-1}}{(1 + z)^{\alpha+1}(1 + u - x - z)^{\alpha-1}}$$

so that it is not straightforward to generate r.v.'s from $f^*(x + z|x)$. Further, $f^*(x + z|x)$ depends on x, which makes it far more difficult to bounding the variance than in the Siegmund case.   □

We will not discuss the r.v. generation issue here, but to resolve the difficulties in bounding the variance, we return to the general Markov chain case. Assume that the importance distribution is a Markov chain with transition probabilities of the form $\widetilde{p}(x, y) = p(x, y)/r(x, y)$, where for each x one would typically try to choose $r(x, y)$ as $c(x)/a(y)$ where $a(y)$ is roughly asymptotically proportional to $h(y)$ and $c(x)$ ensures the normalization $\sum_{y \in E} \widetilde{p}(x, y) = 1$. The estimator for $z = h(x_0)$ is then

$$I(\tau_G < \infty) \prod_{n=1}^{\tau_G} r(X_{n-1}, X_n), \tag{5.6}$$

with the X_n simulated as a Markov chain with $X_0 = x_0$ and transition probabilities $\widetilde{p}(x, y)$. Had one used instead the $p^*(x, y)$, one could be sure that the simulation would terminate, i.e. $\mathbb{P}^*(\tau_G < \infty) = 1$. Given the way the $r(x, y)$ have been chosen, one could hope that also $\widetilde{\mathbb{P}}(\tau_G < \infty) = 1$, but this is a separate problem that we will ignore in the following.

As noted above, a crucial but not easy point is to estimate and bound the variance of Z, or equivalently the second moment vector $\boldsymbol{m}_2$ with elements $m_2(x) = \widetilde{\mathbb{E}}_x Z^2$, $x \notin G$. In the rest of this section, we follow Blanchet & Glynn [175]. The main idea of [175] is to use a Lyapounov function technique, cf. part (iii) of the following result. Define $\boldsymbol{K}$ as the $G^c \times G^c$ matrix with elements $k(x, y) = r(x, y)p(x, y)$, and let $\boldsymbol{\eta}$ be the column vector with elements $\eta(x) = \mathbb{P}_x(X_1 \in G) = \sum_{y \in G} k(x, y)$. Note that $\boldsymbol{m}_2$ and $\boldsymbol{\eta}$ have dimension G^c.

Theorem 5.5 (i) *The vector $\boldsymbol{m}_2$ is the minimal solution to $\boldsymbol{m}_2 = \boldsymbol{\eta} + \boldsymbol{K}\boldsymbol{m}_2$.*
(ii) $\boldsymbol{m}_2 = \sum_{m=0}^{\infty} \boldsymbol{K}^m \boldsymbol{\eta}$.
(iii) *Let $\boldsymbol{k}$ be an G^c-vector such that $\boldsymbol{K}\boldsymbol{k} \leq \boldsymbol{k} - \boldsymbol{\eta}$. Then $\boldsymbol{m}_2 \leq \boldsymbol{k}$.*

Given the potential of Theorem 5.5 and the following Corollary 5.6 for the (in general very difficult) problem of bounding the variance of the estimator (5.6), we give the

Proof. We have

$$\mathbb{E}_{x_0}[Z^2 \mid \tau_G = m] = \sum_{x_1, \ldots, x_{m-1} \notin G, \, x_m \in G} \widetilde{p}_{x_{m-1}, x_m} \prod_{n=1}^{m-1} r(X_{n-1}, X_n)^2 \widetilde{p}_{x_{n-1}, x_n} \, .$$

This is the x_0th element of the vector $\boldsymbol{K}^m \boldsymbol{\eta}$. Summing over m, (ii) follows. (i) is then an easy consequence.

For (iii), we have $\boldsymbol{\eta} \leq \boldsymbol{k} - \boldsymbol{K}\boldsymbol{k}$ and hence $\boldsymbol{K}^m \boldsymbol{\eta} \leq \boldsymbol{K}^m \boldsymbol{k} - \boldsymbol{K}^{m+1} \boldsymbol{k}$ for all m. Thus

$$\sum_{m=0}^{n} \boldsymbol{K}^m \boldsymbol{\eta} \leq \boldsymbol{k} - \boldsymbol{K}^{n+1} \boldsymbol{k} \leq \boldsymbol{k} \, .$$

Letting $n \to \infty$ and using (ii) gives (iii). □

Intuitively, one should choose $a(x)$ as a good approximation to $h(x)$ and then take $k(x)$ of the form $a(x)^2 \widehat{k}(x)$ with $\widehat{k}(x) = O(1)$. This idea is made precise in the following corollary and its proof:

Corollary 5.6 *Assume $r(x,y)$ has the form $c(x)/a(y)$ and that*

$$c(x) \sum_{y \in E} p(x,y)a(y)\widehat{k}(y) \leq \widehat{k}(x)a(x)^2 \qquad (5.7)$$

for some E-vector $\widehat{k}$ and all $y \notin G$. If $\widehat{k}(x) \geq 1$ for all $x \in E$ and $a(y) \geq \kappa > 0$ for all $y \in G$, then $m_2(x) \leq \kappa^{-2}a(x)^2\widehat{k}(x)$ for all $x \notin G$.

Proof. We first note that

$$\eta(x) = \sum_{y \in G} p(x,y)\frac{c(x)}{a(y)} \leq \kappa^{-2}c(x)\sum_{y \in G} p(x,y)a(y)$$

$$\leq \kappa^{-2}c(x)\sum_{y \in G} p(x,y)a(y)\widehat{k}(y). \qquad (5.8)$$

Define $k(x) = \kappa^{-2}a(x)^2\widehat{k}(x)$. Then for $x \notin G$, we have

$$\kappa^{-2}c(x)p(x,y)a(y)\widehat{k}(y) = \kappa^{-2}\widehat{k}(x,y)a(y)^2\widehat{k}(y) = \widehat{k}(x,y)k(y).$$

Thus combining with (5.8), it follows from (5.7) divided by κ^2 that $\boldsymbol{k} \geq \boldsymbol{Kk} + \boldsymbol{\eta}$. Now appeal to Theorem 5.5(iii). $\qquad\square$

In the rest of this section, we assume that $X_n = -u + Y_1 + \cdots + Y_n$ is a random walk with negative drift, subexponential increments and (for simplicity) density $f(z)$, and take $G = (0, \infty)$. The ruin probability is then $\psi(u) = h(-u)$ and τ_G is the ruin time. The start of implementing the h-transform ideas is easy: as for light tails, we have an approximation for $h(v)$ for $v < 0$, now $h(v) \approx a(v) = C\overline{F}_I(-v)$ where F_I is the integrated tail distribution of the increment distribution F, cf. X.3.1. In the representation $r(x,y) = c(x)/a(y)$, it will be convenient to be able to think of $a(y)$ as the tail of a r.v. Z, that we take as the r.v. with

$$\mathbb{P}(Z > z) = \min\left[1, \frac{1}{\mathbb{E}|Y|}\int_z^\infty \overline{F}(s)\,\mathrm{d}s\right].$$

Thus, we take $a(y) = \mathbb{P}(Z > z)$ and have

$$c(x) = \int_{\mathbb{R}} p(x,y)a(y)\,\mathrm{d}y = \mathbb{E}a(x+Y) = \mathbb{P}(Y + Z > -x).$$

The most obvious procedure is now to use the estimator

$$Z = Z(u) = I(\tau_G < \infty)\prod_{n=1}^{\tau_G} \frac{c(X_{n-1})}{a(X_n)}. \qquad (5.9)$$

However, as for the Dupuis-Leder-Wang algorithm one encounters the difficulty that the most obvious choice does not have the desired efficiency properties but needs modification, more precisely to

$$Z = Z(u) = I(\tau_G < \infty) \prod_{n=1}^{\tau_G} \frac{c(X_{n-1} + x^*)}{a(X_n + x^*)}. \tag{5.10}$$

Here $x^* = x^*(\gamma)$ is taken as in the following lemma (recall the definition of the class S^* from p. 302):

Lemma 5.7 *Assume $Y^+ \in S^*$. Then:*
(i) $c(x) - a(x) = o(\overline{F}(-x))$ *as $x \downarrow -\infty$;*
(ii) *given $\gamma \in (0, 1]$, there exists $x^*(\gamma) \leq 0$ such that*

$$\frac{a(x)^2 - c(x)^2}{\overline{F}(-x)c(x)} \geq -\gamma \quad \text{for all } x \leq x^*(\gamma). \tag{5.11}$$

Theorem 5.8 *Assume $Y^+ \in S^*$, let $0 < \gamma < 1$, let x^* be defined as in (5.11) and $d(x^*) = \mathbb{P}(Z > -x^*)$. Then*

$$\limsup_{u \to \infty} \frac{\widetilde{\mathbb{E}}Z(u)^2}{\psi(u)^2} \leq \frac{1}{(1 - \gamma)d(x^*)^2}.$$

The proof of Theorem 5.8 and Lemma 5.7 (which is a crucial step in bounding the variance) are long and technical, although in principle elementary, and will not be reproduced here. We note once more that random variate generation in (5.11) is not a standard problem.

Notes and references A further relevant reference in the setting of the Blanchet-Glynn algorithm is Blanchet, Glynn & Liu [176].

To summarize our discussion of dynamic importance sampling, the method is not straightforward to implement in the heavy-tailed case. The most obvious ideas need modification and tuning to produce efficient algorithms, and these steps may require tedious calculations, cf. Lemma 5.7. Further, bounding the variance is not straightforward at all and random variate generation may present problems.

However, the Blanchet-Glynn algorithm is remarkable by being the first to be efficient for an infinite horizon problem with heavy tails when no alternative representation (say the Pollaczeck-Khinchine geometric sum in the compound Poisson model) is available. In fact, it is shown in Bassamboo, Juneja & Zeevi [140] that no static importance sampling algorithm exists for efficient simulation of the tail of the maximum of a random walk (at least in the regularly varying case).

6   Regenerative simulation

Our starting point is the duality representations in III.3: for many risk processes $\{R_t\}$, there exists a dual process $\{V_t\}$ such that

$$\psi(u, T) \; = \; \mathbb{P}\Big(\inf_{0 \le t \le T} R_t < 0\Big) \; = \; \mathbb{P}(V_T > u),$$

$$\psi(u) \; = \; \mathbb{P}\Big(\inf_{t \ge 0} R_t < 0\Big) \; = \; \mathbb{P}(V_\infty > u), \tag{6.1}$$

where the identity for $\psi(u)$ requires that V_t has a limit in distribution V_∞.

In most of the simulation literature (say in queueing applications), the object of interest is $\{V_t\}$ rather than $\{R_t\}$, and (6.1) is used to study V_∞ by simulating $\{R_t\}$ (for example, the algorithm in Section 3 produces simulation estimates for the tail $\mathbb{P}(W > u)$ of the GI/G/1 waiting time W). However, we believe that there are examples also in risk theory where (6.1) may be useful. One main example is $\{V_t\}$ being regenerative (see A.1): then by Proposition A1.3,

$$\psi(u) \; = \; \mathbb{P}(V_\infty > u) \; = \; \frac{1}{\mathbb{E}\omega} \mathbb{E} \int_0^\omega I(V_t > u)\,\mathrm{d}t \tag{6.2}$$

where ω is the generic cycle for $\{V_t\}$. The method of *regenerative simulation*, which we survey below, provides estimates for $\mathbb{P}(V_\infty > u)$ (and more general expectations $\mathbb{E}g(V_\infty)$). Thus the method provides one answer on how to avoid simulating $\{R_t\}$ for an infinitely long time period.

For details, consider first the case of independent cycles. Simulate a zero-delayed version of $\{V_t\}$ until a large number N of cycles have been completed. For the ith cycle, record $\boldsymbol{Z}^{(i)} = \big(Z_1^{(i)}, Z_2^{(i)}\big)$ where $Z_1^{(i)} = \omega_i$ is the cycle length, $Z_2^{(i)}$ the time during the cycle where $\{V_t\}$ exceeds u and $z_j = \mathbb{E}Z_j^{(i)}$, $j = 1, 2$. Then $\boldsymbol{Z}^{(1)}, \ldots, \boldsymbol{Z}^{(N)}$ are i.i.d. and

$$\mathbb{E}Z_1^{(i)} = z_1 = \mathbb{E}\omega, \quad \mathbb{E}Z_2^{(i)} = z_2 = \mathbb{E}\int_0^\omega I(V_t > u)\,\mathrm{d}t\,.$$

Thus, letting

$$\overline{Z}_1 = \frac{1}{N}\big(Z_1^{(1)} + \cdots + Z_1^{(N)}\big), \quad \overline{Z}_2 = \frac{1}{N}\big(Z_2^{(1)} + \cdots + Z_2^{(N)}\big),$$

$$\widehat{\psi}(u) \; = \; \frac{\overline{Z}_2}{\overline{Z}_1} \; = \; \frac{Z_2^{(1)} + \cdots + Z_2^{(N)}}{Z_1^{(1)} + \cdots + Z_1^{(N)}}\,,$$

the LLN yields $\overline{Z}_1 \overset{\text{a.s.}}{\to} z_1$, $\overline{Z}_2 \overset{\text{a.s.}}{\to} z_2$,

$$\widehat{\psi}(u) \overset{\text{a.s.}}{\to} \frac{z_2}{z_1} = \frac{\mathbb{E} \int_0^\omega I(V_t > u)\, dt}{\mathbb{E}\omega} = \psi(u)$$

as $N \to \infty$. Thus, the regenerative estimator $\widehat{\psi}(u)$ is consistent.

To derive confidence intervals, let $\boldsymbol{\Sigma}$ denote the 2×2 covariance matrix of $\boldsymbol{Z}^{(i)}$. Then

$$\frac{1}{\sqrt{N}}\left(\overline{Z}_1 - z_1, \overline{Z}_2 - z_2\right) \overset{\mathscr{D}}{\to} N_2(0, \boldsymbol{\Sigma}).$$

Therefore, a standard transformation technique (sometimes called the delta method, cf. [79, IV.4]) yields

$$\frac{1}{\sqrt{N}}\left(h\left(\overline{Z}_1, \overline{Z}_2\right) - h\left(z_1, z_2\right)\right) \overset{\mathscr{D}}{\to} N(0, \sigma_h^2)$$

for $h : \mathbb{R}^2 \to \mathbb{R}$ and $\sigma_h^2 = \nabla_h \boldsymbol{\Sigma} \nabla_h'$, $\nabla_h = (\partial h/\partial z_1 \ \ \partial h/\partial z_2)$. Taking $h(z_1, z_2) = z_2/z_1$ yields $\nabla_h = (-z_2/z_1^2 \ \ 1/z_1)$,

$$\frac{1}{\sqrt{N}}\left(\widehat{\psi}(u) - \psi(u)\right) \overset{\mathscr{D}}{\to} N(0, \sigma^2) \tag{6.3}$$

where

$$\sigma^2 = \frac{z_2^2}{z_1^4}\Sigma_{11} + \frac{1}{z_1^2}\Sigma_{22} - 2\frac{z_2}{z_1^3}\Sigma_{12}. \tag{6.4}$$

The natural estimator for $\boldsymbol{\Sigma}$ is the empirical covariance matrix

$$\boldsymbol{S} = \frac{1}{N-1}\sum_{i=1}^{N}\left(\boldsymbol{Z}^{(i)} - \overline{\boldsymbol{Z}}\right)\left(\boldsymbol{Z}^{(i)} - \overline{\boldsymbol{Z}}\right)^{\mathsf{T}}$$

so σ^2 can be estimated by

$$s^2 = \frac{\overline{Z}_2^2}{\overline{Z}_1^4}S_{11} + \frac{1}{\overline{Z}_1^2}S_{22} - 2\frac{\overline{Z}_2}{\overline{Z}_1^3}S_{12} \tag{6.5}$$

and the 95% confidence interval is $\widehat{\psi}(u) \pm 1.96s/\sqrt{N}$.

The regenerative method is not likely to be efficient for large u but rather a brute force one. However, in some situations it may be the only one resolving the infinite horizon problem, say risk processes with a complicated structure of the point process of claim arrivals and heavy-tailed claims. There is potential also for combining it with some variance reduction method.

Notes and references The literature on regenerative simulation is extensive, and we will not attempt a literature survey here.

7 Sensitivity analysis

We return to the problem of IV.9, to evaluate the sensitivity $\psi_\zeta(u) = (d/d\zeta)\,\psi(u)$ where ζ is some parameter governing the risk process. In IV.9, asymptotic estimates were derived using the renewal equation for $\psi(u)$. We here consider simulation algorithms which have the potential of applying to substantially more complex situations.

Before going into the complications of ruin probabilities, consider an extremely simple example, the expectation $z = \mathbb{E}Z$ of a single r.v. Z of the form $Z = \varphi(X)$ where X is a r.v. with distribution depending on a parameter ζ. Here are the ideas of the two main approaches in today's simulation literature:

The score function (SF) method. Let X have a density $f(x,\zeta)$ depending on ζ. Then $z(\zeta) = \int \varphi(x) f(x,\zeta)\,dx$ so that differentiation yields

$$
\begin{aligned}
z_\zeta &= \frac{d}{d\zeta}\int \varphi(x) f(x,\zeta)\,dx = \int \varphi(x)\frac{d}{d\zeta} f(x,\zeta)\,dx \\
&= \int \varphi(x)\frac{(d/d\zeta)f(x,\zeta)}{f(x,\zeta)} f(x,\zeta)\,dx = \mathbb{E}[SZ],
\end{aligned}
$$

where

$$
S = \frac{(d/d\zeta)f(X,\zeta)}{f(X,\zeta)} = \frac{d}{d\zeta}\log f(X,\zeta)
$$

is the score function familiar from statistics. Thus, SZ is an unbiased Monte Carlo estimator of z_ζ.

Infinitesimal perturbation analysis (IPA) uses sample path derivatives. So assume that a r.v. with density $f(x,\zeta)$ can be generated as $h(U,\zeta)$ where U is uniform$(0,1)$. Then $z(\zeta) = \mathbb{E}\varphi\big(h(U,\zeta)\big)$,

$$
z_\zeta = \mathbb{E}\Big[\frac{d}{d\zeta}\varphi\big(h(U,\zeta)\big)\Big] = \mathbb{E}\Big[\varphi'\big(h(U,\zeta)\big)h_\zeta(U,\zeta)\Big],
$$

where $h_\zeta(u,\zeta) = (\partial/\partial\zeta)h(u,\zeta)$. Thus, $\varphi'\big(h(U,\zeta)\big)h_\zeta(U,\zeta)$ is an unbiased Monte Carlo estimator of z_ζ. For example, if $f(x,\zeta) = \zeta e^{-\zeta x}$, one can take $h(U,\zeta) = -\log U/\zeta$, giving $h_\zeta(U,\zeta) = \log U/\zeta^2$.

The derivations of these two estimators are heuristic in that both use an interchange of expectation and differentiation that needs to be justified. For the SF method, this is usually unproblematic and involves some application of dominated convergence. For IPA there are, however, non-pathological examples where sample path derivatives fail to produce estimators with the correct expectation. To see this, just take φ as an indicator function, say $\varphi(x) = I(x > x_0)$

and assume that $h(U, \zeta)$ is increasing in ζ. Then, for some $\zeta_0 = \zeta_0(U)$, $\varphi(h(U, \zeta))$ is 0 for $\zeta < \zeta_0$ and 1 for $\zeta > \zeta_0$ so that the sample path derivative $\varphi'(h(U, \zeta))$ is 0 w.p. one. Thus, IPA will estimate z_ζ by 0 which is obviously not correct. In the setting of ruin probabilities, this phenomenon is particularly unpleasant since indicators occur widely in the CMC estimators. A related difficulty occurs in situations involving the Poisson number N_t of claims: also here the sample path derivative w.r.t. β is 0. The following example demonstrates how the SF method handles this situation.

Example 7.1 Consider the sensitivity $\psi_\beta(u)$ w.r.t. the Poisson rate β in the compound Poisson model. Let $M(u)$ be the number of claims up to the time $\tau(u)$ of ruin (thus, $\tau(u) = T_1 + \cdots + T_{M(u)}$). The likelihood ratio up to $\tau(u)$ for two Poisson processes with rates β, β_0 is

$$\prod_{i=1}^{M(u)} \frac{\beta e^{-\beta T_i}}{\beta_0 e^{-\beta_0 T_i}} I\big(\tau(u) < \infty\big).$$

Taking expectation, differentiating w.r.t. β and letting $\beta_0 = \beta$, we get

$$\begin{aligned}
\psi_\beta(u) &= \mathbb{E}\left[\sum_{i=1}^{M(u)} \left(\frac{1}{\beta} - T_i\right) I\big(\tau(u) < \infty\big)\right] \\
&= \mathbb{E}\left[\left(\frac{M(u)}{\beta} - \tau(u)\right) I\big(\tau(u) < \infty\big)\right].
\end{aligned}$$

To resolve the infinite horizon problem, change the measure to $\mathbb{P}_L$ as when simulating $\psi(u)$. We then arrive at the estimator

$$Z_\beta(u) = \left(\frac{M(u)}{\beta} - \tau(u)\right) e^{-\gamma u} e^{-\gamma \xi(u)}$$

for $\psi_\beta(u)$ (to generate $Z_\beta(u)$, the risk process should be simulated with parameters β_L, B_L).

We recall (Proposition IV.9.4) that $\psi_\beta(u)$ is of the order of magnitude $ue^{-\gamma u}$. Thus, the estimation of $\psi_\beta(u)$ is subject to the same problem concerning relative precision as in rare event simulation. However, since

$$\mathbb{E}_L Z_\beta(u)^2 \leq \left(\frac{M(u)}{\beta} - \tau(u)\right)^2 e^{-2\gamma u} = O(u^2) e^{-2\gamma u},$$

we have

$$\frac{\mathbb{V}ar_L\big(Z_\beta(u)\big)}{z_\beta(u)^2} \sim \frac{O(u^2) e^{-2\gamma u}}{u^2 e^{-2\gamma u}} = O(1)$$

so that in fact the estimator $Z_\beta(u)$ has bounded relative error. □

Remark 7.2 IPA and score functions are not the only ones around. Here are some further alternatives:

Finite differences are simply a stochastic version of numerical differentiation. So, assume Z can be generated as $h(\boldsymbol{X}, \zeta)$ for a suitable random vector $\boldsymbol{X}$. Then the estimate of z_ζ is

$$\frac{h(\boldsymbol{X}, \zeta + h/2) - h(\boldsymbol{X}, \zeta - h/2)}{h} \tag{7.1}$$

(there are several possible variants; this one uses common random numbers and central differences). In many situations, this idea is the simplest one to implement. Its problem is that the estimate is biased. In the limit $h \downarrow 0$, (7.1) becomes the IPA estimator.

The idea of *weak derivatives* is measure-valued differentiation. Suppose that we are interested in the sensitivity of $\mathbb{E}h(\boldsymbol{Y}, X)$ w.r.t. ζ, where X has density $f(x, \zeta)$ with $f'(x, \zeta) = \partial f(x, \zeta)/\partial\zeta$ and $\boldsymbol{Y}$ is a random vector with distribution independent of ζ. Since $\int f'(x, \zeta)\, \mathrm{d}x = 0$, we will typically be able to write $f'(x, \zeta)$ as $k f_+(x, \zeta) - k f_-(x, \zeta)$ where $f_+(x, \zeta)$, $f_-(x, \zeta)$ are probability densities and k a constant. If W_+, W_- are r.v.'s with these densities, we therefore have

$$\begin{aligned}
\frac{\mathrm{d}}{\mathrm{d}\zeta}\mathbb{E}h(\boldsymbol{Y}, X) &= \frac{\mathrm{d}}{\mathrm{d}\zeta}\int \mathbb{E}h(\boldsymbol{Y}, x)\, f(x, \zeta)\, \mathrm{d}x = \int \mathbb{E}h(\boldsymbol{Y}, x)\, f'(x, \zeta)\, \mathrm{d}x \\
&= \mathbb{E}\big[k h(\boldsymbol{Y}, W_+) - k h(\boldsymbol{Y}, W_-)\big],
\end{aligned}$$

so that the desired estimator can be taken as $k h(\boldsymbol{Y}, W_+) - k h(\boldsymbol{Y}, W_-)$. For example, in the Poisson case $f(x, \zeta) = \mathrm{e}^{-\zeta}\zeta^x/x!$ we get

$$f'(x, \zeta) = \mathrm{e}^{-\zeta}\zeta^{x-1}/(x-1)! - \mathrm{e}^{-\zeta}\zeta^x/x! = f(x-1, \zeta) - f(x, \zeta)$$

(with the convention $(-1)/(-1)! = 0$) so that $k = 1$ and we can generate W_+, W_- as $V_+ + 1$, V_- with being V_+, V_- Poisson(ζ).

Finally, in finance (where the sensitivities go under the name *Greeks*) methods based on formulas from *Malliavin calculus* have become popular. □

Notes and references A general survey of simulation methods for evaluating sensitivities is given in Asmussen & Glynn [79]. For topics not treated there in detail, see e.g. Heidergott [456, 457] for weak derivatives, and Fournie *et al.* [368] and Kohatsu-Riga & Montero [549] for the Malliavin approach. A general reference for IPA is Glasserman [417], one for the SF method is Rubinstein & Shapiro [754].

Example 7.1 is from Asmussen & Rubinstein [100] who also work out a number of similar sensitivity estimators, in part for different measures of risk than ruin probabilities, for different models and for the sensitivities w.r.t. different parameters.

There has been much work on resolving the difficulties associated with IPA pointed out above. In the setting of ruin probabilities, a relevant reference is Vázquez-Abad [862].

Chapter XVI

Miscellaneous topics

1 More on discrete-time risk models

There are at least two reasons to consider the discrete-time counterparts of continuous-time risk models: one is that the resulting approximation can be computationally easier to handle, in particular when more complex features like interest, investment, dividends and reinsurance are also included. Secondly one could claim that all events (claims, premium payments etc.) are in practice only observable and/or payable at discrete points in time and so a discrete modeling may be considered closer to reality. However, much of the mathematical elegance and insight is usually lost when replacing continuous-time dynamics by discrete ones. If the claim size distribution is also discrete (which is the case we consider here[1]), then the differential equations from the continuous set-up are replaced by difference equations and therefore the probability of ruin can be calculated recursively for given numerical values of the model parameters. A disadvantage of this approach is that it is usually not possible to track the influence of model parameters on the final result and consequently the qualitative behavior of ruin probabilities. On the other hand, the resulting method for calculating ruin probabilities and related quantities is simple and general and, as we shall see below, some relations and identities of continuous-time models have analogues in the discrete-time set-up.

---

[1]Note that at several places in this book we have already dealt with certain discrete-time models as approximations for continuous-time models, then usually with continuous claim size distributions. Here we focus on the fully discrete model to emphasize the computational alternative for obtaining ruin probabilities that it may offer.

Assume that the discrete-time risk reserve process $\{R_n^{(d)}\}$ is given by

$$R_n^{(d)}(u) = u + n - \sum_{i=1}^{n} X_n, \quad n \in \mathbb{N}, \tag{1.1}$$

where $X_1, X_2, \ldots$ are i.i.d. integer-valued non-negative random variables with probability function $h_k = \mathbb{P}(X_1 = k)$ ($k = 0, \ldots, \infty$) and c.d.f. H. The interpretation is that X_j is the total claim amount paid in year (or time unit) j. The initial capital u is also assumed to be a non-negative integer, so that $\{R_n^{(d)}\}$ is always integer-valued. We impose throughout the net profit condition $\mathbb{E}X_1 < 1$.

Remark 1.1 The model (1.1) is often referred to as the *compound binomial model*. This is justified because in each time interval the total claim size is 0 with probability $h_0 > 0$ and hence $\sum_{i=1}^{n} X_i = \sum_{i=1}^{N_n} Y_i$, where N_n is a binomial$(n, 1 - h_0)$ r.v. and $\mathbb{P}(Y_i = k) = h_k/(1 - h_0)$ for $k = 1, 2, \ldots$. The compound Poisson model then appears as the natural continuous-time limit. In that sense the discrete-time model can also help to sharpen the intuition for the continuous-time set-up. □

Define as usual the claim surplus process $\{S_n^{(d)}\}$ by

$$S_n^{(d)} = u - R_n^{(d)}(u) = \sum_{i=1}^{n} X_n - n.$$

The ruin time for (1.1) is defined as

$$\tau^{(d)}(u) = \min\{n \geq 1 : R_n^{(d)}(u) \leq 0\} = \min\{n \geq 1 : S_n^{(d)} \geq u\}$$

and the ruin probability as

$$\psi^{(d)}(u) = \mathbb{P}\big(\tau^{(d)}(u) < \infty\big) = \mathbb{P}\Big(\max_{n \geq 1} S_n^{(d)} \geq u\Big)$$

(we follow here the tradition of the literature to consider the process ruined already if it reaches level 0, but only for some $n \geq 1$, so $\psi(0) < 1$).

Proposition 1.2 *The ruin probability for the discrete-time risk process* (1.1) *satisfies the recursion*

$$\psi^{(d)}(u) = \sum_{y=0}^{u-1} \big(1 - H(y)\big)\,\psi^{(d)}(u - y) + \sum_{y=u}^{\infty} \big(1 - H(y)\big), \quad u = 1, 2, \ldots, \tag{1.2}$$

with starting value $\psi^{(d)}(0) = \sum_{y=0}^{\infty} \big(1 - H(y)\big) = \mathbb{E}(X_1)$.

This is obviously a discrete analogue of the renewal equation IV.(3.2), and becomes clear at once if one conditions on the value y of the first (weak) ascending ladder point S_{τ_+} of the claim surplus process, where

$$\tau_+ = \tau^{(d)}(0) = \min\{n \geq 1 : S_n^{(d)} \geq 0\}$$

provided it has been shown that

$$g_y^{(d)} = \mathbb{P}(S_{\tau_+} = y, \tau_+ < \infty) = 1 - H(y).$$

That $g_y^{(d)} = 1 - H(y)$ can be proved by adapting the proof of Theorem III.5.1 from continuous to discrete time. This is straightforward, which is intuitively plausible from the fact that the claim surplus processes have the common feature of being downward skipfree with unit drift when no claims occur. Alternatively, one may simply refer to the form of $g_y^{(d)}$ as a known result in random walk theory (e.g. [APQ, Cor. 5.6, p. 236] combined with the connection on [APQ, p. 222] between strong and weak ladder points heights). We shall, however, also present a direct proof that avoids the slightly sophisticated probabilistic ideas of III.5 or [APQ].

Proof. In the first time unit, the premium income is 1 and the risk reserve process will only survive if the total claim amount satisfies $X_1 \leq u$ and will then start anew at the level $u + 1 - X_1$ (note that because of the independence assumption of the total claim amounts $X_1, X_2, \dots$ the process is Markov). Hence

$$
\begin{aligned}
\psi^{(d)}(u) &= \sum_{k=0}^{u} h_k \, \psi^{(d)}(u+1-k) + 1 - H(u), \\
&= \sum_{j=1}^{u+1} h_{u+1-j} \, \psi^{(d)}(j) + 1 - H(u), \quad u = 0, 1, 2, \dots \quad (1.3)
\end{aligned}
$$

From this it follows that for $w = 0, 1, 2, \dots$

$$
\begin{aligned}
\sum_{u=0}^{w} \psi^{(d)}(u) &= \sum_{u=0}^{w} \sum_{j=1}^{u+1} h_{u+1-j} \psi^{(d)}(j) + \sum_{u=0}^{w} \left(1 - H(u)\right), \\
&= \sum_{j=1}^{w+1} \psi^{(d)}(j) \sum_{u=j-1}^{w} h_{u+1-j} + \sum_{u=0}^{w} \left(1 - H(u)\right) \\
&= \sum_{j=1}^{w} \psi^{(d)}(j) H(w+1-j) + \psi^{(d)}(w+1) h_0 + \sum_{u=0}^{w} \left(1 - H(u)\right)
\end{aligned}
$$

or equivalently

$$\psi^{(d)}(w+1)\,h_0 \;=\; \psi^{(d)}(0) + \sum_{j=1}^{w} \psi^{(d)}(j)\big(1 - H(w+1-j)\big) - \sum_{u=0}^{w}\big(1 - H(u)\big).$$

At the same time we can read off from (1.3) that

$$\psi^{(d)}(w+1)\,h_0 \;=\; \psi^{(d)}(w) - \sum_{j=1}^{w} h_{w+1-j}\,\psi^{(d)}(j) - \big(1 - H(w)\big),$$

and equating the last two equations gives

$$\psi^{(d)}(w) \;=\; \psi^{(d)}(0) + \sum_{j=1}^{w} \psi^{(d)}(j)\big(1 - H(w-j)\big) - \sum_{j=0}^{w-1}\big(1 - H(j)\big). \qquad (1.4)$$

On the other hand, with $g_y^{(d)}$ as above, we clearly have

$$\psi^{(d)}(u) \;=\; \sum_{y=0}^{u-1} g_y^{(d)}\psi^{(d)}(u-y) + \sum_{y=u}^{\infty} g_y^{(d)}, \quad u = 1, 2, \ldots$$

and

$$\psi^{(d)}(0) \;=\; \sum_{y=0}^{\infty} g_y^{(d)}.$$

We can hence write for $u = 1, 2, \ldots$

$$\psi^{(d)}(u) \;=\; \sum_{y=0}^{u-1} g_y^{(d)}\psi^{(d)}(u-y) + \psi^{(d)}(0) - \sum_{y=0}^{u-1} g_y^{(d)} \qquad (1.5)$$

$$=\; \sum_{y=1}^{u} g_{u-y}^{(d)}\psi_y^{(d)} + \psi^{(d)}(0) - \sum_{y=0}^{u-1} g_y^{(d)}. \qquad (1.6)$$

Comparing (1.4) and (1.6) now establishes

$$g_y^{(d)} = 1 - H(y), \quad y = 0, 1, 2, \ldots$$

and

$$\psi^{(d)}(0) = \sum_{y=0}^{\infty}\big(1 - H(y)\big) = \mathbb{E}(X_1).$$

Inserting the latter formula in (1.5) now gives (1.2).   □

Remark 1.3 Note the complete analogy of the formulas for $g_y^{(d)}$ and $\psi^{(d)}(0)$ with the ones for the compound Poisson risk model in Chapter IV (this comes as no surprise, since as outlined in Remark 1.1 the latter may be interpreted as the continuous-time limit of the compound binomial model). $\square$

Remark 1.4 The proof of Proposition 1.2 via random walk theory has, however, the advantage of naturally giving the form of $g_y^{(d)}$ for more general upward steps than 1 subject to some rootfinding. More precisely, one can handle the model

$$R_n^{(d)}(u) = u + \sum_{i=1}^{n}(Y_n - X_n), \quad n \in \mathbb{N}, \tag{1.7}$$

where the Y_n are i.i.d. with support in $\{0, \dots, r\}$ for some $r \in \mathbb{N}$. For details, see [APQ, Sect. VIII.5a].

An appealing special case of (1.7) is $Y_n = r > 1$. This allows the claims to be orders smaller than the premium inflow, which may appear more realistic than (1.1). $\square$

Example 1.5 Consider the two-point distribution $h_0 = \theta = 1 - h_2$ for $\theta > 1/2$. Then we are in the situation of the Gambler's ruin problem and the recursion (1.2) indeed corresponds to the one of Proof 1 of Proposition II.2.1 (with $a = \infty$) leading to $\psi^{(d)}(u) = \big((1-\theta)/\theta\big)^u$ as already given in II.(2.3). $\square$

Example 1.6 Assume geometrically distributed claim sizes with $h_0 = p$ and $h_k = (1-p)(1-\alpha)\alpha^{k-1}$ for $k \geq 1$ with $0 < p < 1$ and α such that $\mathbb{E}(X_1) = (1-p)/(1-\alpha) < 1$. In this case the recursion (1.2) reduces, after a little algebra, to $\psi^{(d)}(u+1) = (\alpha/p)\psi^{(d)}(u)$ leading to

$$\psi^{(d)}(u) = \frac{1-p}{1-\alpha}\left(\frac{\alpha}{p}\right)^u.$$

$\square$

Consider now the finite-time ruin probability

$$\psi^{(d)}(u,t) = \mathbb{P}(\tau^{(d)}(u) \leq t), \quad t \in \mathbb{N}.$$

Noting that $\psi^{(d)}(u,1) = 1 - H(u)$, it follows from the Markov property that

$$\psi^{(d)}(u,t) = \psi^{(d)}(u,1) + \sum_{k=0}^{u} h_k \psi^{(d)}(u+1-k,t-1) \quad \text{for all } t = 2,3,\dots \tag{1.8}$$

This bivariate recursion[2] can now be used to recursively calculate $\psi(u, t)$ for fixed integer values of u and t, whenever the claim size probability function h_k ($k = 0, 1, 2, \ldots$) is given. Although this simple recursion is one of the reasons why the discrete-time model also has some popularity for approximating $\psi(u, t)$ of continuous-time models (in particular when adding additional features like interest rates, investment and dividends in the model), one should note that in practical applications it can be very computer-intensive to implement and the result only gives the numerical values with no hold of sensitivities to model assumptions such as claim size parameters.

There is also a natural analogue of the adjustment coefficient for this discrete set-up. Define $\gamma^{(d)}$ as the unique positive root of the equation

$$\mathbb{E} e^{r(X_1 - 1)} = 1$$

if it exists. The reason for this definition is that in this way $\left\{ e^{-\gamma^{(d)} R_n^{(d)}} \right\}$ is a martingale and because $R_n^{(d)} \overset{\text{a.s.}}{\to} \infty$ on $\left\{ \tau^{(d)}(u) = \infty \right\}$, Proposition II.3.1 applies and gives

Proposition 1.7 *Assume that the adjustment coefficient $\gamma^{(d)} > 0$ exists. Then*

$$\psi^{(d)}(u) = \frac{e^{-\gamma^{(d)} u}}{\mathbb{E}\left[\exp\{-\gamma R_{\tau^{(d)}(u)}^{(d)}\} \,\big|\, \tau^{(d)}(u) < \infty \right]}, \quad u \geq 0.$$

In particular, the Lundberg inequality $\psi^{(d)}(u) \leq e^{-\gamma^{(d)} u}$ holds.

Notes and references An early reference for the compound binomial model (1.1) is Gerber [398]. Pollaczeck-Khinchine-type formulas for $\psi(u)$ can be found in Gerber [404] and Shiu [802]. De Vylder & Goovaerts [302] investigate possibilities to speed up the recursive calculation (1.8). In particular they give error bounds for $\psi(u, t)$ if the claim size distribution is truncated. Another representation of the recursion (1.8) is given in Willmot [885]. Other quantities like the time of ruin, the surplus prior to ruin and the deficit at ruin in the compound binomial model are e.g. studied in Cheng, Gerber & Shiu [235], Li & Garrido [588] and Liu & Guo [602]. In the compound binomial model the number of periods until a claim occurs is geometrically distributed with parameter h_0. An extension is to allow for more general distributions for the number of interclaim time periods (which is the discrete-time analogue of the extension of a Poisson process to a renewal process). The resulting *discrete-time Sparre Andersen model* has received some interest recently; for a survey see Li, Lu & Garrido [593].

---

[2]Which is the discrete-time analogue of having an additional partial derivative w.r.t. time t in the integro-differential equation for ψ in the compound Poisson case, cf. Chapter V.

Cossette, Landriault & Marceau [260] extend the compound binomial model by assuming that the indicator random variable of whether X_j is non-zero in period j follows a homogeneous Markov chain, for an extension to a Markov-modulated environment for both this indicator r.v. and the claim size distribution see [261] and Yang, Zhang & Lan [905]. Another dependence structure between subsequent claim sizes is considered in Yuen & Guo [907]. For the effect of dependent claims in a discrete-time model with continuous claim size distribution see e.g. Cossette & Marceau [259], Wu & Yuen [898] and Reinhard & Snoussi [732]. De Kok [286] deals with an inhomogeneous risk model.

Dickson & Waters [318] use the recursions of the discrete-time model to effectively approximate finite-time ruin probabilities in the Cramér-Lundberg model, under additional force of interest see [320] and Brekelmans & De Waegenaere [200]. Egidio dos Reis [339] deals with moments of ruin and recovery times. For stochastic ordering concepts in the discrete framework and resulting ordering of ruin probabilities we refer to Denuit & Lefèvre [295].

As mentioned in the beginning of this section, one may take the viewpoint that observations and actions can in practice only happen at discrete points in time, but the underlying risk model has many computational and qualitative advantages when being continuous time. A possible bridge between these conflicting arguments can be to assume an underlying continuous time model and indeed only observe the risk process (and potential ruin) at discrete times. If these discrete time points are assumed to be random, e.g. exponentially distributed, this still leads to explicit formulas of continuous time flavor. By moving towards Erlang(n) (and hence more peaked) observation times with growing n, one approaches the discrete-time set-up with computational vehicle of continuous time models. This procedure is worked out in [20] and is close in spirit to the Erlangization approach for finite time horizon ruin probabilities as discussed in Section IX.8. For statistical inference issues for a continuous time risk model under discrete observations, see Shimizu [797].

2 The distribution of the aggregate claims

We study the distribution of the aggregate claims $A_t = \sum_1^{N_t} U_i$ at time t, assuming that the U_i are i.i.d. with common distribution B and independent of N_t. In particular, we are interested in estimating $\mathbb{P}(A_t > x)$ for large x. This is a topic of practical importance in the insurance business for assessing the probability of a great loss in a period of length t, say one year. Further, the study is motivated from the formulas in V.2 expressing the finite horizon ruin probabilities in terms of the distribution of A_t.

The main example is N_t being Poisson with rate βt. For notational simplicity, we then take $t = 1$ so that

$$p_n = \mathbb{P}(N = n) = \mathrm{e}^{-\beta} \frac{\beta^n}{n!}. \tag{2.1}$$

However, much of the analysis carries over to more general cases, though we do not always spell this out.

2a   The saddlepoint approximation

We impose the Poisson assumption (2.1) and define $A = A_1$. Then $\mathbb{E}\mathrm{e}^{\alpha A} = \mathrm{e}^{\kappa(\alpha)}$ where $\kappa(\alpha) = \beta(\widehat{B}[\alpha] - 1)$. The exponential family generated by A is given by

$$\mathbb{P}_\theta(A \in \mathrm{d}x) = \mathbb{E}\big[\mathrm{e}^{\theta A - \kappa(\theta)}; A \in \mathrm{d}x\big].$$

In particular,

$$\kappa_\theta(\alpha) = \log \mathbb{E}_\theta \mathrm{e}^{\alpha A} = \kappa(\alpha + \theta) - \kappa(\theta) = \beta_\theta\big(\widehat{B}_\theta[\alpha] - 1\big)$$

where $\beta_\theta = \beta\widehat{B}[\theta]$ and B_θ is the distribution given by

$$B_\theta(\mathrm{d}x) = \frac{\mathrm{e}^{\theta x}}{\widehat{B}[\theta]} B(\mathrm{d}x).$$

This shows that the $\mathbb{P}_\theta$-distribution of A has a similar compound Poisson form as the $\mathbb{P}$-distribution, only with β replaced by β_θ and B by B_θ.

The analysis largely follows Example XIII.1.1. For a given x, we define the saddlepoint $\theta = \theta(x)$ by $\mathbb{E}_\theta A = x$, i.e. $\kappa_\theta'(0) = \kappa'(\theta) = x$.

Proposition 2.1 *Assume that* $\lim_{r \uparrow r^*} \widehat{B}''[r] = \infty$,

$$\lim_{r \uparrow r^*} \frac{\widehat{B}'''[r]}{\big(\widehat{B}''[r]\big)^{3/2}} = 0, \tag{2.2}$$

where $r^* = \sup\{r : \widehat{B}[r] < \infty\}$. *Then as* $x \to \infty$,

$$\mathbb{P}(A > x) \sim \frac{\mathrm{e}^{-\theta x + \kappa(\theta)}}{\theta\sqrt{2\pi\,\beta\,\widehat{B}''[\theta]}}. \tag{2.3}$$

Proof. Since $\mathbb{E}_\theta A = x$, $\mathbb{V}ar_\theta(A) = \kappa''(\theta) = \beta \widehat{B}''[\theta]$, (2.2) implies that the limiting $\mathbb{P}_\theta$-distribution of $(A - x)/\sqrt{\beta \widehat{B}''[\theta]}$ is standard normal. Hence

$$
\begin{aligned}
\mathbb{P}(A > x) &= \mathbb{E}_\theta\left[e^{-\theta A + \kappa(\theta)}; A > x\right] = e^{-\theta x + \kappa(\theta)} \mathbb{E}_\theta\left[e^{-\theta(A-x)}; A > x\right] \\
&\sim e^{-\theta x + \kappa(\theta)} \int_0^\infty e^{-\theta\sqrt{\beta \widehat{B}''[\theta]}y} \frac{1}{\sqrt{2\pi}} e^{-y^2/2}\, dy \\
&= \frac{e^{-\theta x + \kappa(\theta)}}{\theta\sqrt{2\pi\,\beta \widehat{B}''[\theta]}} \int_0^\infty e^{-z} e^{-z^2/(2\theta^2 \beta \widehat{B}''[\theta])}\, dz \\
&\sim \frac{e^{-\theta x + \kappa(\theta)}}{\theta\sqrt{2\pi\,\beta \widehat{B}''[\theta]}} \int_0^\infty e^{-z}\, dz = \frac{e^{-\theta x + \kappa(\theta)}}{\theta\sqrt{2\pi\,\beta \widehat{B}''[\theta]}}.
\end{aligned}
$$

$\square$

It should be noted that the heavy-tailed asymptotics is much more straightforward. In fact, just the same dominated convergence argument as in the proof of Theorem X.2.1 yields:

Proposition 2.2 *If B is subexponential and $\mathbb{E}z^N < \infty$ for some $z > 1$, then $\mathbb{P}(A > x) \sim \mathbb{E}N\, \overline{B}(x)$.*

Notes and references Proposition 2.1 goes all the way back to Esscher [358], and (2.3) is often referred to as the *Esscher approximation*.

The present proof is somewhat heuristical in the CLT steps. For a rigorous proof, some regularity of the density $b(x)$ of B is required. In particular, either of the following is sufficient:

A. b is gamma-like, i.e. bounded with $b(x) \sim y\, c_1 x^{\alpha-1} e^{-\delta x}$.

B. b is log-concave, or, more generally, $b(x) = q(x)e^{-h(x)}$, where $q(x)$ is bounded away from 0 and ∞ and $h(x)$ is convex on an interval of the form $[x_0, x^*)$ where $x^* = \sup\{x : b(x) > 0\}$. Furthermore $\int_0^\infty b(x)^\zeta dx < \infty$ for some $\zeta \in (1, 2)$.

For example, **A** covers the exponential distribution and phase-type distributions, **B** covers distributions with finite support or with a density not too far from e^{-x^α} with $\alpha > 1$. For details, see Embrechts *et al.* [347], Jensen [506] and references therein. For higher-order extensions of the asymptotic behavior in Proposition 2.2 see Albrecher, Hipp & Kortschak [29] and references therein. It is also shown there that the folklore use of the shifted asymptotics $\mathbb{P}(A > x) \approx \beta\, \overline{B}(x - \beta\mu_B)$ for Poisson N can be rigorously justified in the sense that, under mild additional assumptions on B, the shifting $\mathbb{P}(A > x) \sim \mathbb{E}N\, \overline{B}\big(x - \mu_B(\mathbb{E}(N^2)/\mathbb{E}(N) - 1)\big)$ improves the asymptotic accuracy of Proposition 2.2 by an order of magnitude.

Asymptotic results for situations where the tail behavior of N determines the tail behavior of A are given in Asmussen, Klüppelberg & Sigman [87] and Robert & Segers [742].

2b   The NP approximation

In many cases, the distribution of A is approximately normal. For example, under the Poisson assumption (2.1), it holds that $\mathbb{E}A = \beta\mu_B$, $\mathrm{Var}(A) = \beta\mu_B^{(2)}$ and that $(A - \beta\mu_B)/(\beta\mu_B^{(2)})^{1/2}$ has a limiting standard normal distribution as $\beta \to \infty$, leading to

$$\mathbb{P}(A > x) \approx 1 - \Phi\left(\frac{x - \beta\mu_B}{\sqrt{\beta\mu_B^{(2)}}}\right). \tag{2.4}$$

The result to be surveyed below improves upon this and related approximations by taking into account second order terms from the Edgeworth expansion.

Remark 2.3 A word of warning should be said right away: the CLT (and the Edgeworth expansion) can only be expected to provide a good fit in the center of the distribution. Thus, it is quite questionable to use (2.4) and related results for the case of main interest, large x.   □

The (first order) *Edgeworth expansion* states that if the characteristic function $\widehat{g}(u) = \mathbb{E}e^{iuY}$ of a r.v. Y satisfies

$$\widehat{g}(u) \approx e^{-u^2/2}(1 + i\delta u^3), \tag{2.5}$$

where δ is a small parameter, then

$$\mathbb{P}(Y \leq y) \approx \Phi(y) - \delta(1 - y^2)\varphi(y). \tag{2.6}$$

Note as a further warning that the r.h.s. of (2.6) may be negative and is not necessarily an increasing function of y for $|y|$ large.

Heuristically, (2.6) is obtained by noting that by Fourier inversion, the density of Y is

$$\begin{aligned}
g(y) &= \frac{1}{2\pi}\int_{-\infty}^{\infty} e^{-iuy}\widehat{g}(u)\,\mathrm{d}u \\
&\approx \frac{1}{2\pi}\int_{-\infty}^{\infty} e^{-iuy}e^{-u^2/2}(1 + i\delta u^3)\,\mathrm{d}u \\
&= \varphi(y) - \delta(y^3 - 3y)\varphi(y),
\end{aligned}$$

and from this (2.6) follows by integration.

In concrete examples, the CLT for $Y = Y_\delta$ is usually derived via expanding the ch.f. as

$$\widehat{g}(u) = \mathbb{E}e^{iuY} = \exp\left\{iu\kappa_1 - \frac{u^2}{2}\kappa_2 - i\frac{u^3}{3}\kappa_3 + \frac{u^4}{4!}\kappa_4 + \cdots\right\}$$

where $\kappa_1, \kappa_2, \ldots$ are the cumulants; in particular,

$$\kappa_1 = \mathbb{E}Y, \quad \kappa_2 = \mathrm{Var}(Y), \quad \kappa_3 = \mathbb{E}(Y - \mathbb{E}Y)^3.$$

Thus if $\mathbb{E}Y = 0$, $\mathrm{Var}(Y) = 1$ as above, one needs to show that $\kappa_3, \kappa_4, \ldots$ are small. If this holds, one expects the u^3 term to dominate the terms of order $u^4, u^5, \ldots$ so that

$$\widehat{g}(u) \approx \exp\left\{-\frac{u^2}{2} - \mathrm{i}\frac{u^3}{3}\kappa_3\right\} \approx \exp\left\{-\frac{u^2}{2}\right\}\left(1 - \mathrm{i}\frac{u^3}{6}\kappa_3\right)$$

so that we should take $\delta = -\kappa_3/6$ in (2.6).

Rather than with the tail probabilities $\mathbb{P}(A > x)$, the NP (normal power) approximation deals with the quantile $a_{1-\epsilon}$, defined as the solution of $\mathbb{P}(A \le y_{1-\epsilon}) = 1 - \epsilon$. A particular case is $a_{.995}$, which is often used as the VaR (*Value at Risk*) for risk management purposes.

Let $Y = (A - \mathbb{E}A)/\sqrt{\mathrm{Var}(A)}$ and let $y_{1-\epsilon}, z_{1-\epsilon}$ be the $(1-\epsilon)$-quantile in the distribution of Y, resp. the standard normal distribution. If the distribution of Y is close to $N(0,1)$, $y_{1-\epsilon}$ should be close to $z_{1-\epsilon}$ (cf., however, Remark 2.3!), and so as a first approximation we obtain

$$a_{1-\epsilon} = \mathbb{E}A + y_{1-\epsilon}\sqrt{\mathrm{Var}(A)} \approx \mathbb{E}A + z_{1-\epsilon}\sqrt{\mathrm{Var}(A)}. \tag{2.7}$$

A correction term may be computed from (2.6) by noting that the $\Phi(y)$ terms dominate the $\delta(1 - y^2)\varphi(y)$ term. This leads to

$$
\begin{aligned}
1 - \epsilon &\approx \Phi(y_{1-\epsilon}) - \delta(1 - y_{1-\epsilon}^2)\varphi(y_{1-\epsilon}) \\
&\approx \Phi(y_{1-\epsilon}) - \delta(1 - z_{1-\epsilon}^2)\varphi(z_{1-\epsilon}) \\
&\approx \Phi(z_{1-\epsilon}) + (y_{1-\epsilon} - z_{1-\epsilon})\varphi(z_{1-\epsilon}) - \delta(1 - z_{1-\epsilon}^2)\varphi(z_{1-\epsilon}) \\
&= 1 - \epsilon + (y_{1-\epsilon} - z_{1-\epsilon})\varphi(z_{1-\epsilon}) - \delta(1 - z_{1-\epsilon}^2)\varphi(z_{1-\epsilon})
\end{aligned}
$$

which combined with $\delta = -\mathbb{E}Y^3/6$ leads to

$$y_{1-\epsilon} = z_{1-\epsilon} + \frac{1}{6}(z_{1-\epsilon}^2 - 1)\mathbb{E}Y^3.$$

Using $Y = (A - \mathbb{E}A)/\sqrt{\mathrm{Var}(A)}$, this yields the NP approximation

$$a_{1-\epsilon} = \mathbb{E}A + z_{1-\epsilon}(\mathrm{Var}(A))^{1/2} + \frac{1}{6}(z_{1-\epsilon}^2 - 1)\frac{\mathbb{E}(A - \mathbb{E}A)^3}{\mathrm{Var}(A)^{3/2}}. \tag{2.8}$$

Under the Poisson assumption (2.1), the kth cumulant of A is $\beta\mu_B^{(k)}$ and so $\kappa_k = \beta\mu_B^{(2)}/(\beta\mu_B^{(k)})^{k/2}$. In particular, κ_3 is small for large β but dominates $\kappa_4, \kappa_5, \ldots$ as required. We can rewrite (2.8) as

$$a_{1-\epsilon} = \beta\mu_B + z_{1-\epsilon}\big(\beta\mu_B^{(2)}\big)^{1/2} + \frac{1}{6}(z_{1-\epsilon}^2 - 1)\frac{\mu_B^{(3)}}{\sqrt{\beta(\mu_B^{(2)})^3}}. \qquad (2.9)$$

Notes and references We have followed largely Sundt [820]. Another main reference is Daykin *et al.* [279]. Note, however, that [279] distinguishes between the NP and Edgeworth approximations.

2c Panjer's recursion

Consider $A = \sum_{i=1}^{N} U_i$, let $p_n = \mathbb{P}(N = n)$, and assume that there exist constants a, b such that

$$p_n = \left(a + \frac{b}{n}\right)p_{n-1}, \quad n = 1, 2, \ldots. \qquad (2.10)$$

For example, this holds with $a = 0$, $b = \beta$ for the Poisson distribution with rate β since

$$p_n = e^{-\beta}\frac{\beta^n}{n!} = \frac{\beta}{n}e^{-\beta}\frac{\beta^{n-1}}{(n-1)!} = \frac{\beta}{n}p_{n-1}.$$

Proposition 2.4 *Assume that B is concentrated on $\{0, 1, 2, \ldots\}$ and write $g_j = \mathbb{P}(U_i = j)$, $j = 0, 1, 2, \ldots$, $f_j = \mathbb{P}(A = j)$, $j = 0, 1, \ldots$. Then $f_0 = \sum_0^\infty g_0^n p_n$ and*

$$f_j = \frac{1}{1 - ag_0}\sum_{k=1}^{j}\left(a + b\frac{k}{j}\right)g_k f_{j-k}, \quad j = 1, 2, \ldots. \qquad (2.11)$$

In particular, if $g_0 = 0$, then

$$f_0 = p_0, \quad f_j = \sum_{k=1}^{j}\left(a + b\frac{k}{j}\right)g_k f_{j-k}, \quad j = 1, 2, \ldots. \qquad (2.12)$$

Remark 2.5 The crux of Proposition 2.4 is that the algorithm is much faster than the naive method, which would consist in noting that (in the case $g_0 = 0$)

$$f_j = \sum_{n=1}^{j} p_n g_j^{*n} \qquad (2.13)$$

where g^{*n} is the nth convolution power of g, and calculating the g_j^{*n} recursively by $g_j^{*1} = g_j$,

$$g_j^{*n} = \sum_{k=n-1}^{j-1} g_k^{*(n-1)} g_{j-k} . \qquad (2.14)$$

Namely, the complexity (number of arithmetic operations required) is $O(j^3)$ for (2.13), (2.14) but only $O(j^2)$ for Proposition 2.4. $\qquad \square$

Proof of Proposition 2.4. The expression for f_0 is obvious. By symmetry,

$$\mathbb{E}\left[a + b\frac{U_i}{j} \,\middle|\, \sum_{i=1}^{n} U_i = j\right] \qquad (2.15)$$

is independent of $i = 1, \ldots, n$. Since the sum over i is $na + b$, the value of (2.15) is therefore $a + b/n$. Hence by (2.10), (2.13) we get for $j > 0$ that

$$
\begin{aligned}
f_j &= \sum_{n=1}^{\infty} \left(a + \frac{b}{n}\right) p_{n-1} g_j^{*n} \\
&= \sum_{n=1}^{\infty} \mathbb{E}\left[a + b\frac{U_1}{j} \,\middle|\, \sum_{i=1}^{n} U_i = j\right] p_{n-1} g_j^{*n} \\
&= \sum_{n=1}^{\infty} \mathbb{E}\left[a + b\frac{U_1}{j} \,;\, \sum_{i=1}^{n} U_i = j\right] p_{n-1} \\
&= \sum_{n=1}^{\infty} \sum_{k=0}^{j} \left(a + b\frac{k}{j}\right) g_k g_{j-k}^{*(n-1)} p_{n-1} \\
&= \sum_{k=0}^{j} \left(a + b\frac{k}{j}\right) g_k \sum_{n=0}^{\infty} g_{j-k}^{*n} p_n = \sum_{k=0}^{j} \left(a + b\frac{k}{j}\right) g_k f_{j-k} \\
&= a g_0 f_j + \sum_{k=1}^{j} \left(a + b\frac{k}{j}\right) g_k f_{j-k} ,
\end{aligned}
$$

and (2.10) follows. $\qquad \square$

If the distribution B of the U_i is non-lattice, it is natural to use a discrete approximation. To this end, let $U_{i,+}^{(h)}$, $U_{i,-}^{(h)}$ be U_i rounded upwards, resp. downwards, to the nearest multiple of h and let $A_{\pm}^{(h)} = \sum_1^N U_{i,\pm}^{(h)}$. An obvious modification of Proposition 2.4 applies to evaluate the distribution $F_{\pm}^{(h)}$ of $A_{\pm}^{(h)}$, letting $f_{j,\pm}^{(h)} = \mathbb{P}(A_{\pm}^{(h)} = jh)$ and

$$
\begin{aligned}
g_{k,-}^{(h)} &= \mathbb{P}\big(U_{i,-}^{(h)} = kh\big) = B\big((k+1)h\big) - B(kh), \quad k = 0, 1, 2, \ldots, \\
g_{k,+}^{(h)} &= \mathbb{P}\big(U_{i,+}^{(h)} = kh\big) = B(kh) - B\big((k-1)h\big) = g_{k-1,-}, \quad k = 1, 2, \ldots
\end{aligned}
$$

Then the error on the tail probabilities (which can be taken arbitrarily small by choosing h small enough) can be evaluated by

$$
\sum_{j=\lfloor x/h \rfloor}^{\infty} f_{j,-}^{(h)} \;\le\; \mathbb{P}(A \ge x) \;\le\; \sum_{j=\lfloor x/h \rfloor}^{\infty} f_{j,+}^{(h)} .
$$

Further examples (and in fact the only ones, cf. Sundt & Jewell [821]) where (2.10) holds are the binomial distribution and the negative binomial (in particular, geometric) distribution. The geometric case is of particular importance because of the following result which immediately follows from combining Proposition 2.4 and the Pollaczeck-Khinchine representation:

Corollary 2.6 *Consider a compound Poisson risk process with Poisson rate β and claim size distribution B. Then for any $h > 0$, the ruin probability $\psi(u)$ satisfies*

$$
\sum_{j=\lfloor u/h \rfloor}^{\infty} f_{j,-}^{(h)} \;\le\; \psi(u) \;\le\; \sum_{j=\lfloor u/h \rfloor}^{\infty} f_{j,+}^{(h)}, \tag{2.16}
$$

where $f_{j,+}^{(h)}, f_{j,-}^{(h)}$ are given by the recursions

$$
f_{j,+}^{(h)} = \rho \sum_{k=1}^{j} g_k^{(h)} f_{j-k,+}^{(h)}, \quad j = 1, 2, \ldots
$$

$$
f_{j,-}^{(h)} = \frac{\rho}{1 - a g_{0,-}^{(h)}} \sum_{k=1}^{j} g_{k,-}^{(h)} f_{j-k,-}^{(h)}, \quad j = 1, 2, \ldots
$$

starting from $f_{0,+}^{(h)} = 1 - \rho$, $f_{0,-}^{(h)} = (1-\rho)/(1 - \rho g_{0,-}^{(h)})$ and using

$$
\begin{aligned}
g_{k,-}^{(h)} &= B_0\big((k+1)h\big) - B_0(kh) = \frac{1}{\mu_B} \int_{kh}^{(k+1)h} \overline{B}(x)\,\mathrm{d}x, \quad k = 0, 1, 2, \ldots, \\
g_{k,+}^{(h)} &= B_0(kh) - B_0\big((k-1)h\big) = g_{k-1,-}^{(h)}, \quad k = 1, 2, \ldots.
\end{aligned}
$$

Remark 2.7 It is clear that the quotient of the upper and the lower bound in (2.16) tends to 1 for $u \to \infty$ if U_i is long-tailed (i.e. $\overline{B}(x - y)/\overline{B}(x) \to 1$ as $x \to \infty$ for all y, cf. Chapter X). Correspondingly, in numerical implementations one typically observes that the difference between the upper and lower bound in (2.16) gets larger for increasing u, but for long-tailed (in particular subexponential) U_i again tends to zero for still larger u.   □

Notes and references The literature on recursive algorithms related to Panjer's recursion is extensive, see e.g. Dickson [307] and references therein. Recursion formulas for counting distributions that are much more general than the class defined by (2.10) have been studied in the literature. A natural and very general class seems to be counting distributions that satisfy a finite-order homogeneous recursion with polynomial coefficients, see e.g. Wang & Sobrero [873]. For a survey that also covers multivariate extensions, see Sundt & Vernic [824]. Gerhold, Schmock & Warnung [416] provide an improved recursion algorithm; see also Hipp [467] for a speed-up from order $O(j^2)$ to order $O(j)$ for phase-type claim size distributions. In recent years, due to the increasing available computer power the emphasis is gradually shifting towards direct numerical inversion of the moment-generating function of the aggregate claim size. In the context of discrete claim size distributions, Fast Fourier Transform techniques can be quite powerful (see Grübel & Hermesmeier [439, 440] for details and Embrechts & Frei [344] for a recent comparison).

2d   The distribution of dependent sums

Whereas for the results in the previous subsections the independence assumption for the summands was crucial, in practice one will often face situations where information is needed about the tails of sums of dependent random variables. Clearly there are infinitely many possible dependence structures for a fixed set of marginal distributions and one cannot expect a complete picture of how dependence affects the behavior of the distribution tail. Nevertheless certain patterns occur and for the tail behavior it seems natural that only the dependence of the summands in the tail is important. In the sequel we will state some results in this direction (mainly) for the sum of two identically distributed subexponential r.v.'s to illustrate the challenges that occur when dependence enters. For more general results see the references in the Notes at the end of the section.

Let us consider the sum $X_1 + X_2$ of two identically distributed subexponential random variables each with distribution function B. By definition, if X_1 and X_2 are independent, then $\mathbb{P}(X_1 + X_2 > x) \sim 2\overline{B}(x)$ as $x \to \infty$ and a natural question is under which assumptions on the dependence structure of X_1 and X_2 and on B the same asymptotic relation holds true with dependence.

A first rough description of tail dependence between X_1 and X_2 is the so-called *(upper) tail dependence coefficient*

$$\lambda = \lim_{v \to \infty} \mathbb{P}(X_2 > v \,|\, X_1 > v).$$

If $\lambda = 0$, then X_1 and X_2 are called tail-independent. The following simple result extends Proposition X.1.1(a).

Lemma 2.8 $\mathbb{P}(\max(X_1, X_2) > x) \sim (2 - \lambda)\overline{B}(x)$.

Proof.

$$\begin{aligned}
\mathbb{P}(\max(X_1, X_2) > x) &= \mathbb{P}(X_1 > x) + \mathbb{P}(X_2 > x) - \mathbb{P}(X_1 > x, X_2 > x) \\
&= \overline{B}(x) + \overline{B}(x) - \overline{B}(x)\,\mathbb{P}(X_2 > x | X_1 > x).
\end{aligned}$$

$\square$

Let us first collect some results for regularly varying marginals, a case that is quite well understood.

Regularly varying marginal distributions

Proposition 2.9 *Let* $\overline{B}(x) \sim L(x)x^{-\alpha}$ *with* $\alpha > 0$. *Then*

$$\limsup_{x \to \infty} \frac{\mathbb{P}(X_1 + X_2 > x)}{\overline{B}(x)} \leq \begin{cases} \left(\lambda^{\frac{1}{\alpha+1}} + (2 - 2\lambda)^{\frac{1}{\alpha+1}}\right)^{\alpha+1}, & 0 \leq \lambda \leq \frac{2}{3}, \\ 2^{\alpha}(2 - \lambda), & \frac{2}{3} < \lambda \leq 1. \end{cases}$$

Proof. Analogously to the proof of Proposition X.1.4, for any $0 < \delta < 1/2$ we have

$$\mathbb{P}(X_1 + X_2 > x) \leq \mathbb{P}\Big(\{X_1 > (1-\delta)x\} \cup \{X_2 > (1-\delta)x\} \cup (\{X_1 > \delta x\} \cap \{X_2 > \delta x\})\Big)$$

$$\leq 2\overline{B}((1-\delta)x) + \mathbb{P}(X_1 > \delta x, X_2 > \delta x) - 2\mathbb{P}(X_1 > (1-\delta)x, X_2 > (1-\delta)x)$$

so that

$$\begin{aligned}
\limsup_{x \to \infty} \frac{\mathbb{P}(X_1 + X_2 > x)}{\overline{B}(x)} &\leq \limsup_{x \to \infty}\left((2 - 2\lambda)\frac{\overline{B}((1-\delta)x)}{\overline{B}(x)} + \lambda\frac{\overline{B}(\delta x)}{\overline{B}(x)}\right) \\
&= \frac{2 - 2\lambda}{(1-\delta)^{\alpha}} + \frac{\lambda}{\delta^{\alpha}}.
\end{aligned}$$

Within the defined range of δ, this upper bound is minimized for

$$\delta^* = \begin{cases} \dfrac{1}{1 + \left(\frac{2}{\lambda} - 2\right)^{\frac{1}{\alpha+1}}}, & 0 \leq \lambda \leq 2/3 \\ 1/2, & 2/3 < \lambda \leq 1, \end{cases}$$

which yields the result. □

This upper bound is sharp for both independence and comonotone dependence (in the latter case,[3] $\mathbb{P}(X_1 + X_2 > x) \sim 2^\alpha \overline{B}(x)$).

A combination of Lemma 2.8 and Proposition 2.9 immediately shows

Corollary 2.10 *If B has a regularly varying tail and $\lambda = 0$, then $\mathbb{P}(X_1 + X_2 > x) \sim 2\overline{B}(x)$ as $x \to \infty$.*

Hence, for regularly varying distributions, tail independence is already a sufficient criterion to guarantee that the tail of the dependent sum behaves asymptotically as if X_1 and X_2 were independent.

An important and natural subclass of distributions with regularly varying marginals are the ones with multivariate regular variation (for consistency we only state the bivariate case, although the extension to n dimensions is obvious). A vector $\mathbf{X} = (X_1, X_2)$ is regularly varying with index $-\alpha < 0$, if there exists a probability measure S on $\mathbb{S}^1_+$ (the unit sphere in $\mathbb{R}^2$ with respect to the Euclidean norm $|\cdot|$ restricted to the first quadrant) and a function $b(x) \to \infty$ such that

$$b^{-1}(x)\,\mathbb{P}\left(\left(\frac{|\mathbf{X}|}{x}, \frac{\mathbf{X}}{|\mathbf{X}|}\right) \in \cdot\right) \overset{\mathscr{D}}{\to} a\,\nu_\alpha \times S, \qquad (2.17)$$

in the space of positive Radon measures on $\left((\epsilon, \infty] \times \mathbb{S}^1_+\right)$ for all $\epsilon > 0$, where $a > 0$ and $\nu_\alpha(t, \infty] = t^{-\alpha}$, $(t > 0, \alpha > 0)$ (see e.g. Resnick [737]). S is often referred to as the *spectral measure* of $\mathbf{X}$.

The above implies in particular that on every ray from $(0,0)$ into the positive quadrant (the direction of which is governed by S), we have a regularly varying tail with index $-\alpha$. Moreover, the tail of $X_1 + X_2$ is also regularly varying with the same index.

For this specific dependence structure, the asymptotic behavior of the sum can be given explicitly in terms of the spectral measure.

Proposition 2.11 *Assume that $\mathbf{X} = (X_1, X_2)$ is exchangable and regularly varying with index $-\alpha < 0$ and spectral measure S. Then*

$$\mathbb{P}(X_1 + X_2 > x) \sim 2\,\overline{B}(x)\,\frac{\int_0^{\pi/2}(\cos\varphi + \sin\varphi)^\alpha\,S(\mathrm{d}\varphi)}{\int_0^{\pi/2}(\cos^\alpha\varphi + \sin^\alpha\varphi)\,S(\mathrm{d}\varphi)}.$$

---

[3]For $\alpha < 1$ (i.e. infinite mean) this also shows that comonotone dependence does not necessarily provide an upper bound for the tail asymptotics of all possible dependence structures with fixed marginals! Intuitively, if the marginal distribution tail is heavy enough, then the two random sources for a possibility of a large sum caused by one of the summands outweighs the effect of summing two large components from one random source.

Proof. Consider in (2.17) the events $|\mathbf{X}|/x > t$ for $t = \frac{1}{\cos\varphi + \sin\varphi}$ and $t = \frac{1}{\cos\varphi}$, where $\varphi \in [0, \pi/2]$ denotes the angle corresponding to $\mathbf{X}/|\mathbf{X}|$. We then obtain

$$b^{-1}(x)\,\mathbb{P}(X_1 + X_2 > x) \;\to\; a\int_0^{\pi/2} (\cos\varphi + \sin\varphi)^\alpha\, S(\mathrm{d}\varphi)$$

and

$$b^{-1}(x)\,\mathbb{P}(X_1 > x) \;\to\; a\int_0^{\pi/2} \cos^\alpha\varphi\, S(\mathrm{d}\varphi),$$

so that the result follows from $S(\mathrm{d}\varphi) = S(\mathrm{d}(\pi/2 - \varphi))$. $\qquad\square$

It occurs in a number of situations that the risks X_i are independent, but they need to be added with some weights that are not independent. Here is a result the proof of which can be found in Goovaerts *et al.* [424].

Proposition 2.12 *Assume that $X_1, \ldots, X_n$ are i.i.d. r.v.'s with regularly varying tail $\overline{B}(x) \sim L(x)\,x^{-\alpha}$ for some $\alpha > 0$ and let $\theta_1, \ldots, \theta_n$ be dependent non-negative r.v.'s, independent of $X_1, \ldots, X_n$. If there exists some $\delta > 0$ s.t. $\mathbb{E}(\theta_k^{\alpha+\delta}) < \infty$ for $1 \le k \le n$, then*

$$\mathbb{P}\Big(\max_{1\le m\le n}\sum_{k=1}^m \theta_k X_k > x\Big) \sim \mathbb{P}\Big(\sum_{k=1}^n \theta_k X_k > x\Big) \sim \overline{B}(x)\sum_{k=1}^n \mathbb{E}(\theta_k^\alpha).$$

If either

$$0 < \alpha < 1, \quad \sum_{k=1}^\infty \mathbb{E}(\theta_k^{\alpha+\delta}) < \infty, \quad \sum_{k=1}^\infty \mathbb{E}\theta_k^{\alpha-\delta} < \infty \quad \textit{for some } \delta > 0$$

or

$$\alpha \ge 1, \quad \sum_{k=1}^\infty (\mathbb{E}(\theta_k^{\alpha+\delta}))^{1/(\alpha+\delta)} < \infty, \quad \sum_{k=1}^\infty [\mathbb{E}\theta_k^{\alpha-\delta}]^{1/(\alpha+\delta)} < \infty \quad \textit{for some } \delta > 0,$$

then

$$\mathbb{P}\Big(\max_{1\le n\le\infty}\sum_{k=1}^n \theta_k X_k > x\Big) \sim \mathbb{P}\Big(\sum_{k=1}^\infty \theta_k (X_k)^+ > x\Big) \sim \overline{B}(x)\sum_{k=1}^\infty \mathbb{E}(\theta_k^\alpha).$$

Example 2.13 Recall the discrete time risk model with stochastic investment of Section VIII.5. If we choose $\theta_k = A_1^{-1}\cdots A_k^{-1}$ and $X_k = B_k$, then Proposition 2.12 applies for the case of regularly varying insurance risk B_k (with index $-\alpha$). The conditions of the Proposition translate into $\mathbb{E}(A_1^{-\alpha+\delta}) < \infty$

and $\mathbb{E}(A_1^{-\alpha \pm \delta}) < 1$, respectively, for some $\delta > 0$. Under these assumptions, Proposition 2.12 gives the finite-time ruin probability

$$\psi(u, n) = \mathbb{P}(\tau(u) \leq n) \sim \overline{B}(x) \frac{\mathbb{E}(A_1^{-\alpha})\Big(1 - (\mathbb{E}(A_1^{-\alpha}))^n\Big)}{1 - \mathbb{E}(A_1^{-\alpha})}, \quad u \to \infty$$

and the infinite-time ruin probability

$$\psi(u) = \mathbb{P}(\tau(u) < \infty) \sim \overline{B}(x) \frac{\mathbb{E}(A_1^{-\alpha})}{1 - \mathbb{E}(A_1^{-\alpha})}, \quad u \to \infty,$$

which refines Theorem VIII.5.8 for this particular case. $\qquad \square$

Other subexponential marginal distributions

From the proof of Proposition 2.9, it becomes clear that Corollary 2.10 also holds true for any $B \in \mathscr{S}$ with heavier tail than regularly varying. On the other hand, in general the marginal tails cannot be much lighter than regularly varying in order to dominate the 'dependence effect' in the tail of the sum given $\lambda = 0$, as the following result shows.

Proposition 2.14 *If the mean excess function $e(x)$ is self-neglecting, i.e.*

$$\lim_{x \to \infty} \frac{e\big(x + a\,e(x)\big)}{e(x)} = 1 \quad \forall\, a \geq 0, \tag{2.18}$$

and if

$$\inf_{a > 0} \liminf_{x \to \infty} \mathbb{P}\big(X_2 > a\,e(x) \,\big|\, X_1 > x\big) > 0, \tag{2.19}$$

then

$$\liminf_{x \to \infty} \frac{\mathbb{P}(X_1 + X_2 > x)}{\overline{B}(x)} = \infty.$$

Proof. From Proposition X.1.18 we know that the self-neglecting property (2.18) implies

$$\lim_{x \to \infty} \frac{\overline{B}\big(x + a\,e(x)\big)}{\overline{B}(x)} = e^{-a}$$

and we have

$$\frac{\overline{B}(x)}{\overline{B}\big(x - a\,e(x)\big)} \sim \frac{\overline{B}\big(x + a\,e(x)\big)}{\overline{B}\big(x + a\,e(x) - a\,e(x + a\,e(x))\big)}$$

$$\sim \frac{\overline{B}\big(x + a\,e(x)\big)}{\overline{B}(x)}.$$

Together with (2.19) this gives

$$
\begin{aligned}
\mathbb{P}(X_1 + X_2 > x) \;&\geq\; \mathbb{P}\big(X_1 > x - a\,e(x), X_2 > a\,e(x)\big) \\
&=\; \mathbb{P}\big(X_1 > x - a\,e(x)\big)\,\mathbb{P}\big(X_2 > a\,e(x)\,\big|\,X_1 > x - a\,e(x)\big) \\
&\sim\; \mathbb{P}\big(X_1 > x - a\,e(x)\big)\,\mathbb{P}\big(X_2 > a\,e(x)\,\big|\,X_1 > x\big) \\
&\geq\; \varepsilon\,\mathbb{P}\big(X_1 > x - a\,e(x)\big) \\
&\sim\; \varepsilon\,\mathbb{P}(X_1 > x)\,\mathrm{e}^{a}
\end{aligned}
$$

for some $\varepsilon > 0$ and any $a > 0$. Hence

$$
\liminf_{x \to \infty} \frac{\mathbb{P}(X_1 + X_2 > x)}{\overline{B}(x)} \;\geq\; \varepsilon\,\mathrm{e}^{a}
$$

and the latter is unbounded for $a \to \infty$. $\square$

Remark 2.15 Recall from Chapter X that condition (2.18) is satisfied for Weibull and lognormal distributions (more generally, for all subexponential distributions which lie in the maximum domain of attraction of the Gumbel distribution, cf. X.6b). A sufficient condition for (2.19) to hold is

$$
\liminf_{x \to \infty} \mathbb{P}\big(X_2 > e^*(x)\,\big|\,X_1 > x\big) \;>\; 0
$$

for any $e^*(x)$ with $e^*(x)/e(x) \to \infty$. One can show that for all B that satisfy (2.18) there exists a dependence structure such that (2.19) is satisfied (cf. [13]). $\square$

On the other hand, for a particular given dependence structure the tail of the sum may well be asymptotically equivalent to the one of the independent sum. This is illustrated by the following example with lognormal marginals and a Gaussian copula (which is tail independent).

Proposition 2.16 *Let Y_1, Y_2 be bivariate normal with the same mean μ, the same variance σ^2 and covariance $\rho \in [-1, 1)$. Then, for $X_1 = \mathrm{e}^{Y_1}$ and $X_2 = \mathrm{e}^{Y_2}$ one has*

$$
\mathbb{P}(X_1 + X_2 > x) \sim 2\,\mathbb{P}(X_1 > x) \;\sim\; \frac{\sqrt{2/\pi}}{\sigma \log x}\,\exp\big\{-(\log x - \mu)^2/2\sigma^2\big\} .
$$

Proof. Rather than giving a rigorous technical proof (for which we refer to Asmussen & Rojas-Nandaypa [96]), we give here just a short heuristical argument supporting the result. Take $\mu = 0$, $\sigma^2 = 1$, $\rho > 0$ for simplicity. Then we can write

$$
Y_1 \;=\; U + V_1\,, \quad Y_2 \;=\; U + V_2\,,
$$

where U, V_1, V_2 are independent univariate Gaussian with mean zero and variances a^2, b^2, b^2, respectively, where $a^2 + b^2 = 1$, $a^2 = \rho$. Given $U = u$, X_1 and X_2 are independent lognormals with log-variance b^2, so by subexponential limit theory

$$
\begin{aligned}
\mathbb{P}\big(X_1 + X_2 > x \,\big|\, U = u\big) &= \mathbb{P}(\mathrm{e}^{V_1} + \mathrm{e}^{V_2} > x\mathrm{e}^{-u}) \\
&\sim \frac{\sqrt{2/\pi}}{b(\log x - u)} \exp\left\{-(\log x - u)^2/2b^2\right\}.
\end{aligned}
$$

We make the guess

$$
\mathbb{P}(X_1 + X_2 > x) \approx \max_u \frac{1}{a\sqrt{2\pi}} \mathrm{e}^{-u^2/2a^2} \mathbb{P}\big(X_1 + X_2 > x \,\big|\, U = u\big) \qquad (2.20)
$$

and ignore everything not in the exponent and constants. Then we have to find the u minimizing

$$
\frac{u^2}{2a^2} - \frac{u\log x}{b^2} + \frac{u^2}{2b^2}
$$

which (using $a^2 + b^2 = 1$) is easily seen to be $u = a^2 \log x$. Substituting back in (2.20), we get

$$
\begin{aligned}
\mathbb{P}(X_1 + X_2 > x) &\approx \exp\left\{-a^4 \log^2 x/2a^2 - (1 - a^2)^2 \log^2 x/2b^2\right\} \\
&= \exp\left\{-\log^2 x/2\right\} \qquad (2.21)
\end{aligned}
$$

in agreement with the claimed assertion (here $\approx$ is used to indicate asymptotics at a rough level, i.e. rougher than $\sim$ or even logarithmic asymptotics as used in large deviations theory). Note that the argument contains some information on how $X_1 + X_2$ exceeds x: U must be approximately $u = a^2 \log x = \rho \log x$ and either V_1 or V_2 but not both large. Translated back to X_1, X_2, this means that one is larger than x and the other of order $\mathrm{e}^u = x^\rho$. □

We finish this section with a fairly general result of Foss & Richards [367] about conditions under which the tail of the dependent sum asymptotically behaves as the tail for the independent sum. Note that a consequence of Proposition X.1.5 is that for $B \in \mathscr{S}$ there always exists a monotone function $h(x) \nearrow \infty$ with

$$
\lim_{x \to \infty} \overline{B}\big(x - h(x)\big)/\overline{B}(x) = 1. \qquad (2.22)
$$

Theorem 2.17 *Let $B \in \mathscr{S}$. Assume that $X_1, X_2, \ldots$ are positive r.v.'s with c.d.f. B_i in a probability space $(\Omega, \mathscr{F}, \mathbb{P})$ such that for each i, $\overline{B}_i(x) \sim c_i \overline{B}(x)$ (with at least one $c_i \neq 0$ and $\exists c > 0$ and $x_0 > 0$ s.t. $\overline{B}_i(x) \leq c\overline{B}(x)$ for all $x > x_0$). Further*

(i) $X_1, X_2, \ldots$ are conditionally independent given a σ-algebra $\mathcal{G} \subset \mathcal{F}$

(ii) for each i there exists a non-decreasing function $r(x)$ and an increasing collection of sets $J_i(x) \in \mathcal{G}$ with $J_i(x) \to \Omega$ as $x \to \infty$ such that

$$\mathbb{P}(X_i > x \,|\, \mathcal{G}) \, I\big(J_i(x)\big) \;\le\; r(x)\overline{B}(x) I\big(J_i(x)\big) \quad a.s.$$

and such that for a function $h(x)$ that satisfies (2.22), uniformly in i,

1. $\mathbb{P}\big(\overline{J_i}(h(x))\big) = \mathrm{o}\big(\overline{B}(x)\big)$,

2. $r(x)\overline{B}\big(h(x)\big) = \mathrm{o}(1)$,

3. $r(x) \int_{h(x)}^{x-h(x)} \overline{B}(x - y) B(\mathrm{d}y) = \mathrm{o}\big(\overline{B}(x)\big)$,

as $x \to \infty$.

Then for all $n \in \mathbb{N}$, $\mathbb{P}(X_1 + \cdots + X_n > x) \sim n\overline{B}(x)$.

Proof. Consider first $X_1 + X_2$. We have the inequalities

$$\mathbb{P}(X_1 + X_2 > x) \;\le\; \mathbb{P}\big(X_1 > x - h(x)\big) + \mathbb{P}\big(X_2 > x - h(x)\big) \\ + \mathbb{P}\big(h(x) \le X_1 \le x - h(x), X_2 > x - X_1\big)$$

and

$$\mathbb{P}(X_1 + X_2 > x) \;\ge\; \mathbb{P}(X_1 > x) + \mathbb{P}(X_2 > x) - \mathbb{P}(X_1 > x, X_2 > x).$$

Now, if Y is another r.v. with c.d.f. B, independent of X_1, X_2,

$$\mathbb{P}\big(h(x) \le X_1 \le x - h(x), X_2 > x - X_1\big)$$
$$= \mathbb{E}\Big[\mathbb{P}\big(h(x) \le X_1 \le x - h(x), X_2 > x - X_1\big) \,|\, \mathcal{G}\Big]$$
$$= \mathbb{E}\Big(\int_{h(x)}^{x-h(x)} \mathbb{P}(X_1 \in \mathrm{d}y \,|\, \mathcal{G})\, \mathbb{P}(X_2 > x - y \,|\, \mathcal{G})\big[I\big(J_2(x - y)\big) + I\big(\overline{J}_2(x - y)\big)\big]\Big)$$
$$\le r(x)\mathbb{E}\Big(\int_{h(x)}^{x-h(x)} \mathbb{P}(X_1 \in \mathrm{d}y \,|\, \mathcal{G})\, \mathbb{P}(Y > x - y)\Big) + \mathbb{E}\Big(I\big(\overline{J}_2(h(x))\big)\Big)$$
$$= r(x) \int_{h(x)}^{x-h(x)} \mathbb{P}(X_1 \in \mathrm{d}y)\overline{B}(x - y) + \mathrm{o}\big(\overline{B}(x)\big) \;=\; \mathrm{o}\big(\overline{B}(x)\big).$$

At the same time

$$\mathbb{P}(X_1 > x, X_2 > x)$$

$$= \mathbb{E}\Big(\big(\mathbb{P}(X_1 > x, X_2 > x \,|\, \mathscr{G}\big) \big(I(J_2(x)) + I(\overline{J}_2(x))\big)\Big)$$

$$\leq \mathbb{E}\Big(\mathbb{P}(X_1 > x \,|\, \mathscr{G})\,\mathbb{P}(X_2 > x \,|\, \mathscr{G})I(J_2(x))\Big) \;+\; \mathbb{E}I\big(\overline{J}_2(x)\big)$$

$$\leq r(x)\overline{B}(x)\mathbb{P}(X_1 > x) + o\big(\overline{B}(x)\big) \;=\; o\big(\overline{B}(x)\big).$$

Consequently, $\mathbb{P}(X_1 + X_2 > x) \sim \mathbb{P}(X_1 > x) + \mathbb{P}(X_2 > x)$. Since w.l.o.g. $c_1 > 0$,

$$\mathbb{P}(X_1 + X_2 > x) \sim (c_1 + c_2)\overline{B}(x).$$

The result for general n now follows by induction.   □

The following extension of Lemma X.1.8 is proved in [367].

Lemma 2.18 *Under the conditions of Theorem 2.17, for any $\epsilon > 0$ there exist $V(\epsilon) > 0$ and $x_0 = x_0(\epsilon)$ such that for any $x > x_0$ and $n \geq 1$*

$$\mathbb{P}(X_1 + \cdots + X_n > x) \;\leq\; V(\epsilon)(1 + \epsilon)^n \overline{B}(x).$$

This result and Theorem 2.17 together with dominated convergence now gives the following extension of Lemma X.2.2.

Proposition 2.19 *Let K be an independent integer-valued r.v. with $\mathbb{E}z^K < \infty$ for some $z > 1$. Under the assumptions of Theorem 2.17 one then has $\mathbb{P}(X_1 + \cdots + X_K > x) \sim \mathbb{E}(\sum_{i=1}^{K} c_i)\,\overline{B}(x)$.*

In several applications in risk theory, conditionally independent r.v. will be an appropriate description of the dependence structure. However, the challenge in the application of the above result is to identify a σ-algebra $\mathscr{G}$ and a corresponding function $h(x)$ that satisfies the assumptions of Theorem 2.17. See [367] for some worked out examples.

Notes and references   Some general non-asymptotic bounds on $\mathbb{P}(X_1 + \cdots + X_n > x)$ are derived in Denuit, Genest & Marceau [293], Cossette, Denuit & Marceau [257], Mesfioui & Quessy [635] and Embrechts & Puccetti [350] (see also [351] for bounds on functions of multivariate risks). Worst-case scenarios are also studied by Rüschendorf [759]. Parts of the material in this section is from Albrecher, Asmussen & Kortschak [13]. It is of course also possible to represent results on the asymptotic behavior of the sum through conditions on the underlying copula. Quite explicit results for Archimedean copulas can be found in Alink, Loewe & Wuethrich [42], see also [43];

for extensions using multivariate extreme value theory see Barbe, Fougéres & Genest [130] and, with an emphasis on non-identically distributed marginals, Kortschak & Albrecher [556]. For further results on asymptotically independent subexponential risks in the maximum domain of attraction of the Gumbel distribution, see Mitra & Resnick [647] and also Laeven, Goovaerts & Hoedemakers [571] with a view towards actuarial applications.

Tang & Wang [832] extend Proposition 2.12 to random variables with dominated variation. Asymptotic tail probabilities for negatively associated sums of heavy-tailed random variables are investigated in Wang & Tang [871] and Geluk & Ng [392].

3 Principles for premium calculation

The standard setting for discussing premium calculation in the actuarial literature does not involve stochastic processes, but only a single risk $X \geq 0$. By this we mean that X is a r.v. representing the random payment to be made (possibly 0). A premium rule is then a $[0, \infty)$-valued function H of the distribution of X, often written $H(X)$, such that $H(X)$ is the premium to be paid, i.e. the amount for which the company is willing to insure the given risk.

Among the standard premium rules discussed in the literature (not necessarily the same which are used in practice!) are the following:

The net premium principle $H(X) = \mathbb{E}X$ (also called *the equivalence principle*). As follows from the fluctuation theory of r.v.'s with finite mean, this principle will lead to ruin if many independent risks are insured. This motivates the next principle,

The expected value principle $H(X) = (1 + \eta)\mathbb{E}X$ where η is a specified safety loading. For $\eta = 0$, we are back to the net premium principle. A criticism of the expected value principle is that it does not take into account the variability of X. This leads to

The variance principle $H(X) = \mathbb{E}X + \eta \mathbb{V}ar(X)$. A modification (motivated from $\mathbb{E}X$ and $\mathbb{V}ar(X)$ not having the same dimension) is

The standard deviation principle $H(X) = \mathbb{E}X + \eta\sqrt{\mathbb{V}ar(X)}$.

The principle of zero utility. Here $v(x)$ is a given utility function, assumed to be concave and increasing with (w.lo.g) $v(0) = 0$; $v(x)$ represents the utility of a capital of size x. The zero utility principle then means $v(0) = \mathbb{E}v\big(H(X) - X\big)$ or, taking into account the initial reserve u in the portfolio,

$$v(u) \;=\; \mathbb{E}v\big(u + H(X) - X\big). \tag{3.1}$$

By Jensen's inequality, $v\big(u + H(X) - \mathbb{E}X\big) \geq \mathbb{E}v\big(u + H(X) - X\big) = 0$ so that $H(X) \geq \mathbb{E}X$. For $v(x) = x$, we have equality and are back to the net premium principle. There is also an approximate argument leading to the variance principle as follows. Assuming that the Taylor approximation

$$v\big(u + H(X) - X\big) \approx u + v'(u)\big(H(X) - X\big) + \frac{v''(u)}{2}\big(H(X) - X\big)^2$$

is reasonable, taking expectations leads to the quadratic equation

$$v''H(X)^2 + H(X)(2v' - 2v''\mathbb{E}X) + v''\mathbb{E}X^2 - 2v'\mathbb{E}X = 0$$

(with v', v'' evaluated at u) with solution

$$H(X) = \mathbb{E}X - \frac{v'}{v''} \pm \sqrt{\left(\frac{v'}{v''}\right)^2 - \mathbb{V}ar(X)}.$$

Write

$$\left(\frac{v'}{v''}\right)^2 - \mathbb{V}ar(X) = \left(\frac{v'}{v''} - \frac{v''}{2v'}\mathbb{V}ar(X)\right)^2 - \left(\frac{v''}{2v'}\mathbb{V}ar(X)\right)^2.$$

If v''/v' is small, we can ignore the last term. Taking $+\sqrt{\cdot}$ then yields

$$H(X) \approx \mathbb{E}X - \frac{v''(u)}{2v'(u)}\mathbb{V}ar X;$$

since $v''(u) \leq 0$ by concavity, this is approximately the variance principle. The most important special case of the principle of zero utility is

The exponential principle which corresponds to $v(x) = (1 - e^{-ax})/a$ for some $a > 0$. Here the initial capital u cancels out and (3.1) leads to

$$H(X) = \frac{1}{a}\log\mathbb{E}e^{aX}.$$

Since m.g.f.'s are log-concave, it follows that $H_a(X) = H(X)$ is increasing as function of a. Further, $\lim_{a\downarrow 0} H_a(X) = \mathbb{E}X$ (the net premium principle) and, provided $b = \operatorname{ess\,sup} X < \infty$, $\lim_{a\to\infty} H_a(X) = b$ (the premium principle $H(X) = b$ is called the *maximal loss principle* but is clearly not very realistic). In view of this, a is called the *risk aversion*.

Note that in the compound Poisson model, the premium collected for the aggregate risk A_t is pt. Equating this with $H(A_t) = \frac{1}{a}\log\mathbb{E}e^{aA_t}$ leads to the Lundberg equation for a. Hence, the premium principle in the Cramér-Lundberg model can be interpreted as an exponential principle with risk aversion γ, given the adjustment coefficient $\gamma > 0$ exists.

The risk-adjusted premium principle

$$H(X) = \int_0^\infty g\big(\mathbb{P}(X > x)\big)\,\mathrm{d}x$$

for a fixed nondecreasing and left-continuous function $g : [0,1] \to [0,1]$ (also called the *distortion function*) such that $g(0) = 0$ and $g(1) = 1$.

The percentile principle Here one chooses a (small) number α, say 0.05 or 0.01, and determines $H(X)$ by $\mathbb{P}(X \le H(X)) = 1 - \alpha$ (assuming a continuous distribution for simplicity).

Some standard criteria for evaluating the merits of premium rules are

1. $\eta \ge 0$, i.e. $H(X) \ge \mathbb{E}X$.

2. $H(X) \le b$ when b (the ess sup above) is finite

3. $H(X + c) = H(X) + c$ for any constant c

4. $H(X + Y) = H(X) + H(Y)$ when X, Y are independent

5. $H(X) = H\big(H(X|Y)\big)$. For example, if $X = \sum_1^N U_i$ is a random sum with the U_i independent of N, this yields

$$H\left(\sum_1^N U_i\right) = H(H(U)N)$$

(where, of course, $H(U)$ is a constant).

Note that $H(cX) = cH(X)$ is not on the list! Considering the examples above, the net premium principle and the exponential principle can be seen to be the only ones satisfying all five properties. The expected value principle fails to satisfy, e.g., 3), whereas (at least) 4) is violated for the variance principle, the standard deviation principle, and the zero utility principle (unless it is the exponential or net premium principle). For more detail, see e.g. Gerber [398] or Sundt [820].

Notes and references The discussed premium principles are standard and can be found in many texts on insurance mathematics, e.g. Gerber [398], Heilmann [458] and Sundt [820]. For an extensive treatment, see Goovaerts *et al.* [423]. In recent years, the discussion about which criteria H should or should not fulfill in various applications has experienced enormous interest and activity in related finance contexts under the terminology of risk measures, see for instance Pflug & Römisch [697] for an overview. On the insurance side, going from the static pricing framework above towards a dynamic one is considered to be an important step for many situations. Time consistency and market consistency play a crucial role in this context; for some recent developments see e.g. Cheridito, Delbaen & Kupper [238], Jobert & Rogers [507], Malamud, Trubowitz & Wüthrich [624] and Pelsser [690].

4 Reinsurance

Reinsurance means that the company (the *cedent*) insures a part of the risk at another insurance company (the *reinsurer*).

Again, we start by formulating the basic concepts within the framework of a single risk $X \geq 0$. A reinsurance arrangement is then defined in terms of a function $h(x)$ with the property $0 \leq h(x) \leq x$. Here $h(x)$ is the amount of the claim x to be paid by the reinsurer and $x - h(x)$ the amount to be paid by the cedent. The function $x - h(x)$ is referred to as the *retention function*. The most common examples are the following two:

Proportional reinsurance $h(x) = \theta x$ for some $\theta \in (0, 1)$. Also called *quota share* reinsurance.

Stop-loss reinsurance $h(x) = (x - b)^+$ for some $b \in (0, \infty)$, referred to as the *retention limit*. Note that the retention function is $x \wedge b$.

Concerning terminology, note that in the actuarial literature the *stop-loss transform* of $F(x) = \mathbb{P}(X \leq x)$ (or, equivalently, of X), is defined as the function

$$b \quad \rightarrow \quad \mathbb{E}(X - b)^+ \; = \; \int_b^\infty (x - b) F(\mathrm{d}x) \; = \; \int_b^\infty \overline{F}(x) \, \mathrm{d}x.$$

An arrangement closely related to stop-loss reinsurance is excess-of-loss reinsurance, see below.

Stop-loss reinsurance and excess-of-loss reinsurance have a number of nice optimality properties. The first we prove is in terms of maximal utility:

Proposition 4.1 *Let X be a given risk, v a given concave non-decreasing utility function and h a given retention function. Let further b be determined by $\mathbb{E}(X - b)^+ = \mathbb{E}h(X)$. Then for any x,*

$$\mathbb{E}v\big(x - [X - h(X)]\big) \; \leq \; \mathbb{E}v(x - X \wedge b).$$

Remark 4.2 Proposition 4.1 can be interpreted as follows. Assume that the cedent charges a premium $P \geq \mathbb{E}X$ for the risk X and is willing to pay $P_1 < P$

for reinsurance. If the reinsurer applies the expected value principle with safety loading η, this implies that the cedent is looking for retention functions with $\mathbb{E}h(X) = P_2 = P_1/(1 + \eta)$. The expected utility after settling the risk is thus

$$\mathbb{E}v(u + P - P_1 - [X - h(X)])$$

where u is the initial reserve. Letting $x = u + P - P_1$, Proposition 4.1 shows that the stop-loss rule $h(X) = (X - b)^+$ with b chosen such that $\mathbb{E}(X - b)^+ = P_2$ maximizes the expected utility. □

Recall the notions of stochastic ordering from Section IV.8. For the proof of Proposition 4.1, we shall need the following lemma:

Lemma 4.3 (OHLIN'S LEMMA) *Let X_1, X_2 be two risks with the same mean, such that*

$$F_1(x) \leq F_2(x), \quad x < b, \qquad F_1(x) \geq F_2(x), \quad x \geq b$$

for some b where $F_i(x) = \mathbb{P}(X_i \leq x)$. Then $X_1 \prec_{\text{cx}} X_2$.

Proof. Define $\Delta(u) = \mathbb{E}(X_2 - u)^+ - \mathbb{E}(X_1 - u)^+$. Clearly $\Delta(0) = 0$ and $\lim_{u \to \infty} \Delta(u) = 0$. But from the representation $\Delta(u) = \int_u^\infty (F_1(x) - F_2(x))\,dx$, we have under the given assumptions that $\Delta(u)$ increases on $(0, b)$ and decreases on (b, ∞). So $\Delta(u) \geq 0$ for all $u \geq 0$, i.e. $X_1 \prec_{\text{icx}} X_2$. Since $\mathbb{E}[X_1] = \mathbb{E}[X_2]$, this implies $X_1 \prec_{\text{cx}} X_2$. □

Proof of Proposition 4.1. It is easily seen that the assumptions of Ohlin's lemma hold when $X_1 = X \wedge b$, $X_2 = X - h(X)$; in particular, the requirement $\mathbb{E}X_1 = \mathbb{E}X_2$ is then equivalent to $\mathbb{E}(X - b)^+ = \mathbb{E}h(X)$. Now just note that $-v$ is convex. □

We now turn to the case where the risk can be written as

$$X = \sum_{i=1}^N U_i \tag{4.1}$$

with the U_i independent; N may be random but should then be independent of the U_i. Typically, N could be the number of claims in a given period, say a year, and the U_i the corresponding claim sizes. A reinsurance arrangement of the form $h(X)$ as above is called *global*; if instead h is applied to the individual claims so that the reinsurer pays the amount $\sum_{i=1}^N h(U_i)$, the arrangement is called *local*.[4]

---

[4]More generally, one could consider $\sum_{i=1}^N h_i(U_i)$.

The following discussion will focus on maximizing the adjustment coefficient. For a global rule with retention function $h^*(x)$ and a given premium P^* charged for $X - h^*(X)$, the cedent's adjustment coefficient γ^* is determined by

$$1 = \mathbb{E}\exp\{\gamma^*[X - h^*(X) - P^*]\}, \qquad (4.2)$$

for a local rule corresponding to $h(u)$ and premium P for $X - \sum_{i=1}^{N} h(U_i)$, we look instead for the γ solving

$$1 = \mathbb{E}\exp\left\{\gamma\left[\sum_{i=1}^{N}[U_i - h(U_i)] - P\right]\right\} = \mathbb{E}\exp\left\{\gamma\left[X - P - \sum_{i=1}^{N} h(U_i)\right]\right\}. \quad (4.3)$$

This definition of the adjustment coefficients is motivated by considering ruin at a sequence of equally spaced time points, say consecutive years, such that N is the generic number of claims in a year and P, P^* the total premiums charged in a year, and referring to the results of VI.3a. The following result shows that if we compare only arrangements with $P = P^*$, a global rule is preferable to a local one.

Proposition 4.4 *To any local rule with retention function $h(u)$ and any*

$$P \geq \mathbb{E}\left[X - \sum_{i=1}^{N} h(U_i)\right], \qquad (4.4)$$

there is a global rule with retention function $h^*(x)$ such that

$$\mathbb{E}h^*(X) = \mathbb{E}\sum_{i=1}^{N} h(U_i) \qquad (4.5)$$

and $\gamma^* \geq \gamma$.

Proof. Define

$$h^*(x) = \mathbb{E}\left[\sum_{i=1}^{N} h(U_i) \,\middle|\, |X = x\right];$$

then (4.5) holds trivially. Applying the inequality $\mathbb{E}\varphi(Y) \geq \mathbb{E}\varphi(\mathbb{E}(Y|X))$ (with φ convex) to $\varphi(y) = e^{\gamma y}$, $Y = \sum_{i=1}^{N}[U_i - h(U_i)] - P$, we get

$$1 = \mathbb{E}\exp\left\{\gamma\left[\sum_{i=1}^{N}[U_i - h(U_i)] - P\right]\right\} \geq \mathbb{E}\exp\{\gamma[X - h^*(X) - P]\}.$$

But since $\gamma \geq 0$, $\gamma^* \geq 0$ because of (4.4), this implies $\gamma^* \geq \gamma$. $\qquad\square$

Remark 4.5 Because of the independence assumptions, expectations like those in (4.3), (4.4), (4.5) simplify a lot. Assuming for simplicity that the U_i are i.i.d., we get $\mathbb{E}X = \mathbb{E}N \cdot \mathbb{E}U$,

$$\mathbb{E}\Big[X - \sum_{i=1}^{N} h(U_i)\Big] = \mathbb{E}N \cdot \mathbb{E}\big[U - h(U)\big],$$

$$\mathbb{E}\exp\Big\{\gamma\Big[\sum_{i=1}^{N}[U_i - h(U_i)] - P\Big]\Big\} = \mathbb{E}\widehat{C}[\gamma]^N, \qquad (4.6)$$

where $\widehat{C}[\gamma] = \mathbb{E}e^{\gamma(U-h(U))}$, and so on. $\qquad\square$

The arrangement used in practice is, however, as often local as global. Local reinsurance with $h(u) = (u - b)^+$ is referred to as *excess-of-loss reinsurance* and plays a particular role:

Proposition 4.6 *Assume the U_i are i.i.d. Then for any local retention function $u - h(u)$ and any P satisfying (4.4), the excess-of-loss rule $h_1(u) = (u - b)^+$ with b determined by*

$$\mathbb{E}(U - b)^+ = \mathbb{E}h(U) \qquad (4.7)$$

(and the same P) satisfies $\gamma_1 \geq \gamma$.

Proof. As in the proof of Proposition 4.4, it suffices to show that

$$\mathbb{E}\exp\Big\{\gamma\Big[\sum_{i=1}^{N} U_i \wedge b - P\Big]\Big\} \leq 1 = \mathbb{E}\exp\Big\{\gamma\Big[\sum_{i=1}^{N}[U_i - h(U_i)] - P\Big]\Big\},$$

or, appealing to (4.6), that $\widehat{C}_1[\gamma] \leq \widehat{C}[\gamma]$ where $\widehat{C}[\gamma] = \mathbb{E}e^{\gamma(U \wedge b)}$. This follows by taking $X_1 = U \wedge b$, $X_2 = U - h(U)$ (as in the proof of Proposition 4.4) and $g(x) = e^{\gamma x}$ in Ohlin's lemma. $\qquad\square$

Notes and references Reinsurance is a classical topic. The material presented here is standard and can be found in many texts on insurance mathematics, e.g. Bowers *et al.* [195], Heilmann [458] and Sundt [820]. See further Hesselager [461] and Dickson & Waters [319]. The original reference for Ohlin's lemma is Ohlin [671].

An early reference for minimization of the ruin probability through reinsurance in an asymptotic sense by maximizing the adjustment coefficient is Waters [876], see Hald & Schmidli [446], Centeno [224, 225] and Guerra & Centeno [441] for more recent extensions. The identification of optimal reinsurance strategies under various objective functions and constraints is an active field of research, see e.g. Centeno & Simões [226] and Albrecher & Teugels [38] for a recent overview.

For optimal dynamic reinsurance in discrete time, see e.g. Dickson & Waters [321]. Optimal adaptive reinsurance strategies in continuous time are discussed in Chapter XIV.

Appendix

A1   Renewal theory

1a   Renewal processes and the renewal theorem

By a *simple point process* on the line we understand a random collection of time epochs without accumulation points and without multiple points. The mathematical representation is either the ordered set $0 \leq T_0 < T_1 < \ldots$ of epochs or the set $Y_1, Y_2, \ldots$ of interarrival times and the time $Y_0 = T_0$ of the first arrival (that is, $Y_n = T_n - T_{n-1}$). The point process is called a *renewal process* if $Y_0, Y_1, \ldots$ are independent and $Y_1, Y_2, \ldots$ all have the same distribution, denoted by F in the following and referred to as the interarrival distribution; the distribution of Y_0 is called the delay distribution. If $Y_0 = 0$, the renewal process is called *zero-delayed*. The number $\max k : T_{k-1} \leq t$ of renewals in $[0, t]$ is denoted by N_t.

The associated renewal measure U is defined by $U = \sum_0^\infty F^{*n}$ where F^{*n} is the nth convolution power of F. That is, $U(A)$ is the expected number of renewals in $A \subseteq \mathbb{R}$ in a zero-delayed renewal process; note in particular that $U(\{0\}) = 1$.

The *renewal theorem* asserts that $U(dt)$ is close to dt/μ, Lebesgue measure dt normalized by the mean μ of F, when t is large. Technically, some condition is needed: that F is non-lattice, i.e. not concentrated on $\{h, 2h, \ldots\}$ for any $h > 0$. Then *Blackwell's renewal theorem* holds, stating that

$$U(t + a) - U(t) \; \rightarrow \; \frac{a}{\mu}, \quad t \to \infty \qquad (A.1)$$

(here $U(t) = U([0, t])$ so that $U(t+a) - U(t)$ is the expected number of renewals in $(t, t+a]$). If F satisfies the stronger condition of being *spread-out* (F^{*n} is non-singular w.r.t. Lebesgue measure for some $n \geq 1$), then *Stone's decomposition* holds: $U = U_1 + U_2$ where U_1 is a finite measure and $U_2(dt) = u(t)\, dt$ where

517

$u(t)$ has limit $1/\mu$ as $t \to \infty$. Note in particular that F is spread-out if F has a density f.

A weaker (and much easier to prove) statement than Blackwell's renewal theorem is the *elementary renewal theorem*, stating that $U(t)/t \to 1/\mu$. Both results are valid for delayed renewal processes, the statements being

$$\mathbb{E}N(t+a) - \mathbb{E}N(t) \;\to\; \frac{a}{\mu}, \quad \text{resp.} \quad \mathbb{E}\frac{N_t}{t} \;\to\; \frac{1}{\mu}.$$

1b   Renewal equations and the key renewal theorem

The *renewal equation* is the convolution equation

$$Z(u) \;=\; z(u) + \int_0^u Z(u-x)F(\mathrm{d}x), \tag{A.2}$$

where $Z(u)$ is an unknown function of $u \in [0, \infty)$, $z(u)$ a known function, and $F(\mathrm{d}x)$ a known probability measure. Equivalently, in convolution notation $Z = z + F * Z$. Under weak regularity conditions (see [APQ, Ch. IV]), (A.2) has the unique solution $Z = U * z$, i.e.

$$Z(u) = \int_0^u z(x)U(\mathrm{d}x). \tag{A.3}$$

Further, the asymptotic behavior of $Z(u)$ is given by the *key renewal theorem*:

Proposition A1.1 *If F is non-lattice and $z(u)$ is directly Riemann integrable (d.R.i.; see [APQ, Ch. IV]), then*

$$Z(u) \;\to\; \frac{\int_0^\infty z(x)\,\mathrm{d}x}{\mu_F}. \tag{A.4}$$

If F is spread-out, then it suffices for (A.4) *that z is Lebesgue integrable with* $\lim_{x \to \infty} z(x) = 0$.

In IV.9, we shall need the following less standard parallel to the key renewal theorem:

Proposition A1.2 *Assume that Z solves the renewal equation* (A.2), *that $z(u)$ has a limit $z(\infty)$ (say) as $u \to \infty$, and that F has a bounded density.[1] Then*

$$\frac{Z(u)}{u} \to \frac{z(\infty)}{\mu_F}, \quad u \to \infty. \tag{A.5}$$

---

[1] This condition can be weakened considerably, but suffices for the present purposes.

Proof. The condition on F implies that $U(\mathrm{d}x)$ has a bounded density $u(x)$ with limit $1/\mu_F$ as $x \to \infty$. Hence by dominated convergence,

$$\frac{Z(u)}{u} = \frac{1}{u}\int_0^u z(u-x)u(x)\,\mathrm{d}x = \int_0^1 z\big(u(1-t)\big)u(ut)\,\mathrm{d}t$$

$$\to \int_0^1 z(\infty)\cdot\frac{1}{\mu_F}\mathrm{d}t = \frac{z(\infty)}{\mu_F}.$$

$\square$

In risk theory, a basic reason that renewal theory is relevant is the renewal equation III.(3.2) satisfied by the ruin probability for the compound Poisson model. Here the relevant F does not have mass one (F is defective). However, asymptotic properties can easily be obtained from the key renewal equation by an exponential transformation also when $F(\mathrm{d}x)$ does not integrate to one. To this end, multiply (A.2) by $\mathrm{e}^{\gamma x}$ to obtain $\widetilde{Z} = \widetilde{z} + \widetilde{F} * \widetilde{Z}$ where $\widetilde{Z}(x) = \mathrm{e}^{\gamma x}Z(x)$, $\widetilde{z}(x) = \mathrm{e}^{\gamma x}z(x)$, $\widetilde{F}(\mathrm{d}x) = \mathrm{e}^{\gamma x}F(\mathrm{d}x)$. Assuming that γ can be chosen such that $\int_0^\infty \mathrm{e}^{\gamma x}F(\mathrm{d}x) = 1$, i.e. that $\widetilde{F}$ is a probability measure, results from the case $\int_0^\infty F(\mathrm{d}x) = 1$ can then be used to study $\widetilde{Z}$ and thereby Z. This program has been carried out in IV.5a. Note, however, that the existence of γ may fail for heavy-tailed F.

1c Regenerative processes

Let $\{T_n\}$ be a renewal process. A stochastic process $\{X_t\}_{t\geq 0}$ with a general state space E is called *regenerative* w.r.t. $\{T_n\}$ if for any k, the post-T_k process $\{X_{T_k+t}\}_{t\geq 0}$ is independent of $T_0, T_1, \ldots, T_k$ (or, equivalently, of $Y_0, Y_1, \ldots, Y_k$), and its distribution does not depend on k. The distribution F of $Y_1, Y_2, \ldots$ is called the *cycle length distribution* and as before, we let μ denote its mean. We let $\mathbb{P}_0$, $\mathbb{E}_0$ etc. refer to the zero-delayed case.

The simplest case is when $\{X_t\}$ has *i.i.d. cycles*. The kth cycle is defined as $\{X_{T_k+t}\}_{0\leq t < Y_{k+1}}$; this expression is to be interpreted as a random element of the space of all E-valued sequences with finite lifelengths. The property of independent cycles is equivalent to the post-T_k process $\{X_{T_k+t}\}_{t\geq 0}$ being independent of $T_0, T_1, \ldots, T_k$ *and* $\{X_t\}_{0\leq t < T_k}$. For example, this covers discrete Markov chains where we can take the T_n as the instants with $X_t = i$ for some arbitrary but fixed state i, or many queueing processes, where the T_n are the instants where a customer enters an empty system (then cycles = busy cycles). However, the present more general definition is needed to deal with say Harris recurrent Markov chains.

A regenerative process converges in distribution under very mild conditions:

Proposition A1.3 *Consider a regenerative process such that the cycle length distribution is non-lattice with $\mu < \infty$. Then $X_t \overset{\mathscr{D}}{\to} X_\infty$ where the distribution of X_∞ is given by*

$$\mathbb{E}g(X_\infty) = \frac{1}{\mu}\mathbb{E}_0\int_0^{Y_1} g(X_t)\,\mathrm{d}t\,. \tag{A.6}$$

If F is spread-out, then $X_t \to X_\infty$ in total variation.

1d   Cumulative processes

Let $\{T_n\}$ be a renewal process with i.i.d. cycles (we allow a different distribution of the first cycle). Then $\{Z_t\}_{t\geq 0}$ is called *cumulative* w.r.t. $\{T_n\}$ if the processes

$$\{Z_{T_n+t} - Z_{T_n}\}_{0\leq t<Y_{n+1}}$$

are i.i.d. for $n = 1, 2, \ldots$. An example is $Z_t = \int_0^t f(X_s)\,\mathrm{d}s$ where $\{X_t\}$ is regenerative w.r.t. $\{T_n\}$, with i.i.d. cycles. This is the case considered in [APQ, VI.3], but in fact, just the same proof as there carries over to show:

Proposition A1.4 *Let $\{Z_t\}_{t\geq 0}$ be cumulative w.r.t. $\{T_n\}$, assume that $\mu < \infty$ and define $U_n = Z_{T_{n+1}} - Z_{T_n}$. Then:*
(a) *If*

$$\mathbb{E}\sup_{0\leq t<Y_1}|Z_{T_0+t} - Z_{T_0}| < \infty,$$

then $Z_t/t \overset{\mathrm{a.s.}}{\to} \mathbb{E}U_1/\mu$;
(b) *If in addition $\mathbb{V}ar(U_1) < \infty$, then $(Z_t - t\mathbb{E}U_1/\mu)/\sqrt{t}$ has a limiting normal distribution with mean 0 and variance*

$$\mathbb{V}ar(U_1) + \left(\frac{\mathbb{E}U_1}{\mu}\right)^2\mathbb{V}ar(Y_1) - \frac{2\mathbb{E}U_1}{\mu}\mathbb{C}ov(U_1, Y_1)\,.$$

1e   Residual and past lifetime

Consider a renewal process and define $\xi(t)$ as the residual lifetime of the renewal interval straddling t, i.e. $\xi(t) = \inf\{T_k - t : t < T_k\}$, and $\eta(t) = \sup\{t - T_k : t \leq T_k\}$ as the age. Then $\{\xi(t)\}$, $\{\eta(t)\}$ are Markov with state spaces $(0, \infty)$, resp. $[0, \infty)$. If $\mu = \infty$, then $\xi(t) \overset{\mathscr{D}}{\to} \infty$ (i.e. $\mathbb{P}(\xi(t) \leq a) \to 0$ for any $a < \infty$) and $\eta(t) \overset{\mathscr{D}}{\to} \infty$. Otherwise, under the condition of Blackwell's renewal theorem, $\xi(t)$ and $\eta(t)$ both have a limiting stationary distribution F_0 given by the density $\overline{F}(x)/\mu$. We denote the limiting r.v.'s by ξ, η. Then it holds more generally that $(\eta(t), \xi(t)) \overset{\mathscr{D}}{\to} (\eta, \xi)$, and we have:

Theorem A1.5 *Under the condition of Blackwell's renewal theorem, the joint distribution of (η, ξ) is given by the following four equivalent statements:*

(a) $\mathbb{P}(\eta > x, \xi > y) = \dfrac{1}{\mu} \displaystyle\int_{x+y}^{\infty} \overline{F}(z)\,\mathrm{d}z;$

(b) *the joint distribution of (η, ξ) is the same as the distribution of $(VW, (1 - V)W)$ where V, W are independent, V is uniform on $(0,1)$ and W has distribution F_W given by* $\mathrm{d}F_W/\mathrm{d}F(x) = x/\mu_F;$

(c) *the marginal distribution of η is F_0, and the conditional distribution of ξ given $\eta = y$ is the overshoot distribution $F_0^{(y)}$ given by* $\overline{F_0^{(y)}}(z) = \overline{F_0}(y+z)/\overline{F_0}(y);$

(d) *the marginal distribution of ξ is F_0, and the conditional distribution of η given $\xi = z$ is $F_0^{(z)}$.*

The proof of (a) is straightforward by viewing $\{(\eta(t), \xi(t))\}$ as a regenerative process, and the equivalence of (a) with (b)-(d) is an easy exercise.

In V.4, we used:

Proposition A1.6 *Consider a renewal process with $\mu < \infty$. Then $\xi(t)/t \overset{\text{a.s.}}{\to} 0$ and, if in addition $\mathbb{E}Y_0 < \infty$, $\mathbb{E}\xi(t)/t \to 0$.*

Proof. The number N_t of renewals before t satisfies $N_t/t \overset{\text{a.s.}}{\to} \mu$. Hence for t large enough, we can bound $\xi(t)$ by $M(t) = \max\{Y_k : k \le 2t/\mu\}$. Since the maximum M_n of n i.i.d. r.v.'s with finite mean satisfies $M_n/n \overset{\text{a.s.}}{\to} 0$ (Borel-Cantelli), the first statement follows. For the second, assume first the renewal process is zero-delayed. Then $\mathbb{E}_0\xi(t)$ satisfies a renewal equation with $z(t) = \mathbb{E}[Y_1 - t; Y_1 > t]$. Hence

$$\mathbb{E}_0\xi(t) = \int_0^t U(\mathrm{d}y)z(t-y) = \int_0^t U(t-\mathrm{d}y)z(y) \le c \sum_{k=0}^{\lfloor t \rfloor + 1} z(k)$$

where $c = \sup_x U(x+1) - U(x)$ ($c < \infty$ because it is easily seen that $U(x+1) - U(x) \le U(1)$). Since $z(k) \le \mathbb{E}[Y_1; Y_1 > t] \to 0$, the sum is o$(t)$ so that $\mathbb{E}_0\xi(t)/t \to 0$. In the general case, use

$$\mathbb{E}\xi(t)/t = \mathbb{E}[Y_0 - t; Y_0 > 0] + \int_0^t \mathbb{E}_0\xi(t-y)\mathbb{P}(Y_0 \in \mathrm{d}y).$$

$\square$

1f Markov renewal theory

By a *Markov renewal process* we understand a point process where the interarrival times $Y_0, Y_1, Y_2, \ldots$ are not i.i.d. but governed by a Markov chain $\{J_n\}$ (we assume here that the state space E is finite) in the sense that

$$\mathbb{P}(Y_n \leq y \mid \mathscr{J}) = F_{ij}(y) \quad \text{on } \{J_n = i, J_{n+1} = j\}$$

where $\mathscr{J} = \sigma(J_0, J_1, \dots)$ and $(F_{ij})_{i,j \in E}$ is a family of distributions on $(0, \infty)$. A stochastic process $\{X_t\}_{t \geq 0}$ is called *semi-regenerative* w.r.t. the Markov renewal process if for any n, the conditional distribution of $\{X_{T_n + t}\}_{t \geq 0}$ given $Y_0, Y_1, \dots, Y_n, J_0, \dots, J_{n-1}, J_n = i$ is the same as the $\mathbb{P}_i$-distribution of $\{X_t\}_{t \geq 0}$ itself where $\mathbb{P}_i$ refers to the case $J_0 = i$.

A Markov renewal process $\{T_n\}$ contains an imbedded renewal process, namely $\{T_{\omega_k}\}$ where $\{\omega_k\}$ is the sequence of instants ω where $J_\omega = i_0$ for some arbitrary but fixed reference state $i_0 \in E$. The semi-regenerative process is then regenerative w.r.t. $\{T_{\omega_k}\}$. These facts allow many definitions and results to be reduced to ordinary renewal- and regenerative processes. For example, the semi-regenerative process is called non-lattice if $\{T_{\omega_k}\}$ is non-lattice (it is easily seen that this definition does not depend on i). Further:

Proposition A1.7 *Consider a non-lattice semi-regenerative process. Assume that $\mu_j = \mathbb{E}_j Y_0 < \infty$ for all j and that $\{J_n\}$ is irreducible with stationary distribution $(\nu_j)_{j \in E}$. Then $X_t \overset{\mathscr{D}}{\to} X_\infty$ where the distribution of X_∞ is given by*

$$\mathbb{E}g(X_\infty) = \frac{1}{\mu} \sum_{j \in E} \nu_j \mathbb{E}_j \int_0^{Y_0} g(X_t) \, dt$$

where $\mu = \sum_{j \in E} \nu_j \mu_j$.

Notes and references Renewal theory and regenerative processes are treated, e.g., in [APQ], Alsmeyer [45] and Thorisson [850].

A2 Wiener-Hopf factorization

Let F be a distribution which is not concentrated on $(-\infty, 0]$ or $(0, \infty)$. Let $X_1, X_2, \dots$ be i.i.d. with common distribution F, $S_n = X_1 + \cdots + X_n$ the associated random walk, and define

$$\tau_+ = \inf\{n > 0 : S_n > 0\}, \quad \tau_- = \inf\{n > 0 : S_n \leq 0\},$$

$$G_+(x) = \mathbb{P}(S_{\tau_+} \leq x, \tau_+ < \infty), \quad G_-(x) = \mathbb{P}(S_{\tau_-} \leq x, \tau_- < \infty).$$

We call τ_+ (τ_-) the strict ascending (weak descending) ladder epoch and G_+ (G_-) the corresponding ladder height distributions.

Probabilistic Wiener-Hopf theory deals with the relation between F, G_+, G_-, the renewal measures

$$U_+ = \sum_{n=0}^{\infty} G_+^{*n}, \quad U_- = \sum_{n=0}^{\infty} G_-^{*n},$$

and the τ_+- and τ_--pre-occupation measures

$$R_+(A) = \mathbb{E} \sum_{n=0}^{\tau_+ - 1} I(S_n \in A), \quad R_-(A) = \mathbb{E} \sum_{n=0}^{\tau_- - 1} I(S_n \in A).$$

The basic identities are the following:

Theorem A2.1 (a) $F = G_+ + G_- - G_+ * G_-$;
(b) $G_-(A) = \int_0^{\infty} F(A - x)R_-(\mathrm{d}x)$, $A \subseteq (-\infty, 0]$;
(c) $G_+(A) = \int_{-\infty}^0 F(A - x)R_+(\mathrm{d}x)$, $A \subseteq (0, \infty)$;
(d) $R_+ = U_-$; (e) $R_- = U_+$.

Proof. Considering the restrictions of measures to $(-\infty, 0]$ and $(0, \infty)$, we may rewrite (a) as

$$G_-(A) = F(A) + (G_+ * G_-)(A), \quad A \subseteq (-\infty, 0], \qquad (A.7)$$
$$G_+(A) = F(A) + (G_+ * G_-)(A), \quad A \subseteq (0, \infty) \qquad (A.8)$$

(e.g. (A.7) follows since $G_+(A) = 0$ when $A \subseteq (-\infty, 0]$). In (A.7), $F(A)$ is the contribution from the event $\{\tau_- = 1\} = \{X_1 \leq 0\}$. On $\{\tau_- \geq 2\}$, define ω as the time where the pre-τ_- path $S_1, \ldots, S_{\tau_- - 1}$ is at its minimum. More rigorously, we consider the last such time (to make ω unique) so that

$$\{\omega = m, \tau_- = n\} = \{S_j - S_m \geq 0, \ 0 < j < m, \ S_j - S_m > 0, \ m < j < n\}.$$

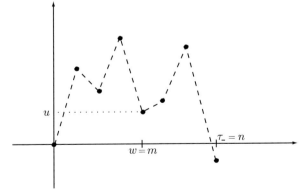

FIGURE A.1

Reversing the time points $0, 1, \ldots, m$ it follows (see Fig. A.1) that

$$\mathbb{P}\big(S_j - S_m \geq 0,\ 0 < j < m,\ S_m \in \mathrm{d}u\big) = \mathbb{P}(\tau_+ = m,\ S_{\tau_+} \in \mathrm{d}u).$$

Also, clearly

$$\mathbb{P}\big(S_j - S_m > 0,\ m < j < n,\ S_n \in A \,\big|\, S_m \in \mathrm{d}u\big) = \mathbb{P}(\tau_- = n - m,\ S_{\tau_-} \in A - \mathrm{d}u)$$

(see again Fig. A.1). It follows that for $n \geq 2$

$$\mathbb{P}(\tau_- = n,\ S_{\tau_-} \in A)$$

$$= \sum_{m=1}^{n-1} \int_0^\infty \mathbb{P}\big(\tau_- = n, \omega = m,\ S_m \in \mathrm{d}u,\ S_{\tau_-} \in A\big)$$

$$= \sum_{m=1}^{n-1} \int_0^\infty \mathbb{P}\big(\tau_+ = m,\ S_{\tau_+} \in \mathrm{d}u\big) \cdot \mathbb{P}(\tau_- = n - m,\ \mathrm{d}S_{\tau_-} \in A - u).$$

Summing over $n = 2, 3, \ldots$ and reversing the order of summation yields

$$\mathbb{P}(\tau_- \geq 2,\ S_{\tau_-} \in A)$$

$$= \int_0^\infty \sum_{m=1}^\infty \mathbb{P}(\tau_+ = m,\ S_{\tau_+} \in \mathrm{d}u) \sum_{n=m+1}^\infty \mathbb{P}\big(S_{\tau_-} = n - m,\ S_{\tau_-} \in A - u\big)$$

$$= \int_0^\infty \mathbb{P}(S_{\tau_+} \in \mathrm{d}u)\mathbb{P}(S_{\tau_-} \in A - \mathrm{d}u)$$

$$= (G_+ * G_-)(A).$$

Collecting terms, (A.7) follows, and the proof of (A.8) is similar.
  (b) follows from

$$G_+(A) = \sum_{n=1}^\infty \mathbb{P}(S_n \in A, \tau_+ = n)$$

$$= \sum_{n=1}^\infty \int_0^\infty \mathbb{P}\big(S_k \leq 0, 0 \leq k < n, S_{n-1} \in \mathrm{d}x, X_n \in A - x\big)$$

$$= \int_0^\infty \sum_{n=1}^\infty F(A - x)\mathbb{P}\big(S_k \leq 0, 0 \leq k < n, S_{n-1} \in \mathrm{d}x\big)$$

$$= \int_0^\infty F(A - x)R_+(\mathrm{d}x),$$

and the proof of (c) is similar. For (d), consider a fixed n and let $X_k^* = X_{n-k+1}$, $S_k^* = X_1^* + \cdots + X_k^* = S_n - S_{n-k}$. Then for $A \subseteq (-\infty, 0]$,

$$
\begin{aligned}
\mathbb{P}(S_n \in A, \tau_+ > n) &= \mathbb{P}\big(S_k \leq 0, 0 \leq k \leq n, S_n \in A\big) \\
&= \mathbb{P}\big(S_n^* \leq S_{n-k}^*, 0 \leq k \leq n, S_n^* \in A\big) \\
&= \mathbb{P}\big(S_n^* \leq S_k^*, 0 \leq k \leq n, S_n^* \in A\big) \\
&= \mathbb{P}\big(S_n \leq S_k, 0 \leq k \leq n, S_n \in A\big)
\end{aligned}
$$

is the probability that n is a weak descending ladder point with $S_n \in A$. Summing over n yields $R_+(A) = U_-(A)$, and the proof of (e) is similar. □

Remark A2.2 In terms of m.g.f.'s, we can rewrite (a) as

$$
1 - \widehat{F}[s] = \big(1 - \widehat{G}_+[s]\big)\big(1 - \widehat{G}_-[s]\big) \tag{A.9}
$$

whenever $\widehat{F}[s]$, $\widehat{G}_+[s]$, $\widehat{G}_-[s]$ are defined at the same time; this always holds on the line $\Re(s) = 0$, and sometimes in a larger strip. Since G_+ is concentrated on $(0, \infty)$, $H_+(s) = 1 - \widehat{G}_+[s]$ is defined and bounded in the half-plane $\{s : \Re(s) < 0\}$ and non-zero in $\{s : \Re(s) < 0\}$ (because $\|G_+\| \leq 1$), and similarly $H_-(s) = 1 - \widehat{G}_-[s]$ is defined and bounded in the half-plane $\{s : \Re(s) \geq 0\}$ and non-zero in $\{s : \Re s > 0\}$. The classical analytical form of the Wiener-Hopf problem is to write $1 - \widehat{F}$ as a product H_+, H_- of functions with such properties. □

Notes and references In its above discrete time version, Wiener-Hopf theory is only used at a few places in this book. However, it serves as model and motivation for a number of results and arguments in continuous time. E.g., the derivation of the form of G_+ for the compound Poisson model (Theorem III.5.1), which is basic for the Pollaczeck-Khinchine formula, is based upon representing G_+ as in (b), and using time-reversion as in (d) to obtain the explicit form of R_+ (Lebesgue measure).

In continuous time, the analogue of a random walk is a process with stationary independent increments (a Lévy process, cf. III.4). In this generality, there is no direct analogue of Theorem A2.1. For example, if $\{S_t\}$ is Brownian motion, then $\tau_+ = \inf\{t > 0 : S_t = 0\}$ is 0 a.s., and G_+, G_- are trivial, being concentrated at 0. Nevertheless, a number of related identities can be derived. An early survey is Bingham [168], and we further refer to Section XI.4d.

Another main extension of the theory deals with Markov dependence. In discrete time, there are direct analogues of Theorem A2.1; see e.g. the survey [57] by the author and the extensive list of references there. Again, such developments motivate the approach in Chapter VII on the Markovian environment model.

The present proof of Theorem A2.1(a) is from Kennedy [529].

A3 Matrix-exponentials

The exponential $e^{\boldsymbol{A}}$ of a $p \times p$ matrix $\boldsymbol{A}$ is defined by the usual series expansion

$$e^{\boldsymbol{A}} = \sum_{n=0}^{\infty} \frac{\boldsymbol{A}^n}{n!}.$$

The series is always convergent because $\boldsymbol{A}^n = \mathrm{O}\bigl(n^k |\lambda|^n\bigr)$ for some integer $k < p$, where λ is the eigenvalue of largest absolute value, $|\lambda| = \max\{|\mu| : \mu \in \mathrm{sp}(\boldsymbol{A})\}$ and $\mathrm{sp}(\boldsymbol{A})$ is the set of all eigenvalues of $\boldsymbol{A}$ (the spectrum).

Some fundamental properties are the following:

$$\mathrm{sp}(e^{\boldsymbol{A}}) = \{e^{\lambda} : \lambda \in \mathrm{sp}(\boldsymbol{A})\} \tag{A.10}$$

$$\frac{\mathrm{d}}{\mathrm{d}t} e^{\boldsymbol{A}t} = \boldsymbol{A} e^{\boldsymbol{A}t} = e^{\boldsymbol{A}t} \boldsymbol{A} \tag{A.11}$$

$$\boldsymbol{A} \int_0^x e^{\boldsymbol{A}t}\,\mathrm{d}t = e^{\boldsymbol{A}x} - \boldsymbol{I} \tag{A.12}$$

$$e^{\boldsymbol{\Delta}^{-1}\boldsymbol{A}\boldsymbol{\Delta}} = \boldsymbol{\Delta}^{-1} e^{\boldsymbol{A}} \boldsymbol{\Delta} \tag{A.13}$$

whenever $\boldsymbol{\Delta}$ is a diagonal matrix with all diagonal elements non-zero.

It is seen from Theorem IX.1.5 that when handling phase-type distributions, one needs to compute matrix-inverses $\boldsymbol{Q}^{-1}$ and matrix-exponentials $e^{\boldsymbol{Q}t}$ (or just $e^{\boldsymbol{Q}}$). Here it is standard to compute matrix-inverses by Gauss-Jordan elimination with full pivoting, whereas there is no similar single established approach in the case of matrix-exponentials. Here are, however, three of the currently most widely used ones:

Example A3.1 (SCALING AND SQUARING) The difficulty in directly applying the series expansion $e^{\boldsymbol{Q}} = \sum_0^{\infty} \boldsymbol{Q}^n/n!$ arises when the elements of $\boldsymbol{Q}$ are large. Then the elements of $\boldsymbol{Q}^n/n!$ do not decrease very rapidly to zero and may contribute a non-negligible amount to $e^{\boldsymbol{Q}}$ even when n is quite large and very many terms of the series may be needed (one may even experience floating point overflow when computing $\boldsymbol{Q}^n$). To circumvent this, write $e^{\boldsymbol{Q}} = (e^{\boldsymbol{K}})^m$ where $\boldsymbol{K} = \boldsymbol{Q}/m$ for some suitable integer m (this is the *scaling* step). Thus, if m is sufficiently large, $\sum_0^{\infty} \boldsymbol{K}^n/n!$ converges rapidly and can be evaluated without problems, and $e^{\boldsymbol{Q}}$ can then be computed as the mth power (by *squaring* if $m = 2$). $\qquad\square$

Example A3.2 (UNIFORMIZATION) Formally, the procedure consists in choosing some suitable $\eta > 0$, letting $\boldsymbol{P} = \boldsymbol{I} + \boldsymbol{Q}/\eta$ and truncating the series in the identity

$$\mathrm{e}^{\boldsymbol{Q}t} = \mathrm{e}^{-\eta t} \sum_{n=0}^{\infty} \frac{\boldsymbol{P}^n (\eta t)^n}{n!} \tag{A.14}$$

which is easily seen to be valid as a consequence of $\mathrm{e}^{\boldsymbol{Q}t} = \mathrm{e}^{\eta(\boldsymbol{P}-\boldsymbol{I})t} = \mathrm{e}^{-\eta t}\mathrm{e}^{\eta \boldsymbol{P}t}$.

The idea which lies behind is *uniformization* of a Markov process $\{X_t\}$, i.e. construction of $\{X_t\}$ by realizing the jump times as a thinning of a Poisson process $\{N_t\}$ with constant intensity η. To this end, assume that $\boldsymbol{Q}$ is the intensity matrix for $\{X_t\}$ and choose η with

$$\eta \geq \max_{i,j} |q_{ij}| = \max_i -q_{ii}. \tag{A.15}$$

Then it is easily checked that $\boldsymbol{P}$ is a transition matrix, and we may consider a new Markov process $\{\widetilde{X}_t\}$ which has jumps governed by $\boldsymbol{P}$ and occurring at epochs of $\{N_t\}$ only (note that since p_{ii} is typically non-zero, some jumps are dummy in the sense that no state transition occurs). However, the intensity matrix $\widetilde{\boldsymbol{Q}}$ is the same as the one $\boldsymbol{Q}$ for $\{X_t\}$ since a jump from i to $j \neq i$ occurs at rate $\widetilde{q}_{ij} = \eta p_{ij} = q_{ij}$. The probabilistic reason that (A.14) holds is therefore that the t-step transition matrix for $\{\widetilde{X}_t\}$ is

$$\mathrm{e}^{\widetilde{\boldsymbol{Q}}t} = \sum_{n=0}^{\infty} \mathrm{e}^{-\eta t} \frac{(\eta t)^n}{n!} \boldsymbol{P}^n$$

(to see this, condition upon the number n of Poisson events in $[0, t]$). $\square$

Example A3.3 (DIFFERENTIAL EQUATIONS) Letting $\boldsymbol{K}_t = \mathrm{e}^{\boldsymbol{Q}t}$, we have $\dot{\boldsymbol{K}} = \boldsymbol{Q}\boldsymbol{K}$ (or $\boldsymbol{K}\boldsymbol{Q}$) which is a system of p^2 linear differential equations which can be solved numerically by standard algorithms (say the Runge-Kutta method) subject to the boundary condition $\boldsymbol{K}_0 = \boldsymbol{I}$.

In practice, what is needed is quite often only $Z_t = \boldsymbol{\pi}\mathrm{e}^{\boldsymbol{Q}t}$ (or $\mathrm{e}^{\boldsymbol{Q}t}\boldsymbol{h}$) with $\boldsymbol{\pi}$ ($\boldsymbol{h}$) a given row (column) vector. One can then reduce to p linear differential equations by noting that $\dot{\boldsymbol{Z}} = \boldsymbol{Z}\boldsymbol{Q}$, $\boldsymbol{Z}_0 = \boldsymbol{\pi}$ ($\dot{\boldsymbol{Z}} = \boldsymbol{Q}\boldsymbol{Z}$, $\boldsymbol{Z}_0 = \boldsymbol{h}$).

The approach is in particular convenient if one wants $\mathrm{e}^{\boldsymbol{Q}t}$ for many different values of t. $\square$

Here is a further method which appears quite appealing at a first sight:

Example A3.4 (DIAGONALIZATION) Assume that $\boldsymbol{Q}$ has diagonal form, i.e. p different eigenvalues $\lambda_1, \ldots, \lambda_p$. Let $\boldsymbol{\nu}_1, \ldots, \boldsymbol{\nu}_p$ be the corresponding left (row) eigenvectors and $h_1, \ldots, h_p$ the corresponding right (column) eigenvectors, $\nu_i Q = \lambda_i \nu_i$, $Q h_i = \lambda_i h_i$. Then $\nu_i h_j = 0$, $i \neq j$, and $\nu_i h_i \neq 0$, and we may adapt some normalization convention ensuring $\nu_i h_i = 1$. Then

$$Q \;=\; \sum_{i=1}^{p} \lambda_i h_i \nu_i \;=\; \sum_{i=1}^{p} \lambda_i h_i \otimes \nu_i, \tag{A.16}$$

$$e^{Qt} \;=\; \sum_{i=1}^{p} e^{\lambda_i t} h_i \nu_i \;=\; \sum_{i=1}^{p} e^{\lambda_i t} h_i \otimes \nu_i. \tag{A.17}$$

Thus, we have an explicit formula for e^{Qt} once the λ_i, ν_i, h_i have been computed; this last step is equivalent to finding a matrix H such that $H^{-1} Q H$ is a diagonal matrix, say $\Delta = (\lambda_i)_{\text{diag}}$, and writing e^{Qt} as

$$e^{Qt} \;=\; H e^{\Delta t} H^{-1} \;=\; H \left(e^{\lambda_i t} \right)_{\text{diag}} H^{-1}. \tag{A.18}$$

Namely, we can take H as the matrix with columns $h_1, \ldots, h_p$.   □

There are, however, two serious drawbacks of this approach:

Numerical instability : If the λ_i are too close, (A.18) contains terms which almost cancel and the loss of digits may be disasterous. The phenomenon occurs not least when the dimension p is large. In view of this phenomenon alone *care should be taken when using diagonalization as a general tool for computing matrix-exponentials.*

Complex calculus : Typically, not all λ_i are real, and we need to have access to software permitting calculations with complex numbers or to perform the cumbersome translation into real and imaginary parts.

Nevertheless, some cases remain where diagonalization may still be appealing.

Example A3.5 If

$$Q = \left(\begin{array}{cc} q_{11} & q_{12} \\ q_{21} & q_{22} \end{array} \right)$$

is 2×2, the eigenvalue, say λ_1, of largest real part is often real (say, under the conditions of the Perron-Frobenius theorem), and hence λ_2 is so because of $\lambda_2 = \text{tr}(Q)$. Everything is nice and explicit here:

$$\lambda_1 = \frac{q_{11} + q_{22} + \sqrt{D}}{2}, \quad \lambda_2 = \frac{q_{11} + q_{22} - \sqrt{D}}{2}, \quad \text{where } D = (q_{11} - q_{22})^2 + 4 q_{12} q_{21}.$$

Write $\boldsymbol{\pi}$ ($= \boldsymbol{\nu}_1$) for the left eigenvector corresponding to λ_1 and $\boldsymbol{k}$ ($= \boldsymbol{h}_1$) for the right eigenvector. Then

$$\boldsymbol{\pi} = (\pi_1 \ \ \pi_2) = a\,(q_{21} \ \ \lambda_1 - q_{11}), \quad \boldsymbol{k} = \begin{pmatrix} k_1 \\ k_2 \end{pmatrix} = b\begin{pmatrix} q_{12} \\ \lambda_1 - q_{11} \end{pmatrix},$$

where a, b are any constants ensuring $\boldsymbol{\pi}\boldsymbol{k} = 1$, i.e.

$$ab\left(q_{12}q_{21} + (\lambda_1 - q_{11})^2\right) = 1.$$

Of course, $\boldsymbol{\nu}_2$ and $\boldsymbol{h}_2$ can be computed in just the same way, replacing λ_1 by λ_2. However, it is easier to note that $\boldsymbol{\pi}\boldsymbol{h}_2 = 0$ and $\boldsymbol{\nu}_2\boldsymbol{k} = 1$ implies

$$\boldsymbol{\nu}_2 = (k_2 \ \ - k_1), \quad \boldsymbol{h}_2 = \begin{pmatrix} \pi_2 \\ -\pi_1 \end{pmatrix}.$$

Thus,

$$\mathrm{e}^{\boldsymbol{Q}t} = \mathrm{e}^{\lambda_1 t}\begin{pmatrix} \pi_1 k_1 & \pi_2 k_1 \\ \pi_1 k_2 & \pi_2 k_2 \end{pmatrix} + \mathrm{e}^{\lambda_2 t}\begin{pmatrix} \pi_2 k_2 & -\pi_2 k_1 \\ -\pi_1 k_2 & \pi_1 k_1 \end{pmatrix}. \tag{A.19}$$

$\square$

Example A3.6 A particular important case arises when

$$\boldsymbol{Q} = \begin{pmatrix} -q_1 & q_1 \\ q_2 & -q_2 \end{pmatrix}$$

is an intensity matrix. Then $\lambda_1 = 0$ and the corresponding left and right eigenvectors are the stationary probability distribution $\boldsymbol{\pi}$ and $\boldsymbol{e}$. The other eigenvalue is $\lambda = \lambda_2 = -q_1 - q_2$, and after some trivial calculus one gets

$$\mathrm{e}^{\boldsymbol{Q}t} = \begin{pmatrix} \pi_1 & \pi_2 \\ \pi_1 & \pi_2 \end{pmatrix} + \mathrm{e}^{\lambda t}\begin{pmatrix} \pi_2 & -\pi_2 \\ -\pi_1 & \pi_1 \end{pmatrix}, \quad \text{where} \tag{A.20}$$

$$\boldsymbol{\pi} = (\pi_1 \ \ \pi_2) = \left(\frac{q_2}{q_1 + q_2} \ \ \frac{q_1}{q_1 + q_2}\right). \tag{A.21}$$

Here the first term is the stationary limit and the second term thus describes the rate of convergence to stationarity. $\square$

Example A3.7 Let

$$\boldsymbol{Q} = \begin{pmatrix} -\dfrac{3}{2} & \dfrac{9}{14} \\[2ex] \dfrac{7}{2} & -\dfrac{11}{2} \end{pmatrix}.$$

Then

$$D = \left(-\frac{3}{2} + \frac{11}{2}\right)^2 + 4\frac{9}{14}\frac{7}{2} = 5^2,$$

$$\lambda_1 = \frac{-3/2 - 11/2 + 5}{2} = -1, \quad \lambda_2 = \frac{-3/2 - 11/2 - 5}{2} = -6,$$

$$1 = ab\left(\frac{9}{14}\frac{7}{2} + (-1 + \frac{3}{2})^2\right) = \frac{5}{2}ab, \quad \boldsymbol{\pi} = a\left(\frac{7}{2} \quad -1 + \frac{3}{2}\right) = a\left(\frac{7}{2} \quad \frac{1}{2}\right),$$

$$\boldsymbol{k} = b\begin{pmatrix} \dfrac{9}{14} \\ -1 + \dfrac{3}{2} \end{pmatrix} = b\begin{pmatrix} \dfrac{9}{14} \\ \dfrac{1}{2} \end{pmatrix},$$

$$\begin{pmatrix} \pi_1 k_1 & \pi_2 k_1 \\ \pi_1 k_2 & \pi_2 k_2 \end{pmatrix} = \frac{2}{5}\begin{pmatrix} \dfrac{9}{10} & \dfrac{9}{70} \\ \dfrac{7}{10} & \dfrac{1}{10} \end{pmatrix},$$

$$e^{\boldsymbol{Q}u} = e^{-u}\begin{pmatrix} \dfrac{9}{10} & \dfrac{9}{10} \\ \dfrac{7}{10} & \dfrac{1}{10} \end{pmatrix} + e^{-6u}\begin{pmatrix} \dfrac{1}{10} & -\dfrac{9}{70} \\ -\dfrac{7}{10} & \dfrac{9}{10} \end{pmatrix}.$$

$\square$

A4 Some linear algebra

4a Generalized inverses

A *generalized inverse* of a matrix A is defined as any matrix $\boldsymbol{A}^-$ satisfying

$$\boldsymbol{A}\boldsymbol{A}^-\boldsymbol{A} = \boldsymbol{A}. \tag{A.22}$$

Note that in this generality it is not assumed that $\boldsymbol{A}$ is necessarily square, but only that dimensions match, and a generalized inverse may not be unique.

Generalized inverses play an important role in statistics. They are most often constructed by imposing some additional properties, for example

$$\boldsymbol{A}\boldsymbol{A}^+\boldsymbol{A} = \boldsymbol{A}, \quad \boldsymbol{A}^+\boldsymbol{A}\boldsymbol{A}^+ = \boldsymbol{A}^+, \quad (\boldsymbol{A}\boldsymbol{A}^+)^\mathsf{T} = \boldsymbol{A}\boldsymbol{A}^+, \quad (\boldsymbol{A}^+\boldsymbol{A})^\mathsf{T} = \boldsymbol{A}^+\boldsymbol{A}. \tag{A.23}$$

A matrix $\boldsymbol{A}^{+}$ satisfying (A.23) is called the *Moore-Penrose inverse* of $\boldsymbol{A}$, and exists and is unique (see for example Rao [727]). E.g., if $\boldsymbol{A}$ is a possibly singular covariance matrix (non-negative definite), then there exists an orthogonal matrix $\boldsymbol{C}$ such that $\boldsymbol{A} = \boldsymbol{C}\boldsymbol{D}\boldsymbol{C}^{\mathsf{T}}$ where

$$
\boldsymbol{D} = \begin{pmatrix}
\lambda_1 & 0 & \cdots & 0 \\
0 & \lambda_2 & \cdots & 0 \\
\vdots & & & \vdots \\
0 & 0 & \cdots & \lambda_p
\end{pmatrix}.
$$

Here we can assume that the λ_i are ordered such that $\lambda_1 > 0, \ldots, \lambda_m > 0$, $\lambda_{m+1} = \ldots = \lambda_p = 0$ where $m \le p$ is the rank of $\boldsymbol{A}$, and can define

$$
\boldsymbol{A}^{+} = \boldsymbol{C} \begin{pmatrix}
\lambda_1^{-1} & 0 & & & 0 \\
& \ddots & & & \\
0 & & \lambda_m^{-1} & 0 & 0 \\
0 & & 0 & 0 & 0 \\
& & & & \ddots & \\
0 & & & & 0
\end{pmatrix} \boldsymbol{C}^{\mathsf{T}}.
$$

In applied probability, one is also faced with singular matrices, most often either an intensity matrix $\boldsymbol{Q}$ or a matrix of the form $\boldsymbol{I} - \boldsymbol{P}$ where $\boldsymbol{P}$ is a transition matrix. Assume that a unique stationary distribution $\boldsymbol{\pi}$ exists. Rather than with generalized inverses, one then works with

$$
\boldsymbol{Q}^{-} = (\boldsymbol{Q} - \boldsymbol{e}\boldsymbol{\pi})^{-1}, \quad (\boldsymbol{I} - \boldsymbol{P})^{-} = (\boldsymbol{I} - \boldsymbol{P} + \boldsymbol{e}\boldsymbol{\pi})^{-1}
$$

(here $(\boldsymbol{I} - \boldsymbol{P} + \boldsymbol{e}\boldsymbol{\pi})^{-1}$ goes under the name *fundamental matrix* of the Markov chain). These matrices are not generalized inverses but act roughly as inverses except that $\boldsymbol{\pi}$ and $\boldsymbol{e}$ play a particular role – e.g.

$$
(\boldsymbol{Q} - \boldsymbol{e}\boldsymbol{\pi})^{-1}\boldsymbol{Q} = \boldsymbol{Q}(\boldsymbol{Q} - \boldsymbol{e}\boldsymbol{\pi})^{-1} = \boldsymbol{I} - \boldsymbol{e}\boldsymbol{\pi}.
$$

Here is a typical result on the role of such matrices in applied probability:

Proposition A4.1 *Let* $\boldsymbol{\Lambda}$ *be an irreducible intensity matrix with stationary row* *vector* $\boldsymbol{\pi}$, *and define* $\boldsymbol{D} = (\boldsymbol{\Lambda} - \boldsymbol{e} \otimes \boldsymbol{\pi})^{-1}$. *Then for some* $b > 0$,

$$\int_0^t e^{\boldsymbol{\Lambda} x}\, \mathrm{d}x \quad = \quad t\boldsymbol{e}\boldsymbol{\pi} + \boldsymbol{D}(e^{\boldsymbol{\Lambda} t} - \boldsymbol{I}) \qquad\qquad\qquad (A.24)$$

$$= \quad t\boldsymbol{e}\boldsymbol{\pi} - \boldsymbol{D} + \mathrm{O}(e^{-bt}), \qquad\qquad\qquad (A.25)$$

$$\int_0^t x e^{\boldsymbol{\Lambda} x}\, \mathrm{d}x \quad = \quad \frac{t^2}{2}\boldsymbol{e}\boldsymbol{\pi} + t(\boldsymbol{D} + \boldsymbol{e}\boldsymbol{\pi}) + \boldsymbol{D}(e^{\boldsymbol{\Lambda} t} - \boldsymbol{I}) - \boldsymbol{D}^2(e^{\boldsymbol{\Lambda} t} - \boldsymbol{I})\ (A.26)$$

$$= \quad \frac{t^2}{2}\boldsymbol{e}\boldsymbol{\pi} + t\boldsymbol{D} - 2\boldsymbol{e}\boldsymbol{\pi} - \boldsymbol{D} + \boldsymbol{D}^2 + \mathrm{O}(e^{-bt}). \qquad (A.27)$$

Proof. Let $\boldsymbol{A}(t)$, $\boldsymbol{B}(t)$ denote the l.h.s. of (A.24), resp. the r.h.s. Then $\boldsymbol{A}(0) = \boldsymbol{B}(0) = 0$,

$$\boldsymbol{B}'(t) \quad = \quad \boldsymbol{e}\boldsymbol{\pi} + \boldsymbol{D}\boldsymbol{\Lambda}e^{\boldsymbol{\Lambda} t} \quad = \quad \boldsymbol{e}\boldsymbol{\pi} + (\boldsymbol{I} - \boldsymbol{e}\boldsymbol{\pi})e^{\boldsymbol{\Lambda} t} \quad = \quad e^{\boldsymbol{\Lambda} t} \quad = \quad \boldsymbol{A}'(t).$$

(A.26) follows by integration by parts:

$$\int_0^t x e^{\boldsymbol{\Lambda} x}\, \mathrm{d}x \quad = \quad \left[x \left\{x\boldsymbol{e}\boldsymbol{\pi} + \boldsymbol{D}(e^{\boldsymbol{\Lambda} x} - \boldsymbol{I})\right\}\right]_0^t - \int_0^t \left\{x\boldsymbol{e}\boldsymbol{\pi} + \boldsymbol{D}(e^{\boldsymbol{\Lambda} x} - \boldsymbol{I})\right\}\, \mathrm{d}x.$$

Finally, the formulas involving $\mathrm{O}(e^{-bt})$ follow by Perron-Frobenius theory, see below.   □

4b   The Kronecker product $\otimes$ and the Kronecker sum $\oplus$

We recall that if $\boldsymbol{A}^{(1)}$ is a $k_1 \times m_1$ and $\boldsymbol{A}^{(2)}$ a $k_2 \times m_2$ matrix, then the Kronecker (tensor) product $\boldsymbol{A}^{(1)} \otimes \boldsymbol{A}^{(2)}$ is the $(k_1 \times k_2) \times (m_1 \times m_2)$ matrix with $(i_1 i_2)(j_1 j_2)$th entry $a_{i_1 j_1}^{(1)} a_{i_2 j_2}^{(2)}$. Equivalently, in block notation $(k_1 = m_1 = 2)$

$$\boldsymbol{A} \otimes \boldsymbol{B} = \begin{pmatrix} a_{11}\boldsymbol{B} & a_{12}\boldsymbol{B} \\ a_{21}\boldsymbol{B} & a_{22}\boldsymbol{B} \end{pmatrix}.$$

Example A4.2 Let $\boldsymbol{\pi}$ be a row vector with m components and $\boldsymbol{h}$ a column vector with k components. Interpreting $\boldsymbol{\pi}, \boldsymbol{h}$ as $1 \times m$ and $k \times 1$ matrices, respectively, it follows that $\boldsymbol{h} \otimes \boldsymbol{\pi}$ is the $k \times m$ matrix with ijth element $h_i \pi_j$. I.e. $\boldsymbol{h} \otimes \boldsymbol{\pi}$ reduces to $\boldsymbol{h}\boldsymbol{\pi}$ in standard matrix notation. Note that $\boldsymbol{h} \otimes \boldsymbol{\pi}$ has rank 1; the rows are proportional to $\boldsymbol{\pi}$, and the columns to $\boldsymbol{h}$, and in fact any rank 1 matrix can be written in this form. For example,

$$\begin{pmatrix} \sqrt{2} \\ \sqrt{3} \end{pmatrix} \otimes \begin{pmatrix} 6 & 7 & 8 \end{pmatrix} = \begin{pmatrix} \sqrt{2} \\ \sqrt{3} \end{pmatrix} \begin{pmatrix} 6 & 7 & 8 \end{pmatrix} = \begin{pmatrix} 6\sqrt{2} & 7\sqrt{2} & 8\sqrt{2} \\ 6\sqrt{3} & 7\sqrt{3} & 8\sqrt{3} \end{pmatrix}.$$

□

Example A4.3 Let

$$A = \begin{pmatrix} 2 & 3 \\ 4 & 5 \end{pmatrix}, \quad B = \begin{pmatrix} \sqrt{6} & \sqrt{7} \\ \sqrt{8} & \sqrt{9} \end{pmatrix}.$$

Then

$$A \otimes B = \begin{pmatrix} 2\sqrt{6} & 2\sqrt{7} & 3\sqrt{6} & 3\sqrt{7} \\ 2\sqrt{8} & 2\sqrt{9} & 3\sqrt{8} & 3\sqrt{9} \\ 4\sqrt{6} & 4\sqrt{7} & 5\sqrt{6} & 5\sqrt{7} \\ 4\sqrt{8} & 4\sqrt{9} & 5\sqrt{8} & 5\sqrt{9} \end{pmatrix}.$$

$\square$

A fundamental formula is

$$(A_1 B_1 C_1) \otimes (A_2 B_2 C_2) = (A_1 \otimes A_2)(B_1 \otimes B_2)(C_1 \otimes C_2). \tag{A.28}$$

In particular, if $A_1 = \nu_1$, $A_2 = \nu_2$ are row vectors and $C_1 = h_1$, $C_2 = h_2$ are column vectors, then $\nu_1 B_1 h_1$ and $\nu_2 B_2 h_2$ are real numbers, and

$$\nu_1 B_1 h_1 \cdot \nu_2 B_2 h_2 = \nu_1 B_1 h_1 \otimes \nu_2 B_2 h_2 = (\nu_1 \otimes \nu_2)(B_1 \otimes B_2)(h_1 \otimes h_2). \tag{A.29}$$

If A and B are both square ($k_1 = m_1$ and $k_2 = m_2$), then the Kronecker sum is defined by

$$A^{(1)} \oplus A^{(2)} = A^{(1)} \otimes I_{k_2} + I_{k_1} \otimes A^{(2)}. \tag{A.30}$$

A crucial property is the fact that the functional equation for the exponential function generalizes to Kronecker notation (note that in contrast $e^{A+B} = e^A e^B$ typically only holds when A and B commute):

Proposition A4.4 $e^{A \oplus B} = e^A \otimes e^B$.

Proof. We shall use the binomial formula

$$(A \oplus B)^\ell = \sum_{k=0}^{\ell} \binom{\ell}{k} A^k \otimes B^{\ell-k}. \tag{A.31}$$

Indeed,

$$(A \oplus B)^\ell = (A \otimes I + I \otimes B)^\ell$$

is the sum of all products of ℓ factors, each of which is $A \otimes I$ or $I \otimes B$; if $A \otimes I$ occurs k times, such a factor is $A^k \otimes B^{\ell-k}$ according to (A.29), and the number of such factors is precisely given by the relevant binomial coefficient.

Using (A.31), it follows that

$$
\begin{aligned}
\mathrm{e}^{A} \otimes \mathrm{e}^{B} &= \left(\sum_{n=0}^{\infty} \frac{A^n}{n!}\right) \otimes \left(\sum_{n=0}^{\infty} \frac{B^n}{n!}\right) = \sum_{\ell=0}^{\infty} \sum_{k=0}^{\ell} \frac{A^k \otimes B^{\ell-k}}{k!(\ell-k)!} \\
&= \sum_{\ell=0}^{\infty} \frac{(A \oplus B)^{\ell}}{\ell!} = \mathrm{e}^{A \oplus B}.
\end{aligned}
$$

$\square$

Remark A4.5 Many of the concepts and results in Kronecker calculus have intuitive illustrations in probabilistic terms. Thus, $P = P^{(1)} \otimes P^{(2)}$ is the transition matrix of the bivariate Markov chain $\{(X_n^{(1)}, X_n^{(2)})\}$, where $\{X_n^{(1)}\}$, $\{X_n^{(2)}\}$ are independent Markov chains with transition matrices $P^{(1)}, P^{(2)}$, and

$$
Q = Q^{(1)} \oplus Q^{(2)} = Q^{(1)} \otimes I + I \otimes Q^{(2)} \tag{A.32}
$$

is the intensity matrix of the bivariate continuous Markov process $\{(Y_t^{(1)}, Y_t^{(2)})\}$, where $\{Y_t^{(1)}\}$, $\{Y_t^{(2)}\}$ are independent Markov processes with intensity matrices $Q^{(1)}, Q^{(2)}$; in the definition (A.32), the first term on the r.h.s. represents transitions in the $\{Y_t^{(1)}\}$ component and the second transitions in the $\{Y_t^{(2)}\}$ component, and the form of the bivariate intensity matrix reflects the fact that due to independence, $\{(Y_t^{(1)}, Y_t^{(2)})\}$ cannot change state in both components at the same time.

A special case of Proposition A4.4 can easily be obtained by probabilistic reasoning along the same lines. Let $P_s, P_s^{(1)}, P_s^{(2)}$ be the s-step transition matrices of $\{Y_t^{(1)}, Y_t^{(2)}\}$, $\{Y_t^{(1)}\}$, resp. $\{Y_t^{(2)}\}$. From what has been said about independent Markov chains, we have $P_s = P_s^{(1)} \otimes P_s^{(2)}$. On the other hand,

$$
P_s = \exp\{sQ\} = \exp\{s(Q^{(1)} \oplus Q^{(2)})\},
$$

$$
P_s^{(1)} = \exp\{sQ^{(1)}\}, \quad P_s^{(2)} = \exp\{sQ^{(2)}big\}.
$$

Taking $s = 1$ for simplicity, $P_s = P_s^{(1)} \otimes P_s^{(2)}$ can therefore be rewritten as

$$
\exp\{Q^{(1)} \oplus Q^{(2)}\} = \exp\{Q^{(1)}\} \otimes \exp\{Q^{(2)}\}.
$$

$\square$

Also the following formula is basic:

Lemma A4.6 *Suppose that $\boldsymbol{A}$ and $\boldsymbol{B}$ are both square such that $\alpha + \beta < 0$ whenever α is an eigenvalue of $\boldsymbol{A}$ and β is an eigenvalue of $\boldsymbol{B}$. Let further $\boldsymbol{\pi}, \boldsymbol{\nu}$ be any row vectors and $\boldsymbol{h}, \boldsymbol{k}$ any column vectors. Then*

$$\int_0^x \boldsymbol{\pi} e^{At} \boldsymbol{h} \cdot \boldsymbol{\nu} e^{Bt} \boldsymbol{k} \, \mathrm{d}t \;=\; (\boldsymbol{\pi} \otimes \boldsymbol{\nu})(\boldsymbol{A} \oplus \boldsymbol{B})^{-1}(e^{A \oplus B \, x} - \boldsymbol{I})(\boldsymbol{h} \otimes \boldsymbol{k}). \qquad (A.33)$$

Proof. According to (A.29), the integrand can be written as

$$(\boldsymbol{\pi} \otimes \boldsymbol{\nu})(e^{At} \otimes e^{Bt})(\boldsymbol{h} \otimes \boldsymbol{k}) \;=\; (\boldsymbol{\pi} \otimes \boldsymbol{\nu})(e^{A \oplus B \, t})(\boldsymbol{h} \otimes \boldsymbol{k}).$$

Now note that the eigenvalues of $\boldsymbol{A} \oplus \boldsymbol{B}$ are of the form $\alpha + \beta$ whenever α is an eigenvalue of $\boldsymbol{A}$ and β is an eigenvalue of $\boldsymbol{B}$, so that by assumption $\boldsymbol{A} \oplus \boldsymbol{B}$ is invertible, and appeal to (A.12). $\qquad \square$

4c The Perron-Frobenius theorem

Let $\boldsymbol{A}$ be a $p \times p$-matrix with non-negative elements. We call $\boldsymbol{A}$ *irreducible* if the pattern of zero and non-zero elements is the same as for an irreducible transition matrix. That is, for each $i, j = 1, \ldots, p$ there should exist $i_0, i_1, \ldots, i_n$ such that $i_0 = i$, $i_n = j$ and $a_{i_{k-1} i_k} > 0$ for $k = 1, \ldots, n$. Similarly, $\boldsymbol{A}$ is called *aperiodic* if the pattern of zero and non-zero elements is the same as for an aperiodic transition matrix.

Here is the Perron-Frobenius theorem, which can be found in a great number of books, see e.g. [APQ, I.6] and references there:

Theorem A4.7 *Let $\boldsymbol{A}$ be a $p \times p$-matrix with non-negative elements. Then:*
(a) The spectral radius $\lambda_0 = \max\{|\lambda| : \lambda \in \mathrm{sp}(\boldsymbol{A})\}$ is itself a strictly positive and simple eigenvalue of $\boldsymbol{A}$, and the corresponding left and right eigenvectors $\boldsymbol{\nu}, \boldsymbol{h}$ can be chosen with strictly positive elements;
(b) if in addition $\boldsymbol{A}$ is aperiodic, then $|\lambda| < \lambda_0$ for all $\lambda \in \mathrm{sp}(\boldsymbol{A})$, and if we normalize $\boldsymbol{\nu}, \boldsymbol{h}$ such that $\boldsymbol{\nu} \boldsymbol{h} = 1$, then

$$\boldsymbol{A}^n \;=\; \lambda_0^n \boldsymbol{h} \boldsymbol{\nu} + \mathrm{O}(\mu^n) \;=\; \lambda_0^n \boldsymbol{h} \otimes \boldsymbol{\nu} + \mathrm{O}(\mu^n) \qquad (A.34)$$

for some $\mu \in (0, \lambda_0)$.

Note that for a transition matrix, we have $\lambda_0 = 1$, $\boldsymbol{h} = \boldsymbol{e}$ and $\boldsymbol{\nu} = \boldsymbol{\pi}$ (the stationary row vector).

The Perron-Frobenius theorem has an analogue for matrices $\boldsymbol{B}$ with properties similar to intensity matrices:

Corollary A4.8 *Let B be an irreducible[2] $p \times p$-matrix with non-negative off-diagonal elements. Then the eigenvalue λ_0 with largest real part is simple and real, and the corresponding left and right eigenvectors ν, h can be chosen with strictly positive elements. Furthermore, if we normalize ν, h such that $\nu h = 1$, then*

$$e^{Bt} = e^{\lambda_0 t} h\nu + O(e^{\mu t}) = e^{\lambda_0 t} h \otimes \nu + O(e^{\mu t}) \tag{A.35}$$

for some $\mu \in (-\infty, \lambda_0)$.

Note that for an intensity matrix, we have $\lambda_0 = 0$, $h = e$ and $\nu = \pi$ (the stationary row vector).

Corollary A4.8 is most often not stated explicitly in textbooks (but see [APQ, II.4d] for intensity matrices!), but is an easy consequence of the Perron-Frobenius theorem. For example, one can consider $A = \eta I + B$ where $\eta > 0$ is so large that all diagonal elements of A are strictly positive (then A is irreducible and aperiodic), relate the eigenvalues of B to those of B via (A.10) and use the formula

$$e^{Bt} = e^{-\eta t} e^{At} = e^{-\eta t} \sum_{n=0}^{\infty} \frac{A^n t^n}{n!}$$

(cf. the analogy of this procedure with uniformization, Example A3.2).

A5   Complements on phase-type distributions

5a   Asymptotic exponentiality

In Proposition IX.1.8, it was shown that under mild conditions the tail of a phase-type distribution B is asymptotically exponential. The next result gives a condition for asymptotic exponentiality, not only in the tail but in the whole distribution. The content is that B is approximately exponential if the exit rates t_i are small compared to the feedback intensities t_{ij} ($i \neq j$). To this end, note that we can write the phase generator T as $Q - (t_i)_{\text{diag}}$ where $Q = T + (t_i)_{\text{diag}}$ is a proper intensity matrix ($Qe = 0$). I.e. the condition is that t is small compared to Q.

Proposition A5.1 *Let Q be a proper irreducible intensity matrix with stationary distribution α, let $t = (t_i)_{i \in E} \neq 0$ have non-negative entries and define $T^{(a)} = aQ - (t_i)_{\text{diag}}$. Then for any β, the phase-type distribution $B^{(a)}$ with representation $\left(\beta, T^{(a)}\right)$ is asymptotically exponential with parameter $t^* = \sum_{i \in E} \alpha_i t_i$ as $a \to \infty$, $\overline{B}^{(a)}(x) \to e^{-t^* x}$.*

---

[2]By this, we mean that the pattern of non-zero off-diagonal elements is the same as for an irreducible intensity matrix.

Proof. Let $\{J_t^{(a)}\}$ be the phase process associated with $B^{(a)}$ and $\zeta^{(a)}$ its life-length, let $\{Y_t^{(a)}\}$ be a Markov process with initial distribution $\boldsymbol{\alpha}$ and intensity matrix $a\boldsymbol{Q}$, and write $Y_t = Y_t^{(1)}$, $\zeta = \zeta^{(1)}$ etc. We can assume that $J_t^{(a)} = Y_t^{(a)}$, $t < \zeta^{(a)}$, and that $Y_t^{(a)} = Y_{at}$ for all t. Let further V be exponential with intensity V and independent of everything else. We can think of $\zeta^{(a)}$ as the first event in an inhomogeneous Poisson process (Cox process) with intensity process $\{t_{Y_v^{(a)}}\}_{v \geq 0}$. Hence we can represent $\zeta^{(a)}$ as

$$
\begin{aligned}
\zeta^{(a)} &= \inf\Big\{t \geq 0 : \int_0^t t_{Y_v^{(a)}}\,\mathrm{d}v = V\Big\} = \inf\Big\{t \geq 0 : \int_0^t t_{Y_{av}}\,\mathrm{d}v = V\Big\} \\
&= \inf\Big\{t \geq 0 : \int_0^{at} t_{Y_v}\,\mathrm{d}v = aV\Big\} = \frac{1}{a}\sigma(aV),
\end{aligned}
$$

where $\sigma(x) = \inf\{t \geq 0 : \int_0^t t_{Y_v}\,\mathrm{d}v = x\}$. By the law of large numbers for Markov processes, $\int_0^t t_{Y_v}\,\mathrm{d}v/t \overset{\text{a.s.}}{\to} t^*$, and this easily yields $\sigma(x)/x \overset{\text{a.s.}}{\to} 1/t^*$. Hence $\zeta^{(a)} \overset{\text{a.s.}}{\to} V/t^*$. $\square$

We shall, in fact, prove a somewhat more general result which was used in the proof of Proposition VII.1.9. In addition to the asymptotic exponentiality, it states that the state, from which the phase process is terminated, has a limit distribution:

Proposition A5.2 $\mathbb{P}_i\big(\zeta^{(a)} > x,\ J_{\zeta^{(a)}-}^{(a)} = i\big) \to \mathrm{e}^{-t^*x} \cdot \dfrac{\alpha_i t_i}{t^*}$.

Proof. Assume first $t_i > 0$ for all i and let $I_x = Y_{\sigma(x)}$. Then $\{I_x\}$ is a Markov process with $I_0 = Y_0$. Conditioning upon whether $\{Y_t\}$ changes state in $[0, \mathrm{d}x/t_i]$ or not, we get

$$
\mathbb{P}_i(I_{\mathrm{d}x} = j) = (1 + q_{ii}\frac{\mathrm{d}x}{t_i})\delta_{ij} + q_{ij}\frac{\mathrm{d}x}{t_i}(1 - \delta_{ij}).
$$

Hence the intensity matrix of $\{I_x\}$ is $(q_{ij}/t_i)_{i,j\in E}$, from which it is easily checked that the limiting stationary distribution is $(\alpha_i t_i/t^*)_{i\in E}$.

Now let $a' \to \infty$ with a in such a way that $a' < a$, $a'/a \to 1$, $a - a' \to \infty$ (e.g. $a' = a - a^\epsilon$ where $0 < \epsilon < 1$). Then $\sigma(a'V)/\sigma(aV) \overset{\text{a.s.}}{\to} 1$. Since

$$
J_{\zeta^{(a)}-}^{(a)} = Y_{\zeta^{(a)}}^{(a)} = Y_{a\zeta^{(a)}} = Y_{\sigma(aV)},
$$

it follows that

$$\mathbb{P}_i\left(\zeta^{(a)} > x, J^{(a)}_{\zeta^{(a)}-} = j\right)$$

$$= \mathbb{P}_i\left(\frac{\sigma(aV)}{a} > x, Y_{\sigma(aV)} = j\right) \sim \mathbb{P}_i\left(\frac{\sigma(a'V)}{a'} > x, Y_{\sigma(aV)} = j\right)$$

$$= \mathbb{E}_i\left[\left(\frac{\sigma(a'V)}{a'} > x\right)\mathbb{P}\left(Y_{\sigma(aV)} = j \mid \mathscr{F}_{\sigma(a'V)}\right)\right]$$

$$\sim \mathbb{E}_i\left[I\left(\frac{\sigma(a'V)}{a'} > x\right) \cdot \frac{\alpha_i t_i}{t^*}\right] \rightarrow e^{-t^* x} \cdot \frac{\alpha_i t_i}{t^*}.$$

Reducing the state space of $\{I_x\}$ to $\{i \in E : t_i > 0\}$, an easy modification of the argument yields finally the result for the case where $t_i = 0$ for one or more i.

$\square$

Notes and references Propositions A5.1 and A5.2 do not appear to be in the literature. However, these results are in the spirit of rare events theory for regenerative processes (e.g. Keilson [523], Gnedenko & Kovalenko [420] and Glasserman & Kou [418]). See also Korolyuk, Penev & Turbin [555].

5b Discrete phase-type distributions

The theory of discrete phase-type distributions is a close parallel of the continuous case, so we shall be brief.

A distribution B on $\{1, 2, \ldots\}$ is said to be *discrete phase-type with representation* $(E, \boldsymbol{P}, \boldsymbol{\alpha})$ if B is the lifelength of a terminating Markov chain (in discrete time) on E which has transition matrix $\boldsymbol{P} = (p_{ij})$ and initial distribution $\boldsymbol{\alpha}$. Then $\boldsymbol{P}$ is substochastic and the vector of exit probabilities is $\boldsymbol{p} = \boldsymbol{e} - \boldsymbol{P}\boldsymbol{e}$.

Example A5.3 As the exponential distribution is the simplest continuous phase-type distribution, so is the geometric distribution, with point probabilities $b_k = (1-p)^{k-1}p$, $k = 1, 2, \ldots$, the simplest discrete phase-type distribution: here E has only one element, and thus the parameter p of the geometric distribution can be identified with the exit probability vector $\boldsymbol{p}$.

$\square$

Example A5.4 *Any discrete distribution B with finite support,* say $b_k = 0$, $k > K$, *is discrete phase-type.* Indeed, let $E = \{1, \ldots, K\}$, $\boldsymbol{\alpha} = \boldsymbol{b} = (b_k)_{k=1,\ldots,K}$ and

$$p_{kj} = \begin{cases} 1 & k > 1, \ j = k - 1, \\ 0 & \text{otherwise,} \end{cases}, \quad p_k = \begin{cases} 1 & k = 1 \\ 0 & k > 1 \end{cases}.$$

$\square$

Theorem A5.5 *Let B be discrete phase-type with representation $(\boldsymbol{P}, \boldsymbol{\alpha})$. Then:*
(a) *The point probabilities are $b_k = \boldsymbol{\alpha} \boldsymbol{P}^{k-1} \boldsymbol{p}$;*
(b) *the generating function $\widehat{b}[z] = \sum_{k=1}^{\infty} z^k b_k$ is $z \boldsymbol{\alpha} (\boldsymbol{I} - z\boldsymbol{P})^{-1} \boldsymbol{p}$;*
(c) *the nth moment $\sum_{k=1}^{\infty} k^n b_k$ is $(-1)^n n! \, \boldsymbol{\alpha} \boldsymbol{P}^{-n} \boldsymbol{p}$.*

5c   Closure properties

Example A5.6 (CONVOLUTIONS)   Let B_1, B_2 be phase-type with representations $(E^{(1)}, \boldsymbol{\alpha}^{(1)}, \boldsymbol{T}^{(1)})$, resp. $(E^{(2)}, \boldsymbol{\alpha}^{(2)}, \boldsymbol{T}^{(2)})$. Then the convolution $B = B_1 * B_2$ is phase-type with representation $(E, \boldsymbol{\alpha}, \boldsymbol{T})$ where $E = E^{(1)} + E^{(2)}$ is the disjoint union of $E^{(1)}$ and $E^{(2)}$, and

$$\alpha_i = \left\{ \begin{array}{ll} \alpha_i^{(1)}, & i \in E^{(1)} \\ 0, & i \in E^{(2)} \end{array} \right. , \quad \boldsymbol{T} = \left(\begin{array}{cc} \boldsymbol{T}^{(1)} & \boldsymbol{t}^{(1)} \boldsymbol{\alpha}^{(2)} \\ 0 & \boldsymbol{T}^{(2)} \end{array} \right) \qquad \text{(A.36)}$$

in block-partitioned notation (where we could also write $\boldsymbol{\alpha}$ as $(\boldsymbol{\alpha}^{(1)} \, 0)$). A reduced phase diagram (omitting transitions within the two blocks) is

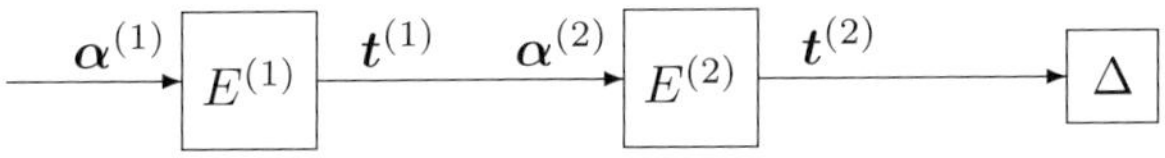

FIGURE A.2

The form of these results is easily recognized if one considers two independent phase processes $\big\{ J_t^{(1)} \big\}$, $\big\{ J_t^{(2)} \big\}$ with lifetimes U_1, resp. U_2, and piece the processes together by

$$J_t = \left\{ \begin{array}{ll} J_t^{(1)}, & 0 \le t < U_1 \\ J_{t-U_1}^{(2)}, & U_1 \le t < U_1 + U_2 \\ \Delta, & t \ge U_1 + U_2. \end{array} \right.$$

Then $\{J_t\}$ has lifetime $U_1 + U_2$, initial distribution $\boldsymbol{\alpha}$ and phase generator $\boldsymbol{T}$.
$\square$

Example A5.7 (THE NEGATIVE BINOMIAL DISTRIBUTION)   The most trivial special case of Example A5.6 is the Erlang distribution E_r which is the convolution of r exponential distributions. The discrete counterpart is the negative binomial distribution with point probabilities

$$b_k = \left(\begin{array}{c} k-1 \\ r-1 \end{array} \right) (1-p)^{k-r} p^r, \quad k = r, r+1, \ldots.$$

This corresponds to a convolution of r geometric distributions with the same parameter p, and hence the negative binomial distribution is discrete phase-type, as is seen by minor modifications of Example A5.6.   □

Example A5.8 (FINITE MIXTURES)   Let B_1, B_2 be phase-type with representations $(E^{(1)}, \boldsymbol{\alpha}^{(1)}, \boldsymbol{T}^{(1)})$, resp. $(E^{(2)}, \boldsymbol{\alpha}^{(2)}, \boldsymbol{T}^{(2)})$. Then the mixture $B = \theta B_1 + (1 - \theta) B_2$ $(0 \leq \theta \leq 1)$ is phase-type with representation $(E, \boldsymbol{\alpha}, \boldsymbol{T})$ where $E = E^{(1)} + E^{(2)}$ is the disjoint union of $E^{(1)}$ and $E^{(2)}$, and

$$\alpha_i = \left\{ \begin{array}{ll} \theta \alpha_i^{(1)}, & i \in E^{(1)} \\ (1 - \theta)\alpha_i^{(2)}, & i \in E^{(2)} \end{array} \right\}, \quad \boldsymbol{T} = \left(\begin{array}{cc} \boldsymbol{T}^{(1)} & 0 \\ 0 & \boldsymbol{T}^{(2)} \end{array} \right) \qquad \text{(A.37)}$$

(in block-partitioned notation, this means that $\boldsymbol{\alpha} = \left(\theta \boldsymbol{\alpha}^{(1)} \quad (1 - \theta)\boldsymbol{\alpha}^{(2)} \right)$). A reduced phase diagram is

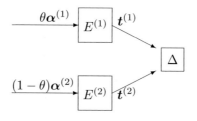

FIGURE A.3

In exactly the same way, a mixture of more than two phase-type distributions is seen to be phase-type. In risk theory, one obvious interpretation of the claim size distribution B to be a mixture is several types of claims.   □

Example A5.9 (INFINITE MIXTURES WITH $\boldsymbol{T}$ FIXED)   Assume that $\boldsymbol{\alpha} = \boldsymbol{\alpha}^{(\alpha)}$ depends on a parameter $\alpha \in A$ whereas E and $\boldsymbol{T}$ are the same for all α. Let $B^{(\alpha)}$ be the corresponding phase-type distribution, and consider $B^{(\nu)} = \int_A B^{(\alpha)} \nu(\mathrm{d}\alpha)$ where ν is a probability measure on A. Then it is trivial to see that $B^{(\nu)}$ is phase-type with representation $\left(\boldsymbol{\alpha}^{(\nu)}, \boldsymbol{T}, E \right)$ where $\boldsymbol{\alpha}^{(\nu)} = \int_A \boldsymbol{\alpha}^{(\alpha)} \nu(\mathrm{d}\alpha)$.   □

Example A5.10 (GEOMETRIC COMPOUNDS)   Let B be phase-type with representation $(E, \boldsymbol{\alpha}, \boldsymbol{T})$ and $C = \sum_{n=1}^{\infty} (1 - \rho)\rho^{n-1} B^{*n}$. Equivalently, if $U_1, U_2, \ldots$ are i.i.d. with common distribution and N is independent of the U_k and geometrically distributed with parameter ρ, $\mathbb{P}(N = n) = (1 - \rho)\rho^{n-1}$, then C is the distribution of $U_1 + \cdots + U_N$. To obtain a phase process for C, we need to restart the phase process for B w.p. ρ at each termination. Thus, a reduced phase diagram is

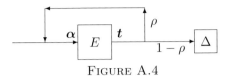

FIGURE A.4

and C is phase-type with representation $(E, \boldsymbol{\alpha}, \boldsymbol{T} + \rho \boldsymbol{t}\boldsymbol{\alpha})$. Minor modifications of the argument show that

1. If U_1 has a different initial vector, say $\boldsymbol{\nu}$, but the same $\boldsymbol{T}$, then $U_1 + \cdots + U_N$ is phase-type with representation $(E, \boldsymbol{\nu}, \boldsymbol{T} + \rho \boldsymbol{t}\boldsymbol{\alpha})$;

2. if B is defective and $N + 1$ is the first n with $U_n = \infty$, then $U_1 + \cdots + U_N$ is zero-modified phase-type with representation $(\boldsymbol{\alpha}, \boldsymbol{T} + \boldsymbol{t}\boldsymbol{\alpha}, E)$. Note that this was exactly the structure of the lifetime of a terminating renewal process, cf. Corollary IX.2.2. □

Example A5.11 (OVERSHOOTS) The overshoot of U over x is defined as the distribution of $(U - x)^+$. It is zero-modified phase-type with representation $(E, \boldsymbol{\alpha}\mathrm{e}^{\boldsymbol{T}x}, \boldsymbol{T})$ if U is phase-type with representation $(E, \boldsymbol{\alpha}, \boldsymbol{T})$. Indeed, if $\{J_t\}$ is a phase process for U, then J_x has distribution $\boldsymbol{\alpha}\mathrm{e}^{\boldsymbol{T}x}$.

If we replace x by a r.v. X independent of U, say with distribution F, it follows by mixing (Example A5.9) that $(U - X)^+$ is zero-modified phase-type with representation $(E, \boldsymbol{\alpha}\widehat{F}[\boldsymbol{T}], \boldsymbol{T})$ where

$$\widehat{F}[\boldsymbol{T}] = \int_0^\infty \mathrm{e}^{\boldsymbol{T}x} F(\mathrm{d}x)$$

is the matrix m.g.f. of F, cf. Proposition IX.1.7. □

Example A5.12 (PHASE-TYPE COMPOUNDS) Let $f_1, f_2, \ldots$ be the point probabilities of a discrete phase-type distribution with representation $(E, \boldsymbol{\alpha}, \boldsymbol{P})$, let B be a continuous phase-type distribution with representation $(F, \boldsymbol{\nu}, \boldsymbol{T})$ and $C = \sum_{n=1}^\infty f_n B^{*n}$. Equivalently, if $U_1, U_2, \ldots$ are i.i.d. with common distribution B and N is independent of the U_k with $\mathbb{P}(N = n) = f_n$, then C is the distribution of $U_1 + \cdots + U_N$. To obtain a phase representation for C, let the phase space be $E \times F = \{ij : i \in E, j \in F\}$, let the initial vector be $\boldsymbol{\alpha} \otimes \boldsymbol{\nu}$ and let the phase generator be $\boldsymbol{I} \otimes \boldsymbol{T} + \boldsymbol{P} \otimes (\boldsymbol{t}\boldsymbol{\alpha})$. □

Example A5.13 (MINIMA AND MAXIMA) Let U_1, U_2 be random variables with distributions B_1, B_2 of phase-type with representations $(E^{(1)}, \boldsymbol{\alpha}^{(1)}, \boldsymbol{T}^{(1)})$, resp. $(E^{(2)}, \boldsymbol{\alpha}^{(2)}, \boldsymbol{T}^{(2)})$. Then *the minimum $U_1 \wedge U_2$ and the maximum $U_1 \vee U_2$ are again phase-type.*

To see this, let $\{J_t^{(1)}\}$, $\{J_t^{(2)}\}$ be independent with lifetimes U_1, resp. U_2. For $U_1 \wedge U_2$, we then let the governing phase process be $\{J_t\} = \{(J_t^{(1)}, J_t^{(2)})\}$, interpreting exit of either of $\{J_t^{(1)}\}$, $\{J_t^{(2)}\}$ as exit of $\{J_t\}$. Thus the representation is

$$(E^{(1)} \times E^{(2)}, \boldsymbol{\alpha}^{(1)} \oplus \boldsymbol{\alpha}^{(2)}, \boldsymbol{T}^{(1)} \oplus \boldsymbol{T}^{(2)}).$$

For $U_1 \vee U_2$, we need to allow $\{J_t^{(2)}\}$ to go on (on $E^{(2)}$) when $\{J_t^{(1)}\}$ exits, and vice versa. Thus the state space is $E^{(1)} \times E^{(2)} \cup E^{(1)} \cup E^{(2)}$, the initial vector is $(\boldsymbol{\alpha}^{(1)} \otimes \boldsymbol{\alpha}^{(2)} \ 0 \ 0)$, and the phase generator is

$$\begin{pmatrix} \boldsymbol{T}^{(1)} \oplus \boldsymbol{T}^{(2)} & \boldsymbol{T}^{(1)} \oplus \boldsymbol{t}^{(2)} & \boldsymbol{t}^{(1)} \oplus \boldsymbol{T}^{(2)} \\ 0 & \boldsymbol{T}^{(1)} & 0 \\ 0 & 0 & \boldsymbol{T}^{(2)} \end{pmatrix}.$$

$\square$

Notes and references The results of the present section are standard, see Neuts [660] (where the proof, however, relies more on matrix algebra than the probabilistic interpretation exploited here).

5d Phase-type approximation

A fundamental property of phase-type distributions is denseness. That is, any distribution B on $(0, \infty)$ can be approximated 'arbitrarily close' by a phase-type distribution $\widetilde{B}$:

Theorem A5.14 *To a given distribution B on $(0, \infty)$, there is a sequence $\{B_n\}$ of phase-type distributions such that $B_n \xrightarrow{\mathcal{D}} B$ as $n \to \infty$.*

Proof. Assume first that B is a one-point distribution, say degenerate at b, and let B_n be the Erlang distribution $E_n(\delta_n)$ with $\delta_n = n/b$. The mean of B_n is $n/\delta_n = b$ and the variance is $n/\delta_n^2 = b^2/n$. Hence it is immediate that $B_n \xrightarrow{\mathcal{D}} B$.

The general case now follows easily from this, the fact that any distribution B can be approximated arbitrarily close by a distribution with finite support, and the closedness of the class of phase-type distributions under the formation of finite mixtures, cf. Example A5.8. Here are the details at two somewhat different levels of abstraction:

(diagonal argument, elementary) Let $\{b_k\}$ be any dense sequence of continuity points for $B(x)$. Then we must find phase-type distributions B_n with $B_n(b_k) \to B(b_k)$ for all k. Now we can find first a sequence $\{D_m\}$

of distributions with finite support such that $D_n(b_k) \to B(b_k)$ for all k as $n \to \infty$. By the diagonal argument (subsequent thinnings), we can assume that $|D_n(b_k) - B(b_k)| \le 1/n$ for $n \ge k$. Let the support of D_n be $\{x_1(n), \ldots, x_{q(n)}(n)\}$, with weight $p_i(n)$ for $x_i(n)$. Then from above,

$$C_{r,n} = \sum_{i=1}^{q(n)} p_i(n) E_r\left(\frac{r}{x_i(n)}\right) \xrightarrow{\mathcal{D}} \sum_{i=1}^{q(n)} p_i(n)\delta_{x_i(n)} = D_n, \quad r \to \infty.$$

Hence we can choose $r(n)$ in such a way that

$$\left|C_{r(n),n}(b_k) - D(b_k)\right| \le \frac{1}{n}, \quad k \le n.$$

Then

$$\left|C_{r(n),n}(b_k) - B(b_k)\right| \le \frac{2}{n}, \quad k \le n,$$

and we can take $B_n = C_{r(n),n}$. $\qquad\square$

(abstract topological) The essence of the argument above is that the closure (w.r.t. the topology for weak convergence) $\overline{\mathscr{P}HT}$ of the class $\mathscr{P}HT$ of phase-type distributions contains all one-point distributions. Since $\mathscr{P}HT$ is closed under the continuous operation of formation of finite mixtures, $\overline{\mathscr{P}HT}$ contains all finite mixtures of one-point distributions, i.e. the class $\mathscr{L}_0$ of all discrete distributions. But $\overline{\mathscr{L}_0}$ is the class $\mathscr{L}$ of all distributions on $[0, \infty)$. Hence $\mathscr{L} \subseteq \overline{\mathscr{P}HT}$ and $\mathscr{L} = \overline{\mathscr{P}HT}$. $\qquad\square$

Theorem A5.14 is fundamental and can motivate phase-type assumptions, say on the claim size distribution B in risk theory, in at least two ways:

insensitivity Suppose we are able to verify a specific result when B is of phase-type say that two functionals $\varphi_1(B)$ and $\varphi_2(B)$ coincide. If $\varphi_1(B)$ and $\varphi_2(B)$ are weakly continuous, then it is immediate that $\varphi_1(B) = \varphi_2(B)$ for *all* distributions B on $[0, \infty)$.

approximation Assume that we can compute a functional $\varphi(B)$ when B is phase-type, and that φ is known to be continuous. For a general B_0, we can then approximate B_0 by a phase-type B, compute $\varphi(B)$ and use this quantity as an approximation to $\varphi(B_0)$. In particular, if information on B_0 is given in terms of observations (i.i.d. replications), one would use the B given by some statistical fitting procedure (see below).

It should be noted, however, that this procedure should be used with care if $\varphi(B)$ is the ruin probability $\psi(u)$ and u is large.

Let $\mathscr{E}$ be the class of functions $f : [0, \infty) \to [0, \infty)$ such that $f(x) = O(e^{\alpha x})$, $x \to \infty$, for some $\alpha < \infty$.

Corollary A5.15 *To a given distribution B on $(0, \infty)$ and any $f_1, f_2, \ldots \in \mathscr{E}$, there is a sequence $\{B_n\}$ of phase-type distributions such that $B_n \xrightarrow{\mathscr{D}} B$ as $n \to \infty$ and $\int_0^\infty f_i(x) B_n(\mathrm{d}x) \to \int_0^\infty f_i(x) B(\mathrm{d}x)$, $i = 1, 2, \ldots$*

Proof. By Fatou's lemma, $B_n \xrightarrow{\mathscr{D}} B$ implies that

$$\liminf_{n \to \infty} \int_0^\infty f_i(x) B_n(\mathrm{d}x) \geq \int_0^\infty f_i(x) B(\mathrm{d}x),$$

for each i, and hence it is sufficient to show that we can obtain

$$\limsup_{n \to \infty} \int_0^\infty f_i(x) B_n(\mathrm{d}x) \leq \int_0^\infty f_i(x) B(\mathrm{d}x), \quad i = 1, 2, \ldots . \tag{A.38}$$

We first show that for each $f \in \mathscr{E}$,

$$B = \delta_z, \quad B_n = E_n\left(\frac{n}{z}\right) \quad \Rightarrow \quad \int_0^\infty f(x) B_n(\mathrm{d}x) \to \int_0^\infty f(x) B(\mathrm{d}x) = f(z). \tag{A.39}$$

Indeed, if $f(x) = e^{\alpha x}$, then

$$\int_0^\infty f(x) B_n(\mathrm{d}x) = \left(\frac{\frac{n}{z}}{\frac{n}{z} - \alpha}\right)^n = \left(\frac{1}{1 - \frac{\alpha z}{n}}\right)^n \to e^{\alpha z} = f(z) = \int_0^\infty f(x) B(\mathrm{d}x),$$

and the case of a general f then follows from the definition of the class $\mathscr{E}$ and a uniform integrability argument.

Now returning to the proof of (A.38), we may assume that in the proof of Theorem A5.14 D_n has been chosen such that

$$\int_0^\infty f_i(x) D_n(\mathrm{d}x) \leq \left(1 + \frac{1}{n}\right) \int_0^\infty f_i(x) B(\mathrm{d}x), \quad i = 1, \ldots, n.$$

By (A.39),

$$\int_0^\infty f_i(x) C_{r,n}(\mathrm{d}x) \to \int_0^\infty f_i(x) D_n(\mathrm{d}x),$$

and hence we may choose $r(n)$ such that

$$\int_0^\infty f(x) C_{r(n),n}(\mathrm{d}x) \leq \left(1 + \frac{1}{n}\right) \int_0^\infty f(x) B(\mathrm{d}x), \quad i = 1, \ldots, n.$$

$\square$

Corollary A5.16 *To a given distribution B on $(0, \infty)$, there is a sequence $\{B_n\}$ of phase-type distributions such that $B_n \xrightarrow{\mathcal{D}} B$ as $n \to \infty$ and all moments converge, $\int_0^\infty x^i B_n(\mathrm{d}x) \to \int_0^\infty x^i B(\mathrm{d}x)$, $i = 1, 2, \ldots$*

In compound Poisson risk processes with arrival intensity β and claim size distribution B satisfying $\beta \mu_b < 1$, the adjustment coefficient $\gamma = \gamma(B, \beta)$ is defined as the unique solution > 0 of $\widehat{B}[\gamma] = 1 + \gamma/\beta$. The adjustment coefficient is a fundamental quantity, and therefore the following result is highly relevant as support for phase-type assumptions in risk theory:

Corollary A5.17 *To a given $\beta > 0$ and a given distribution B on $(0, \infty)$ with $\widehat{B}[\gamma + \epsilon] < \infty$ for some $\epsilon > \gamma = \gamma(B, \beta)$, there is a sequence $\{B_n\}$ of phase-type distributions such that $B_n \xrightarrow{\mathcal{D}} B$ as $n \to \infty$ and $\gamma_n \to \gamma$ where $\gamma_n = \gamma(B_n, \beta)$.*

Proof. Let $f_i(x) = e^{(\gamma + \epsilon_i)x}$ for some sequence $\{\epsilon_i\}$ with $\epsilon_i \in (0, \epsilon)$ and $\epsilon_i \downarrow 0$ as $i \to \infty$. If $\epsilon_i > 0$, then

$$\widehat{B_n}[\gamma + \epsilon_i] \to \widehat{B}[\gamma + \epsilon_i] > 1 + \frac{\gamma}{\beta}$$

implies that $\gamma_n \leq \gamma + \epsilon_i$ for all sufficiently large n. I.e. $\limsup \gamma_n \leq \gamma$. $\liminf \geq$ is proved similarly. $\square$

We state without proof the following result:

Corollary A5.18 *In the setting of Corollary A5.16, one can obtain $\gamma(B_n, \beta) = \gamma$ for all n.*

Notes and references   Theorem A5.14 is classical; the remaining results may be slightly stronger than those given in the literature, but are certainly not unexpected.

5e   Phase-type fitting

As has been mentioned a number of times already, there is substantial advantage in assuming the claim sizes to be phase-type when one wants to compute ruin probabilities. For practical purposes, the problem thus arises of how to fit a phase-type distribution B to a given set of data $\zeta_1, \ldots, \zeta_N$. The present section is a survey of some of the available approaches and software for inplementing this.

We shall formulate the problem in the slightly broader setting of fitting a phase-type distribution B to a given set of data $\zeta_1, \ldots, \zeta_N$ *or* a given distribution B_0. This is motivated in part from the fact that a number of non-phase-type distributions like the lognormal, the loggamma or the Weibull have been argued to provide adequate descriptions of claim size distributions, and in part from the fact that many of the algorithms that we describe below have been formulated within the set-up of fitting distributions. However, from a more conceptual point of view the two sets of problems are hardly different: an equivalent representation of a set of data $\zeta_1, \ldots, \zeta_N$ is the empirical distribution B_e, giving mass $1/N$ to each ζ_i.

Of course, one could argue that the results of the preceding section concerning phase-type approximation contains a solution to our problem: given B_0 (or B_e), we have constructed a sequence $\{B_n\}$ of phase-type distribution such that $B_n \overset{\mathscr{D}}{\to} B_0$, and as fitted distribution we may take B_n for some suitable large n. The problem is that the constructions of $\{B_n\}$ are not economical: the number of phases grows rapidly, and in practice this sets a limitation to the usefulness (the *curse of dimensionality*; we do not want to perform matrix calculus in hundreds or thousands dimensions).

A number of approaches restrict the phase-type distribution to a suitable class of mixtures of Erlang distributions. The earliest such reference is Bux & Herzog [211] who assumed that the Erlang distributions have the same rate parameter, and used a non-linear programming approach. The constraints were the exact fit of the two first moments and the objective function to be minimized involved the deviation of the empirical and fitted c.d.f. at a number of selected points. In a series of papers (e.g. [509]), Johnson & Taaffe considered a mixture of two Erlangs (with different rates) and matched (when possible) the first three moments. Schmickler (the MEDA package; e.g. [767]) has considered an extension of this set-up, where more than two Erlangs are allowed and in addition to the exact matching of the first three moments a more general deviation measure is minimized (e.g. the L_1 distance between the c.d.f.'s).

The characteristics of all of these methods is that even the number of parameters may be low (e.g. three for a mixture of two Erlangs), the number of phases required for a good fit will typically be much larger, and this is what matters when using phase-type distributions as computational vehicle in say renewal theory, risk theory, reliability or queueing theory. It seems therefore a key issue to develop methods allowing for a more general phase diagram, and we next describe two such approaches which also have the feature of being based upon the traditional statistical tool of maximum likelihood.

A method developed by Bobbio and co-workers (see e.g. [179]) restricts attention to *acyclic* phase-type distributions, defined by the absence of loops in the phase diagram. The likelihood function is maximized by a local linearization method allowing to use linear programming techniques.

Asmussen & Nerman [91] implemented maximum likelihood in the full class of phase-type distributions via the EM algorithm; a program package written in C for the SUN workstation or the PC is available as shareware, cf. [476]. The observation is that the statistical problem would be straightforward if the whole $(E_\Delta$-valued) phase process $\big\{\overline{J}_t^{(k)}\big\}_{0\leq t\leq \zeta_k}$ associated with each observation ζ_k was available. In fact, then the estimators would be of simple occurrence-exposure type,

$$\widehat{\alpha}_i = \frac{\sum_{k=1}^N I\big(\overline{J}_0^{(k)} = i\big)}{N}, \quad \widehat{t}_{ij} = \frac{N_{ij}}{T_i}, \quad i \in E,\ j \in E_\Delta,$$

where

$$T_i = \sum_{k=1}^N \int_0^{\zeta_k} I\big(\overline{J}_t^{(k)} = i\big)\,\mathrm{d}t, \quad N_{ij} = \sum_{k=1}^N \sum_{t\in[0,\zeta_k]} I\big(\overline{J}_{t-}^{(k)} = i,\ \overline{J}_t^{(k)} = j\big)$$

(T_i is the total time spent in state i and N_{ij} is the total number of jumps from i to j). The general idea of the EM algorithm ([291]) is to replace such unobserved quantities by the conditional expectation given the observations; since this is parameter-dependent, one is led to an iterative scheme, e.g.

$$t_{jk}^{(n+1)} = \frac{\mathbb{E}_{\boldsymbol{\alpha}^{(n)},\boldsymbol{T}^{(n)}}\big(N_{jk}|\zeta_1,\ldots,\zeta_N\big)}{\mathbb{E}_{\boldsymbol{\alpha}^{(n)},\boldsymbol{T}^{(n)}}\big(T_j|\zeta_1,\ldots,\zeta_N\big)} \quad (j \neq k),$$

and similarly for the $\alpha_i^{(n+1)}$. The crux is the computation of the conditional expectations. E.g., it is easy to see that

$$\mathbb{E}_{\boldsymbol{\alpha}^{(n)},\boldsymbol{T}^{(n)}}\big(T_i|\zeta_1,\ldots,\zeta_N\big) = \sum_{k=1}^N \mathbb{E}_{\boldsymbol{\alpha}^{(n)},\boldsymbol{T}^{(n)}}\left[\int_0^{\zeta_k} I\big(\overline{J}_t^{(k)} = i\big)\,\mathrm{d}t\,\Big|\,\zeta_k\right]$$

$$= \sum_{k=1}^N \frac{\int_0^{\zeta_j} \boldsymbol{\alpha}^{(n)}\mathrm{e}^{\boldsymbol{T}^{(n)}x}\boldsymbol{e}_i \cdot \boldsymbol{e}_i^{\mathsf{T}}\mathrm{e}^{\boldsymbol{T}^{(n)}(\zeta_k-x)}\boldsymbol{t}^{(n)}\,\mathrm{d}x}{\boldsymbol{\alpha}^{(n)}\mathrm{e}^{\boldsymbol{T}^{(n)}\zeta_k}\boldsymbol{t}^{(n)}}$$

and this and similar expressions are then computed by numerical solution of a set of differential equations.

In practice, the methods of [179] and [91] appear to produce almost identical results. Thus, it seems open whether the restriction to the acyclic case is a severe loss of generality.

Yet a third method, implemented in Bladt & Lauritzen [172], is based on Markov chain Monte Carlo where the main computational step is based on simulation.

A6   Tauberian theorems

The following classical results (see e.g. Bingham, Goldie & Teugels [169, Th.1.7.1 and Th.8.1.6] on regularly varying tails are often useful for asymptotic results on ruin probabilities $\psi(u)$ for $u \to \infty$ when some information on the behavior of its Laplace transform $\widehat{\psi}[-s]$ for $s \to 0$ is available.

Theorem A6.1 *Let U be a non-decreasing right-continuous function on $\mathbb{R}$ with $U(x) = 0$ for $x < 0$ and denote by $\widehat{U}[-s] = \int_0^\infty e^{-sx} dU(x)$ its Laplace-Stieltjes transform. If $L(x)$ is a slowly varying function and $c \geq 0, \alpha \geq 0$, then the following two assertions are equivalent:*

(i) $U(x) \sim c\, x^\alpha L(x)/\Gamma(1+\alpha), \quad x \to \infty,$

(ii) $\widehat{U}[-s] \sim cs^{-\alpha} L(1/s), \quad s \downarrow 0.$

Consider next a positive r.v. X with c.d.f. F, Laplace-Stieltjes transform $\widehat{F}[-s] = \int_0^\infty e^{-sx} dF(x)$ and $\mu_n = \mathbb{E}(X^n)$. Denote

$$f_n(s) = (-1)^{n+1}\left(\widehat{F}[-s] - \sum_{j=0}^n \frac{\mu_j(-s)^j}{j!}\right)$$

and

$$g_n(s) = \frac{d^n f_n(s)}{ds^n} = \mu_n - (-1)^n \widehat{F}^{(n)}[-s].$$

In particular, $f_0(s) = g_0(s) = 1 - \widehat{F}[-s]$.

Theorem A6.2 *Let $L(x)$ be a slowly varying function and $\mu_n < \infty$. Write $\alpha = n + \eta$ with $0 \leq \eta \leq 1$. Then the following assertions are equivalent:*

(i) $f_n(s) \sim s^\alpha L(1/s), \quad s \downarrow 0,$

(ii) $g_n(s) \sim \frac{\Gamma(\alpha+1)}{\Gamma(\eta+1)} s^\eta L(1/s), \quad s \downarrow 0,$

(iiii) $\int_x^\infty t^n dF(t) \sim n!\, L(x), \qquad x \to \infty, \text{ when } \eta = 0,$

$\qquad 1 - F(x) \sim \frac{(-1)^n}{\Gamma(1-\alpha)} x^{-\alpha} L(x), \quad x \to \infty, \text{ when } 0 < \eta < 1,$

$\qquad \int_0^x t^{n+1} dF(t) \sim (n+1)!\, L(x), \quad x \to \infty, \text{ when } \eta = 1.$

For $\eta > 0$ a further equivalent statement is

$$(-1)^{n+1} \widehat{F}^{(n+1)}[-s] \sim \frac{\Gamma(\alpha+1)}{\Gamma(\eta)} s^{\eta-1} L(1/s), \quad s \downarrow 0.$$

Bibliography

[1] J. Abate, G.L. Choudhury & W. Whitt (1994) Waiting-time tail probabilities in queues with long-tail service-time distributions. *Queueing Systems* **16**, 311–338.

[2] J. Abate & W. Whitt (1992) The Fourier-series method for inverting transforms of probability distributions. *Queueing Systems* **10**, 5–87.

[3] J. Abate & W. Whitt (1999) Explicit M/G/1 waiting–time distributions for a class of long-tail service time distributions. *Oper. Res. Letters* **25**, 25–31.

[4] M. Abramowitz & I. Stegun (1972) *Handbook of Mathematical Functions* (10th ed.). Dover, New York.

[5] I. Adan and V. Kulkarni (2003) Single-server queue with Markov dependent inter-arrival and service times. *Queueing Systems* **45**, 113–134.

[6] R. Adler (1990) *An Introduction to Continuity, Extrema, and Related Topics for General Gaussian Processes.* Institute of Mathematical Statistics, Hayward.

[7] L. Afonso, A. Egidio dos Reis & H. Waters (2009) Calculating continuous-time ruin probabilities for a large portfolio with varying premiums. *Astin Bulletin* **39**, 117–136.

[8] S. Ahn & A. Badescu (2007) On the analysis of the Gerber-Shiu discounted penalty function for risk processes with Markovian arrivals. *Insurance: Mathematics and Economics* **41**, 234–249.

[9] S. Ahn & V. Ramaswami (2004). Transient analysis of fluid flow models via stochastic coupling to a queue. *Stoch. Models* **20**, 71–101.

[10] S. Ahn & V. Ramaswami (2006). Efficient algorithms for transient analysis of stochastic fluid flow models. *J. Appl. Probab.* **42**, 531–549.

[11] H. Albrecher (2004) Operational Time. *Encyclopedia of Actuarial Science*, 1207–1208. Wiley.

[12] H. Albrecher & S. Asmussen (2006) Ruin probabilities and aggregate claims distributions for shot noise Cox processes. *Scand. Act. J.* **2006**, 86–110.

[13] H. Albrecher, S. Asmussen & D. Kortschak (2006) Tail asymptotics for the sum of two heavy-tailed dependent risks. *Extremes* **9**, 107–130.

[14] H. Albrecher, F. Avram & D. Kortschak (2010). On the efficient evaluation of ruin probabilities for completely monotone claim size distributions. *J. Comp. Appl. Math.* **233**, 2724–2736.

[15] H. Albrecher, A. Badescu & D. Landriault (2008) On the dual risk model with tax payments. *Insurance: Mathematics and Economics* **42**, 1086–1094.

[16] H. Albrecher, S. Borst, O.J. Boxma & J. Resing (2009) The tax identity in risk theory — a simple proof and an extension. *Insurance: Mathematics and Economics* **44**, 304–306.

[17] H. Albrecher, S. Borst, O.J. Boxma & J. Resing (2010) Ruin excursions, the G/G/∞ queue and tax payments in renewal risk models. *Preprint.*

[18] H. Albrecher & O.J. Boxma (2004) A ruin model with dependence between claim sizes and claim intervals. *Insurance: Mathematics and Economics* **35**, 245–254.

[19] H. Albrecher & O.J. Boxma (2005) On the discounted penalty function in a Markov-dependent risk model. *Insurance: Mathematics and Economics* **37**, 650–672.

[20] H. Albrecher, E. Cheung & S. Thonhauser (2010) A randomized approach to analyzing the compound Poisson risk model under periodic observations. *Preprint.*

[21] H. Albrecher, C. Constantinescu, G. Pirsic, G. Regensburger & M. Rosenkranz (2010) An algebraic operator approach to the analysis of Gerber-Shiu functions. *Insurance: Mathematics and Economics* **46**, 42–51.

[22] H. Albrecher, C. Constantinescu & E. Thomann (2010) Asymptotic results in renewal risk models with investments. *Preprint.*

[23] H. Albrecher, H. Gerber & H. Yang (2010) A direct approach to the discounted penalty function. *Preprint.*

[24] H. Albrecher & S. Haas (2010) A numerical approach to ruin models with excess of loss reinsurance and reinstatements. *COMPSTAT 2010*, Springer-Verlag, in press.

[25] H. Albrecher & J. Hartinger (2007) A risk model with multilayer dividend strategy. *North American Actuarial J.* **11**, 43–64.

[26] H. Albrecher, J. Hartinger & S. Thonhauser (2007) On exact solutions for dividend strategies of threshold and linear barrier type in a Sparre Andersen model. *Astin Bulletin* **37**, 203–233.

[27] H. Albrecher, J. Hartinger & R.F. Tichy (2005) On the distribution of dividend payments and the discounted penalty function in a risk model with linear dividend barrier. *Scand. Act. J.* **2005**, 103–126.

[28] H. Albrecher & C. Hipp (2007) Lundberg's risk process with tax. *Blätter DGVFM* **28**, 13–28.

[29] H. Albrecher, C. Hipp & D. Kortschak (2010) Higher-order expansions for compound distributions and ruin probabilities with subexponential claims. *Scand. Act. J.* **2010**, 105–135.

[30] H. Albrecher & R. Kainhofer (2002) Risk theory with a non-linear dividend barrier. *Computing* **68**, 289–311.

[31] H. Albrecher, R. Kainhofer & R.F. Tichy (2003) Simulation methods in ruin models with non-linear dividend barriers. *Math. Comput. Simulation* **62**, 277–287.

[32] H. Albrecher & J. Kantor (2002) Simulation of ruin probabilities for risk processes of Markovian type. *Monte Carlo Methods & Appl.* **8**, 111–127.

[33] H. Albrecher & D. Kortschak (2009) On ruin probability and aggregate claim representations for Pareto claim size distributions. *Insurance: Mathematics and Economics* **45**, 362–373.

[34] H. Albrecher & C. Macci (2008) Large deviation bounds for ruin probability estimators in some risk models with dependence. *Proceedings of the Fourth Int. Workshop on Applied Probability, Compiègne.*

[35] H. Albrecher, J. Renaud & X. Zhou (2008) A Lévy insurance risk process with tax. *J. Appl. Probab.* **45**, 363–375.

[36] H. Albrecher & J.L. Teugels (2006) Exponential behavior in the presence of dependence in risk theory. *J. Appl. Probab.* **43**, 257–273.

[37] H. Albrecher & J.L. Teugels (2006) Asymptotic analysis of a measure of variation. *Th. Probab. Math. Statist.* **74**, 1–9.

[38] H. Albrecher & J.L. Teugels (2010) *Reinsurance: Actuarial and Financial Aspects.* Wiley, to appear.

[39] H. Albrecher, J.L. Teugels & R.F. Tichy (2001) On a gamma series expansion for the time-dependent probability of collective ruin. *Insurance: Mathematics and Economics* **29**, 345–355.

[40] H. Albrecher & S. Thonhauser (2009) Optimality results for dividend problems in insurance. *Rev. R. Acad. Cien. Serie A. Mat.* **103**, 295–320.

[41] H. Albrecher & R.F. Tichy (2000) On the convergence of a solution procedure for a risk model with gamma-distributed claims. *Mitt. Ver. Schweiz. Vers. Math.* **2000**, 115–127.

[42] S. Alink, M. Löwe & M.V. Wüthrich (2004) Diversification of aggregate dependent risks. *Insurance: Mathematics and Economics* **35**, 77–95.

[43] S. Alink, M. Löwe & M.V. Wüthrich (2005) Diversification for general copula dependence. *Stat. Neerl.* **61**, 446–465.

[44] U.T. Alparslan & G. Samorodnitsky (2008) Asymptotic analysis of the ruin with stationary stable steps generated by dissipative flows. *Scand. Act. J.* **2008**, 180–201.

[45] G. Alsmeyer (1991) *Erneuerungstheorie.* Teubner.

[46] R. Ambagaspitiya (2003) Aggregate survival probability of a portfolio with dependent subportfolios. *Insurance: Mathematics and Economics* **32**, 431–443.

[47] R. Ambagaspitiya (2009) Ultimate ruin probability in the Sparre Andersen model with dependent claim sizes and claim occurrence times. *Insurance: Mathematics and Economics* **44**, 464–472.

[48] V. Anantharam (1988) How large delays build up in a $GI/GI/1$ queue. *Queueing Systems* **5**, 345–368.

[49] D. Applebaum (2004) *Lévy Processes and Stochastic Calculus.* Cambridge University Press.

[50] G. Arfwedson (1954) Research in collective risk theory. *Skand. Aktuar Tidskr.* **37**, 191–223.

[51] G. Arfwedson (1955) Research in collective risk theory. The case of equal risk sums. *Skand. Aktuar Tidskr.* **38**, 53–100.

[52] V. Araman & P.W. Glynn (2006) Tail asymptotics for the maximum of perturbed random walk. *Ann. Appl. Probab.* **16**, 1411–1431.

[53] K. Arndt (1984) On the distribution of the supremum of a random walk on a Markov chain. In: *Limit Theorems and Related Problems*, pp. 253-267. Optimizations Software, New York.

[54] S. Asmussen (1982) Conditioned limit theorems relating a random walk to its associate, with applications to risk reserve processes and the GI/G/1 queue. *Adv. Appl. Probab.* **14**, 143-170.

[55] S. Asmussen (1984) Approximations for the probability of ruin within finite time. *Scand. Act. J.* **1984**, 31–57; *ibid.* **1985**, 57.

[56] S. Asmussen (1985) Conjugate processes and the simulation of ruin problems. *Stoch. Proc. Appl.* **20**, 213–229.

[57] S. Asmussen (1989a) Aspects of matrix Wiener-Hopf factorisation in applied probability. *The Mathematical Scientist* **14**, 101–116.

[58] S. Asmussen (1989b) Risk theory in a Markovian environment. *Scand. Act. J.* **1989**, 69–100.

[59] S. Asmussen (1991) Ladder heights and the Markov-modulated M/G/1 queue. *Stoch. Proc. Appl.* **37**, 313–326.

[60] S. Asmussen (1992a) Phase-type representations in random walk and queueing problems. *Ann. Probab.* **20**, 772–789.

[61] S. Asmussen (1992b) Light traffic equivalence in single server queues. *Ann. Appl. Probab.* **2**, 555–574.

[62] S. Asmussen (1995a) Stationary distributions for fluid flow models and Markov–modulated reflected Brownian motion. *Stochastic Models* **11**, 21–49.

[63] S. Asmussen (1995b) Stationary distributions via first passage times. *Advances in Queueing: Models, Methods & Problems* (J. Dshalalow ed.), 79–102. CRC Press, Boca Raton, Florida.

[64] S. Asmussen (1998a) Subexponential asymptotics for stochastic processes: extremal behaviour, stationary distributions and first passage probabilities. *Ann. Appl. Probab.* **8**, 354–374.

[65] S. Asmussen (1998b) Extreme value theory for queues via cycle maxima. *Extremes* **1**, 137–168.

[66] S. Asmussen (1998c) A probabilistic look at the Wiener-Hopf equation. *SIAM Review* **40**, 189–201.

[67] S. Asmussen (1999) On the ruin problem for some adapted premium rules. *Probabilistic Analysis of Rare Events* (V.K. Kalashnikov & A.M. Andronov, eds.), 3–15. Riga Aviation University.

[68] S. Asmussen (2000) Matrix-analytic models and their analysis. *Scand. J. Statist.* **27**, 193–226.

[69] S. Asmussen (2003) *Applied Probability and Queues.* Second Edition. Springer-Verlag.

[70] S. Asmussen, F. Avram & M.R. Pistorius (2004) Russian and American put options under exponential phase-type Lévy models. *Stoch. Proc. Appl.* **109**, 79–111.

[71] S. Asmussen, F. Avram & M. Usabel (2002) Erlangian approximations for ruin probabilities. *Astin Bulletin* **32**, 267–281.

[72] S. Asmussen & K. Binswanger (1997) Simulation of ruin probabilities for subexponential claims. *Astin Bulletin* **27**, 297–318.

[73] S. Asmussen, K. Binswanger & B. Højgaard (2000) Rare events simulation for heavy–tailed distributions. *Bernoulli* **6**, 303–322.

[74] S. Asmussen & M. Bladt (1996) Renewal theory and queueing algorithms for matrix–exponential distributions. *Matrix-Analytic Methods in Stochastic Models* (A.S. Alfa & S. Chakravarty, eds.), 313–341. Marcel Dekker, New York.

[75] S. Asmussen & M. Bladt (1996) Phase–type distributions and risk processes with premiums dependent on the current reserve. *Scand. Act. J.* **1996**, 19–36.

[76] S. Asmussen, L. Fløe Henriksen & C. Klüppelberg (1994) Large claims approximations for risk processes in a Markovian environment. *Stoch. Proc. Appl.* **54**, 29–43.

[77] S. Asmussen, S. Foss & D. Korshunov (2003) Asymptotics for sums of random variables with local subexponential behaviour. *J. Theor. Probab.* **16**, 489–518.

[78] S. Asmussen, A. Frey, T. Rolski & V. Schmidt (1995) Does Markov-modulation increase the risk? *Astin Bulletin* **25**, 49–66.

[79] S. Asmussen & P.W. Glynn (2007) *Stochastic Simulation. Algorithms and Analysis*. Springer-Verlag.

[80] S. Asmussen & B. Højgaard (1996) Finite horizon ruin probabilities for Markov-modulated risk processes with heavy tails. *Th. Random Processes* **2**, 96–107.

[81] S. Asmussen & B. Højgaard (1999) Approximations for finite horizon ruin probabilities in the renewal model. *Scand. Act. J.* **1999**, 106–119.

[82] S. Asmussen, B. Højgaard & M. Taksar (2000) Optimal risk control and dividend distribution policies. Example of excess-of-loss reinsurance for an insurance corporation. *Finance and Stochastics* **4**, 299–324.

[83] S. Asmussen, V. Kalashnikov, D. Konstantinides, C. Klüppelberg & G. Tsitsiashvili (2002) A local limit theorem for random walk maxima with heavy tails. *Statist. Probab. Lett.* **56**, 399–404.

[84] S. Asmussen & O. Kella (2000) A multidimensional martingale for Markov additive processes and its applications. *Adv. Appl. Probab.* **91**, 376–393.

[85] S. Asmussen & O. Kella (2001) Optional stopping of some exponential martingales for Lévy processes with or without reflection. *Stoch. Proc. Appl.* **91**, 47–55.

[86] S. Asmussen & C. Klüppelberg (1996) Large deviations results for subexponential tails, with applications to insurance risk. *Stoch. Proc. Appl.* **64**, 103–125.

[87] S. Asmussen, C. Klüppelberg & K. Sigman (1999) Sampling at subexponential times, with queueing applications. *Stoch. Proc. Appl.* **679**, 265–286.

[88] S. Asmussen & G. Koole (1993). Marked point processes as limits of Markovian arrival streams. *J. Appl. Probab.* **30**, 365–372.

[89] S. Asmussen & D.P. Kroese (2006) Improved algorithms for rare event simulation with heavy tails. *Adv. Appl. Probab.* **38**, 545–558.

[90] S. Asmussen, D. Madan & M. Pistorius (2008) Pricing equity default swaps under an approximation to the CGMY Lévy model. *J. Comp. Finance* **11**, 79–93.

[91] S. Asmussen , O. Nerman & M. Olsson (1996). Fitting phase-type distributions via the EM algorithm. *Scand. J. Statist.* **23**, 419–441.

[92] S. Asmussen & H.M. Nielsen (1995) Ruin probabilities via local adjustment coefficients. *J. Appl. Probab.* **32**, 736–755.

[93] S. Asmussen & C.A. O'Cinneide (2000/01) On the tail of the waiting time in a Markov-modulated M/G/1 queue. *Oper. Res.* **50**, 559–565.

[94] S. Asmussen & C.A. O'Cinneide (2000/2001) Matrix–exponential distributions [Distributions with a rational Laplace transform] *Encyclopedia of Statistical Sciences, Supplementary Volume* (Kotz, Johnson, Read eds.). Wiley.

[95] S. Asmussen & D. Perry (1992) On cycle maxima, first passage problems and extreme value theory for queues. *Stochastic Models* **8**, 421–458.

[96] S. Asmussen & L. Rojas-Nandaypa (2008) Asymptotics of sums of lognormal random variables with Gaussian copula. *Stat. Probab. Lett.* **78**, 2709–2714.

[97] S. Asmussen & T. Rolski (1991) Computational methods in risk theory: a matrix–algorithmic approach. *Insurance: Mathematics and Economics* **10**, 259–274.

[98] S. Asmussen & T. Rolski (1994) Risk theory in a periodic environment: Lundberg's inequality and the Cramér–Lundberg approximation. *Math. Oper. Res.* 410-433.

[99] S. Asmussen & R.Y. Rubinstein (1995) Steady-state rare events simulation in queueing models and its complexity properties. *Advances in Queueing: Models, Methods & Problems* (J. Dshalalow ed.), 429–466. CRC Press, Boca Raton.

[100] S. Asmussen & R.Y. Rubinstein (1999) Sensitivity analysis of insurance risk models. *Management Science* **45**, 1125–1141.

[101] S. Asmussen, H. Schmidli & V. Schmidt (1999) Tail approximations for non-standard risk and queueing processes with subexponential tails. *Adv. Appl. Probab.* **31**, 422–447.

[102] S. Asmussen & V. Schmidt (1993) The ascending ladder height distribution for a class of dependent random walks. *Statistica Neerlandica* **47**, 1–9.

[103] S. Asmussen & V. Schmidt (1995) Ladder height distributions with marks. *Stoch. Proc. Appl.* **58**, 105–119.

[104] S. Asmussen & S. Schock Petersen (1989) Ruin probabilities expressed in terms of storage processes. *Adv. Appl. Probab.* **20**, 913–916.

[105] S. Asmussen & K. Sigman (1996) Monotone stochastic recursions and their duals. *Probab. Th. Eng. Inf. Sc.* **10**, 1–20.

[106] S. Asmussen & M. Taksar (1997) Controlled diffusion models for optimal dividend payout. *Insurance: Mathematics and Economics* **20**, 1–15.

[107] S. Asmussen & J.L. Teugels (1996) Convergence rates for M/G/1 queues and ruin problems with heavy tails. *J. Appl. Probab.* **33**, 1181–1190.

[108] K.B. Athreya & P. Ney (1972) *Branching Processes.* Springer-Verlag.

[109] B. Avanzi (2009) Strategies for dividend distribution: a review. *North American Actuarial J.* **13**, 217–251.

[110] F. Avram, N. Leonenko & L. Rabehasaina (2009) Series expansions for the first passage distribution of Wong-Pearson jump diffusions. *Stoch. Anal. Appl.* **27**, 770–796.

[111] F. Avram, Z. Palmowski & M. Pistorius (2008) Exit problem of a two-dimensional risk process from the quadrant: exact and asymptotic results. *Ann. Appl. Probab.* **18**, 2421–2449.

[112] F. Avram & M. Usabel (2003) Finite time ruin probabilities with one Laplace inversion. *Insurance: Mathematics and Economics* **32**, 371–377.

[113] F. Avram & M. Usabel (2004) Ruin probabilities and deficit for the renewal risk model with phase-type interarrival times. *Astin Bulletin* **34**, 315–332.

[114] F. Avram & M. Usabel (2008) The Gerber-Shiu expected discounted penalty-reward function under an affine jump-diffusion model. *Astin Bulletin* **38**, 461–481.

[115] P. Azcue & N. Muler (2009) Optimal investment strategy to minimize the ruin probability of an insurance company under borrowing constraints. *Insurance: Mathematics and Economics* **44**, 26–34.

[116] A. Badescu, L. Breuer, A. da Silva Soares, G. Latouche, M.-A. Remiche & D. Stanford (2005) Risk processes analyzed as fluid queues. *Scand. Act. J.* **2005**, 127–141.

[117] A.L. Badescu, E. Cheung & D. Landriault (2009) Dependent risk models with bivariate phase-type distributions. *J. Appl. Probab.* **46**, 113-131.

[118] A.L. Badescu, S. Drekic & D. Landriault (2007) On the analysis of a multi-threshold Markovian risk model. *Scand. Act. J.* **2007**, 248–260.

[119] A.L. Badescu & D. Landriault (2009) Applications of fluid flow matrix analytic methods in ruin theory — a review. *Rev. R. Acad. Cien. Serie A. Mat.* **103**, 353–372.

[120] A.L. Badescu & D. Stanford (2005) A generalization of the De Vylder approximation for the probability of ruin. *Econ. Comp. Cybern. Stud. Res.* **40**, 245-265.

[121] B. von Bahr (1974) Ruin probabilities expressed in terms of ladder height distributions. *Scand. Act. J.* **1974**, 190–204.

[122] B. von Bahr (1975) Asymptotic ruin probabilities when exponential moments do not exist. *Scand. Act. J.* **1975**, 6–10.

[123] L. Bai & J. Guo (2008) Optimal proportional reinsurance and investment with multiple risky assets and no-shorting constraint. *Insurance: Mathematics and Economics* **42**, 968–975.

[124] G. Baigger (1988) Limsup-theorems and their applications to risk theory. *Transactions XXIIIth International Congress of Actuaries, Helsinki, I*, 1–21.

[125] C.T.H. Baker (1977) *The Numerical Solution of Integral Equations.* Clarendon Press, Oxford.

[126] A. Baltrūnas (1999) Second order behaviour of ruin probabilities. *Scand. Act. J.* **1999**, 120–133.

[127] A. Baltrūnas (2005) Second order behaviour of ruin probabilities in the case of large claims. *Insurance: Mathematics and Economics* **36**, 485–498.

[128] A. Baltrūnas (2001) Some asymptotic results for transient random walks with applications to insurance risk. *J. Appl. Probab.* **38**, 108–121.

[129] A. Baltrūnas & C. Klüppelberg (2004) Subexponential distributions — large deviations with applications to insurance and queueing models. *Aust. N. Z. J. Stat.* **46**, 145–154.

[130] P. Barbe, A. Fougéres & C. Genest (2006) On the tail behavior of sums of dependent risks. *Astin Bulletin* **36**, 361–373.

[131] P. Barbe, W.P. McCormick & C. Zhang (2007) Asymptotic expansions for distributions of compound sums of random variables with rapidly varying subexponential distribution. *J. Appl. Probab.* **44**, 670–684.

[132] P. Barbe & W.P. McCormick (2009) *Asymptotic expansions for infinite weighted convolutions of heavy tail distributions and applications.* Mem. Amer. Math. Soc.

[133] P. Barbe & W.P. McCormick (2010) An extension of a logarithmic form of Cramér's ruin theorem to some FARIMA and related processes. *Stoch. Proc. Appl.* **120**, 801–828.

[134] P. Barbe & W.P. McCormick (2010) Veraverbeke's theorem at large: on the maximum of some processes with negative drift and heavy tail innovations. *Extremes*, in press.

[135] O.E. Barndorff-Nielsen (1977) Exponentially decreasing distributions for the logarithm of particle size. *Proc. Roy. Soc. London A* **353**, 401–419.

[136] O.E. Barndorff-Nielsen (1978) *Information and Exponential Families in Statistical Theory.* Wiley.

[137] O.E. Barndorff-Nielsen, T. Mikosch & S. Resnick, eds. (2001) *Lévy Processes. Theory and Applications.* Birkhäuser.

[138] O.E. Barndorff-Nielsen & H. Schmidli (1995) Saddlepoint approximations for the probability of ruin in finite time. *Scand. Act. J.* **1995**, 169–186.

[139] O.E. Barndorff-Nielsen & A. Shiryaev (2010) *Change of Time and Change of Measure.* World Scientific.

[140] A. Bassamboo, S. Juneja & A. Zeevi (2006) On the efficiency loss of state-independent importance sampling in the presence of heavy-tails. *Oper. Res. Letters* **34**, 521–531.

[141] N. Bäuerle & A. Blatter (2010) Discrete-time stochastic control of insurance portfolios. *Math. Meth. Oper. Res.*, in press.

[142] N. Bäuerle, A. Blatter & A. Müller (2008) Dependence properties and comparison results for Lévy processes. *Math. Method. Oper. Res.* **67**, 161–186.

[143] N. Bäuerle & R. Grübel (2005) Multivariate counting processes: copulas and beyond. *Astin Bulletin* **35**, 379–408.

[144] N. Bäuerle & R. Grübel (2008) Multivariate risk processes with interacting intensities. *Adv. Appl. Probab.* **40**, 578–601.

[145] E. Bayraktar, K. Moore & V. Young (2008) Minimizing the probability of lifetime ruin under random consumption. *North American Actuarial J.* **12**, 384–400.

[146] E. Bayraktar & V. Young (2007) Minimizing the probability of lifetime ruin under borrowing constraints. *Insurance: Mathematics and Economics* **41**, 196–221.

[147] E. Bayraktar & V. Young (2008) Minimizing the probability of ruin when consumption is ratcheted. *North American Actuarial J.* **12**, 428–442.

[148] E. Bayraktar & V. Young (2008) Maximizing utility of consumption subject to a constraint on the probability of lifetime ruin. *Finance Research Lett.* **5**, 204–212.

[149] N. Bäuerle (2004) Traditional versus non-traditional reinsurance in a dynamic setting. *Scand. Act. J.* **2004**, 355-371.

[150] N.G. Bean, M. Fackrell & P. Taylor (2008) Characterization of matrix-exponential distributions. *Stoch. Models* **24**, 339–363.

[151] J. Beekman (1969) A ruin function approximation. *Trans. Soc. Actuaries* **21**, 41–48.

[152] J. Beekman (1974) *Two Stochastic Processes*. Halsted Press, New York.

[153] J.A. Beekman (1985) A series for infinite time ruin probabilities. *Insurance: Mathematics and Economics* **4**, 129–134.

[154] J. Beirlant, Y. Goegebeur, J. Segers & J.L. Teugels (2004) *Statistics of Extremes*. Wiley.

[155] F. Belzunce, E.M. Ortega, F. Pellerey & J.M. Ruiz (2006) Variability of total claim amounts under dependence between claims severity and number of events. *Insurance: Mathematics and Economics* **38**, 460–468.

[156] V. Bernyk, R.C. Dalang & G. Peskir (2008) The law of the supremum of a stable Lévy process with no negative jumps. *Ann. Probab.* **36**, 1777–1789.

[157] J. Bertoin (1996) *Lévy Processes*. Cambridge University Press.

[158] J. Bertoin & R. Doney (1994). Cramér's estimate for Lévy processes. *Stat. Probab. Lett.* **21**, 363–365.

[159] J. Bertoin, R.A. Doney & R.A. Maller (2008) Passage of Lévy processes across power law boundaries at small times. *Ann. Probab.* **36**, 160–197.

[160] D. Bertsekas & S. Shreve (1978) *Stochastic Optimal Control: the Discrete-Time Case*. Academic Press.

[161] C.J. Beveridge, D.C.M. Dickson & X. Wu (2008) Optimal dividends under reinsurance. *Mitt. Ver. Schweiz. Vers. Math.* **08**, 149–166.

[162] R. Biard, C. Lefèvre & S. Loisel (2008) Impact of correlation crises in risk theory: asymptotics of finite-time ruin probabilities for heavy-tailed claim amounts when some independence and stationarity assumptions are relaxed. *Insurance: Mathematics and Economics* **43**, 412–421.

[163] R. Biard, S. Loisel, C. Macci & N. Veraverbeke (2010) Asymptotic behavior of the finite-time expected time-integrated negative part of some risk processes. *J. Math. Anal. Appl.* **367**, 535–549.

[164] B.M. Bibby & M. Sørensen (2003) Hyperbolic processes in finance. *Handbook of Heavy Tailed Distributions in Finance*. Elsevier, 211–248.

[165] E. Biffis & A. Kyprianou (2010) A note on scale functions and the time value of ruin for Lévy risk processes. *Insurance: Mathematics and Economics* **46**, 85–91.

[166] E. Biffis & M. Morales (2010) On a generalization of the Gerber-Shiu function to path-dependent penalties. *Insurance: Mathematics and Economics* **46**, 92–97.

[167] P. Billingsley (1968) *Convergence of Probability Measures*. Wiley, New York.

[168] N.H. Bingham (1975) Fluctuation theory in continuous time. *Adv. Appl. Probab.* **7**, 705–766.

[169] N.H. Bingham, C.M. Goldie & J.L. Teugels (1987) *Regular Variation*. Cambridge University Press.

[170] T. Björk & J. Grandell (1985) An insensitivity property of the ruin probability. *Scand. Act. J.* **1985**, 148–156.

[171] D. Blackwell (1962) Discrete dynamic programming. *Ann. Math. Statist.* **33**, 719–726.

[172] M. Bladt, A. Gonzalez & S.L. Lauritzen (2003). The estimation of phase-type related functionals using Markov chain Monte Carlo methods. *Scand. Act. J.* **2003**, 280–300.

[173] M. Bladt & M.F. Neuts (2003). Matrix-exponential distributions: calculus and interpretations via flows. *Stoch. Models* **19**, 113–124.

[174] J. Blanchet & P. Glynn (2006) Complete corrected diffusion approximations for the maximum of a random walk. *Ann. Appl. Probab.* **16**, 951–983.

[175] J. Blanchet & P. Glynn (2008) Efficient rare-event simulation for the maximum of heavy-tailed random walks. *Ann. Appl. Probab.* **18**, 1351–1378.

[176] J. Blanchet, P. Glynn & J.C. Liu (2007) Fluid heuristics, Lyapunov bound and efficient importance sampling for a heavy-tailed GI/G/1 queue. *QUESTA* **57**, 99–113.

[177] T. Björk & J. Grandell (1988) Exponential inequalities for ruin probabilities in the Cox case. *Scand. Act. J.* **1988**, 77–111.

[178] P. Bloomfield & D.R. Cox (1972) A low traffic approximation for queues. *J. Appl. Probab.* **9**, 832–840.

[179] A. Bobbio & M. Telek (1994) A benchmark for PH estimation algorithms: results for acyclic PH. *Stoch. Models* **10**, 661–667.

[180] P. Boogaert & V. Crijns (1987) Upper bounds on ruin probabilities in case of negative loadings and positive interest rates. *Insurance: Mathematics and Economics* **6**, 221–232.

[181] P. Boogaert & A. de Waegenaere (1990) Simulation of ruin probabilities. *Insurance: Mathematics and Economics* **9**, 95–99.

[182] A.A. Borovkov (1976) *Asymptotic Methods in Queueing Theory*. Springer-Verlag.

[183] A.A. Borovkov (2009) Insurance with borrowing: first- and second-order approximations. *Adv. Appl. Probab.* **41**, 1141–1160.

[184] A.A. Borovkov & K.A. Borovkov (2002) On probabilities of large deviations for random walks. I. Regularly varying distribution tails. *Theory Prob. Appl.* **46**, 193–213.

[185] A.A. Borovkov & K.A. Borovkov (2008) *Asymptotic Analysis of Random Walks. Heavy-Tailed Distributions*. Cambridge University Press.

[186] A.A. Borovkov & A.A. Mogul'skiĭ (2001) Integro-local limit theorems including large deviations for sums of random vectors. II. *Theory Probab. Appl.* **45**, 3–22.

[187] A.A. Borovkov & A.A. Mogul'skiĭ (2001) Limit theorems in the problem of reaching the boundary by a multidimensional random walk. *Siberian Math. J.* **42**, 245–270.

[188] A.A. Borovkov & A.A. Mogul'skiĭ (2007) On large and superlarge deviations of sums of independent random vectors under the Cramér condition. II. *Theory Probab. Appl.* **51**, 567–594.

[189] K.A. Borovkov & D.C.M. Dickson (2008) On the ruin time distribution for a Sparre Andersen process with exponential claim sizes. *Insurance: Mathematics and Economics* **42**, 1104–1108.

[190] A.A. Borovkov, Z. Palmowski & O.J. Boxma (2003) On the integral workload process of the single server queue. *J. Appl. Probab.* **40**, 200–225.

[191] R.J. Boucherie, O.J. Boxma & K. Sigman (1997) A note on negative customers, GI/G/1 workload, and risk processes. *Probab. Th. Eng. Inf. Sc.* **11**, 305–311.

[192] M. Boudreault, H. Cossette, D. Landriault & E. Marceau (2006) On a risk model with dependence between interclaim arrivals and claim sizes. *Scand. Act. J.* **2006**, 265–285.

[193] O.J. Boxma & J.W. Cohen (1998) The M/G/1 queue with heavy-tailed service time distribution. *IEEE J. Sel. Areas Commun.* **16**, 749–763.

[194] O.J. Boxma & J.W. Cohen (1999) Heavy-traffic analysis for the GI/G/1 queue with heavy-tailed service time distributions. *Queueing Systems* **33**, 177–204.

[195] N.L. Bowers, Jr., H.U. Gerber, J.C. Hickman, D.A. Jones & C.J. Nesbitt (1986) *Actuarial Mathematics.* The Society of Actuaries, Itasca, Illinois.

[196] M. Bratiychuk & D. Derfla (2007) On a modification of the classical risk process. *Insurance: Mathematics and Economics* **41**, 156–162.

[197] P. Bratley, B.L. Fox & L. Schrage (1987) *A Guide to Simulation.* Springer-Verlag.

[198] J. Bregman & C. Klüppelberg (2005) Ruin estimation in multivariate models with Clayton dependence structure. *Scand. Act. J.* **2005**, 462–480.

[199] L. Breiman (1968) *Probability.* Addison-Wesley, Reading.

[200] R. Brekelmans & A. De Waegenaere (2001) Approximating the finite-time ruin probability under interest force. *Insurance: Mathematics and Economics* **29**, 217–229.

[201] L. Breuer (2008) First passage times for Markov additive processes with positive jumps of phase type. *J. Appl. Probab.* **45**, 779–799.

[202] M. Brito, A.C.M. Freitas (2008) Edgeworth expansion for an estimator of the adjustment coefficient. *Insurance: Mathematics and Economics* **43**, 203–208.

[203] P.J. Brockwell, S.I. Resnick & R.L. Tweedie (1982) Storage processes with general release rule and additive inputs. *Adv. Appl. Probab.* **14**, 392–433.

[204] F. Broeck, M.J. Goovaerts & F. De Vylder (1986) Ordering of risks and ruin probabilities. *Insurance: Mathematics and Economics* **5**, 35–39.

[205] M. Brokate, C. Klüppelberg, R. Kostadinova, R. Maller & R.C. Seydel (2008) On the distribution tail of an integrated risk model: a numerical approach. *Insurance: Mathematics and Economics* **42**, 101–106.

[206] S. Browne (1995) Optimal investment policies for a firm with a random risk process: exponential utility and minimizing the probability of ruin. *Math. Oper. Research* **20**, 937–958.

[207] J.A. Bucklew (1990) *Large Deviation Techniques in Decision, Simulation and Estimation.* Wiley, New York.

[208] H. Bühlmann (1970) *Mathematical Methods in Risk Theory.* Springer-Verlag.

[209] D.Y. Burman & D.R. Smith (1983) Asymptotic analysis of a queueing system with bursty traffic. *Bell. Syst. Tech. J.* **62**, 1433–1453.

[210] D.Y. Burman & D.R. Smith (1986) An asymptotic analysis of a queueing system with Markov–modulated arrivals. *Oper. Res.* **34**, 105–119.

[211] W. Bux & U. Herzog (1977) The phase concept: approximations of measured data and performance analysis. *Computer Performance* (K.M. Chandy & M. Reiser eds.), 23–38. North–Holland.

[212] J. Cai (2002) Ruin probabilities with dependent rates of interest. *J. Appl. Probab.* **39**, 312–323.

[213] J. Cai (2004) Ruin probabilities and penalty functions with stochastic rates of interest. *Stoch. Proc. Appl.* **112**, 53–78.

[214] J. Cai & D.C.M. Dickson (2002) On the expected discounted penalty function at ruin of a surplus process with interest. *Insurance: Mathematics and Economics* **30**, 389–404.

[215] J. Cai & D.C.M. Dickson (2003) Upper bounds for ultimate ruin probabilities in the Sparre Andersen model with interest. *Insurance: Mathematics and Economics* **32**, 61–71.

[216] J. Cai, R. Feng & G.E. Willmot (2009) Analysis of the compound Poisson surplus model with liquid reserves, interest and dividends. *Astin Bulletin* **39**, 225–247.

[217] J. Cai, R. Feng & G.E. Willmot (2009) On the total discounted operating costs up to default and its applications. *Adv. Appl. Probab.* **41**, 495–522.

[218] J. Cai & J. Garrido (1998) Aging properties and bounds for ruin probabilities and stop-loss premiums. *Insurance: Mathematics and Economics* **23**, 33–43.

[219] J. Cai & H. Li (2005) Multivariate risk model of phase type. *Insurance: Mathematics and Economics* **36**, 137–152.

[220] J. Cai & C. Xu (2006) On the decomposition of the ruin probability for a jump-diffusion surplus process compounded by a geometric Brownian motion. *North American Actuarial J.* **10**(2), 120–129.

[221] R. Cardoso & H. Waters (2003) Recursive calculation of finite time ruin probabilities under interest force. *Insurance: Mathematics and Economics* **33**, 659–676.

[222] R. Cardoso & H. Waters (2003) Calculation of finite time ruin probabilities for some risk models. *Insurance: Mathematics and Economics* **37**, 197–215.

[223] P. Carr (1998) Randomization and the American put. *Review of Financial Studies* **11**, 597–626.

[224] M.L. Centeno (2002) Measuring the effects of reinsurance by the adjustment coefficient in the Sparre Anderson model. *Insurance: Mathematics and Economics* **30**, 37–49.

[225] M.L. Centeno (2005) Dependent risks and excess of loss reinsurance. *Insurance: Mathematics and Economics* **37**, 229–238.

[226] M.L. Centeno & O. Simões (2009) Optimal reinsurance. *Rev. R. Acad. Cien. Serie A. Mat.* **103**, 387–404.

[227] S. Chadjiconstantinidis & K. Politis (2005) Non-exponential bounds for stop-loss premiums and ruin probabilities. *Scand. Act. J.* **2005**, 335–357.

[228] S. Chadjiconstantinidis & K. Politis (2007) Two-sided bounds for the distribution of the deficit at ruin in the renewal risk model. *Insurance: Mathematics and Economics* **41**, 41–52.

[229] G. Chan & H. Yang (2006) Upper bounds for ruin probability under time series models. *Probab. Th. Eng. Inf. Sc.* **20**, 529–542.

[230] W.S. Chan, H. Yang & L. Zhang (2003) Some results on ruin probabilities in a two-dimensional risk model. *Insurance: Mathematics and Economics* **32**, 345–358.

[231] C.S. Chang, D.D. Yao & T. Zajic (1999) Large deviations, moderate deviations and queues with long-range dependent input. *Adv. Appl. Probab.* **31**, 254–277.

[232] C. Chen & H. Panjer (2009) A bridge from ruin theory to credit risk. *Rev. Quant. Finan. Acc.* **32**, 373–403.

[233] Y. Chen, C. Lee & Y. Sheu (2007) An ODE approach for the expected discounted penalty at ruin in a jump-diffusion model. *Finance Stoch.* **11**, 323–355.

[234] Y. Chen & K. Ng (2007) The ruin probability of the renewal model with constant interest force and negatively dependent heavy-tailed claims. *Insurance: Mathematics and Economics* **40**, 415–423.

[235] S. Cheng, H.U. Gerber & E.S.W. Shiu (2000) Discounted probabilities and ruin theory in the compound binomial model. *Insurance: Mathematics and Economics* **26**, 239–250.

[236] Y. Cheng & J. Pai (2003) On the nth stop-loss transform order of ruin probabilities. *Insurance: Mathematics and Economics* **32**, 51–60.

[237] Y. Cheng & Q. Tang (2003) Moments of the surplus before ruin and the deficit at ruin in the Erlang(2) risk process. *North American Actuarial J.* **7**(1), 1–12.

[238] P. Cheridito, F. Delbaen & M. Kupper (2006) Coherent and convex monetary risk measures for unbounded càdlàg processes. *Fin. Stoch.* **10**, 427–448.

[239] E.C.K. Cheung & D. Landriault (2009) Perturbed MAP risk models with dividend barrier strategies. *J. Appl. Probab.* **46**, 521–541.

[240] E.C.K. Cheung & D. Landriault (2010). Analysis of a generalized penalty function in a semi-Markovian risk model. *North American Actuarial J.*, in press.

[241] E.C.K Cheung, D. Landriault, G.E. Willmot & J. Woo (2010) Structural properties of Gerber-Shiu functions in dependent Sparre Andersen models. *Insurance: Mathematics and Economics* **46**, 117–126.

[242] E.C.K Cheung, D. Landriault, G.E. Willmot & J. Woo (2010) Gerber-Shiu analysis with a generalized penalty function. *Scand. Act. J.*, in press.

[243] Y. Chi (2010) Analysis of the expected discounted penalty function for a general jump-diffusion risk model and applications in finance. *Insurance: Mathematics and Economics* **46**, 385–396.

[244] Y. Chi, S. Jaimungal & S. Lin (2010) An insurance risk model with stochastic volatility. *Insurance: Mathematics and Economics* **46**, 52–67.

[245] S.N. Chiu & C. Yin (2005) Passage times for a spectrally negative Lévy process with applications to risk theory. *Bernoulli* **11**, 511–522.

[246] K.L. Chung (1974) *A Course in Probability Theory* (2nd ed.). Academic Press, New York San Francisco London.

[247] E. Çinlar (1972) Markov additive processes. II. *Z. Wahrscheinlichkeitsth. verw. Geb.* **24**, 93–121.

[248] D.B.H. Cline (1994) Intermediate regular variation and Π variation. *Proc. London Math. Soc.* **68**, 594–616.

[249] J.W. Cohen (1982) *The Single Server Queue* (2nd ed.) North-Holland.

[250] J. Collamore (1996) Hitting probabilities and large deviations. *Ann. Probab.* **24**, 2065–2078.

[251] J. Collamore (2002) Importance sampling techniques for the multi-dimensional ruin problem for general Markov additive sequences of random vectors. *Ann. Appl. Probab.* **12**, 382–421.

[252] J. Collamore (2009) Random recurrence equations and ruin in Markov-dependent stochastic economic environment. *Ann. Appl. Probab.* **19**, 1404–1458.

[253] I. Coulibaly & C. Lefèvre (2008) On a simple quasi-Monte Carlo approach for classical ultimate ruin probabilities. *Insurance: Mathematics and Economics* **42**, 935–942.

[254] C. Constantinescu (2006) *Renewal risk processes with stochastic returns on investments — a unified approach and analysis of the ruin probabilities.* Ph.D. Thesis, Oregon State University.

[255] C. Constantinescu & E. Thomann (2006) Analysis of the ruin probability using Laplace transforms and Karamata Tauberian theorems. *Actuarial Research Clearing House* **39**, 1–10.

[256] P.L. Conti (2005) A nonparametric sequential test with power 1 for the ruin probability in some risk models. *Stat. Probab. Lett.* **72**, 333–343.

[257] H. Cossette, M. Denuit & É. Marceau (2002) Distributional bounds for functions of dependent risks. *Mitt. Ver. Schweiz. Vers. Math.* **2002**, 45–65.

[258] H. Cossette, T. Duchesne & É. Marceau (2000) Modeling catastrophes and their impact on insurance portfolios. *North American Actuarial J.* **7**(4), 1–22.

[259] H. Cossette & É. Marceau (2000) The discrete-time risk model with correlated classes of business. *Insurance: Mathematics and Economics* **26**, 133–149.

[260] H. Cossette, D. Landriault & É. Marceau (2004) Exact expressions and upper bound for ruin probabilities in the compound Markov binomial model. *Insurance: Mathematics and Economics* **34**, 449–466.

[261] H. Cossette, D. Landriault & É. Marceau (2004) Compound binomial risk model in a Markovian environment. *Insurance: Mathematics and Economics* **35**, 425–443.

[262] H. Cossette, É. Marceau & F. Marri (2008) On the compound Poisson risk model with dependence based on a generalized Farlie-Gumbel-Morgenstern copula. *Insurance: Mathematics and Economics* **43**, 444–455.

[263] M. Cottrell, J.C. Fort & G. Malgouyres (1983) Large deviations and rare events in the study of stochastic algorithms. *IEEE Trans. Aut. Control* **AC–28**, 907–920.

[264] D.R. Cox (1955) Use of complex probabilities in the theory of stochastic processes. *Proc. Cambr. Philos. Soc.* **51**, 313–319.

[265] H. Cramér (1930) *On the Mathematical Theory of Risk.* Skandia Jubilee Volume, Stockholm.

[266] H. Cramér (1955) Collective risk theory. *The Jubilee volume of Försäkringsbolaget Skandia*, Stockholm.

[267] K. Croux & N. Veraverbeke (1990). Non-parametric estimators for the probability of ruin. *Insurance: Mathematics and Economics* **9**, 127–130.

[268] M. Csörgő & J. Steinebach (1991) On the estimation of the adjustment coefficient in risk theory via intermediate order statistics. *Insurance: Mathematics and Economics* **10**, 37–50.

[269] M. Csörgő & J.L. Teugels (1990) Empirical Laplace transform and approximation of compound distributions. *J. Appl. Probab.* **27**, 88–101.

[270] D. Daley & T. Rolski (1984) A light traffic approximation for a single server queue. *Math. Oper. Res.* **9**, 624–628.

[271] D. Daley & T. Rolski (1991) Light traffic approximations in queues. *Math. Oper. Res.* **16**, 57–71.

[272] L. Dang, N. Zhu & H. Zhang (2009) Survival probability for a two-dimensional risk model. *Insurance: Mathematics and Economics* **44**, 491–496.

[273] A. Dassios & P. Embrechts (1989) Martingales and insurance risk. *Stoch. Models* **5**, 181–217.

[274] A. Dassios & J. Jang (2003) Pricing of catastrophe reinsurance and derivatives using the Cox process with shot noise intensity. *Finance Stoch.* **7**, 73–95.

[275] A. Dassios & J. Jang (2005) Kalman-Bucy filtering for linear systems driven by the Cox process with shot noise intensity and its application to the pricing of reinsurance contracts. *J. Appl. Probab.* **42**, 93–107.

[276] B. D'Auria, J. Ivanovs, O. Kella & M. Mandjes (2010) First passage process of a Markov additive process, with applications to reflection problems. *Preprint.*

[277] A. Davidson (1946) On the ruin problem of collective risk theory under the assumption of a variable safety loading (in Swedish). Försäkringsmatematiska Studier Tillägnade Filip Lundberg, Stockholm. English version published in *Skand. Aktuar. Tidskr. Suppl.*, 70–83 (1969).

[278] M.H.A. Davis (1993) *Markov Models and Optimization.* Chapman & Hall.

[279] C.D. Daykin, T. Pentikäinen & E. Pesonen (1994) *Practical Risk Theory for Actuaries.* Chapman & Hall.

[280] K. Dębicki (1999) A note on LDP for supremum of Gaussian processes over infinite horizon. *Stat. Probab. Lett.* **44**, 211–220.

[281] K. Dębicki (2002) Ruin probability for Gaussian integrated processes. *Stoch. Proc. Appl.* **98**, 151–174.

[282] K. Dębicki, Z. Michna & T. Rolski (1998) On the supremum from Gaussian processes over infinite horizon. *Probab. Math. Statist.* **18**, 83–100.

[283] B. De Finetti (1957) Su un'Impostazione Alternativa della Teoria Collettiva del Rischio. *Transactions of the 15th Int. Congress of Actuaries*, 433–443.

[284] L. De Haan & A. Ferreira (2006) *Extreme Value Theory. An Introduction.* Birkhäuser.

[285] P. Deheuvels & J. Steinebach (1990) On some alternative estimators of the adjustment coefficient in risk theory. *Scand. Act. J.* **1990**, 135–159.

[286] T.G. De Kok (2003) Ruin probabilities with compounding assets for discrete time finite horizon problems, independent period claim sizes and general premium structure. *Insurance: Mathematics and Economics* **33**, 645–658.

[287] F. Delbaen & J. Haezendonck (1985) Inversed martingales in risk theory. *Insurance: Mathematics and Economics* **4**, 201–206.

[288] F. Delbaen & J. Haezendonck (1987) Classical risk theory in an economic environment. *Insurance: Mathematics and Economics* **6**, 85–116.

[289] A. Dembo, Karlin & O. Zeitouni (1994) Large exceedances for multidimensional Lévy processes. *Ann. Appl. Probab.* **4**, 432–447.

[290] A. Dembo & O. Zeitouni (1992) *Large Deviations Techniques and Applications.* Jones and Bartlett, Boston.

[291] A.P. Dempster, N.M. Laird & D.B. Rubin (1977) Maximum likelihood from incomplete data via the EM algorithm. *J. Roy. Statist. Soc.* **22**, 583–602.

[292] M. Denuit, J. Dhaene, M.J. Goovaerts & R. Kaas (2005) *Actuarial Theory for Dependent Risks*, Wiley.

[293] M. Denuit, C. Genest & É. Marceau (1999) Stochastic bounds on sums of dependent risks. *Insurance: Mathematics and Economics* **25**, 85–104.

[294] M. Denuit, C. Genest & É. Marceau (2002) Criteria for the stochastic ordering of random sums, with actuarial applications. *Scand. Act. J.* **2002**, 3–16.

[295] M. Denuit & C. Lefèvre (1997) Some new classes of stochastic order relations among arithmetic random variables, with applications in actuarial sciences. *Insurance: Mathematics and Economics* **20**, 197–213.

[296] V. Dermitzakis, S.M. Pitts & K. Politis (2010) Lundberg-type bounds and asymptotics for the moments of the time to ruin. *Methodol. Comput Appl. Probab.* **12**, 155–175.

[297] L. Devroye (1986) *Non-Uniform Random Variate Generation.* Springer-Verlag.

[298] F. De Vylder (1977) A new proof of a known result in risk theory. *J. Comp. Appl. Math.* **3**, 227–229.

[299] F. De Vylder (1978) A practical solution to the problem of ultimate ruin probability. *Scand. Act. J.* **1978**, 114–119.

[300] F. De Vylder (1996) *Advanced Risk Theory.* Editions de l'Universite de Bruxelles.

[301] F. De Vylder & M.J. Goovaerts (1984) Bounds for classical ruin probabilities. *Insurance: Mathematics and Economics* **3**, 121–131.

[302] F. De Vylder & M.J. Goovaerts (1988) Recursive calculation of finite–time ruin probabilities. *Insurance: Mathematics and Economics* **7**, 1–7.

[303] F. De Vylder & M.J. Goovaerts (1999) Explicit finite-time and infinite-time ruin probabilities in the continuous case. *Insurance: Mathematics and Economics* **24**, 155–172.

[304] F. De Vylder & M.J. Goovaerts (1999) Inequality extensions of Prabhu's formula in ruin theory. *Insurance: Mathematics and Economics* **24**, 249–271.

[305] D.C.M. Dickson (1992) On the distribution of the surplus prior to ruin. *Insurance: Mathematics and Economics* **11**, 191–207.

[306] D.C.M. Dickson (1994) An upper bound for the probability of ultimate ruin. *Scand. Act. J.* **1994**, 131–138.

[307] D.C.M. Dickson (1995) A review of Panjer's recursion formula and its applications. *British Actuarial J.* **1**, 107–124.

[308] D.C.M. Dickson (1998) Discussion on 'On the time value of ruin'. *North American Actuarial J.* **2**, 74.

[309] D.C.M. Dickson (1999) *Insurance Risk and Ruin*. Cambridge University Press.

[310] D.C.M. Dickson (2008) Some explicit solutions for the joint density of the time of ruin and the deficit at ruin. *Astin Bulletin* **38**, 259–276.

[311] D.C.M. Dickson & A.D. Egidio dos Reis (1996) On the distribution of the duration of negative surplus. *Scand. Act. J.* **1996**, 148–164.

[312] D.C.M. Dickson & J.R. Gray (1984) Exact solutions for ruin probability in the presence of an upper absorbing barrier. *Scand. Act. J.* **1984**, 174–186.

[313] D.C.M. Dickson & J.R. Gray (1984) Approximations to the ruin probability in the presence of an upper absorbing barrier. *Scand. Act. J.* **1984**, 105–115.

[314] D.C.M. Dickson & C. Hipp (1998) Ruin probabilities for Erlang(2) risk processes. *Insurance: Mathematics and Economics* **22**, 251–262.

[315] D.C.M. Dickson & C. Hipp (1999) Ruin problems for phase-type(2) risk processes. *Scand. Act. J.* **2000**, 147–167.

[316] D.C.M. Dickson & C. Hipp (2001) On the time to ruin for Erlang(2) risk processes. *Insurance: Mathematics and Economics* **22**, 251–262.

[317] D.C.M. Dickson & S. Li (2010) Finite time ruin problems for the Erlang(2) risk model. *Insurance: Mathematics and Economics* **46**, 12–18.

[318] D.C.M. Dickson & H.R. Waters (1991) Recursive calculation of survival probabilities. *Astin Bulletin* **21**, 199–221.

[319] D.C.M. Dickson & H.R. Waters (1996) Reinsurance and ruin. *Insurance: Mathematics and Economics* **19**, 61–80.

[320] D.C.M. Dickson & H.R. Waters (1999) Ruin probabilities with compounding. *Insurance: Mathematics and Economics* **25**, 49–62.

[321] D.C.M. Dickson & H.R. Waters (2006) Optimal dynamic reinsurance. *Astin Bulletin* **36**, 415–432.

[322] D.C.M. Dickson & G.E. Willmot (2005) The density of the time to ruin in the classical Poisson risk model. *Astin Bulletin* **35**, 45–60.

[323] T. Dieker (2006) Applications of factorization embeddings for Lévy processes. *Adv. Appl. Probab.* **38**, 768–791.

[324] B. Djehiche (1993) A large deviation estimate for ruin probabilities. *Scand. Act. J.* **1993**, 42–59.

[325] R.A. Doney (1991) Hitting probabilities for spectrally positive Lévy processes. *J. London Math. Soc.* **44**, 566–576.

[326] R.A. Doney & A. Kyprianou (2006) Overshoots and undershoots of Lévy processes. *Ann. Appl. Probab.* **16**, 91–106.

[327] E. van Doorn & G.J.K. Regterschot (1988) Conditional PASTA. *Oper. Res. Lett.* **7**, 229–232.

[328] S. Drekic, D.C.M. Dickson, D.A. Stanford & G.E. Willmot (2004) On the distribution of the deficit at ruin when claims are phase-type. *Scand. Act. J.* **2004**, 105–20.

[329] S. Drekic, J. Stafford & G.E. Willmot (2004) Symbolic calculation of the moments of the time of ruin. *Insurance: Mathematics and Economics* **34**, 109–120.

[330] S. Drekic & G.E. Willmot (2005) On the moments of the time of ruin with applications to phase-type claims. *North American Actuarial J.* **9**(2), 17–30.

[331] N.G. Duffield & N. O'Connell (1995) Large deviations and overflow probabilities for the general single–server queue, with applications. *Math. Proc. Camb. Philos. Soc.* **118**, 363–374.

[332] D. Dufresne (2001) On a general class of risk models. *Austral. Act. J.* **7**, 755–791.

[333] F. Dufresne & H.U. Gerber (1988) The surpluses immediately before and at ruin, and the amount of claim causing ruin. *Insurance: Mathematics and Economics* **7**, 193–199.

[334] F. Dufresne & H.U. Gerber (1991) Risk theory for the compound Poisson process that is perturbed by a diffusion. *Insurance: Mathematics and Economics* **10**, 51–59.

[335] F. Dufresne, H.U. Gerber & E.S.W. Shiu (1991) Risk theory with the Gamma process. *Astin Bulletin* **21**, 177–192.

[336] P. Dupuis, K. Leder & H. Wang (2007) Importance sampling for sums of random variables with regularly varying tails. *ACM TOMACS* **17**(3).

[337] P. Dupuis & H. Wang (2004) Importance sampling, large deviations and differential games. *Stochastics and Stochastics Reports* **76**, 481–508.

[338] E.B. Dynkin (1965) *Markov Processes* **I**. Springer-Verlag.

[339] A.D. Egidio dos Reis (2000) On the moments of ruin and recovery times. *Insurance: Mathematics and Economics* **27**, 331–343.

[340] A.D. Egidio dos Reis (2002) How many claims does it take to get ruined and recovered? *Insurance: Mathematics and Economics* **31**, 235–248.

[341] K.T. Eisele (2008) Recursions for multivariate compound phase variables. *Insurance: Mathematics and Economics* **42**, 65–72.

[342] D.C. Emanuel, J.M. Harrison & A.J. Taylor (1975) A diffusion approximation for the ruin probability with compounding assets. *Scand. Act. J.* **1975**, 37–45.

[343] P. Embrechts (1988). Ruin estimates for large claims. *Insurance: Mathematics and Economics* **7**, 269–274.

[344] P. Embrechts & M. Frei (2009). Panjer recursion versus FFT for compound distributions. *Math. Meth. Oper. Res.* **69**, 497–508.

[345] P. Embrechts, J. Grandell & H. Schmidli (1993) Finite-time Lundberg inequalities in the Cox case. *Scand. Act. J.* **1993**, 17–41.

[346] P. Embrechts, R. Grübel & S.M. Pitts (1993) Some applications of the fast Fourier transform in insurance mathematics. *Stat. Neerl.* **47**, 59–75.

[347] P. Embrechts, J.L. Jensen, M. Maejima & J.L. Teugels (1985). Approximations for compound Poisson and Polya processes. *Adv. Appl. Probab.* **17**, 623–637.

[348] P. Embrechts & T. Mikosch (1991). A bootstrap procedure for estimating the adjustment coefficient. *Insurance: Mathematics and Economics* **10**, 181–190.

[349] P. Embrechts, C. Klüppelberg & T. Mikosch (1997) *Modelling Extremal Events for Finance and Insurance.* Springer-Verlag.

[350] P. Embrechts and G. Puccetti (2006) Bounds for functions of dependent risks. *Finance Stoch.* **10**, 341–352.

[351] P. Embrechts and G. Puccetti (2006) Bounds for functions of multivariate risks. *J. Multivar. Anal.* **97**, 526–547.

[352] P. Embrechts & H. Schmidli (1994) Ruin estimation for a general insurance risk model. *Adv. Appl. Probab.* **26**, 404–422.

[353] P. Embrechts & N. Veraverbeke (1982) Estimates for the probability of ruin with special emphasis on the possibility of large claims. *Insurance: Mathematics and Economics* **1**, 55–72.

[354] P. Embrechts & J.A. Villaseñor (1988) Ruin estimates for large claims. *Insurance: Mathematics and Economics* **7**, 269–274.

[355] F. Enikeeva, V. Kalashnikov & D. Rusaityte (2001) Continuity estimates for ruin probabilities. *Scand. Act. J.* **2001**, 18–39.

[356] A.K. Erlang (1909) Sandsynlighedsregning og telefonsamtaler. *Nyt Tidsskrift for Matematik* **B20**, 33–39. Reprinted 1948 as 'The theory of probabilities and telephone conversations' in *The Life and work of A.K. Erlang* **2**, 131–137. Trans. Danish Academy Tech. Sci.

[357] J.D. Esary, F. Proschan & D.W. Walkup (1967) Association of random variables, with applications. *Ann. Math. Statist.* **38**, 1466–1474.

[358] F. Esscher (1932) On the probability function in the collective theory of risk. *Skand. Akt. Tidsskr.* 175–195.

[359] S. Ethier & T. Kurtz (1986) *Markov Processes. Characterization and Convergence.* Wiley.

[360] M. Fackrell (2009) An alternative characterization of matrix-exponential distributions. *Adv. Appl. Probab.* **41**, 1005–1022.

[361] W. Feller (1966) *An Introduction to Probability Theory and its Applications* **I** (3rd ed.). Wiley.

[362] W. Feller (1971) *An Introduction to Probability Theory and its Applications* **II** (2nd ed.). Wiley.

[363] P.J. Fitzsimmons (1987) On the excursions of Markov processes in classical duality. *Probab. Th. Rel. Fields* **75**, 159–178.

[364] W.H. Fleming & H.M. Soner (1993) *Controlled Markov Processes and Viscosity Solutions*. Springer-Verlag.

[365] S. Foss & D. Korshunov (2007) Lower limits and equivalences for convolution tails. *Ann. Probab.* **35**, 366–383.

[366] S. Foss, Z. Palmowski & S. Zachary (2005) The probability of exceeding a high boundary on a random time interval for a heavy-tailed random walk. *Adv. Appl. Probab.* **15**, 1936–1957.

[367] S. Foss & A. Richards (2010) On sums of conditionally independent subexponential random variables. *Math. Oper. Res.*, in press.

[368] E. Fournié, J.-M. Lasry, J. Lebouchoux, P.-L. Lions & N. Touzi (1999) Applications of Malliavin calculus to Monte Carlo methods in finance. *Finance Stoch.* **3**, 391–412.

[369] N. Frangos, S. Vrontos & A. Yannacopoulos (2007) Ruin probability at a given time for a model with liabilities of the fractional Brownian motion type: A partial differential equation approach. *Scand. Act. J.* **2007**, 285–308.

[370] P. Franken, D. König, U. Arndt & V. Schmidt (1982) *Queues and Point Processes*. Wiley.

[371] E.W. Frees (1986) Nonparametric estimation of the probability of ruin. *Astin Bulletin* **16**, 81–90.

[372] M. Frenz & V. Schmidt (1992) An insensitivity property of ladder height distributions. *J. Appl. Probab.* **29**, 616–624.

[373] A. Frey & V. Schmidt (1996) Taylor series expansion for multivariate characteristics of classical risk processes. *Insurance: Mathematics and Economics* **18**, 1–12.

[374] A. Frolova, Y. Kabanov & S. Pergamenshchikov (2002) In the insurance business risky investments are dangerous. *Fin. Stoch.* **6**, 227–235.

[375] E. Frostig (2003) Ordering ruin probabilities for dependent claim streams. *Insurance: Mathematics and Economics* **32**, 93–114.

[376] E. Frostig (2005) The expected time to ruin in a risk process with constant barrier via martingales. *Insurance: Mathematics and Economics* **37**, 216–228.

[377] E. Frostig (2008) On a risk model with dividend payments perturbed by a Brownian motion — an algorithmic approach. *Astin Bulletin* **38**, 183–206.

[378] E. Frostig & M. Denuit (2009) Ruin probabilities and optimal capital allocation for heterogeneous life annuity portfolios. *Scand. Act. J.* **2009**, 295–305.

[379] C.D. Fuh (1997) Corrected ruin probabilities for ruin probabilities in Markov random walks. *Adv. Appl. Probab.* **29**, 695–712.

[380] C.D. Fuh & T. Lai (1998) Wald's equations, first passage times and moments of ladder variables in Markov random walks. *J. Appl. Probab.* **35**, 566–580.

[381] H. Furrer (1998) Risk processes perturbed by α–stable Levý motion. *Scand. Act. J.* **1998**, 59–74.

[382] H. Furrer (1996) A note on the convergence of the infinite-time ruin probabilities when the weak approximation is α–stable Lévy motion. Unpublished, contained in the author's PhD thesis, ETH Zürich.

[383] H. Furrer, Z. Michna & A. Weron (1997) Stable Lévy motion approximation in collective risk theory. *Insurance: Mathematics and Economics* **20**, 97–114.

[384] H. Furrer & H. Schmidli (1994) Exponential inequalities for ruin probabilities of risk processes perturbed by diffusion. *Insurance: Mathematics and Economics* **15**, 23–36.

[385] J. Gaier & P. Grandits (2004) Ruin probabilities and investment under interest force in the presence of regularly varying tails. *Scand. Act. J.* **2004**, 256–278.

[386] J. Gaier, P. Grandits & W. Schachermayer (2003) Asymptotic ruin probabilities and optimal investment. *Ann. Appl. Probab.* **13**, 1054–1076.

[387] A. Ganesh, C. Macci & G.L. Torrisi (2007) A class of risk processes with reserve-dependent premium rate: sample path large deviations and importance sampling. *Queueing Systems* **55**, 83–94.

[388] F. Gao & J. Yan (2009) Sample path large and moderate deviations for risk model with delayed claims. *Insurance: Mathematics and Economics* **45**, 74–80.

[389] J. Garcia (2005) Explicit solutions for the survival probabilities in the classical risk model. *Astin Bulletin* **35**, 113–130.

[390] J. Garrido (1989) Stochastic differential equations for compounded risk reserves. *Insurance: Mathematics and Economics* **8**, 165–173.

[391] J. Garrido & M. Morales (2006) On the expected discounted penalty function for Lévy risk processes. *North American Actuarial J.* **10**, 196–218.

[392] J.L. Geluk & K. Ng (2006) Tail behavior of negatively associated heavy tailed sums. *J. Appl. Probab.* **43**, 587–593.

[393] J.L. Geluk, L. Peng & C.G. de Vries (2000) Convolutions of heavy-tailed random variables and applications to portfolio diversification and MA(1) time series. *Adv. Appl. Probab.* **32**, 1011–1026.

[394] H.U. Gerber (1969) Entscheidungskriterien für den zusammengesetzten Poisson-Prozess. *Mitt. Ver. Schweiz. Vers. Math.* **69**, 185–227.

[395] H.U. Gerber (1970) An extension of the renewal equation and its application in the collective theory of risk. *Skand. Aktuarietidskrift* **1970**, 205–210.

[396] H.U. Gerber (1971) Der Einfluss von Zins auf die Ruinwahrscheinlichkeit. *Mitt. Ver. Schweiz. Vers. Math.***71**, 63–70.

[397] H.U. Gerber (1973) Martingales in risk theory. *Mitt. Ver. Schweiz. Vers. Math.* **73**, 205–216.

[398] H.U. Gerber (1979) *An Introduction to Mathematical Risk Theory*. S.S. Huebner Foundation Monographs, University of Pennsylvania.

[399] H.U. Gerber (1981) On the probability of ruin in the presence of a linear dividend barrier. *Scand. Act. J.* **1981**, 105–115.

[400] H.U. Gerber (1981) On the probability of ruin in an autoregressive model. *Mitt. Ver. Schweiz. Vers. Math.* **81**, 213–219.

[401] H.U. Gerber (1982) Ruin theory in the linear model. *Insurance: Mathematics and Economics* **1**, 177–184.

[402] H.U. Gerber (1986). *Life Insurance Mathematics*. Springer-Verlag.

[403] H.U. Gerber (1988) Mathematical fun with ruin theory. *Insurance: Mathematics and Economics* **7**, 15–23.

[404] H.U. Gerber (1988) Mathematical fun with the compound binomial process. *Astin Bulletin* **18**, 161–168.

[405] H.U. Gerber, M.J. Goovaerts & R. Kaas (1987) On the probability and severity of ruin. *Astin Bulletin* **17**, 151–163.

[406] H.U. Gerber & B. Landry (1998) On the discounted penalty at ruin in a jump-diffusion and the perpetual put option. *Insurance: Mathematics and Economics* **22**, 263–276.

[407] H.U. Gerber, X.S. Lin & H. Yang (2006) A note on the dividends-penalty identity and the optimal dividend barrier. *Astin Bulletin* **36**, 489–503.

[408] H.U. Gerber & E.S.W. Shiu (1997) The joint distribution of the time of ruin, the surplus immediately before ruin, and the deficit at ruin. *Insurance: Mathematics and Economics* **21**, 129–137.

[409] H.U. Gerber & E.S.W. Shiu (1998) On the time value of ruin. *North American Actuarial J.* **2**, 48–72.

[410] H.U. Gerber & E.S.W. Shiu (1999) From ruin theory to pricing reset guarantees and perpetual put options. *Insurance: Mathematics and Economics* **24**, 3–14.

[411] H.U. Gerber & E.S.W. Shiu (2005) The time value of ruin in a Sparre Andersen model. *North American Actuarial J.* **9**, 49–84.

[412] H.U. Gerber & E.S.W. Shiu (2006) On optimal dividend strategies in the compound Poisson model. *North American Actuarial J.* **10**, 76–93.

[413] H.U. Gerber, E.S.W. Shiu & N. Smith (2008) Methods for estimating the optimal dividend barrier and the probability of ruin. *Insurance: Mathematics and Economics* **42**, 243–254.

[414] H.U. Gerber & H. Yang (2007) Absolute ruin probabilities in a jump-diffusion model with investment. *North American Actuarial J.* **11**(3), 159–169.

[415] H.U. Gerber & H. Yang (2010) Obtaining the dividends-penalty identities by interpretation. *Insurance: Mathematics and Economics*, in press.

[416] S. Gerhold, U. Schmock & R. Warnung (2010) A generalization of Panjer's recursion and numerically stable risk aggregation. *Fin. Stoch.* **14**, 1432–1122.

[417] P. Glasserman (1991) *Gradient Estimation via Perturbation Analysis*. Kluwer.

[418] P. Glasserman & S.-G. Kou (1995) Limits of first passage times to rare sets in regenerative processes. *Ann. Appl. Probab.* **5**, 424–445.

[419] P.W. Glynn & W. Whitt (1994) Logarithmic asymptotics for steady-state tail probabilities in a single-server queue. In *Studies in Applied Probability* (J. Galambos & J. Gani, eds.). *J. Appl. Probab.* **31A**, 131–156.

[420] B.V. Gnedenko & I.N. Kovalenko (1989) *Introduction to Queueing Theory* (2nd ed.). Birkhäuser.

[421] C.M. Goldie (1991) Implicit renewal theory and tails of solutions of random equations. *Ann. Appl. Probab.* **1**, 126–166.

[422] C.M. Goldie & R. Grübel (1996) Perpetuities with thin tails. *Adv. Appl. Probab.* **28**, 463–480.

[423] M.J. Goovaerts, F. de Vylder & J. Haezendonck (1984) *Insurance Premiums*. North-Holland.

[424] M.J. Goovaerts, R. Kaas, R. Laeven, Q. Tang & R. Vernic (2005) The tail probability of discounted sums of Pareto-like losses in insurance. *Scand. Act. J.* **2005**, 446–461.

[425] M.J. Goovaerts, R. Kaas, A.E. van Heerwarden & T. Bauwelinckx (1990) *Effective Actuarial Methods*. North-Holland.

[426] J. Grandell (1977) A class of approximations of ruin probabilities. *Scand. Act. J., Suppl.*, **1977**, 37–52.

[427] J. Grandell (1978) A remark on 'A class of approximations of ruin probabilities'. *Scand. Act. J.* **1978**, 77–78.

[428] J. Grandell (1979) Empirical bounds for ruin probabilities. *Stoch. Proc. Appl.* **8**, 243–255.

[429] J. Grandell (1990) *Aspects of Risk Theory*. Springer-Verlag.

[430] J. Grandell (1992) Finite time ruin probabilities and martingales. *Informatica* **2**, 3–32.

[431] J. Grandell (1997) *Mixed Poisson Processes*. Chapman & Hall.

[432] J. Grandell (2000) Simple approximations of ruin probabilities. *Insurance: Mathematics and Economics* **26**, 157–173.

[433] J. Grandell & C.-O. Segerdahl (1971) A comparison of some approximations of ruin probabilities. *Skand. Aktuar Tidskr.* **54**, 143–158.

[434] P. Grandits (2005) Minimal ruin probabilities and investment under interest force for a class of subexponential distributions. *Scand. Act. J.* **2005**, 401–416.

[435] B. Grigelionis (1992) On Lundberg inequalities in a Markovian environment. *Proc. Winter School on Stochastic Analysis and Appl.*. Akademie-Verlag, Berlin.

[436] B. Grigelionis (1993) Two-sided Lundberg inequalities in a Markovian environment. *Liet. Matem. Rink.* **33**, 30–41.

[437] B. Grigelionis (1996) Lundberg-type stochastic processes. *Probability Theory and Mathematical Statistics* (I.A. Ibragimov, ed.), 167–176.

[438] R. Grübel (1991) G/G/1 via FFT. Statistical Algorithm 265. *Applied Statistics* **40**, 355–365.

[439] R. Grübel & R. Hermesmeier (1999) Computation of compound distributions I: aliasing errors and exponential tilting. *Astin Bulletin* **29**, 197–214.

[440] R. Grübel & R. Hermesmeier (1999) Computation of compound distributions II: Discretization errors and Richardson extrapolation. *Astin Bulletin* **30**, 309–331.

[441] M. Guerra & M.L. Centeno (2008) Optimal reinsurance policy: the adjustment coefficient and the expected utility criteria. *Insurance: Mathematics and Economics* **42**, 529–539.

[442] D.V. Gusak & V.S. Korolyuk (1969) On the joint distribution of a process with stationary increments and its maximum. *Th. Probab. Appl.* **14**, 400–409.

[443] A. Gut (1988) *Stopped Random Walks. Theory and Applications*. Springer-Verlag.

[444] M. Gyllenberg & D. Silvestrov (2000) Cramér-Lundberg approximations for non-linearly perturbed risk processes. *Insurance: Mathematics and Economics* **26**, 75–90.

[445] H. Hadwiger (1940) Über die Wahrscheinlichkeit des Ruins bei einer grossen Zahl von Geschäften. *Archiv für Mathematische Wirtschafts- und Sozialforschung* **6**, 131–135.

[446] M. Hald & H. Schmidli (2004) On the maximization of the adjustment coefficient under proportional reinsurance. *Astin Bulletin* **34**, 75–83.

[447] X. Hao & Q. Tang (2008) A uniform asymptotic estimate for discounted aggregate claims with subexponential tails. *Insurance: Mathematics and Economics* **43**, 116–120.

[448] X. Hao & Q. Tang (2009) Asymptotic ruin probabilities of the Lévy insurance model under periodic taxation. *Astin Bulletin* **39**, 479–494.

[449] J.M. Harrison (1977) Ruin problems with compounding assets. *Stoch. Proc. Appl.* **5**, 67–79.

[450] J.M. Harrison & A.J. Lemoine (1977) Limit theorems for periodic queues. *J. Appl. Probab.* **14**, 566–576.

[451] J.M. Harrison & S.I. Resnick (1976) The stationary distribution and first exit probabilities of a storage process with general release rule. *Math. Oper. Res.* **1**, 347–358.

[452] J.M. Harrison & S.I. Resnick (1977) The recurrence classification of risk and storage processes. *Math. Oper. Res.* **3**, 57–66.

[453] J. Hartinger & D. Kortschak (2007) On the efficiency of the Asmussen-Kroese-estimator and its application to stop-loss transforms. *Blätter DGVFM* **30**, 363–377.

[454] A.E. van Heerwarden (1991) *Ordering of Risks: Theory and Actuarial Applications.* Tinbergen Institute Research Series **20**, Amsterdam.

[455] P. Heidelberger (1995) Fast simulation of rare events in queueing and reliability models. *ACM TOMACS* **6**, 43–85.

[456] B. Heidergott, F.J. Vázquez-Abad, G. Pflug & T. Farenhorst-Yuan (2010) Gradient estimation for discrete event systems by measure-valued differentiation. *ACM TOMACS*, in press.

[457] B. Heidergott, T. Farenhorst-Yuan & F.J. Vázquez-Abad (2010) A perturbation analysis approach to phantom estimators for waiting times in the GI/G/1 queue. *Discrete Event Dynamical Systems*, in press.

[458] W.-R. Heilmann (1987) *Grundbegriffe der Risikotheorie.* Verlag Versicherungswirtschaft e.V., Karlsruhe.

[459] U. Herkenrath (1986) On the estimation of the adjustment coefficient in risk theory by means of stochastic approximation procedures. *Insurance: Mathematics and Economics* **5**, 305–313.

[460] U. Hermann (1965) Ein Approximationssatz für Verteilungen stationärer zufälliger Punktfolgen. *Math. Nachr.* **30**, 377–381.

[461] O. Hesselager (1990) Some results on optimal reinsurance in terms of the adjustment coefficient. *Scand. Act. J.* **1990**, 80–95.

[462] C.C. Heyde & D. Wang (2009) Finite-time ruin probability with an exponential Lévy process investment return and heavy-tailed claims. *Adv. Appl. Probab.* **41**, 206–224.

[463] B.M. Hill (1975) A simple general approach to inference about the tail of a distribution. *Ann. Statist.* **3**, 1163–1174.

[464] C. Hipp (1989a) Efficient estimators for ruin probabilities. *Proc. Fourth Prague Symp. on Asymptotic Statistics* (P. Mandl & M. Huskova, eds.), 259–268.

[465] C. Hipp (1989b) Estimators and bootstrap confidence intervals for ruin probabilities. *Astin Bulletin* **19**, 57–70.

[466] C. Hipp (2004) Stochastic control with application in insurance. *in: Stochastic Methods in Finance*, 127–164. Springer-Verlag.

[467] C. Hipp (2006) Speedy convolution algorithms and Panjer recursions for phase-type distributions. *Insurance: Mathematics and Economics* **38**, 176–188.

[468] C. Hipp & R. Michel (1990) *Risikotheorie: Stochastische Modelle und Statistische Methoden*. Verlag Versicherungswirtschaft e.V., Karlsruhe.

[469] C. Hipp & M. Plum (2000) Optimal investment for insurers. *Insurance: Mathematics and Economics* **27**, 215–228.

[470] C. Hipp & M. Plum (2000) Optimal investment for investors with state dependent income, and for insurers. *Finance Stoch.* **7**, 299–321.

[471] C. Hipp & M. Taksar (2000) Stochastic control for optimal new business. *Insurance: Mathematics and Economics* **26**, 185-192.

[472] C. Hipp & M. Vogt (2000) Optimal dynamic XL reinsurance. *Astin Bulletin* **33**, 193-207.

[473] M.L. Hogan (1986) Comment on corrected diffusion approximations in certain random walk problems. *J. Appl. Probab.* **23**, 89–96.

[474] R.V. Hogg & S.A. Klugman (1984) *Loss Distributions*. Wiley.

[475] L. Horváth & E. Willekens (1986) Estimates for the probability of ruin starting with a large initial reserve. *Insurance: Mathematics and Economics* **5**, 285–293.

[476] O. Häggström, S. Asmussen & O. Nerman (1992) EMPHT — a program for fitting phase-type distributions. *Studies in Statistical Quality Control and Reliability* **1992:4**. Mathematical Statistics, Chalmers University of Technology and the University of Göteborg. Software available from http://home.imf.au.dk/asmus.

[477] T. Höglund (1974) Central limit theorems and statistical inference for Markov chains. *Z. Wahrscheinlichkeitsth. verw. Geb.* **29**, 123–151.

[478] T. Höglund (1990) An asymptotic expression for the probability of ruin within finite time. *Ann. Probab.* **18**, 378–389.

[479] T. Höglund (1991). The ruin problem for finite Markov chains. *Ann. Probab.* **19**, 1298–1310.

[480] B. Højgaard & M. Taksar (1997) Optimal proportional reinsurance policies for diffusion models. *Scand. Act. J.* **1997**, 166–180.

[481] B. Højgaard & M. Taksar (1998) Optimal proportional reinsurance policies for diffusion models with transaction costs. *Insurance: Mathematics and Economics* **22**, 41–51.

[482] F. Hubalek & A. Kyprianou (2010) Old and new examples of scale functions for spectrally negative Lévy processes. *in: Progress in Probability*, Birkhäuser.

[483] J. Huh & A. Kolkiewicz (2008) Computation of multivariate barrier crossing probability and its applications in credit risk models. *North American Actuarial J.* **12**, 263–291.

[484] H. Hult & J. Svensson (2009) On importance sampling with mixtures for random walks with heavy tails. arXiv:math/0909.333v.

[485] H. Hult & F. Lindskog (2010) Heavy-tailed insurance portfolios: buffer capital and ruin probabilities. *Preprint.*

[486] H. Hult & G. Samorodnitsky (2010) Large deviations for point processes based on stationary sequences with heavy tails. *J. Appl. Probab.* **47**, 1–40.

[487] J. Hüsler & V. Piterbarg (1999) Extremes of a certain class of Gaussian processes. *Stoch. Proc. Appl.* **83**, 257–271.

[488] J. Hüsler & V. Piterbarg (2004) On the ruin probability for physical fractional Brownian motion. *Stoch. Proc. Appl.* **113**, 315–332.

[489] J. Hüsler & V. Piterbarg (2008) A limit theorem for the time of ruin in a Gaussian ruin problem. *Stoch. Proc. Appl.* **118**, 2014–2021.

[490] M. Huzak, M. Perman, H. Sikic & Z. Vondracek (2004) Ruin probabilities and decompositions for general perturbed risk processes. *Ann. Appl. Probab.* **14**, 1378–1397.

[491] M. Huzak, M. Perman, H. Sikic & Z. Vondracek (2004) Ruin probabilities for competing claim processes. *J. Appl. Probab.* **41**, 679–690.

[492] D.L. Iglehart (1969) Diffusion approximations in collective risk theory. *J. Appl. Probab.* **6**, 285–292.

[493] Z. Ignatov, V. Kaishev & R. Krachunov (2001) An improved finite-time ruin probability formula and its Mathematica implementation. *Insurance: Mathematics and Economics* **29**, 375–386.

[494] Z. Ignatov & V. Kaishev (2004) A finite-time ruin probability formula for continuous claim size severities. *J. Appl. Probab.* **41**, 570–578.

[495] C. Irgens & J. Paulsen (2004) Optimal control of risk exposure, reinsurance and investments for insurance portfolios. *Insurance: Mathematics and Economics* **35**, 21–51.

[496] V.B. Iversen & L. Staalhagen (1999) Waiting time distributions in M/D/1 queueing systems. *Electronic Lett.* **35**, No. 25.

[497] M. Jacobsen (2003) Martingales and the time to ruin. *Stoch. Proc. Appl.* **107**, 29–51.

[498] M. Jacobsen (2005) The time to ruin for a class of Markov additive risk process with two-sided jumps. *Adv. Appl. Probab.* **37**, 963–992.

[499] D. Jagerman (1985) Certain Volterra integral equations arising in queueing. *Stoch. Models* **1**, 239–256.

[500] D. Jagerman (1991) Analytical and numerical solution of Volterra integral equations with applications to queues. Manuscript, NEC Research Institute, Princeton, N.J.

[501] J. Janssen (1980) Some transient results on the M/SM/1 special semi-Markov model in risk and queueing theories. *Astin Bulletin* **11**, 41–51.

[502] J. Janssen & J.M. Reinhard (1985) Probabilités de ruine pour une classe de modeles de risque semi-Markoviens. *Astin Bulletin* **15**, 123–133.

[503] H. Jasiulewicz (2001) Probability of ruin with variable premium rate in a Markovian environment. *Insurance: Mathematics and Economics* **29**, 291–296.

[504] P.R. Jelenkovic & A.A. Lazar (1998) Subexponential asymptotics of a Markov-modulated random walk with a queueing application. *J. Appl. Probab.* **35**, 325–247.

[505] A. Jensen (1953) *A Distribution Model Applicable to Economics.* Munksgaard, Copenhagen.

[506] J.L. Jensen (1995) *Saddle Point Approximations.* Clarendon Press, Oxford.

[507] A. Jobert & L.G.C. Rogers (2008) Valuations and dynamic convex risk measures. *Math. Finance* **18**, 1–22.

[508] H. Joe (1997) *Multivariate Models and Dependence Concepts.* Chapman & Hall.

[509] M. Johnson & M. Taaffe (1989/90) Matching moments to phase distributions. *Stoch. Models* **5**, 711–743; *ibid.* **6**, 259–281; *ibid.* **6**, 283–306.

[510] S. Juneja (2007) Estimating tail probabilities of heavy tailed distributions with asymptotically zero error. *Queueing Systems* **57**, 115–127.

[511] S. Juneja & P. Shahabuddin (2002) Simulating heavy-tailed processes using delayed hazard rate twisting. *ACM TOMACS* **12**, 94–118.

[512] S. Juneja & P. Shahabuddin (2006) Rare event simulation techniques: and introduction and recent advances. *Handbook on Simulation* (S. Henderson & B. Nelson, eds.), 291–350. North-Holland.

[513] A. Juri (2002) Supermodular order and Lundberg exponents. *Scand. Act. J.* **2002**, 17–36.

[514] R. Kaas & M.J. Goovaerts (1986) General bound on ruin probabilities. *Insurance: Mathematics and Economics* **5**, 165–167.

[515] R. Kaas, M.J. Goovaerts, J. Dhaene & M. Denuit (2003) *Modern Actuarial Risk Theory.* Kluwer.

[516] V. Kalashnikov (1996) Two-sided bounds of ruin probabilities. *Scand. Act. J.* **1996**, 1–18.

[517] V. Kalashnikov (1997) *Geometric Sums: Bounds for Rare Event with Applications.* Kluwer.

[518] V. Kalashnikov (1999) Bounds for ruin probabilities in the presence of large claims and their comparison. *North American Actuarial J.* **3**, 116–129.

[519] V. Kalashnikov & D. Konstantinides (2000) Ruin under interest force and subexponential claims: a simple treatment. *Insurance: Mathematics and Economics* **27**, 145–149.

[520] V. Kalashnikov & R. Norberg (2002) Power tailed ruin probabilities in the presence of risky investments. *Stoch. Proc. Appl.* **98**, 211–228.

[521] E.P.C. Kao (1988) Computing the phase-type renewal and related functions. *Technometrics* **30**, 87–93.

[522] S. Karlin & H.M. Taylor (1981) *A Second Course in Stochastic Processes*. Academic Press, New York.

[523] J. Keilson (1966) A limit theorem for passage times in ergodic regenerative processes. *Ann. Math. Statist.* **37**, 866–870.

[524] J. Keilson & D.M.G. Wishart (1964) A central limit theorem for processes defined on a finite Markov chain. *Proc. Cambridge Philos. Soc.* **60**, 547–567.

[525] J. Keilson & D.M.G. Wishart (1964) Boundary problems for additive processes defined on a finite Markov chain. *Proc. Cambridge Philos. Soc.* **61**, 173–190.

[526] J. Keilson & D.M.G. Wishart (1964). Addenda for processes defined on a finite Markov chain. *Proc. Cambridge Philos. Soc.* **63**, 187–193.

[527] O. Kella & W. Whitt (1992) Useful martingales for stochastic storage processes with Lévy input. *J. Appl. Probab.* **29**, 396–403.

[528] J.H.B. Kemperman (1961) *The Passage Problem for a Markov Chain*. University of Chicago Press.

[529] J. Kennedy (1994) Understanding the Wiener–Hopf factorization for the simple random walk. *J. Appl. Probab.* **31**, 561–563.

[530] D.P. Kerekesha (2004) An exact solution of the risk equation with a step current reserve function. *Th. Probab. Math. Statist.* **69**, 61–66.

[531] H. Kesten (1974) Renewal theory for functionals of a Markov chain with general state space. *Ann. Probab.* **2**, 355–386.

[532] B. Kim & H.S. Kim (2007) Moments of claims in a Markovian environment. *Insurance: Mathematics and Economics* **40**, 485–497.

[533] J.F.C. Kingman (1961) A convexity property of positive matrices. *Quart. J. Math. Oxford* **12**, 283–284.

[534] J.F.C. Kingman (1962) On queues in heavy traffic. *Quart. J. Roy. Statist. Soc.* **B24**, 383–392.

[535] J.F.C. Kingman (1964) A martingale inequality in the theory of queues. *Proc. Camb. Philos. Soc.* **60**, 359–361.

[536] S. Klugman, H. Panjer & G.E. Willmot (2008) *Loss Models* (3rd ed.). Wiley.

[537] C. Klüppelberg (1988) Subexponential distributions and integrated tails. *J. Appl. Probab.* **25**, 132–141.

[538] C. Klüppelberg (1989) Estimation of ruin probabilities by means of hazard rates. *Insurance: Mathematics and Economics* **8**, 279–285.

[539] C. Klüppelberg (1993) Asymptotic ordering of risks and ruin probabilities. *Insurance: Mathematics and Economics* **12**, 259–264.

[540] C. Klüppelberg & R. Kostadinova (2008) Integrated insurance risk models with exponential Lévy investment. *Insurance: Mathematics and Economics* **42**, 560–577.

[541] H.J. Kushner & P. Dupuis (2001) *Numerical Methods for Stochastic Control Problems in Continuous Time* (2nd ed.). Springer-Verlag.

[542] C. Klüppelberg & A. Kyprianou (2006) On extreme ruinous behaviour of Lévy insurance risk processes. *J. Appl. Probab.* **43**, 594–598.

[543] C. Klüppelberg, A. Kyprianou & R. Maller (2004) Ruin probabilities and overshoots for general Lévy insurance risk processes. *Ann. Appl. Probab.* **14**, 1766–1801.

[544] C. Klüppelberg & T. Mikosch (1995) Explosive Poisson shot noise with applications to risk retention. *Bernoulli* **1**, 125–147.

[545] C. Klüppelberg & T. Mikosch (1995) Modelling delay in claim settlement. *Scand. Act. J.* **1995**, 154–168.

[546] C. Klüppelberg & U. Stadtmüller (1998) Ruin probabilities in the presence of heavy–tails and interest rates. *Scand. Act. J.* **1998**, 49–58.

[547] C. Knessel & C. Peters (1994) Exact and asymptotic solutions for the time-dependent problem of collective ruin I. *SIAM J. Appl. Math.* **54**, 1745-1767.

[548] C. Knessel & C. Peters (1996) Exact and asymptotic solutions for the time-dependent problem of collective ruin II. *SIAM J. Appl. Math.* **56**, 1471–1521.

[549] A. Kohatsu-Higa & M. Montero (2004) Malliavin calculus in finance. *Handbook of Numerical and Computational Methods in Finance* (S.T. Rachev, ed.), 111–174. Birkhäuser.

[550] D. Konstantinides & T. Mikosch (2005) Large deviations and ruin probabilities for solutions of stochastic recurrence equations with heavy-tailed innovations. *Ann. Probab.* **33**, 1992–2035.

[551] D. Konstantinides, Q. Tang & G. Tsitsiashvili (2002) Estimates for the ruin probability in the classical risk model with constant interest force in the presence of heavy tail. *Insurance: Mathematics and Economics* **31**, 447–460.

[552] R. Korn, E. Korn & G. Kroisandt (2010) *Monte Carlo Methods in Finance and Insurance.* Chapman & Hall/CRC.

[553] R. Korn & A. Wiese (2008) Optimal investment and bounded ruin probability: constant portfolio strategies and mean-variance analysis. *Astin Bulletin* **38**, 423–440.

[554] V.S. Korolyuk (1975) *Boundary Problems for Compound Poisson Processes.* Naukova Dumka.

[555] V.S. Korolyuk, I.P. Penev & A.F. Turbin (1973) An asymptotic expansion for the absorption time of a Markov chain distribution. *Cybernetica* **4**, 133–135 (in Russian).

[556] D. Kortschak & H. Albrecher (2009a) Asymptotic results for the sum of dependent non-identically distributed random variables. *Methodol. Comput. Appl. Probab.* **11**, 279–306.

[557] D. Kortschak & H. Albrecher (2009b) An asymptotic expansion for the tail of compound sums of Burr distributed random variables. *Stat. Probab. Lett.* **80**, 612–620.

[558] M. Kötter & N. Bäuerle (2008) The periodic risk model with investment. *Insurance: Mathematics and Economics* **42**, 962–967.

[559] S. Kou & H. Wang (2003) First passage time of a jump diffusion process. *Adv. Appl. Probab.* **35**, 504–531.

[560] S. Kou & H. Wang (2004) Option pricing under a double exponential jump diffusion model. *Management Science* **50**, 1178–1192.

[561] U. Küchler & M. Sørensen (1997) *Exponential Families of Stochastic Processes.* Springer-Verlag.

[562] H. Kunita (1976) Absolute continuity of Markov processes. *Seminaire de Probabilties* **X**. Lecture Notes in Mathematics **511**, 44–77. Springer-Verlag.

[563] A. Kuznetsov (2010) Wiener-Hopf factorization and distribution of extrema for a family of Lévy processes. *Ann. Appl. Probab.*, in press.

[564] A. Kyprianou (2006) *Introductory Lectures on Fluctuations of Lévy Processes with Applications.* Springer-Verlag.

[565] A. Kyprianou & R. Loeffen (2008) Refracted Lévy processes. *Ann. Inst. Henri Poincaré* **46**, 24–44.

[566] A.E. Kyprianou & M.R. Pistorius (2003) Perpetual options and Canadization through fluctuation theory. *Ann. Appl. Probab.* **13**, 1097–1098 .

[567] A. Kyprianou & V. Rivero (2008) Special, conjugate and complete scale functions for spectrally negative Lévy processes. *Electr. J. Probab.* **57**, 1672–1701.

[568] A. Kyprianou, V. Rivero & R. Song (2010) Convexity and smoothness of scale functions and de Finetti's control problem. *J. Theor. Probab.* **23**, 547–564.

[569] A. Kyprianou & X. Zhou (2009) General tax structures and the Lévy insurance risk model. *J. Appl. Probab.* **46**, 1146–1156.

[570] C. Labbé & K. Sendova (2008) The expected discounted penalty function under a risk model with stochastic income. *Appl. Math. Comp.* **215**, 1852–1867.

[571] R. Laeven, M.J. Goovaerts & T. Hoedemakers (2005) Some asymptotic results for sums of dependent random variables, with actuarial applications. *Insurance: Mathematics and Economics* **37**, 154–172.

[572] D. Landriault & G.E. Willmot (2008) On the Gerber-Shiu discounted penalty function in the Sparre Andersen model with an arbitrary interclaim time distribution. *Insurance: Mathematics and Economics* **42**, 600–608.

[573] D. Landriault & G.E. Willmot (2009) On the joint distributions of the time to ruin, the surplus prior to ruin, and the deficit at ruin in the classical risk model. *North American Actuarial J.* **13**, 252–270.

[574] G. Latouche & V. Ramaswami (1999) *Introduction to Matrix-Analytic Methods in Stochastic Modelling.* SIAM.

[575] C. Lefèvre & S. Loisel (2009) Finite-time horizon ruin probabilities with discrete, possibly dependent, claim amounts. *Methodol. Comput. Appl. Probab.* **11**, 425–441.

[576] T. Lehtonen & H. Nyrhinen (1992a) Simulating level–crossing probabilities by importance sampling. *Adv. Appl. Probab.* **24**, 858–874.

[577] T. Lehtonen & H. Nyrhinen (1992b) On asymptotically efficient simulation of ruin probabilities in a Markovian environment. *Scand. Act. J.* **1992**, 60–75.

[578] R. Leipus & J. Šiaulys (2007) Asymptotic behaviour of the finite-time ruin probability under subexponential claim sizes. *Insurance: Mathematics and Economics* **40**, 498–508.

[579] A.J. Lemoine (1981) On queues with periodic Poisson input. *J. Appl. Probab.* **18**, 889–900.

[580] A.J. Lemoine (1989) Waiting time and workload in queues with periodic Poisson input. *J. Appl. Probab.* **26**, 390–397.

[581] S. Levendorskii (2004) Pricing of the American put under Lévy processes. *Int. J. Theoret. Appl. Finance* **7**, 303–335.

[582] A. Lewis & E. Mordecki (2008) Wiener-Hopf factorization for Lévy processes having positive jumps with rational transforms. *J. Appl. Probab.* **45**, 118–134.

[583] J. Li, Z. Liu & Q. Tang (2007) On the ruin probability of a bidimensional perturbed risk model. *Insurance: Mathematics and Economics* **41**, 185–195.

[584] S. Li (2003) Discussion on 'Moments of the surplus before ruin and the deficit at ruin in the Erlang(2) risk process'. *North American Actuarial J.* **7**(3), 119–122.

[585] S. Li (2008) Discussion on 'The discounted joint distribution of the surplus prior to ruin and the deficit at ruin in a Sparre Andersen model'. *North American Actuarial J.* **12**, 208–210.

[586] S. Li (2008) The time of recovery and the maximum severity of ruin in a Sparre Andersen model. *North American Actuarial J.* **12**, 413–425.

[587] S. Li & D.C.M. Dickson (2006) The maximum surplus before ruin in an Erlang(n) risk process and related problems. *Insurance: Mathematics and Economics* **38**, 529–539.

[588] S. Li & J. Garrido (2002) On the time value of ruin in the discrete time risk model. *Working Paper 02-18, University Carlos III of Madrid*, 1–28.

[589] S. Li & J. Garrido (2004) On ruin for the Erlang(n) risk process. *Insurance: Mathematics and Economics* **34**, 391–408.

[590] S. Li & J. Garrido (2005) On a general class of renewal risk processes: analysis of the Gerber-Shiu function. *Adv. Appl. Probab.* **37**, 836–856.

[591] S. Li & J. Garrido (2005) The Gerber-Shiu function in a Sparre Andersen risk process perturbed by diffusion. *Scand. Act. J.* **2005**, 161–186.

[592] S. Li & Y. Lu (2008) The decompositions of the discounted penalty functions and dividend-penalty identity in a Markov-modulated risk model. *Astin Bulletin* **38**, 53–71.

[593] S. Li, Y. Lu & J. Garrido (2009) A review of discrete-time risk models. *Rev. R. Acad. Cien. Serie A. Mat.* **103**, 321–337.

[594] X.S. Lin & K.P. Sendova (2008) The compound Poisson risk model with multiple thresholds. *Insurance: Mathematics and Economics* **42**, 617–627.

[595] X.S. Lin & T. Wang (2009) Pricing perpetual American catastrophe put options: a penalty function approach. *Insurance: Mathematics and Economics* **44**, 287–295.

[596] X.S. Lin & G.E. Willmot (2000) The moments of the time of ruin, the surplus before ruin, and the deficit at ruin. *Insurance: Mathematics and Economics* **27**, 19–44.

[597] X.S. Lin, G.E. Willmot & S. Drekic (2003) The classical risk model with a constant dividend barrier: analysis of the Gerber-Shiu discounted penalty function. *Insurance: Mathematics and Economics* **33**, 551–566.

[598] D. Lindley (1952) The theory of queues with a single server. *Proc. Cambr. Philos. Soc.* **48**, 277–289.

[599] F. Lindskog & A.J. McNeil (2003) Common Poisson shock models: applications to insurance and credit risk modelling. *Astin Bulletin* **33**, 209–238.

[600] L. Lipsky (2009) *Queueing Theory — a Linear Algebraic Approach* (2nd ed.). Springer-Verlag.

[601] C.S. Liu & H. Yang (2004) Optimal investment for an insurer to minimize its probability of ruin. *North American Actuarial J.* **8**, 11–31.

[602] S.X. Liu & J.Y. Guo (2006) The discrete risk model revisited. *Methodol. Comput. Appl. Probab.* **8**, 303–313.

[603] Y. Liu & J. Ma (2009) Optimal reinsurance/investment problems for general insurance models. *Ann. Appl. Probab.* **19**, 1495-1528.

[604] R. Loeffen (2008) On optimality of the barrier strategy in de Finetti's dividend problem for spectrally negative Lévy processes. *Ann. Appl. Probab.* **18**, 1669–1680.

[605] R. Loeffen & P. Patie (2010) Absolute ruin in the Ornstein-Uhlenbeck type risk model. *Preprint*.

[606] R. Loeffen & J. Renaud (2010) De Finetti's optimal dividends problem with an affine penalty function at ruin. *Insurance: Mathematics and Economics* **46**, 98–108.

[607] S. Loisel (2005) Differentiation of some functionals of risk processes, and optimal reserve allocation. *J. Appl. Probab.* **42**, 379–392.

[608] S. Loisel, C. Mazza & D. Rullière (2009) Convergence and asymptotic variance of bootstrapped finite-time ruin probabilities with partly shifted risk processes. *Insurance: Mathematics and Economics* **45**, 374–381.

[609] A. Löpker & D. Perry (2010) The idle period of the finite G/M/1 queue with an interpretation in risk theory. *Queueing Syst.* **64**, 395–407.

[610] Y. Lu & S. Li (2005) On the probability of ruin in a Markov-modulated risk model. *Insurance: Mathematics and Economics* **37**, 522–532.

[611] Y. Lu & C.C. Tsai (2007) The expected discounted penalty at ruin for a Markov-modulated risk process perturbed by diffusion. *North American Actuarial J.* **11**, 136–149.

[612] D. Lucantoni (1991) New results on the single server queue with a batch Markovian arrival process. *Stoch. Models* **7**, 1–46.

[613] D. Lucantoni, K.S. Meier-Hellstern & M.F. Neuts (1990) A single server queue with server vacations and a class of non-renewal arrival processes. *Adv. Appl. Probab.* **22**, 676–705.

[614] F. Lundberg (1903) **I** *Approximerad Framställning av Sannolikhetsfunktionen.* **II** *Återförsäkring av Kollektivrisker.* Almqvist & Wiksell, Uppsala.

[615] F. Lundberg (1926) *Försäkringsteknisk Riskutjämning.* F. Englunds Boktryckeri AB, Stockholm.

[616] S. Luo (2008) Ruin minimization for insurers with borrowing constraints. *North American Actuarial J.* **12**, 143–174.

[617] S. Luo & M. Taksar (2010) On absolute ruin minimization under a diffusion approximation model. *Insurance: Mathematics and Economics*, in press.

[618] J. Ma & X. Sun (2003) Ruin probabilities for insurance models involving investments. *Scand. Act. J.* **2003**, 217–237.

[619] C. Macci & L. Petrella (2006) Mixtures of conjugate prior distributions and large deviations for level crossing probabilities. *Sankhya* **68**, 61–89.

[620] C. Macci, G. Stabile & G. Torrisi (2005) Lundberg parameters for non standard risk processes. *Scand. Act. J.* **2005**, 417–432.

[621] C. Macci & G. Torrisi (2004) Asymptotic results for perturbed risk processes with delayed claims. *Insurance: Mathematics and Economics* **34**, 307–320.

[622] D.B. Madan & W. Schoutens (2008) Break on through to the single side. *J. Credit Risk* **4**, Sept. Issue.

[623] A.M. Makowski (1994) On an elementary characterization of the increasing convex order, with an application. *J. Appl. Probab.* **31**, 834–841.

[624] S. Malamud, E. Trubowitz & M. Wüthrich (2008) Market consistent pricing of insurance products. *Astin Bulletin* **38**, 483–526.

[625] V.K. Malinovskii (1994) Corrected normal approximation for the probability of ruin within finite time. *Scand. Act. J.* **1994**, 161–174.

[626] V.K. Malinovskii (1996) Approximation and upper bounds on probabilities of large deviations of ruin within finite time. *Scand. Act. J.* **1996**, 124–147.

[627] V.K. Malinovskii (1998) Non-Poissonian claims arrivals and calculation of the probability of ruin. *Insurance: Mathematics and Economics* **22**, 123–138.

[628] V. Mammitzsch (1986) A note on the adjustment coefficient in ruin theory. *Insurance: Mathematics and Economics* **5**, 147–149.

[629] M. Mandjes (2007) *Large Deviations for Gaussian Queues.* Wiley.

[630] A. Martin-Löf (1983) Entropy estimates for ruin probabilities. *Probability and Mathematical Statistics* (A. Gut & J. Holst eds.), 129–139. Uppsala University.

[631] A. Martin-Löf (1986) Entropy, a useful concept in risk theory. *Scand. Act. J.* **1986**, 223–235.

[632] C. Mazza & D. Rullière (2004) A link between wave governed random motions and ruin processes. *Insurance: Mathematics and Economics* **35**, 205–222.

[633] A. McNeil, R. Frey, & P. Embrechts (2005) *Quantitative Risk Management. Concepts, Techniques and Tools.* Princeton University Press.

[634] K.S. Meier (1984). A fitting algorithm for Markov–modulated Poisson processes having two arrival rates. *Europ. J. Oper. Res.* **29**, 370–377.

[635] M. Mesfioui and J.F. Quessy (2005) Bounds on the value-at-risk for the sum of possibly dependent risks. *Insurance: Mathematics and Economics* **37**, 135–151.

[636] R. Michel (1987) A partial ordering of claim amount distributions and its relation to ruin probabilities in the Poisson model. *Mitt. Ver. Schweiz. Vers. Math.* **1987**, 75–79.

[637] Z. Michna (1998) Self-similar processes in collective risk theory. *J. Appl. Math. Stoch. Anal.* **11**, 429–448.

[638] T. Mikosch (2009) *Non-Life Insurance Mathematics* (2nd ed.). Springer-Verlag.

[639] T. Mikosch & A. Nagaev (2001) Rates in approximations to ruin probabilities for heavy-tailed distributions. *Extremes* **4**, 67–78.

[640] T. Mikosch & G. Samorodnitsky (2000) The supremum of a negative drift random walk with dependent heavy-tailed steps. *Ann. Appl. Probab.* **10**, 1025–1064.

[641] T. Mikosch & G. Samorodnitsky (2000) Ruin probability with claims modeled by a stationary ergodic stable process. *Ann. Probab.* **28**, 1814–1851.

[642] H.D. Miller (1961) A convexity property in the theory of random variables defined on a finite Markov chain. *Ann. Math. Statist.* **32**, 1260–1270.

[643] H.D. Miller (1962) A matrix factorization problem in the theory of random variables defined on a finite Markov chain. *Proc. Cambridge Philos. Soc.* **58**, 268–285.

[644] H.D. Miller (1962) Absorption probabilities for sums of random variables defined on a finite Markov chain. *Proc. Cambridge Philos. Soc.* **58**, 286–298.

[645] R. Ming & W. Wang (2009) On the expected discounted penalty function for risk process with tax. *Stat. Probab. Lett.*, in press.

[646] R. Ming, W. Wang & L. Xiao (2010) On the time value of absolute ruin with tax. *Insurance: Mathematics and Economics* **46**, 67–84.

[647] A. Mitra & S. Resnick (2009) Aggregation of rapidly varying risks and asymptotic independence. *Adv. Appl. Probab.* **41**, 797–828.

[648] M. Miyazawa & V. Schmidt (1993) On ladder height distributions of general risk processes. *Ann. Appl. Probab.* **3**, 763–776.

[649] M. Morales (2006) On the expected discounted penalty function for a perturbed risk process driven by a subordinator. *Insurance: Mathematics and Economics* **40**, 293–301.

[650] M. Morales & W. Schoutens (2003) A risk model driven by a Lévy process. *Appl. Stochastic Models Bus. Ind.* **19**, 147–167.

[651] G.V. Moustakides (1999) Extension of Wald's first lemma to Markov processes. *J. Appl. Probab.* **36**, 48–59.

[652] A. Müller & G.C. Pflug (2001) Asymptotic ruin probabilities for risk processes with dependent increments. *Insurance: Mathematics and Economics* **28**, 381–392.

[653] A. Müller & D. Stoyan (2002) *Comparison Methods for Stochastic Models and Risks*. Wiley.

[654] S.V. Nagaev (1957) Some limit theorems for stationary Markov chains. *Th. Probab. Appl.* **2**, 378–406.

[655] O. Narayan (1998) Exact asymptotic queue length distribution for fractional Brownian motion traffic. *Adv. Perf. Anal.* **1**, 39–63.

[656] R. Nelsen (1999) *An Introduction to Copulas.* Springer-Verlag.

[657] P. Ney & E. Nummelin (1987) Markov additive processes I. Eigenvalue properties and limit theorems. *Ann. Probab.* **15**, 561–592.

[658] M.F. Neuts (1977) A versatile Markovian point process. *J. Appl. Probab.* **16**, 764–779.

[659] M.F. Neuts (1978) Renewal processes of phase type. *Naval Research Logistics Quart.* **25**, 445–454.

[660] M.F. Neuts (1981) *Matrix-Geometric Solutions in Stochastic Models.* Johns Hopkins University Press, Baltimore.

[661] M.F. Neuts (1989) *Structured Stochastic Matrices of the M/G/1 Type and their Applications.* Marcel Dekker.

[662] M.F. Neuts (1992) Models based on the Markovian arrival process. *IEICE Trans. Commun.* **E75-B**, 1255–1265.

[663] J. Neveu (1961) Une generalisation des processus à accroisements indépendantes. *Sem. Math. Abh. Hamburg* **25**, 36–61.

[664] A.C.G. Ng (2005) Discussion of 'The time value of ruin in a Sparre Andersen model'. *North American Actuarial J.* **9**(4), 131–134.

[665] R. Norberg (1990) Risk theory and its statistics environment. *Statistics* **21**, 273–299.

[666] H. Nyrhinen (1998) Rough descriptions of ruin for a general class of surplus processes. *Adv. Appl. Probab.* **30**, 1008–1026.

[667] H. Nyrhinen (1999a) Large deviations for the time of ruin. *J. Appl. Probab.* **36**, 733–746.

[668] H. Nyrhinen (1999b) On the ruin probabilities in a general economic environment. *Stoch. Proc. Appl.* **83**, 319–330.

[669] H. Nyrhinen (2001) Finite and infinite time ruin probabilities in a stochastic economic environment. *Stoch. Proc. Appl.* **92**, 265–285.

[670] C.A. O'Cinneide (1990) Characterization of phase-type distributions. *Stoch. Models* **6**, 1–57.

[671] J. Ohlin (1969) On a class of measures for dispersion with application to optimal insurance. *Astin Bulletin* **5**, 249–266.

[672] B. Øksendal & A. Sulem (2007) *Applied Stochastic Control of Jump Diffusions* (2nd ed.). Springer-Verlag.

[673] A. Olivieri & E. Pitacco (2003) Solvency requirements for pension annuities. *J. Pension Econ. Fin.* **2**, 127–157.

[674] M. Olvera-Cravioto, J.H. Blanchet & P.W. Glynn (2010) On the transition from heavy traffic to heavy tails for the M/G/1 queue: the regularly varying case. *Ann. Appl. Probab.*, in press.

[675] E. Omey & E. Willekens (1986) Second order behaviour of the tail of a subordinated probability distributions. *Stoch. Proc. Appl.* **21**, 339–353.

[676] E. Omey & E. Willekens (1987) Second order behaviour of distributions subordinate to a distribution with finite mean. *Stoch. Models* **3**, 311–342.

[677] A.G. Pakes (1975) On the tail of waiting time distributions. *J. Appl. Probab.* **12**, 555–564.

[678] Z. Palmowski (2002) Lundberg inequalities in a diffusion environment. *Insurance: Mathematics and Economics* **31**, 303–313.

[679] Z. Palmowski & M. Pistorius (2010) Cramér asymptotics for finite time first passage probabilities of general Lévy processes, *Preprint*.

[680] J. Paulsen (1993) Risk theory in a stochastic economic environment. *Stoch. Proc. Appl.* **46**, 327–361.

[681] J. Paulsen (1998) Sharp conditions for certain ruin in a risk process with stochastic return on investments. *Stoch. Proc. Appl.* **75**, 135–148.

[682] J. Paulsen (1998) Ruin theory with compounding assets — a survey. *Insurance: Mathematics and Economics* **22**, 3–16.

[683] J. Paulsen (2002) On Cramér-like asymptotics for risk processes with stochastic return on investments. *Ann. Appl. Probab.* **12**, 1247–1260.

[684] J. Paulsen (2008) Ruin models with investment income. *Probab. Surv.* **5**, 416–434.

[685] J. Paulsen & H.K. Gjessing (1997a) Optimal choice of dividend barriers for a risk process with stochastic return on investments. *Insurance: Mathematics and Economics* **20**, 215–223.

[686] J. Paulsen & H.K. Gjessing (1997b) Present value distributions with applications to ruin theory and stochastic equations. *Stoch. Proc. Appl.* **71**, 123–144.

[687] J. Paulsen & H.K. Gjessing (1997c) Ruin theory with stochastic return on investments. *Adv. Appl. Probab.* **29**, 965–985.

[688] J. Paulsen, J. Kasozi & A. Steigen (2005) A numerical method to find the probability of ultimate ruin in the classical risk model with stochastic return on investments. *Insurance: Mathematics and Economics* **36**, 399–420.

[689] F. Pellerey (1995) On the preservation of some orderings of risks under convolution. *Insurance: Mathematics and Economics* **16**, 23–30.

[690] A. Pelsser (2010) Time-consistent and market-consistent actuarial valuations. *Preprint*.

[691] S. Pergamenshchikov & O. Zeitouny (2006) Ruin probability in the presence of risky investments. *Stoch. Proc. Appl.* **116**, 267–278. *ibid.* **119**, 305–306.

[692] D. Perry, W. Stadje & S. Zacks (2002) First-exit times for compound Poisson processes for some types of positive and negative jumps. *Stoch. Models* **18**, 139–157.

[693] D. Perry, W. Stadje & S. Zacks (2002) Hitting and ruin probabilities for compound Poisson processes and the cycle maximum of the $M/G/1$ queue. *Stoch. Models* **18**, 553–564.

[694] A.A. Pervozvansky (1998) Equation for survival probability in a finite time interval in case of non-zero real interest force. *Insurance: Mathematics and Economics* **23**, 287–295.

[695] S. Schock Petersen (1989) Calculation of ruin probabilities when the premium depends on the current reserve. *Scand. Act. J.* **1989**, 147–159.

[696] D. Pfeifer & J. Nešhlehová (2004) Modeling and generating dependent risk processes for IRM and DFA. *Astin Bulletin* **34**, 333–360.

[697] G.C. Pflug & W. Römisch (2007) *Modeling, Measuring and Managing Risk.* World Scientific.

[698] H. Pham (2009) *Continuous-time Stochastic Control and Optimization with Financial Applications.* Springer-Verlag.

[699] C. Philipson (1968) A review of the collective theory of risk. *Skand. Aktuar. Tidskr.* **61**, 45–68, 117–133.

[700] P. Picard (1994) On some measures of the severity of ruin in the classical Poisson model. *Insurance: Mathematics and Economics* **14**, 107–115.

[701] P. Picard & C. Lefèvre (1997) The probability of ruin in finite time with discrete claim size distribution. *Scand. Act. J.* **1997**, 58–69.

[702] P. Picard & C. Lefèvre (1998) The moments of ruin time in the classical risk model with discrete claim size distribution. *Insurance: Mathematics and Economics* **23**, 157–172.

[703] M. Pistorius (2005) A potential-theoretic review of some exit problems of spectrally negative Lévy processes. *Lecture Notes in Mathematics*, **1857**, 30–41. Springer-Verlag.

[704] M. Pistorius (2006) On maxima and ladder processes for a dense class of Lévy process. *J. Appl. Probab.* **43**, 208–220.

[705] E.J.G. Pitman (1980) Subexponential distribution functions. *J. Austr. Math. Soc.* **29A**, 337–347.

[706] S. Pitts (1994) Nonparametric estimation of compound distributions with applications in insurance. *Ann. Inst. Stat. Math.* **46**, 537–555.

[707] S.M. Pitts, R. Grübel & P. Embrechts (1996) Confidence bounds for the adjustment coefficient. *Adv. Appl. Probab.* **28**, 820–827.

[708] S.M. Pitts & K. Politis (2007) Approximations for the Gerber-Shiu expected discounted penalty function in the compound Poisson risk model. *Adv. Appl. Probab.* **39**, 385–406.

[709] K. Politis (2003) Semiparametric estimation for non-ruin probabilities. *Scand. Act. J.* **2003**, 75–96.

[710] M. Powers (1995) A theory of risk, return and solvency. *Insurance: Mathematics and Economics* **17**, 101–118.

[711] N.U. Prabhu (1961). On the ruin problem of collective risk theory. *Ann. Math. Statist.* **32**, 757–764.

[712] N.U. Prabhu (1965) *Queues and Inventories.* Wiley.

[713] N.U. Prabhu (1980) *Stochastic Storage Processes. Queues, Insurance Risk, and Dams.* Springer-Verlag.

[714] N.U. Prabhu & Zhu (1989) Markov-modulated queueing systems. *Queueing Systems* **5**, 215–246.

[715] W.H. Press, B.F. Flannery, S.A. Teukolsky, & W.T. Vetterling (1986) *Numerical Recipes. The Art of Scientific Computing.* Cambridge University Press.

[716] D. Promislow (1991) The probability of ruin in a process with dependent increments. *Insurance: Mathematics and Economics* **10**, 99–107.

[717] D. Promislow (2005) Minimizing the probability of ruin when claims follow Brownian motion with drift. *North American Actuarial J.* **9**, 109–128.

[718] P. Protter (2004). *Stochastic Integration and Differential Equations* (2nd ed.). Springer-Verlag.

[719] G. Psarrakos (2008) Tail bounds for the distribution of the deficit in the renewal risk model. *Insurance: Mathematics and Economics* **43**, 197–202.

[720] G. Psarrakos & K. Politis (2008) Tail bounds for the joint distribution of the surplus prior to and at ruin. *Insurance: Mathematics and Economics* **42**, 163–176.

[721] R. Pyke (1959) The supremum and infimum of the Poisson process. *Ann. Math. Statist.* **30**, 568–576.

[722] V. Ramaswami (1980) The N/G/1 queue and its detailed analysis. *Adv. Appl. Probab.* **12**, 222–261.

[723] V. Ramaswami, D.G. Woolford & D.A. Stanford (2010) The Erlangization method for Markovian fluid flows. *Ann. Oper. Res.*, in press.

[724] C.M. Ramsay (1984) The asymptotic ruin problem when the healthy and sick periods form an alternating renewal process. *Insurance: Mathematics and Economics* **3**, 139–143.

[725] C.M. Ramsay (2003) A solution to the ruin problem for Pareto distributions. *Insurance: Mathematics and Economics* **33**, 109–116.

[726] C.M. Ramsay (2007) Exact waiting time and queue size distributions for equilibrium M/G/1 queues with Pareto service. *Queueing Syst.* **57**, 147–155.

[727] C.R. Rao (1965) *Linear Statistical Inference and Its Applications.* Wiley.

[728] R. Redheffer (1966) Algebraic properties of certain integral transforms. *Amer. Math. Mon.* **73**, 91–95.

[729] G.J.K. Regterschot & J.H.A. de Smit (1986) The queue M/G/1 with Markov–modulated arrivals and services. *Math. Oper. Res.* **11**, 465–483.

[730] J.M. Reinhard (1984) On a class of semi-Markov risk models obtained as classical risk models in a Markovian environment. *Astin Bulletin* **14**, 23–43.

[731] J.M. Reinhard & M. Snoussi (2000) The probability of ruin in a discrete semi-Markov risk model. *Blätter DGVFM* **24**, 477–490.

[732] J.M. Reinhard & M. Snoussi (2001) On the distribution of the surplus prior to ruin in a discrete-time semi-Markov risk model. *Astin Bulletin* **31**, 255–273.

[733] J. Ren (2005) The expected value of the time of ruin and the moments of the discounted deficit at ruin in the perturbed classical risk process. *Insurance: Mathematics and Economics* **37**, 505–521.

[734] J. Ren (2007) The discounted joint distribution of the surplus prior to ruin and and the deficit at ruin in a Sparre Andersen model. *North American Actuarial J.* **11**(3), 128–136.

[735] J. Ren (2008) A connection between the discounted and non-discounted expected penalty functions in the Sparre Andersen risk model. *Stat. Probab. Lett.* **79**, 324–330.

[736] J. Renaud (2009) The distribution of tax payments in a Lévy insurance risk model with a surplus-dependent taxation structure. *Insurance: Mathematics and Economics* **45**, 242–246.

[737] S. Resnick (1987) *Extreme Values, Regular Variation and Point Processes.* Springer-Verlag.

[738] S.I. Resnick (2007) *Heavy-Tail Phenomena. Probabilistic and Statistical Modeling.* Springer-Verlag.

[739] S. Resnick & G. Samorodnitsky (1997) Performance decay in a single server exponential queueing model with long range dependence. *Oper. Res.* **45**, 235–243.

[740] B. Ripley (1987) *Stochastic Simulation.* Wiley.

[741] C.Y. Robert (2005) Asymptotic probabilities of an exceedance over renewal thresholds with an application to risk theory. *J. Appl. Probab.* **42**, 153–162.

[742] C.Y. Robert & J. Segers (2008) Tails of random sums of a heavy-tailed number of light-tailed terms. *Insurance: Mathematics and Economics* **43**, 85–92.

[743] L.C.G. Rogers (1994) Fluid models in queueing theory and Wiener–Hopf factorisation of Markov chains. *Ann. Appl. Probab.* **4**, 390–413.

[744] L.C.G. Rogers & D. Williams (2000) *Diffusions, Markov Processes, and Martingales.* Cambridge University Press.

[745] T. Rolski (1987) Approximation of periodic queues. *J. Appl. Probab.* **19**, 691–707.

[746] T. Rolski, H. Schmidli, V. Schmidt & J.L. Teugels (1999) *Stochastic Processes for Insurance and Finance.* Wiley.

[747] W. Rongming & L. Haifeng (2002) On the ruin probability under a class of risk processes. *Astin Bulletin* **32**, 81–90.

[748] S.M. Ross (1974) Bounds on the delay distribution in GI/G/1 queues. *J. Appl. Probab.* **11**, 417–421.

[749] H.–J. Rossberg & G. Siegel (1974) Die Bedeutung von Kingmans Integralgleichungen bei der Approximation der stationären Wartezeitverteilung im Modell GI/G/1 mit und ohne Verzögerung beim Beginn einer Beschäftigungsperiode. *Math. Operationsforsch. Statist.* **5**, 687–699.

[750] B. Roynette, P. Vallois & A. Volpi (2007) Asymptotic behavior of the hitting time, overshoot and undershoot for some Lévy processes. *ESAIM Prob. Stat.* **12**, 58–93.

[751] R.Y. Rubinstein (1981) *Simulation and the Monte Carlo Method.* Wiley.

[752] R.Y. Rubinstein & D.P. Kroese (2008) *Simulation and the Monte Carlo Method* (2nd ed.). Wiley.

[753] R.Y. Rubinstein & B. Melamed (1998) *Modern Simulation and Modelling.* Wiley.

[754] R.Y. Rubinstein & A. Shapiro (1993) *Discrete Event Systems: Sensitivity Analysis and Stochastic Optimization via the Score Function Method*. Wiley.

[755] M. Rudemo (1973) Point processes generated by transitions of a Markov chain. *Adv. Appl. Probab.* **5**, 262–286.

[756] H. Rue & L. Held (2005) *Gaussian Random Markov Fields*. Chapman & Hall.

[757] D. Rullière & S. Loisel (2004) Another look at the Picard-Lefèvre formula for finite-time ruin probabilities. *Insurance: Mathematics and Economics* **35**, 187–203.

[758] D. Rullière & S. Loisel (2005) The win-first probability under interest force. *Insurance: Mathematics and Economics* **37**, 421–442.

[759] L. Rüschendorf (2010) Worst case portfolio vectors and diversification effects. *Finance Stoch.*, in press.

[760] T. Rydén (1994) Parameter estimation for Markov modulated Poisson processes. *Stochastic Models* **10**, 795–829.

[761] T. Rydén (1996) An EM algorithm for estimation in Markov-modulated Poisson processes. *Comp. Statist. Data Anal.* **21**, 431–447.

[762] J. Sarkar & A. Sen (2005) A generalized defective renewal equation for the surplus process perturbed by diffusion. *Insurance: Mathematics and Economics* **36**, 421–432.

[763] K. Sato (1999) *Lévy Processes and Infinite Divisibility*. Cambridge University Press.

[764] G. Samorodnitsky & M.L. Taqqu (1994) *Non–Gaussian Stable Processes*. Chapman & Hall.

[765] M. Schäl (2004) On discrete-time dynamic programming in insurance: exponential utility and minimizing the ruin probability. *Scand. Act. J.* **2004**, 189–210.

[766] S. Schlegel (1998) Ruin probabilities in perturbed risk models. *Insurance: Mathematics and Economics* **22**, 93–104.

[767] L. Schmickler (1992) MEDA: mixed Erlang distributions as phase-type representations of empirical distribution functions. *Stoch. Models* **6**, 131–156.

[768] H. Schmidli (1994) Diffusion approximations for a risk process with the possibility of borrowing and interest. *Stoch. Models* **10**, 365–388.

[769] H. Schmidli (1995) Cramér-Lundberg approximations for ruin probabilities of risk processes perturbed by a diffusion. *Insurance: Mathematics and Economics* **16**, 135–149.

[770] H. Schmidli (1996) Martingales and insurance risk. *Lecture Notes of the 8th Summer School on Probability and Mathematical Statistics (Varna)*, 155–188. Science Culture Technology Publishing, Singapore.

[771] H. Schmidli (1997a) Estimation of the Lundberg coefficient for a Markov modulated risk model. *Scand. Act. J.* **1997**, 48–57.

[772] H. Schmidli (1997b) An extension to the renewal theorem and an application in risk theory. *Ann. Appl. Probab.* **7**, 121–133.

[773] H. Schmidli (1999a) On the distribution of the surplus prior and at ruin. *Astin Bulletin* **29**, 227–244.

[774] H. Schmidli (1999b) Perturbed risk processes: a review. *Th. Stoch. Proc.* **5**, 145–165.

[775] H. Schmidli (2001a) Optimal proportional reinsurance policies in a dynamic setting. *Scand. Act. J.* **2001**, 40–68.

[776] H. Schmidli (2001b) Distribution of the first ladder height of a stationary risk process perturbed by α-stable Lévy motion. *Insurance: Mathematics and Economics* **28**, 13–20.

[777] H. Schmidli (2002) On minimizing the ruin probability by investment and reinsurance. *Ann. Appl. Probab.* **12**, 890–907.

[778] H. Schmidli (2005) Discussion of the paper 'The time value of ruin in a Sparre Andersen model'. *North American Actuarial J.* **9**(2), 69–70.

[779] H. Schmidli (2008) *Stochastic Control in Insurance.* Springer-Verlag.

[780] H. Schmidli (2010) On the Gerber-Shiu function and change of measure. *Insurance: Mathematics and Economics* **46**, 3–11.

[781] H. Schmidli (2010) Conditional law of risk processes given that ruin occurs. *Insurance: Mathematics and Economics* **46**, 281–289.

[782] K.D. Schmidt (1996) *Lectures on Risk Theory.* Teubner.

[783] W. Schoutens (2003) *Lévy Processes in Finance.* Wiley.

[784] H.L. Seal (1969) *The Stochastic Theory of a Risk Business.* Wiley.

[785] H.L. Seal (1972) Numerical calculcation of the probability of ruin in the Poisson/Exponential case. *Mitt. Verein Schweiz. Versich. Math.* **72**, 77–100.

[786] H.L. Seal (1972). Risk theory and the single server queue. *Mitt. Verein Schweiz. Versich. Math.* **72**, 171–178.

[787] H.L. Seal (1974) The numerical calculation of $U(w,t)$, the probability of non-ruin in an interval $(0,t)$. *Scand. Act. J.* **1974**, 121–139.

[788] H.L. Seal (1978) *Survival Probabilities.* Wiley.

[789] G.A.F. Seber (1984) *Multivariate Observations.* Wiley.

[790] C.-O. Segerdahl (1942) Über einige Risikotheoretische Fragestellungen. *Skand. Aktuar Tidsskr.* **61**, 43–83.

[791] C.-O. Segerdahl (1955) When does ruin occur in the collective theory of risk? *Skand. Aktuar Tidsskr.* **1955**, 22–36.

[792] C.-O. Segerdahl (1959) A survey of results in the collective theory of risk. *Probability and statistics — the Harald Cramér volume* (U. Grenander, ed), 279–299. Almqvist & Wiksell, Stockholm.

[793] B. Sengupta (1989) Markov processes whose steady-state distribution is matrix-geometric with an application to the GI/PH/1 queue. *Adv. Appl. Probab.* **21**, 159–180.

[794] B. Sengupta (1990) The semi-Markov queue: theory and applications. *Stoch. Models* **6**, 383–413.

[795] M. Shaked & J.G. Shantikumar (1993) *Stochastic Orders and Their Applications.* Academic Press.

[796] X. Shen, Z. Lin & Y. Zhang (2009) Uniform estimate for maximum of randomly weighted sums with applications to ruin theory. *Methodol. Comput. Appl. Probab.* **11**, 669–685.

[797] Y. Shimizu (2009) A new aspect of a risk process and its statistical inference. *Insurance: Mathematics and Economics* **44**, 70–77.

[798] E.S. Shtatland (1966) On the distribution of the maximum of a process with independent increments. *Th. Probab. Appl.* **11**, 483–487.

[799] A. Shwartz & A. Weiss (1995) *Large Deviations for Performance Analysis*. Chapman & Hall.

[800] E.S.W. Shiu (1987) Convolution of uniform distributions and ruin probability. *Scand. Act. J.* **1987**, 191–197.

[801] E.S.W. Shiu (1989) Ruin probability by operational calculus. *Insurance: Mathematics and Economics* **8**, 243–249.

[802] E.S.W. Shiu (1989) The probability of eventual ruin in the compound binomial model. *Astin Bulletin* **19**, 179–190.

[803] J. Siaulys & R. Asanaviciute (2006). On the Gerber-Shiu discounted penalty function for subexponential claims. *Lithuanian Math. J.* **46**, 487–493.

[804] T. Siegl & R.F. Tichy (1996) Lösungsmethoden eines Risikomodells bei exponentiell fallender Schadensverteilung. *Mitt. Ver. Schweiz. Vers. Math.* **96**, 85–118.

[805] T. Siegl & R.F. Tichy (1999) A process with stochastic claim frequency and a linear dividend barrier. *Insurance: Mathematics and Economics* **24**, 51–65.

[806] D. Siegmund (1975) The time until ruin in collective risk theory. *Mitteil. Verein Schweiz Versich. Math.* **75**, 157–166.

[807] D. Siegmund (1976) Importance sampling in the Monte Carlo study of sequential tests. *Ann. Statist.* **4**, 673–684.

[808] D. Siegmund (1976) The equivalence of absorbing and reflecting barrier problems for stochastically monotone Markov processes. *Ann. Probab.* **4**, 914–924.

[809] D. Siegmund (1979) Corrected diffusion approximations in certain random walk problems. *Adv. Appl. Probab.* **11**, 701–719.

[810] D. Siegmund (1985) *Sequential Analysis*. Springer-Verlag.

[811] K. Sigman (1992) Light traffic for workload in queues. *Queueing Systems* **11**, 429–442.

[812] K. Sigman (1994) *Stationary Marked Point Processes: An Intuitive Approach*. Chapman & Hall.

[813] E. Slud & C. Hoesman (1989) Moderate- and large-deviations probabilities in actuarial risk theory. *Adv. Appl. Probab.* **21**, 725–741.

[814] W.L. Smith (1953) Distribution of queueing times. *Proc. Cambridge Philos. Soc.* **49**, 449–461.

[815] M. Song, Q. Meng, R. Wu & J. Ren (2010) The Gerber-Shiu discounted penalty function in the risk process with phase-type interclaim times. *Appl. Math. Comp.* **216**, 523–531.

[816] E. Sparre Andersen (1957) On the collective theory of risk in the case of contagion between the claims. *Transactions XVth International Congress of Actuaries, New York, II*, 219–229.

[817] D.A. Stanford & K.J. Stroiński (1994) Recursive method for computing finite–time ruin probabilities for phase–distributed claim sizes. *Astin Bulletin* **24**, 235–254.

[818] E. Straub (1988) *Non-Life Insurance Mathematics.* Springer-Verlag.

[819] L. Sun & H. Yang (2004) On the joint distributions of surplus immediately before ruin and the deficit at ruin for Erlang(2) risk processes. *Insurance: Mathematics and Economics* **34**, 121–125.

[820] B. Sundt (1993) *An Introduction to Non-Life Insurance Mathematics* (3rd ed.). Verlag Versicherungswirtschaft e.V., Karlsruhe.

[821] B. Sundt & W.S. Jewell (1981) Further results on recursive evaluation of compound distributions. *Astin Bulletin* **12**, 27–39.

[822] B. Sundt & J.L. Teugels (1995) Ruin estimates under interest force. *Insurance: Mathematics and Economics* **16**, 7–22.

[823] B. Sundt & J.L. Teugels (1997) The adjustment coefficient in ruin estimates under interest force. *Insurance: Mathematics and Economics* **19**, 85–94.

[824] B. Sundt & R. Vernic (2009) *Recursions for Convolutions and Compound Distributions with Insurance Applications.* Springer-Verlag.

[825] R. Suri (1989) Perturbation analysis: the state of the art and research issues explained via the GI/G/1 queue. *Proceedings of the IEEE* **77**, 114–137.

[826] S. Täcklind (1942) Sur le risque dans les jeux inequitables. *Skand. Aktuar. Tidskr.* **1942**, 1–42.

[827] L. Takács (1966) *Combinatorial Methods in the Theory of Stochastic Processes.* Wiley.

[828] Q. Tang (2004) Asymptotics for the finite time ruin probability in the renewal model with consistent variation. *Stoch. Models* **20**, 281–297.

[829] Q. Tang (2005) The finite-time ruin probability of the compound Poisson model with constant interest force. *J. Appl. Probab.* **42**, 608–619.

[830] Q. Tang & G. Tsitsiashvili (2003) Precise estimates for the ruin probability in finite horizon in a discrete-time model with heavy-tailed insurance and financial risks. *Stoch. Proc. Appl.* **108**, 299–325.

[831] Q. Tang & G. Tsitsiashvili (2004) Finite- and infinite-time ruin probabilities in the presence of stochastic returns on investments. *Adv. Appl. Probab.* **36**, 1278–1299.

[832] Q. Tang & D. Wang (2006) Tail probabilities of randomly weighted sums of random variables with dominated variation. *Stoch. Models* **22**, 253–272.

[833] Q. Tang, G. Wang & K.C. Yuen (2010) Uniform tail asymptotics for the stochastic present value of aggregate claims in the renewal risk model. *Insurance: Mathematics and Economics* **46**, 362–370.

[834] Q. Tang & L. Wei (2010) Asymptotic aspects of the Gerber-Shiu function in the renewal risk model using Wiener-Hopf factorization and convolution equivalence. *Insurance: Mathematics and Economics* **46**, 19–31.

[835] M. Taqqu, W. Willinger & R. Sherman (1997) Proof for a fundamental result in self-similar modeling. *Comp. Comm. Rev.* **27**, 5–23.

[836] G.C. Taylor (1976) Use of differential and integral inequalities to bound ruin and queueing probabilities. *Scand. Act. J.* **1976**, 197–208.

[837] G.C. Taylor (1978) Representation and explicit calculation of finite–time ruin probabilities. *Scand. Act. J.* **1978**, 1–18.

[838] G.C. Taylor (1979) Probability of ruin under inflationary conditions or under experience ratings. *Astin Bulletin* **16**, 149–162.

[839] G.C. Taylor (1979) Probability of ruin with variable premium rate. *Scand. Act. J.* **1980**, 57–76.

[840] G.C. Taylor (1986) *Claims Reserving in Non-Life Insurance.* North-Holland.

[841] J.L. Teugels (1982) Estimation of ruin probabilities. *Insurance: Mathematics and Economics* **1**, 163–175.

[842] J.L. Teugels & N. Veraverbeke (1973) Cramér-type estimates for the probability of ruin, *C.O.R.E. Discussion Paper No 7316.*

[843] S. Thonhauser & H. Albrecher (2007) Dividend maximization under consideration of the time value of ruin. *Insurance: Mathematics and Economics* **41**, 163–184.

[844] O. Thorin (1974) On the asymptotic behaviour of the ruin probability when the epochs of claims form a renewal process. *Scand. Act. J.* **1974**, 81–99.

[845] O. Thorin (1977). Ruin probabilities prepared for numerical calculations. *Scand. Act. J.* **1977**, 65–102.

[846] O. Thorin (1982) Probabilities of ruin. *Scand. Act. J.* **1982**, 65–102.

[847] O. Thorin (1986) Ruin probabilities when the claim amounts are gamma distributed. Unpublished manuscript.

[848] O. Thorin & N. Wikstad (1977) Numerical evaluation of ruin probabilities for a finite period. *Astin Bulletin* **7**, 137–153.

[849] O. Thorin & N. Wikstad (1977) Calculation of ruin probabilities when the claim distribution is lognormal. *Astin Bulletin* **9**, 231-246.

[850] H. Thorisson (2000) *Coupling, Stationarity and Regeneration.* Springer-Verlag.

[851] R.F. Tichy (1984) Über eine zahlentheoretische Methode zur numerischen Integration und zur Behandlung von Integralgleichungen. *Österreich. Akad. Wiss. Math.-Natur. Kl. Sitzungsber. II* **193**, 329–358.

[852] H. Tijms (1986) *Stochastic Modelling and Analysis: A Computational Approach.* Wiley.

[853] J. Trufin, H. Albrecher & M. Denuit (2009) Impact of underwriting cycles on the solvency of an insurance company. *North American Actuarial J.* **13**, 385–403.

[854] J. Trufin, H. Albrecher & M. Denuit (2010) Properties of a risk measure derived from ruin theory. *The Geneva Risk and Insurance Review*, in press.

[855] J. Trufin, H. Albrecher & M. Denuit (2010) Ruin problems under IBNR dynamics. *Appl. Stochastic Models Bus. Ind.*, in press.

[856] C.C. Tsai (2009) On the ordering of ruin probabilities for the surplus process perturbed by diffusion. *Scand. Act. J.* **2009**, 187–204.

[857] C.C. Tsai & L. Sun (2004) On the discounted distribution functions for the Erlang(2) risk process. *Insurance: Mathematics and Economics* **35**, 5–19.

[858] C.C. Tsai & G.E. Willmot (2002) A generalized defective renewal equation for the surplus process perturbed by diffusion. *Insurance: Mathematics and Economics* **30**, 51–66.

[859] M. Usabel (1999) A note on the Taylor series expansions for multivariate characteristics of classical risk processes. *Insurance: Mathematics and Economics* **25**, 37–47.

[860] M. Usabel (1999) Practical approximations for multivariate characteristics of risk processes. *Insurance: Mathematics and Economics* **25**, 397–413.

[861] M. Van Wouve, F. De Vylder & M.J. Goovaerts (1983) The influence of reinsurance limits on infinite time ruin probabilities. In: *Premium Calculation in Insurance* (F. De Vylder, M.J. Goovaerts, J. Haezendonck eds.). Reidel, Dordrecht.

[862] F. Vázquez-Abad (1999) RPA pathwise derivative estimation of ruin probabilities. *Insurance: Mathematics and Economics* **26**, 269–288.

[863] N. Veraverbeke (1993) Asymptotic estimates for the probability of ruin in a Poisson model with diffusion. *Insurance: Mathematics and Economics* **13**, 57–62.

[864] N. Veraverbeke and J.L. Teugels (1975/76) The exponential rate of convergence of the distribution of the maximum of a random walk. *J. Appl. Probab.* **12**, 279–288; ibid. **13**, 733–740.

[865] N. Veraverbeke and J.L. Teugels (1975) Regular speed of convergence for the maximum of a random walk. *Colloq. Math. Soc. Janos Bolyai* **11**, 399–406.

[866] C. Wagner (2001) A note on ruin in a two state Markov model. *Astin Bulletin* **32**, 349–358.

[867] R. Wang, L. Xu & D. Yao (2007) Ruin problems with stochastic premium stochastic return on investments. *Front. Math. China* **2**, 467–490.

[868] S. Wang, C. Zhang & G. Wang (2010) A constant interest risk model with tax payments. *Stoch. Models*, in press.

[869] A. Wald (1947) *Sequential Analysis*. Wiley.

[870] V. Wallace (1969) The solution of quasi birth and death processes arising from multiple access computer systems. Unpublished Ph.D. thesis, University of Michigan.

[871] D. Wang & Q. Tang (2004) Maxima of sums and random sums for negatively associated random variables with heavy tails. *Stat. Probab. Lett.* **68**, 287–295.

[872] G. Wang & R. Wu (2008) The expected discounted penalty function for the perturbed compound Poisson risk process with constant interest. *Insurance: Mathematics and Economics* **42**, 59–64.

[873] S. Wang & M. Sobrero (1994) Further results on Hesselager's recursive procedure for calculation of some compound distributions. *Astin Bulletin* **24**, 161–168.

[874] Z. Wang, J. Xia & L. Zhang (2007) Optimal investment for an insurer: The martingale approach. *Insurance: Mathematics and Economics* **40**, 322–334.

[875] H.R. Waters (1983) Probability of ruin for a risk process with claims cost inflation. *Scand. Act. J.* **1983**, 148–164.

[876] H.R. Waters (1983) Some mathematical aspects of reinsurance. *Insurance: Mathematics and Economics* **2**, 17–26.

[877] J. Wei, H. Yang & R. Wang (2010) On the Markov-modulated insurance risk model with tax. *Blätter DGVFM* **31**, 65–78.

[878] L. Wei (2009) Ruin probability in the presence of interest earnings and tax payments. *Insurance: Mathematics and Economics* **45**, 133–138.

[879] L. Wei (2009) Ruin probability of the renewal model with risky investment and large claims. *Science in China, Ser. A: Mathematics* **52**, 1539–1545.

[880] L. Wei & H. Yang (2004) Explicit expressions for the ruin probabilities of Erlang risk processes with Pareto individual claim distributions. *Acta Math. Appl. Sinica* **20**, 495–506.

[881] C. Weng, Y. Zhang & K.S. Tan (2009) Ruin probabilities in a discrete time risk model with dependent risks of heavy tail. *Scand. Act. J.* **2009**, 205–218.

[882] W. Whitt (1989) An interpolation approximation for the mean workload in a GI/G/1 queue. *Oper. Res.* **37**, 936–952.

[883] W. Whitt (2002) *Stochastic Process Limits.* Springer-Verlag.

[884] W. Willinger, M. Taqqu, R. Sherman & D. Wilson (1995) Self-similarity in high-speed traffic: analysis and modeling of ethernet traffic measurements. *Stat. Sci.* **10**, 67–85.

[885] G.E. Willmot (1994) Ruin probabilities in the compound binomial model. *Insurance: Mathematics and Economics* **12**, 133–142.

[886] G.E. Willmot (1994) Refinements and distributional generalizations of Lundberg's inequality. *Insurance: Mathematics and Economics* **15**, 49–63.

[887] G.E. Willmot (2004) A note on a class of delayed renewal risk processes. *Insurance: Mathematics and Economics* **34**, 251–257.

[888] G.E. Willmot, J. Cai & X. Lin (2001) Lundberg inequalities for renewal equations. *Adv. Appl. Probab.* **33**, 674–689.

[889] G.E. Willmot & D.C.M. Dickson (2003) The Gerber-Shiu discounted penalty function in the stationary renewal risk model. *Insurance: Mathematics and Economics* **32**, 403–411.

[890] G.E. Willmot, D.C.M. Dickson, S. Drekic & D. Stanford (2004) The deficit at ruin in the stationary renewal risk model. *Scand. Act. J.* **2004**, 241–255.

[891] G.E. Willmot & X. Lin (1994) Lundberg bounds on the tails of compound distributions. *J. Appl. Probab.* **31**, 743–756.

[892] G.E. Willmot & X. Lin (1998) Exact and approximate properties of the distribution of surplus before and after ruin. *Insurance: Mathematics and Economics* **23**, 91–110.

[893] G.E. Willmot & J.K. Woo (2007) On the class of Erlang mixtures with risk theoretic applications. *North American Actuarial J.* **11**(2), 99–115.

[894] R.W. Wolff (1990) *Stochastic Modeling and the Theory of Queues.* Prentice–Hall.

[895] R. Wu, Y. Lu & Y. Fang (2007) On the Gerber-Shiu discounted penalty function for the ordinary renewal risk model with constant interest. *North American Actuarial J.* **11**(2), 119–134.

[896] R. Wu, G. Wang & L. Wei (2003) Joint distributions of some actuarial random vectors containing the time of ruin. *Insurance: Mathematics and Economics* **33**, 147–161.

[897] R. Wu, G. Wang & C. Zhang (2005) On a joint distribution for the risk process with constant interest force. *Insurance: Mathematics and Economics* **36**, 365–374.

[898] X. Wu & K.C. Yuen (2003) A discrete-time risk model with interaction between classes of business. *Insurance: Mathematics and Economics* **33**, 117–133.

[899] H. Yang (1998) Non-exponential bounds for ruin probability with interest effect included. *Scand. Act. J.* **1998**, 66–79.

[900] H. Yang (2003) Ruin theory in a financial corporation model with credit risk. *Insurance: Mathematics and Economics* **33**, 135–145.

[901] H. Yang & L. Zhang (2001) The joint distribution of the surplus immediately before ruin and the deficit at ruin under interest force. *North American Actuarial J.* **5**(3), 92–103.

[902] H. Yang & L. Zhang (2005) Optimal investment for insurer with jump-diffusion risk process. *Insurance: Mathematics and Economics* **37**, 615–634.

[903] H. Yang & L. Zhang (2006) Ruin problems for a discrete time risk model with random interest rate. *Math. Meth. Oper. Res.* **63**, 287–299.

[904] H. Yang & Z. Zhang (2009) Discounted penalty functions and dividend payments in Sparre Andersen model with multi-layer dividend strategy. *Insurance: Mathematics and Economics* **42**, 984–991.

[905] H. Yang, Z. Zhang & C. Lan (2009) Ruin problems in a discrete Markov risk model. *Stat. Probab. Lett.* **79**, 21–28.

[906] G. Yin, Y.J. Liu & H. Yang (2006) Bounds of ruin probability for regime-switching models using time scale separation. *Scand. Act. J.* **2006**, 111–127.

[907] K.C. Yuen & J. Guo (2001) Ruin probabilities for time-correlated claims in the compound binomial model. *Insurance: Mathematics and Economics* **29**, 47–57.

[908] K.C. Yuen, J. Guo & K.W. Ng (2005) On ultimate ruin in a delayed-claims risk model. *J. Appl. Probab.* **42**, 163–174.

[909] K.C. Yuen, J. Guo & X. Wu (2002) On a correlated aggregate claims model with Poisson and Erlang risk processes. *Insurance: Mathematics and Economics* **31** 205–214.

[910] K.C. Yuen, J. Guo & X. Wu (2005) On the first time of ruin in the bivariate compound Poisson model. *Insurance: Mathematics and Economics* **38**, 298–308.

[911] K. Yuen, G. Wang (2005) Some ruin problems for a risk process with stochastic interest. *North American Actuarial J.* **9**(3), 129–142.

[912] K. Yuen, G. Wang & W. Li (2007) The Gerber-Shiu expected discounted penalty function for risk processes with interest and a constant dividend barrier. *Insurance: Mathematics and Economics* **40**, 104–112.

[913] X. Zhang & T.K. Siu (2009) Optimal investment and reinsurance of an insurer with model uncertainty. *Insurance: Mathematics and Economics* **45**, 81–88.

[914] Z. Zhang & H. Yang (2010) A generalized penalty function in the Sparre Andersen risk model with two-sided jumps. *Stat. Probab. Lett.* **80**, 597–607.

[915] Z. Zhang, H. Yang & S. Li (2010) The perturbed compound Poisson risk model with two-sided jumps. *J. Comp. Appl. Math.* **233**, 1773–1784.

[916] Z. Zhang, K.C. Yuen & W.K. Li (2007) A time-series risk model with constant interest for dependent classes of business. *Insurance: Mathematics and Economics* **41**, 32–40.

[917] M. Zhou & J. Cai (2009) A perturbed risk model with dependence between premium rates and claim sizes. *Insurance: Mathematics and Economics* **45**, 382–392.

[918] X. Zhou (2005) On a classical risk model with a constant dividend barrier. *North American Actuarial J.* **9**(4), 95–108.

[919] X. Zhou (2006) Classical risk model with a multi-layer premium rate. *Actuarial Research Clearing House* **41**, 1–10.

[920] X. Zhou (2007) Exit problems for spectrally negative Lévy processes reflected at either the supremum or infimum. *J. Appl. Probab.* **44**, 1012–1030.

[921] J. Zhu & H. Yang (2009) On differentiability of ruin functions under Markov-modulated models. *Stoch. Proc. Appl.* **119**, 1673–1695.

[922] V.M. Zolotarev (1964) The first passage time of a level and the behavior at infinity for a class of processes with independent increments. *Th. Probab. Appl.* **9**, 653–661.

[923] B. Zwart, S. Borst & K. Dębicki (2005) Subexponential asymptotics of hybrid fluid and ruin models. *Ann. Appl. Probab.* **15**, 500–517.

Index